THE ROUTLEDGE HANDBOOK OF NATURE BASED TOURISM DEVELOPMENT

This handbook offers a comprehensive overview of the themes and concepts related to nature-based tourism development. Providing interdisciplinary insights from leading researchers, academics, and practitioners across the globe, it delivers a critical and timely contribution to the knowledge around nature-based tourism.

Nature-based tourism is currently the fastest-growing tourism sector globally and for many destinations, the most significant tourism segment. Organized into five parts, this handbook provides contemporary and cutting-edge perspectives on core topics and explores their linkages. It considers, among others, various natural settings and natural attractions where nature-based tourism can be exercised, including: protected and conserved areas, islands, and mountains; the emerging themes shaping the contemporary nature-based tourism development, including ethics, Sustainable Development Goals, the COVID-19 crisis, over-tourism, climate change, resilience; and new approaches toward the visitor management and low-impact experience design, including regenerative and transformative tourism, destination stewardship and pro-environmental behaviour. Part I introduces the concept of nature-based tourism and the emerging challenges in the field. Part II explores the key components in the management and planning of nature-based tourism development. In Part III the handbook focuses on visitor experience design and management and Part IV highlights the impacts of nature-based tourism. Part V examines the future of nature-based tourism and possible solutions to mitigate associated challenges in the field.

The handbook offers a valuable contribution with a systematic outlook of the phenomenon of nature-based tourism and critical perspectives on key concepts, policy, and practice. It shares current knowledge, innovative tools, and sustainable solutions with substantial evidence and societal impact. The book will appeal to students, researchers, and professionals in the fields of tourism, human geography, leisure studies, business studies, and sociology.

Ante Mandić, PhD, is an Assistant Professor of Sustainable Tourism at the Department of Tourism and Economy, Faculty of Economics, Business and Tourism, University of Split, Croatia. He is also an affiliated faculty at the Human Dimensions of Natural Resources Department, Warner College of Natural Resources, Colorado State University. He serves as a manager of the SmartCulTour Living Lab and an Associate Editor of the *Journal of Ecotourism*. He is a member of the Executive Committee and the coordinator of Knowledge Development at the International Union for Conservation of Nature – World Commission for Protected Areas TAPAS.

His teaching predominantly focuses on sustainable tourism development. In his research, he pays attention to heritage destinations, particularly protected and conserved natural areas. He has published a number of academic papers in his field in respectful international journals as well as book chapters and books. Dr Mandić has extensive experience working as a researcher or educator on international projects founded by INTERREG MEDITERRANEAN (Shape-Tourism; BluTour System; Co-Evolve; DESTIMED (educator) and also the HORIZON 2020 program (SmartCulTour)).

Sandeep K. Walia, PhD, is presently working as Professor and Deputy Dean in the School of Hotel Management and Tourism at Lovely Professional University, Jalandhar, Punjab, India. He is an alumnus of Himachal Pradesh University Shimla and IGNOU, New Delhi. Dr. Sandeep K. Walia is associated as Visiting Faculty to the School of Tourism, Hospitality and Event Management, Universiti Utara Malaysia Kedah (a QS-ranked university), Malaysia, and Himalayan Institutes of Alternative Learning (an upcoming Mountain University) in Leh, Laddakh, as a Curriculum Development/Board of Studies Expert and Visiting Faculty. His areas of interest are community-based tourism, tourism marketing, and cultural tourism. He has worked as Program Director/Paper Coordinator for the subject *Aviation Industry – Ticketing and Frontier Formalities* and developed E-content under the MHRD-Sponsored UGC Project titled EPG Pathshala for the online learning at the Postgraduate Level. He has also acted as a faculty coordinator for *Peer to Peer: Facebook Global Digital Challenge Spring 2017*. Dr Sandeep K. Walia is Founder Editor-in-Chief of the *Tourism and Hospitality Research Review* (International Journal of LPU), and the Managing Editor and Editor for the journals *Tourism Dimensions* and *Tourism Spectrum* (both international refereed journals). He has authored 9 books and 15 research papers in various Scopus and ABDC Journals.

THE ROUTLEDGE HANDBOOK OF NATURE BASED TOURISM DEVELOPMENT

Edited by Ante Mandić and Sandeep K. Walia

Routledge
Taylor & Francis Group
LONDON AND NEW YORK

Cover image: © Getty Images

First published 2023
by Routledge
4 Park Square, Milton Park, Abingdon, Oxon OX14 4RN

and by Routledge
605 Third Avenue, New York, NY 10158

Routledge is an imprint of the Taylor & Francis Group, an informa business

British Library Cataloguing-in-Publication Data
A catalogue record for this book is available from the British Library

ISBN: 978-1-032-13770-4 (hbk)
ISBN: 978-1-032-13775-9 (pbk)
ISBN: 978-1-003-23074-8 (ebk)

DOI: 10.4324/9781003230748

Typeset in Bembo
by SPi Technologies India Pvt Ltd (Straive)

Dedicated to our Parents and Family

CONTENTS

FIGURES

TABLES

MAP

CONTRIBUTORS

Vedat Acar is currently an Assistant Professor at the Department of Tourism Guiding, Faculty of Tourism, Aydın Adnan Menderes University, Turkey. He started his academic career as a research assistant at the same university in 2012. Prior to joining the university, he had worked as a tourist guide in guided tours for some years. He has also been a licensed tourist guide of the Turkish Ministry of Culture and Tourism since 2009. He is now conducting some research on tourist behaviors through using the participant observation method. He is one of the Editors-in-Chief of the *Journal of Qualitative Tourist Guiding Research* (TURNAD).

Karl Agius is a Visiting Lecturer at the Institute for Tourism, Travel and Culture and the Islands and Small States Institute at the University of Malta, where he lectures on island tourism, special interest tourism, research methods and environmental issues in the hospitality industry. His research interests include protected area tourism and archipelagos. His doctoral research studied the marine ecotourism potential of central Mediterranean Islands. Recent publications include 'Mitigating seasonality patterns in an archipelago: the role of ecotourism' (*Maritime Studies*) and '"Splendid isolation": Embracing islandness in a global pandemic' (*Island Studies Journal*).

Julia N. Albrecht is a Senior Lecturer in Tourism at the Department of Tourism, University of Otago, New Zealand. Her research interests include tourism and destination management, visitor management, both with a focus on nature-based tourism. Julia's work has been published in leading journals such as the *Journal of Sustainable Tourism*, the *Annals of Tourism Research, Tourism Management*, the *Journal of Cleaner Production*, and the *International Journal of Tourism Research*. Recent projects investigated the management of New Zealand's ecosanctuaries, the implementation of sustainability in destination management, and visitor pledges as possible means to achieve desirable visitor behaviour. Julia is a co-editor of the *Journal of Outdoor Recreation and Tourism* as well as being as on the Editorial Board of the *Journal of Hospitality and Tourism Research*. Julia is the sole editor of the books *Visitor Management in Tourism Destinations* (2017) and *Managing Visitor Experiences in Nature-based Tourism* (2021).

Michal Apollo is an Assistant Professor at the Institute of Earth Sciences, University of Silesia in Katowice, Poland, a Fellow of Yale University's Global Justice Program, New Haven, USA, a Visiting Scholar at Hainan University-Arizona State University Joint International Tourism

College, China, and a Fellow at the Center for Tourism Research, Wakayama University, Japan. Michal's areas of expertise are tourism management, consumer behaviours as well as environmental and socio-economic issues. Currently, he is working on a concept of sustainable use of environmental and human resources. He is also an enthusiastic, traveller, diver, mountaineer (he has climbed Mt Fuji three times), ultra-runner, photographer, science popularizer, and an NGOs activist. He is an author or co-author of over 50 journal articles and three books on mountain tourism: *Environmental Impacts of Mountaineering: A Conceptual Framework* (2021), *Mountaineering Tourism: A Critical Perspective* (2022, Routledge), and *Mountaineering Adventure Tourism and Local Communities: Social, Environmental and Economic Interactions* (2022). Michal is a Board Member of Directors of Academics Stand Against Poverty, USA.

Gül Erkol Bayram is currently an Associate Professor in the School of Tourism and Hotel Management, Department of Tour Guiding, University of Sinop, Sinop, Turkey. Dr. Erkol Bayram has worked as an internal trainer and teacher in the arena of tourism management. She is a professional tour guide in Turkey. Her doctorate, on Tourism Management, is from Sakarya University, Turkey. Her core subjects are organizational behaviour, tourism management, tour guiding, tourism policy and planning, and women's studies. She has published many book chapters in the international arena, including with the following publishers: IGI GLOBAL, CABI Routledge, Emerald Publishing UK, and also with Apple Academic Press as an editor. She has also been invited as keynote, speaker, panel member to conferences and gatherings in a number of different countries.

Georges Bellos is a Master's Degree Holder in Geological and Archaeological Sciences, an MBA in Management, with an emphasis on economic development and tourism management, and educational leadership. He is the holder of a Teaching Diploma in Entrepreneurial Skills Training, and was awarded a Master's degree in Educational Leadership, with an emphasis on Sustainable Development, while he continues to pursue his studies in tourism education.

Ella Björn is a project manager/project planner at the Faculty of Art and Design, University of Lapland in Finland. She holds the degree of Master's of Social Sciences (Tourism Research) and currently works for the *Active Arctic* project (funded by the European Regional Development Fund) and the *SmartCulTour* project (Smart Cultural Tourism as a Driver of Sustainable Development of European Regions), an EU-funded research project in the frame of Horizon 2020. Her research interests are in service design and tourism research.

Huong T. Bui is a Professor at the College of Asia Pacific Studies, Ritsumeikan Asia Pacific University (APU), Japan. Prior to her career in academia, Huong worked in the tourism industry and was a consultant for tourism development projects in South-east Asia for the World Bank, the Japan International Cooperation Agency (JICA) and the Netherlands Development Organization (SNV). Her research expertise revolves around conflict resolution, crisis management, sustainable resource management and adaptive resilience of tourism destinations. She has published more than 50 journal articles and book chapters and edited three books on tourism in Asia: *Tourism and Development in Southeast Asia* (2020, Routledge), *Nature-based Tourism in Asia Mountain's Protected Area* (2021), and *Vietnam Tourism: Policies and Practices* (2022).

Emre Cantekinler graduated from the Department of Tourism Administration at Bogazici University. His interest in environmental subjects was shaped by a series of environmental tourism courses he took during his early university years. He worked on rural tourism projects in two classes and wanted to contribute to the reduction of the adverse effects of tourism on the environment and culture. He took part in research on tourists' cultural and environmental impact on destinations in the last year of his university. His career in the tourism industry began during his early university years. He is increasingly interested in corporate social responsibility and is becoming familiar with corporate life with the aim of combining his desire to reduce the adverse effects of tourism with corporate goals.

Hrvoje Carić is presently working as Research Associate at Institute for tourism, Zagreb, Croatia. He has done his B.A. degree in Economics and Environmental sciences (double major) from Concordia College, Mn, USA. M. Sc. in Environmental Management and Policy (IIIEE, Lund, Sweden) PhD in the field of Environmental/Ecological Economics in tourism related pollution and biodiversity. His focused research areas are Environmental Indicators, Carrying Capacity and Nature Conservation, Cost-Benefit Analysis, and Risk Assessment.

Samantha Chaperon is Principal Lecturer in Tourism and Events at the University of Greenwich, in the United Kingdom. Her research interests include tourism in islands and archipelagos, governance, tourism policy, core–periphery relations, and public–private partnerships for tourism. Samantha is the Book Reviews Editor for the *International Journal of Island Research* and a member of the Scientific Committee of the Observatory on Tourism for Island Economies.

Emine Cihangir is an Assistant Professor in the Tourism Department at Van Yuzuncu Yıl in Turkey. In addition to her Bachelor's and Master's degrees, she has a PhD in Corporate Professional Management from Marmara University in Istanbul. She has professional experience in hotel management in collaboration of the tourism projects in eastern part of Turkey. She has published numerous papers on a range of aspects of sustainable tourism, such as gendering, female entrepreneurship, city branding, collaborative branding, education for sustainable development, and sustainable tourism in national and international journals. Apart from papers, she has already edited four books in the field of tourism. She is also co-editor of the *Geographies, Planning & Tourism Studies* journal.

Kübra Cihangir-Çamur is an Associate Professor. After graduating from METU as a city and regional planner, she started her academic profession at Ankara's Gazi University in the Faculty of Architecture. She has a Master's degree from METU and a PhD from Ankara University, which focuses on urban planning, policy and legal issues, urban geography, tourism planning, policy, urban and rural conservation, site management and tourism entrepreneurship. She was the planner member of the "Cultural Assets Protection Board" on behalf of the Ministry of Culture and Tourism between 2009 and 2018. Her recent studies mainly focus on a conceptual framework of modernism-postmodernism, tourism entrepreneurship, and the legislative framework of planning. She has published in several national and international journals as well as producing plans and programmes for the protection and sustainable development of urban and rural areas and touristic destinations in Turkey. She is also co-editor of the *Geographies, Planning & Tourism Studies* journal.

Ian Convery has spent the last 25 years working on understanding societal interactions with, connections to, and perceptions of, the 'natural world'. His current interests are focused on public engagement with species reintroductions and rewilding, and he is the Principal Investigator on the HLF-funded 'Back on our Map' (BOOM) multi-species reintroduction project in South

Cumbria. He established and co-chairs the IUCN CEM Rewilding Thematic Group and has been a member of the IUCN World Commission for Protected Areas since 2016. Ian is a Fellow of the Royal Geographical Society and is a Director of the Lifescapes Project conservation charity, which owns and manages the Natural Capital Laboratory in collaboration with AECOM. He has worked in Armenia, Azerbaijan, Bolivia, Ethiopia, Georgia, Kazakhstan, Kyrgyzstan, Mozambique, and Tanzania.

Kellen Copeland is a PhD candidate, 'ABD' status, of the Environmental Science Graduate Program (ESGP) at Oregon State University. His work centers on wildlife–human relations, environmental anthropology, wildlife ecotourism, and social entrepreneurship. His dissertation focuses on the 'salmon–sea lion–human complex' in the Columbia River Basin and unpacks how tourism encounters, natural resource management, and biocultural identity shape the lives of nonhuman species.

Shelby Copeland is a Master of Tourism Management student at Colorado State University. She is also a research assistant in the Multispecies Livelihoods Lab, advised by Bastian Thomsen. Shelby is interested in human–wildlife systems, rewilding, and posthumanism, pushing passed the anthropocentric lens that saturates current sustainable development practices. Shelby has worked in wolf reintroduction, analyzing wolf legislation in the seven states reintroducing the grey wolf. Much of this research occurred in conjunction to the introduction of Proposition 114 in Colorado, a statute stipulating the reintroduction of grey wolves to the territory. She was able to hear from multiple stakeholders and community members regarding the initiative, propelling her to further investigate wildlife–human complexes. She hopes to pursue law school and finish in Animal Law, a relatively new field aiming to give animal 'personhood' status to eliminate 'culling' tactics to control wildlife populations. She is a fierce advocate for wildlife sovereignty, land back initiatives, climate change advocacy, and community aid.

Selen Dağ graduated from the Department of Tourism Administration at Bogazici University. After graduation, she started working as an executive assistant. She is interested in the environment, sustainability, and social subjects such as inequalities, discrimination, and diversity. Her interest in environmental sustainability emerged during her university life, especially after witnessing tourism's role in environmental issues. She focused on rural tourism and aimed to reveal travelers' awareness and expectations. She is eager to continue her education with Green Business Management and Sustainability in future and is planning to participate in postgraduation programmes in Canada.

Ivana Damnjanović, PhD, is an Associate Professor and Vice-Dean for International Cooperation at the Faculty of Health and Business Studies, Singidunum University, Serbia. Her academic and professional interests include: a spectrum of topics in sustainable tourism (protected areas in particular); community wellbeing; the regenerative and transformational potential of tourist experience, including storytelling, interpretation, and educational aspects; and wellness through recreation, adventure, and nature-based tourism.

She has been an active Global Sustainable Tourism Council member, a Transformational Travel Council Herald, a Planet Happiness Ambassador, IUCN WCPA TAPAS Group wider network member, a Future of Tourism Coalition signatory, and a local faculty and NGO projects' leader.

Focused on future generations of sustainable tourism professionals, she facilitates learning experiences through creating enhanced and tailor-made students' educational opportunities at home and in the international context. She enjoys working with students with diverse cultural backgrounds (teaching in Serbia and across Europe since 2005). As the faculty Vice-Dean for international cooperation, she enables students to create their international stories through exchange projects.

Ayşe Deliormanli is Lecturer in the Department of Transportation Services in Beykent University Vocational School, She is also a PhD student in the, Department of Tourism Management, Eskişehir Osmangazi University. She has publications in the areas of tourism management and alternative tourism. She has lectured on travel and transportation, civil aviation and reservations courses.

Max Duggan is a PhD student in the Human Dimensions of Natural Resources department of Colorado State University. He is a member of the Multispecies Livelihood Lab and has contributed to research focused on wildlife reintroduction, rehabilitation, coexistence, and ecotourism.

Ebru Düşmezkalender is working as Associate Professor in the Department of Tourism Management, within the Faculty of Tourism at Eskişehir Osmangazi University in Turkey. Her research areas are tourism, tourism management, alternative tourism, and organizational behavior in tourism. She has many publication in international journals and book chapters in the tourism field, along with being involved in a range of projects.

Hilde Nikoline Hambro Dybsand is a postdoctoral researcher at The Norwegian University of Life Sciences. Her research interests include nature-based tourism, wildlife tourism, participant experiences, human–wildlife interactions, and virtual tourism experiences. Nikoline recently completed her PhD project on participant experiences in wildlife watching tourism, in which she investigated which elements were most important to wildlife tourists, and how these elements could contribute to the desired outcomes and/or reduce the negative impacts of wildlife watching tourism. In her current position, she is involved in the Norwegian research project *Marine Ecosystem Accounting (MAREA) for integrated coastal planning in the Oslofjord*. She is also teaching courses on tourism, nature-based tourism, nature interpretation and methods in social sciences.

Dorthe Eide is Professor in Organization and Management at Nord University Business School, Norway. She holds a PhD from the University of Tromsø, within organization and management, studying knowing, learning, and innovation from a practice-based approach in small and medium-sized hotels. Her main research contexts are different experience sectors particularly related to nature, culture, and meals. The main research fields are within management, innovation and value co-creation, by topics such as the design, quality and consumption of experiences, networks, labs, and the co-creation of innovation and research. Increasingly, she is also investigating sustainability and innovation with regard to organizations and local communities. She manages the research group Marketing, Management and Innovation of Experiences (MMIE), teaches at all levels and has managed several research and innovation projects. She has co-edited books on *Meaning, power and media*, and *Innovation in tourism industries*, and guest-edited a special issue on 'Network and network driven innovation in tourism' for the *Scandinavian Journal of Hospitality and Tourism*, as well as having been a chair of several conference tracks. She has also published contributions in various books and journals.

Günay Erol currently works as an Associate Professor in the Tourism Faculty at Nevşehir Hacı Bektaş Veli University in Turkey. He has experience in different tourism businesses and he has various certificates relating to his competency in ticketing and reservation and online reservation. His research areas are in the areas of tourism and hotel management, tourist behaviors and mobility, organizational behavior in tourism, social media and tourism, and tourism and gastronomy. He has published many contributions in international journals and he has published many book chapters and projects in the general field of tourism. Dr. Erol has also carried out studies into the use of local cuisine in tourism. He continues to teach in areas such as alternative tourism, tourism marketing, food and beverage management, tourism and travel, and vocational foreign language at the Ürgüp SET Vocational School and Tourism Faculty at Nevşehir Hacı Bektaş Veli University. Additionally, he also instructs postgraduate sectoral seminars and field studies courses in local cuisine.

Bengi Ertuna is a Professor in the Department of Tourism Administration at Bogazici University. Her research interests include corporate social responsibility, business engagement with development, and sustainable tourism. She has also joined various tourism development projects, particularly those focusing on assessing and monitoring impacts on community wellbeing.

David A. Fennell researches mainly in the areas of ecotourism, tourism ethics, and moral issues tied to the use of animals in the tourism industry, and sustainability. A major thrust of his research involves the use of theory from other disciplines (such as biology and philosophy) to gain traction on many of tourism's most persistent issues and problems. Fennell is the founding Editor-in-Chief of the *Journal of Ecotourism*.

Samuel R. Fennell is a PhD Student at Colorado State University in the Human Dimensions of Natural Resources department. He is currently a member of the Multispecies Livelihoods Lab, which is led by his primary advisor Dr. Bastian Thomsen. His current research interests revolve around the issues of tourism, animal ethics, and animal rights and welfare with an anthropological and philosophical lens.

Sonia Ferrari is an Associate Professor of Tourism Marketing and Place Marketing at the Department of Business and Law Science, University of Calabria, Italy. At the university, where she has been a researcher since 1993, she has taught on the subjects of event marketing, service management, and tourism management. Her main research interests are: tourism and place marketing, roots tourism, spa and wellness tourism, tourism in protected areas and natural parks, event marketing, service marketing and management, and sustainable tourism. She is a Member of the World Commission on Protected Areas WCPA of IUCN (International Union for Conservation of Nature) since 2016.

Peter Fredman is a Professor of Nature-based Tourism at the Mid-Sweden University and the Norwegian University of Life Sciences. His research spans over several aspects of outdoor recreation participation, visitor management, and sustainability in the nature-based tourism domain. Much of his research focuses on the Nordic context, where tourism and recreation in nature is influenced by open access and the 'friluftsliv' (outdoor life) tradition. Peter has been the leader of several large projects, including the Norwegian BIOTOUR project, Tourism in the new bioeconomy: From place-based resources to value added experiences, and the Swedish Mistra Sport & Outdoors research and co-creation program for sustainable sports and outdoor recreation.

Francesc Fusté-Forné is a Professor and Researcher at the Department of Business, University of Girona. He holds a PhD in Tourism (University of Girona) and a PhD in Communication (Ramon Llull University) and his research is focused on food and rural marketing and tourism. In particular, he has studied the connections between authenticity, food heritages and identities, landscapes and landscapers, regional development, rural activities, street food and tourist experiences. He also conducts applied research on the role of gastronomy in relation to mass media and as a driver of social changes.

Karen Guillén has been the Academic Coordinator of the Tourism Administration Program at Universidad San Ignacio de Loyola since 2019. She has a Master's degree in International Tourism from the University of Lleida, Catalonia, specializing Tourism Promotion and Marketing. Karen also has a specialization in Data Sciences and Big Data from the Massachusetts Institute of Technology (MIT), certification in Business Intelligence from the Centrum PUCP, specialization in Cloud Computing by Google and also an accreditation about Post-Crisis Hospitality Management from the University of South Florida (USF). She has experience in both the public and private sectors of tourism management, and also in hospitality management, and had been working in many international brands, such as IHG Intercontinental Hotel Groups and Marriott. She is also an undergraduate thesis advisor in the Tourism Administration Career and is dedicated to the development of research in the tourism sector, vinculated Tourism with Data Science in some methodological models.

Vegard Gundersen is a Senior Researcher at the Norwegian Institute of Nature Research. His research often concerns the interface between human use and ecology, including both qualitative and quantitative research methods within fields such as recreational monitoring, human disturbance and impact, environmental perception, relational values, vulnerable analyses' carrying capacity, visitor management and multiple-use forestry. Studies of local participation in planning and management processes have focused particular interest on the field of outdoor recreation in urban areas and the human dimension of mountain areas.

Marco Haid is an Assistant Professor at the Division for Management in Health and Sports Tourism at the University UMIT Tirol in Hall. His research interests lie, in particular, in destination management, nature-based tourism, and sustainable product development in tourism. His work has been published in the *Journal of Sustainable Tourism*, *Journal of Cleaner Production*, and other high-ranking journals. He serves as a reviewer for various tourism journals and is a lecturer in educational initiatives for tourism practitioners.

Viana Hassan is actively involved in academic and research work in tourism and cultural management in Lebanon and Malta. She received her PhD in Cultural Heritage from Saint Joseph University, Beirut. Her thesis was entitled 'Medical Tourism in Lebanon'. She has more than 17 years of tourism experience in a range of related travel and tourism fields. She is currently a visiting lecturer at the University of Malta, Research Student Supervisor at ITS, a Lecturer at a Lebanese University and a Senior Business Lecturer in the Scholars School Systems UK.

Burcin Kalabay Hatipoglu, PhD, is a research associate at the School of Business, Canberra, and Industrial Relations Research Group (IRRG), and an adjunct lecturer at the School of Business, the University of New South Wales, Sydney (UNSW). She has worked on regional planning and development for 15 years, collaborating with small businesses and communities in less advantaged areas. Her research focuses on the challenges facing humanity: gender equality, migration, sustainable consumption, and production.

Asif Hussain, PhD, is the founding director of Sustainability and Resilience Institute | Tourism Advisory New Zealand. Hussain is a dedicated entrepreneur, social worker, and philanthropist, and facilitated numerous community development projects at an international level. His research focus is on regenerative tourism, resilience, indigenous studies, infrastructure development, community development, and sustainability.

Glen Hvenegaard is a Professor of Environmental Science at the University of Alberta's Augustana Campus in Camrose, Alberta. Glen's research examines human interactions with nature, with a focus on interpretation, parks, birds, ecotourism, and rural sustainability. Glen is a member of the World Commission on Protected Areas (and the Tourism and Protected Areas Working Group), a Killam Laureate, and a fellow with LEAD International (Leadership for Environment and Development). He is co-editor of *Parks and Protected Areas: Mobilizing Knowledge for Effective Decision-Making* (2021), *Tourism and Visitor Management in Protected Areas: Guidelines for Sustainability* (2018; 5 translations) and *Taking the Next Steps: Sustainability Planning and Collaboration in Rural Canada* (2016).

Nicolaia Iaffaldano is a Researcher of Economics and Business Management at the Ionic Department in the Legal and Economic System of Mediterranean, University of Bari, Italy. She has taught marketing, economics and management of business networks in the course of studies in Economics and Business Administration, and Cultural Heritage Sciences for Tourism and Environment at the University of Bari. Her main research interests are management and marketing of tourist destinations, service management and marketing, tourism in protected areas and natural parks, sustainable tourism, and the relationship between business strategy and local development, with particular attention to sustainability.

Sara V. Iversen is a postdoctoral researcher at Aarhus University in Denmark. She started her research journey in Conservation Biology from the University of Cumbria, UK and continued into a PhD in Landscape Ecology (Lancaster University, UK). Upon completion, Sara took up a position with Leeds University as a field assistant where she got the opportunity to further develop an interest for landscape approaches to conservation and a postdoctoral position at Newcastle University, focusing on the social element of natural resource management. Her diverse research activities include fieldwork in the Carpathian mountains for a brown bear project, genetic analysis and fieldwork of Elasmobranchs in Gambia and taxonomy analysis of butterflies using genetic applications in Scotland. She is currently in a postdoctoral position within the Center for Sustainable Landscapes under Global Change at Aarhus University, where she investigates social systems and conflict mitigation in land use, natural resource management and protected areas.

Frank S. Jensen is Professor with Special Responsibilities for Outdoor Recreation at the University of Copenhagen. He has a broad knowledge of outdoor recreation research, and his research interest includes values, attitudes, and behavior in relation to nature and wildlife management. Frank has an extensive experience in conducting projects in Denmark regarding monitoring the use of nature areas for recreation, and assessment of the general populations' preferences for different forest environments. In recent years, his research has contributed substantially with background information for the development of the first Danish national outdoor recreation policy. Frank has been active in Scandinavian projects (e.g. Visitor Monitoring Methods in the Nordic and Baltic Countries), and European co-operations like COST-actions (e.g. E33-Forrec) INTEREG-projects (PROGRESS) and EU-funded research projects (e.g. Multifor, Eforwood and DIABOLO).

Thomas E. Jones is Professor of Environment & Development Studies at Ritsumeikan Asia Pacific University (APU) in Kyushu, Japan. His research interests include nature-based tourism, environmental policy, and Protected Area management. Originally from the UK, he first came to Japan in the last century as an international student. After completing his PhD at the University of Tokyo, he worked as a consultant for the municipal government carrying out visitor surveys in the Japanese Alps. Jones has also been lucky enough to join a survey of Mount Fuji climbers every summer from 2008 to 2017. Recent projects have been related to wildlife tourism (the Snow Monkeys & Rabbit Island); heritage planning (core and buffer zones in Hoi An Ancient Town); and sustainability (student perceptions of the SDGs). Together with these two co-authors, the same trio recently co-edited a book entitled *Nature-Based Tourism in Asia's Mountainous Protected Areas: A Trans-regional Review of Peaks and Parks* (2021).

Sıla Karacaoğlu is Associate Professor in the Bilecik Şeyh Edebali University Faculty of Applied Sciences, Department of Tourism Guidance. She received her PhD degree from the Mersin University in tourism management in 2017. In her doctoral dissertation, she investigates "Community Perceptions, Attitudes and Support for Community Based Tourism: The Case of Misi Village". Her research interests are tourism marketing, sustainable tourism, cultural heritage, and tourist guiding.

Jeetesh Kumar is Senior Lecturer in the School of Hospitality, Tourism and Events, Taylor's University, Malaysia. His doctorate in Hospitality and Tourism is from the Taylor's University, with research into the economic impacts of business events in Malaysia. He has two postgraduate degrees: a Professional Master's in Hospitality Management and International Tourism from the University of Toulouse, France and the other in Business Administration (MBA – Marketing) from Hamdard University, Pakistan. His research areas include economic impacts, economic modelling, MICE, medical tourism, and behavioral studies. He has worked on consultancy and research projects at the national level and authored more than 35 publications, including research articles and book chapters. Jeetesh is an active member of several national and international associations, conference and journals.

Suneel Kumar is an Associate Professor at the Department of Commerce, Shaheed Bhagat Singh College, University of Delhi, India. His area of specialization is in the area of tourism and marketing. He has published more than two dozen research papers in international and national reputed journals, including *Journal of Organizational Change Management, Open Geosciences, Empirical Economic Letters*. He has a teaching experience of more than a decade and specializes in marketing, accounting and financial management.

Maria Lexhagen is a Professor of Tourism Studies at Mid-Sweden University and the research center ETOUR. She is engaged in teaching undergraduate and graduate students in tourism as well as the supervision of PhD students. Her work focuses on tourism marketing, consumer behavior, tourism management, and destination development, with a specific interest in digitalization aspects. Recent and ongoing research covers areas such as fans and destinations in popular culture tourism, business intelligence in tourism, field- and lab experimental research on climate-friendly food choice and climate-smart solutions for tourism businesses and destinations, destination branding, and customer perceived value. She has acted as a project leader in many research projects related to digitalization and destination development and her work is published in international scientific journals as well as books.

Hong Li is currently a postdoctoral research fellow affiliated with the Department of Informatics at the University of Oslo. Hong holds the degree of Doctor of Arts with distinction in the field of Human–Computer Interaction from the University of Lapland. She has developed a multicultural and multidisciplinary background by pursuing her education in China, Japan, the UK, Finland, and Norway. Her research interests are inherently wide-ranging and expanding, including health care, e-health, remote care, digital infrastructures, information systems, user studies, user experience research, participatory design, and service design. Hong is the lead author for this book chapter, which was written during her time at the University of Lapland, where she worked as a senior researcher in the *Co-Stars Service Design Research Group* and was the project coordinator for the *SmartCulTour* (Smart Cultural Tourism as a Driver of Sustainable Development of European Regions) project, an EU-funded research project in the frame of Horizon 2020.

Frank Lindberg is Professor of Marketing at Nord University Business School, Norway. He earned his PhD in 2001 at Copenhagen Business School, Denmark. Since then, he has worked as Associate Professor at University of Nordland and University of Gothenburg, and Vice-Dean and Dean at Bodø Graduate School of Business. Lindberg has been Visiting Scholar at Copenhagen Business School, University of California, Berkeley, USA and at the University of Southern Denmark. His research covers areas such as the dynamics of branding, tourism markets and consumption, and recently with a focus on tensions and challenges of tourist consumption communities – especially related to Consumer Culture Theory. Lindberg has published articles in various journals, such as the *Journal of Consumer Behaviour, Consumption Markets & Culture, Marketing Theory*, the *Journal of Marketing Management, Annals of Tourism Research, Current Issues in Tourism*, the *Scandinavian Journal of Hospitality and Tourism*, the *Journal of Gastronomy and Tourism*, the *Journal of Hospitality Marketing & Management* and the *Journal of Sustainable Tourism.*

Kristin Løseth works as an Associate Professor at the Western Norway University of Applied Sciences (HVL). In her PhD from Aalborg University, she examined knowledge development and innovation in the context of nature-based tourism businesses. Her research interests include product development and cooperation in nature-based tourism, knowledge development at the intersection between leisure and industry, and how to integrate issues of sustainability into tourism educations. She is program coordinator and teaches at a bachelor program in nature-based tourism at HVL. Løseth has recently been part of the Norwegian research project BIOTOUR: *From place-based natural resources to value added experiences: Tourism in the new bioeconomy*, supported by the Research Council of Norway. The aim of BIOTOUR was to research and disseminate key conditions for the future development of nature-based tourism in Norway and thereby contribute to business innovation, community resilience, and sustainable use of resources.

Goda Lukoseviciute is a PhD candidate in Tourism Management at the Faculty of Economics, University of Algarve, Portugal. She has a MSc in Environmental Engineering. As a researcher, she is a member of the Research Centre for Tourism, Sustainability and Well-being (CinTurs).

Her research interest includes tourism economics, sustainable tourism management, and nature-based recreation.

Christine Lundberg is a Professor of Tourism Management at the University of Stavanger, Norway. Her research primarily focusses on tourism and events in the wake of popular culture expressions such as films, TV shows, literature, and music and she co-founded the international research network POPCULTOUR 2010.

Monika Lüthje works as a Senior Lecturer in Tourism Research at the Faculty of Social Sciences at the University of Lapland, Finland. She holds the degree of Doctor of Social Sciences (Tourism Research). Her research has focused on host and guest experiences of tourism from a cultural studies perspective. Her current research interests are cultural sensitivity and participatory and practice-based approaches as ways to develop Arctic tourism more respectful toward Indigenous and other local cultures. She was the Responsible Leader/PI of the ARCTISEN (Culturally Sensitive Tourism in the Arctic) project and works currently as a Post-Doctoral Researcher in several other projects.

Ante Mandić is currently working as an Assistant Professor in the field of sustainable tourism at the Department of Tourism and Economy, Faculty of Economics, Business and Tourism, University of Split, Croatia. His teaching predominantly focuses on sustainable tourism development. In his research, he pays attention to heritage destinations, particularly protected natural areas. He has published a number of academic papers in his field in respectful international journals, including, among others, *Tourism Management Perspectives*, the *Journal of Outdoor Recreation and Tourism*, the *Journal of Ecotourism* (Taylor & Francis), *Environment Systems and Decisions*, *Environment Development and Sustainability*, the *Journal of Hospitality*, and *Tourism Technology* as well as book chapters. He recently published a book *Mediterranean Protected Areas: Challenges and Solutions*, and is the co-editor of *The Routledge Handbook of Nature Based Tourism Development* (Routledge), and "*Cultural Sustainable Tourism*,. Dr. Mandić has extensive experience working as a researcher or educator on international projects founded by INTERREG MEDITERRANEAN (ShapeTourism; BluTour System; Co-Evolve; DESTIMED (educator)) and the HORIZON 2020 programme (SmartCulTour). He serves as a manager of SmartCulTour Living Lab and an Associate Editor of the *Journal of Ecotourism* He is a member of the Executive Committee and coordinator of Knowledge Development at the IUCN WCPA TAPAS.

John T. Mgonja is a Senior Lecturer in Tourism and Recreation at the Department of Tourism and Recreation, College of Forestry, Wildlife and Tourism, Sokoine University of Agriculture, Morogoro, Tanzania. His main research interests span from nature-based tourism and community-based tourism to food tourism.

Satu Miettinen is a Dean and a Professor of Service Design at the Faculty of Art and Design, University of Lapland in Finland. She has also worked as a Professor of Applied Art and Design during the years 2011–2016. She is also leading the service design research group "Co-Stars" at the University of Lapland. She has published widely on service design and arts-based research. She is also an experience research lead (Principal Investigator) in several international research consortiums.

Zovko Mira, PhD, was employed at the Croatian Institute of Public Health from 1997 to 2006, and at the Croatian Agency for Environmental and Nature Protection and the Ministry of Economy and Sustainable Development from 2006 to 2021. She was a member of the European Information and Observation Network (EIONET), editor and writer of the State of the Environment Report in the Republic of Croatia, an associate in drafting the European State of Environment Report (SOER; EEA) and a member of Steering Committee on the Future of Global Environment Outlook (GEO; UNEP). She is the author of 14 scientific papers, and a member of several associations and project teams in the field of biomonitoring, environmental and nature protection, sustainability of tourism and environmental economics. She is a member of expert bodies that evaluate pre-certification documentation and conducts workshops and trainings related to the implementation of environmental management systems (EU Ecolabel, ISO 14001:2015, ISO/TS 13811:2015, EMAS), the circular economy and waste management. Today, she is an employee of the Ministry of Tourism and Sport where she actively works on raising the level of tourism sustainability.

Priyakrushna Mohanty (PK) is an Assistant Professor at the Department of Tourism, School of Professional Studies, Garden City University, Bengaluru. He is a former UGC Senior Research Fellow at the Department of Tourism Studies, Pondicherry University, India. PK is also an awardee of the prestigious Travel Corporation (India) Gold Medal for his outstanding performance in Master's Degree in Tourism Studies from Pondicherry University, India. He also holds a Master's Degree in Finance along with three PG Diploma Degrees in Rural Development, Research Methodology, and Teaching Skills. PK served the Indian Railway Catering and Tourism Corporation Ltd. for two years, following which he was recruited as a Guest Faculty in the Department of Tourism Studies, Pondicherry University. He has published more than 25 papers in both international and national journals and edited books. He has presented more than 25 papers in both international and national conferences and has been invited by a number of national and international institutes as guest speaker. Currently, he is serving as the editor for five book projects with a range of academic publishers. PK is passionate about a range of academic areas: research methodology, tourism sustainability, sustainable livelihood, events tourism, and e-tourism, along with gender issues in tourism development.

G. Nagarjuna is an Assistant Professor in the Department of Tourism Management, School of Business and Management, Bannerghatta Campus, Christ (Deemed to be University). Before starting his academic career, he worked in Holiday Bliss as an executive in tour operations. Later, he joined Christ University as a Research Assistant for the Major Research Project on "An Evaluation of Eco and Sustainable Tourism Practices of Selected Resorts in Karnataka". He holds a doctoral degree in the sustainable tourism practices of Karnataka. He started his full-time academic career by joining the Department of Tourism Studies in Indian Academy Degree College, and later served as an assistant professor at Mount Carmel College, Bangalore. He has presented research and conceptual papers in seminars and conferences at the national and international levels. He has published articles in acclaimed journals and also participated in research workshops. He was one of the organizing committee members for a one-day national seminar on 'Tourism and Community Development'. Apart from his research and academic interest, he is also interested in theatrical performance.

David Newsome is Emeritus Associate Professor in the Environmental and Conservation Sciences Group, Murdoch University, Perth, Western Australia. Research interests focus on nature-based tourism with a particular emphasis on the environmental impacts of recreation and tourism, the sustainability of tourism in national parks and nature reserves, the evaluation of the quality of ecotourism operations, geological tourism, and wildlife tourism. David is a member of the IUCN World Commission on Protected Areas and has a particular interest in undertaking projects that will aid in the development of sustainable tourism in the Asian region. His publications include more than 150 journal articles, book chapters and reports and has also the books *Natural Area Tourism: Ecology, Impacts and Management* and *Wildlife Tourism*. David is also co-editor of *Geotourism, Geotourism: the Tourism of Geology and Landscape* and *Global Geotourism Perspectives*. A series of books that have helped set the scene for the emergence of geotourism as a distinct discipline within the area of natural area tourism.

Agnes M. Nowaczek is a Professor at the School of Hospitality, Tourism and Sport at Niagara College, Canada. Outside of academia, Dr. Nowaczek is an Ecotourism Consultant and TICO Certified Travel Counsellor, aiming to create deeper connections between travelers and nature. She has completed projects in Poland, Peru, Costa Rica, and North America, all centered around sustainable tourism and ecotourism. Her professional and academic interests focus on visitor behavior and management, experiential education and transformational visitor experiences, policy planning, program design, and evaluation. Dr. Nowaczek is currently pursuing several projects related to her earlier work on experiential education, as well as new projects examining the authenticity of ecotourism activities and building regional identity and resilience. She is passionate about incorporating holistic worldviews, such as Indigenous perspectives, into education and research. To support her discipline, Dr. Nowaczek serves as the Associate Editor for the *Journal of Ecotourism*, and as the Co-Chair of the North American Ecotourism Network (NAEN), which is the NA branch of the Global Ecotourism Network (GEN).

Burcu Ormanlıoğlu is a student in the Department of Tourism Administration at Bogazici University.

Thomas Panagopoulos has an MSc in Renewable Natural Resources and a PhD in Forestry and Natural Environment. He has been the Department Head of Landscape Architecture at the University of Algarve, Portugal and has acted as Principal Investigator in a total of approved funding of over 8 million euros. This is a result of his research strategy that crosses many disciplinary boundaries to create a holistic transdisciplinary approach to science, and his multicultural background in fostering research at the international level.

Luís Nobre Perreira is Professor of Applied Quantitative Methods at the School of Management, Hospitality and Tourism of the University of Algarve. He is Vice-President of the Research Centre for Tourism, Sustainability and Well-being. He is also active in conducting funded research projects in the field of tourism. Professor Luis Pereira holds a PhD degree in Quantitative Methods Applied to Economics and Management. His research interests include tourism demand modelling and forecasting, decision support systems for tourism, sustainable tourism, revenue management, and dynamic pricing.

Priyan Perera is a Professor in Forestry and Environmental Science at the Department of Forestry and Environmental Science, University of Sri Jayewardenepura, Sri Lanka, with over 15 years of experience in research, teaching, extension, and outreach in the fields of environmental management and forestry. Priyan earned his PhD (Forestry) and his Master's degrees in Forestry and Environmental Management from the Louisiana State University, Baton Rouge LA, USA.

Priyan is interested in interdisciplinary research related to environmental management and sustainability. His research interests are mainly in the interface between natural sciences and social sciences and related to sustainable tourism, wildlife tourism in protected areas, recreation ecology, human dimensions in wildlife management, and environmental sustainability. He is also an Honorary Research Fellow in Environmental and Conservation Sciences, Murdoch University, Australia.

Neethiahnanthan Ari Ragavan has more than 27 years of academic experience and holds a Doctorate degree from France. He is highly involved in hospitality and tourism education in Malaysia and the region. In November 2013, he initiated the setting up of the ASEAN Tourism Research Network, now aptly known as ASEAN Tourism Research Association (ATRA), a regional cooperation among universities in ASEAN countries for greater research and education collaboration on issues related to hospitality and tourism. He is currently the President of the ASEAN Tourism Research Association (ATRA) and has been active in various international boards such as the Pacific Asia Travel Association (PATA). His areas of research include higher education models and system, hotel and tourism development and innovation; and food and culture. Presently, he is the Executive Dean, Faculty of Social Sciences and Leisure Management overseeing five schools and one institute. One of these schools is currently ranked Top 20 in the world by QS World University Rankings by subject, which is the School of Hospitality, Tourism & Events. It is internationally benchmarked by the United Nations World Tourism Organization (UNWTO) TedQual certification.

Pablo Romo is a Professor of Sustainability for the College of Business Studies at Universidad San Ignacio de Loyola since 2017. He has a Master's degree in Ecotourism from the National Agrarian University La Molina and one in Social Investment from Pacifico University. He graduated in Business Management from Universidad San Ignacio de Loyola. He has experience in public and private organizations in different sectors like education, early childhood, conservation, and ecotourism. He is also the founder of Ikam Expeditions, an ecotourism operator in the Peruvian Amazon.

Geetanjali Sageena works as an Assistant Professor of Environmental Studies at Keshav Mahavidyalaya University of Delhi (India). She received her doctorate degree from Evolutionary Biology Laboratory from Zoology Department, University of Delhi (India). Dr. Sageena has a keen interest in environment-related issues with simultaneous responsible teaching. She has published her research findings in various international and national journals of academic repute. She has given several invited/plenary talks at conferences. She has received some recognition for her research career. Her research interests are focused on sustainable development, environmental studies, toxicity, behaviors and other related aspects.

Klas Sandell is a Professor Emeritus in Human Geography at Karlstad University. With a start in human ecology, views of nature, and development issues, his later research spans over several aspects of outdoor recreation, nature-based tourism and outdoor teaching with a special interest in environmental education and public access. Particular focus has been on the historical aspects of outdoor recreation and the right of public access in Sweden, including landscape relations and the Nordic tradition of "Friluftsliv" (lit. free-air-life) in the light of the need for a more sustainable development. Education material in these areas have also been an important theme, involving history, current trends and education, but also concrete outdoor teaching practice.

Carolin Seiferth is an MSc candidate in Sustainable Destination Development at Uppsala University, Sweden, and Education for Sustainable Development at Gothenburg University, Sweden. She is passionate about sustainable tourism, particularly about creating an authentic traveler experience and building resilience in nature-based destinations. At this early point in her academic career, Seiferth has already reached an understanding of industry experience, including her six-month internship with Cayuga Collection in Monteverde, Costa Rica. As an alumna of Niagara College, Canada, she has served as co-lead of an international field study to Costa Rica and a guest speaker on the topic of organizational resilience. Since 2021, Seiferth has been working as a research assistant at Stockholm Resilience Centre, Sweden. Her interdisciplinary research focuses on sustainability transformations, senses of place and stewardship practices, as well as nature-based tourism destinations. Currently, she is part of 'The Stockholm Hub on Environment, Climate and Security' which explores how worldwide biodiversity loss might lead to conflict and crises. She is an active member of the German Tourism Research Group.

Serçin Dilem Şenkal is a student in the Department of Tourism Administration at Bogazici University. Her main academic interests are social, economic, and environmental inequalities, sustainability, Buen Vivir, and human wellbeing. In addition to departmental classes on tourism, she also attended several courses on sustainability, sustainable development, social ecology, and designing a multidimensional poverty index. She is eager to search for a sustainable version of tourism and other businesses and transform them.

Mehmet Şeremet has a PhD from Plymouth University on GIS education in the UK and Turkey. His current post is that of Reader in Human Geography at Van Yuzuncu Yil University in Turkey. He has published several papers in international journals, including *JGHE*, *Geography*, *IRGEE*, and *Geojournal* and chapters in collections from international publishers such as Springer, Taylor & Francis, and Emerald. He leads sustainable tourism-oriented projects in different parts of Turkey, including the Black Sea Region, and the Aegean and West Mediterranean Regions. He is also the co-editor of the *Turkish Geographical Review* and *Geographies, Planning & Tourism Studies* journals. His recent research includes critical subjects in both human geography and geographical education, such as GIS education, environmental education, tourism policy and planning, gender and sustainability, feminist geography, and critical GIS.

Shekhar is a full-time Research Scholar in the Faculty of Management Studies, University of Delhi, India. His area is specialization is strategic management and tourism marketing. He has published papers in several reputed journals, including the *Journal of Organizational Change Management, Open Geosciences*, the *Journal of Family Business Management*. He has also served as a visiting faculty at several colleges in University of Delhi. He has more than three years of teaching experience and specializes in organization behavior and strategic management.

Valerie A. Sheppard researches in the areas of tourism ethics, resilience, and sustainability. She has a special interest in the ethical considerations related to animals used in the tourism experience. She is a field notes editor for the *Journal of Ecotourism*.

Tuija Sievänen is a senior scientist who worked at Natural Resources Institute Finland until her retirement in 2020. Her research interests are in outdoor recreation and nature-based tourism studies. Her major achievements include the inventory system for monitoring outdoor recreation demand and the development of outdoor recreation statistics for Finland.

N. Sneha is an Assistant Professor Department of Tourism Management, School of Business & Management, Christ (Deemed to be University), Bannerghatta Main Road, Hulimavu, Bengaluru. Sneha has completed her Master's Degree in Tourism from Christ University, Bengaluru. She worked with leading corporate companies in the travel domain before starting her academic journey. She has been involved in key roles as an IQAC member, coordinator for MOOC, associate coordinator, custodian of examinations, placement coordinator, principal investigator of management-funded research project and organizer of various conferences, inter-collegiate tourism fests and workshops. She was awarded the doctorate in Tourism Management by Madurai Kamaraj University. She is a member of the Board of Studies at various institutions and also a member of the ITHC, TAAI and SIHC. With her academic and industrial experience, she has widely participated and published in various national and international seminars. She is a frequent invitee as a Resource Person for scholarly talks and panel discussions. Her research interests involve topics in the domain of social media, customer relationship management, responsible tourism, tour operations, and tourism marketing.

Marios Sotiriadis is Professor in Tourism Marketing at Ningbo University, China. Formerly, he was Professor at the Hellenic Mediterranean University, Greece; Tutor at the Hellenic Open University, Greece; Visiting Professor, University of South Africa; and Module Leader and Online Tutor, Global MBA, Queen Mary University of London. He received his PhD in Tourism Marketing from the University of Nice Sophia-Antipolis, Nice, France. He is the author of nine books, seven manuals and five study guides. He has undertaken a variety of research and consultancy projects for both public and private organizations within the tourism industry. His research and writing interests include in the areas of tourism destinations and business marketing. His articles have been published by international journals and he has also made many presentations at conferences.

Sandra Sotomayor is a researcher and teacher for the Tourism Management Department at Universidad San Ignacio de Loyola. She has a Master´s degree in Parks, Recreation and Tourism from the University of Missouri and a Bachelor's degree in Marketing from Universidad Peruana de Ciencias Aplicadas. She conducts research on different tourism-related topics, mainly focusing on the impacts of tourism in rural and nature settings. Her main research interests include sustainable tourism development, destination image and promotion, and the tourism sector workforce, topics that she examines using qualitative and quantitative approaches. She has participated in various local and international conferences, both in the United States and Latin America. She also has extensive experience in tourism development gained in the public tourism sector.

Anna Spenceley is an independent consultant who works on sustainable tourism issues. She has been analyzing the impact of COVID-19 on the tourism sector in protected areas for the EU, UNEP, UNESCO and Luc Hoffmann Institute, and has compiled a blog post of over 1500 materials on "COVID-19 and sustainable tourism" for public use. She is a signatory to both the "Future of Tourism" and "Tourism Declares a Climate Emergency". Anna is Chair of the IUCN World Commission on Protected Areas (WCPA) Tourism and Protected Areas Specialist Group (TAPAS Group), is on the Board of the Global Sustainable Tourism Council, and sits on the Independent Advisory Panel of Travalyst. She is editor of the *Handbook for sustainable tourism practitioners* from Edward Elgar, and co-author of *Private sector tourism in conservation areas in Africa* from CABI.

Abdullah Tanrisevdi is currently a Professor in Tourism Marketing, Research Methods and Special Interest Tourism at the Department of Tourism Guiding, Faculty of Tourism, Aydın Adnan Menderes University, Turkey. He visited Oklahoma State University (OSU) – Center for Hospitality and Tourism Research (CHTR) as a visiting scholar in 2011 and received his post-doc from the University of South Florida in 2015. He is the Editor-in-Chief of the *Adnan Menderes University Journal of Travel and Tourism Research.*

Bastian Thomsen is an Assistant Professor of Human Dimensions of Natural Resources at Colorado State University, a Postdoctoral Affiliate of the School of Anthropology & Museum Ethnography at the University of Oxford, and an Affiliate Faculty in the Environmental Sciences Graduate Program at Oregon State University. His work centers on wildlife–human relations, environmental anthropology, wildlife ecotourism, and social entrepreneurship.

Marco Valeri is Senior Lecturer in Organizational Behavior at the Faculty of Economics, Niccolò Cusano University, Italy. He holds a PhD in Strategic Management and Organizational Behavior, is a member of several editorial boards for international tourism journals and co-editor of several international handbooks on entrepreneurship and tourism and hospitality management.

Naomi van der Velden is a plant community ecologist with interests in spatio-temporal dynamics in natural ecosystems, including forests and alpine vegetation, and the ecology and productivity of agroecosystems. Her more recent research has included strong elements of citizen science and participatory research in building collaborative networks and investigating resilient food-production systems. Dr. van der Velden holds a PhD in Forest Ecology from the University of Leeds. She has most recently worked as a Senior Researcher in Agroecology at the Permaculture Association (Britain) and as a Senior Lecturer in Ecology and Sustainability in the National School of Forestry & Centre for Wildlife Conservation at the University of Cumbria.

Odd Inge Vistad is a Senior Research Scientist in the Human Dimension Department at the Norwegian Institute for Nature Research. His research has focused on the use and management of natural resources, especially of National Parks and other Protected Areas, but also on resource vulnerability and sustainable development. A key interest is outdoor recreation and nature-based tourism – the historical and cultural aspects, social and ecological conflicts, public health, modernization and commercialization. He has primarily worked in Norway (including Svalbard), but also in Sweden and in the Baltic states.

Marković Vukadin Izidora is employed at the Institute for Tourism in Zagreb since 2009. She is also lecturing at both the VERN University and Edward Bernays University, and she has vast experience of workshops, focus groups, seminars and educational programs. As an expert in sustainable tourism, she participated in a number of European projects concerning spatial development of tourism, as well as in a large number of masterplans of tourism development at national, regional, and local levels. Izidora has participated, or is participating, in a number of international projects, highlighting the role of Interreg Project Manager, *Destimed* +. As part of the work on CROSTO, she actively focused on data collection and the calculation of indicators. Izidora has conducted numerous research studies as part of her scientific research work, and it is important to point out that she is the leader of the scientific project *Development of Integral Indicators and Carrying Capacity Model for PAs*. She has published 40 scientific papers, dominated by works focused on tourism management in Protected Areas and the measurement of sustainable tourism indicators.

Sandra Wall-Reinius is Associate Professor of Human Geography at the Mid-Sweden University, Sweden, where she is also researcher at the European Tourism Research Institute (ETOUR). Her research interests relate to sustainability and responsibility in tourism, tourism–community relations, and nature conservation studies. Much of her research focuses on nature-based tourism and outdoor recreation in the Swedish mountains, as well as land-use conflicts.

Daniel Wolf-Watz holds a PhD in Human Geography and is an affiliated researcher at Mid-Sweden University. In his research he focuses mainly on outdoor recreation and nature-based tourism, and she has a special interest in the relationship between outdoor recreation and environmental concern.

Ziyi Yan is a graduate student in the Department of Tourism Management at the University of Ningbo, China and the University of Angers, France. She studied in the Department of Agriculture and Forestry Economics and Management and the Department of English of Sichuan Agricultural University. To date, she has published two journal papers. She has participated in various international conferences and scientific research projects in the tourism industry. Her research and writing interests include rural tourism and tourism destination marketing.

Nuray Yıldız graduated from the Department of Tourism Guiding, Aydın Adnan Menderes University,. During her education, she participated in the Erasmus+ Programme at Sannio University in Italy. Upon graduating from the university, she worked for two years in the operation and reservation departments of Puerto Travel in Kuşadası. Additionally, she had several years of experience as a Spanish and English translator in the hunting tours. She is currently a PhD student at the Department of Tourism Management. She has published a number of research papers and book chapters in the field of tourist guiding since 2021 and she is now conducting some research on tourism marketing and special interest tourism.

ACKNOWLEDGEMENTS

To acknowledge, write and express gratitude seems the toughest task to us. *The Routledge Handbook of Nature Based Tourism Development* would have remained a dream of our life without the support of all these people to whom we would like to express and extend my gratitude and appreciation in this page.

We treat this section as to place our sincere gratitude to Faye Leerink, Commissioning Editor, for her consistent help and guidance to complete this handbook. We are also thankful to Prachi Priyanka, Editorial Assistant of Taylor & Francis, for her kind support. We would also like to express our word of appreciation for all other members of Taylor & Francis' editorial, production and marketing teams for the successful execution of the *Handbook*.

We are thankful to all the contributors who have contributed the masterpiece of their authentic, unmatched research to this handbook. We are highly thankful for their outright support throughout the book project.

We do not have the words to express our debt of gratitude to our parents, family and friends for having faith in us and for generating in us a persistent interest for pursuing academic research.

We have learnt that 'academic research' could not be accomplished alone. Thanks to all of you who made this handbook come true.

Ante Mandić & Sandeep K. Walia

INTRODUCTION AND ORGANIZATION OF THE BOOK

Ante Mandić and Sandeep K. Walia

Nature-based tourism (NBT) is most often defined as a type of tourism that relies on experiences directly related to natural attractions, often involving activities in protected natural and wilderness areas. It is also discussed in terms of enjoying wilderness or relatively undisturbed natural areas such as lakes, islands, mountains, and valleys and engaging in outdoor activities like hiking, skiing, trekking, and swimming (Fredman et al., 2012; Fredman & Tyrväinen, 2010). Although specific progress toward understanding what NBT entails has been made, scholars have still not reached a consensus on its definition. Instead, the concept is often used as a synonym for, among others, ecotourism, adventure tourism, and wilderness tourism. Within the Millennium Ecosystem Assessment (2005), NBT has been acknowledged as a cultural service through the lens of ecosystem services, the non-material benefits people obtain from the ecosystem through spiritual enrichment, cognitive development, reflection, recreation, and aesthetic experiences.

The fundamental paradigm underlying the development of contemporary NBT is sustainable development (SD). However, considering that it often involves activities in the wilderness, protected and conserved areas, in reality, well-planned and organised NBT is most often exercised as, and aligned with, ecotourism which could be seen as both tourism development conception aligned with SD principles and nature-dependent niche tourism forms. This means that the visitor activities and experiences dominantly shaped within natural environments centre around conservation, sustainability, responsibility, environmental education, wellbeing, ethics, and benefits for local communities as critical tenets.

From the visitor experience design perspective, the fundamental challenges in studying NBT relate to revealing the consumers and the producers within this complex system; differentiating regular tourists from NB tourists and NBT-specific activities from services offered in the destination; understanding visitors' needs and motives; creating and communicating value through transformative and memorable visitor experiences; and ultimately capturing the value from customers through satisfaction and commitments to destination. From the destination management perspective, the critical challenges are related to mobilising destination stewardship, promoting sustainable tourism as a driver of nature conservation, and fostering the resilience of often tourism-dependent communities (Mandić & McCool, 2022).

NBT is currently the fastest-growing tourism sector globally and the most significant tourism segment for many destinations (WTTC, 2019; World Bank, 2021). The demand for

DOI: 10.4324/9781003230748-1

transformative, memorable and authentic nature-based experiences is high and will probably continue to grow in the years to come. Many NB destinations faced increasing visitation amid the current pandemic, since activities in natural settings are now recognised as beneficial to human health and wellbeing (Spenceley et al., 2021; White et al., 2019). Nature is one of the main reasons to go on holiday for many European tourists (World Tourism Organization, 2018), while sustainable NBT has significant potential to generate funds for conservation and shape people's attitudes toward the environment (Spenceley, 2021).

The reports on the rapid growth of NBT add significant weight to the economic case for biodiversity conservation but seem to contradict widespread concerns that people are becoming increasingly isolated from nature. As more and more people live in urban areas, getting in touch with nature provides ultimately different holiday experiences. Immersion in nature indeed provides unique experiences that differ from daily life. Simultaneously, such activities exert many pressures on the very resources on which they rely, thus implying a need for careful planning and management (Mandić & Petrić, 2021). Understanding the recent developments calls for more research focused on the challenges NBT faces and a holistic, cross-disciplinary approach to broad geographical representation across the topic. The knowledge basis for NBT is still quite fragmented, lacks a system approach, which would involve studies examining both the demand and supply aspects of NBT, destination and tourism business perspectives through the lens of triple-bottom-line impact assessments, stakeholder participation, changing demand, and recent challenges such as the COVID-19 crisis, over-tourism, climate change, regenerative and transformative tourism, and resilience.

These gaps and our passion for NBT were the dominant arguments for delivering this Handbook. Our overall aim was to extend recent writing on NBT development to deliver a comprehensive and systematic outlook of the phenomenon and, in doing so, to comment critically on vital concepts within both policy and practice. While preparing the manuscript, we encouraged collaboration across and beyond disciplines, contributions from all over the world that demonstrate the engagement of academics, field experts, and communities dealing with various NBT planning, development, and management challenges. The resulting Handbook, featuring contributions of leading academicians from all continents, is helpful to both researchers and practitioners. It serves as media to share current knowledge, innovative tools, and sustainable solutions with substantial evidence and societal impact and goes well beyond a simple standard practice description.

The Handbook, comprising of 34 theoretical, empirical and practical contributions, deals with critical concepts and emerging themes (Part I), the management and planning of NBT development (Part II), visitor experiences (Part III), the impacts (Part IV), and the future of NBT (Part V).

The first part includes seven chapters. It begins with the study by Newsome and Perera (Chapter 1) reflecting on NBT before, during, and after the COVID-19 pandemic. Adapting the system approach, the authors discuss the need for a new direction in NBT, the collective responsibilities of public and private sectors, and the transition to a more sustainable path in the post-pandemic era. Next comes the chapter by Sheppard and Fennell (Chapter 2), advocating for animal justice in tourism, laying the foundation for understanding how this goal could be achieved, and assessing the opportunities to make it a reality. Ertuna et al. (Chapter 3) reflect on the role of the responsibility of travellers and NBT service providers to progress toward sustainable development in rural communities. Their discussion, focuses on visitor experience attributes, motivations and sustainability perception, providing a roadmap toward the more sustainable tourism development in rural areas in Turkey. In Chapter 4, Iversen et al. relied on photovisualisation of critical woodland scenarios in an upland landscape in the United

Kingdom to investigate visitor preferences for woodland levels and estimate the economic value of NBT and the impacts of the increase of woodland cover on visitor activities. The section further continues with the contribution of Løseth (Chapter 5), investigating the value chain, cooperation and knowledge transfer between the mountain guides in Norway. Løseth demonstrates how informal cooperation within the niche is critical for efficient knowledge transfer by demonstrating the ambivalence of the actors toward cooperation with the wider tourism industry. Hvenegaard (Chapter 6) focuses on the alignment of the objectives, strategies, and outcomes of ecotourism, wildlife festivals, and sustainability in Canada. A higher-level organisational analysis demonstrated a low level of the alignment of objectives, strategies, and outcomes due to a lack of coherent planning processes and revealed how wildlife festivals could support efforts toward improving sustainability. The final contribution in this part, by Ferrari and Iaffaldano (Chapter 7), delivers a discussion of the role of the stakeholders in the destination renovation and repositioning process in southern Italy. Particular attention is given to the discussion of how the growth of social capital could support the alignment of NBT with the principles of SD.

The second part includes eight chapters. Although the strengths of most of the protected areas lie in their accumulated knowledge of conservation biology, biodiversity interpretation, and visitor flow management, they often struggle with environmental risk assessments, spatial planning, use, and conservation monitoring. To address this gap, Carić (Chapter 8) proposed an audit framework enabling more integrated, informed, and efficient management. Next, Freedman et al. discussed various approaches for foresight in outdoor recreation and tourism in nature in Nordic countries, including Finland, Sweeden, Denmark, and Norway. The authors acknowledge the contribution of the traditional foresight techniques and call for the development of new ones to address the contemporary challenges in society and understand outdoor recreation better. Hussain and Fuste-Forne (Chapter 10) demonstrate the power of connectivity to transform an undisturbed natural ecosystem and isolated communities in the Gilgit-Baltistan region between India and Pakistan into a tourism-dependent economy. Following this, Lukoseviciute et al. (Chapter 11) focus on the role of digital technologies in trail-related recreation management. The content analysis reveals the potential of the technologies to investigate visitor satisfaction along with destination attributes, motivations, and values, providing safe trail-related recreation and monitoring environmental trail conditions. In the following chapter, Li et al. explore the methodological layering of art-based methods to engage critical stakeholders in the Utsjoki indigenous community in north Finland. The authors argue that integrating art-based and service design methods can serve as an instrument to foster stakeholder engagement and describe how to achieve a shared understanding of sustainable tourism development in this isolated community. The contribution of Hambro Dysband and Fredman (Chapter 13) deals with the role of the wildlife watching tourism guides in the provision of sustainable wildlife experiences. Using Cohen's framework as a foundation, the authors compare the musk ox safaris in Dovrefjell, Norway, to Polar bear tourism activities in Svalbard, Norway and Churchill, Canada. They reveal uncertainty and encounter as two innovative components of these wildlife experiences. Following this, the contribution of Bayram (Chapter 14) focuses on the roles that tour guides need to undertake within the NBT activities. The author reflects on, among other things, leadership, mediation, and resource management roles. Finally in this part, Sotiriadis and Yan (Chapter 15) reflect on critical factors for successful business networking in Zhejiang Province in China, providing insight and guidance for implementing and managing a collaborative networking strategy in nature-based tourism and an integrated management approach to attain desirable outcomes.

Part III contains nine chapters. Eide and Lindberg (Chapter 16) reflect on customer interactions, value co-creation, and experience design. They discuss how guided sea kayaking is not only about nature activity and sensing nature but also about connecting with residents and local culture through interactions and storytelling. In the following chapter, Sotomayor et al. (Chapter 17) investigate the transformative birdwatching experiences in the Peruvian Amazon. The contribution holds impactful managerial implications and provides the roadmap for the transformative birdwatching experience design in different settings. Jones et al. (Chapter 18) investigate the past and present Mount Fuji climbers to re-design experiences that use historical elements of pilgrimage to moderate contemporary peak-bagging trends. The analysis confirms how contemporary climbers recognise the sacred rites of their pilgrim predecessors instead of seeking to summit quickly without following Shugendo customs. The authors propose a 120 km walk from central Tokyo to the Sengen Shrine at Fuji's northern foot as a counter-strategy. Next, in Chapter 19, Nowaczek and Seiferth extend and adapt the Guests' Site-adapted Sustainability Tours Experience Model to explore the inclusive visitor experiences and equity in sustainable tourism in Costa Rica. Agius and Chaperon (Chapter 20) reflect on the co-creation of visitor experiences in small central Mediterranean islands. The authors address the intervening concepts such as motivation, experience design, visitor satisfaction, and loyalty and discuss the managerial implications. In Chapter 21, Damnjanović introduces Airsoft as a novel adventure nature-based activity. The author tends to decompose these innovative visitor experiences and investigate this unique visitor segment's motivations and preferences. In Chapter 22, Mgonja focused on the relationship between tourist demographic characteristics, wildlife species diversity preferences, and willingness to revisit northern protected areas in Tanzania. The analysis reveals, among others, the different willingness to revisit and the requirements regarding the wildlife species diversity in visited parks between males and females. In the following chapter, Karacaoglu reviews the literature on travel characteristics, typologies, and motivation of nature-based tourists, emphasising the visitor segmentation as the foundation of the contemporary NBT marketing. In the final chapter of this theme, Sneha and Nagarjuna (Chapter 24) reflect on experiential tourism and NBT trends in India from the marketing perspective and provide innovative marketing solutions for experiential tourism planning and development.

Part IV consists of six chapters. Zovko et al. (Chapter 25) focus on monitoring the sustainability of NBT development. The authors identified climate change, efficient resource utilisation, and the strengthening of the local communities as three critical contemporary themes within protected areas. Having done so, they then evaluated the usability of the established framework of indicators to monitor these phenomena. Kumar et al. (Chapter 26) employed a system approach to discuss the sustainable mountain tourism development in the foothills of the Himalaya. The authors employed total interpretative structural modelling to identify the factors affecting the sustainability of NBT development and potential adaptive strategies. The lessons learned from this analysis could be helpful in similar mountain destinations. In the following chapter, Acar et al. reflect on the behaviour of hunting tourists in the Konya Province in Turkey. The study is built on practical observations and phenomenology and reveals the complexity of visitor behaviour and the consequences of various acts and choices in destinations. Hassan and Bellos (Chapter 28) deliver a first insight into the impacts associated with the development of hiking tourism in Moukhtara, Lebanon. The authors provide the guidelines for policymakers in this tourism destination in the initial phase of the tourism area life cycle. In the following chapter, Erol et al. reflect on NBT growth and the development of accommodation facilities. The analysis centres around destination space and sustainability, providing insight into the visitor choices, behaviours, and preferences. In the final contribution to this theme, Kumar et al. (Chapter 30) deliver a synthesis of current knowledge on a phenomenon to explore the linkages between nature-based tourism and conservation.

The fifth part includes four chapters. The contribution of Spenceley (Chapter 31) synthesises the materials published over the past year on COVID-19 and NBT. The author summarises the impacts of the pandemic on the sector, and the strategies adopted by operators and destinations to adapt to the crisis, tools, and resources to help adjust the management and operations of NBT and finally highlights interventions required in light of the pandemic. The following contribution of Seremet et al. (Chapter 32) evaluated the Turkey NBT policy, planning and management. The authors provide a critical account of how Turkey incorporated NBT development into the overall tourism development agenda and how they tailored policies to design experiences to meet continuously evolving tourism demand. In the following chapter, Albrecht and Haid introduce a framework for sustainable tourism product development for high-quality NB visitor experiences. The authors proposed a Triple Diamond Design process explicitly featuring sustainability as a concern in the development phase, which could be adapted to suit protected and conserved areas. In the final chapter of the Handbook, Fennell et al. (Chapter 34) present a multi-species livelihoods ethical framework that positions non-human animals in a light that necessarily places them into a standing of equal consideration to that of humans. The authors reflect on the complexities and disadvantages amended to current tourism policy, degrowth management, and education and provide critical insights into how community-based tourism can improve upon a collective consideration of local human and non-human life.

References

Fredman, P., & Tyrväinen, L. (2010). Frontiers in nature-based tourism. *Scandinavian Journal of Hospitality and Tourism*, 10(3), 177–189. https://doi.org/10.1080/15022250.2010.502365

Fredman, P., Wall-Reinius, S., & Grundén, A. (2012). The nature of nature in nature-based tourism. *Scandinavian Journal of Hospitality and Tourism*, 12(4), 289–309. https://doi.org/10.1080/15022250.2012.752893

Mandić, Ante, & McCool, S. (2022). A critical review and assessment of the last 15 years of experience design research in a nature-based tourism context. *Journal of Ecotourism*. https://doi.org/10.1080/14724049.2022.2099877

Mandić, Ante, & Petrić, L. (2021). *Mediterranean Protected Areas in the Era of Overtourism*. Springer, Cham. https://doi.org/10.1007/978-3-030-69193-6

Millennium Ecosystem Assessment. (2005). Ecosystems and human well-being: Current state and trends, Volume 1. https://www.millenniumassessment.org/en/index.html

Spenceley, A. (2021). The future of nature-based tourism: Impacts of COVID-19 and paths to sustainability. 53. https://luchoffmanninstitute.org/future-nature-based-tourism-report/

Spenceley, A., McCool, S., Newsome, D., Báez, A., Barborak, J. R., Blye, C., Bricker, K., Cahyadi, H. S., Halpenny, E., Hvenegaard, G., King, D. M., Hvenegaard, G., Leung, Y., Mandić, A., Naidoo, R., Rüede, D., Sano, J., Sarhan, M., Santamaria, V., Beraldo Sousa, T., & Zschiegner, A. K. (2021). Tourism in protected and conserved areas amid the COVID-19 PANDEMIC. *Parks – The International Journal of Protected Areas and Conservation*, *27*(March), 103–118.

White, M. P., Alcock, I., Grellier, J., Wheeler, B. W., Hartig, T., Warber, S. L., Bone, A., Depledge, M. H., & Fleming, L. E. (2019). Spending at least 120 minutes a week in nature is associated with good health and wellbeing. *Scientific Reports*, 9(1), 1–11. https://doi.org/10.1038/s41598-019-44097-3

World Bank. (2021). Banking on protected Areas. https://www.worldbank.org/en/topic/environment/publication/banking-on-protected-areas-promoting-sustainable-protected-area-tourism-to-benefit-local-communities

World Tourism Organization. (2018). European union tourism trends. In *European Union Tourism Trends*. https://doi.org/10.18111/9789284419470

WTTC. (2019). The economic impact of global wildlife tourism travel & tourism as an economic tool for the protection of wildlife. www.wttc.Org, August. https://travesiasdigital.com/wp-content/uploads/2019/08/The-Economic-Impact-of-Global-Wildlife-Tourism-Final-19.pdf

PART I

Nature-based tourism

Concepts and emerging challenges

1

NATURE-BASED TOURISM

Before, during, and after COVID-19

David Newsome and Priyan Perera

Introduction and context

The importance of tourism as an industry has grown rapidly over the last 25 years and was predicted to reach more than 1.8 billion by 2030 (Holden, 2016; UNTWO, 2017). In a recent compilation of data, Spenceley (2021) reported that there were 1.4 billion international arrivals in 2018. Moreover, in terms of the nature-based tourism component it is estimated that protected areas receive 8 billion visits annually, highlighting the importance of the nature-based tourism sector within the overall global tourism industry. The WTTC (2019) indicates that in 2018 the wildlife tourism industry employed 8.8 million people and contributed US$29.3 billion to GDP in Africa alone. The increasing interest in wildlife tourism in the global context has previously been recognised by Balmford et al. (2009) and Curtin and Kragh (2014). There is no doubt that engaging in nature-based tourism has brought joy and provided rich memories for millions of people. Moreover, the benefits of wildlife tourism are increasingly acknowledged as a tool for the funding of conservation programmes as well as providing livelihoods for local communities that need income and services (Buckley, 2003, 2010; Newsome and Hassell, 2014).

In conjunction with the heightened attention given to nature-based tourism is a widening visitor demographic which is fostered by competitively priced ease of travel, profiling of nature attractions, increased marketing campaigns and sharing of experiences on social media (Newsome and Hughes, 2018; da Mota and Pickering, 2021). The present-day visitor to natural areas sits on a spectrum ranging from the very informed, environmentally aware, and specialist traveller through to a novice, uninformed person who might only be interested in a selfie or a close-up photograph of a wild animal. The situation can be complicated by an increasing suite of people who are seeking entertainment, adventure, sport, challenge or simply 'doing something different' in an outdoor setting (Newsome et al., 2011; Newsome, 2014; Newsome and Hughes, 2018). At the same time the mental and physical health benefits of people 'being in nature' are also being increasingly recognised (Kaplan and Kaplan, 1989; Keniger et al., 2013; Buckley et al., 2019; Coventry et al., 2021; Lackey et al., 2021). Moreover, the loss of natural landscapes driven by urbanization has increased people's need for green spaces and the demand for nature-based tourism (Frost et al., 2014; Elmahdy et al., 2017; Margaryan and Fredman, 2017; Weaver and Lawton, 2017).

DOI: 10.4324/9781003230748-3

There is a need for infrastructure such as roads and airstrips to gain access to the many protected areas in which nature-based tourism takes place. Successful tourism requires accommodation and service providers and there are commercial interests which may or may not promote conservation, environmental awareness, and appropriate visitor behaviour (Newsome et al., 2013; Newsome and Hughes, 2017). Recreation and tourism in natural areas also requires staffing, resourcing, facilities, and adequate protected area management effectiveness to manage impacts. At the same time, securing tangible community benefits for local/host communities has become an integral aspect of sustainable nature-based tourism, and community-based nature-based tourism models are becoming increasingly popular tools in rural development (Spenceley and Goodwin, 2007; Jaafar et al., 2013; Bello et al., 2016).

The high demand for nature-based tourism experiences can lead to large numbers of people being on-site at the same time, leading to disturbance, traffic congestion, and crowding (for example, Newsome et al., 2012; Newsome, 2013, Prakash et al., 2019). Uncontrolled numbers of visitors can lead to over-tourism scenarios where management have difficulty in managing impacts and visitor satisfaction is compromised (Newsome et al., 2019; Prakash et al., 2019). As a response, many contemporary biocentric management and monitoring approaches especially in higher-order protected areas tend to view visitors as an inherent threat, which has resulted is suboptimal sustainability outcomes and escalated conflicting interests between visitors and park management (Weaver, 2006; Perera et al., 2015). In this light, Weaver and Lawton (2017) have highlighted the necessity of a paradigm shift in protected area management models that reposition visitors as an inherent opportunity, and restructure visitor management in protected areas in line with visitor motivations and mobilization for mass participation in on-site park enhancement activities. Hence, the context provided here demonstrates the complex situation in which nature-based tourism occurs, and this includes visitor demand and interests, the need for adequate conservation of natural resources, planning and management of potential negative impacts.

The situation in 2021 has been brought to the fore by the COVID-19 pandemic which had shut down global tourism by March 2020. What became apparent during the pandemic was how much protected areas relied on tourism for conservation outcomes (Spenceley et al., 2021). Alongside this situation the implications of a lack of tourism putting wild places at risk of degradation arose. Here lies the conundrum regarding the benefits versus the impacts of tourism in the natural environment. The purpose of this chapter, therefore, is to further describe the situation with nature-based tourism before the COVID-19 lockdown, consider what happened during the lockdown, and then consider what the future of nature-based tourism might be like in a post COVID-19 world.

Nature-based tourism before COVID-19

Many texts and articles have been written as to what nature-based tourism comprises reflecting its importance as a sub-category in the overall tourism system and consequently receiving much attention from academics and practitioners (for example, see Patterson, 2007; Buckley, 2009; Fennell, 2014; Fredman, 2021). The positioning of nature-based tourism as a sustainable form of tourism and its reliance on protected areas has additionally been highlighted by Spenceley (2014), Holden (2016), and Blumstein et al. (2017). Often used interchangeably with the term 'ecotourism' the 'gold standard' approach in the practice of nature-based tourism comprises five core principles, namely: it is environmentally based; it is sustainable in its approach; it engages with local communities; it has a strong educative focus; and it fosters visitor satisfaction (Newsome et al., 2013). There are many examples of good nature-based tourism practices around the world (for example, Buckley, 2003, 2010). An example of an excellent nature-based tourism experience is the Mt Lesueur Wildflower Trail, a plant biodiversity hotspot in Western Australia, which is characterized and supported by a protected environment, managed access with a small management footprint, trail planning and design, interpretation at designated day use areas along with multi-stakeholder participation. In terms of visitor experience, Mt Lesueur National Park provides natural experiences/authenticity, controlled access to natural sites, facilities to enhance the experience and the opportunity to learn about nature (Newsome et al. 2013).

Many authors agree that best practice nature-based tourism includes protected areas that are free from natural resource exploitation and have adequate funding (e.g. Leverington et al., 2010). Pertinent literature also emphasizes that management have the capacity to manage the tourism which is taking place (Schulze et al., 2018; Spenceley et al., 2021). Newsome et al. (2013) go on to explore this relationship with other types of outdoor activity and suggest that the core of nature-based tourism consists of ecotourism, wildlife tourism, and geotourism (Table 1.1). Not included in the definition of nature-based tourism presented here is consumptive hunting tourism, fishing, zoo tourism, adventure racing, sporting activities that take place in natural areas and outdoor recreational activities such as surfing, rock climbing, caving, dirt bike challenges, outdoor concerts, and endurance challenge events. Outdoor activities such as these reflect the complex visitor demographic mentioned earlier (Newsome and Hughes, 2018).

Table 1.1 Inter-related and extended components of nature-based tourism

Tourism in the natural environment	*Core definition/criteria*	*Example and geographic scope**
Nature-based Tourism	Tourism with a focus on all expressions of the natural environment fostering positive engagement, sustainable livelihoods, environmental awareness and the conservation of nature	Ecotourism Wildlife tourism Geotourism Astro-tourism
Ecotourism	Attractions should be nature-based, Visitor interactions should be focused on education. Experience and product management should follow principles and practices associated with ecological, socio-cultural, and economic sustainability	Ecotourism involving hiking, camping and patronage of eco-tours and lodges Wildlife tourism Geotourism Wildflower Tourism Entomo-tourism Last-chance Tourism
Geotourism	Focus is on the geology and landscape promoting tourism to geosites	Patronage of geoparks Volcano tourism (Galapagos Islands Ecuador, Philippines, Indonesia, USA) Show caves (USA, Malaysia, Vietnam) Scenic attractions (China, USA, Australia, New Zealand, Antarctica) Fossil sites (Australia, South Korea, China) Key geosites (USA, UK, Australia)

(*Continued*)

Table 1.1 (Continued)

Tourism in the natural environment	*Core definition/criteria*	*Example and geographic scope**
Wildlife Tourism	The desire for visitors to see, photograph and potentially interact with wild animals in the natural environment	Bird watching (Global) Animal migrations (Canada, Kenya, Tanzania) Wildlife destinations (Madagascar, South Georgia UK) Marine wildlife (Australia, Antarctica, South Africa) Iconic species (Uganda, Malaysia, Norway, Ecuador)
Wildflower Tourism	Focus is on seeing and photographing flora and especially displays wildflowers in the natural environment	Trees and forests (USA, Madagascar, Australia, Brazil) Orchids (Madagascar, Australia) Pitcher plants and Rafflesia (Malaysia) Biodiversity hotspots (South Africa and Australia)
Mycotourism	Focus is on fungi (future emphasis is on fungi in the wild)	Europe
Entomo-tourism	Focus is on insects (future emphasis is on insects in the wild)	Fireflies, Malaysia Butterflies (Europe, Taiwan)
Astro-tourism	Focus is on night sky phenomena and star gazing	USA, Canada, Australia, Norway, Iceland
Last-chance Tourism	Tourism that is centred around vanishing/diminishing wildlife, natural phenomena, and natural attractions (tourist motivation has environmental elements and such tourism may help to profile the plight of 'last-chance' destinations as well as fostering conservation actions)	Polar bear (Canada and Spitsbergen) Tiger (India) Indri (Madagascar Endangered birds (global) Komodo Dragon (Indonesia) Great white sharks (South Africa) Glaciers (South America, Greenland)

* List not exhaustive but indicative and designed to indicate the global scope.

The reason for clarifying this position is important because human impacts from ostensibly benign activities in and around natural areas and when viewing wildlife are real (Newsome et al., 2005, 2013; Wolf et al., 2019). Additional problems include conflicts between different types of visitors and fostering inappropriate attitudes towards the environment when contrasting uses are taking place in a natural setting (Newsome, 2013, 2014). Moreover, Shulze et al. (2018) report that recreation and tourism occupy high positions on a list of current and future threats to terrestrial protected areas. Problems surrounding nature-based tourism in recent years include rapid tourism development, rapid increases in visitation and degraded environmental conditions associated with tourism (Newsome et al., 2019; Prakash et al., 2019). Moreover, reduced environmental values and pressures on facilities and services of many popular protected areas due to overcrowding and 'over-tourism' have been documented in literature (Dudley et al., 2018; Gale et al., 2019). The situation can be compounded if protected areas, which tourism relies upon, are impacted by the removal of vegetation and poaching. Protected areas and geoparks in Indonesia, for example, are subject to heavy visitation and degraded environmental conditions associated with tourism. Adjacent and competing land uses impact on geoparks and management authorities are compromised by inadequate funding for conservation and reserve management (Cahyadi and Newsome, 2021). Conflicting interests of peripheral communities alongside exclusion of local inhabitants in the planning and management of protected areas have also been recognised as a threat to sustainable tourism development in protected areas (Dudley and Stolton, 2018).

Given this situation, there has been much research conducted attempting to understand and document negative environmental impacts and in the promotion of sustainable tourism (for example, see accounts and summaries in: Buckley, 2004; Higginbottom, 2004; Newsome et al., 2005; Newsome et al., 2013; Leung et al., 2018; Dowling and Newsome, 2018; Scott et al., 2021). Situations that mitigate good practice are overcrowding, the inability of management to enforce regulations, lack of educational and interpretive services, lack or poorly maintained facilities and an absence of codes of conduct. Planning and management of the putative social and environmental impacts of nature-based tourism are many and vary according to environmental conditions, the wildlife context, and conditions under which the specific tourism activity takes place (Newsome et al., 2013; Holden and Fennell, 2013; Hall et al., 2015). As indicated previously, managing nature-based tourism requires staffing, resources, and expertise. Various planning approaches and management strategies are indicated in Tables 1.2 and 1.3. At very popular destinations, where visitation is high, management can be extensive in some cases leading to a significant management footprint (Newsome et al., 2012).

Dudley et al. (2018) lists a hundred research priorities which are structured under four broad themes of management, ecology, governance and social (including political and economic issues) to set future agendas of protected area research, with tourism management in protected areas being identified as a key priority. Hughes et al. (2018) posit that researching tourism in protected areas is a complex matter requiring input and expertise from many disciplines such

Table 1.2 Planning frameworks applied in nature-based tourism

Framework	*Identifies impacts*	*Requires identification of indicators and standards that are used to measure management effectiveness*	*Requires management actions and monitoring*
Carrying capacity	Yes	Can be applied	Probably unreliable because of difficulties in specific impact identification
Recreation opportunity spectrum	Yes	No	To some degree
Limits of acceptable change	Yes	Yes	Yes
Visitor impact management	Yes	Yes	Yes
Visitor activity management process	Yes	To some degree	To some degree
Visitor experience and resource protection	Yes	To some degree	Yes

Derived from Newsome et al. (2013) and Hughes et al. (2018).

Table 1.3 Management actions applied in nature-based tourism

Action	*Specific types*
Visitor communication and education	Printed and electronic communication
	Fixed panel signage
	Ranger presence
	Visitor centres
Site access control	Managed entry points
	Access road planning and design
	Trail planning and design
	No access areas
Site hardening	Boardwalks
	Viewing platforms
	Barriers
Regulating visitors	Controls on visitor numbers
	Managing group size
	Managing length of stay
	Deterrence and enforcement
Managing the tourism industry	Codes of conduct and guidelines
	Certification
	Environmental management systems
	Regulation via licencing
	Regulation via leases

Derived from Newsome et al. (2013) and Hughes et al. (2018)

as environmental policy, tourism planning, protected area management, recreation ecology, wildlife management, conservation biology, environmental technology, socio-cultural analysis, psychology of human behaviour, geospatial analysis, marketing, and tourism economics. Further to this, because there are many stakeholders and actors in the nature-based tourism industry, resolving the best way forward is potentially fraught with disagreement and as part of a political decision-making situation is subject to differing points of view. For example, in a consideration of the impacts of whale watching, recent correspondence by Palazzo (2021) is somewhat critical of the academic research that leads, and may lead to, restrictions on ecotourism activities. At the same time Palazzo (2021) acknowledges that harassment of wildlife, off-trail trampling and crowding in ecotourism situations are problems but is also of the opinion: "that's not what academic zealots are deriding; it's legitimate wildlife watching activities with no proven harm". Furthermore, he calls for a halt to ecotourism being treated as an impact source and for it to be recognised more as a tool for conservation. The mention of 'academic zealots' is interesting implying that such tourism impact research overstates the case and has the intention of restricting or even stopping 'legitimate' and harmless ecotourism activities. While Palazzo (2021) mentions that whale shark viewing at Oslob in the Philippines has turned direct exploitation for food into an ecotourism situation he fails to acknowledge the poor quality of the Oslob operation, due to over-tourism, as compared to a scientifically informed and regulated best practice management scenario as at Exmouth in Western Australia (Mau, 2008; Snow, 2020). However, it is also worth acknowledging that Palazzo's (2021) views are somewhat supported by disagreement in the scientific community regarding whether ecological impacts, such as provisioning, are negatively impacting whale sharks at such high-profile tourism attractions as in the Philippines (see Meekan and Lowe, 2019).

Notwithstanding the concerns raised above. the previously mentioned research effort has gone a long way in providing data that can be used in policy development and in the guidance of tourism management actions. The conservation, public engagement, and local community benefits of nature-based tourism are very important and maintaining research that can detect negative impacts and reduce damage is vital for the sustainability of the nature-based tourism industry. Coupled with this is there is a growing need for adequate protected area management that is focussed on maintaining natural ecosystem processes and wildlife populations (Leverington et al., 2010; Newsome et al., 2013). The task is further complicated, depending on where you are in world, by increasing human populations, landscape-level pressures such as increasing urbanization, agricultural encroachment, and climate change (UNSDG, 2019; Attenborough, 2020). Given the massive interest, large numbers of people involved, increasing diversity of natural attractions, worsening state of the natural world and potential impacts of unregulated/poorly managed tourism, it could be said that 'nature-based tourism' is at a pivotal stage. Recognising all the benefits that tourism in the natural environment can bring to humanity in the form of enjoyment and psychological benefits, coupled with the scope for mitigation of land degradation, pollution and conserving nature, we stand at a turning point and this turning point has been catalysed and brought into prominence by COVID-19.

Nature-based tourism during COVID-19

The onset of the COVID-19 pandemic resulted in widespread travel restrictions and border closures and brought a halt to nature-based tourism activities around the world (Newsome, 2020; EU, 2021). Millions of jobs connected with the nature tourism industry were either lost or 'put on hold' and valuable funding for local communities and conservation programmes seriously depleted (Spenceley et al., 2021). The positive and negative effects of the pandemic for nature-based tourism and on the protected areas, on which such tourism relies have been further discussed by Buckley (2021), Spenceley (2021), Smith et al. (2021), Primack and Terry (2021) and Bates et al. (2021).

Positive aspects of the pandemic for the environment that have been reported include reduced pollution and greenhouse gas emissions, increased wildlife sightings and reduced tourism pressure on wildlife and natural resources (Newsome, 2020; Manenti et al., 2020; Khan et al., 2021). Such impacts, however, were only likely to be of a temporary nature because as the pandemic progressed and governments worked out COVID safety and recovery protocols, basic industrial activity was always going to resume. As discussed later, tourism recovery was slower but commenced in the domestic format while international borders remained closed.

Widely reported negative impacts included economic losses for the ecotourism industry, loss of tourism revenues for management, conservation, and environmental monitoring, as well as reduced funding and COVID-19 induced staff health issues further hindering management capacity (Spenceley et al., 2021). The closure of protected areas in South America has been witness to encroachment by local communities seeking food and forest products and to 'grab' land (Praeli, 2020). Vale et al. (2021) further reported declines in environmental regulation and enforcement in Brazil, but these problems have been reported from many countries as reflected in increased rates of deforestation, increased habitat encroachment and poaching of wildlife around the World (Rondeau et al., 2020).

The pandemic further highlighted the strong connection of people and natural areas that underpin nature-based tourism, exposing both vulnerabilities and opportunities. For instance, Smith et al. (2021) reported a drop in visitation by 96 percent in South Africa's National Parks as a direct impact of short-term lockdowns. This was accompanied by a tourism revenue drop by 90 percent, which highlighted the risk and fragility of those parks being reliant on one main income stream. Similar observations have been reported by McGinlay (2020) and Mandić (2021) for protected areas of Europe.

Considering the before-mentioned pre-COVID-19-related negative impacts, Newsome (2020) posited that the time in right to reflect on any harm done to the environment and visitor satisfaction in situations where, under pre-COVID, there was poor quality, sub-optimal and crowded nature-based tourism situations. Newsome (2020) went on to consider two divergent post-COVID-19 pathways. The first considered that if there was a widespread return to 'business as usual' circumstances, in some cases destinations further damaged by COVID-19-related lockdown impacts, we could be faced with a new baseline of degraded tourism attractions. Alternatively, if we took the best of what was taking place prior to COVID-19 and learnt from the damage caused by a 'lack' of tourism and its consequent impacts in natural areas, then we would be able to realize a new awareness, thus fostering a sustainable future. The next section, however, explores what happened in some nature-based tourism situations following the relaxation of lockdown conditions.

Relaxation of lockdown and increased tourism activity in the natural environment

Earlier in this chapter the popularity and significance of nature-based tourism was established. However, the COVID-19 pandemic brought a halt to both domestic and international travel. Following on from this is the increasing evidence that being 'in nature' is good for our physical and mental health. Continuing this theme further is the recommendation that outdoor recreation is a tangible way for people to counter pandemic lockdown-induced stress (e.g. Brooks et al., 2020; Pouso et al., 2021). In a Norwegian study, Venter et al. (2021) reported substantial increases in outdoor activity in forests and protected areas both during and then after lockdowns were lifted. Moreover, observations from the USA, Germany and Indonesia show a rebound in domestic tourism in protected areas following the relaxation of restrictions (Spenceley et al., 2021).

Increases in domestic tourism around the world have been witness to environmentally damaging behaviours (e.g. McGinlay et al., 2020; Cahyadi and Newsome, 2021). Anecdotal evidence from the Shark Bay World Heritage Area of Western Australia indicates a dramatic increase in domestic tourism with visitation running at 2.5 times higher than normal in 2021. More importantly, Shark Bay staff report that there is also a greater number of people who have little regard for the environment which is concomitant with a rise of inappropriate behaviour and damage. In Europe, McGinlay et al. (2020) identify new demographic profiles, overcrowding, conflicts between different user groups and problematic behaviour as significant challenges.

In a rapid assessment of the impacts of domestic tourism on selected geoparks in Indonesia, Cahyadi and Newsome (2021) describe over-tourism impacts in geoparks across the archipelago. Pre-pandemic tourism impact scenarios included congestion, littering, air pollution from vehicle emissions, accidental fire damage to natural vegetation, disruption of wildlife and no opportunity to undertake site protection and restoration. While all these impacts substantially declined or were absent during lockdown conditions, all of them were re-established once restrictions were lifted and the domestic tourism cohort could travel again.

Accordingly, as it stands at this point in time, there appears to be a return to 'business as usual' or there has been insufficient time for decision-making authorities and protected area managers to reflect on problematical pre-COVID impact scenarios. Hence the caution raised by Newsome (2020) along with the question: What will nature-based tourism look like after COVID-19?

What could nature-based tourism look like after COVID-19?

The call for 'building back better' was well established before the post-lockdown domestic tourism rebound that is evident today had gained momentum. As stated *previously*, *Newsome* (2020) called for reflection as to what were good and bad forms of nature-based tourism before the pandemic shut down borders and restricted the movement of people. Spenceley et al. (2021) emphasized that 'building back better' did not mean returning to pre-pandemic conditions but envisioned sustainable tourism that particularly addressed biodiversity loss and climate change. Buckley (2021) also remained cautious as to whether much-needed changes that aim at mitigating negative scenarios would occur. Buckley (2021) also pointed out that the pandemic provided recreation ecologists with an opportunity to document, monitor, and assess the implications of reduced tourism. Examples of this include investigating tourism sourced disturbance to wildlife and the impact of roads and traffic in protected areas. Such an opportunity additionally helps in mediating the dilemma mentioned earlier where some practitioners articulate concerns about research potentially restricting 'benign' wildlife tourism operations (e.g. Meekan and Lowe, 2019; Palazzo, 2021).

A recent review of wildlife tourism practices in the Asia Pacific region provides some useful guiding criteria as to how nature-based tourism could be implemented in the future and in the global context. These criteria include conservation, habitat restoration, and conservation focussed research; attention to animal welfare and respect for wildlife; provision of interpretation; local community engagement and best practice management such as adherence to codes of conduct and relevant legislation (Scott et al., 2021). Attendant to this is for decision-making authorities and practitioners to have a clear understanding and acknowledgment of a valid and accepted definition of ecotourism and its application in the nature-based tourism industry. In exploring a readily understandable definition, Newsome et al. (2013) proposed that ecotourism, and by extension nature-based tourism, should have a strong focus on the natural environment, be educational, be sustainably managed, involve local community participation, and promote visitor satisfaction. Regarding this last point, visitor satisfaction is linked to visitor education and environmental management. This in turn fosters informed and appropriate attitudes towards wildlife, addressing over-tourism scenarios and acknowledging good management practices. Educated and informed visitors are then in a position to be able to judge the efficacy of conservation and tourism management practices in any given situation (e.g. see Prakash et al., 2019).

The pandemic further highlighted the need for local communities that are reliant on tourism in natural areas to enhance their resilience to mitigate the vulnerability and risks posed by such global catastrophes as well as the emerging impacts of climate change (Gabriel-Campos et al., 2021). The role of protected areas in climate change adaptation (Mansourian et al., 2009), the impact of tourism on climate change, and the growing sensitivity of tourists to such issues have now been established (Hares et al., 2010; Scott et al., 2012). Hence, the strategies for 'building back better' require implementation of appropriate risk management measures to counteract negative impacts on communities from scenarios that can affect economies reliant on nature-based tourism. A more holistic approach is needed as the effects of such hazards are often interlinked. For instance, Sharma et al. (2021) outlined four key factors for building resilience in the tourism industry: government response, technology innovation, local belongingness, and consumer and employee confidence. Strategies such as diversifying tour packages and utilizing local resources and products have been proven to be successful amidst the recent pandemic (Noorashid and Chin, 2021).

Optimal post-COVID-19 nature-based tourism policy development will require reflection, knowledge, commitment, funding, staffing and training. Attendant to this is the need

Table 1.4 Managing tourism in the natural environment

Access management
Provision of facilities and infrastructure
Site management/hardening
Education
Policing and compliance enforcement
Zoning and dispersion
An increase in the number of available sites
Controls on visitor numbers
Limits of acceptable change to determine capacity limits and site closures
Demarketing
Economic instruments
Site/whole section/park closures

Adapted from Newsome et al. (2013).

for adequate and effective protected area management. All actors will need to be at the table, including politicians, regional and national tourism planning agencies, protected area managers, conservation groups and other non-governmental organizations and academics who can feed their research into the conversation. It will be important to anticipate an increase in demand once international borders are lifted. As stated previously, the domestic tourism rebound in Europe and Indonesia has brought problems. Identification and applicability of the options available to us are equally important (Tables 1.3 and 1.4).

McGinlay et al. (2020) emphasize some lessons learned from the impacts associated with the lifting of pandemic lockdowns. Management of visitor numbers is the key to developing sustainable tourism models. They advise that avoidance of overcrowded conditions can be achieved via visitor education, spatial planning and the promotion of tourism that does not involve large visitor numbers.

Conclusion

In the global context, nature-based tourism has been growing rapidly and was predicted to be part of an ever-increasing global travel phenomenon by 2030. For many destinations the increase in visitor numbers, in combination with a more complex visitor demographic, has given rise to a mix of attitudes and expectations especially where more generalized outdoor recreation blurs into nature-based tourism activities. In recent times, the value of 'being in nature' has brought mental and physical health benefits to the fore. Such benefits appear to have become more widely recognised under COVID-19 restrictions and lockdowns, where being outdoors and experiencing nature have been seen as relief from isolation and confinement and as a means of mitigating pandemic induced stress.

The COVID-19 pandemic brought an abrupt halt to global tourism activity and visits to nature-based tourism destinations, such as national parks and geoparks, went into catastrophic decline. Under lockdown conditions the absence of visitors resulted in declines in economic activity that impacted millions of people around the world. At the same time, in the absence of tourism activity pollution levels declined, protected areas were relieved from heavy visitor pressure, and increased sightings of wildlife were recorded. On the other hand, park and wildlife management agencies in many countries became increasingly faced with difficulties posed by illegal activities such as poaching and a general decline in 'on the ground' conservation programmes.

The post-lockdown period has seen domestic tourism in many countries undergo an explosive growth since the lifting of restrictions. This is partly because people are still largely restricted in the international travel sphere and being outdoors is seen as a healthy way to spend leisure time. The future of tourism after COVID-19 has been widely debated and there are calls for a more sustainable approach that counters heavy visitor pressure, congestion, and environmental damage. However, as it stands at the moment, a lot more work needs to be done to achieve such goals.

References

Attenborough, D. (2020). *A life on our planet.* Witness Books, London.

Balmford, A., Beresford, J., Green, J., Naidoo, R., Walpole, M., & Manica, A. (2009). A global perspective on trends in nature-based tourism. *PLoS biology*, 7(6), e1000144.

Bates, A. E., Primack, R. B., Biggar, B. S., Bird, T. J., Clinton, M. E., Command, R. J., & Parmelee, J. R. (2021). Global COVID-19 lockdown highlights humans as both threats and custodians of the environment. *Biological Conservation, 263*, 109175.

Bello, F. G., Carr, N., & Lovelock, B. (2016). Community participation framework for protected area-based tourism planning. *Tourism Planning & Development, 13*(4), 469–485.

Blumstein, D., Geffroy, B., Samia, D. S. M., & Bessa, E. (2017). *Ecotourism's promise and peril: A biological evaluation.* Springer International Publishing, Cham, Switzerland, 185 p.

Brooks, S. K., Webster, R. K., Smith, L. E., Woodland, L., Wessely, S., Greenberg, N., & Rubin, G. J. (2020). The psychological impact of quarantine and how to reduce it: Rapid review of the evidence. *The Lancet, 395*(10227), 912–920. https://doi.org/10.1016/S0140-6736(20)30460-8

Buckley, R. (2003). *Case studies in ecotourism.* CABI Publishing, Oxford, UK.

Buckley, R. (2004). *Environmental impacts of ecotourism.* CABI Publishing, Oxford, UK.

Buckley, R. (2009). *Ecotourism: Principles and practices.* CABI Publishing, Oxford, UK.

Buckley, R. (2010). *Conservation tourism.* CABI Publishing, Oxford, UK.

Buckley, R. (2021). Pandemic travel restrictions provide a test of net ecological effects of ecotourism and new research opportunities. *Journal of Travel Research, 60*(7), 1612–1614.

Buckley, R., Brough, P., Hague, L., Chauvenet, A., Fleming, C., Roche, E., Sofija, E., & Harris, N. (2019). Economic value of protected areas via visitor mental health. *Nature Communications, 10*(1), 1–10.

Cahyadi, H. S., & Newsome, D. (2021). The post COVID-19 tourism dilemma for geoparks in Indonesia. *International Journal of Geoheritage and Parks.* https://doi.org/10.1016/j.ijgeop.2021.02.003

Coventry, P. A., Brown, J. E., Pervin, J., Brabyn, S., Pateman, R., Breedvelt, J., … & White, P. L. (2021). Nature-based outdoor activities for mental and physical health: Systematic review and meta-analysis. *SSM-Population Health, 16*, 1–14.

Curtin, S., & Kragh, G. (2014). Wildlife tourism: Reconnecting people with nature. *Human Dimensions of Wildlife, 19*(6), 545–554.

da Mota, V. T., & Pickering, C. (2021). Assessing the popularity of urban beaches using metadata from social media images as a rapid tool for coastal management. *Ocean & Coastal Management, 203*, 1–11.

Dowling, R, & Newsome, D. (Eds.). (2018). *Handbook of geotourism.* Edward Elgar Publishing, Cheltenham, UK.

Dudley, N., Hockings, M., Stolton, S., Amend, T., Badola, R., Bianco, M., … & Zhang, Y. (2018). Priorities for protected area research. *PARKS*, 24(1), 35–50. https://doi.org/10.2305/IUCN.CH.2018.PARKS-24-1ND.en

Dudley, N., & Stolton, S. (2018). Protected areas: Challenges and responses for the coming decade. In: *Equilibrium Research Dialogue*: 1, Bristol, UK.

Elmahdy, Y. M., Haukeland, J. V., & Fredman, P. (2017). *Tourism megatrends, a literature review focused on nature-based tourism (MINA fagrapport 42).* Norwegian University of Life Sciences, ISSN: 2535-2806. https://nmbu.brage.unit.no/nmbu-xmlui/bitstream/handle/11250/2648159/mif42.pdf?sequence=1&isAllowed=y

EU (2021). European commission, directorate-general for international partnerships. In: A. Spenceley (Ed.), *Tourism and visitation to protected areas amid COVID-19: Guidance for protected area authorities and managers.* Publications Office of the European Union, Brussels, Belgium. https://data.europa.eu/doi/10.2841/772

Fennell, D. A. (2014). *Ecotourism*. Routledge, London and New York.

Fredman, P. (Ed.). (2021). *Nordic perspectives on nature-based tourism: From place-based resources to value-added experiences*. Edward Elgar Publishing, Cheltenham, UK.

Frost, W., Laing, J., & Beeton, S. (2014). The future of nature-based tourism in the Asia-Pacific region. *Journal of Travel Research*, *53*(6), 721–732.

Gabriel-Campos, E., Werner-Masters, K., Cordova-Buiza, F., & Paucar-Caceres, A. (2021). Community eco-tourism in rural Peru: Resilience and adaptive capacities to the Covid-19 pandemic and climate change. *Journal of Hospitality and Tourism Management*, *48*, 416–427.

Gale, T., Ednie, A., & Beeftink, K. (2019). Worldviews, levels of consciousness, and the evolution of planning paradigms in protected areas. *Journal of Sustainable Tourism*, *27*(11), 1609–1633.

Hall, C. M., Gossling, S., & Scott, D. (Eds.). (2015). *The Routledge handbook of tourism and sustainability*. Routledge.

Hares, A., Dickinson, J., & Wilkes, K. (2010). Climate change and the air travel decisions of UK tourists. *Journal of Transport Geography*, *18*(3), 466–473.

Higginbottom, K. (Ed.). (2004). *Wildlife tourism: Impacts, management and planning*. Common Ground Publishing, Australia.

Holden, A. (2016). *Environment and tourism*. Routledge, London and New York.

Holden, A., & Fennell, D. A. (Eds.). (2013). *The Routledge handbook of tourism and the environment*. London and New York: Routledge.

Hughes, M., Newsome, D., & Rodger, K. (2018). Researching the nexus of protected natural area tourism management: Milestones and future directions. In: C. Cooper, S. Volo, W. Garner, & N. Scott (Eds.), *Sage handbook on tourism management: Theories, concepts and disciplinary approaches to tourism*. pp. 448–458. Sage, London and Singapore.

Jaafar, M., Kayat, K., Tangit, T. M., & Yacob, M. F. (2013). Nature-based rural tourism and its economic benefits: A case study of Kinabalu National Park. *Worldwide Hospitality and Tourism Themes*, 5(4), 342–352.

Kaplan, R., & Kaplan, S. (1989). *The experience of nature: A psychological perspective*. Cambridge University Press, New York, USA.

Keniger, L. E., K. J. Gaston, K. N. Irvine, & R. A. Fuller. (2013). What are the Benefits of Interacting with Nature? *International Journal of Environmental Research and Public Health*, *10*, 913–935.

Khan, I., Shah, D., & Shah, S. S. (2021). COVID-19 pandemic and its positive impacts on environment: An updated review. *International Journal of Environmental Science and Technology*, *18*(2), 521–530.

Lackey, N. Q., Tysor, D. A., McNay, G. D., Joyner, L., Baker, K. H., & Hodge, C. (2021). Mental health benefits of nature-based recreation: A systematic review. *Annals of Leisure Research*, *24*(3), 379–393.

Leung, Y. F., Spenceley, A., Hvenegaard, G., & Buckley, R. (2018). *Tourism and visitor management in protected areas: Guidelines for sustainability* (Vol. 27). IUCN, Gland, Switzerland.

Leverington, F., Costa, K. L., Pavese, H., Lisle, A., & Hockings, M. (2010). A global analysis of protected area management effectiveness. *Environmental Management*, *46*(5), 685–698.

Mandić, A. (2021). Protected area management effectiveness and COVID-19: The case of Plitvice Lakes National Park, Croatia. *Journal of Outdoor Recreation and Tourism*, *6*(1), 1–27.

Manenti, R., Mori, E., Di Canio, V., Mercurio, S., Picone, M., Caffi, M., Brambilla, M., Ficetola, G., & Rubolini, D. (2020). The good, the bad and the ugly of COVID-19 lockdown effects on wildlife conservation: Insights from the first European locked down country. *Biological Conservation*, *249*, 108728.

Mansourian, S., Belokurov, A., & Stephenson, P. J. (2009). The role of forest protected areas in adaptation to climate change. *Unasylva*, *60*(231–232), 63–69.

Margaryan, L., & Fredman, P. (2017). Bridging outdoor recreation and nature-based tourism in a commercial context: Insights from the Swedish service providers. *Journal of Outdoor Recreation and Tourism*, *17*, 84–92.

Mau, R. (2008). Managing for conservation and recreation: The Ningaloo whale shark experience. *Journal of Ecotourism*, 7(2–3), 213–225.

McGinlay, J., Gkoumas, V., Holtvoeth, J., Fuertes, R. F. A., Bazhenova, E., Benzoni, A., Botsch, K., Martel, C., Sánchez, C., Cervera, I., Chaminade, G., Doerstel, J.,García, C., Jones, A., Lammertz, M., Lotman, K., Odar, M., Pastor, T., Ritchie, C., Santi, S., Smolej, M., Rico, F.,Waterman, H., Zwijacz-Kozic, T., Kontoleon, A., Dimitrakopoulos, P., & Jones, N. (2020). The impact of COVID-19 on the management of European protected areas and policy implications. *Forests*, *11*, 1214. https://doi.org/10.3390/f11111214

Meekan, M., & Lowe, J. (2019). Oslob whale sharks–Preconceived ideas about provisioning. *Tourism Management*, *75*, 630–631.

Newsome, D. (2013). An 'ecotourist's recent experience in Sri Lanka. *Journal of Ecotourism*, 12, 210–220.

Newsome, D. (2014). Appropriate policy development and research needs in response to adventure racing in protected areas. *Biological Conservation*, 171, 259–269.

Newsome, D. (2020). The collapse of tourism and its impact on wildlife tourism destinations. *Journal of Tourism Futures*. https://doi.org/10.1108/JTF-04-2020-0053

Newsome, D., Dowling, R., & Leung, Y-F. (2012). The nature and management of geotourism: A case study of two established iconic geotourism destinations *Tourism Management Perspectives*, *2–3*, 19–27.

Newsome, D., Dowling, R., & Moore, S. (2005). *Wildlife tourism*. Channel View Publications, Clevedon, UK.

Newsome, D., & Hassell, S. (2014). Tourism and conservation in Madagascar: The importance of Andasibe National Park. *Koedoe*, 56, 1–8.

Newsome, D., & Hughes, M. (2017). Jurassic World as a contemporary wildlife tourism theme park allegory. *Current Issues in Tourism*, 20, 1311–1319.

Newsome, D., & Hughes, M. (2018). The contemporary conservation reserve visitor phenomenon! *Biodivers Conserv*, *27*, 521–529.

Newsome, D., Lacroix, C., & Pickering, C. (2011). Adventure racing events in Australia: Context, assessment and implications for protected area management. *Australian Geographer*, *42*(4): 403–418.

Newsome, D., Moore, S. A., & Dowling, R. K. (2013). *Natural area tourism*. Channel View Publications, Bristol, UK.

Newsome, D., Rodger, K., Pearce, J., & Chan, J. (2019). Visitor satisfaction with a key wildlife tourism destination within the context of a damaged landscape. *Current Issues in Tourism*, *22*, 729–746.

Noorashid, N., & Chin, W. L. (2021). Coping with Covid-19: The resilience and transformation of community-based tourism in Brunei Darussalam. *Sustainability*, *13*(15), 8618. https://doi.org/10.3390/su13158618

Palazzo, J. T. (2021). A whale watching me. *Live Encounters Magazine*. https://liveencounters.net/2021-le-mag/07-july-2021/jose-truda-palazzo-jr-a-whale-watching-me/ (accessed 2/10/21).

Patterson, C. (2007). *The business of ecotourism: The complete guide for nature and culture-based tourism operators*. Trafford Publishing, Bloomington, USA.

Perera, P., Senevirathna, M. C., & Vlosky, R. P. (2015). Recreationist perspectives, attitudes, and perceptions towards national park management in Sri Lanka. *Tourism: An International Interdisciplinary Journal*, *63*(4), 497–514.

Pouso, S., Borja, Á., Fleming, L. E., Gomez-Baggethun, E., White, M. P., & Uyarra, M. C. (2021). Contact with blue-green spaces during the COVID-19 pandemic lockdown beneficial for mental health. *Science of The Total Environment*, *756*, 143984. https://doi.org/10.1016/j.scitotenv.2020.143984

Praeli, Y. S. (2020). For South America's wilderness areas, COVID-19 brings risk and respite, *Mongabay*. https://news.mongabay.com/2020/08/for-south-americas-wilderness-areas-covid-19-brings-risk-and-respite/ (accessed 25/10/21).

Prakash, S. L., Perera, P., Newsome, D., Kusuminda, T., & Walker, O. (2019). Reasons for visitor dissatisfaction with wildlife tourism experiences at highly visited national parks in Sri Lanka. *Journal of Outdoor Recreation and Tourism*, *25*, 102–112.

Primack, R. B., & Terry, C. (2021). New social trails made during the pandemic increase fragmentation of an urban protected area. *Biological Conservation*, *255*, 108993.

Rondeau, D., Perry, B., &Grimard, F. (2020). The consequences of COVID-19 and other disasters for wildlife and biodiversity. *Environmental and Resource Economics*, *76*(4), 945–961.

Schulze, K., Knights, K., Coad, L., Geldmann, J., Leverington, F., Eassom, A., Marr, M., Butchart, S. H., Hockings, M., & Burgess, N. D. (2018). An assessment of threats to terrestrial protected areas. *Conservation Letters*, 11(3), e12435. https://doi.org/10.1111/conl.12435

Scott, D., Gössling, S., & Hall, C. M. (2012). International tourism and climate change. *Wiley Interdisciplinary Reviews: Climate Change*, *3*(3), 213–232.

Scott, N., Green, R., & Newsome, D. (2021). Sustainable development of wildlife tourism in Asia and the Pacific – Maintaining Good Practice. *Chinese Journal of Ecotourism*, *11*(5), 1–11. Chinese Academy of Sciences, Beijing, China.

Sharma, G. D., Thomas, A., & Paul, J. (2021). Reviving tourism industry post-COVID-19: A resilience-based framework. *Tourism Management Perspectives*, *37*, 100786. https://doi.org/10.1016/j.tmp.2020.100786

Smith, M. K. S., Smit, I. P., Swemmer, L. K., Mokhatla, M. M., Freitag, S., Roux, D. J., & Dziba, L. (2021). Sustainability of protected areas: Vulnerabilities and opportunities as revealed by COVID-19 in a national park management agency. *Biological Conservation, 255*, 108985.

Snow, S. (2020). *Controlling a whale shark mass tourism destination: New study investigates crowding and tourist expectations*. LAMAVE Press Release.

Spenceley, A. (2014). Tourism and protected areas: A growing nexus of challenge and opportunity. *Koedoe, 56*(2), 1–2.

Spenceley, A. (2021). *The future of nature-based tourism*. Luc Hoffman Institute/WWF, Basel, Switzerland.

Spenceley, A., & Goodwin, H. (2007). Nature-based tourism and poverty alleviation: Impacts of private sector and parastatal enterprises in and around Kruger National Park, South Africa. *Current Issues in Tourism, 10*(2–3), 255–277.

Spenceley, A., McCool, S., Newsome, D., Báez, A., Barborak, J., Blye, C-J., Bricker, K., Cahyadi, H., Corrigan, K., Halpenny, E., Hvenegaard, G., Malleret King, D., Leung, Y-F., Mandić, A., Naidoo, R., Rüede, D., Sano, J., Sarhan, M., Santamaria, V., Sousa, T., & Zschiegner, A. (2021). Tourism in protected and conserved areas amid the Covid-19 pandemic. *PARKS, 27*, 108–118.

UNSDG. (2019). UN Report: Nature's dangerous decline' unprecedented'; species extinction rates accelerating. https://www.un.org/sustainabledevelopment/blog/2019/05/nature-decline-unprecedented-report/ (accessed Sept 3, 2021).

UNTWO. (2017). *Tourism highlights*. World Tourism Association, Madrid, Spain.

Vale, M. M., Berenguer, E., de Menezes, M. A., de Castro, E. B. V., de Siqueira, L. P., & Rita de Cássia, Q. P. (2021). The COVID-19 pandemic as an opportunity to weaken environmental protection in Brazil. *Biological Conservation, 255*, 108994.

Venter, Z. S., Barton, D. N., Gundersen, V., Figari, H., & Nowell, M. S. (2021). Back to nature: Norwegians sustain increased recreational use of urban green space months after the COVID-19 outbreak. *Landscape and Urban Planning, 214*, 104175.

Weaver, D. (2006). *Sustainable tourism: Theory and practice*. Butterworth-Heinemann, London.

Weaver, D. B., & Lawton, L. J. (2017). A new visitation paradigm for protected areas. *Tourism Management, 60*, 140–146.

Wolf, I. D., Croft, D. B., & Green, R. J. (2019). Nature conservation and nature-based tourism: A paradox? *Environments, 6*(9), 104.

WTTC. (2019). *The economic impact of global wildlife tourism: Travel and tourism as an economic tool for the protection of wildlife*. World Travel & Tourism Council. Oxford Economics, UK.

2

THE MOST VULNERABLE AND LEAST HEARD

Advocating for animal justice in tourism

Valerie A. Sheppard and David A. Fennell

Introduction

We are in the midst of an emerging transformation in tourism scholarship and practice. This transformation is centred on concern over the needs and interests of animals, which are thoroughly infused into the fabric of tourism on many scales. This moral turn (Caton, 2012), rather like an *interspecies* turn, is likely a function of the continued evolution in tourism studies research. For example, a call for papers in 2020 on the topic of impacts on the welfare of wildlife from tourism, along with ethical issues, by guest editors of the *Journal of Sustainable Tourism* yielded 35 abstracts. Even five years ago, this breadth of work would not have emerged. This turn is also a result of the many troubling practices that continue to compromise the welfare of animals used in tourism in so many contexts, including captive animals, animal workers, animals used for sport, and so on (Fennell, 2012a). We witness poor human–animal interaction practices regularly on the worldwide web and through social media. This reality has prompted some large travel organizations, like TripAdvisor, to announce it will no longer generate revenue from or promote attractions that contribute to the captivity of cetaceans (TripAdvisor, 2019).

These trends in tourism are no doubt nested within a broader set of issues around global challenges facing the tourism industry, such as climate change, environmental degradation, habitat and biodiversity loss. The time is upon us, most would argue, because our current trajectory, unbridled, will mean serious irreversible change that we will have trouble adjusting to now and in the future (McKibben, 2019). On the other hand, we also see acts of human kindness towards animals that have been placed in stressful and life-threatening situations. For example, the catastrophic fires in 2019–20 in Australia prompted organizations and individuals to rescue and care for displaced and injured animals on what appears to be an unprecedented level, given the magnitude of animal deaths (estimated to be more than 1 billion) by January 2020 (ABC News, 2020).

The concern for animal interests is thus an important part of an *expanding consciousness* platform that is comprised of two essential ingredients as the tourism agenda moves forward: sustainability and ethics (Fennell, 2018). In turn, this demands that we develop a solid moral

DOI: 10.4324/9781003230748-4

underlay to conventional practices that ought to be challenged in the present day. Animal advocacy and sanctuary organizations are one way to construct such a moral underlay, by putting the expanding human consciousness into practice. Consequently, this chapter employs case study methods to understand how animal advocacy and sanctuary organizations define and apply principles of justice, specifically as they relate to the care of and/or advocacy for animals, including those used in tourism-related ventures. Understanding the extent to which achieving justice for animals used in tourism is possible, and assessing the opportunities to make this a reality, is an important starting point on the road to measuring the extent to which achieving justice for all animals.

Review of literature

The genesis of pro-animal research in tourism can be traced back to the turn of the present century; however, the history of human care and concern for animals is evidenced in all major time periods and as far back as pre-Christian times, according to Niven (1967). Examples in tourism that have an explicit focus on one type of ethics or another include work by Hughes (2001) on the UK dolphin industry, Shani and Pizam (2008) on an ethics for animal-based attractions; Burns, Macbeth, and Moore (2011) on ecocentric ethics and wildlife tourism; and a comprehensive treatment of tourism, animals and ethics by Fennell (2012a). More recently, research has focused on animals as food and the ethics around these practices (Kline, 2018), as well as the unethical use of wildlife in tourism (Moorehouse, D'Cruze, & Macdonald, 2017). The use of several different theories of animal ethics, such as rights, welfare, utilitarianism, ecocentrism, and ecofeminism, have been advantageous in demonstrating the broad terrain in which different tourism–animal practices have been investigated (see Fennell & Sheppard, 2011, 2021; Sheppard & Fennell, 2019).

The American philosopher Tom Regan sought to develop a moral theory as it relates to humans' treatment of animals. Regan (2004, p. 232) states, "There is no single interpretation or theory of justice." Rather, justice is likened to an umbrella under which lie many other concepts, such as fairness, collaboration, gratitude, trust, reciprocity, retribution, anger, indignation, punishment, forgiveness and reconciliation among animals (Pierce & Bekoff, 2012). Achieving justice for animals is constrained because of a Cartesian backdrop that continues to have ramifications in the present day. These shortcomings include claims that animals lack consciousness, rationality, sentience, and intentionality; including the fact that animals are unable to sign contracts, reciprocate within society, fail to understand good versus bad, and act merely on instincts (Rawls, 1971; Rowlands, 1997; Garner, 2012).

Regan writes that moral agents are the recipients of direct moral duties and concern because they are in the possession of select capabilities (noted above), while moral patients are to be afforded only indirect duties because they are not in the possession of these capabilities. In more detail, Regan (2004) argues that moral agents are:

> Individuals who have a variety of sophisticated abilities, including in particular the ability to bring impartial moral principles to bear on the determination of what, all considered, morally ought to be done and, having made this determination, to freely choose or fail to choose to act as morality, as they conceive it, requires. Because moral agents have these abilities, it is fair to hold them morally accountable for what they do … they must also bear the moral responsibility of doing (or not doing it).
>
> (pp. 151–152)

By contrast, moral patients:

> … lack the prerequisites that would enable them to control their own behavior in ways that would make them morally accountable for what they do. A moral patient lacks the ability to formulate, let alone bring to bear, moral principles in deliberating about which one among a number of possible acts it would be right or proper to perform.
>
> (Regan, 2004, p. 152)

Regan further argues that we ought to have duties to moral patients because they have inherent value on the basis of consciousness, intentionality, and sentience; they are, therefore, subjects-of-a-life. Respect is a central feature of Regan's beliefs on inherent value, which flows from principles of formal justice. Egalitarianism is important in the way animals deserve the same level of respect equally, while non-perfectionism is also an important characteristic because one individual should not to be viewed as more important than any other on the basis of the aforementioned characteristics. Furthermore, justice demands that we should not use others as resources for our own ends, and that we should be compelled, as a duty, to help others who are the recipients of injustice from others (see also Fennell, 2012b). This latter point corresponds to Nozick's (1974) work on positive and negative rights.

Nozick (1974) argued that a positive right necessitates action, and this action may be of a legal or a moral basis. Positive rights mean that we ought to assist another in attaining their goals. For example, we help a farm animal to better health by enlisting the help of a veterinarian. By contrast, a negative right refers to the fact that others must allow an agent to pursue their ends in a climate of noninterference, i.e., the right not to be subjected to the actions of other groups or individuals. For example, governments have a duty to ensure proper welfare for citizens (positive right). But this duty may infringe upon the rights of others of the citizenry who must help pay for this added welfare through the raising of taxes – the right not to have more money taken away by government (negative right). The importance of positive and negative rights has been emphasized in the human rights literature. Negative rights were common in earlier interpretations of human rights (first-generation rights), whilst positive rights (second-generation), i.e., the imperative to act, are more closely aligned with more recent interpretations of social human rights and entitlements (Hirschl, 2000).

Shooster (2015) has taken up the cause of positive rights in animal ethics research by arguing that while governments have taken care of the rights of people through a range of different welfare programs around food, shelter and health, as noted above, the same luxury ought to be extended to animals that share many of the same needs as and capabilities of humans: shelter, food, reproduction, habitat, clean air and water, and so on. These "expansive positive legal rights," according to Shooster, suggest that governments must be legally obliged to use all resources at hand to make the lives of wild animals better, even if these rights are only aspirational at this point in time.

Overall, the literature review reveals that there is a lack of focus on moving from theory to practice, particularly as it relates to animals used in tourism. The research that is the subject of this chapter seeks to begin addressing this gap in the literature. In the following sections we describe our methods employed to understand how animal advocacy and sanctuary organizations define and apply the principles associated with justice, specifically as it relates to their care of and/or advocacy for animals, including those used in tourism-related ventures and experiences.

Methods

This research employed an exploratory case study approach. According to Yin (2003, p. 1), case studies are appropriate when the phenomenon under study has "real-life" contemporary context and have been used liberally in tourism research in a number of contexts. Examples include ecotourism (Sanson, 1994), pro-poor tourism (Butler, Curran, & O'Gorman, 2013), and volunteer tourism (Broad, 2008). Consequently, a case study approach was a suitable research strategy for understanding the contextual conditions associated with this study's research objectives. For these purposes, the unit of analysis is three organizations, from various parts of the world. Each provides sanctuary to, and/or advocate for, animals including those used in tourism-related operations and experiences.

A total of seven organizations involved in animal welfare, either directly (by providing sanctuary), indirectly (advocating), or a combination of both, were identified through website research, in combination with knowledge previously existing within the purview of the researchers; however, the researchers had not previously worked with or visited these organizations. Potential participants were chosen and contacted solely based upon: 1) their internet presence and their use of the word "justice," or "just," or the antonyms of these words, "injustice" or "unjust," on their websites; and, 2) either a direct or indirect connection to tourism. A variety of websites were utilized, including ResponsibleTravel.com and Charitywatch.org, to help refine the search. Of the seven organizations located in the internet search, only three organizations agreed to participate (see Table 2.1). The low sample size is directly related to the fact that few organizations met the two criteria for inclusion in the study.

After securing each organization's agreement to participate, data collection took place via an emailed questionnaire, directed at a senior-level person. Three questionnaire variations were designed, with questions targeted at each organization. To achieve the exploratory objectives the questionnaire employed open-ended questions, focused on understanding: 1) how the organization defines and works toward promoting and/or achieving justice for animals; 2) academic and non-academic sources used, if any, by the organization to assist it with understanding what is meant by justice for animals; 3) certification programs the organization is involved with, if any; 4) the importance tourism plays in the organization's operations (e.g. conservation, education, or revenue generation); and 5) what the organization perceives is required to enable further progress as it relates to justice for animals.

Table 2.1 Sample organizational breakdown

Organization Name	*Geographical Location Represented*	*Animal welfare (tourism) focus*
1. Lilongwe Wildlife Trust	Malawi, Africa	Direct link to tourism
2. Edgar's Mission Farm Sanctuary	Victoria, Australia	Direct link to tourism
3. Animal Justice Canada Legislative Fund	Toronto, Canada	Indirect link to tourism (organization depends upon tourism)

Data analysis and results

Lilongwe Wildlife Trust (LWT), Malawi, East Africa: analysis

LWT (founded in 2008) is a non-profit organization situated in Malawi, a landlocked country in East Africa. Malawi, with a population of 16.3 million, encompasses a landmass of 118,484 square kilometres (Government of the Republic of Malawi, 2013). Malawi is characterized by its natural beauty and friendly citizens; however, the country is challenged by poverty, deforestation, and wildlife crime (Lilongwe Wildlife Trust, n.d.). It is considered a centre for the transit of illicit wildlife products, which threatens the extinction of some of the nation's wildlife, specifically elephants (Lilongwe Wildlife Trust, n.d.). A World Travel & Tourism Council (2018) report indicates that tourism made a direct contribution to Malawi's gross domestic product of USD $221.5 billion in 2017 and 3 percent of total employment in the country.

LWT, with a staff of 90 spread out across all three regions of Malawi, outlines four visions: 1) wildlife rescue and welfare ("Every wild animal in Malawi is free from human-inflicted suffering"); 2) conservation justice ("Malawi is free from wildlife and forest crime"); 3) education and advocacy ("Everyone in Malawi respects and protects wildlife"); and 4) biodiversity research ("Evidence-based management is integral to conservation") (see https://www.lilongwewildlife.org/about/about-lwt/). LWT works both nationwide and across borders, investing more than $2 million in conservation projects in Malawi in 2017, responding to conservation challenges and partnering with organizations it believes can help it deliver results (see https://www.lilongwewildlife.org/about/about-lwt/).

LWT's *Conversation Justice* project is focused on preventing the trafficking and poaching of wildlife by working collaboratively with local Malawi government officials and "like-minded organizations" to implement enforcement initiatives to "stop the illegal killing and trade of wildlife" (see https://www.lilongwewildlife.org/programmes/wildlife-law-enforcement/). Actions LWT have undertaken include co-authoring the *Illegal Wildlife Trade (IWT) Review*, which examined the drivers of IWT in Malawi and made recommendations to strengthen the country's wildlife enforcement and its criminal justice system.

LWT also operates an animal sanctuary, and permits visitors to tour its sanctuary. It was awarded a "Responsible Tourism Award" for conservation of wildlife and habitats in 2011 and has received an 'Outstanding Contribution to Tourism' award from the Malawi Tourist Board.

Findings

LWT defines justice for animals as ensuring they are "free from suffering," which includes "living in the wild wherever possible." The organization has undertaken many on-the-ground projects in order to "contribute to wildlife justice," including operating Malawi's only wildlife sanctuary, establishing a wildlife investigations unit, a wildlife emergency response unit, and a wildlife release program. LWT also operates an animal welfare educational program delivered in local communities and hosts school visits to the sanctuary.

LTW works with government officials to develop guidelines, policy, and legislation. For example, the organization has been involved in developing legislation known as the *Wildlife Act*, and it also developed *Captive Care Guidelines*, for captive animals. When asked specifically which sources, academic or non-academic, the organization utilized to help it understand what justice for animals means, LWT acknowledged it "could do better" in this area. The respondent representing the organization stated, "We are developing [a] technical centre of excellence."

LWT also works toward promoting and enabling justice for animals through its membership in a variety of wildlife-focused organizations and accreditation programs. For example, LWT is a member of the Pan African Sanctuary Alliance (PASA), a coalition of wildlife centres and sanctuaries. It is an accredited/verified organization through the Global Federation of Animal Sanctuaries (GFWA) and was the first organization to receive PAW accreditation, through the Born Free Foundation. It is the only accredited wildlife rescue and rehabilitation sanctuary operating in Malawi. When asked specifically why certification is important to the organization, the organization's respondent stated that certification is "a sign to the outside world that we are adhering to the best possible standards." She added that certification is also "an internal look to ensure we are always striving for operational excellence."

Another vehicle for promoting and enabling justice for wildlife is LWT's membership in national as well as international programs. It is an approved member of the Species Survival Network (SSN) and is the SSN representative for Malawi, working to promote and enforce the *Convention on International Trade in Endangered Species of Wildlife Fauna and Flora* (CITES). LWT is a founding member of the *United for Rangers*, providing support for rangers and working collaboratively on illegal wildlife investigations, and a founding member of the *International Conservation Caucus Foundation* (ICCF), Malawi's Conservation Council. ICCF Malawi supports Malawi's *Parliamentary Conservation Council* (MPCC). LWT also works toward promoting justice for animals by raising the profile of Malawi's wildlife, nationally and internationally, and has been recognized for its contributions to the advancement of tourism in Malawi. For example, it was awarded the 2011 "Responsible Tourism Award" in the *Conservation of Wildlife and Habitats* category. It has also received other recognitions and tourism-related awards for its work.

LWT described the overall importance of tourism (through its volunteer program, educational courses, and day visitors) to the organization in terms of generating approximately 75 percent of the overall revenue for its wildlife sanctuary. LWT runs a fee-based volunteer-education program that provides individuals from around the world the opportunity to take courses and/or work as a sanctuary or research volunteer in Malawi. The program enables volunteers to experience Africa through the various on-the-ground programs it offers.

Edgar's Mission Farm Sanctuary, Victoria, Australia: analysis

Edgar's Mission (founded in 2003 by Pam Ahern) is a non-profit organization, situated in Victoria, Australia. Victoria, the country's smallest mainland state, is the second most populated (population 6.4 million as of 2017 and estimated to reach 6.5 by June of 2020) (Population Australia, 2019). It is located in the southeastern area of Australia. Visit Victoria, the country's destination marketing board, describes the county's diversity of attractions as encompassing beaches and coast lines, national parks and reserves, parks and gardens, natural springs, coastal and alpine villages, wildlife, zoos and farms, heritage attractions, food and wine, architecture and design (Visit Victoria, n.d.).

Edgar's Mission focuses on rescuing and providing sanctuary to mostly farm animals in need (Edgar's Mission, 2019) and it depends upon tourists (~450 per month) for its survival. The organization is in the process of laying "the groundwork to become accredited with the Global Federation of Animal Sanctuaries," but reports that it is "a small organization in a period of growth," and is struggling with time constraints. It currently houses over 450 animals, and encompasses 153 acres of land in Victoria, Australia, approximately 60 kilometres north of Melbourne. The organization's vision is "for a humane and just world for all." Additionally, the organization focuses on education, outreach, advocacy, community enrichment, and farm tours.

Findings

Edgar's Mission defines justice for animals as "fair and just treatment." This was further explained as:

> … extending kindness to (non-human) animals, a group of beings we consider the most vulnerable and least heard among us … righting the wrongs our society has afforded them … encouraging a broader application of kindness and justice for all animals.

Edgar's Mission promotes justice for animals by telling "the stories of the animals as individuals," as it believes "learning and witnessing the animal as individual is far more powerful and engaging than telling the story of 100 sheep." Visitors to the sanctuary learn about a sheep named Molly Brown, the sole survivor of a truck rollover. The intention is "to expand [visitors'] circle of compassion to include all animals" and "to put a face to the millions of sheep who are slaughtered every year in this country." There are parallels for telling stories of animals as individuals in the animal ethics literature, as some scholars believe that zoos, for example, have an obligation to develop programs on interspecies friendship in inducing a higher level of concern for animals (Levin, 2015).

The organization works in other ways to promote justice for animals. For example, visitors to the sanctuary are permitted to "meet the animals [which will] always be on the animal's terms." The organization provides supporting materials to encourage people to make "kinder choices in their life, from plant-based eating," to "cruelty-free products," to "reducing their environmental footprint." It also lobbies governments and decision-makers on behalf of farmed animals," which includes "an active humane education program" and communication through "social media platforms." For example, the organization has been working for years to obtain an exemption from ear-tagging the animals in its care, on the grounds that the practice is cruel (due to health and safety risks). Rather, the organization microchips the animals in its care, which it claims is "an acceptable and legal means of identification of domesticated pets."

The organization states justice for animals is achieved because they are enabled to live their lives in full at the sanctuary. This includes providing for both the physical and psychological needs of the animals at the sanctuary. For example, on a physical level, Edgar's Mission provides rehabilitation and husbandry services and infrastructure (i.e. scratching posts, shade sails, pig showers, a purpose-built mountain for goats, etc.). On a psychological level, they provide "hope" in "hopeless" situations (some animals arrive to the sanctuary in dire condition, both physically and psychologically). In hopeless situations, they ensure that animals who have reached the final life stage, "where a life worth living for them is no longer possible," are provided with a peaceful passage "surround[ed] by love and kindness."

Edgar's Mission considers tourism "vital to our work, as it allows us to spread our message of kindness for all animals exponentially." In its experience, the organization has discovered that visitors often "become inspired" or "further inspired" (in the case of those who already provide support) to financially support the "work and mission," after their visits to the sanctuary. Edgar's Mission is situated in a renowned tourist area, which was a prime consideration for the sanctuary's chosen location in the Macedon Ranges. In addition, the sanctuary recently established three "Tiny Houses," on site, in an effort to attract more visitors. Edgar's Mission states that their "aim is twofold." First, it allows "people to experience our sanctuary and all its magic," and secondly, it "create[s] an income source."

When asked "What is needed to make further progress on enabling justice for animals," the organization stated that "getting a seat at the table" will enable it to make further progress on enabling justice for animals. Specifically, the organization sees the need to establish "ourselves as a reliable and credible source of information." In order to do this the organization recognizes the need for:

> … more opportunities to speak on behalf of animals. Increased awareness about what is happening to animals, matched by how people can stop animal suffering and cruelty they may either be complicit in (i.e. by not buying the products of animal suffering) or be able to have influence in changing behaviours.
>
> (i.e. acts of cruelty and injustice in foreign countries by writing letters, raising awareness)

Increased funding would also enable the organization to "create bigger and more far-reaching media campaigns and potentially expand our landholding."

Overall, the organization reports feeling "greatly buoyed by the ever-increasing amount of media stories and interest in animal matters, from their sentience, the treatment of farmed animals, to plant-based meals and alternatives to animal-based products." In an earlier question, the organization expanded upon these thoughts stating:

> Throughout history the ethical progress of societies has been measured by their ability to embrace those once considered different (color of skin, religion followed, gender), we believe that society's treatment of animals is our next great social justice movement – a belief we are comforted to know we are not alone in having.

As it relates to the COVID-19 pandemic, the organization spoke about the impacts at both the societal and organizational level. For example, Edgar's Mission reported its concern that lockdowns and the resulting financial hardship could result in many uncared-for animals within the general population; however, at the same time they perceived that "some people [appeared] more attentive to their animals and open to adopting animals." Edgar's Mission also spoke about the logistical difficulties for authorities investigating animal cruelty reports during periods of lockdown and stated concern for "lower levels of donations for not-for-profits "working towards a more just world for animals…"

From an organizational perspective, the lockdowns prevented the organization from receiving tour visitors, as well as volunteers who assisted with various activities, including tours. While the organization "offered live virtual tours on Facebook," this approach was perceived to be "not as impactful as physical visits in changing the hearts and minds of people." In addition to the loss of tour revenue, the organization reported that it was not eligible for governmental COVID-19 assistance programs. Edgar's Mission learned from the pandemic experience that it must "remain adaptable" particularly in terms of utilizing various outreach channels "to ensure than animals' voices are heard."

Animal Justice, Canada: analysis

Animal Justice Canada Legislative Fund (Animal Justice) is a non-profit, federally incorporated organization, based in Toronto, Canada. The organization is not directly involved in the rescue or care of animals. Its mission is to "protect animals from cruelty and neglect" in Canada, which takes place mostly at the legal level (Animal Justice, 2022). The organization was founded by

Toronto business lawyer and political campaigner, Nicholas dePencier Wright. Including its current executive director, Camille Labchuck, the organization is comprised mostly of lawyers. They "lead the legal fight for animals in Canada, working for strong new animal protection legislation, vigorous enforcement of existing laws, and fighting for animals in courtrooms across the country" (personal communication of 6 January 2020). The organization seeks to instigate meaningful policy changes through its work and also fights to protect freedom of speech as it relates to animal advocates.

On a policy level, Animal Justice has had many successes. Recent examples include a campaign to hold the Canadian dairy industry accountable for misleading advertising claiming Canadian dairy products contained no growth hormones. In another example, Canada's House of Commons approved recommendations put forth by Animal Justice related to amending Bill C-84, the country's animal anti-cruelty legislation. The changes included making all sexual abuse of animals illegal, including a restriction from owning or residing with animals by those convicted of bestiality. Animal Justice also proposed the repeal of the provision requiring that birds rescued from cockfighting be automatically euthanized. Recovered birds will be individually assessed for their ability to live out the rest of their lives in an animal sanctuary. Bill C-84 received Royal Assent on June 21, 2019. Most recently, Animal Justice filed a complaint against Marineland in Niagara Falls, Canada, as it continues to use "dolphins in illegal performances" (Daily Hive News, 2021).

Findings

While the organization reports that it does not "have a formal policy or definition internally," it defines justice for animals as "support[ing] a world where animals are valued as individuals, and not [on a basis] of what they provide to humans." In understanding what justice means for animals, the organization stated that: "Our views are primarily influenced by philosophical writings, including books and articles by leading animal rights philosophers." Overall, the organization works to promote and or achieve justice for animals by working …

> … to align Canada's outdated animal protection laws with our moral values, both by promoting strong new animal protection legislation and progressive court judgments. We work to expose abusive industries for their mistreatment of animals, and ensure animal abusers are prosecuted.
>
> The organization states that what is needed to make further justice for animals, the organization stated that there is an urgent need for … more funding to compete with the industries that abuse them [meaning animals] and spend significant resources fighting laws that would improve legal protections for animals. Funding would also allow capacity building and greater training, to build a strong workforce of animal advocates committed to executing strategy to improve legal protections for animals.

As it relates to the question regarding whether or not the organization was working toward certification by national or international animal ethics certifying bodies, the organization reported that it was not, adding that while it is not aware of any animal ethics certifying bodies, it did "not see the advantage of this."

Initially, COVID-19 made it challenging for the organization "to do substantial legislative work," as the public, governments, and the media were "focused on COVID, and less interested in hearing about animal protection issues." At the same time, "governments bailed out industries that are harmful to animals, including farms and zoos. Farmers killed off a lot of animals for economic inconvenience, through brutal means like ventilation shutdown, due to supply chain disruptions." Overall, the organization does feel optimistic that "the pandemic may lead to policy action to address the exploitation of animals that contribute to the emergence of novel viruses in humans, including the wildlife trade and farming." The organization reported that it qualified for wage subsidies for "several months," and had access to loans, a percentage of which were forgiven.

Discussion and implications for the tourism industry

The analysis reveals some interesting perspectives regarding how the various organizations define justice for animals. Specifically, the definitions reveal some common language. Although a variety of words were used to describe justice, they could be classified as being similar in intent. For example, the words "kindness" and "fairness" were mentioned, as well as phrases that include "freedom from suffering," "living in the wild wherever possible," and "righting [past] wrongs." We link this finding back to the animal justice research undertaken by Fennell and Sheppard (2021), and suggest that the participant organizations recognize the need for and desire to enable deep justice for animals. Deep justice for animals is defined in terms of living wild, free, and not directly manipulated or controlled by humans. However, the findings also suggest that the organizations recognize that this is not always possible. In such cases, they are focused on enabling and or providing an intermediate level of justice for animals, where the focus is on prioritizing the interests of animals over humans. This latter point is evidenced by the fact that both sanctuary organizations allow human–animal interactions, as appropriate for the individual animals, and as instigated by the animals. In other words, such human–animal interactions are not forced as they would be in organizations functioning at shallow or no justice levels.

Interestingly, two of the organizations specifically referred to valuing animals as individuals. Their focus on valuing animals as individuals provides further evidence of a deep justice perspective, where the focus is on *the wellbeing of individual animals*. Edgar's Mission, one of the two organizations that specifically referred to valuing animals as individuals (the other being Animal Justice), contends that such an approach is more "powerful," than taking a collective approach. Edgar's Mission believes that valuing animals on an individual basis has the potential to "expand the circle of compassion to all animals," not just ones that we might traditionally consider worthy of compassion (i.e. domestic pets). This is particularly the case as it relates to farm animals, which is particularly challenging to the tourism industry. Animals not only serve as a source of entertainment and free labour as it relates to tourism, they are also consumed by tourists.

Overall, all three organizations share many common approaches to promoting and achieving justice for animals. For example, all work in varying capacities at the governance level, whether it be with the aim of influencing policy (LWT, Edgar's Mission) or legislative changes (Animal Justice) which may lead to justice or a deeper level of justice for animals (Fennell & Sheppard, 2021). Influencing policy change can include activities such as creating greater public awareness by providing facility tours (e.g., LWT, Edgar's Mission) (see Section 5.3 re tourism), seeking out public speaking engagements, and working with other organizations (e.g. LTW's collaboration with Africa Geographic) to create awareness, with the aim of influencing policy and legislative change.

LWT and Edgar's Mission both spoke about providing justice to animals directly through the existence of their sanctuaries. Each described the various ways in which they seek to provide justice to animals, which, in addition to providing sanctuary, included collaborating with governments, policymakers, and other strategic organizations. In the case of LWT, it also collaborates on enforcement, emergency responses, and release programs. Both spoke about the role they play in educating the community and beyond as it relates to animal welfare. On the other hand, Animal Justice works indirectly by working to "promot[e] strong new animal protection legislation and progressive court judgments." They are focused on "align[ing] Canada's outdated animal protection laws with our moral values," and "expos[ing] abusive industries for their mistreatment of animals and ensure animal abusers are prosecuted."

Overall, the activities of all three organizations seek to pressure governments to acknowledge and accept their legal obligations to improve the lives of animals. An expansive positive legal rights perspective, as advocated by Shooster (2015), would extend justice to animals as it relates to ensuring adequate care in the areas of shelter, food, reproduction, habitat, clean air and water, etc. While on the surface it would appear that these aspects could easily be legislated, governments appear reluctant to go down this path, particularly from a tourism perspective. As observed by Sheppard and Fennell (2019), given tourism's connection to gastronomy and the fact that animals are not merely a source of entertainment, they are also consumed by tourists (and in some cases fed to other animals as a source of entertainment) it seems unlikely that we will see such legislation in the near future (see https://www.nationalgeographic.com/news/2017/06/wildlife-watch-china-donkey-tigers-zoo/).

Two of the three organizations indicated a desire to have knowledge of and access to academic and non-academic sources to assist with their understanding of what is meant by justice for animals. This suggests the need for those involved in related academic research to disseminate their research findings beyond the narrow world of academia, to those who are actually working in animal sanctuaries and those fighting for justice for animals. The same could be said for tourism-related operations that involve the use of animals as part of the tourism experience. Two of the organizations spoke about the negative impacts of the COVID-19 pandemic; however, both felt optimistic that this might provide an opportunity for making progress on animal justice.

The analysis also reveals some valuable implications from a tourism management perspective in terms of two findings: 1) the importance of prioritizing the interests of animals over humans, such that interactions are instigated by the animals; and 2) the role organizations can and, we would argue, should play in creating awareness amongst their visitors as to the power they (meaning the visitors) hold in influencing policy change at the governmental level. For example, organizations should consider how they can better design the visitor experience, such that animals are able to initiate interactions with human. They should also consider how to use their influence amongst visitors to create awareness of visitors' power to influence positive policy change (e.g. marketing materials, speaking points during tours). The COVID pandemic, ongoing at the time of publication, presents opportunities for the tourism industry to refocus its recovery efforts on a enabling a more sustainable, just, and resilient tourism industry, where enabling and promoting justice for animals used in tourism is a primary focus.

Limitations and future research

This study is limited by the small number of case studies, which were specifically focused on understanding how animal advocacy and sanctuary organizations define and apply the principles of justice presented here. Future studies could expand upon these case studies, and also include organizations that do not mention the words justice/injustice or just/unjust. We contend that understanding how tourism organizations define and apply the principles of justice, if at all, will provide foundational knowledge that could then be utilized to craft literacy and education programs for tourists and operators. Similarly, future research could focus on an information gathering process which could then be utilized to create an online library-resource centre for use by a variety of tourism stakeholders (e.g. governments, operators, educators, tourists, destination marketing boards). An emerging area of research is related to COVID-19 and the need to understand the opportunities and challenges COVID-19 presents in making progress on animal justice. We conclude that, as a next step, it is critical to examine tourism organizations who directly use animals either as beasts of burden, entertainment, trophies, or some other manner, in order to understand how they define justice and apply the principles of justice in their operations.

Conclusion

As justice continues to advance in the realm of tourism studies and practice, we expect to see an emerging climate where the social costs associated with the lack of justice is more closely examined, related not only to human animals, but also to non-human animals. To this point, we assume that those offenders – both brokers and tourists – will increasingly have to answer to a public that demands a higher level of regard for those who are unable to speak for themselves (Shooster, 2015). Animal advocacy and sanctuary organizations will play an increasingly important role as it relates to seeking and providing justice for animals. Indeed, we argue that the proper surrogate to take up the cause for animal rights is organizations, such as those highlighted in this research. They may have the best ability to make progress, given the reluctance, unwillingness, and/or inability of governments to secure expansive legal rights for animals. We conclude by pointing the finger back at ourselves, acknowledging that academics studying in this field, must do a better job ensuring animal advocacy and sanctuary organizations have the resources they require for advancing the animal justice agenda. And we argue that the time has come for a dedicated animal justice organization to push the justice agenda forward in tourism studies and practice.

Declaration of interest

None for both authors. All organizations identified in this study have given their permission to be so identified. The authors also wish to thank the organizations who participated in this study, as well as the reviewers who provided valuable comments and suggested edits. We are grateful to all for their time and input.

References

ABC News. (2020). *More than 1 billion animals estimated dead in Australia wildfires, expert says*. Retrieved from https://www.abcactionnews.com/news/world/more-than-1-billion-animals-estimated-dead-in-australia-wildfires-expert-says

Animal Justice. (2022). *Our mission*. Retrieved from https://animaljustice.ca/about-us

Broad, S. (2008). Living the Thai life—a case study of volunteer tourism at the Gibbon Rehabilitation Project, Thailand. *Tourism Recreation Research*, *18*(3), 63–72.

Burns, G. L., MacBeth, J., & Moore, S. (2011). Should dingoes die? Principles for engaging ecocentric ethics in wildlife tourism management. *Journal of Ecotourism*, *10*(3), 179–196.

Butler, R. W., Curran, R., & O'Gorman, K. D. (2013). Pro-poor tourism in a first world urban setting: Case study of Glasgow Govan. *International Journal of Tourism Research*, *15*, 443–457.

Caton, K. (2012). Taking the moral turn in tourism studies. *Annals of Tourism Research*, *39*(4), 1906–1928.

Daily Hive News. (2021, October 6). *Complaint filed against Marineland over dolphin performances*. Retrieved from https://dailyhive.com/toronto/complaint-filed-against-marineland-dolphin-performances

Edgar's Mission Farm Sanctuary. (2019). *About us*. Retrieved from https://www.edgarsmission.org.au/about-us/

Fennell, D. A. (2012a). *Tourism and animal ethics*. London: Routledge.

Fennell, D. A. (2012b). Tourism and animal rights. *Tourism Recreation Research*, *37*(2), 157–166.

Fennell, D. A. (2018). Sustainability ethics in tourism: The imperative next imperative, *Tourism Recreation Research*, *44*(1), 117–130. doi:10.1080/02508281.2018.1548090

Fennell, D. A., & Sheppard, V. A. (2011). Canada's 2010 Olympic & Paralympic Winter Games' legacy: Applying an ethical lens to the post-games' sled dog cull. *Journal of Ecotourism*, *10*(3), 197–213. doi:10.1080/14724049.2011.617452

Fennell, D. A., & Sheppard, V. A. (2021). Tourism, animals and the scales of justice. *Journal of Sustainable Tourism*, *29*(2–3), 314–335. doi:10.1080/09669582.2020.1768263

Garner, R. (2012). Rawls, animals and justice: New literature, same response. *Res Publica*, *18*, 159–172.

Government of the Republic of Malawi. (2013). *Government of Malawi*. Retrieved from http://www.malawi.gov.mw/

Hirschl, R. (2000). "Negative" rights vs. "positive" entitlements: A comparative study of judicial interpretations of rights in an emerging neo-liberal economic order. *Human Rights Quarterly*, *22*(4), 1060–1098.

Hughes, P. (2001). Animals, values and tourism – structural shifts in UK dolphin tourism provision. *Tourism Management*, *22*, 321–329.

Kline, C. (2018) (Ed.). *Tourism experiences and animal consumption: Contested values, morality and ethics*. Abingdon, Oxon, UK: Routledge.

Leading-Animal-Welfare-Policy-With-A-Commitment-To-End-Whale-Dolphin-Captivity. Visit Victoria. (n.d.). *Victoria's Regions*. Retrieved from https://www.visitvictoria.com/regions

Levin, A. (2015). Zoo animals as specimens, animals as friends: The life and death of Marius the giraffe. *Environmental Philosophy*, *12*(1), 21–44.

Lilongwe Wildlife Trust (n.d.). *About Lilongwe Wildlife Trust*. Retrieved from https://www.lilongwewildlife.org/about/about-lwt/

McKibben, B. (2019). *Falter: Has the human game begun to play itself out?* New York: Henry Holt and Company.

Moorehouse, T., D'Cruze, N. C., & Macdonald, D. W. (2017). Unethical use of wildlife in tourism: What's the problem, who is responsible, and what can be done? *Journal of Sustainable Tourism*, *25*(4), 505–516.

Niven, C. D. (1967). *History of the human movement*. London: Johnson.

Nozick, R. (1974). *Anarchy, state and utopia*. Oxford: Basil Blackwell.

Pierce, J., & Bekoff, M. (2012). Wild justice redux: What we know about social justice in animals and why it matters. *Social Justice Research*, *25*(2), 122–139.

Population Australia. (2019). Population of Victoria 2020. Retrieved from http://www.population.net.au/population-of-victoria/

Rawls, J. (1971). *A theory of justice*. Oxford: Oxford University *Press*.

Regan, T. (2004). *The case for animal rights*. Los Angeles: University of California Press.

Rowlands, M. (1997). Contractarianism and animal rights. *Journal of Applied Philosophy*, *14*, 235–247.

Sanson, L. (1994). An ecotourism case study in sub-Antarctic islands. *Annals of Tourism Research*, *21*(2), 344–354.

Shani, A. and Pizam, A. (2008). Towards an ethical framework for animal-based attractions. *International Journal of Contemporary Hospitality Management*, *20*(6), 679–693.

Sheppard, V. A., & Fennell, D. A. (2019). Progress in tourism public sector policy: Toward an ethic for non-human animals. *Tourism Management*, *73*(August), 134–142. doi:10.1016/j.tourman.2018.11.017

Shooster, J. (2015). Justice for all: Including animal rights in social justice activism. *The Harbinger*, *40*(39), 39–44.

TripAdvisor. (2019, October 2). TripAdvisor updates industry-leading animal welfare policy with a commitment to end whale & dolphin captivity. Retrieved from https://tripadvisor.mediaroom.com/2019-10-02-TripAdvisor-Updates-Industry-

World Travel & Tourism Council. (2018). *Travel and Tourism Economic Impact 2018 Malawi*. Retrieved from https://www.wttc.org/economic-impact/country-analysis/country-reports/#undefined

Yin, R. K. (2003). *Case study research: Design and methods* (3rd ed.). Thousand Oaks, CA: Sage Publications.

3

A CRITICAL INQUIRY INTO THE SUSTAINABILITY OF NBT

Responsibility of the traveler

Bengi Ertuna, Burcin Kalabay Hatipoglu, Emre Cantekinler, Selen Dağ, Burcu Ormanlıoğlu and Serçin Dilem Şenkal

Introduction

Tourism has emerged as a policy tool for development in rural areas, characterized by rich natural resources and limited economic development opportunities. There is also a current rise in tourist interest in natural and rural areas, especially following the Covid-19 pandemic (UNWTO, 2020). However, the increased number of travelers creates various sustainability challenges in these areas with fragile socioeconomic structures, making travelers a critical stakeholder category, shaping nature-based tourism (NBT) through their preferences and dynamic interaction with other stakeholders and tourism providers.

Even before the pandemic, policies for developing tourism in rural areas have been advocated by international organizations for spreading tourism activity to mitigate the negative consequences of over-tourism (Galvani et al., 2020; UNWTO, 2020). Following the international trends, policies for applying tourism activity and the current rise of tourist interest in rural areas have been accompanied by plans to develop new NBT destinations in Turkey. For example, the General Directorate of Forestry in Turkey has recently announced an Ecotourism Action Plan, which included management plans to develop ecotourism infrastructure for 110 new destinations (The Republic of Turkey, Ministry of Agriculture and Forestry, 2021). All these developments raised concerns over the sustainability of increased NBT activity in rural areas, impacts on the communities and the environment, and the responsibility of the tourism providers and tourists. Tourist motivations and preferences and the dynamics of their interaction with the environment and communities make responsible tourists an essential element of the sustainability of tourism in rural areas, including NBT. In this regard, studies that focus on "sustainable tourists" may contribute to identifying ways of building tourist awareness and creating sustainable travel alternatives (Seeler et al., 2021, p. 5).

The sustainable tourist has been the focus of many studies. A strand of these studies aimed to identify sustainable tourist definitions, which extract several elements of sustainable tourism, including transformative travel experience, green behaviour, going into the culture of the destination, protecting the environment, and consuming less (Juvan and Dolnicar, 2014; Seeler et al., 2021). In their construct for conceptualizing and measuring sustainable behaviour among tourists, Lee et al. (2013) define a sustainable tourist as a person that "respects local culture, conserves natural environment, and reduces interference of local environment"

DOI: 10.4324/9781003230748-5

(p. 457). These stated aspects of sustainable tourist definitions are expected to build a tourism activity that would generate positive contributions and minimize the harm to the fragile communities and environment in rural areas. Based on this expectation, we aim to question the congruence between sustainability perceptions and NBT travel preferences of the travelers to identify sustainability gaps for NBT. In doing so, using a sample of NBT travelers in Turkey, we first question the travel motivations and experience preferences of nature-based tourists. We then critically evaluate the consequences of nature-based travelers' preferences and relate these with their sustainability perceptions to identify sustainability gaps potential threats to the future of NBT. We suggest that it is essential to identify these threats and offer solutions for tourism planners to maintain and grow the natural resources to benefit the present and future communities and travelers.

Travelers and nature-based tourism

Consumption of tourism is a complex, dynamic and multi-dimensional process (Sharpley, 2018, p. 125). The tourists' values, perceptions, preferences, and motivations also evolve through time and influence their travel planning decisions (Luo and Deng, 2008). These decisions have significant ramifications on how tourism development will affect the communities and environment in different contexts, especially in fragile natural areas (Budeanu et al., 2016). Consequently, the role and responsibility of the traveler become critical for the sustainability of NBT.

Understanding tourists' preferences and behaviour is significant for product development and services management (Adam et al., 2019) and sustainably developing nature-based tourism. When tourists possess environmental beliefs and concerns, their attitude (Hultman et al., 2015; Teeroovengadum, 2019) and intentions for NBT are positively affected (Pham and Khanh, 2021). Yet, evidence from the field suggests that tourists' attitudes and motivations are heterogeneous for visiting natural areas (Luo and Deng, 2008). For example, Luo and Deng (2008) regard being close to nature, learning about nature, and escaping from routine and issues with the cities as motivational factors for pursuing nature-based attractions. Moreover, Carvache-Franco et al. (2019) state that the prime motivations of ecotourists are self-development, building relationships, and ego-defensive action. Further, Kastenholz et al. (2018) note that besides the appeal of recreational activities, some visitors are also attracted by an opportunity to interact with local nature, culture, and communities. And for some, enjoyment of nature also involves engagement with the wildlife and endemic plants (Zhang et al., 2021). Furthermore, tourists can drive multiple values from a tourism experience (Breiby et al., 2020). Overall, the studies infer that leisurely activities, self-development, and social factors are reasons for travelers visiting nature-based destinations (Adam et al., 2019).

Many studies have aimed at developing a typology based on the environmental commitment of tourists and their support for tourism's role in development (Buffa, 2015; Kastenholz et al., 2018; Zhang et al., 2021). They identified different segments of tourists based on motivational and behavioural factors, such as ecotourists, nature-based tourists, responsible tourists, and green tourists (even different shades of green). While identification of these different segments provides a nuanced understanding of the attributes that may influence sustainability implications, broader definitions include all these segments, with a general interest in ecotourism but differing degrees of environmental commitment and different motivations for participating in NBT activity, and form the basis of classifying and understanding the ecotourism market (Ruhanen, 2019). Consequently, sustainability implications can be expected to vary depending on these differences embodied in broader definitions. For example, lack of awareness on sustainability implications of their tourism activity, lower levels of environmental commitment, or a motive for engaging in NBT for acquiring social status may influence sustainability implications of tourists' behaviour. In addition, the literature identifies inconsistencies between attitudes and behaviours of ecotourists. For example, Juvan and Dolnicar (2014) identify an attitude–behaviour gap in sustainable tourism and investigate this gap's reasons. Mkono and Hughes (2020) focus on "eco-guilt and eco-shame" in the context of tourism consumption, arising from the discrepancy between tourists' awareness of the environmental impacts of their tourism activity. Whatever the reasons are for the lack of tourists' sustainable behaviour, the cultural transformation of the traveler is advocated for sustainable tourism implications (Budeanu and Emtairah, 2014). Such a transformation entails multiple aspects of the tourism activity and experience.

In her study of post-consumption reflections of travelers, Ruhanen (2019) identified seven themes for NBT experience preferences and ecotourism products. These themes include "staff, nature, facilities, food, activities, learning and experience" (p. 112). While the "staff" theme facilitated satisfaction with the experience, the "facilities" theme included quality of accommodation and amenities as the second most frequently emerging theme in traveler evaluations, reflecting the importance attributed to "high-quality standards and access to well-functioning modern amenities" (p. 113). In addition, her findings reveal a dissatisfaction of some travelers with higher prices and cost of eco-friendly accommodation. Along the same lines, income generated for host communities and effects on local culture was picked up only by a few customers. Depending on the analysis of travelers' evaluations and these themes, Ruhanen (2019) concluded that service and experience-related aspects are more important to travelers than aspects related to sustainability.

An independent review on the economics of biodiversity recently suggested that we place natural capital at the centre of our economies and advise decision-makers to shift consumer preferences through behavioural change and pricing (Dasgupta, 2021). Thus, changes in travel patterns can have favorable results on reducing the pressures on biodiversity and contribute to the building of the natural capital. Despite the importance of the role of the traveler for sustainability transformations, there is scarce scholarly work highlighting their responsibility (Seeler et al., 2021). We aim to fill this gap and identify threats to sustainability in NBT by analyzing the relationship between travelers' sustainability perceptions and their preferences on NBT experience attributes as indicated by the conceptual framework presented in Figure 3.1.

Methods of the study

This research aimed to explore possible threats to NBT through travelers' influences. Using a mixed-methods research design, opinions and insights of travel agencies were obtained through qualitative techniques, while travelers' preferences were collected through quantitative methods. The exploratory sequential approach to data collection and analysis moved in two phases (Molina-Azorin and Font, 2016). The qualitative data helped develop the questionnaire, narrow the focus of our variables, and enable local adaptation. After the questionnaire administration, the results of both sections were interpreted together. Other researchers have also used mixed-methods and sequential design to study NBT and sustainability (Spenceley, 2005). The exploratory sequential design is used for developing and testing new instruments (Creswell et al., 2008, p. 182). The use of qualitative methods and quantitative results in this study enabled the researchers to include the insights of travel agencies for explaining how travelers decide and behave in nature-based settings while supporting the investigation results.

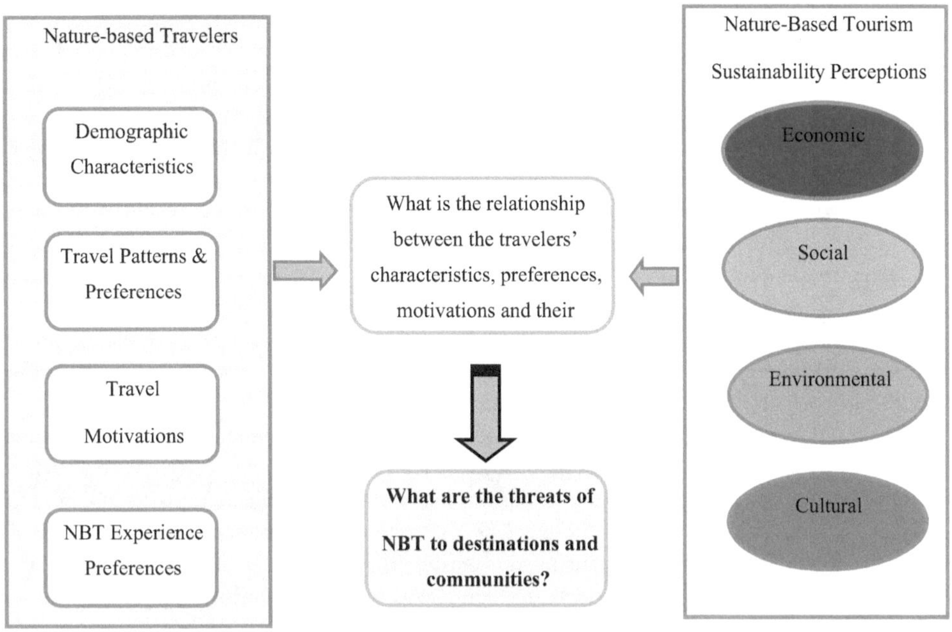

Figure 3.1 Conceptual model.

Data collection

We focused our data collection on three NBT travel agencies in Turkey. These agencies organize small-group, experience-based tours to various nature-based destinations in the country. As required by our study's purpose and research design, the agencies have social media presence and display an explicit statement of #responsible travel. They plan and organize socially, culturally, and environmentally responsible tours into these destinations and communities.

We first conducted semi-structured interviews with the owners/managers during March 2021 and one group interview with all three managers on April 15, 2021. Due to the pandemic restrictions, interviews were held online, recorded with participants' permission, and transcribed. In the discussions, we focused on the motives, preferences, and sustainability understanding of the travelers and the dyadic relationship between the travelers and customers in shaping the outcomes of NBT. After analyzing the results of these interviews, we adapted and localized the dimensions we identified from the literature and structured our questionnaire (32 questions) for the online survey.

Based on our conceptual model, the variables used in this study are as follows: Demographic variables include respondents' gender, age, education, income level, and city of residence (see Table 3.1 for categories of demographic variables). "Travel patterns and preferences" consist of annual travel frequency, annual rural travel frequency, rural travel during the pandemic, period of stay in rural travels, preferred type of accommodation during rural travels, and minimum requirements from accommodation (see Table 3.2 for operationalization of variables of travel patterns and preferences and Table 3.2. for preferences for accommodation). Binary variables (yes/no) on various motivations (Luo and Deng, 2008; Carvache-Franco et al., 2019) for travel to rural areas were used to identify respondents' travel motivations (see Table 3.4 for motivation categories). To identify NBT experience preferences, a 5-point Likert scale (1 = strongly disagree to 5 = strongly disagree) reflecting agreement with statements on personal preferences relating to the dimensions of transportation, food & beverage, safety & security, prices, host community interaction, and activity is used. Statements reflecting preferences on each of these dimensions are identified from the literature (Esfandiar et al., 2019; Ruhanen, 2019; Breiby et al., 2020) were reduced and adapted in the group interview with the company manager/ owners. The resulting statements used for each of these dimensions are available in Table 3.5, and the notation (R) is used to indicate reverse statements according to sustainable tourist definitions (Lee et al., 2013; Juvan and Dolnicar, 2014; Seeler et al., 2021). Lastly, NBT sustainability perceptions are identified using a 5-point Likert scale (1 = strongly disagree to 5 = strongly disagree) was used on statements reflecting sustainability perceptions of the respondents on the economic environmental, social, and cultural dimensions of sustainability. Statements identified from the literature (Panzer-Krause, 2020) on these dimensions were adapted, as shown in Table3.6, in which reverse coding is indicated by the notation (R).

We administered the survey on a sample of customers following the social media accounts of these three companies. Similar methodologies were used in recent research on ecotourism, in which data were derived from digital forums and platforms (Ruhanen, 2019; Mkono and Hughes, 2020). Our survey was announced to the followers and posted on the social media accounts of the three companies during the period May 26-June 4, 2021. We arrived at a sample of 231 travelers who responded and completed the survey during this period.

Data analysis

Based on the survey results, we critically evaluated the relation between travel patterns and NBT experience choices with sustainability perceptions of the travelers. The SPSS Statistics Version 27 was used for data analysis. First, we identified the NBT travelers' demographic characteristics, travel motivations, travel patterns and preferences, and NBT experience preferences as frequency results of survey questions. We then critically evaluated travel patterns and preferences and NBT experience preferences, considering the attributes of sustainable tourists identified from the literature. Second, using hierarchical regression analysis, we relate the NBT experience preferences to different dimensions of NBT sustainability perceptions of the travelers, investigating the role of travel motivations while controlling for demographic characteristics. In doing so, we specify four models, taking each dimension of sustainability perceptions (economic, environmental, social, and cultural dimensions) as the dependent variables and dimensions of NBT experience preferences (transportation, food, and beverage, safety, and security, prices, host community interaction, and activity) and travel motivations as independent variables, while controlling for demographic characteristics of the travelers.

Results

Our sample consists of 231 respondents from the followers of the social media accounts of three NBT travel companies in Turkey. Almost 75 percent of our respondents are female, and nearly half of the sample is above 40 (mean age 40.38 and median age 39). Respondents are highly educated (56.7 percent with undergraduate and 31.6 percent with a graduate degree). The profile of our sample confirms with previous studies that NBT travelers are "older, better educated, and better off than general travelers" (Buffa, 2015 p. 14045). The demographic characteristics of our sample are presented in Table 3.1.

Respondents of our study are frequent travelers, including travels to rural areas. Almost 60 percent of our respondents travel more than four times a year, while 51.3 percent travel to

Table 3.1 Sample Characteristics

Demographic Characteristics	#	%
Gender		
Female	170	74.9
Male	57	25.1
Age		
18–29 years old	41	18
30–39 years old	80	35.1
40–49 years old	56	24.6
50 and above	51	22.4
Education		
High School	27	11.7
University	131	56.7
Graduate	73	31.6

Source: Compiled by authors

Table 3.2 Travel Patterns

Travel pattern and preferences	#	%
Annual travel frequency		
Once	11	4.8
Twice	39	16.9
Three times	44	19.0
Four times and more	137	59.3
Annual rural travel frequency		
None	19	8.2
Once	65	28.1
Twice	46	19.9
Three times and more	119	51.3
Rural travel during the pandemic		
Traveled to rural destinations	158	68.4
Plans to travel next year	212	91.8
Period of stay in rural travels		
Daily trips to rural areas	21	9.1
One to two days	59	25.5
Three to five days	103	44.6
One week and more	48	20.8

Source: Compiled by authors

rural areas three or more times annually. During the COVID-19 pandemic, their travel patterns continued; most of the respondents (68.4 percent) traveled to rural destinations, and almost all (91.8 percent) had plans to travel to rural areas next year. In their travels to rural areas, they mostly stay for three to five days; only 9.1 percent prefer daily trips, while 20.8 percent stay one week or more in the rural areas. Travel frequency results are presented in Table 3.2.

The most preferred types of accommodation are bungalows, hotels, and guesthouses (25.5, 23.4, and 22.1 percent, respectively). Tent or caravan is preferred for accommodation by 17.7

percent, and staying in a local home is the preferred accommodation type for only 11.3 percent of the respondents. Most of the respondents' minimum requirements include running water/hot water and electricity in the accommodation facilities. More than half of the respondents (51.5 percent) state private bathroom as a necessity, while internet/Wi-Fi, air conditioning, and television are stated among minimum requirements 61, 30, and 10 respondents, respectively. Several travelers said that their minimum requirements from accommodation facilities differ when they travel with their children.

Our findings on accommodation facilities support the conclusions of Ruhanen (2019) regarding the importance of modern amenities for travelers' NBT experience. Some travelers even require amenities that might have negative environmental impacts (such as air conditioners) or need replacing local amenities with modern ones (such as replacing traditional furnaces for heating with radiators). These demands of the travelers also resonate in the host community. One of the owners/managers of NBT travel companies reported that one guesthouse replaced the traditional furnaces with modern electric radiators upon the travelers' requests. Thus, the everyday comforts were necessary for the visitors, as confirmed in our research, supporting previous research (Ruhanen, 2019). Results of respondents' preferences on accommodation facilities and amenities are presented in Table 3.3.

Interest in nature is the primary motivation for NBT for most respondents (87 percent). On the other hand, discovering new places and getting away from city life are the second and third most frequently cited motivations for NBT travels, reflecting the motivations of more occasional ecotourists (Ruhanen, 2019). Interest in a rural lifestyle and experiencing local food and culture are shared by almost 35 percent of the respondents as motivations for travel, aligning with the definitions of ecotourism (UNWTO, 2018) and sustainable tourists (Lee et al., 2013). While a considerable number of respondents (29 percent) report feeling safe from COVID-19 risk in nature among the reasons for participating in NBT, lower prices in natural areas are stated as a motivation by 9.1 percent of respondents. These two motivations also hint at the presence of more occasional NBT travelers. Most of the motivations expressed by the respondents confirm the definition of ecotourism by the UNWTO (2018), which suggests that "the main motivation of the tourist is the observation and appreciation of nature as well as the traditional cultures prevailing in natural areas" (Ruhanen, 2019, p. 111). Statistics relating to motivations for NBT travel are presented in Table 3.4.

Table 3.3 Preferences for Accommodation

Preferred accommodation type # %		
Bungalow	59	25.5
Hotel	54	23.4
Guesthouse	51	22.1
Tent or caravan	41	17.7
Local Home	26	11.3
Minimum requirements for accommodation		
Running water/hot water	170	73.6
Electricity	158	68.4
Private bathroom	119	51.5
Internet/Wi-Fi	61	26.4
Air conditioning	30	13.0
Television	10	4.3

Source: Compiled by authors

Table 3.4 Motivations for NBT Travel

Motivation	#	%
To be in nature	201	87.0
To discover new places	143	61.9
To get away from city life	139	60.2
Interest in a rural lifestyle	82	35.5
To experience local food and culture	80	34.6
To feel safe from Covid-19 risk	67	29.0
To meditate in nature	31	13.4
Lower prices	21	9.1

Source: Compiled by authors

Sustainability perceptions of travelers do not statistically differ based on their motivations, except the motivations of experiencing local food and culture and lower prices. According to univariate statistics on the differences between motivations and sustainability perceptions, there is a statistically significant difference between the sustainability perception scores of travelers that seek local experiences and those that do not, on dimensions of environmental ($p = 0.048$) and social ($p = 0.000$) dimensions of sustainability. Those travelers who have a motivation to experience local food and culture have higher scores relating to their perceptions of sustainability of NBT on environmental (with a mean of 4.050 vs. 3.825 out of 5) and social (with a mean of 4.154 vs 3.823) dimensions. However, this motivation does not significantly differentiate the travelers on the economic and cultural dimensions of sustainability. On the other hand, lower prices as a motivating factor for NBT leads to a significant difference in NBT sustainability perceptions of travelers only on the economic dimension. Those who are motivated by lower prices have statistically lower ($p = 0.016$) economic sustainability perception (mean score 3.381) as compared to those that are not stating lower prices as their motivations for NBT (mean score 3.712). Consequently, seeking local experiences emerges as a significant motivation that is associated with travelers' sustainability perceptions.

Our results on respondents' preferences on different dimensions of their NBT experience reflect mixed implications on the sustainability of NBT as they vary concerning their inclination for personal benefits vs. benefits for sustainability. They generally prefer to drive to the destination and expect a developed transportation system, reflecting a preference for ease of access. Almost unanimously, travelers indicate a preference for local food and restaurants. Consistent with this finding, their preference for packaged food and expectations for food according to special dietary requirements (e.g., vegan, lactose-free, gluten-free) are low. They also do not look for or prefer chain restaurant brands.

On the other hand, most travelers expect to have health services and security personnel at the destination. Furthermore, they mostly require COVID-19 safety certificates. The responses on transportation, food and beverage, and safety and security indicate the importance of service quality for travelers and a preference for local food. Yet travelers expect lower prices for food and beverage, free entry to natural parks, and lower price levels in NBT areas than city destinations. Travelers' expectations for lower prices in the NBT experience hinder sustainability benefits but align with the dissatisfaction of ecotourists due to premium prices, as similarly reported by Ruhanen (2019).

In their NBT experience, travelers would like to be treated as guests, not customers, in their interaction with the host community. They prefer the host community not to impose lifestyle restrictions (e.g., restrictions on alcohol and dress code), a finding that may indicate

Table 3.5 Preferences on NBT Experience Attributes

Score on 5-point Likert scale (1 = strongly disagree to 5 = strongly agree)	*Mean*	*Std. Dev*
Transportation		
Prefer to drive to the destination	3.77	1.28
Expect a developed transportation system	3.05	1.26
Food and beverage		
Prefer local food and restaurants	4.77	0.54
Prefer packaged food - R*	2.23	1.15
Expect special dietary requirements to be satisfied - R	2.25	1.28
Prefer chain restaurant brands - R	1.32	0.68
Safety and security		
Require Covid-19 safety certificates	3.40	1.39
Expect to have nearby health services	3.84	1.11
Expect to have security personnel in the destination	3.12	1.29
Prices		
Expect lower prices than city destinations - R	3.54	1.16
Expect free natural park entry - R	3.61	1.44
Expect lower food and beverage prices - R	3.63	1.22
Host community interaction		
Prefer no restrictions from the local community - R	4.41	0.98
Willing to comply with local cycles	3.97	1.06
Prefer a local guide	4.08	1.12
Would like to be treated as a guest, not a customer	4.54	0.79
Activity		
Would like to participate in local activities	4.00	1.12
Would like to participate in sports activities	4.55	0.76
Would like to take courses	3.72	1.29

* R represents reverse coded statements in further analysis for calculating dimension scores concerning implications on sustainability.

a possible conflict with local culture. Conversely, they show a general willingness to comply with the community's local life cycles (e.g., waking up or mealtimes). Travelers also display an interest in the learning dimension of NBT experience, as indicated by their preference for a local guide. In the activity component of the NBT experience, most of the travelers would like to participate in sports activities; they are generally willing to participate in local activities or take courses (e.g., yoga, cooking), reflecting the importance of learning experiences in NBT and making the NBT experience more meaningful (Breiby et al., 2020). Agreement scores (mean and standard deviation) of travelers with the statements reflecting their preferences on NBT experience attributes are presented in Table 3.5.

NBT experience dimensions are not correlated with each other except the relation between preferences of safety & security and transportation (correlation coefficient (r) = 0.385, p = 0.000); safety and security and food & beverage (r = −0.240, p = 0.000); and the negative correlation between preferences for price and transportation (r = −0.292, p = 0.000); price and safety and security (r = −0.288, p = 0.000). There is also a positive relationship between activity and host community interaction (r= 0.190, p = 0.004). Absence of a significant relationship between the other dimensions on a univariate level suggests that these dimensions relate to different aspects of the NBT experience.

We investigate the sustainability perceptions of travelers relating to NBT, using statements reflecting economic, environmental, social, and cultural dimensions of sustainability. The findings reveal differences in sustainability perceptions about different dimensions and mixed signals within dimensions of sustainability. In the economic dimension of sustainability, while most travelers strongly agree that NBT generates income for the locals, the majority also agree that NBT provides inexpensive travel options, resulting in a low contribution to local income. On the environmental dimension, there is no consensus on the sustainability impact. Overall, travelers neither agree nor disagree with the statement that NBT depletes natural resources. The finding resonates with the debated question on whose responsibility is to take care of the environment (Mkono and Hughes, 2020). On the other hand, travelers mostly agree that NBT provides nature-friendly accommodation.

On the social dimension, travelers' perceptions of sustainability of NBT are primarily favorable. They agree that NBT decreases migration from rural areas and provides opportunities for local women. However, some perceive a negative impact of NBT on authenticity and social fabric. In the cultural dimension, sustainability perceptions are also mostly positive, indicating a general agreement that NBT helps visibility of cultural assets to preserve local recipes, but with a possible threat of changing local culture. Sustainability perception scores (mean and standard deviation) on statements reflecting different dimensions of sustainability is presented in Table 3.6.

When we investigate the relations between different dimensions of sustainability perceptions of travelers at the univariate level, we observe that the economic sustainability perceptions score is not related to other dimensions. On the other hand, there is a significant relationship between environmental and social sustainability ($r = 0.441$; $p = 0.000$); environmental and cultural sustainability ($r = 0.394$, $p = 0.000$); social and cultural sustainability ($r = 0.394$, $p = 0.000$) perceptions of travelers about NBT. It is worth noting that travelers have consistent perceptions relating to environmental, social, and cultural dimensions while their perceptions of the economic sustainability of NBT diverge from others.

Table 3.6 Sustainability Perceptions

Score on 5-point Likert scale (1 = strongly disagree to 5 = strongly agree)	*Mean*	*Std.*
Nature-based tourism		
Economic		
Generates income for locals	4.77	0.517
Provides inexpensive travel option – R	3.40	1.130
Environmental		
Depletes natural resources – R	2.47	1.295
Provides nature-friendly accommodation	4.28	0.900
Social		
Provides opportunities for local women	4.29	0.926
Destroys authenticity and social fabric – R	2.81	1.180
Decreases migration from rural areas	4.33	0.954
Cultural		
Changes local culture – R	3.36	1.226
Preserves local recipes	4.52	0.739
Provides visibility of cultural assets	4.79	0.505

R represents reverse coded statements in further analysis for calculating dimension scores concerning implications on sustainability.

We analyze the relation between sustainability perceptions, NBT experience preferences, and motivations using hierarchical regression analysis. There is a significant relationship between different NBT experience preferences and sustainability perceptions of travelers considering the economic ($p < .001$), social ($p < .01$), and cultural ($p < .05$) dimensions of sustainability, while there is no statistically significant relationship between environmental sustainability perceptions and NBT experience preferences of travelers. The absence of a relationship between the experience preferences and perceptions on the environmental sustainability of NBT may indicate travelers' lack of awareness of the environmental consequences of NBT activity. We also investigated the role of motivations in this relationship by introducing the motivation of seeking local experiences in the second step of the hierarchical regression. We also found a significant additional explanatory power of motivation for the social dimension of sustainability perceptions. In other words, seeking local experiences improved the significance of the relationship between social sustainability perceptions and experience preferences.

Perceptions of the travelers about the economic sustainability of NBT are significantly related to the experience preferences related to price ($p < .001$). As the expectations that prices related to NBT experience products/services increase, the perceptions of travelers on the economic sustainability of NBT increase. Awareness and acceptance of premium prices willingness to pay for (entry fee for natural parks) NBT increase economic sustainability perceptions on NBT. There is also a significant relationship ($p < .05$) between economic sustainability perceptions and preferences on safety and security dimensions, including the presence of health facilities and security personnel, which can increase economic development in the region.

Sustainability perceptions on the social dimension are significantly related to preference scores on host community interaction and activity dimensions. As the scores on preferences for interaction with locals and participating in local activities increase, social sustainability perceptions of NBT increase. In addition, there is also a significant and positive influence of the control variable of age. As the traveler's age increases, their perception of the social sustainability of NBT also increases. Results of this model indicate the importance of interactions between the travelers and host communities and learning opportunities for the sustainability of NBT on the social dimension.

On the cultural dimension, there is a positive relationship between preferences relating to food and beverage and safety and security and social sustainability of NBT, but the relationship is slightly significant ($p < .05$). As scores on preferences for local food increase and packaged food and chain restaurants decrease, perceptions on sustainability implications of NBT on the cultural dimension increase (Table 3.7).

Discussion

The choices made by travelers regarding tourism services and products have the power to determine the pace of change toward sustainable development in destinations. To our knowledge, there is scarce research addressing the role of the traveler for sustainability transformations, thus providing insights on the demand side of nature-based tourism (NBT) (Seeler et al., 2021). Our study adds to previous research by identifying threats of NBT to communities and destinations and grouping the implications under four sustainability dimensions (Figure 3.2).

First, for the economic dimension, our findings on the price dimension of the NBT experience support previous research (Ruhanen, 2019) and indicate pressure on the pricing of NBT services and products. In general, travelers expect relatively lower prices and are unwilling to pay for NBT services and products (e.g., entry fees for natural parks). Building on the work of Hultman et al. (2015), we can assume their lack of awareness of travelers' adverse impact

Table 3.7 Relationship between NBT Preferences, Motivations, and Sustainability Perceptions

Sustainability Perceptions:	*Economic*		*Environmental*		*Social*		*Cultural*	
Constant	3.159***	3.112***	1.882*	1.986*	2.390***	2.621***	2.958***	2.994***
Gender	−0.035	−0.035	0.119	0.120	−0.034	−0.032	−0.084	−0.084
Age	−0.004	−0.004	0.015**	0.015**	0.010**	0.009**	0.005	0.005
Transportation	−0.073	−0.076	0.006	0.014	0.020	0.037	−0.074	−0.071
Food and beverage	0.006	0.015	0.144	0.126	−0.071	−0.112	0.163*	0.156*
Safety and security	0.093*	0.094**	0.005	0.005	0.016	0.014	0.080*	0.080*
Prices	0.238***	0.238***	0.003	0.002	−0.031	−0.034	−0.018	−0.018
Host community interaction	−0.015	−0.011	0.114	0.105	0.206**	0.187*	0.047	0.044
Activity	0.035	0.041	0.083	0.069	0.171**	0.141*	0.055	0.050
Seeking local experience		−0.052		0.116		0.258**		0.040
Adjusted R2	0.135	0.133	0.028	0.028	0.078	0.107	0.044	0.041
F-value	5.364***	4.797***	1.801	1.704	3.355**	3.965***	2.276*	2.049*
R2 change		0.001		0.004		0.032		0.001
F-value change		0.385		0.935		7.973**		0.289

$^{*}p < 0.05$.
$^{**}p < 0.01$.
$^{***}p < 0.001$.

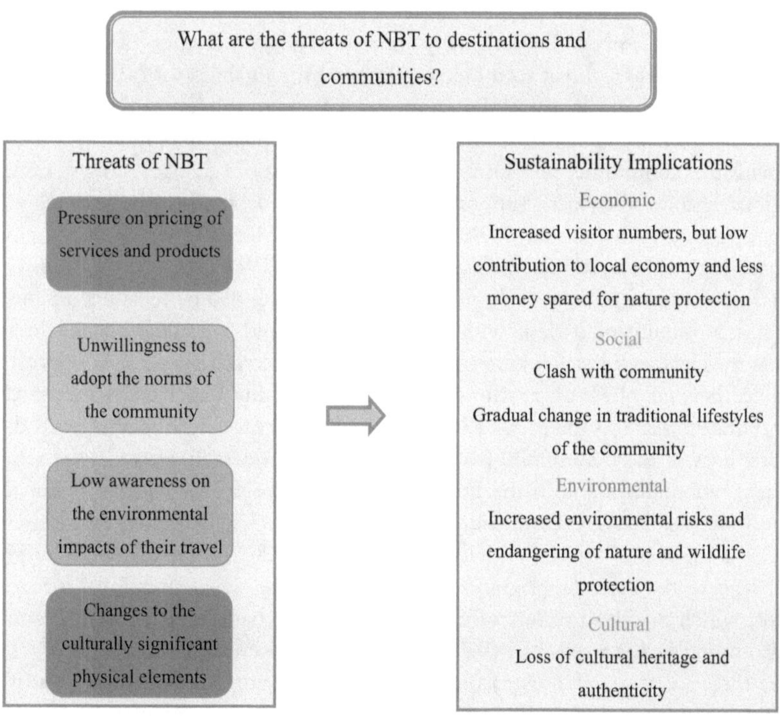

Figure 3.2 Threats to NBT and Sustainability Implications.

on the environment as one explanation for their unwillingness to pay the premium process. Unfortunately, their attitude toward the pricing of these services has significant consequences for the community and the destinations. Increased visitor numbers in these fragile natural landscapes coupled with pressure on prices suggest a low contribution to local economies and fewer resources available for nature protection, which emerge from our findings as to the major threat to the sustainability of NBT.

Besides supporting the economic livelihood of a community, tourism activity should also provide funds for the conservation of wildlife and protected areas (Fennell, 2021; Hatipoglu et al., 2021; Stronza et al., 2019). Creating awareness about the need to pay for NBT services and products would contribute to economic sustainability and generate funds for conservation efforts. In addition, premium prices would also help limit the number of travelers in NBT settings, thus lowering the risk of overcrowding and its harms.

Second, our findings on the environmental sustainability dimension built on the observed discrepancy by Mkono and Hughes (2020) and Juvan and Dolnicar (2014) between travelers' awareness of the environmental impacts of their tourism activity and consumption patterns. In our sample, travelers gave high importance to the enjoyment of nature, escape from city life, and discovering new places. However, they don't seem to realize that escape has its environmental consequences, such as the transportation out of the city centre to rural areas and its contribution to climate change. Thus, they display low awareness of the travelers' environmental impacts on nature. Secondly, a contradiction arises as their experience preferences (e.g., amenities for accommodation) conflict with their intentions to be in natural settings. Our finding, perhaps, confirms Ruhanen's (2019) conclusion that the service- and experience-related aspects are more important to travelers than aspects related to sustainability. Or some travelers are not knowledgeable about the implications of their experience choices. No matter what the reasons are for this contradiction between their attitudes and behaviours, the preference of travelers to maintain amenities relating to modern daily life reflects a low awareness of their environmental impacts and emerges as a threat to the environmental sustainability of NBT.

Third, our findings on the social dimension contribute to the self-development and education aspects of NBT (Fennell, 2021). Travelers' experience preferences indicate a willingness to interact with the community and join in the local activities (e.g., harvesting tea, milking the animals, joining local festivities), supporting importance attributed to the learning experience in previous studies (Ruhanen, 2019; Breiby et al., 2020). Moreover, those seeking local experiences have higher social sustainability perceptions of NBT. However, our findings also point to a critical social threat. Travelers do not want to be constrained by local norms and customs, presenting a possibility of a clash with the community and a gradual change in traditional lifestyles as travelers seek local experiences. While these interactions provide opportunities for sustainability beyond NBT destinations through the learning experiences for travelers, they also incorporate threats to the social fabric and traditional lifestyles together with distractions to the daily lives of the communities as the number of travelers increases in the area.

Similarly, our findings add to the literature on tourism risks for cultural sustainability. No matter how different travelers' motivations to natural areas can be, one common element in their travels is the search for authenticity (Panzer-Krause, 2020). Yet their traveler preferences imply changes to the culturally significant physical elements (e.g., traditional furnaces, quilts, and baths), which may lead to loss of cultural heritage and authenticity, as host communities try to accommodate the needs of travelers in the face of increased competition.

All the threats that we identify in our study draw attention to the critical role of travelers. However, confirming prior research (Ruhanen, 2019), travelers seem to be driven by personal benefits rather than benefits to sustainability. This outcome brings us to question the mediating role that needs to be fulfilled by other stakeholders (e.g., travel agents, travel planners, community leaders, destination planners) in designing the NBT services and products, considering the threats identified because of traveler preferences and perceptions. For example, we see a need to identify culturally significant physical elements that should be preserved to conserve cultural heritage and authenticity, despite travelers' demands for having their modern daily amenities in rural NBT areas. In this case, there might also be a need for translating the demand of travelers to local communities, as communities might be misinterpreting the demands from customers.

Moreover, local communities seem mostly unaware of the treasures (e.g., rural landscapes, traditional ways of preservation, and cooking techniques) that they have and the value these might represent to the travelers. Travel agents specialized in NBT, and rural specifics might play a mediating role. The niche travel agency managers frequently mention intervening in changes initiated by the communities to satisfy demand, such as increasing capacity by adding annexes to their buildings and changing traditional furnaces with electrical heating devices. The managers also report intervening to stop travelers from disturbing the local daily rituals. The role of these mediating agents should be supported by training on the sustainability of NBT offered on regional and local levels or certifications for NBT travel agencies.

Conclusions

To reduce the pressures on biodiversity, change should happen in the existing consumption patterns as well as in the supply side of tourism (Dasgupta, 2021). As such, the findings of our study have implications for tourism planners and regulators in nature-based destinations. We recognize that behavioural change of the traveler for minimizing the negative impacts of their travel is slow (Juvan and Dolnicar, 2014). However, service providers can create change by enhancing their service design. Our study offers opportunities for engraving awareness-raising components into nature-based activities. Tourism planners can adopt a learning-based experience design to educate travelers on nature and the ecosystem (Rong-Da Liang et al., 2020; Hatipoglu et al., 2021; Seeler et al., 2021), satisfying travelers' self-development needs. For example, the NBT travelers in our sample are an educated older age group; they are interested in experiencing the local culture and are open to learning. Enriched NBT experiences with the learning and reflection components would create value for these travelers and at the same time support them in their individual journey toward sustainability transitions. In those regions with different visitor profiles, the learning-based experiences can be altered for younger people or families with children. Furthermore, regulators can also contribute via improved regulations and interventions in high-risk areas that need protection. Thus, natural capital can be protected and grown through a mix of market and legislative mechanisms (Dasgupta, 2021).

Our main aim in this study was to critically discuss travelers' responsibility as agents of change for sustainable development (Budeanu et al., 2016). We identify travelers' conception of relatively lower prices on NBT services and products as one of the major threats to economic sustainability. We partially relate this threat to travelers' ignorance of NBT's environmental impacts and the necessity to use a portion of the tourism income for conserving nature and wildlife. Additionally, the lack of knowledge on their preferences' short- and long-term impacts extended to the social and cultural domains. Thus, in this study, by unveiling threats to the sustainability of NBT arising from travelers' experience preferences, motivations, and sustainability perceptions, we demonstrated the role of travelers in responsible travel and contributed to the literature on sustainable development.

Our study captured an educated and older female group that traveled to nature-based destinations frequently and stayed there for longer than a day. However, the experience preferences of other demographic groups and daily travelers could differ from our sample. Therefore, we call for further research on different segments of travelers, such as families with children, millennials, and extending the sustainability impacts of NBT beyond destination.

References

Adam, I., Adongo, C. A., & Amuquandoh, F. E. (2019). A structural decompositional analysis of eco-visitors' motivations, satisfaction and post-purchase behaviour. *Journal of Ecotourism, 18*(1), 60–81.

Breiby, M. A., Eva Duedahl, E., Øian H., & Ericsson, B. (2020). Exploring sustainable experiences in tourism, *Scandinavian Journal of Hospitality and Tourism, 20*(4), 335–351.

Budeanu, A., & Emtairah, T. (2014). A fresh look into tourist consumption: Is there hope for sustainability? In C. Weeden, & K. Boluk (eds.), *Managing Ethical Consumption in Tourism, Critical Studies in Tourism, Business and Management Series*. Routledge, New York. (ISBN: 978-0-415-71676-5).

Budeanu, A., Miller, G., Moscardo, G., & Ooi, C-S. (2016). Sustainable tourism, progress, challenges and opportunities: An introduction, *Journal of Cleaner Production, 111*(B), 285–294.

Buffa, F. (2015). Young tourists and sustainability. Profiles, attitudes, and implications for destination strategies. *Sustainability*, 7(10), 14042–14062.

Carvache-Franco, M., Segarra-Oña, M., & Carrascosa López, C. (2019). Motivation's analysis in ecotourism through an empirical application: Segmentation, characteristics and motivations of the consumer. *Geo Journal of Tourism and Geosites, 24*(1), 60–73.

Creswell, J., Clark, J., Gutmann, V., & Hanson, W., 2008. Advanced mixed methods research designs. In: J. Clark, & J. Creswell (eds.), *The Mixed Methods Reader*. Sage Publications, Thousand Oaks CA.

Dasgupta, P. (2021) *The Economics of Biodiversity: The Dasgupta Report*. HM Treasury, London.

Esfandiar, K., Pearce, J., & Dowling, R. (2019). Personal norms and pro-environmental binning behaviour of visitors in national parks: The development of a conceptual framework, *Tourism Recreation Research, 44*(2), 163–177.

Fennell, D. A. (2021). Introduction. In D. A. Fennell (ed.), *Routledge Handbook of Ecotourism*. Routledge, New York.

Galvani, A., Lew, A. A., & Perez, M. S. (2020). COVID-19 is expanding global consciousness and the sustainability of travel and tourism. *Tourism Geographies, 22*(3), 567–576.

Hatipoglu, B. K., Denizci, F. C., & Imamoğlu, T. (2021). Biodiversity conservation through an agroecotourism project: The case of Ovacık Village, Turkey. In D. A. Fennell (ed.), *Routledge Handbook of Ecotourism* (pp. 331–343). Routledge, New York.

Hultman, M., Kazeminia, A., & Ghasemi, V. (2015). Intention to visit and willingness to pay premium for ecotourism: The impact of attitude, materialism, and motivation. *Journal of Business Research, 68*(9), 1854–1861.

Juvan, E., & Dolnicar, S. (2014). The attitude–behaviour gap in sustainable tourism. *Annals of Tourism Research, 48*, 76–95.

Kastenholz, E., Eusébio, C., & Carneiro, M. J. (2018). Segmenting the rural tourist market by sustainable travel behaviour: Insights from village visitors in Portugal. *Journal of Destination Marketing & Management, 10*, 132–142.

Lee, T. H., Jan, F., & Yang, C. (2013). Conceptualizing and measuring environmentally responsible behaviors from the perspective of community-based tourists. *Tourism Management, 36*, 454–468.

Luo, Y., & Deng, J. (2008). The New Environmental Paradigm and nature-based tourism motivation. *Journal of Travel Research, 46*(4), 392–402.

Mkono, M., & Hughes, K. (2020). Eco-guilt and eco-shame in tourism consumption contexts: Understanding the triggers and responses. *Journal of Sustainable Tourism, 28*(8), 1223–1244.

Molina-Azorín, J. F., & Font, X. (2016). Mixed methods in sustainable tourism research: An analysis of prevalence, designs and application in JOST (2005–2014). *Journal of Sustainable Tourism, 24*(4), 549–573.

Panzer-Krause, S. (2020) The lost rural idyll? Tourists' attitudes towards sustainability and their influence on the production of rural space at a rural tourism hotspot in Northern Ireland, *Journal of Rural Studies, 80*, 235–243.

Pham, H. S. T., & Khanh, C. N. T. (2021). Ecotourism intention: The roles of environmental concern, time perspective and destination image. *Tourism Review, 76*(5), 1141–1153.

Rong-Da Liang, A., Hsiao, T. Y., Chen, D. J., & Lin, J. H. (2020). Agritourism: Experience design, activities, and revisit intention. *Tourism Review*, *76*(5), 1180–1195.

Ruhanen, L. (2019). The prominence of eco in ecotourism experiences: An analysis of post-purchase online reviews. *Journal of Hospitality and Tourism Management*, *39*, 110–116.

Seeler, S., Zacher, D., Pechlaner, H., & Thees, H. (2021). Tourists as reflexive agents of change: Proposing a conceptual framework towards sustainable consumption. *Scandinavian Journal of Hospitality and Tourism*, 1–19. DOI:10.1080/15022250.2021.1974543

Sharpley, R. (2018). *Tourism, Tourists, and Society*, 5th Edition, Routledge, Abington.

Spenceley, A. (2005). Nature-based tourism and environmental sustainability in South Africa. *Journal of Sustainable Tourism*, *13*(2), 136–170.

Stronza, A. L., Hunt, C. A., & Fitzgerald, L. A. (2019). Ecotourism for Conservation? *Annual Review of Environment and Resources*, *44*, 229–253.

Teeroovengadum, V. (2019), Environmental identity and ecotourism behaviours: An examination of the direct and indirect effects. *Tourism Review*, *74*(2), 280–292.

The Republic of Turkey, Ministry of Agriculture and Forestry (2021). Ecotourism Action Plan 2021–2025. https://www.ogm.gov.tr/tr/e-kutuphane-sitesi/Yayinlar/ECOTOUR%C4%B0SM%20ACT%C4%B0ON%20PLAN%202021-2025.pdf

United Nations World Tourism Organization (UNWTO) (2018). Ecotourism and protected areas. http://sdt.unwto.org/content/ecotourism-and-protected-areas

United Nations World Tourism Organization (UNWTO) (2020). *UNWTO Recommendations on Tourism and Rural Development – A Guide to Making Tourism an Effective Tool for Rural Development*. UNWTO, Madrid.

Zhang, G., Higham, J., & Albrecht, J. N. (2021). Ecological restoration and visitor experiences: Insights informed by environmental philosophy. *Journal of Sustainable Tourism*, 1–19.

4

A CASE STORY OF THE CONNECTION BETWEEN WOODLANDS, LANDSCAPE, AND NATURE-BASED RECREATION IN UPLAND ENGLAND

Sara V. Iversen, Naomi van der Velden and Ian Convery

"You can't plant new woodlands here, the tourists will stop coming."

Introduction

In recent years, the world has increasingly experienced extreme weather events such as severe flooding, storms and droughts, all of which have had devastating effects on society (IPCC, 2021). This has led to an increased focus on the role landscapes can play in mitigating the outcomes of climate change (Locatelli et al., 2015; Norton et al., 2015; Hunsberger et al., 2017; IPCC, 2021). Valuing natural systems and related societal benefits has become an important area of policy development (at global and national scales), with a particular focus on natural capital accounting, ecosystem service assessment and nature-based solutions (Zank et al., 2016; Costanza et al., 2017). There is increased political focus by governments to initiate and implement natural resource management policies to improve the environment, mitigate climate change and to protect biodiversity (e.g. UNFCCC, 2021), but also to recognize the positive human health benefits landscapes provide (Stratford et al., 2017).

Afforestation, when following recommendations for implementation (Di Sacco et al., 2021), is often seen as a strategy that can meet multiple social and environmental objectives. Creating new woodlands and increasing woodland cover is beneficial for improving services such as climate change mitigation (Reed et al., 2009; Stratford et al., 2017), flood protection and water quality (Broadmeadow & Nisbet, 2010) as well as having economic benefits on a local scale (Hardaker et al., 2021), particularly for cultural services, such as nature-based recreation (NBR). Significant levels of tree planting are required, however, to achieve the targets set by the UK Committee on Climate Change (UKCCC, 2019), who recommend afforestation rates of between 30,000 and 50,000 hectares per annum in order to reach net-zero emissions by 2050. Meeting these targets would increase woodland cover from the current average of 13 percent of UK land area to 17–19 percent, delivering carbon sequestration in the range of 16.2–27.5

DOI: 10.4324/9781003230748-6

Mt CO_2 equivalent (UKCCC, 2019). These ambitious targets (albeit within the realms of past planting rates, would also likely lead to significant landscape change in some areas.

The COVID-19 pandemic has highlighted how important natural areas are to society, providing opportunities for physical exercise and enhancing mental wellbeing. During the pandemic, many countries have experienced an increase of visitor numbers to natural parks and protected areas, and a corresponding increase in demand for nature-based recreational (NBR) activities (Derks et al., 2020; Landry et al., 2021; Spennemann & Whitsed, 2021). NBR is a specific form of tourism that can be defined as the activities people may leisurely engage with in natural areas, and includes hill walking, fishing, cycling, running, wildlife viewing and horse riding (Fredman & Tyrväinen, 2010). There is good evidence that participating in NBR can have a positive impact – both directly and indirectly – on the health and wellbeing of people (Sherman et al., 2005; Brown & Bell, 2007; Li et al., 2011; Ward & Aspinall, 2011; Mapes, 2012). NBR locations can also be perceived as a refuge from urban life as a place to heal (Stenbacka, 2001; Cloke, 2003) or by those seeking a perceived rural idyll (Daugstad, 1999; Christou et al., 2018). NBR is especially important to the tourism sector in UK upland areas (Cumbria Tourism, 2019). In the English county of Cumbria, which predominantly consist of uplands, NBR accounts for 68 percent of visitors and the tourism sector as a whole contributed £3.13 billion to the local economy in 2019 (Cumbria Tourism, 2019).

Alterations of landscapes, for whatever reason and in whatever form, can provoke highly emotional responses, and there is significant scope for conflict when trying to achieve multiple objectives across a landscape (Reed et al., 2009). The increased interest in NBR as a driver of landscape change and (local) economic development must be carefully balanced with regional and national strategies that will significantly alter landscapes, such as afforestation. In a previous work we discussed how changes in upland land use (woodland planting) on a large landscape scale within a Cumbrian National Park, driven by changes in environmental policy and stakeholder interest, can affect ecosystem services and cultural values (Iversen, 2019). Stakeholder participation is a critical factor in influencing the uptake, implementation and success, or otherwise, of any scheme (Quick & Bryson, 2016). Local stakeholders often have differing views on how their landscapes should be managed and for what purposes, and this risks causing conflict and can act as a barrier for action (Bendtsen et al., 2021). In particular, stakeholder emotions and values have been suggested to be a potential barrier to tree planting (Duesberg et al., 2013) yet this has been an underrepresented area of focus within woodland creation research (Buijs & Lawrence, 2013).

We therefore aim to examine the contribution of combined quantitative and qualitative methods in assessing cultural ecosystem services, such as NBR, and the social dimension of multiple stakeholder perspectives by using a case study example of specific landscape changes. We aim to identify local stakeholder concerns and areas of conflict from a qualitative perspective and combine it with an empirical approach of economic valuation of NBT. In Iversen et al. (2021) and Iversen et al. (2022), we explored the topic in detail focusing on economic evaluation and visitor perceptions in one study and the subjective perspectives of stakeholders in another. Here, we combine the two studies and, by doing so, we hope to provide insight and understandings that are directly relevant for both NBT and natural resource management.

The study is situated around a case study of increasing woodland cover in an upland landscape in a National Park in upland England. The question of if and how increased woodland cover influences NBR visitors' visiting patterns thus forms the first part of our study, whereby we carry out an economic evaluation of NBT within the study area and visitor perspectives on the proposed land use changes (photovisualization and survey work). This is followed by an in-depth exploration of perspectives on tree-planting in the uplands from a wide range of local stakeholders, explored via the use of Q-methodology in the second part of this study. Together these approaches provide a detailed case study for the land management sector (e.g. conservationists, farmers, foresters, estate managers, etc.).

Methods

Introducing the case study

The UK contains 15 National Parks (IUCN category V) and eleven of these are in upland areas (IUCN, 2021). They have, both historically and increasingly, been landscapes sustaining important nature-based recreation activities and all have their own defining characteristics (Frost & Hall, 2009). The purpose and management of the UK's National Parks are, however, under scrutiny. The 2019 Glover Report notes that the National Park system in England is 'fragmented, sometimes marginalised and often misunderstood'. And further, 'are currently unable to fulfil their statutory purpose to 'conserve and enhance the natural beauty, wildlife and cultural heritage' of their areas. This is a damning finding, and the usual category V cultural landscapes argument seems hopelessly out of date given the current (and well documented) state of wildlife loss in the UK (Carver & Fisher, 2018). This perspective is important when discussing woodland creation within Cumbria, particularly in a post-COP26 context, but also in relation to NBR and its connection to the landscape.

The situation in England is complicated further by National Character Areas (NCA). The NCA of the Howgill Fells covers an area of 10,360 ha. It is situated in Cumbria, in the north-west of England and lies within the boundaries of the Yorkshire Dales National Park. The Howgill Fells are broadly representative of the upland regions of Cumbria in that they are rural, isolated, strongly influenced by hill farming with a strong cultural identity and fairly homogenous socio-demographic characteristics (Natural England, 2014). The area also has an identified objective of increasing woodland, and a large amount of tree planting is either being considered or has already taken place.

The objective of increasing woodland in upland areas, both within the study site and in Cumbria as a whole, has generated concerns amongst local stakeholders (Iversen, 2019). This is embedded in stakeholders' subjective differing views and concerns of changes ahead in terms of changes to agro-environment schemes, as well as firmly rooted in feelings and emotions. For example, there is a strong (mostly local) view that increased woodland cover could negatively impact tourism. In the early stages of our pilot study, local stakeholders would often voice the opinion of; "You can't plant woodlands here, the tourists will stop coming." Tourism is important to the area, with annual visitor numbers of approximately 460,000 a year and 68 percent of these visiting for NBR (Cumbria Tourism, 2015). Also, in other areas of Europe, where agriculture has declined, tourism has become the principal income generator (Butler et al., 1998; Garrod et al., 2005; Kneafsey, 2000) and can additionally result in improved socioeconomic wellbeing, higher employment growth rates and incomes. (Reeder & Brown, 2005; Cumbria Tourism, 2019; Streifeneder & Dax, 2020). A visitor survey by Cumbria Tourism (2015) showed that 96 percent of visitors felt that the NCA 'is a good place for outdoor activities'. The top three key reasons for visiting were: 'because of the physical scenery and landscape' (61 percent), 'because of the atmospheric characteristics of the area – peaceful, relaxing, beautiful etc.' (40 percent) and 'been before' (37 percent), followed by 'undertaking a specific activity' (19 percent). It is therefore perhaps understandable that any proposals to change the landscape (and negatively affect tourism) should cause concern amongst local stakeholders.

Assessing the economic value of nature-based recreation using TESSA

Generally, it is a common, yet challenging, problem to measure cultural ecosystem services (Daniel et al., 2012; Kenter et al., 2015). Even measuring more quantifiable ecosystem services site-specifically is difficult and often carried out via modelling approaches (Zank et al., 2016). Given that a cultural service is likely to be heavily influenced by locality (Convery & O'Brien, 2012), site-specific assessments are vital, if they are to be meaningful. Tourism data on different levels are important for natural resource and tourism management but comparisons between sites in different geographical regions can be difficult (Leung et al., 2018). Some regional protocols are developed for guidance on how to monitor and estimate tourism in nature areas, which enables comparability to a higher extent, such as the coordinated effort of producing a visitor monitoring manual between the Nordic and Baltic countries (Kajala, 2007). The ecosystem service toolkit TESSA (Toolkit for Ecosystem Service Site-based Assessment) – Figure 4.1 – offers guidance on economic estimation of the impacts on NBR when changing from one land use or management to another (Bradbury et al., 2021) by using on-site data collection via surveys (Soe Zin et al., 2019; Jones et al., 2020) often accompanied by use of graphical media (Hegetschweiler et al., 2017).

To assess the question of whether or not planting trees would actually decrease NBR tourism numbers to the case study area, we conducted a survey using in-situ non-probability convenience intercepts sampling (Chhetri et al., 2004; den Breejen, 2007; Newing, 2011; Kim & Weiler, 2013). Following the guidelines by the TESSA Toolkit (Peh et al., 2013), this was combined with a photovisualization approach (Kim & Weiler, 2013), which is commonly used and recommended in studies where attitudes towards changing landscapes are explored (Chhetri et al., 2004; den Breejen, 2007; Soliva & Hunziker, 2009). This allows the participants to get a sense for a more detailed experience and sense of being within or immersed in the landscape under the different scenarios (Lange, 2001; Orland et al., 2001; Soliva & Hunziker, 2009).

We surveyed a total of 493 NBR visitors to the area and determined their preference for certain woodland levels, ranging from 10 percent to 100 percent. The survey results were then combined with economic visitor data from Cumbria Tourism (2015), and a local Tourist Information Centre (see Iversen et al., 2021 for more details). From these combined data we both determined the economic value of NBR tourism under different woodland level scenarios, and also assessed the extent to which NBR visitor patterns would change, if woodland cover were to increase.

Assessing local stakeholder perspectives on nature-based recreation

These results were combined with a qualitative assessment on key stakeholders' perspectives on how they thought woodland creation would impact nature-based recreation. Individual perspectives, emotions, and the finding of areas of consensus on a conflicted topic, can only be explored by the use of methods that can investigate subjectivity rather than objective views. Psychology approaches developed to better understand and identify underlying values and emotions underpinning stakeholder behaviour (Hall, 2008; Walder & Kantelhardt, 2018) or perspectives on complex and conflict-intense case studies (Chamberlain et al., 2012) are increasingly being used in land use conflict situations. One such method – Q-methodology – has been successfully used to explore the breadth of human subjectivity from which to develop more effective management strategies, including human subjectivity in environmental conflict situations (Bredin et al., 2015; Dempsey, 2021), and the methodology is increasingly

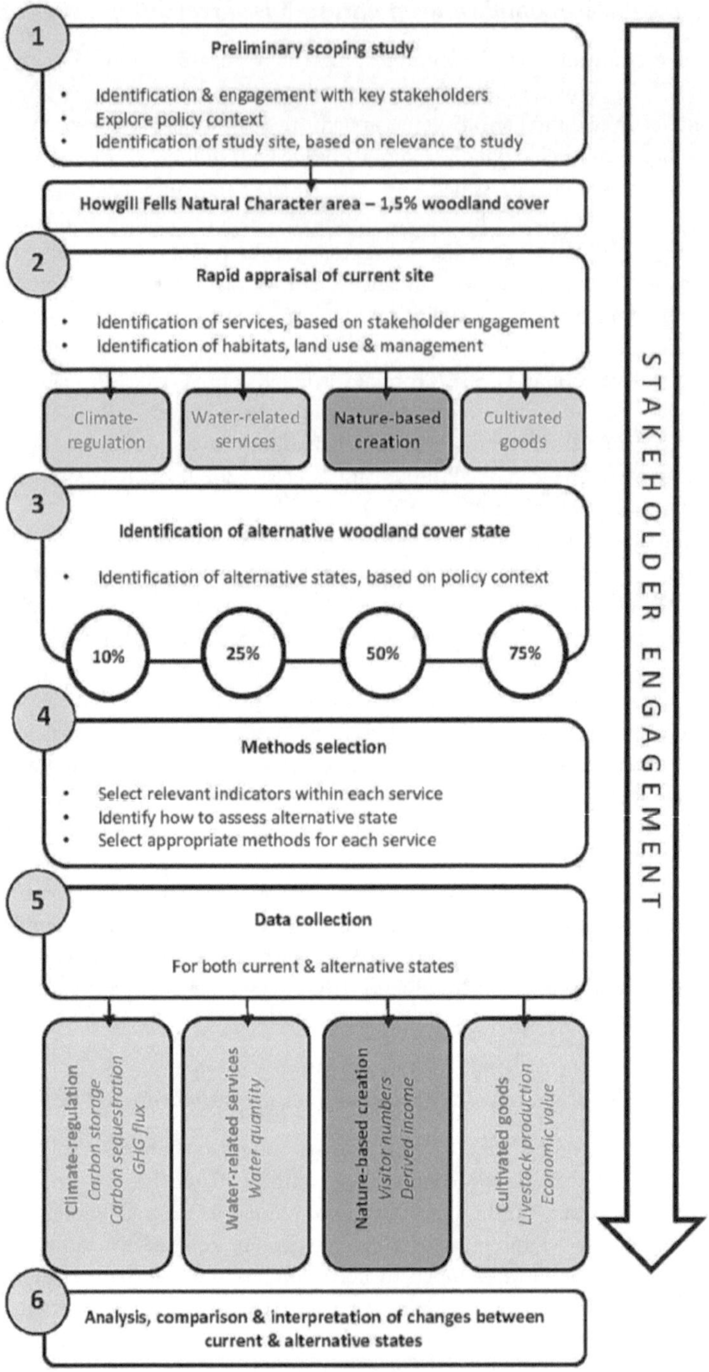

Figure 4.1 TESSA ecosystem service assessment outline and systematic approach to the assessment of four ecosystem services delivered by the study site. Each service will be assessed for both the current and alternative states and a comparison of the potential impacts of the change in land use will be explored. NBR assessment focus in dark grey. Developed from Peh et al., (2013).

gaining popularity across disciplines (Haslam & McGarty, 2014; Zabala et al., 2018). We used Q-methodology for our case study and invited sixty local stakeholders to take part in a Q interview. Selected participants were all relevant to woodland creation in the study area. These included farm or woodland creation advisors (9), NGOs (8), business or interest groups (9), general public with common rights (8), council (2), landowner (2), and farmers who had already planted (11) or were considering planting (11). From this group, three main shared viewpoints (factors) were identified. In brief, Q-methodology combines both quantitative and qualitative data collection and analysis through statistical discourse analysis, using an application of factor analysis tools (Stephenson, 1935; Brown, 1996). Following the Q-methodology approach, a selection of 47 unique statements representing all opinions on the subject of woodland creation in this upland area was created. Participants were then asked to evaluate and rank each statement on a ranked grid, depending on their level of agreement or disagreement and how much importance they attributed to the statement. From these individual "Q-sorts" we were able to gain detailed insights into the range and strength of opinions concerning woodland creation. We were able to create a typology of participants based on their respective views, beliefs and related influencing factors. Further details of how the Q-methodology was used in this research can be found in Iversen et al. (2022) and for more general information on the approach in conservation research see Zabala et al., 2018).

Results

Assessing the economic value of nature-based recreation

Figure 4.2 indicates that the majority of visitors believed that changes in woodland levels would not have an impact on the likelihood of them visiting again. However, there was a difference between visiting patterns and landscape preferences. In comparison, the majority of visitors had a preference for 25–50 percent woodland (Figure 4.3) levels, but visiting patterns would not decrease, until 75 pecent woodland levels were to occur. Actual extant woodland levels, for comparison, were 1.5 percent.

Assessing local stakeholder perspectives on nature-based recreation

Perspectives of local stakeholders on the topic of NBR gave three main viewpoints; 1 – 'Not enough is done to protect the environment', 2 – 'Changing the landscape is changing us', 3 – let's not let our emotions get in the way – seeing the bigger picture'. These are considered in more detail below.

Viewpoint 1 – "Not enough is done to protect the environment"

Twenty-one participants (35 percent) are significantly associated with this viewpoint. They are from a wide range of backgrounds, such as NGOs, interest groups, general public, public houses and environmental advisors. The majority of them do not make a living from the land directly, but have an interest or connection in other ways, such as in advisory or specific interest roles, and they all have a strong interest in the area. Four farmers were included, all of which are farming as part of an environmentally sensitive farming scheme.

This is a group of people very much driven by intrinsic values and who appear to feel that not enough is done to protect the environment. Woodlands for environmental benefits, such as wildlife and biodiversity, are very important to them and they believe that more should

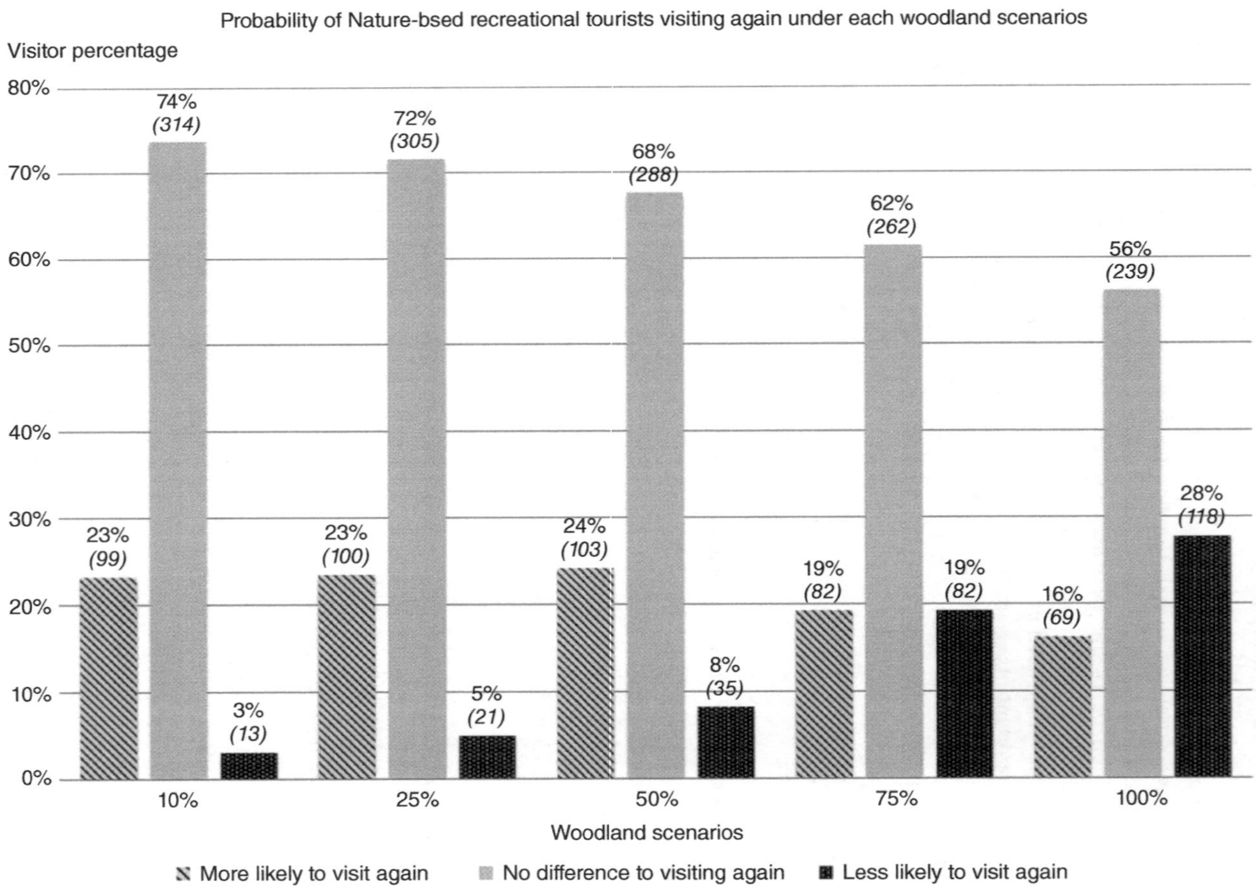

Figure 4.2 Visitors' preference for percentage of woodland cover under each of the woodland scenarios, as well as the percentage with no preference at all.

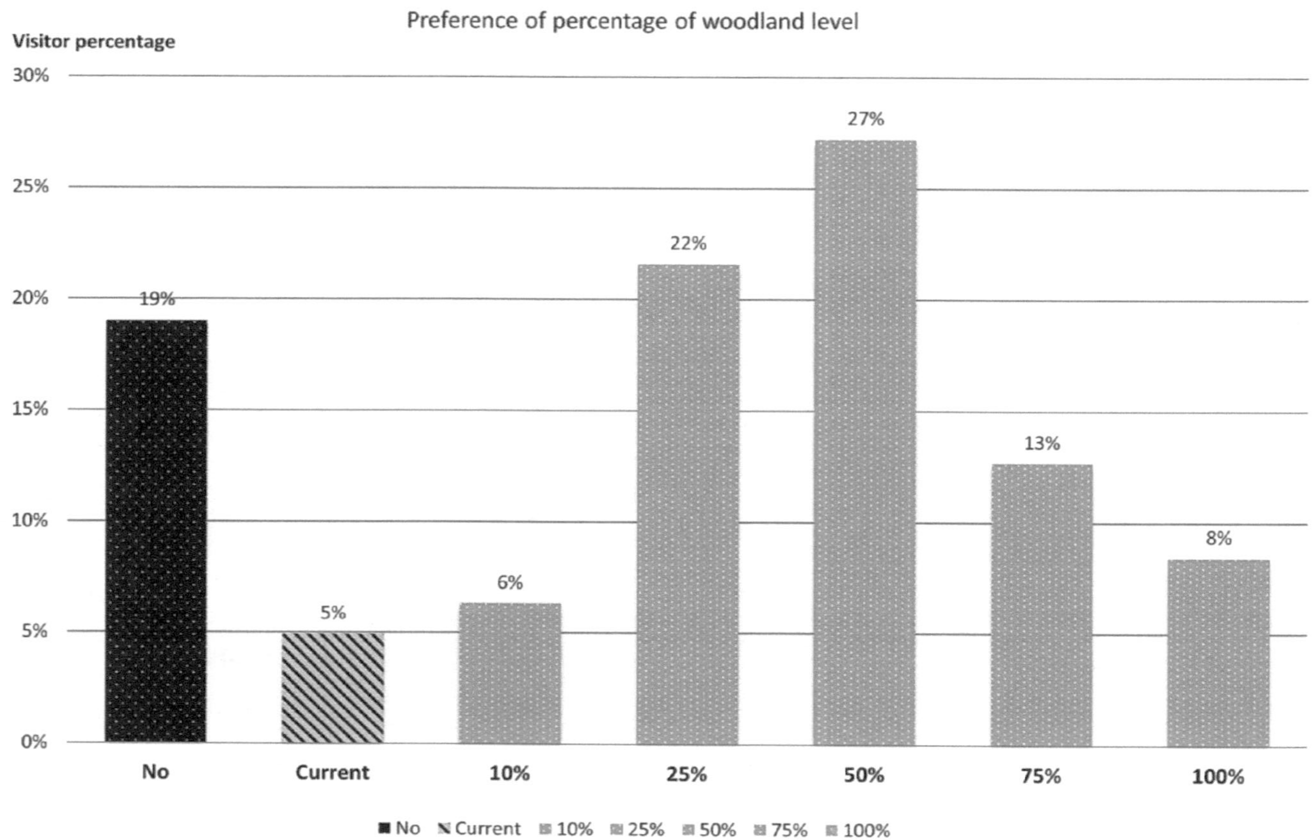

Figure 4.3 Visitors' preference for percentage of woodland cover under each of the woodland scenarios, as well as the percentage with no preference at all.

be done towards environmental protection within the study site. Moreover, they also believe that woodlands would be beneficial not only to the environment but also to people making a living from the land. And that these societal benefits could be achieved via woodland-related economic benefits to local businesses, that come from nature-based recreation tourism or the concept of payments for ecosystem services. Some of the benefits of woodlands, such as physical and mental benefits to health, creating a sense of wilderness and the general encouragement of outdoor leisure activity are perceived as already in existence within the study site. This group of participants also believe that the aesthetics of the landscape is important and therefore new woodlands have to be planted in tune with the landscape. They view landscape changes as acceptable and woodlands are not seen as having a negative impact on the identity and local cultural heritage.

Viewpoint 2 – "Changing the landscape is changing us"

Seventeen people (28 percent) are associated with this viewpoint. With the exception of two people, all are actively hill farming and reside in the area or are retired farmers. The two who are not from a farming background are, through their profession and personal interest, strongly interested in the protection of the landscape.

This perspective can be characterized by a strong sense of place, heritage culture and a way of life. What the area and landscape delivers in its current state (pastoral farming) is seen to be of bigger value, both for the local community but also for society as a whole. The notion that aesthetics and nature-based recreational activities would benefit if more woodlands were created is regarded with strong contempt. Nature-based recreational activities are perceived as already being plentiful and the current landscape is well functioning for the delivery of this. That such values are of more, or equal, importance than what it should deliver in terms of agricultural produce is met with strong disapproval.

Viewpoint 3 – "Let's not let our emotions get in the way – seeing the bigger picture"

Seventeen people (28 percent) are significantly associated with this viewpoint. Participants consist of a variety of forestry, landscape and farming-related advisors and businesses, as well as farmers. All of them have currently, or in the past, had an economic experience with woodland creation, either in an advisory role, business or directly. Besides the people in advisory roles, all of them have a business or are diversifying their farm business with additional incomes related to the land and area.

This group are driven by economic considerations (e.g. tourism is seen as an important income stream), with a sustainability focus (resource security) and an interest in future multi-functional woodlands, alongside a view that more should be done to protect the environment. There is a preference for evidence-based decision-making (and corresponding rejection of emotionally driven evidence-poor arguments).

The value of interrelation and interdisciplinary approaches

The evaluation of the visitor responses to increased tree cover scenarios alongside this deeper investigation of stakeholder perspectives allows for a balanced approach to landscape change and woodland creation. Viewpoints 1 and 3 were broadly supportive of tree-planting, although the first was driven largely by intrinsic values and feelings, and the latter by more logical arguments and an economic context. For Viewpoint 3, particularly, a clearer understanding of the extent to which tree-planting will impact the visitor experience and likelihood of returning can help to determine where tree-planting efforts could be focused to increase visitor numbers, and the overall extent of woodland coverage to aim for in a particular location to meet desired objectives.

However, it is in Viewpoint 2 that the importance and value of the interrelated nature of this study becomes clearest. "If we change the way the landscape looks, tourists will stop coming", was identified from the Q-methodology study as a strong concern for some of these stakeholders. This concern is directly addressed by the results of the TESSA impact assessment, which showed that no negative impact to the tourism economy is likely if woodland levels are kept under 75 percent.

TESSA as an ecosystem service assessment tool has thus been useful in identifying and quantifying site-specific changes to NBR. However, to become implementable, meaningful and useful, the findings have benefitted from being integrated with a qualitative understanding of stakeholder perspectives. Using new interdisciplinary approaches also has the benefit of exploring the unknown, which may yield surprising results. For stakeholders within viewpoint 2, for example, planting trees in the uplands of Cumbria is based on emotions and feelings rather than evidence. These feelings are dominated by frustration and concern for the changes ahead and (in their view) what the land ought to be used for, i.e. pastoral farming. This insight was not identified by the TESSA assessment, but via the Q-methodology analysis, and this combined approach offers a much deeper understanding of this complex topic. This demonstrates the value of an interrelated and interdisciplinary study. Our research identified an important issue from one area of data collection (TESSA), and addressed this through another data collection approach (Q-methodology). This is important in that it enables results to be more useful and targeted for local landscape planning and also facilitates a better understanding of the complexity in natural resource management.

In Cumbria, local tourist boards and planning authorities have expressed concerns that increasing woodland cover could reduce tourism revenue (Iversen, 2020); our research suggests that such concerns are largely unfounded, at least for woodland cover at or below 75 percent. On a local scale, our results provide useful evidence to a range of stakeholders involved in managing the Cumbrian landscape (including, but not limited to, estate managers, forestry organizations, the conservation sector, farmers, statutory agencies and the tourism sector), and can help support future decision-making regarding tree-planting and woodland creation within the NCA.

As always, however, further research is required. There is a clear need to better understand how and why people vary in their attitudes towards woodland creation to enable effective real-world change. Such studies need to be context- and location-specific, and we recommend combined approaches including the use of Q-methodology, which enables the rich complexity of all stakeholders' perspectives to be captured and valued. Linked to this, understanding key alignments and groupings of participants taken into account and allows for subjective opinions to be individually raised and shared equally amongst participants, without differing viewpoints causing antagonistic conflict. The understanding of the key stakeholder groupings, such as the three identified here, could be beneficially used in ensuring equal stakeholder representation, with a focus on not only organizational connections, when establishing future working groups and consultation activities. On a broader scale, our study adds to the national and international knowledge base of the relationship between the upland landscape, woodlands and cultural ecosystem services, such as NBR, and can inform further land management and policy decision-making. As a caveat, however, our findings are specific to the Cumbrian uplands and may not be applicable to areas with a distinctively different landscape character.

Concluding comments

The implications of this research are that any attempt to engage with local stakeholders regarding land use change must target both the 'heart and the head'. Simply providing scientific evidence regarding the need for more woodland cover and how it will impact NBR, will not succeed unless there is also emotional/psychological engagement appropriate to stakeholder concerns. As a recent *Guardian* article has reported "conservationists often think they can convert people to their view using a science-based approach, but this tends to alienate people and makes them do the opposite of what you want" (Weston, 2021). Some local stakeholders had raised a concern that increasing woodland cover would stop tourists visiting the study site, and negatively impact the cultural heritage of the area. Others, with a nature conservation agenda, strongly opposed that and felt that more woodland would be beneficial both for people and nature. This mindset of 'they are wrong and I am right' alienates people and can increase conflict, often resulting in less sustainable outcomes. If you exclude the people with whom you need to collaborate because of either differing viewpoints or a failure to recognize an existing spectrum of views, then necessary and significant progress towards key local and global environmental and cultural goals will never be achieved.

References

Bendtsen, E. B., Clausen, L. P. W. and Hansen, S. F. (2021). A review of the state-of-the-art for stakeholder analysis with regard to environmental management and regulation. *Journal of Environmental Management*, 279, 111–773.

Bradbury, R. B., Butchart, S. H., Fisher, B., Hughes, F. M., Ingwall-King, L., MacDonald, M. A., Merriman, J. C., Peh, K. S. H., Pellier, A. S., Thomas, D. H. and Trevelyan, R., (2021). The economic consequences of conserving or restoring sites for nature. *Nature Sustainability*, 4(7), 602–608.

Bredin, Y. K., Lindhjem. H., van Dijk, J., Linnell, J. D. C. (2015). Mapping value plurality towards ecosystem services in the case of Norwegian wildlife management: A Q analysis. *Ecological Economics*, 118, 198–206.

Broadmeadow, S. and Nisbet, T. (2010). *Opportunity mapping for woodland creation to reduce diffuse sediment and phosphate pollution in the lake district*. Farnham, Surrey, UK.

Brown, S. R. (1996). Q methodology and qualitative research. *Qualitative Health Research*, 6(4), 561–567.

Brown, T. and Bell, M. (2007). Off the couch and on the move: Global public health and the medicalisation of nature. *Social Science & Medicine*, 64(6), 1343–1354.

Buijs, A. and Lawrence A. (2013). Emotional conflicts in rational forestry: Towards a research agenda for understanding emotions in environmental conflicts. *Forest Policy and Economics*, 33, 104–111.

Butler, R., C. Hall and J. Jenkins. (1998). *Tourism and recreation in rural areas*. Chichester: Wiley.

Carver, S. and Fisher, M., (2018). Reviewing England's National Parks: An opportunity for rewilding? *ECOS*, 39, 3. https://www.ecos.org.uk/8984-2/

Chamberlain, E. C., Rutherford, M. B. and Gibeau, M. L. (2012). Human perspectives and conservation of grizzly bears in Banff National Park, Canada. *Conservation Biology*, 26, 420–431.

Chhetri, P., Arrowsmith, C. and Jackson, M. (2004). Determining hiking experiences in nature-based tourist destinations. *Tourism Management*, 25, 31–43.

Christou, P., Farmaki, A. and Evangelou, G. (2018). Nurturing nostalgia: A response from rural tourism stakeholders. *Tourism Management*, 69, 42–51.

Cloke, P. J. ed. (2003). *Country visions*. Pearson Education.

Convery, I. and O'Brien, V. (2012). Cultural landscape and sense of place: Community and tourism representations of the Barossa. In *Making sense of a place – multidisciplinary perspectives*. Ed. Convery, I., Corsane, G. and Davis, P. The Boydell Press, Woodbridge, UK.

Costanza, R., De Groot, R., Braat, L., Kubiszewski, I., Fioramonti, L., Sutton, P. … and Grasso, M. (2017). Twenty years of ecosystem services: How far have we come and how far do we still need to go?. *Ecosystem Services*, 28, 1–16.

Cumbria Tourism. (2015). Cumbria Visitor Survey 2015. Report by QA Research.

Cumbria Tourism. (2019). Tourism in Cumbria 2019: Key facts and Trends. Report by QA Research.

Daniel, C. D., Muhar, A., Arnberger, A., Aznar, O., Boyd, J. W. and Chan, K. M. A. (2012). Contributions of cultural services to the ecosystem services agenda. *Proceedings of the National Academy of Sciences of the United States of America*, 19(23), 8812–8819.

Daugstad, K. (1999). Mellom romantikk og realisme. Om seterlandskapet som ideal og realitet. PhD dissertation in geography. Trondheim: Norwegian University of Science and Technology.

Dempsey, B., (2021). Understanding conflicting views in conservation: An analysis of England. *Land Use Policy*, 104, 105362.

den Breejen, L. D. (2007). The experiences of long-distance walking: A case study of the West Highland Way in Scotland. *Tourism Management*, 28, 1417–1427.

Derks, J., Giessen, L. and Winkel, G., (2020). COVID-19-induced visitor boom reveals the importance of forests as critical infrastructure. *Forest Policy and Economics*, 118, 102253.

Di Sacco, A., Hardwick, K. A., Blakesley, D., Brancalion, P. H., Breman, E., Cecilio Rebola, L., Chomba, S., Dixon, K., Elliott, S., Ruyonga, G. and Shaw, K., (2021). Ten golden rules for reforestation to optimize carbon sequestration, biodiversity recovery and livelihood benefits. *Global Change Biology*, 27(7), 1328–1348.

Duesberg, S., O'Connor, D. and Dhubháin, Á. N. (2013). To plant or not to plant—Irish farmers' goals and values with regard to afforestation. *Land Use Policy*, 32, 155–164.

Fredman, P. and Tyrväinen, L. (2010). Frontiers in nature-based tourism. *Scandinavian Journal of Hospitality and Tourism*, 10(3), 177–189.

Frost, W. and Hall, C. M. (Eds.). (2009). *Tourism and national parks: International perspectives on development, histories, and change* (Vol. 14). Routledge, London.

Garrod, B., R. Wornell and R. Younell. (2005). Re-conceptualising rural resources as countryside capital: The case of rural tourism. *Journal of Rural Studies* 22, 117–128.

Glover, J. (2019). Landscapes Review: Final Report. defra.gov.uk

Hall, C. (2008). Identifying farmer attitudes towards genetically modified (GM) crops in Scotland: Are they pro- or anti-GM? *Geoforum*, 39, 204–212.

Hardaker, A., Pagella, T. and Rayment, M., (2021). Ecosystem service and dis-service impacts of increasing tree cover on agricultural land by land-sparing and land-sharing in the Welsh uplands. *Ecosystem Services*, 48, 101253.

Haslam, S. A. and McGarty, C. (2014). *Research methods and statistics in psychology*. Sage, California.

Hegetschweiler, K. T., de Vries, S., Arnberger, A., Bell, S., Brennan, M., Siter, N., Olafsson, A. S., Voigt, A. and Hunziker, M., 2017. Linking demand and supply factors in identifying cultural ecosystem services of urban green infrastructures: A review of European studies. *Urban Forestry & Urban Greening*, 21, 48–59.

Hunsberger, C., Corbera, E., Borras Jr, S. M., Franco, J. C., Woods, K., Work, C., de la Rosa, R., Eang, V., Herre, R., Kham, S. S. and Park, C. (2017). Climate change mitigation, land grabbing and conflict: Towards a landscape-based and collaborative action research agenda. *Canadian Journal of Development Studies/Revue canadienne d'études du développement*, 38(3), 305–324.

IPCC. (2021). *Climate change 2021: The physical science basis. Contribution of working group i to the sixth assessment report of the intergovernmental panel on climate change* [Masson-Delmotte, V., P. Zhai, A. Pirani, S. L. Connors, C. Péan, S. Berger, N. Caud, Y. Chen, L. Goldfarb, M. I. Gomis, M. Huang, K. Leitzell, E. Lonnoy, J. B. R. Matthews, T. K. Maycock, T. Waterfield, O. Yelekçi, R. Yu and B. Zhou (eds.)]. Cambridge University Press. In Press.

IUCN (2021). https://www.iucn.org/theme/protected-areas/about/protected-area-categories. Accessed 11/11/2021.

Iversen, S. (2019). Impact & Impacts and perspectives of woodland creation in upland Cumbria, UK. Lancaster University (United Kingdom).

Iversen, S. (2020). Impact & Perspectives of Woodland Creation in Upland Cumbria, UK. Ph.D thesis from Lancaster University, UK.

Iversen, S. V., Holt, C., van der Velden, N., Mansfield, L., Convery, I., Kjeldsen, C. and Thorsøe, M. H. (2021). Impacts of woodland planting on nature-based recreational tourism in upland England – a case study. DOI: 10.32942/osf.io/ur6qz

Iversen, S. V., van der Convery, I, Velden, N., Mansfield, L. and Holt, C. D. S. (2022). Why understanding stakeholder perspectives and emotions is important in upland woodland creation – a case study from Cumbria. *UK. Land Use Policy*, 114, 105929.

Jones, L., Holland, R. A., Ball, J., Sykes, T., Taylor, G., Ingwall-King, L. and Peh, K. (2020). A place-based participatory mapping approach for assessing cultural ecosystem services in urban green space. *People and Nature*, 2(1), 123–137.

Kajala, L. (2007). *Visitor monitoring in nature areas: A manual based on experiences from the Nordic and Baltic countries*. Nordic Council of Ministers.

Kenter, J. O., O'Brien, L., Hockley, N., Ravenscroft, N., Fazey, I., Irvine, K. N., Reed, M. S., Christie, M., Brady, E. and Bryce, R. (2015). What are shared and social values of ecosystems? *Ecological Economics*, 111, 86–99.

Kim, A. K. and Weiler, B. (2013). Visitors' attitudes towards responsible fossil collecting behavior: An environmental attitude-based segmentation approach. *Tourism Management*, 36, 602–612.

Kneafsey, M. (2000). Tourism, place identities and social relations in the european rural periphery. *European Urban and Regional Studies*, 7, 35–50.

Landry, C. E., Bergstrom, J., Salazar, J. and Turner, D., (2021). How has the COVID-19 pandemic affected outdoor recreation in the US? A revealed preference approach. *Applied Economic Perspectives and Policy*, 43(1), 443–457.

Lange, E. (2001). The limits of realism: Perceptions of virtual landscapes. *Landscape and Urban Planning*, 54, 163–182.

Leung, Y. F., Spenceley, A., Hvenegaard, G. and Buckley, R. (2018). *Tourism and visitor management in protected areas: Guidelines for sustainability* (Vol. 27). Gland, Switzerland: IUCN.

Li, Q., Otsuka, T., Kobayashi, M., Wakayama, Y., Inagaki, H. and Katsumata, M. (2011). Acute effects of walking in forest environments on cardiovascular and metabolic parameters. *European Journal of Applied Physiology*, 111(11), 2845–2853.

Locatelli, B., Catterall, C. P., Imbach, P., Kumar, C., Lasco, R., Marín-Spiotta, E., Mercer, B., Powers, J. S., Schwartz, N. and Uriarte, M., (2015). Tropical reforestation and climate change: Beyond carbon. *Restoration Ecology*, 23(4), 337–343.

Mapes, N. (2012). Have you been down to the woods today? Working with older people. *Community Care Policy & Practice*, 16(1), 7–16.

Natural England (2014). *Mapping values: The vital nature of our uplands – an atlas linking environment and people (NE209)*. Published by Natural England, UK.

Newing (2011). *Conducting research in conservation – a social science perspectives*. Routledge, Oxon, UK.

Norton, B. A., Coutts, A. M., Livesley, S. J., Harris, R. J., Hunter, A. M. and Williams, N. S., (2015). Planning for cooler cities: A framework to prioritise green infrastructure to mitigate high temperatures in urban landscapes. *Landscape and Urban Planning*, 134, 127–138.

Orland, B., Budthimedhee, K. and Uutsitalo, J. (2001). Considering virtual worlds as representations of landscape realities and as tools for landscape planning. *Landscape and Urban Planning*, 54, 139–148.

Peh, K. S. H., Balmford, A., Bradbury, R. B., Brown, C., Butchart, S. H., Hughes, F. M., Stattersfield, A., Thomas, D. H., Walpole, M., Bayliss, J. and Gowing, D., (2013). TESSA: A toolkit for rapid assessment of ecosystem services at sites of biodiversity conservation importance. *Ecosystem Services*, 5, 51–57.

Quick, K. S. and Bryson, J. M., (2016). Public participation. In *Handbook on theories of governance*. Edward Elgar Publishing, Cheltenham.

Reed, M. S., Bonn, A., Slee, W., Beharry-Borg, N., Birch, J., Brown, I., Burt, T. P., Chapman, D., Chapman, P. J., Clay, G. D., Cornell, S. J., Fraser, E., Glass, J. H., Holden, J., Hodgson, J. A., Hubacek, K., Irvine, B., Jin, N., Kirkby, M. J., Kunin, W. E., Moore, O., Moseley, D., Prell, C., Price, M. F., Quinn, C. H., Redpath, S., Reid, C., Stagl, S., Stringer, L. C., Termansen, M., Thorp, S., Towers, W. and Worrall, F. (2009). The future of the uplands. *Land Use Policy*, 26, S204–S216.

Reeder, R. J. and Brown, D. M. (2005). Recreation, tourism and rural well-being. United States Department of Agriculture. Economic Research Reports number 7.

Sherman, S. A., Varni, J. W., Ulrich, R. S. and Malcarne, V. L. (2005). Post-occupancy evaluation of healing gardens in a pediatric cancer center. *Landscape and Urban Planning*, 73,167–183.

Soe Zin, W., Suzuki, A., Peh, K. S. H. and Gasparatos, A. (2019). Economic value of cultural ecosystem services from recreation in popa mountain national park, Myanmar: A comparison of two rapid valuation techniques. *Landscape*, 8(12), 194.

Soliva, R. and Hunziker, M. 2009. Beyond the visual dimension: Using ideal type narratives to analyse people's assessments of landscape scenarios. *Land Use Policy*, 26(2), 284–294.

Spennemann, D. H. and Whitsed, R., (2021). The impact of COVID-19 on the Australian outdoor recreation industry from the perspective of practitioners. *Journal of Outdoor Recreation and Tourism*, 100445.

Stenbacka, C., (2001). Qualitative research requires quality concepts of its own. *Management Decision* 1.

Stephenson, W. (1935). Technique of factor analysis. *Nature*, 136(3434), 297–297.

Stratford, C., Miller, J., House, A., Old, G., Acreman, M., Dueñas-Lopez, M. A., Nisbet, T., Newman, J., Burgess-Gamble, L., Chappell, N. and Clarke, S., (2017). Do trees in UK-relevant river catchments influence fluvial flood peaks. *A Systematic Review*, 1. Wallingford, UK, NERC/Centre for Ecology & Hydrology, 46 pp. (CEH Project no. NEC06063.

Streifeneder, T. and Dax, T. (2020). Agritourism in Europe: Enabling Factors and Current Developments of Sustainable On-Farm Tourism in Rural Areas. In Global Opportunities and Challenges for Rural and Mountain Tourism.

UKCCC (UK Committee on Climate Change (UKCCC). (2019). Net Zero Technical Report https://www.theccc.org.uk/publication/net-zero-technical-report/. Accessed 28/11/2021.

UNFCCC. (2021). Communication of long-term strategies. United Nations Framework Convention on Climate Change, https://unfccc.int/process/the-paris-agreement/long-term-strategies. Accessed 28/11/2021.

Walder, P. and Kantelhardt, J. (2018). The environmental behaviour of farmers – capturing the diversity of perspectives with a Q methodological approach. *Ecological Economics*, 143, 55–63.

Ward, T. C. and Aspinall, P. A. 2011. Natural environments and their impact on activity, health, and quality of life. *Applied Psychology. Health and Well-Being* 3(3), 230–260.

Weston, P. 2021. Planting a vision: Why the secret to rewilding success is about people, not trees. *The Guardian*, https://www.theguardian.com/environment/2021/sep/27/planting-a-vision-why-the-secret-to-rewilding-success-is-about-people-not-trees-aoe. Accessed: 26/11/21.

Zabala, A., Sandbrook, C. and Mukherjee, N. (2018). When and how to use Q methodology to understand perspectives in conservation research. *Conservation Biology* 32, 1185–1194.

Zank, B., Bagstad, K. J., Voigt, B. and Villa, F., (2016). Modeling the effects of urban expansion on natural capital stocks and ecosystem service flows: A case study in the Puget Sound, Washington, USA. *Landscape and Urban Planning*, 149, 31–42.

5

COOPERATION IN NICHES OF THE NATURE-BASED TOURISM INDUSTRY

The case of mountain guiding in Norway

Kristin Løseth

Introduction: the suppliers of nature-based tourism products and the need for collaboration

Nature-based tourism is considered an important contributor to rural diversification and development. While the nature-based tourism industry is growing overall, the suppliers of commercial nature-based tourism products and experiences have been found to be small, located in rural areas, and often operated partially seasonally (Stensland, Forbord, Fossgard & Løseth, 2021). Research on small tourism businesses has identified challenges in developing knowledge tied to the domination of small and medium-sized businesses, high levels of labour turnover, seasonality, rapid changes of ownership and the lack of dedicated career ladders (Hjalager, 2002).

These challenges, which have also been identified specifically in adventure tourism businesses (Løseth, 2014), lead to a lack of absorbative capacity (Cohen & Levinthal, 1990: 128) in small tourism businesses. The term refers to how "the ability to evaluate and utilize outside knowledge largely is a function of prior related knowledge". Hoarau (2014) argues that many tourism organizations may lack both the experience and the knowledge expertise to recognize, absorb, and use transferred knowledge. Collaboration is a natural way of combating such knowledge and resource constraints. Collaboration on the production side of the tourism industry refers to the interaction between businesses as well as between businesses and other relevant stakeholders, such as public bodies, that contribute and facilitate such interactions (Hoegh-Guldberg, 2018: 35). The nature of the tourism experience favors cooperation as the total tourism experience often depends on the coordinated services provided by several businesses (Hjalager, 2002: 470). In a nature-based tourism context, the providers of activities and attractions are highly dependent on tourism infrastructure such as restaurants, accommodation, and transport to be able to achieve value creation (Stensland et al., 2021: 158). Other effects of cooperation in a nature-based tourism context have been identified as product improvements and innovations, increased internal capabilities and competencies, and access to wider networks (Forbord & Sivertsvik, 2021). Network collaborations have been found both to facilitate access to tangible and intangible resources (Saxena, 2005) and to enhance performance (Novelli, Schmitz & Spencer, 2006).

Still, the small size of the businesses means limited resources for cooperation, particularly at bigger scales such as tourism networks (Fuglesang & Eide, 2013). The fragmentation of the

 DOI: 10.4324/9781003230748-7

industry may also halt cooperation, as it consists of entities of different scales from different areas, that hinders communication and knowledge transfer (Czernek, 2017; Hjalager, 2002). The heterogeneity of the industry may thus act as a break on cooperation and knowledge transfer. Even if the businesses are situated in geographical proximity, they may have little in common in terms of markets, knowledge needs, business aims, and visions for the industry at large.

In this complex and diverse industry, the methodological approach singles out and explores cooperation in a tourism niche. A few studies suggest that businesses with similar products have the potential for more specific learning through cooperation, and more direct imitation (Weidenfeld, Williams & Butler, 2010: 610). Collaborations between businesses with similar products enhance exploitative knowledge transfers (Sørensen, 2007; Weidenfeld et al., 2010), meaning that deeper learning in a particular field takes place through the refinement and development of existing competencies (Sørensen, 2007: 25).

This study will follow up on the findings above, by looking at the pattern of cooperation in mountain guide businesses in Norway that provide similar ski and mountaineering products. The businesses provide products for narrow markets consisting of individuals with a keen interest in skiing and/or mountaineering. In a niche market tourists can be identified as having similar specialized needs or interests and by a strong desire for the specific products offered (Robinson & Novelli, 2011: 5). The studies mentioned above suggest that knowledge transfer is facilitated between actors that provide similar products, but little is known about *how* businesses within a niche cooperate and how the niche cooperates with the tourism industry at large. Here, cooperation in the niche of mountain guiding in Norway is explored through the following research question:

What characterizes cooperation within the niche of mountain guiding, and how do actors of the niche cooperate with actors from other parts of the tourism industry?

The research has managerial implications as it aims to make the actors more conscious of patterns of cooperation as well as some of the pitfalls of cooperating mainly with the likeminded. The study has policy implications as differences in knowledge needs and patterns of cooperation between tourism niches and the general industry are made visible, pointing to where measures can be taken to support communication and knowledge transfer.

In the remainder of this chapter, key theoretical issues concerning cooperation and knowledge development in tourism are examined, before narrowing down to look at the specificities of nature-based tourism niches. There then follows the methodological consideration of the study, a presentation and discussion of the key findings, and, finally, an assessment of theoretical, managerial, and policy implications.

Cooperation and knowledge development in a tourism context

While cooperation has many functions in the development of experience products, the transfer and development of knowledge resources are increasingly seen as vital (Saxena, 2005; Hjalager, 2002). Since the turn of the century there has been a growing scholarly interest in knowledge management in tourism (see Cooper, 2015; Hall & Williams, 2008; Hjalager, 2002; Shaw & Williams, 2009 for theoretical reviews). The literature suggests that particularly the development and use of *external* knowledge new to the business is important for business development (Hjalager, 2002; Shaw & Williams, 2009; Weidenfeld et al., 2010). External knowledge enters a business in different ways, and Polanyi's (2000; orig. 1966) typology of explicit and tacit knowledge has been particularly useful to understand how to access, develop, and transfer knowledge within and between organizations. His divisions imply that while explicit knowledge is easily accessible through the internet, work manuals, or in the daily flow of information, other competencies require practical know-how that is more tacit in nature, and which requires close interaction to be transferred from one person to another (Nonaka & Takeuschi, 1995). Due to the dependency of spatial proximity and interaction, practical know-how has been described as more "sticky", or place-dependent, than explicit knowledge (Gertler, 2003). Cooper (2018: 508) argues that the bulk of knowledge in tourism is tacit and that this knowledge is a core competency of the organization "*as it both facilitates strategic advantage and allows differentiation from rivals*".

The close interaction needed for the transfer of tacit knowledge requires a level of trust between the actors (Czernek & Czakon, 2016). Hjalager (2009: 470) argues that there is little mutual trust among tourism businesses, as they first and foremost see themselves as competitors, rather than colleagues. This view has been challenged by the more dynamic concept of 'coopetition' (Czernek & Czakon, 2016; Van der Zee & Vanneste, 2015; Wank & Krakover, 2008) which acknowledges that competition and collaboration often go hand in hand.

Research on cooperation in the tourism industry has explored the topic from different theoretical and methodological perspectives. Three perspectives are dealt with here, as they can all contribute to a more nuanced understanding of cooperation in niche tourism. The first approach is broad and emphasizes how tourism businesses are embedded in social networks consisting of actors as nodes with connections between them (Elvekrok, Veflen, Scholderer & Sørensen, 2022; Sørensen, 2007; Van der Zee & Vanneste, 2015; Zach & Racherla, 2011). These networks vary in density. Sørensen (2007: 25) describes how dense networks, characterized by strong relations and high levels of trust, support knowledge exploitation. More sparse networks support knowledge exploration and the search for new knowledge, as the number of connections increase, and the variety of information is maximized. Sørensen argues that an aim must be to find a good balance between knowledge exploitation and knowledge exploration.

The second approach is narrower and examines how cooperation is influenced by geographical distance. Cooperation can be both local and over vast distances (Asheim & Isaksen, 2002; Bathelt, Malmberg & Maskell, 2004). Theories of agglomeration highlight the knowledge benefits of being spatially close as it facilitates close interaction and learning, particularly regarding know-how (Boschma, 2005; Gertler, 2003). The organization of tourism businesses in destinations can be seen as one such spatial arrangement. While the availability of different services at a destination benefits the tourists, difficulties of developing and maintaining cooperating activities at the destination level have been identified, due to low levels of professionalism (Pikkemaat & Weiermair, 2007), differences in motives (Ateljevic & Doorne, 2000; Shaw & Williams, 2004) and fear of imitation (Hjalager, 2002). Tourism actors prefer to cooperate with partners that are similar in terms of resources, governance structures, and business vision (Zach & Racherla, 2011: 105). These findings are supported by Boschma (2005), who argues that geographical proximity facilitates cooperation but is no guarantee for knowledge transfer to take place: Equally important is the sharing of a similar knowledge base and expertise (cognitive proximity), trust-based social relationships (social proximity), the organizational arrangement of relations (organizational proximity), and the sharing of institutional rules, habits and values (institutional proximity).

The third theoretical perspective has its origin in the work of Etienne Wenger on communities of practice (1998) and is less concerned with cooperation per se, but more with how learning is socially situated. Wenger (1998: 241) described communities of practice as social learning systems, characterized by a sense of joint enterprise, mutual engagement, and a shared repertoire of communal resources such as routines, tools, language, stories etc. Knowledge is shared and transferred through practice and is accumulated in the experiences and know-how of those engaged in a given practice (Hoarau, 2014; Høegh-Guldberg, 2018). Actors within a niche of nature-based tourism produce similar products and have thus evolved into similar practices. This may explain findings of more specific learning and direct imitation between tourism businesses with similar products (Weidenfeld, Williams & Butler, 2010: 610). Brown and Duguid (1991) warn that knowledge transfer within a community of practice may be constrained in the long run by the shared worldview of its members. An important role in this respect are the 'boundary spanners' who move between different communities of practice.

Taken together, the three perspectives illustrate how cooperation can be understood through multiple levels of interpretation. The number of strong and weak ties a business has in its surroundings say something about its potential for cooperation and the different types of information and support it can gain from this network. The geographical perspective highlights the importance of geographical proximity for cooperation, but also bring in how other types of proximity facilitates cooperation and knowledge transfer. Finally, the community of practice perspective brings attention to the knowledge transfer taking place as people work together in a joint enterprise. In the following section, the three perspectives are used to illuminate distinct features of cooperation in niches of nature-based tourism.

Cooperation and knowledge transfer in nature-based tourism

The nature-based tourism industry can be seen as a macro-niche, which can be further segmented into micro-niches that have common markets, common resource utilization, and knowledge needs (Robinson & Novelli, 2011). Empirical research on cooperation and knowledge development in niches of the nature-based tourism industry is scarce, with a few important exceptions (Fadda, 2020; Hoarau, 2014; Helgadóttir & Sigurðardóttir, 2008; Shaw & Williams, 2004).

A particular characteristic of many nature-based tourism niches is the close link to specific leisure activities. The products have developed out of traditional leisure activities such as hiking, climbing, surfing, or sea kayaking through gradual processes of commercialization and commoditization (Beedie & Hudson, 2003; Cloke & Perkins, 2002; Varley, 2006). The knowledge needed to take part in these activities are transferred to a commercial setting. Not surprisingly, the close relation between niches in nature-based tourism and specific leisure activities has been found to affect know-how, entrepreneurial motivation, and entrepreneurial networks (Løseth, 2014, 2018; Shaw & Williams, 2004).

Sociologist Robert A. Stebbins did research on people's leisure time behaviour and identified how some people became so engaged in their leisure time activity that it influenced their identity, what they did in their spare time, where they chose to live, and even their career pursuits. Stebbins' described this engagement as serious leisure[1] (1982). Important in terms of cooperation, serious leisure also includes a social dimension, and Stebbins describes the social worlds of participants in serious leisure as having *"unique sets of norms, values, beliefs, styles, moral principles, performance standards, and similar shared representations"* (Stebbins, 1999: 71). The participants thus not only share the same knowledge base and expertise, what Boschma (2005) calls cognitive proximity, but also social proximity as the actors shares common norms and values. The similarity facilitates knowledge transfer and cooperation, supported by research findings indicating that actors from niches in nature-based tourism find it easy to cooperate with people from the same niche (Fadda, 2020; Løseth, 2014; Shaw & Williams, 2004).

As knowledge is tied to the practice of the particular nature-based activity, information or know-how may also be transferred from actors outside the tourism industry who possess relevant expertise. An example of this from a nature-based tourism context is a study by Hoarau and Kline (2014: 59). They describe, from a community of practice perspective, how business actors and researchers work together and develop knowledge in cooperation around the tourism product of whale watching. Through common practice they contribute to the *knowledge domain*, a central feature of a community of practice (Wenger, 1998).

Developing a knowledge domain means increasing the business' absorbative capacity (Cohen & Levinthal, 1990). Cooper (2018: 511) argues that many tourism organizations lack absorbative capacity as they "may lack both experience and ability expertise to use transferred knowledge", due to high levels of unskilled labour and labour turnover. From her study of whale watching businesses, Hoarau (2014: 147) suggests that the whale watching firms do have the ability to absorb and utilize knowledge from researchers and consultants due to their close interaction in shared practice. Niche knowledge developed through shared practice may thus mean that the businesses have an enhanced ability to recognize, absorb, and utilize external knowledge within their specific knowledge domain. Clear knowledge domains centred around niches in nature-based tourism has also been described in horse-based tourism (Helgadóttir & Sigurðardóttir, 2008) and surfing (Fadda, 2020; Shaw & Williams, 2004).

Both Wenger's community of practice perspective and Boschma's emphasis on different types of proximities can thus provide valuable theoretical tools to explore cooperation and knowledge transfer in niches of the nature-based tourism industry. Knowledge is developed in practice among actors with a lot in common in terms of knowledge, norms, and values. As knowledge is centred around the particular nature-based tourism activity, or practice, all actors with relevant expertise can contribute to knowledge development within the niche. Stebbins' concept of serious leisure draws the theoretical attention particularly to the knowledge development taking place at borders between leisure activity and commercial niche activity. The theoretical perspectives indicate a potential for cooperation within the niche of mountain, but as the businesses within the niche also compete for the same tourists, it is uncertain whether this potential is utilized or downplayed due to competition. It is also uncertain how actors from the niche cooperate with actors *outside* the niche, and thus how the niche is connected to the wider tourism industry.

Context: Mountain guiding in Norway

The skilled guide is essential for the nature-based tourism experience (Farkić, Filep & Taylor, 2020). Rokenes, Schumann and Rose (2015) claims that tourists on more technically demanding activities depend on their guide in another way than the "normal" tour guide, both due to the high emphasis on safety but also for finding the best snow or the best ski or climbing routes. In the context of mountain guiding specifically, Hough Mackenzie and Kerr (2013: 4) argue that

> guiding requires technical skills and personal responsibility (e.g. for clients, safety, trip planning, logistics), specialized qualifications or training requirements, extended exposure to clients and other guides, and may necessitate quickly learning and acclimatizing to novel outdoor environments and operational systems.

Guiding tourists to majestic peaks has long historic roots in Norway. The first explorers of Norwegian mountains were foreigners who needed practical assistance and local knowledge from the mid-1800s onwards (Horgen, 2016). The international mountaineers had the expert knowledge, but needed support from Norwegian farmers with local knowledge of the terrain and weather patterns.

The first organization of patented guides in Norway was a public initiative back in 1826 at the Jostedal glacier. Apart from this local initiative, the organization of mountain guiding in Norway has mainly been a voluntary affair where a few organizations has played a central role (Eikje et al., 2017). In 1890, The Norwegian Trekking Association established their own mountain guide services which lasted up until the 1970s.

While the economic significance of commercial mountaineering was still small, there was a growing emphasis on the formal qualifications of those leading commercial groups in Norway throughout the 1970s and 80s (Dahl, 2009). While the teaching of mountaineering skills previously had taken place informally, the demand placed on instructors and course content became so important that Norsk Fjellsportforum (Norwegian Mountain Forum, NF) was established, by different climbing and glacier associations.

The growing emphasis on formal qualifications did not lead to increasing regulations in the Norwegian mountains. No certifications are formally required to guide commercially in Norway (Eikje et al., 2017). Nortind is the Norwegian representative of the International Federation of Mountain Guides Associations (IFMGA) and follows their standards in certifying mountain guides. Nortind aims to protect their profession through stronger regulation, but has to date had insufficient impact. This means that in theory anyone can be paid to guide tourists into the mountains, but in practice the commercial actor will have to follow the safety norms of the niche, if he or she is to avoid prosecution in case of an accident.

Today, mountain guiding in Norway mainly takes place in the mountain areas of Lofoten, Sognefjellet, Romsdalen, Sunnmøre and Lyngen. The product usually entails long days or multiple days in small groups, with close interaction between guide and tourists. Given the small guide-to-tourist ratio, mountain guide products can be considered a high-end product. The main products are avalanche, climbing, and glacier courses, and guided trips in steep terrain winter and summer.

Methodological considerations

The aim of this particular data collection, as part of a larger Norwegian Research Council project, BIOTOUR, was to examine how a niche tourism product evolves over time, and the interview guide was developed for this particular purpose. This book chapter is written based on parts of the BIOTOUR material, as the interviews revealed a pattern of cooperation worth further exploration. The ad hoc character of this approach clearly influences the quality of the findings. By being transparent and critical throughout we still argue that the findings can contribute fruitfully as an explorative study on cooperation in tourism niches.

A qualitative approach based on in-depth interviews was chosen to examine what cooperation means to the actors of the tourism niche. The research has a phenomenological position as it seeks to understand and describe the social phenomena of collaboration from the actors' own perspectives (Kvale & Brinkman, 2009: 26).

The sampling of subjects was conducted purposefully. Initial theoretical inquiry into the concept of niche tourism was further elaborated upon in the context of mountain guiding in Norway and discussed in a pilot interview with a key knowledgeable informant (Patton, 2015: 284). This led to an operationalization of the mountain guiding niche as: *Tourism businesses that offer nature-based tourism activities in steep terrain, either winter ski touring (on a complex snowpack) and/or summer mountaineering (that involves the use climbing rope).*

The initial sampling was based on the operationalization above, and a list of relevant businesses was generated both based on internet search and information from the key interview.

As there has been some tension in the world of commercial mountaineering in Norway as to who should be allowed to guide in steep terrain (see Eikje et al., 2017; Løseth & Varley, 2021), we included both guides with IFMGA qualifications, and guides with NF qualifications. There was only one female among the informants, reflecting that most mountain guides who work full time with guiding in Norway today are still male.

As the research question of the original study was concerned with how the niche has evolved over time, we made sure the sample included both guides and/or entrepreneurs who had been in the business over several decades, and guides/entrepreneurs who more recently had entered the niche. The niche could also be experienced to evolve differently in different mountain regions. To explore whether regionally difference mattered, four different mountain regions were chosen: Sunnmøre, Romsdalen, Sogn and Lyngen. The sampling was thus theoretically driven (Miles & Huberman, 1994: 29).

A semi-structured interview guide was used, which needed to be slightly modified to fit the different geographical contexts. The interview guide was developed for the initial study on how niches evolve. The main themes covered that have relevance for this research are on the motivation and way into the work as mountain guide, including what competencies the guides had as they entered the industry. Then, questions on who they cooperate with, both within and outside of the niche, and how this affected their business. They were also asked to elaborate on who had been important for the development of not only their business, but also of the niche of mountaineering. The findings are presented thematically following the interview guide.

12 information-rich interviews were conducted: eleven face-to-face and one via video call. By the last interviews little new information was gained as the level of saturation had been met. This level tends to be reached earlier, when the sample group is homogeneous and the research theme is quite narrow (Tjora, 2017: 143), as was the case here. The interviews were 50–70 minutes in length, all recorded digitally and fully transcribed.

The interviews were then coded. According to Kvale and Brinkmann (2009: 202), coding can be both concept and data-driven. Here, the analysis started with concept-driven coding, where the codes had been developed in advance through consulting relevant literature. This was then followed up by codes developed through readings of the material. The coding was not meant for quantification but made it easier to see relations and repeating patterns in the material. The coded interviews were then used in a thematic analysis centred around the main themes of the interview guide.

The role of the researcher is crucial in qualitative work and one factor that may have influenced the interview situation was my personal knowledge of the niche. I am married to a mountain guide who has worked in the niche full time for 12 years, based in one of the regions of this study but also working in other mountain regions in Norway. This means that I know the niche well through his occupation, and I personally knew five of the interview subjects. Thagaard (2013: 206) argues that knowing the field from within can be both a strength and a weakness. The issues and situation described can be easier to understand, but it can also make the researcher disregard information that is different from one's own opinion.

The fact that the interview guide was developed for a different research question influences the results of the study by making the findings "thinner" than if the interview guide had been fully focused on cooperation. The themes are well covered, but more specific questions would add valuable nuances to the findings.

For transparency, the citations are given "tags" so that the reader can see how the data material is being used without sacrificing anonymity (Tjora, 2017: 249). The interviews have been tagged according to geographical regions, R 1–4, and interview number in that region: N 1–4. The second interview in region 2 will thus be tagged with (R2, N2).

Findings and analysis

For the purpose of clarity, the section is divided into three parts, with the first describing cooperation within the niche, the second examining cooperation between actors within the niche and actors from other parts of the tourism industry, and the third part describing boundary spanning, where knowledge is transferred between the niche and other fields of knowledge.

Cooperation within the niche

The first part of the interviews contained questions on *why* the interviewees chose to become mountain guides, and about their way into the niche. All interviewees referred to leisure time engagement in the activities of skiing and mountaineering, in combination with different type of nature-based tourism guide work, as exemplified by this quote:

> I wanted to work in the mountains, and I saw that I had enough experience with climbing and skiing at a decent level, to enter and complete (the UIAFM education). It was really just a natural development based on all the trips, guiding and instruction I had done.
>
> (R4, N2)

The combination of engagement for a particular leisure activity, leading to summer jobs in a landscape where they can be immersed in their favorite activity, suggest the relevance of Stebbins' (1982) concept *serious leisure*. The interviewees described how guiding work fit with their desired lifestyle: *"Common to us all was that we were young, and we wanted to climb. (...) it was a place where we could come and work hard in short periods and then go off on a climbing trip"* (R4, N1).

Previous studies have pointed out how the high levels of engagement that characterize serious leisure can facilitate the development of specific *know-how* relevant to a specific nature-based tourism niche (Løseth, 2018; Shaw & Williams, 2004). By being involved in specific nature-based activities over several years, both in leisure time, in part-time jobs, and through instructor and guide courses, the mountain guides have acquired quite specialized knowledge. They thus have absorbative capacity (Cohen & Levinthal, 1990) in the field, in contrast to previous findings where lack of knowledge resources has been identified as a barrier to knowledge transfer in tourism (Czernek, 2017).

Literature on serious leisure describe how people identify with a leisure activity and with the community of likeminded practitioners that forms around the activity (Løseth, 2018; Stebbins, 1982). The theme of identity was recurring in the data set, as the interviewees identified themselves with the practice of being and working in the outdoors. In addition, the interviews suggest that those who had been through the IFMGA qualifications saw this three-year process as an important part of their identity as mountain guides:

> I got to know people, which was very good. And then I felt more professionally confident, both regarding my value base, but also regarding techniques and the craft. "This is how we do it" in our practice. It provided a good sense of confidence.
>
> (R2, I2)

The materials suggest informal cooperation between actors within the niche, particularly between businesses within the same region. The trust-based ties and cooperation contrast with the situation described by Hjalager (2002) where relations between tourism actors are dominated by mistrust and competition. The fluid character of the cooperation can be illustrated by the following quotes from two different regions:

> I think we always have had this friendly, professional relationship, where we cooperate and send guests to each other if we can. We have always done it this way; there has never been competition.
>
> (R3, N2)

> A few of us who guide in this area think quite similar about how we work, how we want to run things. So we have in a way cooperated. Each has his own business, and then we use each other when we need help, or we send our guests on. Very little hassle.
>
> (R4, N1)

In busy periods more guides are needed, and due to the cognitive, social and organizational proximity (Boschma, 2005) guides can work for others as freelancers both locally and in other mountain regions nationally and abroad. There has been little research on how labour mobility affects knowledge transfer in tourism (Hall & Williams, 2008: 77; Shaw & Williams, 2009: 330); however, Hjalager (2002) suggests that it can be valuable as long as the reception capacity for new ideas are favorable. This study suggests that work mobility in a niche with competent employees, and thus absorbative capacity within the field, facilitates knowledge transfer between different mountain regions.

Exceptions to this pattern of cooperation and mobility was found between actors who had the IFMGA qualification and those who had an alternative background. According to this material, actors without the IFMGA qualification either decided not to cooperate with IFMGA guides due to differences in guiding and business approach (R2, N1), or found it difficult to cooperate with IFMGA guides:

> With the IFMGA guides, they are a bit "this is how it should be". And it is hard for those of us who do not have that qualification to get access (to the cooperation).
>
> (R3, N1)

The reluctancy to cooperate is not surprising as the IFMGA businesses share institutional rules, habits, and values (Boschma, 2005), and as their union, Nortind, has worked for decades to protect their profession through stronger regulation (Eikje, Horgen, & Arnegård, 2017). The tension can also be related to Aldrich and Fiol's (1994) term *organizational legitimacy*, as businesses in a developing niche tries to define and establish industry standards. Similar struggles for legitimacy have been identified in other niches of the nature-based tourism industry (Løseth, 2014).

With similar products offered to the same niche tourism markets, there is an obvious element of competition between the businesses. Competition was still downplayed, in line with (Fadda, 202) in the interviews. Only one interviewee mentions the regional rivalry: *"Because we (the guide businesses in the region) all want to have the guide directly to us. And perhaps the guest sends a request to several, and then you have to be … to reply first"* (R4, N3).

Little emphasis on competition may be due to the prosperity of the niche, where there is enough work for all in the busy seasons. The situation of simultaneous competition and cooperation has previously been described in the tourism literature as coompetition (Czernek & Czakon, 2016; Van der Zee & Vanneste, 2015; Wank & Krakover, 2008).

The findings of a common practice and knowledge domain, informal cooperation and an identity tied to practice suggests that knowledge development in the niche can be understood through the community of practice perspective (Wenger, 1998). Two factors challenge the use of the concept in this context. First, trust is also a condition of a community of practice (Wenger, 1998). While trustful relationships were described in the niche, the data were not sufficient to say that this was the case overall. The businesses are in competition, and particularly the difference in qualification, led to mistrust among the two groupings. Secondly, one of the critiques of the communities of practice perspectives is the difficulties of assigning boundaries to the community concerning size and spatial reach. The businesses of this study were operating in different mountain regions, and while some cooperated closely within the region, the interviews cannot justify the concept within the niche as a whole.

Cooperation with tourism actors outside the niche

Tourists on guided trips also need complementary services such as food and accommodation (Hjalager, 2002; Stensland et al., 2021), and providers of this type of services were frequently brought up as important cooperation partners outside the niche. Informal cooperation dominated, but some also had exclusives deals:

> We are first and foremost a guide business, so that makes us seek cooperation outside that is win–win, good for us and good for our cooperation partners. They (their business partner) had the economy to build cabins that we, afterwards, could use by renting exclusively from them.
>
> (R3, N2)

More common was informal cooperation with several accommodation providers. The absence of formal deals may reflect that their tourists come in small groups and have different preferences when it comes to accommodation. Keeping several opportunities open may thus be a better bet as "*we have all kinds of customers, in a wide price range*" (R2, N1). Complementary, dyadic relationships were thus found to be valued and common as a way to provide customers with the services they need during their mountain experiences.

There was more ambivalence tied to participation in bigger tourism networks such as regional or destination networks, or different types of tourism development projects. Positive outcomes of tourism projects were mentioned, and one entrepreneur praised the regional tourism organization for their work to promote the region:

> I think they have succeeded well in promoting Norway as a place for activities (…) a lot is taking place out on the big arenas that we don't see here.
>
> (R2, N1)

More common, though, was that the marketing and communication efforts of the general tourism industry was experienced as less useful for the narrow niche of mountain guiding:

> We have tried over time to have sort of a cooperation with tourists offices, but I believe we barely have received one guest through that system.
>
> (R4, N1)

As small businesses they must give priority to the marketing efforts that clearly have an effect. The tourists searching for a mountain guide product have quite clear expectations for experiences and tend to contact the businesses directly. The niche quality of the mountain guiding product is also something that the guides want to protect: *"It is perhaps not always our wish, to be drawn into cooperation, big volumes and those kinds of things"* (R3, N2).

From a safety perspective as well as a quality of experience perspective, their products are not suited for larger groups. Bigger tourism network projects aiming for mass tourism development are thus met with doubts; *"It is almost a threat for us, influencing the quality of what we do. We want small groups"* (R2, N2).

The difference in target markets thus creates an ambivalence to broader cooperation projects. Another difference that challenged cooperation is what Sørensen & Grindsted, 2021) term the actors' "sustainability DNA". Differences in how one conceptualizes sustainable nature-based tourism limits tourism development. This was particularly evident in one mountain region where a bigger nature-based tourism project was opposed by two guides (R4, N1; R4, N2). As one guide says, he would "*rather develop things more slowly and expand more holistically*" (R4, I1). Sustainability is an important issue for entrepreneurs in nature-based tourism (Margaryan, Fredman & Stensland, 2020), but as understandings and values related to sustainability varies, it may challenge cooperation. The commonalities, or proximities (Boschma, 2005), found in terms of knowledge, values, and organization within the niche, is thus lacking in in the niche' relations to the wider industry.

Knowledge fields and boundary spanners

When asked about people who had been important for business' development, several actors not directly part of the tourism industry was brought up, such as customers (R4, I2), friends engaged in the ski industry as filmmakers or photographers (R2, I1), arrangers of ski and hiking festivals (R1, I1; R2, I2; R:4, I2). Common to these supporting actors is their personal interest in skiing and/ or mountaineering. As one interview states: "*The people who really want me to do well are the guests that love to go skiing with me, and among them you'll find a lot of resourceful people with different backgrounds"* (R3, N2).

The common engagement and knowledge base facilitate knowledge flows (Boschma, 2005). Customers, friends, and arrangers of festivals can be boundary spanners, between the mountain guiding niche and relevant knowledge fields. Common to these external actors is personal engagement in the knowledge domain and knowledge about the niche of mountain guiding. The broader knowledge field also led to different side projects for the guides. Four guides (R2, N2; R3, N2; R4, N1; R4, N2) had small part-time positions as avalanche observers for the Norwegian water resources and energy directorate (NVE). Here, their competency is used to report on local snow and avalanche conditions for the national avalanche forecast. Three of the interviewees had written ski guidebooks covering their local region (R2, N2; R3, N1; R4, N2), and two either have worked (R3, N1) or currently work (R1,I) as avalanche advisers. Niche knowledge is thus used in other contexts with overlapping knowledge fields, often including academic research.

Previous studies have found that tourism businesses do not actively employ academic research (Czernik, 2017; Shaw & Williams, 2009). In the mountain guide niche, theoretical models from avalanche research were actively utilized both in safety routines on guided trips, and as part of their avalanche courses. The following quote illustrates how research is integrated in the daily safety routines:

> There is a guide from the US who has written a good article in Avalanche Review, about the routines which should be carried out before, during and after the trip. So, we've tried to implement this in how we work.
>
> (R2, I1)

The findings are in line with the findings of Hoarau (2014) that tourism businesses which have developed absorbative capacity within a knowledge field have the ability to absorb and utilize academic research from overlapping fields of knowledge. The common knowledge base ensures cognitive and social proximity which facilitates communication and knowledge transfer. Hoarau (2014: 147) argues that the small size and "family culture" of the nature-based tourism businesses facilitate close interactions and thus knowledge transfer. However, this study finds that it is identification with the knowledge field, more than the small size and "family culture" of the business that facilitates knowledge transfer.

Conclusions

This chapter has explored, with evidence from the niche of mountain guiding, patterns of cooperation among businesses who provides similar products. The findings support Weidenfeld et al. (2010) and Sørensen (2007) in that product similarity enhances knowledge transfer, in this case particularly through guide mobility. Actors within the niche of mountain guiding identify strongly with their practice, and have close, but informal collaborations with other businesses offering similar products. Their common knowledge domain is the practice of leading groups in steep terrain, winter and summer, and knowledge is developed and spread among actors through shared practice. The identified lack of cooperation and knowledge sharing between actors who have the IFMGA qualification and those who have alternative qualifications is argued to be a barrier to knowledge development within the niche.

The close connection between mountaineering as leisure and work make room for boundary spanners who provide important input for the community. Particularly repeat customers, and actors with knowledge in the related fields of avalanche research, mountain medicine and equipment providers, were bringing important knowledge into the niche. In line with Hoarau (2014), the findings suggest that the experience and education of the guides provides an absorbative capacity to recognize and make use of external knowledge.

While there was close contact between the businesses offering similar products, in a form of coompetition, collaboration with other parts of the tourism industry varied more. The smaller businesses, in particular, had little formalized collaborations outside the niche. A repeating reason for this was that the bigger tourism projects aimed for other markets and had limited understanding for their niche product. Several actors also opposed nature-based tourism development projects that involve large number of tourists, as it was against their views on sustainable tourism (Sørensen & Grindsted, 2021).

The findings contribute to a more nuanced understanding of cooperation and knowledge development in the nature-based tourism industry. In niches of the industry, specialized knowledge is being developed and transferred both between actors in the niche, and through the role of boundary spanners. The identification of such specialized knowledge communities is particularly valuable in the context of rural tourism, challenged with seasonality and high labour turnover.

The theoretical emphasis on practice is valuable in examining knowledge development in a tourism niche, and a natural avenue for future research would be to examine in depth the relevant actors and the knowledge transfer processes taking place in tourism niches through a community of practice perspective (Wenger, 1998). In addition, Boschma's (2005) different types of proximities added nuances to our understandings of what facilitated and constrained cooperation, and is recommended as theoretical framework for future studies on cooperation in the heterogeneous tourism industry.

Managerially, the study can contribute by making actors within the nature-based tourism more aware of the cooperative dynamics of which they are part. Actors within the niche may reflect on how their emphasis on the practice of mountain guiding affects overall cooperative activity. For actors outside the niche the study may lead to a greater understanding of the knowledge base of the mountain guide product, and thus contribute to a better foundation for cooperation.

From a policy perspective, more emphasis should be placed on supporting knowledgeable milieus by facilitating interaction between relevant actors in tourism niches, including boundary spanners. In the mountain guide niche, different seminars and conferences on avalanche safety have been popular among both commercial and non-commercial actors and contributed to knowledge transfer. Similar initiatives thematically centred on practice-relevant knowledge can be efficient avenues for knowledge development. Following Sørensen (2007), this type of knowledge facilitation will support knowledge exploitation, leading to product improvements.

Brown and Duguid (1991) warns that knowledge transfer within a community of likeminded will be constrained in the long run by the shared worldview of its members. Knowledge exploration can lessen this tendency and, from a policy perspective, this can be achieved through regional tourism projects involving actors from different parts of the industry. The material highlights the importance of recognizing the specificities of the tourism products and markets in such projects. To achieve cooperative activity different actors need to have an in-depth understanding of each other's products, aims and visions.

While an ideal aim in terms of business growth and innovation is a balance between exploitation and exploration (Sørensen, 2007), an important lesson learned from this material is that growth is not necessarily a central aim among actors of the niche. Cooperative activity can be equally explained through the guides' and/or entrepreneurs' lifestyle motives and the aim of providing safe and high-quality experiences.

Note

1 Stebbins defines serious leisure as: "the systematic pursuit of an amateur, hobbyist, or volunteer activity, that is sufficiently rewarding despite the costs, such that the participants find a career there in the acquisition and expression of a combination of its special skills, knowledge, and experience (Stebbins, 1997: 17). In Stebbbins writings 'serious' signifies earnestness, sincerity, importance and carefulness.

References

Aldrich, H. and Fiol, M. C. (1994). Fools rush in: The institutional context of industry creation. *Academy of Management Review*, 19(4), 645–670.

Ateljevic, L. and Doorne, S. (2000). "Staying within the fence": Lifestyle entrepreneurship in tourism. *Journal of Sustainable Tourism*, 8(5), 378–392.

Beedie, P and Hudson, S. (2003). 'Emergence of mountain-based adventure tourism', *Annals of Tourism Research*, 30(3), 625–643.

Boschma, R. A. (2005). Proximity and innovation: A critical assessment. *Journal of the Regional Studies Association*, 39(1), 61–74.

Cloke, P. and Perkins, H. C. (2002). Commodification and adventure in New Zealand tourism. *Current Issues in Tourism*, 5(6), 521–549.

Cooper, C. (2018). Managing tourism knowledge: A review. *Tourism Review*, 73(4), 507–519.

Czernek, K. and Czakon, W. (2016). Trust-building processes in tourist coopetion: The case of a Polish region. *Tourism Management*, 52(2016), 380–394.

Eikje, C. A., Horgen, A. and Arnegård, J. (2017). The organizing and regulation of mountain guiding in Scandinavia 1820–2016, with a glance at the Alps. *Sport in Society*, 22(4), 555–572.

Elvekrok, I. Veflen, N., Scholderer, J. and Sørensen, B. T. (2022). Effects of network relations on destination development and business results. *Tourism Management*, 88, 1–10.

Fadda, N. (2020). Entrepreneurial behaviours and managerial approach of lifestyle entreprenerurs in surf tourism: An exploratory study. *Journal of Sport and Tourism*, 24(1), 53–77.

Forbord, M. and Sivertsvik, R. M. (2021). The importance of interactions and networks in the nature-based tourism industry. Chapter 11 in: Fredman, P. and Haukeland, J. V. (Eds.). *Nordic perspectives on nature-based tourism: From place-based resources to value-added experiences*. Cheltenham: Edwar Elgar Publishing.

Gertler, M. S. (2003). Tacit knowledge and the economic geography of context, or the undefinable tacitness of being (there). *Journal of Economic Geography*, 3, 75–99.

Hall, C. M. and Williams, A. M. (2008). *Tourism and innovation*. London: Routledge.

Helgadóttir, G. and Sigurðardóttir, I. (2008). Horse-based tourism: Community, quality, and disinterest in economic value. *Scandinavian Journal of Hospitality and Tourism*, 8(2), 1–17.

Hjalager, A.-M. (2002). Repairing innovation defectiveness in tourism. *Tourism Management*, 23(5), 465–474.

Hoarau, H. (2014). Knowledge acquisition and assimilation in tourism- innovation processes. *Scandinavian Journal of Hospitality and Tourism*, 14(2), 135–151. DOI: 10.1080/15022250.2014.887609

Hoarau, H. and Kline, C. (2014). Science and industry: Sharing knowledge for innovation. *Annals of Tourism Research*, 46, 44–61.

Høegh-Guldberg, O. (2018). *Dynamics of innovation network practices in tourism*. Doctoral dissertation, Inland Norway University of Applied Sciences.

Horgen, A (2016). Respons på fare i naturen, holdninger til sikkerhet i friluftslivet- et historisk perspektiv med utgangspunkt i norsk fjellsport. in: Horgen, A., Fasting, M. L., Lundhaug, T., Magnussen, L. I. and Østrem, K. (Eds.). *Ute! Friluftsliv-pedagogiske, historiske og sosiologiske perspektiver*. Bergen: Fagbokforlaget, 211–234.

Hough Mackenzie, S. and Kerr, J. H. (2013). Stress and emotions at work: An adventure tourism guide's experience. *Tourism Management*, 36, 3–14.

Løseth, K. (2014). Adventure tourism – Exploring relations between knowledge and innovation. PhD dissertation. Aalborg University.

Løseth, K. (2018). Knowledge development in adventure tourism businesses – the influence of serious leisure. *Annals of Leisure Research*, 21(5), 575–591. DOI: 10.1080/11745398.2017.1406808

Løseth, K. and Varley, P. (2021). Commercial mountaineering, Norwegian friluftsliv and the gradual march of commodification. Chapter 13 in: Fredman, P. and Haukeland, J. V. (Eds.). *Nordic perspectives on naturebased tourism: From place-based resources to value-added experiences*. Cheltenham: Edwar Elgar publishing.

Margaryan, L., Fredman, P. and Stensland, S (2020). Lifestyle entrepreneurs as agents of degrowth. Chapter 3 in: Hall, C. M., Lundmark, L. and Zhang, J. J. (Eds.). *Degrowth and tourism: New perspectives on tourism entrepreneurship, destinations, and policy*. London: Routeledge.

Miles, M. B. and Huberman, A. M. (1994). *Qualitative data analysis*. Thousand Oaks: Sage.

Patton, M. Q. (2015) *Qualitative research & evaluation methods* (4th ed.). Thousand Oaks: SAGE Publications.

Pikkemaat, B. and Weiermair, K. (2007). Innovation through cooperation in destinastions: First results of an empirical study in Austria (2007). *Anatolia*, 18(1), 67–83. DOI: 10.1080/13032917.2007.9687036

Polanyi, M. (2000). *Den tause dimensjonen. En innføring til taus kunnskap*. (Norwegian edition. Originally 1966). Oslo: Spatacus Forlag.

Robinson, M. and Novelli, M. (2011). Introction. in: Novelli, M. (Ed.). *Niche tourism: Contemporary issues, trends and cases* (2nd ed.). London: Routledge, 1–11.

Rokenes, A., Schumann, S. and Rose, J. (2015). The art of guiding in nature-based adventure tourism-how guides can create value and positive experiences on mountain bike and backcountry ski tours. *Scandinavian Journal of Hospitality and Tourism*, 15(1), 62–82. DOI: 10.1080/15022250.2015.1061733

Shaw, G. and Williams, A. (2004). From lifestyle consumption to lifestyle production: Changing patterns of tourism entrepreneurship. Chapter 7 in: Thomas, R. (Ed.). *Small firms in tourism: International perspectives*. Oxford: Elsevier.

Shaw, G. and Williams, A. (2009). Knowledge transfer and management in tourism organizations: An emerging research agenda. *Tourism Management*, 30, 325–335.

Sørensen, F. (2007). The geographies of social networks and innovation in tourism. *Tourism Geographies*, 9(1), 22–48. DOI: 10.1080/14616680601092857

Sørensen, F. and Grindsted, T. S. (2021). Sustainability approaches and nature tourism development. *Annals of Tourism Research*, 91, 103307repe

Stebbins, R. A. (1982). Serious leisure.: A conceptual statement. *Pacific Sociological review*, 25, 251–272.

Stensland, S., Forbord, M. Fossgard, K. and Løseth, K. (2021). Characteristics of nature-based tourism firms. Chapter 10 in: Fredman, P. and Haukeland, J. V. (Eds.). *Nordic perspectives on nature-based tourism: From place-based resources to value-added experiences*. Cheltenham: Edwar Elgar Publishing.

Thagaard, T. (2013). *Systematikk og innlevelse. En innføring i kvalitativ metode* (4th ed.). Bergen: Fagbokforlaget.

Tjora, A. (2017). *Kvalitative forskningsmetoder i praksis* (3rd ed.). Oslo: Gyldendal Academisk.

Van der Zee, E. and Vanneste, D. (2015). Tourism networks unravelled; a review of the literature on networks in tourism management studies. *Tourism Management Perspectives*, 15, 46–56.

Varley, P. (2006). Confecting adventure and playing with meaning: The adventure commodification continuum. *Journal of Sport and Tourism*, 11(2), 173–194.

Wenger, E. (1998). *Communities of practice: Learning, meaning, and identity*. Cambridge: Cambridge University Press.

Zach, F. and Racherla, P. (2011). Assessing the value of collaborations in tourism networks: A case study of Elkhart County, Indiana. *Journal of Travel & Tourism Marketing*, 28(1), 97–110. DOI: 10.1080/10548408.2011.535446

6

ECOTOURISM, WILDLIFE FESTIVALS, AND SUSTAINABILITY

Are objectives, strategies, and outcomes aligned?

Glen Hvenegaard

Introduction

Nature-based tourism activities provide visitors with opportunities to interact with the physical environment in a natural setting (Fredman & Tyrväinen, 2010). Each location and activity has its own objectives, which may be social, cultural, political, behavioural, or environmental in nature (Mair & Weber, 2019). However, specifying the target objectives is key to developing supportive strategies and achieving outcomes relevant to the target objectives (Murdick et al., 2001). The lack of clear objectives reduces an organization's ability to develop coherent strategies and achieve any relevant outcomes (Murdick et al., 2001).

Ecotourism activities are widespread globally, depend on natural environments, and strive to achieve objectives of learning and sustainability (Weaver & Lawton, 2007). Wildlife festivals, as a potential subset of ecotourism (Lawton, 2009), have objectives that are similar to ecotourism, are offered around the world, and have widespread appeal and attendance (Hvenegaard & Manaloor, 2007). However, research is needed on the clarity and appropriateness of objectives for wildlife festivals and if festival organizers achieve those objectives. In addition, there is little research to examine how wildlife festivals can achieve more sustainability objectives embedded in ecotourism principles (Lawton & Weaver, 2010).

The alignment of goals within an organization helps it achieve its target outcomes (Ayers, 2015). Without clearly defined and consistent goals, goal displacement often occurs; if an organization's goals are incongruent within its staff or between its goals and strategies, it is difficult to reach outcomes that match the original objectives (Barber & Taylor, 1990). Therefore, this study examines the adherence of wildlife festivals in Canada to ecotourism principles and tests for alignment among festival objectives, strategies, and outcomes.

Literature review

Objectives, strategies, and outcomes

The organizational literature defines objectives or goals as "those ends which the organization seeks to achieve by its existence and operations" (Glueck, 1980, p. 35). Objectives are important

DOI: 10.4324/9781003230748-8

to define an organization or event in its environment, to coordinate decisions, and to assess performance (Glueck, 1980). Organizations often pursue multiple objectives; emphases for those objectives vary over time, by individuals in the organization, by similar organizations, and by how those objectives are implemented (Glueck, 1980; Bode et al., 2010; Dayer et al., 2016). Objectives arise by background environmental factors (e.g., external regulations, changing societal expectations), available resources in the organization, and values of the organizers (Glueck, 1980). Organizations can state objectives at various levels, including project goals, purposes, outputs, and inputs (Baccarini, 1999).

Without clearly defined and consistent objectives for an organization, it is difficult to determine, evaluate, and reach those objectives (Barber & Taylor, 1990). Moreover, ambiguous or non-existent goals can impede performance (Chun & Rainey, 2006). However, clearly defined and consistent objectives can naturally lead to targeted strategies. Therefore, it is important to avoid confusing goals with actions, tactics, or strategies (Enright, 2016).

Strategies (or practices or actions) are the means used to accomplish the objectives of an organization (Glueck, 1980, p. 35). For example, festival strategies are based on an organization's mandate (or mission) and unique situation, and then often gathered within a strategic plan (Getz, 1997). Subsequently, organizations implement strategies as resources and opportunities become available. Ideally, strategies should connect directly to an organization's original objectives (Wintermantel & Mattimore, 1997) and be measurable in terms of impact (Enright, 2016).

An outcome is a restatement of the objectives that can be measured (Heilman & Kennedy-Phillips, 2011). However, evaluating success is constrained by unclear objectives, ineffective information management, varying periods, scarcity of resources for evaluation, and lack of incentives for evaluation (Kapos et al., 2008). Even with the challenges of defining and measuring outcomes, various models can help evaluate performance (Schalock & Bonham, 2003). Organizations can evaluate programs according to a standard of performance, focus on the organization or the individual, and focus on outcomes and measures (Schalock & Bonham, 2003). The organization's original objectives should connect logically to measurable outcomes (Carlsen & Andersson, 2011). In addition, outcomes should connected practically to the strategies employed to achieve these objectives (Hede, 2008).

Ecotourism and wildlife festivals

Because strategic development in any business sector is complex, the sequence of setting objectives, implementing strategies, and measuring outcomes is difficult (Murdick et al., 2001). Organizations need more research to examine this sequence in the context of ecotourism (Cobbinah, 2015). The key characteristics of ecotourism include tourism in natural settings, a focus on learning, and support for sustainability (Cobbinah, 2015). Thus, the key goals of ecotourism are to use natural environments, provide environmental education, promote conservation, and encourage local development (Fennell, 2001; Weaver, 2001, 2005). Put another way, ecotourism is characterized by nature appreciation and learning in natural settings, with management strategies that promote economic, socio-cultural, and economic sustainability (Ross & Wall, 1999; Weaver & Lawton, 2007). Ecotourism activities often include, but are not restricted to, bird watching, whale watching, nature study, and wildlife festivals, the latter of which is the focus of this study (Weaver, 2001; Hvenegaard & Manaloor, 2007; Lawton, 2009).

In general, festivals are annual public celebrations of local features of interest which last a short period and take place in a small region (Getz, 1997). Festivals have grown in the number of offerings, variety, and attendance (Getz, 2007). Typical festival objectives are to generate local economic impacts, provide recreational opportunities, enhance a community image, develop a local sense of community, and raise concern for local features of interest (Mayfield & Crompton, 1995; Walo et al., 1996; Janiskee & Drews, 1998; Derrett, 2003; Hvenegaard & Manaloor, 2004).

Wildlife festivals usually focus on local natural wildlife features, are facilitated by volunteers, attract mostly local and regional visitors, and offer a variety of social, recreational, and educational activities. Wildlife festivals in this chapter encompass any wildlife, birding, or nature-related festivals. Each year in North America, hundreds of wildlife festivals are offered (DiGregorio, 2002; Birdwatcher's Digest, 2019). In addition to the festival objectives listed above, wildlife festivals often promote the conservation of local wildlife (Polson, 1993; Romero & Stangel, 1996; Hvenegaard, 2011), which would be consistent with the sustainability objectives for ecotourism (Fennell, 2001; Weaver, 2005). Thus, wildlife festivals can be considered as a subset of ecotourism because they have similar activities and goals (Slotkin, 2003), but the pattern of wildlife festivals following ecotourism principles is superficial for birding festivals (Lawton & Weaver, 2010).

Purpose

Several investigators have urged research on the actual impacts of ecotourism and wildlife festivals. Hvenegaard et al. (2013) urge more evaluation of outcomes in the context of festival objectives. Mair and Weber (2019) call for more festival research based on conceptual frameworks; in this study, the principles of ecotourism serve as a framework for analysis. Moreover, there is a growing interest in examining sustainability in the context of festivals (Yuan, 2013; Holmes et al., 2015). Last, Wardle et al. (2018) call for more examination of conservation outcomes, rather than conservation strategies.

Therefore, the goals of this research are twofold. First, this chapter examines the extent to which the objectives, strategies, and outcomes of wildlife festivals are consistent with ecotourism principles. Second, it examines the consistency among festival objectives, strategies, and outcomes in order to promote, facilitate, and achieve sustainability objectives inherent with ecotourism. There are four hypotheses:

- The objectives, strategies, and outcomes of wildlife festivals in Canada are consistent with ecotourism principles
- There is alignment between wildlife festival objectives and outcomes
- There is alignment between wildlife festival objectives and strategies
- There is alignment between wildlife festival strategies and outcomes

This research contributes to the literature and practice because it critically evaluates the claim that wildlife festivals are ecotourism events and evaluates their contributions to education and sustainability. Moreover, this research goes beyond describing festivals and objectives, but assesses the relationships among objectives, strategies, and outcomes. These results will provide wildlife festival organizers insights about planning, executing, and evaluating festivals for sustainability outcomes.

Methods

Sampling was restricted to wildlife festivals held in Canada (Table 6.1). To assemble the list of wildlife festivals, we consulted the websites of many national, provincial, and territorial tourism agencies and conservation or natural history organizations. We also consulted wildlife

Table 6.1 Origin of wildlife festivals in Canada by province or territory

Province or territory	*% of 2009 population (n = 90)*	*% of 2009–10 sample (n = 54)*	*% of 2015 sample (n = 40)*
British Columbia	33.3	24.1	25.0
Alberta	14.4	20.4	22.5
Saskatchewan	5.6	9.3	7.5
Manitoba	4.4	5.6	5.0
Ontario	22.2	29.6	30.0
Quebec	5.6	1.9	0
New Brunswick	0	0	0
Nova Scotia	3.3	3.7	2.5
Prince Edward Island	0	0	0
Newfoundland & Labrador	2.2	1.9	2.5
Yukon Territory	8.9	3.7	5.0
Northwest Territories	0	0	0
Nunavut	0	0	0

Source: Complied by Researchers through various sources

festival inventories produced by organizations such as the Nature Canada, National Fish and Wildlife Federation, American Birding Association, and the Bird Watcher's Digest. Finally, we conducted detailed internet searches using key words such as festival, fest, celebration, jamboree, and extravaganza.

We designed a questionnaire to obtain key information from organizers of these wildlife festivals. First, based on Lawton (2009), for each festival, organizers provided the name, year of origin, month of offering, focus species, setting (public protected area, private, land, indoor setting, or water setting), recent attendance (numerical estimate of most recent event), attendance trends (higher, same, or lower than the previous offering), attendee origins (local, regional, province, country), attendee characteristics (serious, casual, or non-naturalists/birders), promotion methods, association with ecotourism initiatives, and the profit/non-profit status.

Second, we asked organizers to indicate the priority level (0 = not an objective, 1 = secondary objective, or 2 = primary objective) given to 25 wildlife festival objectives that have potential links to the objectives of ecotourism. We supplemented Lawton's (2009) list of 21 objectives – the only comprehensive study of wildlife festival objectives – with two more related to nature and two related to economic and social sustainability. The 25 objectives broadly covered the ecotourism objectives; seven related to nature-based criteria, seven to learning, seven to environmental sustainability, and four to economic and social sustainability (Table 6.2).

Third, we asked organizers whether they had participated in any of 17 strategies relevant to wildlife festivals and ecotourism criteria (0 = no, 1 = yes). Starting with a short list of strategies from Lawton and Weaver (2010), we added strategies adapted from reviews of the ecotourism and festival literature and from personal participation in past festivals. The strategies were broad enough to apply to the majority of festivals and included one general approach, one related to nature-based criteria, one to learning, eleven to environmental sustainability, and three to economic and social sustainability (Table 6.2).

From July 2009 to July 2010, we emailed, mailed, or directly gave questionnaires to organizers of all the inventoried festivals. In order to increase the response rate, email messages and phone calls were used to remind organizers to provide responses (one month after initial contact, and then periodically after that).

Table 6.2 Festival objectives, strategies, and outcomes by ecotourism principle

Category	*Objectives ("to....")*	*Strategies*	*Outcomes*
General		Monitoring progress toward achieving festival objectives	None
Nature	Immerse attendees in the natural environment See as many species of wildlife as possible Expose attendees to local ecosystems Give attendees opportunities to observe featured wildlife in the field Give attendees opportunities to observe featured wildlife in captive settings Feature non-living specimens of focus wildlife Give attendees opportunities to improve naturalist and birdwatching skill	Use a public protected area	None
Learning	Get children interested in nature, wildlife, and/or birds Get adults interested in nature, wildlife, and/or birds Give attendees opportunities to learn about the featured wildlife Provide learning opportunities that are entertaining Give attendees opportunities to learn about local culture and/or history Teach attendees about environmental problems that affecting featured wildlife Expose attendees to environmental issues in general	Educate and encourage visitors about environmental protection	Increased knowledge by visitors about the target species
Environmental Sustainability	Attract new members for environmental, nature, and/or birdwatching clubs	Research effects of festival activities on local wildlife species	Increased population of the target species

(*Continued*)

Table 6.2 (Continued)

Category	*Objectives ("to....")*	*Strategies*	*Outcomes*
	Observe featured wildlife in its natural setting in a non-disruptive way	Research tourists' attitudes toward wildlife conservation	Improved attitudes of visitors toward target species
	Make attendees more environmentally aware in their everyday behaviour	Allocate some festival attendance fees to protect wildlife habitats	More visitors helping target species and its habitat
	Give attendees opportunities to volunteer to protect focus wildlife and habitat	Collect donations from organizations to protect wildlife habitats	Improved management of the target species and its habitat
	Raise money to protect the target wildlife	Collect donations from festival participants to protect wildlife habitats	Increased area of protected habitat of the target species
	Raise money to purchase wildlife habitat	Research population trends of local wildlife species	
	Raise money to protect the natural environment more generally	Research health of wildlife habitats in the festival area	
		Conserve energy	
		Recycle materials	
		Re-use signage, banners, and other materials from year to year	
		Minimize waste	
Economic and Social Sustainability	Link with other segments of the local tourism industry	Promote links between local economic benefits and festival activities	Increased economic impact in the local community
	Create employment opportunities for local people	Research local residents' attitudes toward wildlife conservation	Improved attitude of local residents toward target species
	Help the economy of the local community	Reimburse landowners affected by crop damage from problem wildlife	More local residents helping target species and its habitat
	Publicize activities of organizations that are sponsoring this event		

Evaluations of a festival's impact on habitat quality and wildlife populations in the festival's immediate area are rare, but could include assessments from professionals who are familiar with the festivals' activities, such as organizers, scientists, resource managers, and representatives from environmental nongovernmental organizations. Since it is unlikely to obtain data to measure all conservation outcomes (e.g., habitat and population studies before and after a festival), this part of the study relied on perceptions of festival organizers. Moreover, the impact of specific strategies may not be evident immediately; in fact, some impacts might not be discernible until several years have passed. Therefore, in June, 2015, we emailed a follow-up questionnaire to the

2009 respondents to ask about the perceived level of impact that the festival had over the past six years on potential conservation benefits for the target species of the festival (0 = no impact, 1 = possible impact, 2 = definite impact). The outcomes reflected the previous objectives and strategies, including one outcome related to learning, five to environmental sustainability, and three to economic and social sustainability (Table 6.2). Again, these outcomes were general enough to apply to most festivals. In addition, we asked questions about whether the festival was still running, the year the festival was most recently offered, estimated attendance, and attendance trends. We used a similar pattern of reminder messages to increase the response rate.

To explore the relationship among the conservation objectives, mechanisms, and outcomes, we segmented the festivals. First, based on the level of priority for the festival objectives, we used k-means cluster analysis, with a maximum of ten iterations, using initial cluster centres, and excluding cases pairwise. We used Cronbach's alpha to test for reliability. We analyzed results for 2, 3, 4, and 5 clusters, but choose the number of clusters based on cluster means, consistency, cluster size, and interpretability (Mirkin, 2011). To test for differences among clusters, we used one-way analyses of variance and independent samples t-tests. To examine potential relationships among variables, we used Pearson's correlation tests. We relied on an alpha value of $p < 0.05$.

Results

Festival characteristics

The inventory produced a population of 90 wildlife festivals, of which 54 festival organizers (or 60%) responded in 2010. Comparing the host province or territory of the festivals in the population with the festivals in the study, there was no noticeable non-response bias (Table 6.1).

Festivals in the sample began as early as 1969 and as late as 2010. At the time of the first survey, the mean age of the festivals was 12 years. The festivals were concentrated in the spring and summer months, with 3% of festivals in January, 3% in February, 5% in March, 10% in April, 40% in May, 13% in June, 3% in July, 10% in August, 8% in September, 5% in October, 3% in November, and 0% in December. The focus of the festivals varied across species groups, with 17% on general nature and wildlife, 61% on birds (28% in general plus 33% on a specific species or group), 9% on fish, 2% on amphibians, 2% on invertebrates, and 9% on plants. No festivals focused on mammals.

Regarding the origin of visitors, 42% of organizers said that most visitors came from the local area, 39% came from the regional area, 17% came from the rest of the province, and 2% came from beyond the province or from other countries. When asked about the composition of most visitors at their festival, 4% of organizers said they were casual naturalists or birders, 19% said they were non-naturalists and birders, none said they were serious naturalists or birders, and 77% said they were a mixture of all three. Regarding affiliation with ecotourism, 24% of organizers labelled the festival as an ecotourism event. Moreover, 56% of the festivals made connections with other tourism initiatives. Last, 87% were part of non-profit organizations and 13% were part of for-profit organizations.

By 2015, nine festivals had stopped operating. Of the 45 remaining festivals, we received responses from 40 (89% response rate). The distribution by province stayed consistent, indicating no selection bias on this criterion (Table 6.1). Attendance at the festivals in 2015 was lower than in 2009 (mean of 2,751 versus 3,165). This pattern was affirmed by respondents: 35% of organizers said that attendance was lower than the previous year, 53% said about the same, and 13% said higher.

Festival objectives

One festival did not provide complete enough responses about each of the 25 objectives, resulting in a final tally of 53 festivals. Festival objectives rated ≥ 1.5 included immerse attendees in the natural environment; expose attendees to local ecosystems; give attendees an opportunity to observe featured wildlife in the field; get children interested in nature, wildlife, and/or birds; get adults interested in nature, wildlife, and/or birds; give attendees opportunities to learn about the featured wildlife; provide learning opportunities that are entertaining; and observe featured wildlife in its natural setting in a non-disruptive way (Table 6.3). Objectives rated ≤ 0.5 included featuring non-living specimens of focus wildlife, raising money to purchase wildlife habitat, raising money to protect the natural environment more generally, and creating employment opportunities for local people.

While the age of a festival was not correlated with priority given to any of the objectives, attendance was inversely correlated with three objectives, including immerse attendees in the natural environment ($r = -0.288$, $p = 0.040$), expose attendees to local ecosystems ($r = -0.365$, $p = 0.010$), and opportunities to observe featured wildlife in the field ($r = -0.454$, $p = 0.001$).

For those festivals promoted as ecotourism, the priorities were higher for the following objectives: observe wildlife in a non-disruptive way (1.88 vs 1.44, $t = -1.828$, $p = 0.037$); teaching about environmental problems of featured wildlife ($t = -1.700$, $p = 0.048$); raise money to protected target wildlife (1.54 vs 1.13, $t = -1.735$, $p = 0.045$), raise money for wildlife habitat (0.38 vs 0.13, $t = -1.716$, $p = 0.046$); and raise money for general environmental protection (0.75 vs 0.21, $t = -2.545$, $p = 0.007$).

Festival clusters

We selected a 3-cluster solution to ensure a sufficient number of festivals in each category for later analyses, because there were clear differences in objectives between clusters (only three objectives were not statistically different among clusters), and because the results could be interpreted in the context of the study objectives. The Cronbach's alpha was 0.80, which is adequate for reliability (Vaske, 2008).

Cluster 1 contained 32% of the sample, and mean priority scores for 21 of 25 objectives were higher than one or both of the other clusters (Table 6.3). Thus, we describe this cluster as *comprehensive*, given its high priority for most festival objectives, as festivals in this cluster try to 'do it all.' Cluster 2 contained 32% of the sample. Festivals in this cluster rated 16 of 25 objectives lower than one or both of the other clusters, but rated 3 of 25 objectives (i.e., observe featured wildlife in captive settings, get children interested, and get adults interested) higher than one or both of the other clusters. Thus, we describe this cluster as *close interactions*, because festivals promote opportunities to closely interact with and learn from wildlife featured at the festival. Cluster 3 contained 36% of the sample. Festivals in this cluster rated 17 of 25 objectives lower than one or both of the other clusters, but rated 4 of 25 objectives (i.e., immerse attendees in nature, expose attendees to local ecosystems, give opportunities to observe wildlife in the field, and observe featured wildlife in its natural setting in a non-disruptive way) higher than one or both of the other clusters. Thus, we describe this group as *nature interactions*, given the focus on outdoor, natural, and minimal impact recreational opportunities. There were no associations between clusters and age, attendance, attendance trends, or promoting a festival as an ecotourism event.

Table 6.3 Clusters by level of importance for wildlife festival objectives (horizontal lines separate different types of objectives). Bolded items indicate the highest score among clusters; italicized items indicate the lowest score among clusters

Festival Objective ("to...")	*Mean*	*1 (n = 17)*	*2 (n = 17)*	*3 (n = 19)*	*F*	*p*
Immerse attendees in the natural environment	1.53	**1.65**[b]	*1.06*[a]	**1.84**[b]	9.540	<0.001
See as many species of wildlife as possible	0.88	1.20	0.76	0.71	1.941	0.155
Expose attendees to local ecosystems	1.59	**1.88**[b]	*1.44*[a]	**1.72**[b]	5.156	0.009
Give attendees opportunities to observe featured wildlife in the field	1.75	**1.88**[b]	*1.44*[a]	**1.89**[b]	3.934	0.026
Give attendees opportunities to observe featured wildlife in captive settings	0.52	**0.69**[a]	**0.88**[a]	*0.00*[b]	6.287	0.004
Feature non-living specimens of focus wildlife	0.41	**0.75**[a]	0.44[a,b]	*0.06*[b]	5.733	0.006
Give attendees opportunities to improve naturalist and birdwatching skill	1.20	**1.75**[b]	*0.88*[a]	*1.00*[a]	7.086	0.002
Get children interested in nature, wildlife, and/or birds	1.52	**1.63**[a]	**1.82**[a]	*1.16*[b]	5.533	0.007
Get adults interested in nature, wildlife, and/or birds	1.78	1.88[a,b]	**2.00**[a]	*1.53*[b]	4.899	0.012
Give attendees opportunities to learn about the featured wildlife	1.82	**2.00**[a]	1.88[a,b]	*1.61*[b]	3.312	0.045
Provide learning opportunities that are entertaining	1.53	**1.94**[b]	*1.29*[a]	*1.37*[a]	5.711	0.006
Give attendees opportunities to learn about local culture and/or history	0.58	**0.94**[b]	*0.41*[a]	*0.39*[a]	3.245	0.047
Teach attendees about environmental problems affecting featured wildlife	1.23	**1.71**[b]	*0.94*[a]	*1.05*[a]	5.769	0.006
Expose attendees to environmental issues in general	0.75	**1.29**[b]	*0.35*[a]	*0.63*[a]	9.136	<0.001
Attract new members for environmental, nature, and/or bird watching clubs	0.63	**1.35**[b]	*0.25*[a]	*0.32*[a]	15.257	<0.001
Observe featured wildlife in its natural setting in a non-disruptive way	1.53	**1.75**[b]	*1.12*[a]	**1.72**[b]	5.655	0.006
Make attendees more environmentally aware in their everyday behaviour	0.86	**1.53**[b]	*0.75*[a]	*0.33*[a]	12.177	<0.001
Give attendees opportunities to volunteer to protect focus wildlife and habitat	0.69	**1.29**[b]	*0.38*[a]	*0.42*[a]	11.278	<0.001
Raise money to protect the target wildlife	0.55	0.75	0.29	0.61	1.761	0.183
Raise money to purchase wildlife habitat	0.19	0.18	0.00	0.37	2.804	0.070
Raise money to protect the natural environment more generally	0.33	**0.69**[b]	*0.06*[a]	0.26[a,b]	3.917	0.027
Link with other segments of the local tourism industry	0.56	**1.06**[b]	*0.19*[a]	*0.42*[a]	6.956	0.002
Create employment opportunities for local people	0.27	**0.81**[b]	*0.00*[a]	*0.27*[a]	10.564	<0.001
Help the economy of the local community	0.69	**1.13**[a]	0.56[a,b]	*0.42*[b]	4.251	0.020
Publicize activities of organizations that are sponsoring this event	0.70	**1.19**[b]	*0.50*[a]	*0.44*[a]	6.322	0.004

a Clusters with different superscripts have means that are significantly different.

b Clusters with different superscripts have means that are significantly different.

Festival strategies

Strategies employed by two-thirds or more of the festivals included using public protected areas, re-using festival materials, recycling, and providing educational activities (Table 6.4). Strategies employed by one-third or fewer of the festivals were allocating fees for habitat protection, collecting donations, reimbursing landowners, and research on festival impacts, health of wildlife habitats, tourists' attitudes, and residents' attitudes.

There were differences in means among clusters for only 4 of 17 festival strategies. Cluster 1 rated these strategies highest: research on tourists' attitudes towards wildlife, research population trends for local wildlife species, conserve energy, and promote links between economic benefits and festivals. The means for cluster 2 were lowest for the same four strategies, and the means for cluster 3 were lowest for research on population trends of local wildlife species and conserve energy.

Table 6.4 Employment of wildlife festival strategies by clusters (horizontal lines separate different types of strategies). Bolded items indicate the highest score among clusters; italicized items indicate the lowest score among clusters

Strategy	*Mean*	*1*	*2*	*3*	*F*	*p*
Monitor progress toward achieving your festival objectives above	0.42	0.59	0.24	0.42	2.244	0.117
Use a public protected area	0.72	053	0.82	0.79	2.258	0.115
Educational activities to encourage environmental protection	0.68	0.82	0.53	0.68	1.702	0.193
Research the effects of festival activities on local wildlife species	0.06	0.12	0.00	0.05	1.089	0.344
Research tourists' attitudes toward wildlife conservation	0.09	**0.24**[b]	*0.00*[a]	0.05[a,b]	3.258	0.047
Allocate some festival attendance fees to protect wildlife habitats	0.13	0.25	0.06	0.11	1.398	0.257
Collect donations from nearby organizations to protect wildlife habitats	0.15	0.18	0.06	0.21	0.848	0.434
Collect donations from festival participants to protect wildlife habitats	0.28	0.24	0.18	0.42	1.461	0.242
Research population trends for local wildlife species	0.36	**0.65**[b]	*0.18*[a]	*0.26*[a]	5.359	0.008
Research the health of wildlife habitats in the festival area	0.19	0.35	0.06	0.16	2.597	0.085
Conserve energy	0.34	**0.71**[b]	*0.18*[a]	*0.16*[a]	9.853	<0.001
Recycle	0.87	0.95	0.82	0.87	0.794	0.458
Re-use signage, banners, and other materials	0.94	1.00	0.88	0.95	1.089	0.344
Minimize waste	0.60	0.71	0.47	0.63	1.012	0.371
Promote links between local economic benefits and festival activities	0.23	**0.41**[b]	*0.00*[a]	0.26[a,b]	4.746	0.013
Research local residents' attitudes toward wildlife conservation	0.06	0.06	0.06	0.05	0.004	0.996
Reimburse landowners who are affected by crop damage from problem wildlife	0.02	0.06	0.00	0.00	1.061	0.354

a Clusters with different superscripts have means that are significantly different.

b Clusters with different superscripts have means that are significantly different.

There was no correlation between the age of festivals and any festival strategies. Attendance was positively correlated with only collect donations for wildlife habitats ($r = 0.406$, $p = 0.003$). For those festivals promoted as ecotourism, the following strategies were more commonly employed: monitor progress toward festival objectives (0.69 vs. 0.34, $t = -2.296$, $p = 0.013$), allocate attendance fees for wildlife habitats (0.31 vs. 0.08, $t = -2.211$, $p = 0.016$), research population trends of local wildlife species (0.79 vs. 0.26, $t = -2.150$, $p = 0.018$), research effects of festivals on local wildlife species (0.62 vs. 0.29, $t = -1.796$, $p = 0.039$), research tourists' attitudes toward wildlife conservation (0.38 vs. 0.00, $t = -4.967$, $p < 0.001$), research local residents' attitudes toward wildlife conservation (0.15 vs. 0.02, $t = -1.796$, $p = 0.039$), promote links between local economic benefits and festival activities (0.54 vs. 0.12, $t = -3.418$, $p = 0.001$).

Festival outcomes

Outcomes with mean scores ≥ 1.5 were increased knowledge among visitors about target species and improved attitudes of visitors toward target species (Table 6.5). The outcomes with the lowest mean scores (though not < 0.5) were an increase in area of protected habitat of the target species and an increase in population of the target species. There were no significant differences in the means of outcomes among clusters.

There were no correlations between age or attendance and any of the outcomes. Festivals promoted as ecotourism rated higher the outcomes for increase in the populations of the target species (1.00 vs. 0.34, $t = -2.803$, $p = 0.004$) and improved management of the target species and its habitat (1.55 vs. 0.72, $t = -3.322$, $p < 0.001$) than festivals not promoted as ecotourism.

Alignment of festival objectives, strategies, and outcomes

We conducted correlation tests for the many unique combinations of objectives, strategies, and outcomes. The r-values for significant correlations ranged from 0.282 to 0.681. There were only two negative correlations: 1) between the objective of giving opportunities to learn

Table 6.5 Level of achievement of wildlife festival outcomes by clusters (horizontal lines separate different types of outcomes)

Outcome	*Mean*	*1*	*2*	*3*	*F*	*p*
Increased knowledge among visitors about the target species	1.87	1.92	1.77	1.93	0.893	0.418
Increased population of the target species	0.54	0.42	0.46	0.71	0.651	0.527
Improved attitudes of visitors toward target species	1.67	1.83	1.46	1.71	2.117	0.135
More visitors helping target species and its habitat	1.26	1.42	1.00	1.36	1.936	0.159
Improved management of the target species and its habitat	0.95	1.00	0.62	1.21	2.069	0.141
Increased area of protected habitat of the target species	0.54	0.75	0.31	0.57	1.362	0.269
Increased economic impact in the local community	1.26	1.42	1.08	1.29	0.498	0.612
Improved attitude of local residents toward target species	1.41	1.50	1.31	1.43	0.282	0.756
More local residents helping target species and its habitat	1.15	1.33	1.08	1.07	0.956	0.394

about featured wildlife and the outcome of raising money for wildlife habitat (r = −0.361, p = 0.009); and 2) between the objective of getting adults interested in nature, wildlife, and birds and the outcome of raising money for wildlife habitat (r = −0.421, p = 0.002). We summarized the rate of correlations for the various related groupings of objectives, strategies, and objectives. Table 6.6 indicates the number of statistically significant positive correlations compared to the total number of possible positive correlations (in bold), broken down by the categories of general (only for strategies), nature-based, learning, environmental sustainability, and economic and social sustainability. High percentages and percentages higher than those in the corresponding columns and rows would indicate better alignment.

Comparing festival objectives with strategies, the percentage of correlations ranged from 75% (economic/social sustainability objectives with general strategies) to 0% (top of Table 6.6). The mean rates of correlation were 23% for related objectives and strategies, compared to 13% for unrelated objectives and strategies (15% rate for all possible combinations). All but one of the cells comparing objectives to their related strategies had the highest rate of correlations within its column and row, indicating a modest but consistent degree of alignment between objectives and strategies.

Table 6.6 Summary of significant positive correlations among festival objectives, strategies, and outcomes (**bold** values describe the actual versus potential number of correlations among items in related pairs of objectives, strategies, and outcomes)

Combination	*Objectives/ strategies/ outcomes*	*General*	*Nature-based*	*Learning*	*Environmental sustainability*	*Economic and social sustainability*
Objectives with strategies	General	**None**	None	None	None	None
	Nature-based	0/7 = 0%	**2/7 = 29%**	0/7 = 0%	8/77 = 10%	2/21 = 10%
	Learning	0/7 = 0%	1/7 = 14%	**2/7 = 29%**	8/77 = 10%	3/21 = 14%
	Environmental sustainability	2/7 = 29%	1/7 = 14%	2/7 = 29%	**17/77 = 22%**	2/21 = 10%
	Economic and social sustainability	3/4 = 75%	0/4 = 0%	0/4 = 0%	9/44 = 20%	**3/12 = 25%**
Objectives with outcomes	General	**None**	None	None	None	None
	Nature-based	None	**None**	0/7 = 0%	1/35 = 3%	3/21 = 14%
	Learning	None	None	**0/7 = 0%**	1/35 = 3%	5/21 = 24%
	Environmental sustainability	None	None	0/7 = 0%	**9/35 = 26%**	1/21 = 5%
	Economic and social sustainability	None	None	0/4 = 0%	5/20 = 25%	**0/12 = 0%**
Strategies with outcomes	General	**None**	None	0/1 = 0%	5/5 = 100%	0/3 = 0%
	Nature-based	None	**None**	1/1 = 100%	2/5 = 40%	1/3 = 33%
	Learning	None	None	**0/1 = 0%**	0/5 = 0%	0/3 = 0%
	Environmental sustainability	None	None	0/11 = 0%	**7/55 = 13%**	7/33 = 21%
	Economic and social sustainability	None	None	0/3 = 0%	5/15 = 33%	**1/9 = 11%**

Comparing festival objectives with outcomes, the percentage of correlations ranged from 26% (environmental sustainability objectives with environmental sustainability outcomes) to 0% (middle of Table 6.6). The mean rates of correlation were 17% for related objectives and outcomes, compared to 9% for unrelated objectives and outcomes (11% rate for all possible combinations). Among related objectives and outcomes, the rate of correlation was highest only for environmental sustainability objectives and outcomes, and was 0% for the other combinations of related objectives and outcomes (learning and economic/social sustainability).

Comparing festival strategies and outcomes, the percentage of correlations ranged from 100% (general strategies with environmental sustainability outcomes) to 0% (bottom of Table 6.6). The mean rates of correlation were 12% for related strategies and outcomes, compared to 24% for unrelated strategies and outcomes (19% rate for all possible combinations). The rates of correlation for related combinations of strategies and outcomes were low, compared to unrelated combinations of strategies and outcomes. However, the general strategy of monitoring for progress had a 100% rate of correlation with achieving environmental sustainability outcomes.

Discussion

This study sought to examine the level of consistency of wildlife festival objectives, strategies, and outcomes with ecotourism principles. For background, only 24% of this study's organizers labelled their festival as an ecotourism event, compared to 38% of birding festivals in the USA in in 2006 (Lawton, 2009). The primary emphases of festival objectives were on experiencing the natural environment and providing learning opportunities, which are both key ecotourism principles (Cobbinah, 2015). However, festival organizers placed much lower emphasis on objectives related to environmental, economic, or social sustainability, which are also key ecotourism principles (Weaver, 2005).

Similar to the festival objectives above, the employment of festival strategies focused on experiencing the natural environment, learning, and two of eleven items related to environmental sustainability. There was relatively less emphasis placed on strategies related to economic and social sustainability, as well as the other nine aspects of environmental sustainability.

Finally, the achievement of festival outcomes was very limited, and focused on learning and one item related to environmental sustainability (i.e., improved attitude of visitors toward target species). There was less achievement on other environmental sustainability outcomes related to the habitat and populations of the target species.

Overall, festivals in this study were more likely to emphasize nature and learning in their objectives, strategies, and outcomes than environmental or economic/social sustainability. Thus, there is only partial support for the first hypothesis of the objectives, strategies, and outcomes of wildlife festivals in Canada being consistent with ecotourism principles. This support varies by festival cluster. There is strong support for ecotourism principles by the *comprehensive* cluster regarding objectives; this support continues partially to the employment of relevant strategies, but this support tapers off regarding the achievement of festival outcomes. The *close interactions* and *nature interactions* clusters both rated most festival objectives lower than the *comprehensive* cluster. The *closer interactions* cluster rated lowest on four strategies, three of which related to environmental sustainability.

The wildlife festival literature often reports on objectives such as economic impacts (Hvenegaard & Manaloor, 2004), visitor satisfaction (Kim et al., 2008), educational benefits and demand (Slotkin, 2005; Lee et al., 2010), and the broader promotion of regional ecotourism (Slotkin, 2003). Only a few studies have evaluated wildlife festivals according to ecotourism principles. Webb et al. (2013) found that the Kirkwood Wildlife Festival in South Africa succeeded on economic and social sustainability, but urge greater achievement in the area of environmental sustainability. Lawton (2009, p. 265) found that birding festivals in the USA satisfied the "basic ecotourism criteria of nature-based attractions, learning outcomes from interactions with those attractions, and also nondisruptive interactions as per the sustainability agenda," but support for ecotourism principles is neither comprehensive nor rigorous.

This study also tested for consistency among festival objectives, strategies, and outcomes. In support of the second hypothesis, there was some support for alignment between matching festival objectives and outcomes, but the alignment was limited to positive correlations among only 22–29% of the matching options. With that small level of alignment between objectives and related strategies, it will be difficult to obtain tangible outcomes. As such, Ganesan and Paturi (2009) highlight the need to track and monitor objectives and strategies over time to improve decision-making.

There was less support for the third hypothesis about alignment between festival objectives and strategies. There were no correlations between objectives and outcomes for learning or between objectives and outcomes for economic and social sustainability. As well, only 26% of the objectives for environmental sustainability correlated with the outcomes for environmental sustainability. Unsurprisingly, there is less support for this hypothesis than the prior one because the intervening step of developing and implementing strategies can subtly shift away from achieving target outcomes. Ongoing evaluation is required to determine whether objectives and strategies stay in alignment or deviate from each other (Graham, 2011).

Last, there was little support for the last hypothesis about alignment between festival strategies and outcomes. There were no correlations between objectives and outcomes for learning and only one correlation between objectives and outcomes for economic and social sustainability. Surprisingly, there were correlations for all five possible combinations between general strategies (i.e., monitoring) and environmental sustainability outcomes. Richardson (2005) differentiates between strategy formulation (converting objectives into strategies) and strategy execution (converting strategies into outcomes), and encourages careful analyses in both.

Among the limitations for this study, statistical analyses were limited by the small sample size and missing values for some festivals. In addition, the diversity of festivals necessitated asking general questions about festival objectives, strategies, and outcomes. Specific questions relevant to each festival would have allowed for greater precision in evaluating the links among objectives, strategies, and outcomes. Moreover, the level to which festivals meets ecotourism principles is limited by the types of data gathered (Webb et al., 2013). Last, some respondents may not have been the most knowledgeable about answering questions about their festivals.

In terms of further research, studies in countries with more festivals would aid in allowing for broader generalizations. It would also be helpful to examine the alignment of festival objectives, strategies, and outcomes for a range of other festivals with varying emphases (e.g., food and sport). Moreover, to assess outcomes more accurately, it would be helpful to follow up with environmental and community experts (Gursoy et al., 2004). As well, it would be useful to determine if these outcomes last over time and are consistent among regions. As others have investigated, it is important to determine objectives and outcomes assessed at various scales (Beaumont, 2011; Getz, 2008). Finally, the field needs more research on which links, among the various combinations of objectives, strategies, and outcomes, are most coherent. Once those links are established, it would be helpful to conduct in-depth interviews with festival organizers to determine particular enablers and barriers regarding successful identification, implementation, and evaluation of festival strategies (Lawton & Weaver, 2010).

Conclusions

In summary, there was limited identification by wildlife festivals as ecotourism events, but such identification did not vary among festival clusters. Wildlife festivals in Canada placed high emphasis on nature-based and learning objectives, and low emphasis on sustainability objectives. Moreover, that emphasis declined when evaluating festival strategies and outcomes.

The low level of alignment of objectives, strategies, and outcomes is of concern and reflects a lack of coherent planning processes. Festival organizers can do a better job of achieving target outcomes in several ways. First, organizers should rigorously measure festival outcomes, since objectives and strategies on their own, while important, do not directly result in tangible conservation outcomes (Kapos et al., 2009). Second, organizers should collaboratively identify and agree on objectives that have the support of all stakeholders. Clear objectives will help develop purposeful strategies, which, if implemented successfully, will help produce outcomes that are more tangible (Holmes et al., 2015). Third, organizers should clearly link objectives to strategies and outcomes so that the logic and procedures are evidence-based and apparent to all stakeholders. Festival organizers could employ marketing strategies to promote the festival's objectives and support clear communications. Last, organizations should clearly communicate objectives and strategies (Murdick et al., 2001) within and beyond the organization to increase support. As suggested by Lawton (2009), the establishment of a system-wide wildlife festival organization could assist festivals to access existing ecotourism resources, develop clear objectives, support relationships with external organizations, and share experiences about implementing realistic strategies and achieving tangible outcomes. Such an organization could encourage festival organizations to adopt relevant sustainability strategies and even operate an eco-certification scheme (Lawton & Weaver, 2010).

This study contributes to sustainability in nature-based tourism and ecotourism by offering a method of higher-level organizational analysis that critically reflects on the relationships among goals, strategies, and outcomes. In light of global sustainability challenges and environmental change, wildlife festivals can support efforts towards improved sustainability.

Acknowledgements

This work was supported by the Social Sciences and Humanities Research Council of Canada (grant 410-2009-2171), the University of Alberta Killam Fund, the Augustana Faculty Research Fund, and the University of Alberta Research Experience Fund. Thanks to colleagues L. Lawton and V. Manaloor for advice and to J. Rintoul, N. Tiitmamer Kur, K. Brager, and A. Kaiser for help with data collection.

Disclosure statement

No potential conflict of interest was reported by the author.

References

Ayers, R. S. (2015). Aligning individual and organizational performance: Goal alignment in federal government agency performance appraisal programs. *Public Personnel Management, 44*(2), 169–191.

Baccarini, D. (1999). The logical framework method for defining project success. *Project Management Journal, 30*(4), 25–32.

Barber, W. E., & Taylor, J. N. (1990). The importance of goals, objectives, and values in the fisheries management process and oganization: A review. *North American Journal of Fisheries Management, 10*(4), 365–373.

Beaumont, N. (2011). The third criterion of ecotourism: Are ecotourists more concerned about sustainability than other tourists? *Journal of Ecotourism*, *10*(2), 135–148.

Bird Watcher's Digest (2019). *The birders' directory: Travel.* Bird Watcher's Digest. https://www.birdwatchersdigest.com/bwdsite/the-birders-directory-travel-2019-edition.php

Bode, M., Probert, W., Turner, W. R., Wilson, K. A., & Venter, O. (2010). Conservation planning with multiple organizations and objectives. *Conservation Biology*, *25*(2), 295–304.

Carlsen, J., & Andersson, T. D. (2011). Strategic SWOT analysis of public, private and not-for-profit festival organisations. *International Journal of Event and Festival Management*, *2*(1), 83–97.

Chun, Y., & Rainey, H. (2006). Consequences of goal ambiguity in public organizations. In G. Boyne, K. Meier, L. O'Toole, Jr., & R. Walker (Eds.), *Public service performance: Perspectives on measurement and management* (pp. 92–112). Cambridge University Press.

Cobbinah, P. B. (2015). Contextualising the meaning of ecotourism. *Tourism Management Perspectives*, *16*, 179–189.

Dayer, A. A., Rodewald, A. D., Stedman, R. C., Cosbar, E. A., & Wood, E. M. (2016). Wildlife conservation and private protected areas: The discrepancy between Land Trust mission statements and their perceptions. *Environmental Management*, *58*(2), 359–364.

Derrett, R. (2003). Making sense of how festivals demonstrate a community's sense of place. *Event Management*, *8*, 49–58.

DiGregorio, L. (2002). Birding festivals beckon. *Birding*, *34*(1), 77.

Enright, D. (2016). Goals: The definition of success. Accessed Oct 26 2021 from https://www.donenright.com/the-definition-of-success/

Fennell, D. A. (2001). A content analysis of ecotourism definitions. *Current Issues in Tourism*, *4*, 403–421.

Fredman, P., & Tyrväinen, L. (2010) Frontiers in nature-based tourism. *Scandinavian Journal of Hospitality and Tourism*, *10*(3), 177–189.

Ganesan, E., & Paturi, R. (2009). Key performance indicators framework - A method to track business objectives, link business Strategy to processes and detail importance of key performance indicators in enterprise business architecture. *AMCIS 2009 Proceedings*, 736. https://aisel.aisnet.org/amcis2009/736

Getz, D. (1997). *Event management & event tourism*. Cognizant Communication Cooperation.

Getz, D. (2007). *Event studies: Theory, research and policy for planned events*. Elsevier.

Getz, D. (2008). Even tourism: Definition, evolution, and research. *Progress in Tourism Management*, *29*, 403–428.

Glueck, W. F. (1980). *Business policy and strategic management*. McGraw-Hill.

Graham, A. (2011). The objectives and outcomes of airport privatisation. *Airport Management*, *1*(1), 3–14.

Gursoy, D., Kim, K., & Uysal, M. (2004). Perceived impacts of festivals and special events by organizers: An extension and validation. *Tourism Management*, *25*(2), 171–181.

Hede, A.-M. (2008). Food and wine festivals: Stakeholders, long-term outcomes and strategies for success. In C. M. Hall, & L. Sharples (Eds.), *Food and wine festivals and events around the world* (pp. 85–100). Butterworth-Heinemann.

Heilman, S. C., & Kennedy-Phillips, L. (2011). Making assessment easier with the organizational effectiveness model. *About Campus*, *15*(6), 29–32.

Holmes, K., Hughes, M., Mair, J., & Carlsen, J. (2015). *Events and sustainability*. Routledge.

Hvenegaard, G. T. (2011). Potential conservation benefits of wildlife festivals. *Event Management*, *15*(4), 373–386.

Hvenegaard, G.T., T. Delamere, R.H. Lemelin, K. Brager, & A. Auger. (2013). Insect festivals: Celebrating insects and fostering human-insect encounters. In H. Lemelin (Ed.), *The management of insects in recreation and tourism* (pp. 198–216). Cambridge University Press.

Hvenegaard, G. T., & Manaloor, V. (2004). Current and potential expenditure patterns of birding festival participants. *Recent Advances and Research Updates*, *5*(1), 47–60.

Hvenegaard, G. T., & Manaloor, V. (2007). A comparative approach to analyzing local expenditures and visitor profiles of two wildlife festivals. *Event Management*, *10*, 231–239.

Janiskee, R. L., & Drews, P. L. (1998). Rural festivals and community reimaging. In R. Butler, C. M. Hall, & J. Jenkins (Eds.), *Tourism and recreation in rural areas* (pp. 157–175). John Wiley & Sons.

Kapos, V., Balmford, A., Aveling, R., Bubb, P., Carey, P., Entwistle, A., Hopkins, J., Mulliken, T., Safford, R., Stattersfield, A., Walpole, M., & Manica, A. (2008). Calibrating conservation: New tools for measuring success. *Conservation Letters*, *1*(4), 155–164.

Kapos, V., Balmford, A., Aveling, R., Bubb, P., Carey, P., Entwistle, A., Hopkins, J., Mulliken, T., Safford, R., Stattersfield, A., Walpole, M., & Manica, A. (2009). Outcomes, not implementation, predict conservation success. *Oryx*, *43*(3), 336–342.

Kim, Y., Kim, S. S., & Agrusa, J. (2008). An investigation into the procedures involved in creating the Hampyeong Butterfly Festival as an ecotourism resource, successful factors, and evaluation. *Asia Pacific Journal of Tourism Research, 13*(4), 357–377.

Lawton, L. J. (2009). Birding festivals, sustainability, and ecotourism: An ambiguous relationship. *Journal of Travel Research, 48*(2), 259–267.

Lawton, L. J., & Weaver, D. B. (2010). Normative and innovative sustainable resource management at birding festivals. *Tourism Management, 31*(4), 527–536.

Lee, C.-K., Lee, J.-H., Kim, T.-K., & Mjelde, J. W. (2010). Preferences and willingness to pay for bird-watching tour and interpretive services using a choice experiment. *Journal of Sustainable Tourism, 18*(5), 695–708.

Mair, J., & Weber, K. (2019). Event and festival research: A review and research directions. *International Journal of Event and Festival Management, 10*(3), 209–216.

Mayfield, T. L., & Crompton, J. L. (1995). Development of an instrument for identifying community reasons for staging a festival. *Journal of Travel Research, 33*, 37–44.

Mirkin, B. (2011). Choosing the number of clusters. *Data Mining and Knowledge Discovery, 1*, 252–260.

Murdick, R. G., Moor, R. C., Babson, H. C., & Tomlinson, W. H. (2001). *Business policy and strategy: An action guide*. 6 th ed., St. Lucie Press.

Polson, J. (1993). Flocking effect: Small towns learn to love birders who descend on "hot-spots". *American Birds, 47*(5), 1062–1070.

Richardson, J. E. (2005). *The business model: An integrative framework for strategy execution*. SSRN: https://ssrn.com/abstract=932998 or http://dx.doi.org/10.2139/ssrn.932998

Romero, A., & Stangel, P. (1996). Festival birding. *Wildbird*, (January), 34–39.

Ross, S., & Wall, G. (1999). Ecotourism: Towards congruence between theory and practice. *Tourism Management, 20*, 123–132.

Schalock, R. L., & Bonham, G. S. (2003). Measuring outcomes and managing for results. *Evaluation and Program Planning, 26*(3), 229–235.

Slotkin, M. H. (2003). Ecotourism in practice: Birding & wildlife festivals. In G. Nelson, & I. Hronszky (Eds.), *How can science support environmental protection?* (pp. 57–69). Arisztotelesz Publishing.

Slotkin, M. H. (2005). Educational partnerships, sustainability, and ecotourism project development. In G. Nelson, & I. Hronszky (Eds.), *Environmental studies: Implications for sustainability* (pp. 81–89). Arisztotelész Publishing.

Vaske, J. J. (2008). *Survey research and analysis: Applications in parks, recreation and human dimensions*. Venture Publishing.

Walo, M., Bull, A., & Breen, H. (1996). Achieving economic benefits at local events: A case study of a local sports event. *Festival Management & Event Tourism, 4*, 95–106.

Wardle, C., Buckley, R., Shakeela, A., & Castley, J. G. (2018). Ecotourism's contributions to conservation: Analysing patterns in published studies. *Journal of Ecotourism, 20*(2), 99–129.

Weaver, D. (2001). *Ecotourism*. John Wiley & Sons.

Weaver, D. B. (2005). Comprehensive and minimalist dimensions of ecotourism. *Annals of Tourism Research, 32*, 439–455.

Weaver, D. B., & Lawton, L. J. (2007). Twenty years on: The state of contemporary ecotourism research. *Tourism Management, 28*, 1168–1179.

Webb, N., Speyers, T., & Goosen, N. (2013). Short-term festivals as examples of eco-tourism? An evaluation of the Kirkwood Wildlife festival, South Africa. In N. Kozak, & M. Kozak (Eds.), *Tourism research: An interdisciplinary perspective* (pp. 157–171). Scholars Publishing.

Wintermantel, R. E., & Mattimore, K. L. (1997). In the changing world of human resources: Matching measures to mission. *Human Resource Management, 36*(3), 337–342.

Yuan, Y. Y. (2013). Adding environmental sustainability to the management of event tourism. *International Journal of Culture, Tourism and Hospitality Research, 7*(2), 175–183.

7

THE ROLE OF THE LOCAL COMMUNITY IN THE REPOSITIONING PROCESS OF TARANTO AS A SUSTAINABLE TOURIST DESTINATION

Sonia Ferrari and Nicolaia Iaffaldano

Introduction

The Apulian city of Taranto is known worldwide as a symbol of industrial pollution due to the presence of Europe's largest steel works. The plant is the subject of legal battles and citizen protests because of its pollution and the consequent increase in numerous diseases in the area.

Because of this bulky presence, today, Taranto is considered a city affected by severe environmental deterioration. This determines a negative judgement of the city and is the motivation of new urban policies aiming to renovate the local image through an effective urban marketing strategy. The focus of this strategy must be on raising the local community's awareness about environmental sustainability and favouring the conservation, exploitation, and enhancement of natural resources and biodiversity (Iaffaldano & Ferrari, 2021a, 2021b).

In recent years, local policymakers have been pursuing city image repositioning through the reconversion of the main productive and entrepreneurial activities towards the development of a nature-based tourism sector. The City Council and Apulia Region strategies seem to be converging in this aim to preserve local fragile and valuable natural areas and give Taranto a new image as a sustainable tourist destination. The core of this strategy is the establishment of a Marine Protected Area, which the Environment Ministry is now examining.

This is the beginning of an urban renovation towards a new image, in close continuity with an economic reconversion, to achieve the dual aim of protecting the environment and pursuing sustainable economic diversification. All plans revolve around the idea of developing sustainable tourism, rediscovering shellfish farming, and exploiting the sea, which is home to various protected species, particularly cetaceans.

However, this process must consider the multiple interests that need to be brought together, including those of the Military Navy, the fishing industry and, above all, the local community. According to the literature, in most cases, residents' involvement is necessary to define effective governance models and ensure the sustainability of any project.

This chapter aims to focus on the role of the various stakeholders in the urban renovation and repositioning process. The qualitative research is carried out through in-depth interviews

 DOI: 10.4324/9781003230748-9

with different categories of stakeholders to investigate how the growth of social capital can foster the sustainable development of a destination through nature-based tourism.

Literature review

City image, positioning and urban social sustainability

Today, cities compete in many fields, such as productivity, culture, heritage, infrastructure, and sustainability, to attract skills, capital, and, ultimately, increase people's welfare and quality of life (UNHABITAT, 2018). In addition, international place competition is constantly increasing due to numerous factors: the globalization of markets, the spread of innovative information and communication technologies, the tertiarization of productive activities – with the subsequent transformation of society into a reality based on services and information – the increased mobility of people, goods, enterprises, services and technologies – with the resultant delocalization of enterprises in suburban areas – the occupational crisis that spreads over all the main European cities, the progressive loss of importance of city centres and the consequential development of the belt, spread of city lifestyle to extra-urban contexts, not to mention gentrification.

However, in an economic and environmental context in crisis, the current challenge is dealing with issues such as quality of life, sociocultural and environmental sustainability, landscape and new forms of local development. The phenomena of urban crisis and undifferentiated images, which reduce the international competitiveness of some cities, demand strategies aiming at relaunching cities and towns in terms of place and tourism marketing (Golfetto, 2000; Martinotti, 1982). One possibility is given by repositioning urban images through the exploitation of their natural resources, respecting social and environmental sustainability and local history and traditions (Van den Berg et al., 1998; Robertson & Guerrier, 1998; Tarsitano et al., 2021). In this sense, tourist resources have a preeminent role in realizing effective policies of market positioning. This is due to the relevance that such elements of the urban offer have at social, cultural and economic levels today.

Cities can reach a long-lasting competitive advantage through a unique image thanks to marketing strategies based on effective branding and positioning. Since the last century, many European cities have initiated renewal processes, and one of the most prominent policy trends that has emerged is that of the city image (Kolb, 2006). Since the 1980s, in Europe, many old industrial cities in crisis, such as Bilbao, Glasgow, Torino, Liverpool, and Manchester (Law, 1992), had to change their image to launch an urban renovation process, repositioning themselves in the market as tourist destinations (Kavaratzis & Ashworth, 2005).

Currently, as increasingly important tourist destinations, cities are in growing competition with one another (Ashworth & Page, 2011; Ferrari, 2012), attempting to differentiate themselves in terms of their image and tourist offerings (Zukin, 2009), emphasizing their uniqueness and employing strategies such as city branding, flagship projects, and events to become more competitive (Ferrari & Guala, 2015). Regarding this, it is worth pointing out how, nowadays, cities can find the source of their competitive advantages in their own unique and inimitable historical endowment of local resources and specificities (Caroli, 2013). It enables them to realize very effective forms of differentiation, thanks to hardly imitable peculiarities based on their vocation. Even in the long run, these forms of differentiation can strengthen monopolistic power due to their uniqueness. Such factors can impact city image and positioning and become pivotal points of force.

In Italy, urban regeneration began after the Second World War, following the economic boom and industrial development. It is possible to identify three different phases of the urban regeneration process in the country. The first one began in the 1970s and was characterized by a renewed interest in urban identities that led to the redevelopment of historical centres and the relocation of industrial activities. The second phase occurred between the 1980s and 1990s; it consisted of interventions on the emptied areas through experiments that integrated complex programs with development and sustainability objectives. In this phase, large population flows moved from small towns towards high-density areas, thereby leading to the birth of residential districts, which often suffered building speculation. This phenomenon led to an impoverishment of communities' quality of life and situations of social disease. The third phase began in the 1990s and is still ongoing. It aims to respond to the critical issues that arose in the previous phase by promoting social inclusion, new urban spaces and easier access to services and innovation (Messina, 2018). In this last phase, greater attention to social and environmental sustainability is necessary to achieve valid and lasting results through the direct involvement of local community members.

Social sustainability is a multi-faceted concept that intersects with many others, such as quality of life, wellbeing, security, participation, human rights, and empowerment (Dempsey et al., 2011; Rashed & Mahmoud, 2018). Urban social sustainability is defined by Yiftachel and Hedgcock (1993: 140) as: "the continuing ability of a city to function as a long-term viable setting for human interaction, communication and cultural development". Social sustainability is a central issue in urban renovation plans. In fact, a successful renewal project improves the life opportunities of the local community and promotes social cohesion and the sense of belonging in the area (Arthurson, 1998; Danchev, 2005). Because social capital plays a certain role in this process, sustainability in urban renewal projects can only be achieved by involving local residents (Edgar & Taylor, 2000; Fordham, 1995; Ho et al., 2012).

Urban social sustainability is linked to the collective aspects of social life. In order to explore it, the following dimensions must be considered (Dempsey et al., 2011): equity (of access to critical services, resources and opportunities); social interaction/social networks in the community; participation in collective groups and networks in the community; community stability; pride/sense of place, and safety and security.

The city of Taranto

Today, Taranto, an ancient Greek and then Roman colony, has 189,366 inhabitants. In southern Italy, the city overlooks the Gulf of Taranto, enclosed between Apulia and Calabria. Hosting an important port that is strategically positioned along the main routes between East and West, it is the ideal hub for commercial traffic between Europe and the rest of the world. Also known as the *city of the two seas* (the two seas overlooking Taranto are called Mar Piccolo and Mar Grande), Taranto stands out not only for its great cultural heritage, consisting of numerous archaeological sites, necropolises and hypogea, but also for the large variety of its marine resources and beaches. In particular, Taranto boasts an ancient tradition in mussel farming, which emerged by chance in the fifteenth century. The city is one of the largest exporters of mussels in the world. This activity takes on a symbolic and identity value, and farmers often face many difficulties for environmental reasons related to the criticalities of pollution and climate change. However, today the "black mussel" of Taranto is one the most controlled shellfish in Italy and aims to obtain the recognition of *Slow Food presidium* thanks to a specific disciplinary action and the use of new biodegradable fishnets. The sea resource seems to be more and more the core of an urban marketing strategy for Taranto.

The image of the city has been severely damaged over the years by the existence of an Ilva steel works, the largest such plant in Italy. In July 1997, the Italian Council of Ministers declared it an "area at high risk of environmental crisis", and an investigation was opened. Some years later, the plant managers were declared guilty of having spread a large quantity of mineral dust from the deposits near the steel plant and of not having prevented that from spreading. Since then, public opinion and policymakers have been increasingly aware of the health and environmental emergency in the area, which was been evidenced by the heavy pollution found in air, soil, surface and ground waters, both in the Ilva site and in the nearby inhabited urban areas (Lucifora et al., 2015).

A coordinated approach to tourism marketing, through the promotion of natural resources and sites and fish and mussel farming products, might both stimulate this renovation process and respect the traditional vocation of Taranto. In fact, the city history and image are focused on dolphins, the beloved aquatic animals that have always lived in the gulf and have also been the protagonists of numerous legends about the city's birth. Two important innovations are going in this direction: the creation of a Blue Oasis, an instrument for the realization of a Cetacean Sanctuary in the Gulf of Taranto; and the establishment of Mar Piccolo Regional Park in the area in front of Taranto.

Research aim and methodology

This research aims to study the process of urban image renovation of the city of Taranto. The study's main objective is to understand the role of the different stakeholders, with specific attention to the local community's awareness about the whole process and level of involvement in the various phases of the process.

Regarding the methodology, a qualitative approach was adopted because it provides flexibility and adaptability to the research aims and is more interactive, in-depth and sensitive when studying human behaviours, attitudes, ways of thinking and motivations (Carson et al., 2001; Silverman, 1998). For this reason, 25 in-depth, face-to-face, semi-structured interviews were conducted, each lasting approximately 40–50 minutes. The respondents belonged to various categories of stakeholders: city and regional policymakers, representatives of the local community, representatives of cultural, sports, environmental and social promotion associations, people belonging to entrepreneurial and commercial associations as well as to universities and other research centres, operators of local production sectors, port and naval authorities. The sample was composed as follows: city and regional policymakers, 12 percent of the sample interviewees are representatives of local community and 4 percent representatives of regional authorities, 32 percent representatives of cultural, sports, environmental and social promotion associations, 8 percent of entrepreneurial and commercial associations, and 16 percent of universities and other research centres, 20 percent operators of local production sectors, and 8 percent port and naval authorities.

All interviews were recorded and transcribed on the same day to ensure minimal loss of information. A subsequent step was to structure meanings by using narrative and leaving participants' stories as they were told. The narrative provides more in-depth information about participants' ideas and attitudes (Saunders et al., 2009).

Data analysis followed a deductive coding approach. The data were managed by hand through a thematic analysis. The four main themes identified are as follows:

- *the ongoing urban renovation process;*
- *the role of the public institutions involved;*
- *the position and the attitude of private subjects;*
- *the residents' behaviour and their level of involvement in the different phases.*

Findings

The ongoing urban renovation process

The Apulian city of Taranto is known worldwide as a symbol of industrial pollution due to the presence of Europe's largest steel plant. Because of this bulky presence, today Taranto is considered a city affected by severe pollution and environmental deterioration.

In recent decades, the city found itself caught between the problems of unemployment, environmental pollution and health due to the crisis of its steel plant. However, the signs of rebirth are already apparent.

In recent years, many have been working hard to change this negative image, focusing on the extraordinary natural resources of the city and its territory, trying at the same time to respect and enhance the traditional vocation of Taranto.

The city can become an important sustainable tourism destination thanks to many elements: the sea, the quality of the countryside and its products, the culture and the beauty of the landscape. Together with the enhancement of the historical heritage and the numerous measures of urban renovation, there will also be the realization of some important sports centres of attraction within the extraordinary scenery of the city of the two seas. For this reason, there works are currently in progress to improve the historical centre, whose three-thousand-year historical stratification is evident, to promote some sites of great scenic beauty, like Mar Piccolo and Cheradi Islands, and to establish a Marine Protected Area that will safeguard cetaceans and biodiversity; all these initiatives could give a new impetus to cultural, commercial, accommodation and craft activities. In this way, the diversification of economic industries and positive effects on residents' quality of life will be fostered.

The urban renovation process in progress is carefully described by various categories of stakeholders.

The role of the public institutions involved

For some years now, all the local institutions have seriously faced the problem of getting out of the industrial monoculture of Taranto. They are trying to diversify the development prospects, that is, to favour a whole series of activities and projects primarily related to the sea.

In the last 60 years, the Ilva plant has turned Taranto into a city of steel. Because of the crisis of its steel factory, the city found itself overwhelmed by problems. However, at present there are evident signs of its rebirth and the attention that the city, regional, and national policymakers are giving to Taranto. These signs result from legal battles, city protests, a strong popular awareness about industrial pollution, and the consequent increase of numerous diseases in the area.

The city councillor says that the municipal administration has triggered a synergistic process where it has set the objective of becoming the leader of a project involving other political bodies, such as the Region, the national Government and the European Union. He points out:

> We have defined the strategic plan "Ecosistema Taranto. "It starts from a cultural approach that is different from what the history of this city has produced, namely the great disasters of the local economic, social and also cultural system. The vision was to move from an" egocentric system "to an" eco-centric system", i.e., a system where the main actors of the territory come together, work in synergy so that we can work on the construction of a sustainable and circular economy based on what were and are the peculiarities of the territory. For this reason, we have defined a broad strategic plan ranging from urban regeneration to local community participation to the urban regeneration process through an integrated economic system.

Moreover, the city councillor responsible for the city's urban plan states that some of the renovations of the old city involve noble palaces that will be converted into hotels. The city councillor has already got the renovation project of at least 5–6 large hotels that will occupy noble palaces in the old city centre. There is a strong demand from private subjects who want to acquire a series of noble palaces and turn them into luxury hotels. Today in the old town, prices are still very competitive considering that some of these buildings need significant restoration. All the buildings that are to be renewed face the sea, which is why most hotels will directly connect to the private marina; here, there is a project to create numerous yacht berths, thus implementing the harbour tourist business on both Mar Grande and Mar Piccolo.

A marine biology university researcher in the survey says that it is important to make the wider public aware of the degradation of Mar Piccolo. Mar Piccolo has always been the beating heart of the city of Taranto for its marine biodiversity; today, however, it is polluted. The ongoing remediation purpose in the Primo Seno (first inlet) is to return it to a healthy and productive state; this is the objective of the university. He points out that Mar Piccolo was not polluted solely by the huge steel industry and the maritime military arsenal. The real cause that led to the degradation of Mar Piccolo was the carelessness of the locals due to a loss of cultural ties. For this reason, Mar Piccolo, which used to be the environment where oysters and mussels were bred, is now polluted by a variety of waste, including car carcasses, batteries, tyres, cans, plastic bottles and fishing nets that marine biologists have retrieved from the seabed. The researcher proposes that a newly strengthened bond with the marine environment should start again in the schools, involving teachers and students. He says,

> Mar Piccolo is in a state of decay because we did not feel it as our home. This city needs belonging and civic sense, but above all respect for the common good, because only in this way it will be possible to overcome the pollution problem, which is actually much more complex than it seems. The local community and the school must have an active role; otherwise, you will never win this challenge in Taranto.

According to the statement of a Taranto CNR researcher interviewed, the Apulian agri-food product that most represents this city is "Taranto mussel"; this is a resilient speciality that can satisfy a large percentage of the market demand. Mussel farming is sustainable from both an economic and a social point of view as the production of Taranto mussels is among the first in Italy for quantity (yielding about 30–40,000 tons/year) and organoleptic quality; it supports a direct workforce, which before the 2011 environmental crisis was estimated to be about 800 units, very similar in size to that of the Eni Refinery in Taranto. Mussel farming is considered to be environmentally sustainable since a medium-sized mussel can filter about 4–5 litres of water per hour under normal conditions, retaining 90 percent of the particles contained in it. The way mussels feed causes them to absorb and accumulate any contaminants present in water. If mussels are forced to filter out contaminated water, they can grow less until they die. "Mussels, therefore, have no vocation for polluted sites," says the researcher of the Water Research Institute of the CNR.

In Taranto Mar Piccolo, following the environmental emergencies of 2011–12, a whole series of measures have been taken to protect consumers thanks to the vigilant activity of the Taranto CNR. In the Primo Seno, it is impossible to fatten the commercial-sized mussels beyond 28 February of each year. For this reason, the product must be moved from the Primo Seno (first inlet) to the Secondo Seno (second inlet) by that date (Figure 7.1). Currently, the farms occupy the Secondo Seno of Mar Piccolo, where it is possible to carry out the entire production cycle; in the Primo Seno, for sanitary reasons, it is possible to collect the seed but not to fatten the

Figure 7.1 The map of Taranto.

Source: ACTORS ITALIA (2016). Rapporto sulla città di Taranto e il Museo Nazionale Archeologico di Taranto, p. 9, OECD publishing.

adult mussels in the last 3–4 months of the breeding cycle. Therefore, the CNR researcher states that the alternative future of Taranto could start from relaunching mussel farming.

A representative of the port of Taranto says that the Municipality is starting the procedure for the institution of a protected Marine Area in the Gulf of Taranto, which is surely a positive fact as far as environmental protection and marine fauna are concerned, as well as a vehicle for territorial promotion. "The important thing is that the various interests should be reconciled," say both the port representative and that of an environmental association, the WWF. The representative of WWF adds: "Any initiative of protection and preservation of that stretch of coast in the Gulf of Taranto should be obviously compatible with the other existing activities such as military manoeuvres and maritime or cruise traffic." A contrary opinion is expressed by a representative of the Taranto sea sports association, who considers that the arrival of cruise ships in the port of Taranto could conflict with the valorization of the protected marine resources. Moreover, he says that Taranto has already become the destination of some cruise ships, like those belonging to "MSC Cruises". Unfortunately, even if large cruise ships do not generate a form of sustainable tourism, they are a new reality in Taranto.

In addition, the city councillor explains:

> What we did as a Municipality, involving different actors of the territory such as the University, the CNR, the port authority, and the Navy, is to ensure a mapping of the territory, and above all the prospect of creating some Oases, not as if they were inaccessible reserves closed to human activities, but as integral parts of the city.

The areas where the life of cetaceans is prevalent have been identified, and, consequently, so have the possible routes of commercial and cruise ships, partly following the model that has been experienced in northern Europe for years: what have been created are not closed marine reserves but integrated activities for humans, flora and fauna existing on the territory.

The representative of the port of Taranto adds:

> Genoa has already got a protected marine area which has not minimally affected the increase of cruise ships in its harbour. So, fixing the boundaries of a protected area does not necessarily affect the routes and lines of merchant and cruise ships, and it depends on how that delimitation is done: the boundaries found for the Protected Marine Area certainly do not intersect the merchant shipping routes.

A contrary opinion has been expressed by a representative of a commercial association concerning the presence of cruise ships in Taranto. He says:

> If we talk about tourist development in Taranto, I think we have to discuss an economic system that can remain constant over time. I think that cruise ship tourism cannot be a driving force for the development of the tourist dimension of Taranto because it represents a sort of hit-and-run tourism: it is not able to involve the port infrastructures, the ferry service to and from Cheradi Islands, the facilities structuring on the islands of San Pietro and San Paolo, the logistics of Mar Piccolo, and to make mussel farming a form of experiential tourism. We need to develop a tourist system that is stable and can last over time and ensure income to the operators of the sector.

The position and attitude of private subjects

A representative of an entrepreneurial association states that they are defining a different development model for Taranto, linked to the city's sustainable development and its peculiarities. Likewise, a representative of a commercial association says:

> In my opinion, the first thing that must be done is to revitalize the city, because it is a dead city as it has been brought on its knees. Taranto people just can't understand how the dilemma of shutting down or not shutting down the Ilva steel plant will be solved. Today the Ilva is a "hanging sword over our heads" that does not allow us to do and say anything else because you know it is there but cannot get what its destiny will be. Moreover, to my mind, this city has not got ready for the so-called urban renovation process. Indeed, it's true that we complain about shutting down the Ilva plant, but if we assume to shut it down for a moment, can we say what we have built as an alternative? Nothing! What is the system we want to create in order to grow? In my opinion, it is the tourism sector, where all the actors must think in the same way, that is in terms of system, to create an efficient system of hospitality, reception services, hotels, restaurants, transport.

An operator of the local fishery production sector is convinced that the typical consumption of fish products can give the territory an identity dimension, fostering tourist development. He points out:

> The real problem, which acts as a barrier to a hypothesis of this kind, is the lack of logistics because we fishermen have no available areas to use. What I mean is that we have not got the area logistics that can offer a touristic route to those who come to Taranto and make a stop at its historical and picturesque mussel farms. We have missed the opportunity to reclaim the identity of the fishing sector. For example, let us consider the fish market in Taranto: it is located in a marginal area on the coastline of the Tamburi neighbourhood, and many people do not even know how to get there. I was puzzled when I saw the presentation of the renovation project of the Tamburi neighbourhood coastline: it is part of the project" Ecosistema Taranto" created by the Municipality, where Tamburi is the northern part of the coastline of Mar Piccolo Primo Seno. However, the project does not include the entire market area devoted to the fishing sector: it remains there as it is, without any thoroughfare, without any renovation project, and therefore this situation emphasizes how the existing identity root is not valued. For this reason, if this logistical surface isn't created, it will be difficult to think of mussel farmers welcoming tourists, maybe offering them the typical mussel "impepata" (fresh mussels opened in a hot pan and seasoned with oil, garlic and pepper) that they can taste on the spot.

An opposite opinion was expressed by an operator of the local mussel farming production sector. He says:

> I accompanied tourists to see the charming mussel farms in Mar Piccolo and made them taste our mussels. Moreover, through a story I created, I told them the previous scenario of the place and the types of mussel and oyster farms that were made there. I also talked to them about older businesses existing here: the production of sea-silk from a protected species of shell called "pinna nobilis" used to make fabrics and the production of purple from the murex shells used to colour fabrics. All that was very linked to our local identity. This ongoing process of urban renovation and repositioning of the city will be important for us, mussel farmers, because the activities related to what we do will be enhanced, namely, to make mussel farming a form of experiential tourism.

A representative of an environmental association expressed a similar opinion. He says:

> If I had to design a sustainable development strategic plan, I would focus on some tourism segments: nature tourism in the Mar Piccolo, tourism linked to mussel farming and tourism related to cetaceans, namely, the dolphins in the Gulf of Taranto. These are three realities that already exist, and they only need to be better organized.

This interviewee states that the Region authorities have recently established the Regional Natural Park in the Secondo Seno of Mar Piccolo, while in the Primo Seno, reclamations are ongoing. Therefore, tourism is undoubtedly linked to nature tours. In the first place, it is necessary to provide protection, so WWF carries out actual actions against hunters who kill any type of protected avifauna: herons, egrets, flamingos. Many birds, unfortunately, have been hunted; but now, however, thanks to the establishment of the park, hunting is forbidden. However, there (Mar Piccolo), it will be necessary to rebuild habits and provide both the reclamation and the renovation of natural environments. In the second place, it will be necessary to trace the naturalistic routes that WWF has been organizing for many years, canoeing or kayaking, and finally promoting birdwatching. He adds that the ongoing urban renovation and repositioning process will be important for WWF because it will enhance the activities linked to its mission. At present, it is not the only environmental association ensuring naturalistic routes, as there are other associations such as "Marevivo" and "Legambiente" that are rooted in the territory.

A representative of a social promotion association expresses the same opinion:

> This urban renovation and repositioning process will be important for our schools because the activities related to what we do will be promoted. We decided to preserve the ancient craft of naval carpentry that was disappearing. In our opinion, this can be a good way to revitalize the city. At present we are the first and only ones to do this type of activity, and many operators have contacted us, and the city mayor told us that our association, "Officina Maremosso", will be the fulcrum of this sector, aiming at recreating the image of Taranto as a seaside town. For this reason, we try to use the ancient techniques for the maintenance, the construction and the restoration of wooden boats and to involve young people in social projects. We, therefore, pass on the old notions to young people living difficult moments: for example, children at risk of crime and the so-called NEET young people, those "Not in Education, Employment or Training". So, we try to help those who are in a state of uncertain future, offering them a good alternative life: by passing on these notions to them, we hope they will be the new generation, the repository of knowledge that would otherwise be lost.

The residents' behaviour and their level of involvement in the different phases

On the theme of Taranto's renovation, one of the interviewed citizens says that the local community has no awareness about the whole urban regeneration process in progress. She/he underlines that they are sceptical and that some of them are note even aware of the process at all. She/he adds that if local administrators want to design a city development plan, they should do it seriously and try to be in tune with the reality, thus dealing with mussel farming, logistics, products/services quality and city brand. Otherwise, citizens will become disenchanted and sceptical towards the institutions and will no longer believe in what they say.

Another inhabitant of Taranto holds a different view, stating that perhaps the big steel plant will be closed, but not due to a decision on the part of the government, the municipality or the citizens, but rather because the market will consider that Ilva is no longer useful. For this reason, the city of Taranto might have a very different image in the future. She/he also raises doubts about the local community's ability to accept the new resources, tools, and proposals that will emerge. According to the interviewee, today, the local community is not able to accept the demands that are coming because their cultural level is low for several reasons: many cancer patients, Ilva's problems, social defeats, the removal of the best people to work elsewhere. In addition, there are also a number of subjective and personal reasons: their lack of professional fulfilment and frustration, because many young people have no jobs and no plans for the future. Consequently, there is the need for a series of intervention projects to face juvenile delinquency and early school-leaving gravity on the structuring of the social fabric.

To avoid early school-leaving, there are parish recreation centres (in Taranto there are about twenty in total), and an excellent social fabric has been created. The parish attended by another interviewee has realized an event of sporting, educational and pedagogical value, called the big Mundialito Escuela, which is well-known all over Italy and is at its 37th edition. The event consists of a five-a-side football tournament between boys and girls and is organized in schools. It is not a tournament for any young people, but for students who are in regular attendance at school: it helps avoid early school-leaving. Sport is held in high esteem by the parish priest. He is convinced that sporting culture does belong to humankind's desire not to be spectators but protagonists in life. Moreover, playing a team game has a significant educational value.

According to one of the locals, the citizens of Taranto are aware of the urban regeneration process and that natural resources can be exploited. She/he argues that locals know that the "Ecosistema Taranto" project can be seriously developed, even if there might be pockets of resistance. These should be shaken up, in order to make them understand that urban renovation is necessary, even if such change is not easy to achieve. All that happens because most of the population still works in Ilva. However, since this is the only possibility for development in Taranto, and there are no others, it should also be pursued thanks to much funding provided for sustainable development projects.

Discussion and conclusions

In the past few years, the city of Taranto has been facing many important social, environmental and economic problems. The way that has been chosen to obtain an urban image reposition and solve many of these problems is promoting sustainable development and strengthening local identity. The interviews show that the ongoing renovation process in Taranto is a strong and well-established one. In addition, the majority of the interviewed public and private subjects share this opinion, and the different initiatives undertaken involve the whole community. It seems that they will lead to optimum results in the medium term, thus allowing to reach a new position and image for Taranto.

All the interviewees agree that it is necessary to carry out an urban renewal process for Taranto through a large project of repositioning and diversification of the productive activities of the city. The city should achieve a new image as a "seaside city", which over the years it has forgotten and pushed away. The process has already been started and is based on the strategic plan "Ecosistema Taranto;" it is focused on an ecocentric system that should activate a strong sustainable development in terms of circular economy and nature-based tourism. All this must be done through the involvement of the local community and this seems to be clear to all interviewees, who wish for the greater involvement of residents, also through the region's schools.

The idea of enhancing natural resources, primarily marine ones, is shared by all those interviewed. The city must feel to a greater extent than nature is its own resource to be respected and valued and that sustainable development can be achieved through nature-based tourism that could cancel the environmental problems of the past. It should help reposition the image of Taranto. Local stakeholders should focus on an identity and a city brand that must be based on the natural resources of which the sea is rich: dolphins, the historical symbol of Taranto, and mussels, a very important local production activity.

However, some believe that the redevelopment and the enhancement of natural resources, in particular of the cetaceans present in the area, can be limited to some well-defined territories, favouring other activities outside of them, such as cruise tourism, the fishing sector and the mussel and oyster farms. These last activities can become tourist attractions if well managed and organized for visitors. Others, instead, think that the process should be broader and avoid production activities that could damage the natural environment, such as cruise tourism.

Of course, there are still many unresolved issues and ongoing debates, especially among residents, who, in some cases, are not totally aware of the whole process. Challenging actions must be taken to inform and involve the local population, today discouraged and disillusioned by previous failed initiatives and by the serious economic and environmental situation.

In conclusion, the study of this city renovation process is very interesting because every local productive sector, many local and national public bodies and the whole community are involved in a considerable effort to completely change in a few years not only Taranto's image but also its daily life.

As for the limitations and the future research lines, undoubtedly the process of urban renewal in Taranto, which is presently in its earliest stages, must be constantly monitored over the years. This will allow understanding if the actions undertaken and the plan drawn up were effective and in which direction the idea of urban repositioning is evolving. It will also make it possible to know if all the stakeholders, especially the local residents, have been involved in the process, at what level and with what results. In the authors' opinion, the repositioning process currently underway in Taranto is a very relevant and particularly significant one. In the future, it could represent an example for other territories with similar problems and deserves to be studied carefully, not only in the initial phase but also during its evolution in the future years.

References

ACTORS Italia. (2016). Attrattori culturali per il turismo e l'occupazione nelle regioni del sud Italia. In *Rapporto sulla città di Taranto e il Museo Nazionale Archeologico di Taranto*, OECD Publishing, 1–28.

Arthurson, K. (1998). Redevelopment of public housing estates: The Australian experience. *Urban Policy and Research*, 16(1), 35–46.

Ashworth, G. & Page, S.J. (2011). Urban tourism research: Recent progress and current paradoxes. *Tourism Management*, 32(1), 1–15. https://doi.org/10.1016/j.tourman.2010.02.002

Caroli, M. G. (2013). *Il marketing territoriale. Strategie per la competitività sostenibile del territorio*. Milano: Franco Angeli.

Carson, D., Gilmore, A., Perry, C. & Gronhaug, K. (2001). *Qualitative marketing research*, London: Sage Publications.

Danchev A. (2005). Social capital influence on sustainability of development (case study of Bulgaria). *Sustainable Development*, 13(1), 25–37.

Dempsey, N., Bramley, G., Power, S. & Brown, C. (2011). The social dimension of sustainable development: Defining urban social sustainability. *Sustainable Development*, 19(5), 289–300.

Edgar, B. & Taylor, J. (2000). Housing. In Roberts, P. & Sykes, H. (eds.) *Urban regeneration: A handbook*. London: Sage.

Ferrari, S. (2012). *Marketing del turismo. Consumatori, imprese e destinazioni nel nuovo millennio*. Padova: Cedam.

Ferrari, S. & Guala, C. (2015). Mega-events and their legacy: Image and tourism in Genoa, Turin and Milan. *Leisure Studies*, 1–19. https://doi.org/10.1080/02614367.2015.1037788

Fordham, G. (1995). *Made to last*. York: Joseph Rowntre Foundation.

Golfetto, F. (2000). Problemi aperti per il marketing delle città. In E. Valdani & F. Ancarani (eds.) *Strategie di marketing del territorio. Generare valore per le imprese e i territori nell'economia della conoscenza*. Milano: Egea.

Ho, D. C. W., Yau, Y., Law, C. K., Poon, S. W., Yip, H. K. & Liusman, E. (2012). Social sustainability in urban renewal: An assessment of community aspirations. *Urbani Izziv*, 23(1), 125–139.

Iaffaldano, N. & Ferrari, S. (2021a). Applying sustainability indicators in the analysis and segmentation of tourist demand – Implications for marine visitor experiences on the Tremiti Islands, Italy. In J. N. Albrecht (ed.) *Managing visitor experiences in nature-based tourism*, Wallingford, UK: CAB International, 192–204.

Iaffaldano, N. & Ferrari, S. (2021b). Research, education and tourism as place marketing tools: The case of The Jonian Dolphin Conservation in Taranto, Italy. In A. Mandić & L. Petrić (eds.) *Mediterranean protected areas in an era of overtourism: Challenges and solutions*. London: Springer Nature, 281–296.

Kavaratzis, M. & Ashworth, G. J. (2005). City branding: An effective assertion of identity or a transitory marketing trick? *Tijdschrift voor economische en sociale geografie*, 96(5), 506–514.

Kolb, B.M. (2006). *Tourism marketing for cities and towns*. Oxford: Butterworth-Heinemann.

Law, C. M. (1992). Urban tourism and its contribution to economic regeneration. *Urban Studies*, 29(3–4), 599–618. https://doi.org/10.1080/00420989220080581

Lucifora, A., Bianco, F. & Vagliasindi, G. M., (2015). *Environmental crime and corporate miscompliance: Case study on the ILVA steel plant in Italy*. Study in the framework of the EFFACE research project, University of Catania.

Martinotti, G. (ed.) (1982). *La città difficile*. Milano: Franco Angeli.

Messina, E. (2018). L'arte per la rigenerazione urbana: rischi e impatti sociali verso la definizione di nuovi modelli di trasformazione dei sistemi insediativi. *Life Safety and Security*, 6(3), 124–126.

Rashed, R. & Mahmoud, R. (2018). Urban Social Sustainability in the Era of Digital Technology: The Case of NEOM, the World's Future Global Hub across Three Countries. In *The International conference of Sustainable Design of the Built Environment SDBE18*.

Robertson, M. & Guerrier, Y. (1998). Events as entrepreneurial displays: Seville, Barcelona and Madrid. In D. Tyler, Y. Guerrier & M. Robertson (eds.) *Managing tourism in cities. Policy, process and practice*. Chichester: John Wiley & Sons, 215–228.

Saunders, M., Lewis, P. & Thornhill, A. (2009). *Research methods for business student*. Edinburgh Gate, Harlow, UK: Pearson Education.

Silverman, D. (1998). Qualitative research: Meaning or practices? *Information Systems Journal*, 8(1), 3–20.

Tarsitano, E., Rosa, A. G., Posca, C., Petruzzi, G., Mundo, M. & Colao, M. (2021). A sustainable urban regeneration project to protect biodiversity. *Urban Ecosystems*, 24(4), 827–844.

UNHABITAT. (2018). *The global urban competitiveness report 2017–2018 – housing prices: Changing world cities*.

Van den Berg, L., Van der Borg, J. & Van de Meer, J. (1998). *Urban tourism*. Aldershot: Ashgate.

Yiftachel, O. & Hedgcock, D. (1993). Urban social sustainability: The planning of an Australian city. *Cities*, 10(2), 139–157.

Zukin, S. (2009). *Destination culture: How globalisation makes all cities look the same*. Centre for Human and Global Studies at Trinity College, Inaugural Working Paper Series, 1(1), 2–26.

PART II

Management and planning of nature-based tourism development

8

FURTHERING PROTECTED AREA TOURISM AND VISITOR MANAGEMENT AND PLANNING

Synthesis study of Croatian experiences with proposals for improving the framework

Hrvoje Carić

Introduction

The last decade of Mediterranean tourism has been marked by a variety of trends and challenges. The growth and diversification of tourism activities have impacted protected areas (PAs), from the perspective of visitor management and tourism development. As income generation is the primary focus in emerging and developing countries such as Croatia, the formal nature conservation authorities and PAs initially prioritized financial empowerment and over time discussed and partially adopted visitor management tools based on the practical literature. These include: Management Effectiveness Tracking Tool (Stolton et al. 2007), Best Practice Protected Area Guidelines (Thomas & Middleton 2003), and visitor-tourism planning policies (Eagles et al. 2014).

Based on the experience gained and the literature mentioned, the Croatian nature conservation authorities have developed a guide for PAs (HAOP 2018). However, there are still significant challenges related to environmental risk assessment, spatial planning and use, and conservation monitoring. In practice, the management of PA monitors various activities and indicators, but the synthesis of horizontal and vertical aspects is rarely carried out as there is no practice or guidance that clearly links the current objectives and those desired by the above-mentioned guides and frameworks, namely: robust nature conservation (bio-indicators); quality use of space (intensities, flows, thresholds/crowding); and sustainable income generation prioritizing nature.

In addition, the authorities have not managed to fully implement visitor management in PAs, leaving gaps in the functional framework (for more details, see the country report, EC 2019). For this reason, the work presented here aims to use the existing models and experiences, but also the specificities of a PA such as the National Park (NP) Krka, to structure and develop the existing elements in order to form a VMF that is better adapted to the needs of nature conservation and visitors.

DOI: 10.4324/9781003230748-11

A step-by-step approach has been used with clear arguments to inform and underpin the process. In particular, it hopefully contributes to the necessary and often-neglected understanding of the use of space, both in planning and in practice (Mrđa et al. 2014). Since planning and its implementation with continuous improvement is the biggest challenge, the focus should always be on the system that can sustain the continuous improvement. The urgency of a system-based approach is undeniable and necessary given the significant pressures of seasonality, pollution, climate change and inappropriate tourism use of heritage.

This understanding allows for a higher-quality analysis of risks and leads to more effective design of management priorities and actions. The methods described in this text use the study and step-by-step approach to uncover and respond to the lack of operationalization of the management guides commonly used in Croatia (Leung et al. 2018; HAOP 2018), in particular the ROS – Recreation Opportunity Spectrum (USFS 1982; Driver et al. 1987) and LAC – Limits of Acceptable Change (Stankey et al. 1985) frameworks, but also other relevant literature that informs possible solutions and improvements (Alexander 2008; Baud-Bovy & Lawson 1998; Eagles et al. 2002; EUROPARC Federation 2012; IVUMC 2016; Manning 2010; Manning et al. 2011; Manning & Anderson 2012; McCool et al. 2007; Pröbstl et al. 2014; UNWTO 2004). Therefore, the work presented here aims to further existing approaches in PA tourism planning and visitation management using the following steps:

a. Analysis of legislative-management frameworks
b. Zoning – defining visitation spatial elements of VFM
c. VMF risks
d. Integration.

The argument presented here is that the operationalization of the VFM concept should evolve further and take into account: the specificities of the PA sites, the impacts of related legislative frameworks, national governance and management practises. The recent experience in one of Croatia's leading national parks has been used to demonstrate what is inherent in most protected areas with similar economic and socio-political conditions in similar settings around the Mediterranean.

Study site description

National Park Krka (NP Krka) is a 109 km^2 karst phenomenon with seven travertine waterfalls of the Krka River and is one of the most visited national parks in Croatia and south-eastern Europe. The growth in visitor numbers exceeded the nine million mark per year, before being halted by the COVID pandemic. Most visitors are concentrated during the summer and in the 0.3 km^2 area of the attractive Skradinski Buk waterfall. As only 5 percent of visitors explore other sites, NP Krka have initiated a number of measures to improve visitor management. A study on visitor management in Krka National Park was conducted in 2019 by leading Croatian institutions. Two national institutes – the Institute for Tourism and the Ruđer Bošković Institute – and the expert association ADIPA, led by the spatial planning and architecture company Grgurević and Partners, produced the insights used in this text (Grgurević & partneri ltd, et al. 2019).

Analysis of legislative-management frameworks

The analysis of spatial use begins with understanding the legal framework and analyzing its implementation in PA management practice. The complexity of legislative framework relevant to PA can be initially illustrated simply by listing the relevant areas of law: spatial planning, nature conservation, environmental protection, tourism, cultural heritage, transport, energy etc. However, the main problem that the author observed first-hand in Croatia, and which is probably also observed in similar countries and regions, concerns the understanding of the hierarchy and interplay of the legal frameworks mentioned above. Most conservation experts misconceive that nature conservation laws are above other laws. The fact is that they inform other frameworks, has certain operational and planning implications, and is subject to the spatial planning system, as the latter is the overarching system with direct implications for executive development. All this could be attributed to the fact that Croatia is a country still in a long transitional phase towards effective good governance. This has not yet been achieved in environmental protection and nature conservation, as the European Commission (EC) report states (EC 2019).

Gaining a full understanding of interrelationships and hierarchies can be challenging in systems that have frequent and often significant changes in both legal and operational spheres. Nature conservation operational frameworks in Croatia have experienced frequent changes. The key institution – the State Institute for Nature Conservation – was merged into the Environmental Protection Agency and then both of these agencies lost their independent status through being incorporated into the ministry, which has frequently changed its name and purpose over the last two decades by merging and/or dissolving different governance areas such as energy, economy, water management, spatial planning, construction, regional development, etc. The legal framework has gone through an intense period of adaptation to the European Union (EU) framework during the accession phase and is now struggling to reach the required level of governance in some areas.

The basic legislative framework relevant to PAs consists of nature conservation laws and by-laws that feed into the planning and administrative acts of PA, which are informed with expert or scientific studies. The planning documents observed in the case NP Krka are nature conservation documents (the PA management plan and related periodic reports) and nine spatial planning documents (the PA, local and regional administrative units linked to the PA). About 30 expert or scientific studies, formal reports, working documents and databases were produced in NP Krka to support the management and planning of PA.

The methodology of the integrated analysis included desk research of the above documents and then workshops, meetings and interviews with key institutional experts and stakeholders to identify and review the status and current challenges in the implementation of both legislative and executive (planning and management) documents. Additional focus was placed on the planned capital visitor infrastructure projects with a total value of €25 million. All of the above led to a consensus with key experts and stakeholders that visitor management goals need to be revisited, calibrated and improved using related frameworks, such as IUCN (Leung et al. 2018) and IVUMC (2016), management and stakeholder inputs (internal and indigenous knowledge) and spatial tools (visitation system mapping). This was done as part of the methodology described, enabling the revision of the PA's Visitation Management Framework (VMF) as shown below.

The reason for drawing on the IUCN (Leung et al. 2018) and IVUMC (2016) methodologies in revising the Visitation Management Framework was because of their credibility, clarity and overarching scope. The insight from other relevant practices and approaches was also included (McCool et al. 2007; Baud-Bovy & Lawson 1998; Eagles et al. 2002; EUROPARC Federation 2012; Alexander 2008; Manning 2010; Manning et al. 2011; Manning & Anderson 2012; Pröbstl et al. 2014; UNWTO 2004).

Intensive consultations with PA institutional experts and stakeholders continued until a common understanding of the key issues was reached. All the above issues fed into the revision,

Table 8.1 An example of selected Visitation Management Framework revisions

1) Principle	*2) Description*	*3) Action*
Monitoring is the key	Monitoring must be cost-efficient.	Improve public involvement and educating visitors by encouraging their participation in the Monitoring.
Effective management depends on the objectives and key PA values	Goals in the management plan should be linked to desired results	Ensure that the plans management include clear and appropriate conservation objectives, with public participation (informing and education)
Proactive tourism planning and visitation planning and management	Policies and management actions that can be linked to PA values have better chances of effective implementation.	Provide opportunities for visitors to gain knowledge on PA heritage and values through information and interpretation programs.
Fluctuations in market demand and interests are unavoidable	Pressure, intensity usage and expectations vary (for example, impact of the buffer zone tourism activities versus conservation thresholds)	Use zoning to directly influence various recreational options.
Limiting use levels (visits) is only one of the many management options	Many variables, besides usage levels (number of visitors), affect the visitor use of an area (e.g. behavior, movements, group size, seasonality and activities).	Develop diversified educational, interpretation programs, as well as use/behavior restrictions. Identification of indicators – especially bioindicators.
Interested groups and stakeholders should be informed and/or involved	All management decisions affect certain individuals and groups.	Make sure all relevant groups are identified. Involve stakeholders in key issues' deliberation.
Communication is key to increasing knowledge on sustainability	Communicating monitoring of tourism impacts protection/ preservation and community well-being	Develop communication protocols

which was structured according to the following hierarchy: 1) Principle, 2) Description, 3) Action. Table 8.1 illustrates some of the issues and wording used in the revision.

Zoning – defining visitation spatial elements of VFM

The PAs' visitor management consists of various and diverse processes, elements and their interrelationships. To understand this complexity and define the VFM, the building blocks of the spatial elements of the framework need to be identified through auditing and analysis. This process consists of:

I. the formal zoning and existing visitor management practice;
II. the available information on spatio-temporal distribution of visitors;
III. defining and mapping the visitation spatial elements.

I. **Formal zoning and existing visitor management practice**

Zoning of protected areas is one of the basic tools for the planning and management of PA. It establishes existing and future conservation measures and land uses – ranging from a level where almost no human impact is allowed to zones where change is allowed. These degrees do not imply the value of the area, but reflect the need for management and use to preserve the overall biological, geological, and landscape diversity. The zoning of the protected area is defined by two documents: the spatial plan for the special area and the management plan of the PA. Although the zoning of the spatial plan is hierarchically above the zoning of PA, coordination is not only recommended by the nature conservation authorities (HAOP 2016, p. 73) but also considered essential, as described below (Figure 8.1).

In reviewing promotional materials, it is easy to quickly identify the concept of PA's use of space and visitor management practise. There is often an attempt to take the pressure off the main attraction by encouraging the use of other areas. In our case, the promotional materials and online platforms clearly show that the management of PA promotes the whole PA and not only the main attractions. This preferential use of PA should be reconciled with the spatial planning and conservation zoning mentioned above (Figure 8.2).

First row – the first map represents the entire park with the surrounding area and shows routes and attractions (50 × 80 km); the second map is the visitor infrastructure

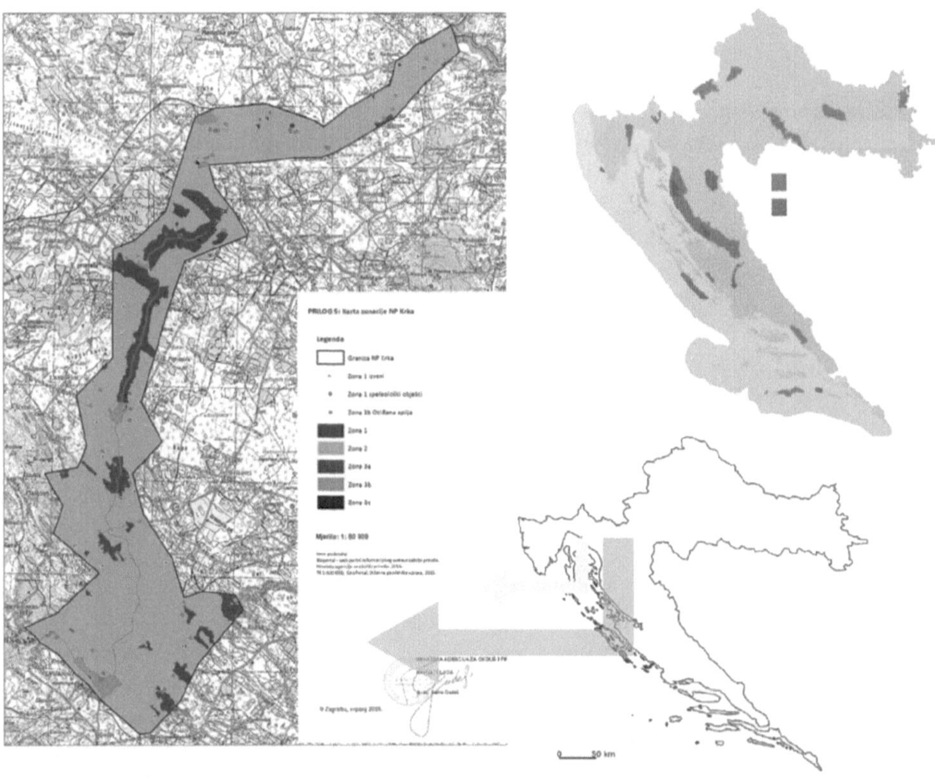

Figure 8.1 Map Reduced cartographic representation of the formal management zoning.

Source: HAOP 2016.

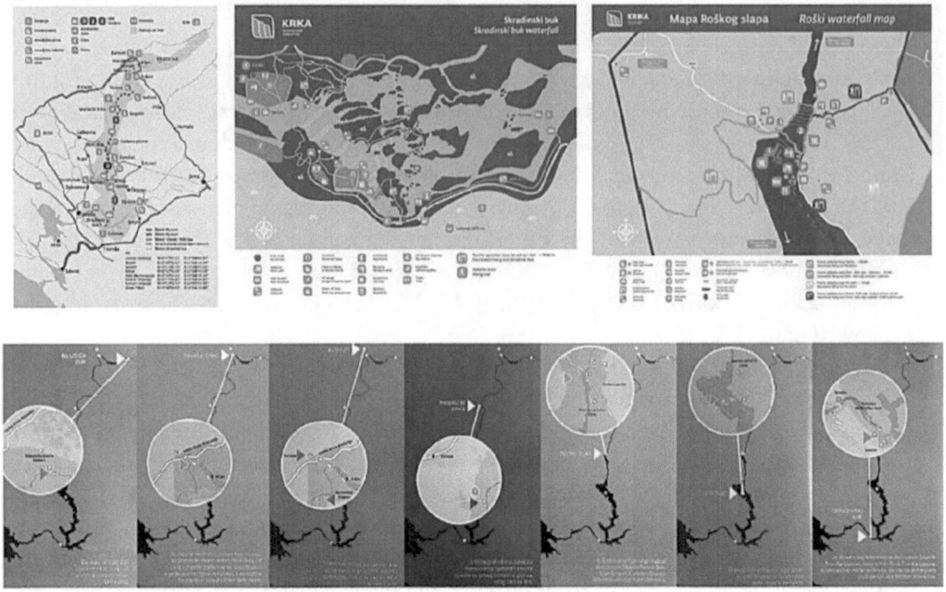

Figure 8.2 Collage - made from chosen excerpts from NP Krka's various promotional materials - discussed through workshops with key institutional experts.

Source: NP Krka, Grgurević i partneri et al. 2019.

of the main attraction Skradinski Buk/Waterfalls (5 × 4 km); the third map is the visitor map of the secondary attraction Roški slap/Waterfalls in the middle of the park, which is envisioned to serve as a "magnet" to relieve the main attraction (0.5 × 0.5 km). Second row – a sequence of seven visitor areas (each including multiple attractions and experiences), starting with the northern, less visited area to the southernmost and the main attraction Skradinski Buk, which attracts 85 percent of visitors.

II. **The available information on the spatio-temporal distribution of visitors**

To see how the above plays out in practice, the information on spatio-temporal distribution of visitors is needed. The input for the spatio-temporal distribution is based on the statistics of the PA entrances, as this was the only data available, so only a general estimate was possible. The mapping clearly showed that a higher resolution is required in the areas with the largest number of visitors. This means that a daily or even real-time insight into the spatial distribution of visitors is needed, especially for the main attraction – the Skradinski Buk area.

III. **Defining and mapping the visitation spatial elements**

Monitoring and assessment of existing use was carried out through fieldwork to verify the above points and improve understanding of the framework for visiting PA, more specifically all the elements that make it up. The fieldwork consisted of reviewing sites to identify and map details of the spatial elements of current use. It was also important to distinguish how the elements relate to the areas outside PA (buffers). The aim was to obtain sufficient input to propose coherent Visitation Management Areas (VMAs), as new units of the VMF, that best meet both conservation and visitor objectives.

The spatial elements that emerged from the analysis are divided into spatial units and linear themes: *Spatial Units* consist of interconnected visitor attractions (areas and places) that form a distinct entity whereas *Linear Themes* are transportation routes (pedestrians, cyclists, boats) that connect different places (attractions) and which can also be an attraction themselves (nature trail, boat ride). The complexity of planning and managing the visitation system is further influenced by the fact that some of the significant spatial units and line themes are wholly or partially outside the PA boundaries. For each of the spatial elements (units and lines) there are individual management issues such as monitoring the spatio-temporal distribution and defining the carrying capacity.

The formal PA Management Zoning (HAOP 2016) was thoroughly reviewed based on a desk analysis of relevant documents and auditing of current use. The participating spatial planning expert (M.J.) derived the VMAs (polygons), which represent both logical management-monitoring units and coherent visitor experiences. The VMAs were further reviewed by NP Krka's institutional experts dealing with operational and planning issues and subsequently confirmed as plausible. This process is shown in the map below – formal zoning, analysis and audit of current use, and final proposal of seven VMAs.

Due to the complexity of nature conservation issues and visitor framework elements of Krka NP, the mapping method was often used for analysis and assessment. This approach made it possible to consider all relevant management aspects, as well as current and predicted use – resulting in seven VMAs that served as the basis for the integration part of this process. However, before the synthesis, the risk needed to be understood (Figure 8.3).

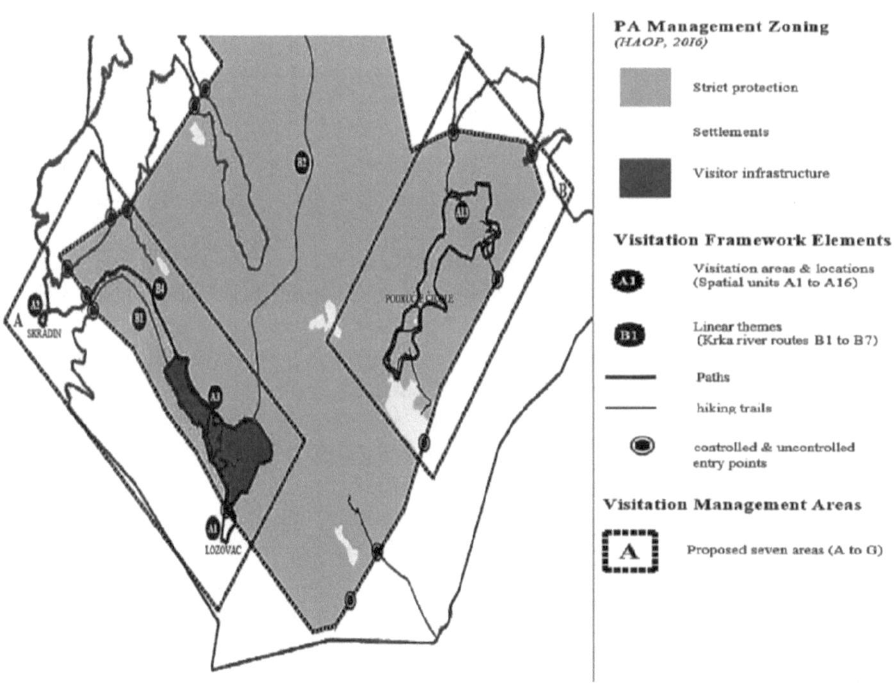

Figure 8.3 Map defining and formulating the Visitor Management Areas based on formal PA management zoning, analysis and audit that disclosed Visitation Framework Elements and proposed Visitation Management Areas.

Source: Grgurević i partneri et al. 2019.

VMF risks

This section uses findings of expert assessments of anthropogenic impacts, carrying capacity analysis, and visitor satisfaction research to disclose existing risks and gaps in the conservation with regard to VMF. Declared potential threats to the biological and landscape diversity are structured in the management plan of NPK according to the following topics: water supply, wastewater and conservation of water resources, solid waste management, roads and railways, forests and forest land, hydropower plants, speleological facilities, agriculture, traditional, hunting grounds around NP and fishing (NP Krka 2011). However, PA management decided to gain more understanding on other risks related to VMF by applying the carrying capacity concept.

The PA visitor carrying capacity helps to define key management challenges by combining information on: nature conservation, visitor experience and economic costs and benefits – including negative externalities (for detailed description see Carić & Jakelić 2018; Carić & Klarić 2011; Mrđa et al. 2014). From the perspective of visitation management, there are three key time spans for which carrying capacity is analyzed: momentary/instantaneous, daily, and annually (for detailed description see Manning 2007; Manning et al. 2011). The momentary carrying capacity is the maximum number of visitors that can stay in a given area at the same time or in a given time frame (usually hourly). Daily carrying capacity is the maximum number of visitors that can visit a given area in one day, and annual carrying capacity is the maximum number of visitors that can visit a given area over the course of a year. While the momentary carrying capacity of an area depends mainly on the characteristics and constraints of an area, the daily and annual carrying capacity depend more on the characteristics of the visit (attractiveness, market demand and trends, special interests, length of stay, etc.). The momentary carrying capacity is crucial for calculating the peak load of an area and thus provides information on constraints and limitations (Carić & Jakelić 2018).

The basic spatial units for the carrying capacity of a complex system are smaller spatial elements (areas, locations, linear themes). The monitoring, analysis and calculation of carrying capacity can only function optimally if the bottom-up logic is applied – from smaller spatial elements to the larger ones (VMAs).

There are four components of carrying capacity – economic, physical (or logistical), ecological and sociological – which have been considered in the research activities on NP Krka over the last decade (2010–20). In this paragraph a brief overview of the components is conducted.

The economic component refers to the basic calculation of the difference between the income from visitors (tickets, promotional materials, etc.) and the costs of maintaining visitor infrastructure and logistics used by visitors (trails, transport, engagement of the staff and the like). The smaller the difference, the greater the possibility that the economic component limits the carrying capacity. A negative financial balance may be justified if it results from preventing the degradation of natural resources, e.g. by keeping visitors away from sensitive areas (in visitor or education centers), or from the need to invest in the priority concerns of local communities. The physical or logistical component shows the maximum number of visitors that can occupy a space. For example, the fewer parking spaces or transport options there are, the smaller the physical component, i.e. the more likely it is placing constraints on the carrying capacity. In the case of a protected area, the ecological component is of particular importance: the number of visitors must not affect nature beyond the acceptable level established by expertly based (informed) management decisions, i.e. the planning documents of the managing body. If visitors do not have a significant direct impact on nature in a given area, or if impacts are mitigated (e.g. erosion control through wooden paths), the potential of the ecological component to limit carrying capacity is low. The impact of visitors on nature can generally be reduced by investing in infrastructure and logistics (trail construction, monitoring, wastewater treatment, etc.), which directs the analysis to the economic component. The sociological component is generally based on visitor satisfaction. The results of the visitor survey in Krka National Park indicate that this satisfaction is closely related to overcrowding, which means that the sociological component is determined by the willingness to accept different degrees of overcrowding (Polytechnic of Šibenik and NP Krka 2019). All components refer to the minimum principle, which states that the capacity of an area is determined by the component that indicates the minimum value of the number of visitors. Furthermore, when calculating the momentary carrying capacity of a PA it is not sufficient to sum up the capacities of all micro locations within the area, but two additional factors must also be taken into account:

1. the distribution of visitors within the area; and
2. the cost–benefit analysis (minimising the impact on nature).

The distribution of visitors within the observed area is important, especially if there is a microsite that attracts and holds a significant number of visitors compared to other microsites in the area. Then this microsite does not contribute with its full momentary capacity to the total capacity of the area, but only to the part of the capacity when the area is under peak load.

The feasibility of using a protected area or site, when determined by a cost–benefit analysis, is important because it simultaneously reinforces the principles of minimising impacts on nature (primarily habitat degradation) and maximising benefits to visitors (revenue, education, interpretation and other aspects of the visitor experience). Put simply, it is about answering the question: Is habitat degradation justified by the benefit acquired? Any impact on natural resources that might result from visitor activities, however minor, can only be justified by a clear (detailed) justification of the objectives in terms of interpretation, education or general positive conservation impacts.

The characteristic of the Krka National Park is that the relatively small area of Skradinski Buk, as a well-known international attraction, generates the largest share and a significant surplus of revenue (entrance fees and other income) compared to expenditure (maintenance, logistics, etc.), which can be used for conservation and other heritage protection measures, monitoring, research, interpretation, education, etc.

The expert interviews conducted as part of the project revealed that the following aspects need to be taken into account when considering new visitor infrastructure: clear definition of visitation objectives, assessment of costs and benefits, conservation risks, carrying capacity calculations, monitoring plan and local population interaction.

Carrying capacity findings informing VMF risks

JU NP Krka has carried out a number of projects with the aim of developing a methodology and calculating the carrying capacity of areas, localities, microlocations and the park as a whole. In this respect, the Ruđer Bošković Institute has carried out most of the research activities, starting with the project that initiated the methodology for maximizing nature conservation during the visitation (project title Environmental carrying capacity for visitors to NP Krka2012/2013 IRB). The project has catalogued and calculated the impacts for each area of the park. The result is that the carrying capacity of the whole park is 4620 visitors, which is only slightly higher than the capacity of the main visitor area Skradinski Buk (about 4400 visitors). The sociological component (visitor crowding) proved to be a limiting component of the capacity of the Skradinski Buk area, and the bridge near the waterfalls proved to be a critical microlocation, i.e. a bottleneck. The calculated carrying capacity of the Skradinski Buk area was exceeded on several days during the high season in 2012, and as the number of visitors increased (25 percent over a five-year period), so did the frequency and volume of proposed thresholds. Mitigation attempts to divert the flow of visitors to less visited areas and locations led to the degradation of nature in these areas. This indicated the need to improve the cost–benefit understanding of such measures. Finally, recommendations were made to increase the capacity of the Skradinski Buk area. The project "Strategy to increase the carrying capacity of the environment for visitors to the Krka National Park: evaluation of ecosystem services" 2015 (IRB 2016) had examined the context in which further capacity increase can be planned, based on: (i) the park's impact on the regional economy; and (ii) the positive and negative aspects of the visitor experience in park areas (with highly developed infrastructure as the main factor). The project estimates an economic impact of 1.5 billion kuna per year and underlines the extraordinary responsibility it has towards the local community and the region, not only as a guardian of natural resources but also as a driver of tourism development. As the region suffers from severe hinterland depopulation due to a lack of employment opportunities and intense coastal summer tourism, management challenges become even more complex (Table 8.2).

ACCTA (Assessment of carrying capacity for tourists in nature protected areas) was a 2014–2017 project financed by the Croatian National Science Foundation (HRZZ). The project's aim was to define visitor satisfaction scaling (crowding perception) as the priority input for the carrying capacity (CC) thresholds. The method used the 2015 data and assessed the Skradinski Buk threshold to be 4,000 visitors instantaneously for a level of satisfaction at which less than 50 percent of visitors are disturbed by the crowding. With 7,500 visitors crowding at the same time, the model estimates that approximately 75 percent of people will be annoyed by the number and density of people which is about half of the daily visitors. The projects and studies of CC show that the total capacity of the park is practically equal to the capacity of the Skradinski Buk area, with smaller contributions from Roški slap (see Figure 8.5). Capacity may depend on parameters such as the time of day and season (e.g. due to the daily or seasonal migration of animals, the meteorological situation, etc.). Taking into account the way Krka National Park is visited and visitors directed, CC was calculated for peak visits. The research carried out also produced a number of concrete proposals for increasing carrying capacity, aimed at efforts to increase the capacity of the Skradinski Buk area as the main factor in the overall capacity of the park. In 2017 and 2018, the management has introduced a number of measures to increase physical capacity, the most important of which are:

- Reconstruction of infrastructure (widening of the footbridge at Skradinski Buk),
- Modification of visitor flows (introducing bus transfers, directing traffic, one-way traffic and circular route),
- Changing the timing of visitor arrivals (cheaper tickets at less busy times, introducing info displays, limiting the maximum number of visitors staying in the Skradinski Buk area while offering alternatives through tour programmes).

Table 8.2 Integrated Overview of a Visitor Management Area with analytical elements and columns designed to be filled in (unshaded fields to be filled in).

Area identification (Conservation zoning information, etc.) …………	*Condition – Existing*	*Condition – Planned*
Describe the purpose of the visit Why PA presents, promotes, develops, maintains the VMA or specific location in the Visitor Management Framework? In the context of protected areas, the term visit – in the sense of using a protected area for visiting (visitor use) – refers to human presence in a protected area for recreational purposes, including education, interpretation, inspiration and physical and mental health (IVUMC 2016). It is thus a matter of how managing authorities define the purpose of management activities in relation to visiting (E.g. "The purpose for which the management opens an area for visiting is …", "The managing authority wishes to present the values of … and to educate about … …".	………… *Excursion, recreation, informing, education, etc.*	………… *Excursion, recreation, informing, education, etc.*
What are the key messages and heritage values conveyed in this Visitor Management Area? (Mission of a PA is to educate, inform, "enlighten", engage … visitors)	*Existing messages (describe)*: …………	*Planned messages (describe)*: …………
Level ("intensity") of communication (pay attention to whether it refers to information on site, on internet, to the work of guides/trainers, and on what microlocation)	*Existing communication (chose)*: a. *basic (information),* b. *medium (educating),* c. *Advanced (participation)*	*Planned communication (chose)*: a. *basic (information),* b. *medium (educating),* c. *Advanced (participation)*
Describe how you will achieve your intended visitation purpose, and the level/intensity of communication: *e.g. draw up an interpretation and communication plan, standardize messages on heritage values, refine/harmonize the marketing – branding with the planned messages and purpose, organize visits in accordance with the interpretation/communication plan, i.e. establish appropriate content programs for visitors, arrange a specific (specify which/what) visitor infrastructure and/or "logistics", educate tour guides,*		

(*Continued*)

Table 8.2 (Continued)

Area identification (Conservation zoning information, etc.) …………	*Condition – Existing*	*Condition – Planned*
Monitoring the spatio-temporal distribution of visitors Basic monitoring: - measurements and data of: hourly entries/exits and/or only total daily data (E.g. entries = number of tickets sold) - number by (micro) locality/ annual, monthly, daily, momentary (instant), - duration of stay - local and current conditions? - tools: manual count, surveys, automatic counters (passes, tickets sold …), Wi-Fi data analysis, GPS applications, statistics on the use of audio guides… Advanced monitoring: - modes of arrival, use of different ways to move about PA, traffic counting - use of offered visitor programs/ content, - spatio-temporal distribution, real-time concentrations; - visitor structure (demographics), motivation, visitor satisfaction, crowding perception, - environmental impacts - the impact of the visit on the local community and the economy of the functional region	*Existing monitoring (describe):…………* *Current monitoring (circle):* a. *satisfies fully* b. *satisfies in part* c. *it does not exist or does not satisfy*	*Planned monitoring (describe):…………*
Monitoring visitor satisfaction a. Crowd perception b. Other (motivation, satisfaction with experience, absorption of “enlightenment” messages …)	*Existing monitoring (describe):…………* *Current monitoring (circle):* a. *satisfies fully* b. *satisfies in part* c. *it does not exist or does not satisfy*	*Planned monitoring (describe):…………* *(eg on-site surveys*

(*Continued*)

Table 8.2 (Continued)

Area identification (Conservation zoning information, etc.) …………	*Condition – Existing*	*Condition – Planned*
Monitoring of desirable and undesirable visitor activities Tools/resources: conservation reports, alerts, application establishment, phone…	*Existing monitoring(describe):…………* *(e.g. site inspection reports)* *Current monitoring (circle):* a. *satisfies fully* b. *satisfies in part* c. *it does not exist or does not satisfy*	*Planned monitoring(describe):…………* *(e.g. via an smartphone App, GSM tracking, counters …)*
Biodiversity monitoring (species, habitats) Examples of elements of existing and planned monitoring: Monitoring of individual habitats and/or species, flow monitoring, microclimatic indicators monitoring, chemical and MB analysis… Regular or occasional scientific surveys of habitats, censuses of populations and species, etc.; relevant indicators to address the state of nature…	*Existing monitoring (describe):…………* *Current monitoring (circle):* a. *satisfies fully* b. *satisfies in part* a. *it does not exist or does not satisfy*	*Planned monitoring (describe):…………*
Visitation threats to biodiversity Examples of elements of an existing, possible, anticipated threat: Noise, vibration, light pollution, changes in microclimatic factors, waste, spread of invasive species, trampling, collection… Erosion, wild animals displacement (specify which and why); 'Planned' threats are those expected to occur after the completion of planned development activities in the assessed or neighboring zones/areas (E.g. construction of a new trail would increase attendance; removal of transit lines would increase pressure on other locations)	*Existing threats:*	*Possible/anticipated threats (describe:* *(due to the implementation of plans, projects, activities in the NPK or in the zone of influence)*

(*Continued*)

Table 8.2 (Continued)

Area identification (Conservation zoning information, etc.) …………	*Condition – Existing*	*Condition – Planned*
Corrective measures for existing and anticipated threats Examples of existing and planned corrections: Elimination or restriction of visits, reduction of groups, temporal or seasonal restriction, placement of warning or educational signs, more frequent monitoring, penalties, physical barriers, modification or removal of infrastructure…	*Current corrections (describe):…………* *Current monitoring (circle):* a. *satisfies fully* b. *satisfies in part* c. *it does not exist or does not satisfy*	*Planned corrections:…………*
Describe how you intend to improve monitoring with regard to existing and possible/ anticipated visiting threats *(e.g. expand, establish monitoring of the following aspect…key chemical-biological parameters, noise, light pollution, destruction of the land cover, erosion…)*		

Integration

During the activities in the NP Krka, the need to integrate all the above knowledge, information and findings became clear. For each Visitor Management Area (see section on "Zoning – defining visitation spatial elements of VFM") a transparent management tool had to be developed that would provide a comprehensive overview. To this end, the integrated Visitor Management Areas matrix (see Table 8.2) was developed to join all the findings from the steps described earlier, but also to provide a review framework that allows for continuous improvement and efficient auditing of the individual area and the VMF as a whole. For that purpose, it is crucial to have a clear understanding of the status of each VMF element in the existing-current and expected-long term perspective (the Condition columns). The VMF analytical elements are:

- interpretation infrastructure and communication practices (visit purpose, key messages, communication intensity),
- monitoring visitor's spatio-temporal distribution, satisfaction and activities,
- biodiversity monitoring, assessing threats and proposing mitigation.

By filling in unshaded areas of the above matrix an auditor (preferably third party) can precisely place information that can lead to concrete actions to improve visitor management. Ideally, benefits are maximized and costs minimized. For example, in considering the establishment of controlled swimming areas for visitors at the 'upper' (near the car park) and/ or 'lower' (at the pedestrian bridge) part of the trails of the Skradinski Buk area (see the above illustrations), the Integrated Overview has highlighted some important considerations. Firstly, swimming at the bridge is downstream from the main conservation phenomena (Sedra waterfalls) and is therefore safer for both visitors and the waterfalls ecosystem. Secondly, CC standards should be considered to disclose use limits with system visitor flows and activities monitoring is urgently needed. Another notable example of Integrated Overview: The strategy of diversifying the offer and dispersing visitors was implemented with the construction of the path to the Manojlovac Waterfall (see previous maps and illustrations). The Integrated Overview shows that this measure has increased accessibility and thus significantly increased the stay at this site. As a result, an important observation was made – the increase in interference with wildlife (especially otters). In the end, the analysis concludes that biomonitoring needs to be strengthened with education and control of visitor behavior in this potentially high-risk segment.

Discussion

These two examples illustrate common PA problem-solving to address pressures using the Integrated Overview Matrix, which underpins the continuous improvement of conservation objectives in individual Visitor Management Areas. However, when all areas are analyzed together on the level of the framework, additional important topics appear. As mentioned earlier, the area has suffered from massive depopulation and the interests of the local community are paramount in finding an answer to how to create more direct economic benefits in a sustainable manner.

These two common examples are presented here to illustrate how problem response is made more efficient through the Integrated Overview. The logic behind this is to enforce continuous improvement of conservation objectives in Visitor Management Areas. When all areas are analyzed together at the framework level, other important issues come to light. As mentioned earlier, the region is largely depopulated and the interests of local people are brought into focus by asking how they can benefit more directly and sustainably from tourism.

NP Krka have started initiatives in this direction. A shepherd with 300 sheep has been granted a concession for grazing on the park's karst pastures – an example of cooperation and the promotion of family farms that can potentially offer tasting of local products, accommodation in rural households and other content. At the same time, practises like these prevent unwanted forest succession that decrease biodiversity, while the development promoted is outside the usual mass tourism and within the framework of green, rural, agro, gastro and ecological tourism. In this way, new NP Krka management objectives are met or existing ones advanced, consistent with the win–win values advocated by programs such as UNESCO MAB. Namely the objectives relevant to this are: sustainable human coexistence in PA; biodiversity conservation: rocky pastures, agrobiodiversity and organic farming; landscape protection; traditional heritage; positive interaction of NPK and resident population; extension of biodiversity practices to PA neighbouring areas.

Conclusion

The milestones of Aichi Target 11 of the Convention on Biological Diversity state that effective and equitable management should be carried out using participatory and science-based site planning processes that include clear biodiversity objectives, targets, management strategies and monitoring and evaluation protocols (CBD 2010). The implementation of the targets is supported by an extensive literature that, among other things, suggests methods for managing protected areas, discusses negative impacts in detail and uses case studies to illustrate the interaction between visitors and nature.

However, management implications remain vague when it comes to systemic and continuous improvement. Therefore, the work presented here aims to further the discourse on integrated visitor management. By using a study approach, the intention was threefold: to illustrate practical applicability; to enable replication of ideas in places with similar conditions (such as the Mediterranean coast and hinterland); and to promote discussion among experts and scholars on the topic of visitor management.

The framework presented here is based on collected expert information from a range of research disciplines, aiming at more informed and integrative approaches that consider spatial and temporal distribution of visitors, visitor attitudes and activities, biodiversity monitoring, environmental risks, local and regional community aspects and income generation. As described, this approach, in the form of an audit framework, can guarantee more effective implementation of the desired conservation objectives by building on the existing formal and informal methods and systems of a PA.

The methods involved four steps or groups of analyses and measures, which are described below. The first step is the analysis of legislative-management frameworks, which allows for a revision based on the best practices described in selected relevant literature. In order to improve the VMF, the polygons were designed based on the existing zoning and the actual use of the space, i.e. the spatio-temporal distribution of visitors in practise. Visitor Management Areas (VMAs) were then formed as units for further analysis and auditing VMF risks through the use of expert assessments on anthropogenic impacts, carrying capacity analysis and visitor satisfaction. When considering microlocations two additional factors must also be taken into account: 1. the distribution of visitors within the area; and 2. the cost–benefit analysis that prioritizes minimising impacts on nature. The final step was integration, which took the form of a matrix integrating all the previous results based on the analytical elements related to visitors (activities conducted, distribution in time and space, monitoring for threat mitigation and information/knowledge offered).

COVID has introduced new challenges and highlighted some of the gaps in visitor management discussed here. As described in the study (McGinlay et al. 2020), social distancing, lockdowns and other measures have created a demand for PA visits, resulting in additional overcrowding and more intensive use of remote areas, which in turn has led to additional habitat disturbance, tension and conflict. The study emphasises that the first policy recommendation for the long-term management of European protected areas is careful spatial planning. This is in line with the findings from years of experience presented here – spatial use and planning are fundamental to VMF. In addition, COVID has introduced some management innovations that are actually consistent with much of the pre-pandemic advice from experts and consultants that has rarely been put into practise. Spenceley et al. (2021) highlight some of these innovations in their research: reservation requirements for high visitation sites, time-limited access requirements, social distancing education, a broader understanding of the benefits of spending time in nature, etc. However, the main long-term strategic direction remains economic resilience based on sustainable indigenous knowledge and practises, with concepts such as UNESCO MAB and agrobiodiversity playing an important role. Going back to basics – monitoring is fundamental for effective day-to-day and crisis management, especially when quick decisions need to be made, but also because investments in the post-pandemic period are likely to shift to "sparsely populated, nature preserved destinations" (Petrić & Mandić 2021), making informed based management and planning even more significant.

Acknowledgements

Marko Jakelić from Grgurevic & Partners Ltd. was not only the bedrock of not only the 2019 project activities presented here, but has also generated ideas used in the methodology. Special thanks also to Roman Ozimec and Tin Klanjšćek.

References

Alexander, M., (2008). *Management Planning for nature Conservation, A Theoretical Basis and Practical Guide*. Amsterdam: Springer Netherlands.

Baud-Bovy, M., Lawson, F. R. (1998), *Tourism and Recreation: Handbook of Planning and Design*. Architectural Press.

Carić, H., Jakelić, M. (2018). Informed Visitor Management in Protected Areas (original title: Informiranoupravljanjeposjećivanjem u zaštićenimpodručjima). *Prostor: A Scholarly Journal of Architecture and Urban Planning*, 26(1 (55)), 146–155. https://doi.org/10.31522/p.26.1(55).11

Carić, H., Klarić, Z. (2011). Istraživanjaprihvatnogkapaciteta – međunarodna i hrvatskaiskustva. Poglavlje u znanstvenojedicijiinstituta za turizam, knjiga br. 5 "Izazoviupravljanjaturizmom". Institut za turizam, Zagreb.

CBD-Convention on Biological Diversity. (2010). Strategic Plan 2011–2020. Aichi Biodiversity targets. Target 11. Retrieved January 05, 2022, from www.cbd.int/sp/targets/

Driver, B. L., Brown, G. H., Stankey, T. G., Gregoire, P. J. (1987). The ROS Planning System: Evolution, Basic Concepts, and Research Needed. *Leisure Sciences*, 9(3), 201–212.

Eagles, P. F. J., Coburn, J., Swartman, B. (2014). Plan quality and plan detail of visitor and tourismpolicies in Ontario Provincial Park management plans, *Journal of Outdoor Recreation and Tourism* 7–8, 44–54.

Eagles, P. F. J., McCool, S. F., Haynes, C. D. A. (2002). *Sustainable Tourism in Protected Areas: Guidelines for Planning and Management*. Switzerland and Cambridge, UK: IUCN Gland. xv + 183pp.

EC (European Commission). Directorate-General for Environment. (2019). *Commission Staff Working Document: The EU Environmental Implementation Review 2019. Country Report – CROATIA*. SWD (2019) 114 final. Brussels, 4.4.2019.

EUROPARC Federation. (2012). *Practical, Profitable, Protected: A Starter Guide to Developing Sustainable Tourism in Protected Areas*. Latvia: European Centre for Eco and Agro Tourism in partnership with the EUROPARC Federation.

Grgurević & Partneri Ltd., Institute for Tourism, Institute Ruđer Bošković, ADIPA. (2019). *Visitor management study the National Park Krka* (original title: Studija upravljanja posjetiteljima NP Krka).

HAOP – Hrvatska Agencija za Okoliš i Prirodu (Croatian Environment and Nature Agency). (2016). Nature Conservation Expert Assessment for the Krka National Park Spatial Plan (original title: Stručnapodlogazaštiteprirode zaprostorni plan područjaposebnihobilježja Nacionalni park Krka").

HAOP – Hrvatska Agencija za Okoliš i Prirodu (Croatian Environment and Nature Agency). (2018). *Guidelines for planning the management of protected areas and/or ecological network areas*. (original title: Smjernice za planiranjeupravljanjazaštićenimpodručjima i/ilipodručjimaekološkemreže.) Ministarstvozaštiteokoliša i energetike/Ministry of Environmental Protection and Energy and Uinted Nations Development Programme Croatia.

IRB – Institute RuđerBošković. (2016). *Project Strategy to increase the carrying capacity of the environment for visitors to the Krka National Park: evaluation of ecosystem services* (original title: ProjektStrategijapovećanjaprihvatnogkapacitetaokoliša za posjetitelje NP Krka:vrednovanjeuslugaeko sustava). Public Institution NP Krka.

IVUMC - Interagency Visitor Use Management Council. (2016). *Visitor Use Management Framework: A Guide to Providing Sustainable Outdoor Recreation*; Edition One, July 2016,

Leung, Yu-Fai, Spenceley, Anna, Hvenegaard, Glen, Buckley, Ralf (eds.) (2018). *Tourism and Visitor Management in Protected Areas: Guidelines for sustainability*. Best Practice Protected Area Guidelines Series No. 27, Gland, Switzerland: IUCN.

Manning, R. (2007.) *Parks and Carrying Capacity: Commons Without Tragedy*. Washington, USA: Island Press.

Manning, R., Valliere, W., Anderson, L., Stansfield McCown, R., Pettengill, P., Reigner, N., Lawson, S., Newman, P., Budruk, M., Laven, D., Hallo, J., Park, L., Bacon, J., Abbe, D., van Riper, C., & Goonan, K. (2011). Defining, measuring, monitoring, and managing the sustainability of parks for outdoor recreation. *Journal of Park and Recreation Administration*. [Online] 29:3

Manning, R. E. (2010). *Studies in Outdoor Recreation: Search and Research for Satisfaction*. Corvallis, OR: Oregon State University Press.

Manning, R. E., Anderson, L. E. (2012). *Managing Outdoor Recreation: Case Studies in the National Parks*. Cambridge, MA: CABI.

McCool, S. F., Clark, R. N., Stankey, G. H. (2007). *An assessment of frameworks useful for public land recreation planning*. Gen. Tech Rep. PNW-GTR-705. Portland, OR: U.S. Department of Agriculture, Forest Service, Pacific Northwest Research Station.

McGinlay, J., Gkoumas, V., Holtvoeth, J., Fuertes, R. F. A., Bazhenova, E., Benzoni, A., Botsch, K., Martel, C. C., Sánchez, C. C., Cervera, I., Chaminade, G., Doerstel, J., García, C. J. F., Jones, A., Lammertz, M., Lotman, K., Odar, M., Pastor, T., Ritchie, C., Santi, S., Smolej, M., Rico, F. S., Waterman, H., Zwijacz-Kozica, T., Kontoleon, A., Dimitrakopoulos, P. G., Jones, N. (2020). The impact of COVID-19 on the management of European protected areas and policy implications. *Forests*, 11(11), 1214. https://doi.org/10.3390/f11111214

Mrđa, A., Carić, H., Bojanić Obad Šćitaroci, B. (2014). Significance of the concept of tourist carrying capacity for spatial planning - Previous research, concept development and methodological approaches (original title: Značajkonceptaturističkenosivosti za prostornoplaniranje - Dosadašnjaistraživanja, razvojkoncepta i metodološkipristupi). *Prostor: A Scholarly Journal of Architecture and Urban Planning* (1330-0652) 22 (2014), 2; 212–227.

NP Krka. (2011). *Management Plan of the National Park Krka* (original title: Plan upravljanja NP Krka), Chapter 2.8. Threaths, pp. 61–62).

Petrić, L., Mandić, A. (2021). Governance and Management of Protected Natural Areas in the Era of Overtourism. *Mediterranean Protected Areas in the Era of Overtourism / Mandić, Ante*; Petrić, Lidija (ur.). Switzerland: Springer International Publishing, 2021. pages. 15–44. https://doi.org/10.1007/978-3-030-69193-6_2

Polytechnic of Šibenik and National Park Krka (2019). *Structure and motivation of visitors in the National Park Krka*. Original title: Struktura i stavoviposjetitelja NP Krka.

Pröbstl, U., Wirth, V., Elands, B. H. M., Bell, S. (Eds.) (2014). *Management of Recreation and Nature Based Tourism in European Forests*. Berlin: Springer.

Spenceley, A., McCool, S., Newsome, D., Báez, A., Barborak, J. R., Blye, C-J, Bricker, K., Sigit Cahyadi, H., Corrigan, K., Halpenny, E., Hvenegaard, G., Malleret King, D., Leung, Y-F, Mandić, A., Naidoo, R., Rüede, D., Sano, J., Sarhan, M., Santamaria, V., Beraldo Sousa, T., Zschiegner, A-K (2021). Tourism in protected and conserved areas amid the COVID-19 pandemic. *Parks*, 27, 103–118.

Stankey, G. H., Cole, R. C., Lucas, M. E., Petersen, S. S., Frissell, D. N. (1985). The Limits of Acceptable Change (LAC) System for Wilderness Planning. Gen Tech Rep INT-176. U.S. Department of Agriculture, U.S. Forest Service, Intermountain Forest and Range Experiment Station, Ogden, UT.

Stolton, S, Hockings, M, Dudley, N, MacKinnon, K, Whitten, T, Leverington, F (2007). *Management Effectiveness Tracking Tool. Reporting Progress at Protected Area Sites: Second Edition.* WWF International, ISBN: 978-2-88085-281-8.

Thomas, L., Middleton, J. (2003).*Guidelines for Management Planning of Protected Areas. Best Practice Protected Area Guidelines Series No. 10.* World Commission on Protected Areas (WCPA), IUCN – The World Conservation Union.

UNWTO – United Nations World Tourism Organisation. (2004). *Tourism Congestion Management at Natural and Cultural Sites.* ISBN 92-844-0763-X World Tourism Organization Calle Capitán Haya, 4228020 Madrid, Spain.

USFS – U.S. Forest Service. (1982). ROS Users Guide. US Department of Agriculture, US Forest Service, Washington.

9

APPROACHES TO FORESIGHT RECREATION AND TOURISM IN NATURE

Experiences from the Nordic Region

Peter Fredman, Tuija Sievänen, Frank S. Jensen, Vegard Gundersen, Sandra Wall-Reinius, Maria Lexhagen, Christine Lundberg, Klas Sandell, Odd Inge Vistad and Daniel Wolf-Watz

Introduction

Understanding the future is inherently important for making the right management decisions, informing policymakers and overall contributing to a sustainable development of tourism. The COVID-19 pandemic that hit the world in 2020 illustrates this profoundly. While international tourism (a significant part of which is nature-orientated) came to an abrupt stop, local and regional demand for recreation in nature boomed and was sustained in countries where such activities were unrestricted (Venter et al., 2020, 2021). Although possible impacts from pandemics on tourism are visible in the literature, both COVID-19 and the subsequent interest in outdoor recreation caught most of us by surprise.

Changes to tourism and outdoor recreation is nothing new, but the first two decades of the 21st century show some remarkable trends before the pandemic hit us (Sandell & Sörlin, 2008; Elmahdy et al., 2017). New and better equipment and services are taking people further and faster into the woods, onto the waters and up mountains. Better-off people are travelling further to experience more in a shorter time. Natural areas are increasingly available through mobile technology. Older people are becoming more active and youngsters are picking up new recreation behaviours, increasingly channelled through social media. But none of this happens in isolation. Trends in outdoor recreational participation reflect numerous drivers of change, such as urbanisation, mobility, income, cultural diversity, aging populations, new technologies, climate change, and political uncertainty (cf. Dwyer et al., 2009; Manning, 2011; Elmahdy et al., 2017; Neuvonen et al., 2018; Fredman & Haukeland, 2021).

It is, however, not possible to study the future empirically as such, since there are no data to be analysed which measures the time ahead of us. Looking at current data on participation in outdoor recreation, most measures are cross-sectional studies from national statistics or data collected in projects without any intentional future follow-ups. To the extent that such data are collected repeatedly over time, it can provide useful information on trends as a basis to *forecast* outdoor recreation, using primarily quantitative analytical methods. If we want to understand future recreational patterns, we need to go beyond this type of mechanist approach and apply complementary methods which can *foresight* the future more comprehensively.

DOI: 10.4324/9781003230748-12

The aim of this chapter is to discuss approaches to foresight recreation in nature and we do this based on experience from research in the Nordic region. The purpose is twofold: (i) to review different monitoring approaches used for trend analyses as a mean to forecast participation in outdoor recreation activities; and (ii) to discuss supplementary approaches to do foresights of outdoor recreation for the benefit of future policies and management decisions. We focus on the Nordic context because of the long-standing traditions of outdoor recreation in this part of the world (Sandell & Sörlin, 2008; Fredman & Haukeland, 2021). We make the presentation based on selected studies undertaken in Finland, Sweden, Denmark, and Norway with the objective of supporting future research and decision-making within outdoor recreation and nature-based tourism.

Critical to our study is the difference between forecast and foresight. While *forecasting* is an estimation of the future in a specific research area by means of scientific methodology, *foresight* is the overall process of creating an understanding and appreciation of information generated by looking ahead (Cuhls, 2003). The European Travel Commission (2021) describes forecasting as "a technique that uses data as input to make informed estimates that are predictive in determining the direction of future trends." FAO (2014) defined a foresight as "a systematic, participatory and multi-disciplinary approach to explore mid to long-term futures and drivers of change." Rather than predicting based on numbers, a foresight implies understanding different alternative futures.

Outdoor recreation in the Nordic context

The Nordic countries share much of the historical and socio-political background on how people relate to, and take recreation in, nature. Climatic conditions in the Nordic region can sometimes be challenging (short days, low temperatures, rainy summers) but they also provide diverse outdoor recreation opportunities across the summer, autumn, winter, and spring seasons. For many people, rural society is only one or two generations back and a large proportion of the now increasingly urbanised population still has strong connections to the countryside and farm cultures through, say, a second home. Nowhere in the world is second-home ownership as common as in the Nordic region (Müller, 2021) and such cabins may be seen as 'gates to nature' that strengthen the contact with nature and offer a place where many outdoor recreational skills are introduced and performed. Another important feature for recreational opportunities in the Nordic outdoors is the *Right of Public Access* which, along with certain obligations, provides good access to most natural areas in Sweden, Norway, and Finland and, to a more limited extent, in Denmark (Kaltenborn et al., 2001; Sandell & Fredman, 2010; Øian et al., 2018).

There is also a discussion of whether there are special features of the Nordic outdoor tradition, characterised by a life in the outdoors away from civilisation and endorsing pedagogical, ethical, and environmental values (cf. Henderson & Vikander, 2007; Sandell & Sörlin, 2008). A study of hikers in Norway may serve as an illustration. Svarstad (2010) invited readers of the Norwegian Trekking Association (DNT) membership magazine and ten regional newspapers to write about 'what is a good hike for you'. Based on the 80 texts submitted, grounded theory analyses revealed three categories of meaning constructions: 1) recreation (hiking as an aspect of a harmonious understanding of contemporary society and everyday life); 2) simple outdoor discourse (hiking as a way of living out critique of the modern society); and 3) belonging category (hiking as a means of establishing a sense of belonging and continuity of one's own life in relation to earlier, traditional ways of living). The second category brings up the Nordic tradition of 'a simple life in nature, with simple means' – advocated under the Nordic term 'friluftsliv', which can be approximated to 'outdoor recreation'.

Most countries in the Nordic region also have a long-standing reputation of promoting outdoor recreation to benefit and provide value to people and societies. As such, it has been closely associated with cultural identity and nature awareness, the development of the welfare state during the 20th century, and regional development through tourism in more recent decades (Gössling & Hultman, 2006; Sandell & Sörlin, 2008). Recent national policies and development programmes for the promotion of outdoor recreation in Sweden, Denmark, Norway, and Finland aim to support these traditions and give future generations the same opportunities (Swedish Government Writ., 2012/13:51; Naturstyrelsen, 2015; Meld. St. 18, 2016; Tyrväinen et al., 2018). However, to do this, it is critical to have knowledge of both current recreational patterns and projected future trends. If knowledge of the present status is poor, then proposals for future actions become less meaningful. Moreover, if information about possible future outcomes is missing, it is difficult to identify the most relevant changes that planning and management agencies should address.

Monitoring outdoor recreation

Measurement and monitoring outdoor recreation in the Nordic region involve two main approaches to collecting data: on-site or population surveys. While on-site studies include people who actually visit the site or area in question, population surveys with different geographical scope will target both visitors and non-visitors to recreation areas. Population surveys at the national level can contribute to national management strategies and policymaking, including non-users of nature-based recreation (Sievänen et al., 2009). On-site surveys, on the other hand, provide detailed insights into area-specific issues, which are essential for regional or local long-term management strategies and policy development (Kajala et al., 2007). Further advantages include their capability to survey international tourists, which are generally not included in national population studies.

On-site monitoring has typically relied on quantitative questionnaire-based methodology (Kajala et al., 2007; Légaré & Haider, 2008; Sievänen et al., 2009; Wall-Reinius, 2010; Ankre et al., 2016; Hansen, 2016). Common approaches include guided questionnaires (staff assisting the visitor with the questionnaire), self-guided questionnaires and two-step approaches with contact cards combined with a more comprehensive follow-up questionnaire distributed after the visit (postal or web-based survey based on email addresses). However, digital technology is now driving much of the development in this field. Data from mobile devices, big data from the Internet and crowdsourced data from social media and apps provide new opportunities as to where and how visitors may be monitored (Guerrero et al., 2016; Pickering et al., 2018). So far, however, the combination of on-site contact cards at trailheads and follow-up surveys have proven the most robust and common approach in the Nordic context, especially in remote locations (Kajala et al., 2007; Fredman et al., 2009; Miljødirektoratet, 2015).

In population studies, data typically come from a sample of the target population in question collected by telephone, postal or face-to-face methods. Finland, Sweden, Denmark, and Norway all measure participation in outdoor recreation as part of their official statistics, but many studies are also affiliated with various research and development projects. As population studies are increasingly challenged by decreasing response rates and high costs (Vaske, 2008; Czajka & Beyler, 2016), there is a growing interest in alternative approaches, such as online panel surveys. While the reliability and validity of such approaches is a subject of debate, they are cost-effective and the survey instrument becomes more flexible compared with traditional telephone or postal surveys (Callegaro et al., 2014).

Trends in participation

Both the above approaches to collecting outdoor recreation data are also useful in studying trends in participation. A trend is a general direction in which something is developing or changing, or how people are behaving. Furthermore, in the outdoor recreation context, trend projections are perhaps the most obvious approach to predicting the future based on previous behaviour (Gartner & Lime, 2000; Légaré & Haider, 2008; White et al., 2016). However, one frequent problem is the lack of data collected consistently over time. This is due to methodological considerations, short-term projects, an absence of long-term policy plans for recreational use, and/or a lack of organisational continuity (Sievänen et al., 2009; Ankre & Wall-Reinius, 2010; Ankre et al., 2016). The Nordic countries provide several examples of both on-site and population studies collecting cross-sectional data at different points in time to identify participation trends in outdoor recreation (cf. Vistad, 1995; Jensen, 2003; Jensen & Koch, 2004; Odden, 2008; Sievänen & Neuvonen, 2011; Wall-Reinius & Bäck, 2011; Dervo et al., 2014; Fredman & Wikström, 2015; Fredman et al., 2014; Fredman et al., 2016). Table 9.1 provides a summary of the main trends identified from these sources. Many of these trends (such as more diverse activity patterns, more physical activities, shorter visits, and decreases in extractive activities) occur in several countries, while other trends seem more distinct. Note that these data reflect trends before the COVID-19 pandemic. Changes observed in all the

Table 9.1 Summary of main trends in outdoor recreation participation in Finland, Sweden, Denmark and Norway

Land	*Method*	*Main trends*
Finland	National population surveys/statistics	- Higher participation rates, but decrease in participation frequency - Participation diversified to more activities - Decrease in fishing, berry- and mushroom picking - Older people are more active
	On-site surveys	- Number of visits to nationals parks are increasing
Sweden	National population surveys/statistics	- Participation in "forest strolls" relatively stable over time - Older people more active (especially females) - Increase in "physical activity" in nature - Decrease in frequent visits to nature - Decrease in berry picking, increase in mushroom picking - Increase in downhill skiing, more 'adventurous' and physical demanding activities
	On-site surveys	- Changing hiking patterns and more long hikes in mountain areas - Demand for more services in nature - Less visitors in some urban proximate forests
Denmark	National population surveys/statistics	- Earlier increase in number of forest visits have stabilized - Increase in proportion of non-visitors - Increase in the number of visits to the forest alone - Decrease in duration of forest visits and distance to the visited forest - More forest visits are connected with physical exercise - More diverse activity patterns
	On-site surveys	- Increase in number of forest visits - Increase in number of visits to forests around the Copenhagen Metropolitan Area and the (western) coastal forest plantations

(*Continued*)

Table 9.1 (Continued)

Land	*Method*	*Main trends*
Norway	National population surveys/statistics	- Long term increase in participation has shifted to decrease - Short walks is still the most popular activity - Increased diversity and specialization of activities - Increases in long distance hikes, biking/MTB, boating and kayak - Skiing increasingly diversified - Decreases in berry- and mushroom picking
	On-site surveys (national parks)	- Increase in one day hikes, and especially short trips (<4 h) - Decrease in multiple-day hikes, and in length of several day hikes - Visitors from more nationalities - Increased use of designated recreational infrastructure - More visitors prefer designated recreational facilities - A European visitor segment seeking 'peace and quiet'

(data provided through personal communication with key researchers in each country before the COVID-19 pandemic)

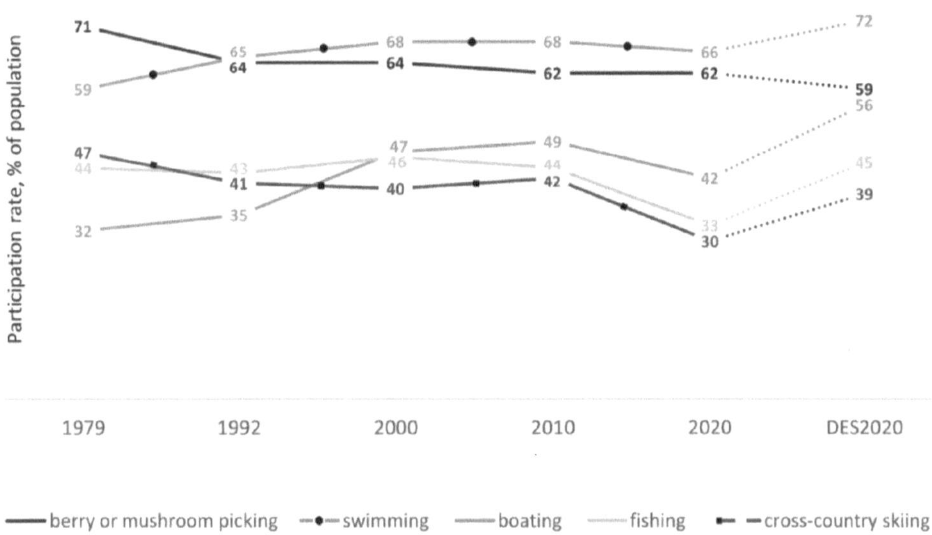

Figure 9.1 Trends and prognosis of participation of outdoor recreation activities in Finland.

Source: Luke unpublished.

countries during the pandemic reveal significant increases in outdoor recreational participation, as described further in the Discussion section.

Data on trends, such as those summarised above, provide researchers with opportunities to forecast outdoor recreational participation. In the Nordic context, an example from Finland can illustrate this (Figure 9.1). Participation rates in five outdoor activities (berry and mushroom picking, swimming in rivers or lakes, boating, fishing and cross-country skiing) are measured at five points in time from late 1970 to 2020. The figure shows that two of the activities (swimming and berry- and mushroom picking) have stayed approximately constant over the last three decades. Two activities (fishing and cross-country skiing) have decreased steadily. Participation rates for 2020 were also predicted based on the existing data in 1979–2010, using Brown's linear Double Exponential Smoothing (DES) method (Chen et al., 2003). The DES2020 points show the prognosis is not very reliable when compared to the actual figures for 2020 in Figure 9.1. Hence, there is a need for supplementary methods, to make better and more insightful predictions of future participation in outdoor recreation.

Increased diversity

One important observation from the trends reported above is the overall pattern that outdoor recreation seems to be becoming increasingly diverse. Participation is diversified into more activities (and sub-activities) at different places and attracting a broader range of people. However, it is difficult to make robust generalisations about this because diversity has many dimensions (Manning, 2011). One example of how outdoor recreation activities may diversify into different sub-activities is alpine skiing. Looking back a few decades, skiing was more or less, done in three different ways: backcountry (tour) skiing; cross-country (track) skiing; and downhill skiing in designated ski areas (Fredman & Heberlein, 2003). Today, one may easily list ten different types of skiing: jibbing, classic, skating, randonee, Telemark, splitboard, sitski, carving, rail, and snowboard. They all require special equipment, places, skills and knowledge (Pröbstl-Haider et al., 2019). From a research perspective, it is a challenge when people participating in these activities do not answer 'yes' to the question as to whether they have participated in 'skiing', making trend analyses for this activity difficult. We also observe similar diversification trends in other recreational categories, such as the use of different types of recreational trails and water-based recreation (cf. Godtman Kling et al., 2017; Kaae et al., 2018; Vistad, 2021).

Foresight outdoor recreation

While increased diversity and greater complexity in the demand for outdoor recreation calls for more in-depth studies, it also challenges research about outdoor recreation as we used to do it. The above Finnish forecasting example is based on the assumptions that population structures are stable over time, that recreational behaviour is not shifting (i.e. that behaviour is reflecting the same interests and values over time), and that the recreational conditions are unchanged. Such assumptions do, of course, put severe restrictions on this approach. If we do not know *why* a trend is increasing, decreasing, or levelling off, it is difficult to know what will happen when external factors, such as population structures or environmental conditions, change. Hence, forecasting through extrapolating trend data is a technocratic and reductionist approach that does not offer much space for deeper understanding of the changes taking place. These kind of data is useful as baseline information to find out "what has happened and what might happen if…", but to *foresight* outdoor recreation, we need approaches that support and broaden our understanding of the future.

In the following subsections, we review a selection of approaches to foresight outdoor recreation that supplement the forecasting of trend data, as described above. While extrapolating trends certainly serves its purpose, we believe the methods described here can provide more in-depth information to better contextualise, understand, and interpret trend data. The authors consider them the most useful approaches, based on their experiences of using them in empirical research into outdoor recreation and nature-based tourism over the last three decades.

Cohort studies

Few people exhibit consistent participation patterns in outdoor recreation across their lifetime. There are obvious reasons to believe that people's characteristics (attitudes, preferences, physical ability, and/or behaviour) will change with age and different life stages. However,

Table 9.2 Participation "at least once" in different outdoor recreation activities while visiting the mountain region of Sweden during 2013. Percent

	AGE 15–25 years	*AGE > 25 years*
Mountain biking*	14.3	5.4
Fishing	22.2	21.9
XC-skiing	36.5	26.0
Downhill skiing	54.0	45.9
Snowmobiling*	28.6	16.1
Nature studies, birdwatching	11.1	12.8

* Significant difference between groups (Chi sq. $p<0.05$)

it is also reasonable to expect that a group of young people raised under similar conditions (cohort) will keep some of their characteristics throughout life. Studying such cohorts may be one approach to foresight (Alm et al., 2012; Vilhelmson et al., 2017) that is also applicable to the context of outdoor recreation (Kelly & Warnick, 1999). As long as demographic tendencies and migrations of the population are stable, long-term trends may be identified. This approach may then be used to foresight outdoor recreation demand and better understand the long-term behaviour of tourism demand. One factor that goes against the cohort method is the aforementioned fact that many external factors will change with time (climate, economy, technology) regardless of which generation a person belongs to. Hence, a 'static' cohort view of change is not likely to provide a complete picture of the future and the cohort approach is more applicable to analysing changes associated with younger generations (Wolch et al., 2011). Research into recreational patterns in the Swedish mountain region provides an illustration of a cohort analysis (Fredman et al., 2016). Table 9.2 shows the numbers for participation 'at least once' among two age groups of Swedes visiting the mountain region during 2013 (age 15–25 and over 25). If the younger group maintains their recreational behaviour over time, we may expect a future increase in mountain biking and snowmobiling, as participation is significantly different between the two groups for these activities.

Predictions by the public

Perhaps one of the more obvious approaches to doing foresights is to ask a sample of the population we are interested in about future behaviour. This implies an indirect measuring tool that supports predictions of the short-term future development of different activities. A Danish national survey may serve as an example. Jacobsen et al. (2014) collected data on households' actual expenditures in relation to domestic outdoor recreation activities during the previous year (2013), as well as expected expenditures for the coming year (2014). Figure 9.2 shows the respondents' expected total consumption in DKK (millions) related to the expected frequency of participation in 17 specific activities (transport, catering, participation fees and other frequency-related consumption costs) for that coming year, as well as percentage changes in relation to the actual consumption for the same activities in the previous year (2013). An increase in net consumption (≈frequency) was expected for all activities, with the exceptions of wind and kite surfing, sailing and riding. The total estimated consumption related to the expected frequency in 2014 corresponds to an increase of 4.2 percent in relation to the consumption in the previous year. A major advantage of this approach is that information is collected directly from the 'source' itself: the actual/potential outdoor recreationist (although the measurement in

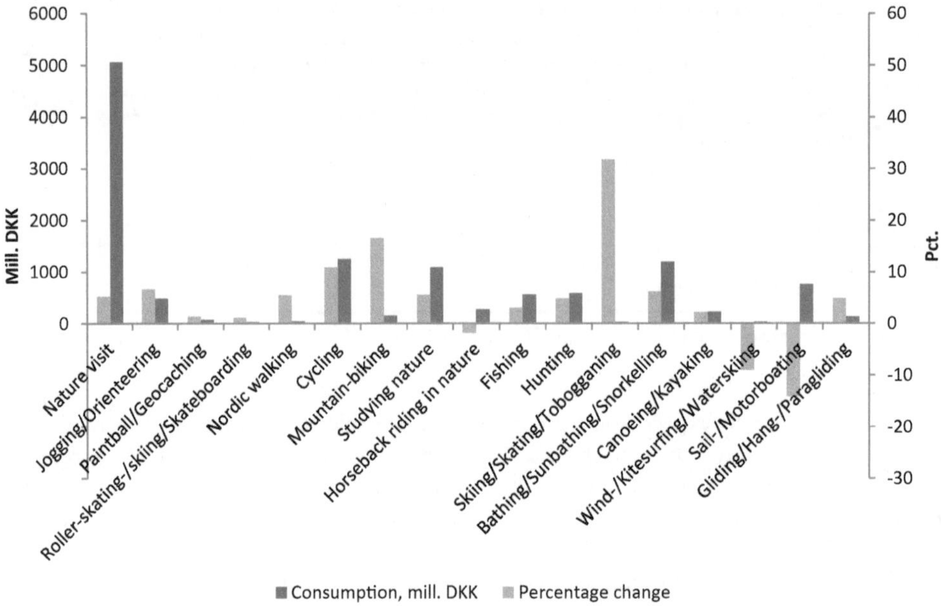

Figure 9.2 Expected total consumption in millions DKK (2014) related to the expected frequency of participation in 17 specific activities (e.g. transport, catering, participation fees and other frequency-related consumption costs), and percentage changes in relation to the actual consumption in the previous year (2013).

Source: Jacobsen et al. (2014).

the example above is indirect by focusing on economic activity). A possible disadvantage is that participants in consumer surveys sometimes exaggerate their answers due to self-selection bias ('over-participation' by the more active recreationists) and that answers to hypothetical questions (about the future) are generally more unreliable than answers to questions about past behavior.

Predictions by experts

Another technique for predicting the future is to use groups or panels of experts. This implies 'in-depth' enquiries with people having extensive knowledge and/or practical experience of the subject under investigation. One frequently used approach goes under the label 'Delphi', named after the tradition of asking the Oracle of Delphi in Ancient Greece about the future. This approach assumes that the future is increasingly less dependent on the past but increasingly more so on the "will of the agents of the present" (Landeta, 2006). The Delphi method implies that experts participate in two or more rounds of surveys and are given the possibility to adjust their answers based on the input from the other experts participating. This way, the method produces one or more 'collective views' about the future. The advantage is that predictions about the future are more accurate when coming from a structured group of individuals than those coming from unstructured groups. The Delphi method is very flexible and can be used in a wide range of situations, but the recommendation is to focus on "one possible future" (European Travel Commission, 2021). The disadvantage of using the Delphi method might be that experts would mainly think 'inside the box' and be instinctively doubtful of new perspectives that challenge their own knowledge and competence. There is also a risk that experts have personal economic interests in a specific trend direction, which may bias the results.

Numerous Delphi studies are reported in the tourism literature, including studies of future challenges for global tourism (von Bergner & Lohmann, 2014), ecotourism (Donohoe, 2011), and nature-based tourism (Konu, 2015). One recent application in the Nordic context focuses on trends in nature-based tourism (Haukeland et al., 2021). This study included experts from five countries representing ten different sectors (including public institutions, destination companies, tour operators, media, non-profit and youth organisations). Sustainability, responsible travel, health and wellbeing, pure nature environments, digitally augmented experiences, soft adventures, and guided services were among the most highly ranked trends.

Experts vs. the public

While the public may predict the future at an 'individual level' with some precision, they may not know about the broader trends in the outdoor recreation sector to the same extent as the experts. Therefore, systematically combining the two approaches described above (predictions by the public and experts) implies yet another approach to foresight outdoor recreational participation (Fredman et al., 2016). Impacts from megatrends, technical development, new materials, forthcoming rules and regulations are issues that often matter which the public may omit in their predictions (Elmahdy et al., 2017). The solution is then to also direct questions to a group of experts selected for their knowledge on the topic under investigation. By asking identical questions to both experts and the public, combining data from the two sources can help to better foresight outdoor recreation. While this represents a combination of two established approaches, we argue that a systematic qualitative analysis based on quantitative and/or qualitative data as input can provide robust predictions of the future.

One illustration of this combined approach is from the Swedish mountain region, a major tourism destination for domestic tourism (Fredman et al., 2016). In this case, participants attending a conference on mountain research (in other words, experts) were asked to answer two questions about their predictions with respect to changes in recreational activities in the mountain region (activities that were increasing and those that were decreasing). The same set of questions was also included in a national survey targeted to the Swedish public, addressing recreation in the mountain region. The results show some remarkable differences between experts and the public (Figure 9.3). For example, compared to the public, many more of the experts think snowmobiling will increase in the future while backpacking, berry and mushroom picking are among the activities the public thinks will increase more than the experts do. Hence, taking the differing views of the two groups into consideration, a more insightful and accurate foresight is possible.

Scenario modeling and future tables

Scenarios are sophisticated descriptions of possible futures and can be either quantitative through statistical modeling (cf. forecasting), or qualitative by using peoples' knowledge, thinking and understanding in a systematic way (Cuhls, 2003; Pouta et al., 2009). The European Travel Commission (2021) describes scenario planning as an approach in which alternative future scenarios are systematically constructed based on information from a wide variety of sources: 'An approach that seeks to understand and predict the forces that push and pull the future in different directions.' There is no strict method for developing scenarios – both explorative and normative approaches are common. In qualitative scenarios, there are usually several different components involved in both the processes and outcomes.

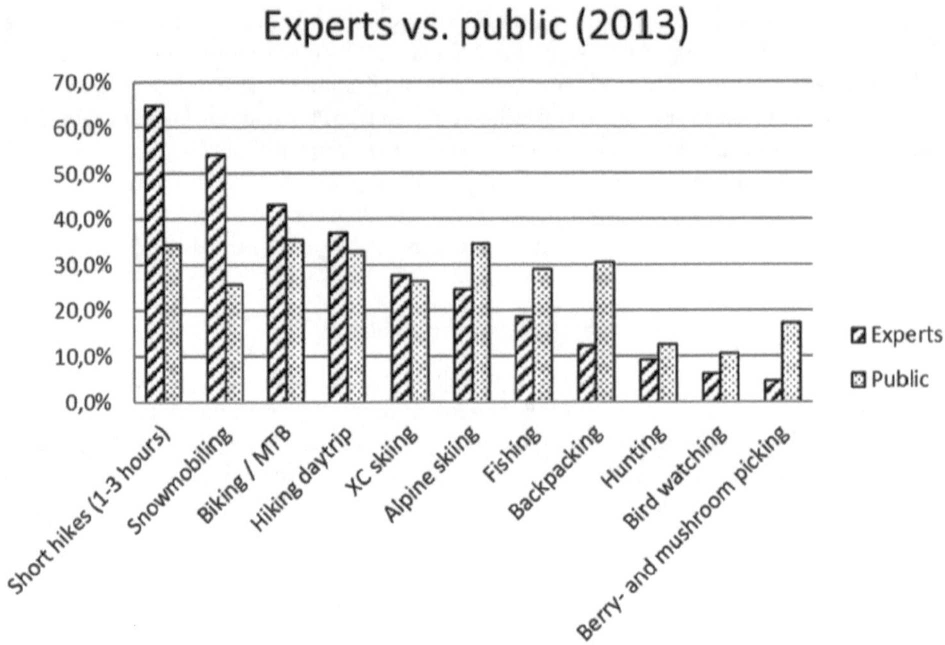

Figure 9.3 Experts vs. public. Predictions of outdoor recreation activities that will increase in the Swedish mountain region in the following 10 years.

Source: Fredman et al. (2016).

One approach to doing scenarios, applied in the Nordic context, is the 'future table'. In this case, the starting point may be a seminar or workshop, with participants of particular relevance to the topic. For example, it may be a selected group of outdoor recreation specialists from different agencies, companies, or other organisations in this sector. In a brainstorming session, the workshop discusses themes selected by the organiser of the seminar. The organiser then develops the first *future table* based on the brainstorming content. In the next step, participants add descriptions of possible *future conditions* according to the agreed themes or sectors in the future table. The outcome of the second phase may be to create *future paths*, i.e. how to reach future outcomes from the present. Finally, workshop participants evaluate *alternative futures*, including strategic and practical conclusions (Seppälä, 2013). The text output from such workshops may also be analysed through various discourse analyses. The typical output from a future table is 'alternative futures' (scenarios), one or more of which may be described as a systematic storyline or arrays of alternative futures. Hence, the future table method provides a 'big picture' of the direction of changes in outdoor recreation, based on expert input.

One example of the future table method to foresight outdoor recreation is from Finland, and is organised around three workshops for experts in the field (Neuvonen et al., 2018). The brainstorming phase of the process was done beforehand by an online mind-mapping tool which provided a set of driving forces for the discussion. The workshops created nine different alternative futures, of which eight were positive and desired and one was a dystopia (Figure 9.4). The process itself received positive feedback from participants and the experts assessed its output as very useful.

Figure 9.4 An example of future tables used to develop alternative futures as the outcome. Here is a table of driving forces for an alternative future called 'Warehouse of outdoor recreation'.

Source: Neuvonen et al. (2018).

Ethnographic online approaches

Given the considerable technical development during the last couple of decades, researchers aiming at understanding the future can find valuable sources of information online. We therefore add research methods that apply ethnographic approaches such as cyber-ethnography, virtual ethnography and netnography to the set of tools providing insights into future trends in outdoor recreation. These methods offer researchers a platform to observe and collect data from the Internet on the behaviour of individuals and groups. Not only do these approaches complement outdoor recreational data outside of traditional, physical field sites, they also offer researchers access to unobtrusive and authentic descriptions of individuals' behaviours and preferences (Del Fresno & López-Pelaez, 2014; Kozinets, 2015; Olafsson et al., 2021).

Observational ethnographies (very similar to the methods employed in traditional ethnography) have been used when researchers take on the role of online non-participant observers or 'lurkers' as well as researchers actively participating in online communities (Kozinets, 2006). These participative and immersive techniques are contrasted with quantitative research methods, such as datamining of web usage and social media (Fuchs et al., 2014) and social network analysis (Casanueva et al., 2016). The study of online content, patterns and behaviours exhibited by virtual communities provide researchers with in-depth knowledge on the characteristics of groups and individuals and their decision-making processes that may be informative with respect to future trends (Sharma et al., 2018).

Foresight through such ethnographic approaches means access to contemporary social worlds, people's activities, and their online interactions. By collecting online empirical data such as photos, video, or text, interpretative insights (carefully focusing on and analysing what is publicly available on the Internet) can inform awareness and understanding of future trends. For example, Instagram photos and their associated hashtags (#) were used to identify the most frequently occurring user-generated content published by visitors to the Swedish mountain region (Fredman et al., 2016). Compared with methods that use stakeholder views on future trends (such as predictions by experts and the public), ethnographic online research methods provide insights into individuals' everyday lives and decision-making processes that are collected in an environment not fabricated by researchers. This can enable a more authentic understanding of the future and future trends, at least among those sharing their experiences on the Internet. Online ethnography also has the advantage of accessible data that does not require costly fieldwork and which can account for more respondents compared to qualitative interview studies.

Crowdsourced data

Yet another emerging field of study in tourism and outdoor recreation is the use of big data (cf. Höpken et al., 2019, 2021). While the potential for using big data as a replacement for other data sources in traditional quantitative approaches may seem unlimited, there are many intriguing methodological considerations (Heikinheimo et al., 2020). In traditional survey terms, the dimensions measured include defining the target population, describing the sampling frame (including geographical scope) and the sample (such as non-probability), as well as identifying the sample size and other metrics (mode, scales and demographics such as age groups). The character of traditional data differs largely from big data (Barton et al., 2021). Such crowdsourced data is typically semi-structured, more difficult to manage, and needs special data management tools (Venter et al., 2020). The fact that datasets are large is not, in itself, that important and has no value unless one can analyse the data successfully and identify trends across many different datasets. In principle, it is the same assessment of validity and reliability for crowdsourced data as in traditional datasets, but it can be much more difficult to determine the data quality (Venter et al., 2021).

Several big data sources and methods that are also appropriate for assessing outdoor recreation are critically dependent on availability, accuracy, cost and privacy protection (Guilarte & Quintáns, 2019; Heikinheimo et al., 2020). Reliable use of big data requires the combination and validation of data sources that can compensate for individual deficiencies. As such, they can often work in tandem with the approaches to foresight outdoor recreation, as described in this chapter. However, efficient use of big data often requires collaboration between sector authorities on investing in method development, data platform solutions, data acquisition, analysis and application to policy, planning and visitor management.

Looking into the crystal ball

The approaches to foresight outdoor recreation presented in this chapter all have their pros and cons. There is no standard way to understand the future and the COVID-19 pandemic has disclosed some of the difficulties in controlling large-scale unforeseen future events that overwhelm the more successive trends that we want to foresight. A brief outlook in that respect shows that participation in nature-based recreation has increased significantly in the Nordic region during the pandemic but that changes are uneven across populations and their drivers multifaceted (Skriver-Hansen et al., 2022; Selvaag et al., 2021; Venter et al., 2020, 2021; Præstholm et al., 2020). Such circumstances call for more insightful approaches to studying the future through foresights. In Table 9.3, we have summarised core features of the methods reviewed for this chapter which may provide further guidance to evaluate the different approaches.

The aim of our study is to present approaches to foresight recreation in nature, to aid better future policy and management decisions. Through this approach, we want to reveal more

Table 9.3 Summary of foresight methods

	Quantitative trend data	*Scale-Scope: Small/detail vs large/overall*	*Evaluation by Experts-Public*	*Level of estimation*	*Pros/cons*
Cohort studies	Yes	Both	Public	Individual	External factors can change with time; Suitable for young generations.
Predictions by the public	No	Both, but mainly large/overall	Public	Individual	Good for short-term forecast and overall patterns; Low response rates.
Predictions by experts	No	Large/overall	Experts	Group	In-depth knowledge on specific topic; Risk of "expert bias".
Experts vs. the public	May be included	Both	Combined	Group	Robust "mixed-method" approach combining qualitative and quantitative data.
Scenario modeling and future tables	May be included	Both, but mainly large/overall	Combined	Group	Flexible method; Can describe alternative futures.
Ethnographic online approaches	No	Large/overall	Public	Individual and Group	Large data at low cost; Difficult to do generalizations.
Crowdsourced data	Yes	Both	Public	Individual and Group	Large dataset at low cost; Concerns over privacy and data regulations.

awareness of the future and show how scholars may make robust predictions, independent of what kind of trends they face. For that reason, we argue in favour of more in-depth methods such as scenario modeling and future tables to deal with alternative trends. Models that mix different methods, such as experts vs. the public, will be more robust in making foresights, especially if they combine quantitative and qualitative data. Cohort studies and predictions by the public or experts alone connect with a single source of information and their predictive power becomes less reliable. Ethnographic online and big-data approaches are often faster and cheaper. They overcome geographical limitation much more easily than most of the other methods presented (which include extensive fieldwork). They may easily include large numbers of people and retroactively trace historic data, as opposed to studying only the current situation. However, sampling error is an issue that may cause uncertainties when generalising to groups outside the sample. Dealing with large amounts of data often requires certain analytical skills (see section on crowdsourced data), which applies more or less to all the methods presented.

Implementing knowledge based on trends also depends on geographical scale and whether it is done at the local, regional, or national level (Yuan & Fredman, 2008). The importance of national trends is indisputable when it concerns serving national strategies and policies (Meld. St. 18, 2016; Naturstyrelsen, 2015; Swedish Government Writ., 2012/13:51; Tyrväinen et al., 2018). At the local level, in, say, a national park or urban-proximate recreational forest, the need for more area-specific knowledge is imperative to increasing visitor satisfaction, reducing user conflicts and managing visitors in a way that has less negative impact on the protection of natural resources (Gundersen et al., 2019). There are also important connections between the geographical levels. For example, national trends regarding outdoor recreational participation are also needed at the local and regional levels to put visitor information in context. For example, how a particular national park is suitable for certain activities with respect to uses that are increasing or decreasing on a national scale.

The Nordic tradition of national surveys and on-site visitor monitoring at outdoor locations has proven effective, but contemporary changes in society and technology are challenging our familiar methods and call for more knowledge of a wide range of methodologies. Researchers must maintain a critical view of the methodology used in trend studies, as many errors are associated with the different methods. Validity (in the sense of consistence and correlation between construct and measurement) is a common challenge for surveys of outdoor activities. This is primarily because the terminology used may be interpreted differently among respondents and because the respondents are participating in multiple activities and have multiple motives for visiting. The target population, geographical scope, and sampling frame varies greatly between surveys and estimates of coverage error, sampling error, non-response error, and adjustment error are rarely reported (Kajala et al., 2007; Fredman et al., 2009). Hence, we recommend researchers continue standardising survey methodologies so that the data are comparable in space and over time. Information based on systematic monitoring is the backbone of forecasting and further developments in foresight. The current situation shows that many methodologies and methods are employed but very few are implemented consistently over time to allow comparison and longitudinal research and thus help identify change. Longitudinal studies often focus on single topics, such as activity participation. This further limits their usefulness and capacity to predict outdoor recreation in a wider sense. Hence, a weak empirical basis to support and inform decision-making calls for alternative approaches to understanding the future of outdoor recreation.

Acknowledgement

This chapter is one result of a long-term collaboration between researchers in Finland, Sweden, Denmark, and Norway. Significant contributions to the progress of this paper have been received from several research agencies; the Magnificent Mountain Research Program funded by the Swedish Environmental Protection Agency, the BIOTOUR project funded by the Norwegian Research Council, Natural Resources Institute Finland, and the Knowledge-based Tourism Development Project funded by the European Regional Development Fund.

References

Alm, S., Palme, J. & Westholm, E. (2012). *Att utforska framtiden: valda perspektiv*. Stockholm: Dialogos.

Ankre, R., Fredman, P. & Lindhagen, A. (2016). Managers' Experiences of Visitor Monitoring in Swedish Outdoor Recreational Areas. *Journal of Outdoor Recreation and Tourism*, 14, 35–40. DOI:10.1016/j.jort.2016.04.008

Ankre, R. & Wall-Reinius, S. (2010). Methodological Perspectives: the Application of Self-registration Cards in the Swedish Coasts and Mountains. *Scandinavian Journal of Hospitality and Tourism*, 10(1), 74–80.

Barton, D.N., Gundersen, V. & Venter, Z.S. (2021). *The use of Big Data in programs for stimulating physical activity – State of the art and suggestions for indicators in Norway*. NINA Report 1937. Norwegian Institute for Nature Research.

Callegaro, M., Baker, R., Bethlehem, J., Göritz, A.S., Krosnick, J.A. & Lavrakas, P.J. (Eds.) (2014). *Online Panel Research. A Data Quality Perspective*. John Wiley & Sons, 477 p.

Casanueva, C., Gallego, Á. & García-Sánchez, M.R. (2016). Current Issues in Method and Practice – Social Network Analysis in Tourism. *Current Issues in Tourism*, 19(12), 1190–1209.

Chen, R.J.C., Bloomfield, P. & Fu J.S. (2003). An Evaluation of Alternative Forecasting Methods to Recreation Visitation. *Journal of Leisure Research*, 35(4), 441–454.

Cuhls, K. (2003). From Forecasting to Foresight Processes—New Participative Foresight Activities in Germany. *Journal of Forecasting*, 22, 93–111.

Czajka, J.L. & Beyler, A. (2016). Background Paper. Declining Response Rates in Federal Surveys: Trends and Implications. *Mathematica Policy Research*, Washington, DC. 54 pp + appendix.

Del Fresno, M. & López-Pelaez, A. (2014). Social Work and Netnography: The Case of Spain and Generic Drugs. *Qualitative Social Work*, 13(1), 85–107.

Dervo, B.K., Skår, M., Köhler, B., Øian, H., Vistad, O.I., Andersen, O. & Gundersen, V. (2014). *Friluftsliv i Norge anno 2014 – status ogutfordringer* - NINA Rapport 1073. 98 s + appendices.

Donohoe, H.M. (2011). A Delphi Toolkit for Ecotourism Research. *Journal of Ecotourism*, 10(1), 1–20.

Dwyer, L., Edwards, D., Mistilis, N., Roman, C. & Scott, N. (2009). Destination and Enterprise Management for a Tourism Future. *Tourism Management*, 30(1), 63–74.

Elmahdy, Y.E., Haukeland, J.V. & Fredman, P. (2017). *Tourism megatrends, a literature review focused on nature-based tourism*. Norwegian University of Life Sciences, MINA fagrapport 42. 74 pp.

European Travel Commission (2021). *Handbook on Tourism Forecasting Methodologies*. ETC Market Intelligence Report, Brussels, Belgium. 97 p.

FAO (2014). *A Glossary of Terms commonly used in Futures Studies*. http://www.fao.org/docs/eims/upload/315951/Glossary%20of%20Terms.pdf [Accessed 10 March 2016).

Fredman, P. & Haukeland, J.V. (Eds.) (2021). *Nordic Perspectives on Nature-based Tourism. From Place-based Resources to Value-added Experiences*. Edward Elgar Publishing, Cheltenham, UK, 328 p.

Fredman, P. & Heberlein, T.A. (2003). Changes in Skiing and Snowmobiling in Swedish Mountains. *Annals of Tourism Research*, 30(2), 485–488.

Fredman, P., Romild, U., Emmelin, L. & Yuan, M. (2009). Non-Compliance with On-Site Data Collection in Outdoor Recreation Monitoring. *Visitor Studies*, 12(2), 164–181.

Fredman, P., Stenseke, M. & Sandell, K. (eds.) (2014). *Friluftsliv i förändring: Studier från svenska upplevelse-landskap* (Outdoor Recreation in Change: Studies from Recreation Landscapes of Sweden). Carlsson bokförlag, Stockholm.

Fredman, P. & Wikström, D. (2015). *Besök och besökare i Fulufjällets nationalpark sommaren 2014 (med jämförelser åren 2001 och 2003)*. Etour Report 2015:5, Östersund, Sweden.

Fredman, P., Wolf-Watz, D., Sandell, K., Wall-Reinius, S., Lexhagen, M., Lundberg, C. & Ankre, R. (2016). *Dagens miljömål och framtidens fjällupplevelser – Iakttagelser av aktivitetsmönster, landskaps- relationer och kommunikationsformer*. Mittuniversitetet, ETOUR, rapport 2016, 3.

Fuchs, M., Höpken, W. & Lexhagen, M (2014). Big Data Analytics for Knowledge Generation in Tourism Destinations – A Case from Sweden. *Journal of Destination Marketing and Management*, 3(4), 198–209.

Gartner, W.C. & Lime, D.W. (2000). *Trends in Outdoor Recreation, Leisure and Tourism*. CABI Publishing. 458 s.

Godtman Kling, K., Fredman, P. & Wall Reinius, S. (2017). Trails for Tourism and Outdoor Recreation: A Systematic Literature Review. *TOURISM*, 65(4), 488–508.

Gössling, S. & Hultman, J. (eds.) (2006). *Ecotourism in Scandinavia: Lessons in Theory and Practice*, CABI, Wallingford.

Guerrero, P., Møller, M.S., Olafsson, A.S. & Snizek, B. (2016). Revealing Cultural Ecosystem Services Through Instagram Images: The Potential of Social Media Volunteered Geographic Information for Urban Green Infrastructure Planning and Governance. *Urban Planning*, 1(2), 1–17. DOI:10.17645/up.v1i2.609

Guilarte, Y.P. & Quintáns, B.D. (2019). Using Big Data to Measure Tourist Sustainability: Myth or Reality? *Sustainability*, 11(20), 5641.

Gundersen, V., Vistad, O.I., Panzacchi, M., Strand, O. & Van Moorter, B. (2019). Large-Scale Segregation of Tourists and Wild Reindeer in Three Norwegian National Parks: Management Implications. *Tourism Management*, 75, 22–33. DOI:10.1016/j.tourman.2019.04.017

Hansen, A.S. (2016). *Understanding recreational landscapes. Developing a knowledge base on outdoor recreation monitoring in Swedish coastal and marine areas*. Ph. D. Thesis. Human Geography Series B, no 127. University of Gothenburg, Department of Economy and Society.

Haukeland, J.V., Fredman, P., Siegrist, D., Tyrväinen, L., Lindberg, K. & Elmahdy, Y.M. (2021). Trends in nature-based tourism. In: Fredman, P. & Haukeland, J.V. (Eds.), *Nordic Perspectives on Nature-based Tourism: From Place-Based Resources To Value-Added Experiences*. Edward Elgar Publishing, Cheltenham, UK, p. 328.

Heikinheimo, V., Tenkanen, H., Bergroth, C., Järv, O., Hiippala, T. & Toivonen, T. (2020). Understanding the Use of Urban Green Spaces from User-Generated Geographic Information. *Landscape and Urban Planning*, 201, 103845.

Henderson, B. & Vikander, N. (eds.) (2007). *Nature First: Outdoor Life the Friluftsliv Way*. Natural Heritage Books, Toronto.

Höpken, W., Eberle, T., Fuchs, M. & Lexhagen, M. (2019). Google Trends Data for Analysing Tourists' Online Search Behaviour and Improving Demand Forecasting: The Case of Åre, Sweden. *Information Technology & Tourism*, 21(1), 45–62.

Höpken, W., Eberle, T., Fuchs, M. & Lexhagen, M. (2021). Improving Tourist Arrival Prediction: A Big Data and Artificial Neural Network Approach. *Journal of Travel Research*, 60(5), 998–1017.

Jacobsen, L.B., Jensen, F.S., Bakhtiari, F. & Thorsen, B.J. (2014). *Friluftslivets nationaløkonomiskefodaftryk*. IFRO Rapport 229. Institut for Fødevare- ogRessourceøkonomi, Københavns Universitet. 51 pp + Appendix.

Jensen, F.S. (2003). *Friluftsliv i 592 skoveog andre naturområder*. Skovbrugsserien nr. 32. Forskningscentret for Skov& Landskab. Hørsholm, Denmark. 335 pp.

Jensen, F.S. & Koch, N.E. (2004). Twenty-five years of forest recreation research in Denmark and its influence on forest policy. *Scandinavian Journal of Forest Research*, 19(Suppl. 4), 93–102.

Kaae, B.C., Olafsson, A.S. & Draux, H. (2018). *Blåt friluftsliv i Danmark*. IGN Rapport 2018. Institut for Geovidenskabog Naturforvaltning, Københavns Universitet. 174 p.

Kajala, L., Almik, A., Dahl, R., Dikšaitė, L, Erkkonen, J., Fredman, P., Jensen, F.S., Karoles, K., Sievänen, T., Skov-Petersen, H., Vistad, O.I. & Wallsten, P. (2007). *Visitor Monitoring in Nature Areas – A Manual Based on Experiences from the Nordic and Baltic Countries* Edited by: Kajala, L., Almik, A., Dahl, R., Dikšaitė, L, Erkkonen, J., Fredman, P., Jensen, F.S., Karoles, K., Sievänen, T., Skov-Petersen, H., Vistad, O.I. & Wallsten, P. Tema Nord 534. Copenhagen: Nordic Council of Ministers, Stockholm.

Kaltenborn, B., Haaland, H. & Sandell, K. (2001). The Public Right of Access – Some Challenges to Sustainable Tourism Development in Scandinavia. *Journal of Sustainable Tourism*, 9, 417–433.

Kelly, J.R. & Warnick, R.B. (1999). *Recreation Trends and Markets: The 21st Century*. Sagamore Publishing, Champagne, IL. 199 p.

Konu, H. (2015). Developing Nature-Based Tourism Products with Customers by Utilising the Delphi Method. *Tourism Management Perspectives*, 14, 42–54.

Kozinets, R.V. (2006). Netnography 2.0. In: Belk, Russell W. (Ed.), *Handbook of Qualitative Research Methods in Marketing*. Edward Elgar Publishing, Cheltenham, UN and Northampton, MA, pp. 129–142.

Kozinets, R.V. (2015). *Netnography: Redefined*. Toronto: Sage Publications Ltd.

Landeta, J. (2006). Current Validity of the Delphi Method in Social Sciences. *Technological Forecasting & Social Change*, 73(5), 467–482.

Légaré, A-M. & Haider, W. (2008). Trend Analysis of Motivation-Based Clusters at the Chilkoot Trail National Historic Site of Canada. *Leisure Sciences*, 30, 158–176.

Manning, R.E. (2011). *Studies in Outdoor Recreation. Search and Research for Satisfaction*. Oregon State University Press.

Meld. St. 18. (2016). *Friluftsliv. Natur som kildetilhelseog livskvalitet*. Meldingtil Stortinget (2015–2016), Den Kongelige Klima- og Miljödepartementet, Oslo, p. 118.

Miljødirektoratet (2015). *Planleggingoggjennomføring av brukerundersøkelser i norske nasjonalparkerogøvrigeverneområder*. Veileder, M-416, 16 p.

Müller, D.K. (2021). 20 Years of Nordic Second-Home Tourism Research: A Review and Future Research Agenda. *Scandinavian Journal of Hospitality and Tourism*, 21(1), 91–101.

Naturstyrelsen, (2015). *Alle tidersfriluftsliv. Redegørelse om Danmarks friluftspolitik*. Miljøministeriet, 36 p.

Neuvonen, M., Riala, M., Nummelin, T., Sievänen, T. & Tuulentie, S. (2018). Future Perspectives on Outdoor Recreation in Finland. *Leisure/Loisir*, 42(4), 365–388. DOI:10.1080/14927713.2019.1581991

Odden, A. (2008). *Hvaskjer med norsk friluftsliv? Enstudieavutviklingstrekkinorskfriluftsliv 1970–2004 [What is happening in Norwegian outdoor recreation: A study of trends in Norwegian outdoor recreation 1970–2004]* (Doctoral dissertation). Trondheim, Norway: Gegrafiskinstitt, NTNU.

Øian, H., Fredman, P. Sandell, K., Sæþórsdóttir, A.D., Tyrväinen, L. & Jensen, F.S. (2018). *Tourism, Nature and Sustainability: A Review of Policy Instruments in the Nordic Countries*. Nordic Council of Ministers, TemaNord, Copenhagen, 534, 99 p.

Olafsson, A.S., Møller, M.S., Mattijssen, T., Gulsrud, N.M., Breman, B. & Buijs, A. (2021). Social Media and Experiences of Nature: Towards a Plurality of Senses of Place. In Di Masso, A., Raymond, C.M., Williams, D.R., Manzo, L.C., & von Wirth, T. (Eds.), *Changing Senses of Place: Navigating Global Challenges*. Cambridge University Press, pp. 271–284.

Pickering, C., Rossi, S.D., Hernando, A. & Barros, A. (2018). Current Knowledge and Future Research Directions for the Monitoring and Management of Visitors in Recreational and Protected Areas. *Journal of Outdoor Recreation and Tourism*, 21, 10–18.

Pouta, E., Neuvonen, M. & Sievänen, T. (2009). Participation in Cross-country Skiing in Finland under Climate Change: Application of Multiple Hierarchy Stratification Perspective. *Journal of Leisure Research*, 41(1), 91–108.

Præstholm, S., Olafsson, A.S., Liu, Y., Gentin, S., Kaae, B.C. & Møller, M.S. (2020). *Udendørsaktiviteter under Covid-19 (forår 2020) – Resultaterfraborgerpanelets online brugerkortlægningafbynatur i Bispebjerg*. Institut for GeovidenskabogNaturforvaltning, Københavns Universitet. IGN Rapport Nr. December 2020.

Pröbstl-Haider, U., Richins, H. & Türk, S. (2019). *Winter Tourism – Trends and Challenges*. CAB International, Wallingford, Oxfordshire. 548 p.

Sandell, K. & Fredman, P. (2010). The Right of Public Access: Opportunity or Obstacle for Nature Tourism in Sweden? *Scandinavian Journal of Hospitality and Tourism*, 10(3), 291–309.

Sandell, K. & Sörlin, S. (2008). *Friluftshistoria – från 'härdandefriluftslif' till ekoturismochmiljöpedagogik: Temani det svenskafriluftslivetshistoria* (The History of Outdoor Recreation – From'the Hardy Outdoor Life' to Ecotourism and Environmental Education: Themes in Swedish Outdoor History). Carlsson bokförlag, Stockholm.

Selvaag, S.K., Wold, L.C., Gundersen, V., Keller, R. & Vistad, O.I. (2021). Domestic Travel in Times of Crisis – What Happened at Nature Attractions During the Covid-Summer of 2020? *Utmark*, 1, 20–36.

Seppälä, Y. (2013). Tulevaisuustaulukkomenetelmä – Sovelluksenavanhustenhuolto. [Future table method – an application for eldercare]. In: Kuusi, O., Bergman, T. & Salminen, H (Eds.), *Mitentutkimmetulevaisuuksia? [How to study futures]*. Acta Futura Fennica 5. Finnish Society for Future Studies, pp. 137–153.

Sharma, R., Ahuja, V. & Alavi, S. (2018). The Future Scope of Netnography and Social Network Analysis in the Field of Marketing. *Journal of internet commerce*, 17(1), 26–45.

Sievänen, T., Arnberger, A., Dehez, J. & Jensen, F.S. (2009). Monitoring of Forest Recreation Demand. In Bell, S., Simpson, M., Tyrvainen, L., Sievänen, T. & Pröbstl, U. (Eds.), *European Forest Recreation and Tourism. A Handbook.* Taylor & Francis Group, London and New York, pp. 105–133.

Sievänen, T. & Neuvonen, M. (2011). *Luonnonvirkistyskäyttö 2010.*[Outdoor recreation 2010]. *Metlantyöraportteja* 212, 190.

Skriver-Hansen, A., Wolf-Watz, D., Fredman, P. & Beery, T. (2022). Outdoor Recreation in Sweden During and after the Covid-19 Pandemic – Management and Policy Implications, *Journal of Environmental Planning and Management,* DOI:10.1080/09640568.2022.2029736

Svarstad, H. (2010). Why Hiking? Rationality and Reflexivity Within Three Categories of Meaning Construction. *Journal of Leisure Research,* 42(1), 91–110, DOI:10.1080/00222216.2010.11950196

Swedish Government Writ. 2012/13:51. *Målförfriluftspolitiken.* Regeringskansliet, Stockholm.

Tyrväinen, L. Sievänen, T., Konu, H., Tuohino, A., Aapala, K. & Ojala, O. (2018). *Mitenkehittääluonnonvirkistys-jamatkailukäyttöäSuomessa? (New ways to develop tourism and recreation on forest and water areas in Finland).* Valtioneuvostonselvitys- jatutkimustoiminta (Publications of the Government's analysis, assessment and research activities). 2018, 2:8. 32 p. http://tietokayttoon.fi/julkaisu?pubid=24101

Vaske, J.J. (2008). *Survey Research and Analysis: Applications in Parks, Recreation, and Human Dimensions.* Venture Publishing, Pennsylvania, USA. 635 pp.

Venter, Z.S., Barton, D., Gundersen, V., Figari, H. & Nowell, M. (2020). Urban nature in a time of crisis: recreational use of green space increases during the COVID-19 outbreak in Oslo, Norway. *Environmental Research Letters.* DOI:10.1088/1748-9326/abb396

Venter, Z.S., Barton, D., Gundersen, V., Figari, H. & Nowell, M. (2021). Back to Nature: Norwegians Sustain Increased Recreational Use of Urban Green Space Months after the COVID-19 Outbreak. *Landscape and Urban Planning..* DOI:10.1016/j.landurbplan.2021.104175

Vilhelmson, B., Elldér, E. & Thuli, E. (2017). What Did We Do When the Internet Wasn't Around? Variation in Free-Time Activities Among Three Young-Adult Cohorts from 1990/1991, 2000/2001, and 2010/2011. *New Media and Society,* 20(8), 2898–2916.

Vistad, O.I. (1995). *I skogen og i skolten - Einanalyse av friluftsliv, miljøoppleving, påverknadogforvaltning i Femundsmarka, med jamføringartil Rogen og Långfjället.* (In the forest and in the head. An analysis of outdoor recreation, recreational experiences, recreational impacts and management in Femundsmarka, compared to Rogen and Långfjället. PhD Thesis, University of Trondheim, Norway.

Vistad, O.I. (2021). Stieniendring? Om det godes paradoks. In Flø, B.E. & Flemseter, F. (Eds.), *Utmark i endring.* Cappelen Damm Akademisk, Oslo.

von Bergner, N.M. & Lohmann, M. (2014). Future Challenges for Global Tourism. *Journal of Travel Research,* 53(4), 420–432.

Wall-Reinius, S. (2010). Researching Tourists in the Outdoors – Challenges and Experiences from Protected Areas in Sweden. In Hall, M.C. (Ed.), *Fieldwork in Tourism: Methods, Issues and Reflections,* London: Routledge, pp. 232–239.

Wall-Reinius, S. & Bäck, L. (2011). Changes in Visitor Demand: Inter-Year Comparisons of Swedish Hikers' Characteristics, Preferences and Experiences. *Scandinavian Journal of Hospitality and Tourism,* 11(Sup1), 38–53.

White, E.M., Bowker, J.M., Askew, A.E., Langner, L.L., Arnold, J.R. & English, D.B.K. (2016). *Federal outdoor recreation trends: effects on economic opportunities.* General Technical Report - Pacific Northwest Research Station, USDA Forest Service 2016 No.PNW-GTR-945.

Wolch, J., Jerrett, M., Reynolds, K., McConnell, R., Chang, R., Dahmann, N., Brady, N., Gilliland, F., Su, J.G. & Berhaned, K. (2011). Childhood Obesity and Proximity to Urban Parks and Recreational Resources: A Longitudinal Cohort Study. *Health Place,* 17(1), 207–214.

Yuan, M. & Fredman, P. (2008). A Call for a Broad Spatial Under-standing of Outdoor Recreation Use. In Raschi, A. & Trampeti, S. (Eds.), *Management for Protection and Sustainable Development. Proceedings of the 4th International Conference on Monitoring and Management of Visitor Flows in Recreational and Protected Areas, MMV.* Montecatini Terme, Italy, 14–19 October, 2008.

10

THE IMPACT OF TRANSPORT INFRASTRUCTURE DEVELOPMENT ON NATURE-BASED TOURISM

The case of the isolated communities of Gilgit-Baltistan, Pakistan

Asif Hussain and Francesc Fusté-Forné

Introduction and context

This chapter explores the importance of physical mobility in terms of access to resources and availability which alters livelihood outcomes by improving resilience and reducing vulnerabilities (Currie et al., 2007; Hussain, 2017; Wachs & Kumagai, 1973). In particular, 'Livelihood Framework for Transport Infrastructure Development and Tourism' (Hussain, 2019) is used to understand the complexity of livelihood systems in terms of policy development. The framework suggests that livelihood capitals are transformed as a result of transport infrastructure development in relation to tourism at both the macro and micro levels. Drawing from a qualitative design which is based on 98 interviews, the chapter focuses on the relationship between transport infrastructure development and nature-based tourism.

Gilgit-Baltistan is a strategically disputed region between India and Pakistan located in the north-east of Pakistan. Until the construction of the Karakoram Highway (1958–1978), the traditional livelihood system of the region was based on subsistence farming and barter trade (Hussain, 2017, 2019). Gilgit-Baltistan is a culturally unique area where oral folktales were used to preserve history; the whole region was like an ethnographic museum and cultural traits were preserved because of the extreme limited exchanges with the rest of the world (Hussain, 2017, 2019). Improved access and mobility due to road infrastructure development has resulted in the arrival of international and domestic tourists impacting the traditional livelihood systems.

This region was geographically isolated until the construction of an all-weather Karakoram Highway between 1958 and 1978. China's Belt and Road Initiative project aims to improve connectivity between 69 Eurasian countries by investing in infrastructure development, including the Karakoram Highway (CPEC-LTP, 2017; DSA, 2017; Hussain, 2019; PwC, 2018). Gilgit-Baltistan was selected to understand the consequences of road infrastructure development. It is located at 35°54'59.99" N 74°17'60.00" E connecting East to West through road networks and it covers a geographic area of 72,496 square kilometres (IFAD, 2016; World

DOI: 10.4324/9781003230748-13

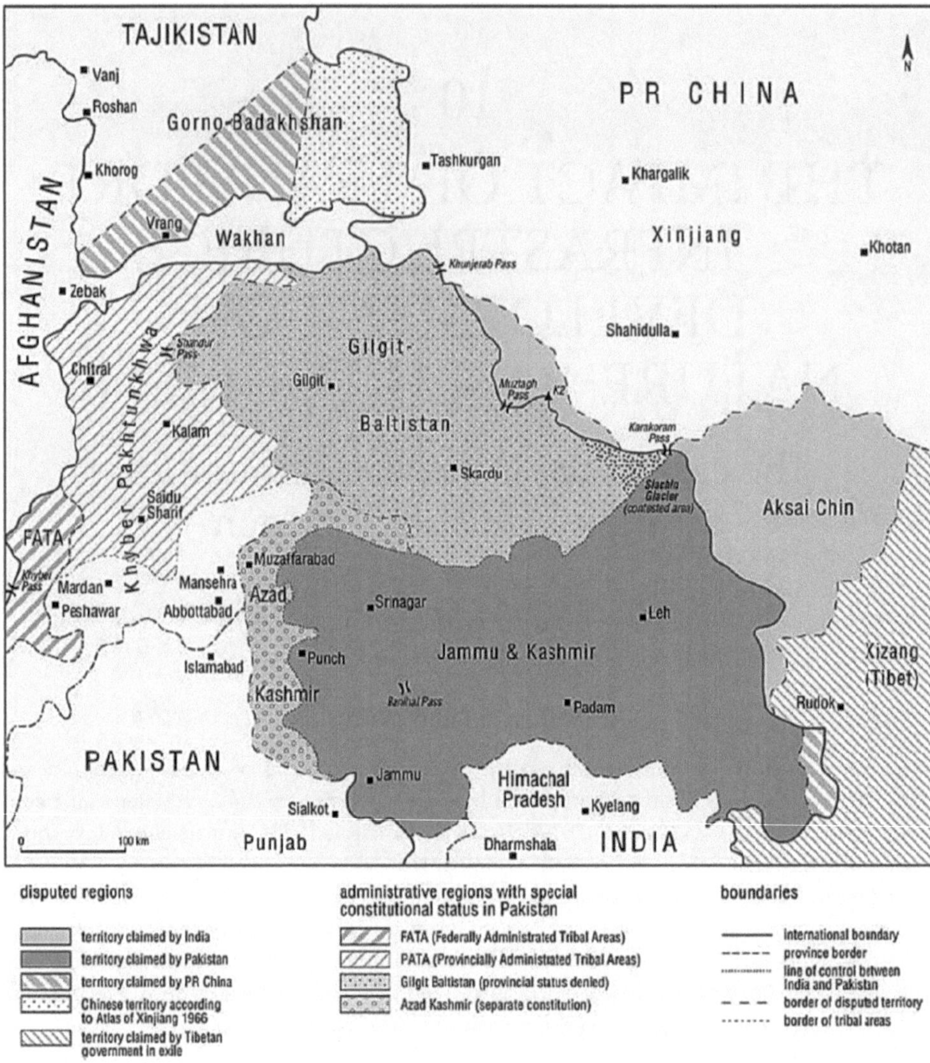

Figure 10.1 Map showing disputed regions in Central Asia including Gilgit-Baltistan.

Bank, 2011) with an estimated population of 1.8 million (Burki, 2015). The region borders the Uyghur autonomous region of Xinjiang in the North-East (China), disputed Jammu and Kashmir to the South and South-East (the Indian-administered part of Kashmir), disputed Azad Jammu and Kashmir to the South (the Pakistan-administered part of Kashmir), Khyber Pakhtunkhwa to the West (province of Pakistan), and the Wakhan corridor to the North-West (part of the Badakhshan province of Afghanistan) as shown in Figure 10.1.

The region of Gilgit-Baltistan is the home for indigenous people with unique characteristics in terms of their history, language and cultural norms (Hussain, 2019). The region of Gilgit-Baltistan acts like a multi-ethnic and diverse ecosystem (Jensen, 2007). The indigenous cultures and activities are centuries-old (Arif, 2001; Dani, 2001). Numerous areological sites in Gilgit-Baltistan point toward Tibetans, Mughals, Dards and Persians (Arora, 1940; Dani, 2001; Murphy, 2011; Tarar, 1991; Wallace, 1996; World Bank, 2011). Recent DNA tests have shown that the haplogroup of the people of Gilgit (*matulfae*[1]) and Chilas (*matulfae*) "settled in the region approximately 30–50,000 years ago respectively and are the ancestors of the indigenous people of North and South America who migrated across the Bering Strait Land Bridge" (Hussain, 2019, pp. 39–40).

People of Gilgit-Baltistan used to practice Tibetan Buddhism and Bon (a form of ancient Tibetan religion) before the advent of Islam around 700 years ago (Hussain, 2019). Traditionally, the indigenous people of Gilgit-Baltistan relied on subsistence farming, livestock and barter trade as the main livelihood (AKRSP, 2007; MacDonald, 1996; World Bank, 2011). Because of the increased tourism inflow, people are diversifying towards tourism (Ali & Benjaminsen, 2004; Hussain et al., 2017; Khan, 2013; Kreutzmann, 1991, 2004, 2006, 2013).

The natural landscape, rich history, heritage and archaeological sites are the main tourist attractions in the area (Hussain, 2019; Kreutzmann, 1991; Owen-Edmunds, 2013; World Bank, 2011). The Great Trigonometric Survey (a project aimed to measure the Indian Subcontinent between 1802–1871 for demarcation purposes) of British India played a critical role in the exploration of the hidden mountain valleys of Gilgit-Baltistan, including an attempt to measure the highest mountains on the Indian Subcontinent (Gill, 2001). The survey led to the attempts to numerous attempts to summit K2 (second highest peak) in 1902 and 1909 led by the Duke of Abruzzi, who declared K2 'unconquerable' (Hussain, 2019). The successful summit of K2 in 1954 acted as a snowball and from that time numerous adventure seekers (mountaineers) began to visit Gilgit-Baltistan (ESCAP, 1993; Jensen, 2007; MacDonald, 1996; World Bank, 2011). Because of the security concerns since 9/11, however, there has been a significant decline in the number of international tourists (AKRSP, 2007; DoTGB, 2014; Hussain et al., 2017; Owen-Edmunds, 2013); however, due to improvements in road infrastructure domestic tourism increased dramatically (Hussain, 2017, 2019).

The geostrategic location of Gilgit-Baltistan attracted road infrastructure development. Chinese President Xin Jinping announced the 'Belt and Road Initiative' in 2013 (FMPRC, 2013); four years later, in 2017, this mega project was embedded into the Chinese constitution (Reuters, 2017). The 'Belt and Road Initiative' aims to create an economic corridor between 69 Eurasian countries by investing in infrastructure development (PwC, 2018). The 'Belt and Road Initiative' is 'The 21st Century Maritime Silk Road', with six corridors (PwC, 2018). An estimated $1.2 to 1.3 trillion will have been spent on the project by 2027, with China acting as an economic pivot (Chatzky & McBride, 2019). The China–Pakistan Economic Corridor, passing through the disputed region of Gilgit-Baltistan, is one aspect of this mega project (CPEC-LTP, 2017; DSA, 2017; Hussain, 2019; Post, 2017). China Pakistan Economic Corridor provides the shortest trade route for China to access the Persian Gulf (Chowdhury, 2013; Hussain, 2019). This shortest route that connects West to East gives a comparative advantage to Gilgit-Baltistan and also visibility as a tourism destination.

Description of the case study area

In understanding tourism systems, connectivity is a key factor for the economic and social development of communities in remote and inaccessible areas. While infrastructure brings hope, it also translates into a change in the local environment that alters the attitudes of residents, for example concerning entrepreneurship (Hussain, 2017, 2019). This is the case of Gilgit-Baltistan, a strategically located, mountainous, and culturally rich disputed region in Central Asia which is administered by Pakistan. The traditional livelihood systems of the indigenous mountain communities were based on subsistence farming (Ehlers & Kreutzmann, 2000; Hussain et al., 2017; MacDonald, 1998; Seong et al., 2009). Gilgit-Baltistan was geographically isolated until the construction of road infrastructure between 1958 and 1978 for geopolitical reasons (Hussain, 2017, 2019). Traditionally, history and knowledge were preserved and passed on in the form of oral folk tales (Hussain, 2019). Sökefeld (1997, p. 83) described this isolated region as an "ethnographic museum where archaic cultural traits are preserved because [of] extremely limited exchange with surrounding countries existed." Indigenous communities relied heavily on local resources and bartered trade (Ehlers & Kreutzmann, 2000; Hussain et al., 2017; Seong et al., 2009). The trade was slow and primarily local in nature because of difficulties in achieving access to and from the region and the seasonal impairment caused by the presence of high-altitude mountain passes (Hussain, 2019). Despite limited interaction with the outside world, the indigenous communities experienced a sustainable lifestyle (Hussain et al., 2017).

The all-weather road infrastructure removed the centuries-old isolated status of indigenous mountain communities as visitors from all over the world came to explore this region (Hussain, 2019; Kreutzmann, 1991). As a result, local communities diversified their livelihood from subsistence farming toward tourism (Khan, 2012; Owen-Edmunds, 2013; World Bank, 2011). Over the course of the next four decades, the traditional subsistence farming has diminished and the communities of Gilgit-Baltistan are experiencing a range of human, social, natural, physical, financial, and institutional changes (Hussain et al., 2017). Gilgit town has become an economic hub and an agent of change at both micro and macro levels (Hussain, 2019). Since the construction of the Karakoram Highway, the population of Gilgit has grown significantly – from 4000 in 1961 (Kreutzmann, 1991) to 216,000 in 1998 (GoP, 2018) – and the indigenous people of Gilgit town "who call themselves '*matulfae*', those who prepare the soil of Gilgit, the sons of soil ... are a minority of Gilgit's population today" (Sökefeld, 1997, p. 84). The new dependency on the monetary economy, as opposed to barter trade, has resulted in livelihood diversification from subsistence farming toward the services sector and people are now dependent on consumer goods imported from outside Gilgit-Baltistan (Pakistan and China). The single road which connects Gilgit-Baltistan with the rest of the world is vulnerable to natural and manmade disasters, which increase the region's vulnerability due to uncertainty in the supply chain (Hussain et al., 2017).

Transport infrastructure development and tourism

Physical mobility, resource allocation, and availability are the fundamental factors that ensure sustainable and resilient livelihoods. The notion of reduced vulnerability is determined by improved income sources, social stability, and protection of natural resource allocation and access, which are fundamental to the improvement of wellbeing (DFID, 1999; Scoones, 2009; Shen, 2009). Guaranteed physical mobility and access to livelihood capitals, such as human, social, natural, physical, financial and institutional, are central for sustained and resilient livelihoods (Currie, Stanley, & Stanley, 2007; Hussain, 2019; Kenyon, Lyons, & Rafferty, 2002; Wachs & Kumagai, 1973). Transport infrastructure such as roads, both directly and indirectly, impact physical mobility and resource allocation by providing access to resources (Estache, 2008; Olsson, 2009; Peters, 2000). Road infrastructure reduces logistic costs, results in increased regional and national integration (Hussain, 2019; Khadaroo & Seetanah, 2008; Prideaux, 2000), facilitates the flow of goods, and generates income opportunities (Bourguignon & Pleskovic, 2008; Ghosh & De, 1998; Kumaraswamy & Zhang, 2001).

The response of individuals and the community's response to livelihood shocks and stress shape livelihood outcomes (Ellis, 1998, 2000; Hussain, 2017, 2019; Mbaiwa & Sakuze, 2009; Wang, Zhang, & Liu, 2010). Geographically isolated communities rely fundamentally on cultural memory, traditions, and belief systems to ensure sustainable and resilient livelihood systems characterised by short-term and long-term survival strategies (Béné et al., 2014; Bradtmöller, Grimm, & Riel-Salvatore, 2017; Lew, 2014; Rosen et al., 2012). Indigenous nations and cultures across the globe used collaborative, joint planning, and a diverse skillset that enabled resilient crisis management. It is argued that sustainability resists change whereas resilience embraces change and adapts proactively (Derissen, Quaas, & Baumgärtner, 2011; Espiner et al., 2017; Strickland-Munro, Allison, & Moore, 2010).

Transport infrastructure is an important component of development (Ghosh & De, 1998). It was concluded in *The Annual Bank Conference on Development Economics* (2007) that reduction in transport logistic costs is fundamental to securing sustainable economic development (Bourguignon & Pleskovic, 2008). The World Bank estimated that the "poorest countries need to spend 9 percent of their GDP on the operation, maintenance, and expansion of their infrastructure to reach Millennium Development Goals" (Estache, 2008, p. 47). This is because transport infrastructure increases investment, facilitates the flow of goods and services, and generates employment and income (Ghosh & De, 1998; Kumaraswamy & Zhang, 2001). It is argued that the economic return on transport infrastructure such as roads is up to 200% (Canning & Bennathan, 2000; Estache, Briceno, & Shafik, 2004) because roads provide local solutions and facilitates livelihood outcomes by ensuring mobility (Estache, 2008). Mobility, in turn, provides access to resources such as market, education, employment, and healthcare (Church, Frost, & Sullivan, 2000; Currie et al., 2007; Hussain et al., 2017; Kenyon et al., 2002). Road infrastructure cannot be treated in isolation (Doorn, 1986; Prideaux, 1993) as the elements of the transport system, such as operations, modes and interactions of transport infrastructure, are interlinked (Prideaux, 2000).

Tourism influences and shapes livelihood strategies in a destination (Ashley, 2000; Cohen & Cohen, 2012; Hussain, 2019; Tao & Wall, 2009). Tourists must travel to the destination to consume the tourism product (Benson & Whitehead, 1985; Leiper, 1992). The tourism system comprises tourists, generating regions, transit routes, destination regions, and the tourism industry (Leiper, 1992, 2004). The transport infrastructure is a significant component of tourism systems as it plays a critical role in connecting tourists from tourist-generating regions to a destination (Force, 2003; Khadaroo & Seetanah, 2008; Tang & Rochananond, 1990). The transport system is the foundation for tourism development, and transport infrastructure is the precondition for the travel industry (Prideaux, 2000). The production and consumption of tourism happen in a destination to create tourism experiences. Therefore, destination development provides the foundation for domestic and international tourists (Prideaux, 2000). The tourism industry is involved not only with the movement of people but also with significant impacts on human, social, natural, physical, financial, and institutional capitals (Hussain, 2019). While transport infrastructure development improves mobility it also poses the risk of the spread of diseases such as COVID-19. It is evident from the spread of COVID-19 that human mobility is the main cause for the spreading of the virus (Hussain et al., 2021; Zheng et al., 2021). Because of the present-day rapid mobility, the velocity of spread of such viruses has increased and this has significantly impacted global mobility and tourism (Gössling et al., 2020; Lapointe, 2020; Sigala, 2020; UNWTO, 2020). As a consequence, recent debates have also discussed the relevance of sustainable mobilities in post-pandemic tourism recovery (Fusté-Forné & Michael, 2021).

Methods

Qualitative research methods such as in-depth interviews were used to collect empirical data (Bloch, 1996; Glaser, 1992; Jennings, 2005) between December 2015 and April 2016 in

Gilgit-Baltistan. Interviews are the preferred tools for this study because of their conversational mood where researchers remain open-minded in terms of structure and direction of the interview (Glaser & Strauss, 1967; Heron, 1981; Riley, 1996). Gatekeepers in the case study area were identified before entering the field and helped locate potential participants, building trust and relationships to collect rich information (Hussain, 2019; Jennings, 2005; Rogers, 1980). The snowball sampling technique was used to select participants based on their direct and indirect involvement with tourism (Cooper & Schindler, 2008; Tashakkori & Teddlie, 1998). In total, a sample of 98 participants were interviewed in Urdu, Shina, Balti, and English. The interviews were translated into English by the first author. The average interview duration was between 45 and 60 minutes. All interviews were recorded with the help of two recording devices. The primary themes of the interview included livelihood assets, livelihood activities, vulnerabilities, institutional arrangements, transport infrastructure development, tourism, and livelihood outcomes. Interviews were transcribed into English and classified based on themes, open codes, and focus codes. Similar arguments from multiple participants were attributed in the bracket. The data were analysed using qualitative data analysis software, MAXQDA version 12, and manual thematic analysis. Participant names were coded.

Results

Access to Gilgit-Baltistan and Tourist Visitation

The strategic geographical location of Gilgit-Baltistan influenced transport infrastructure development in the region. Access to Gilgit-Baltistan was improved in multiple phases, starting from improvement in access via the formation of mule tracks, airport development and via roading infrastructure development within Gilgit-Baltistan and with the outside world. It was the improvement in roading infrastructure development (1958–1978) that led to a significant increase in the number of tourists. Before road infrastructure development, the majority of the tourists were international adventure seekers (P85, P84, P89, P84, P58, P13, P85, P12). According to the participants, the peak tourist inflow (international) was between 1985 and the late 1990s. There are no reliable official statistics; however, the trend was expressed in a conference (Hussain, 2016) and in the *Economic Development Report* by the World Bank (World Bank, 2011). Before road infrastructure development (the Karakoram Highway), the majority of the tourists who visited Gilgit-Baltistan were mountaineers as shown in Figure 10.2. Participants unanimously argue the strong correlation between road infrastructure development and tourism inflow (Hussain, 2019).

In this sense, after the opening of the Karakoram Highway in the 1980s, around 150 mountain expeditions used to visit Gilgit-Baltistan as compared to the time before the road infrastructure (P85, P1, P6, P2, P7,P69, P74, P70, P78, P71, P30, P24, P37, P34, P45, P65, P33, P39). An increase in tourism led to the creation of many tourism-related jobs and gradually tourism become the main source of livelihood of those villages which offer mountain expeditions, especially in the Braldu valley in Skardu and the Hushe valley in Ghanche district. This phenomenon was reported by several participants (P95, P90, P85, P88, P90, P98, P51, P55, P59, P65, P44, P36, P46, P39, P41, P42, P50). The labour market opportunities and monetary incentives attracted people toward tourism and "many made tourism a profession by specialising in various roles, such as high-altitude portering, guiding, tour operating, hotel and restaurant operating" (Hussain, 2019, pp. 140–141).

Livelihood diversification strategies in Gilgit-Baltistan are risk-minimising strategies that often revolve around the resources available at the time (P98, P88, P85, P86, P90, P92, P62,

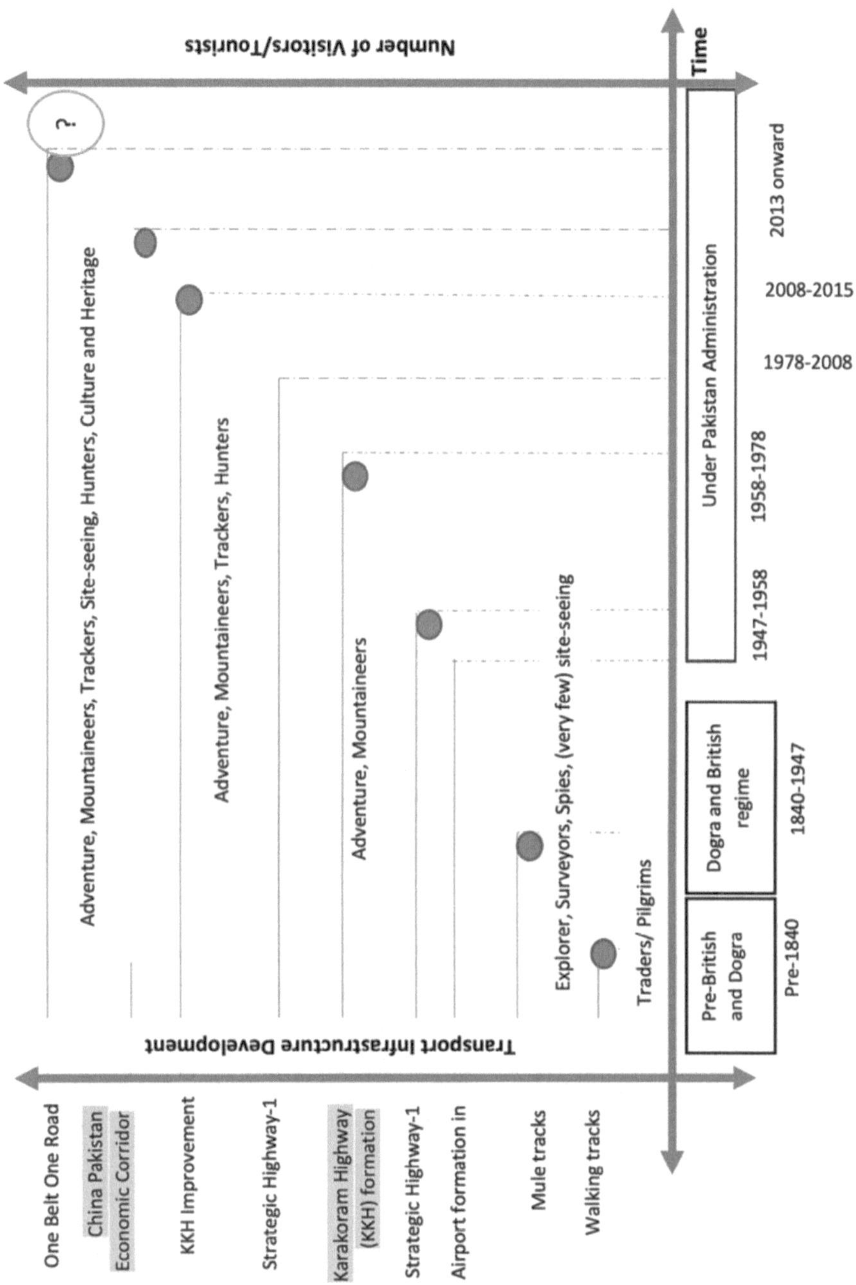

Figure 10.2 Improvement in access to Gilgit-Baltistan throughout history and visitors/tourist's inflow.

Source: Hussain (2019, p. 165).

P12, P56, P81, P96). Because of road infrastructure development, people, at both the macro and the micro levels, have diversified from farm to off-farm-related livelihood activities (P1, P7, P3, P8, P30). Off-farm livelihood strategies are mainly concentrated in the tourism sector (P87, P88, P90, P12, P92, P94, P12, P93). Participants argue that in some places almost 90% of people rely directly on tourism-related livelihood activities (P83, P84, P85, P86, P87, P88, P90, P91, P92,P93, P94, P95, P96). For example,

> people got involved with tourism because of poverty, lack of employment opportunities and the benefit of instant payments by tourists ... months of hard work in the field [subsistence farming] were replaced with a few hours of a labouring job for tourists [mainly as porters], which would enable them to buy groceries from the market... opportunities provided by transport infrastructure development and tourism led to an introduction of a monetary economy as opposed to barter trade in the past.
>
> (Hussain, 2019, p. 141)

Post-road livelihood diversification strategies

The improvement in road infrastructure development and the opportunities offered by tourism attracted numerous people to diversify their livelihood toward tourism. Today, livelihood activities in the region of Gilgit-Baltistan constitute tourism, non-tourism, farm and non-farm-related activities, as shown in Figure 10.3. The main livelihood of the people of the Braldu valley in Skardu and the Hushe valley of Ghanche (study valleys) mainly relied on tourism as a livelihood activity. It was reported that

> tourism-related diversification has disintegrated the traditional livelihood practices. People now wait for the tourist season and do not perform livelihood chores as before (P84, P89, P94, P98). Others argued that because of easy access and availability of food items in the market people did not work hard in the fields nor develop barren land for agriculture (P83, P24, P81). This showed an increased dependency on tourism which has increased the workload of women, children, and the elderlies as they have to perform household chores, farm-related activities (watering the field) and looking after livestock while male members were busy with tourists (P98, P90). After the roading infrastructure and an increase in tourism numbers, the trend of keeping livestock has also changed. According to household heads, people used to keep a lot of livestock, but now the trend is a decline (P84, P86, P83, P89, P90, P91, P96). For instance, in Hushe participants claimed that because of tourism they had lost almost fifty percent of their pasture land as they cannot send their female members and children to pastureland in summer (which was a tradition) (P88, P96, P87). This was because local people cannot trust tourists, porters, and outsiders and there have been many incidents where people have drunk milk off their animals, stolen vegetables, burned firewood and so on (P96, P83).
>
> (Hussain, 2019, p. 151)

This stage is similar to the consolidation stage of Tourism Life Cycle (Buttler, 1980) as the discontent from tourism activities is evident amongst locals.

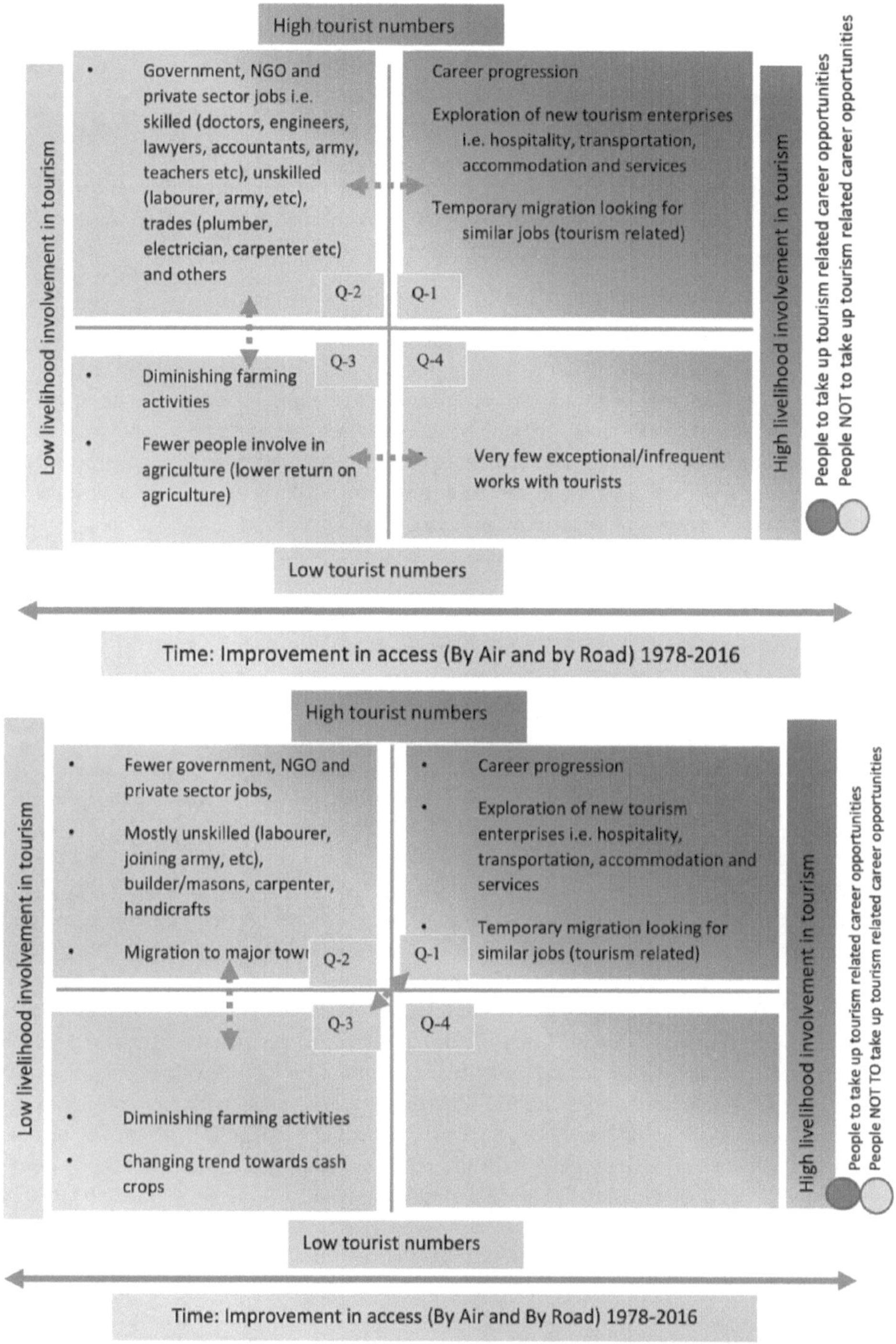

Figure 10.3 Post-Road Livelihood Diversification Strategies. 1) At Macro (above) 2) At Micro Levels (below).

Source: Hussain (2019, p. 173).

It was also reported in Hushe valley that animals have become feeble because of intense grazing in one area (Hushe). However, this was not reported in other valleys, because their pastureland was far away from tourist tracks. Infect people have bought more livestock because of the income from the tourists. A community leader explained that

> tourism-related activities have not affected our pastureland as the route to K2 is not via our pasture… infect, because of the income from tourism people have bought more livestock and the number of livestock-keeping has increased, and people often sell livestock to tourists and in the market in Skardu.
>
> (Hussain, 2019, p. 152)

The improvement in roading infrastructure increased both skilled, unskilled and services offered by both government and the private sector. It was noticed that because of road infrastructure development and tourist flow, subsistence farming activities have diminished at macro levels (Gilgit-Baltistan levels) because of a diminishing rate of return, as shown in Figure 10.3. Today, the services and tourism sector is considered as the main source of livelihood as opposed to subsistence farming in the past (pre-road). "Many of the people involved in agriculture have diversified their farming practices from cereal to cash crops such as potatoes and fruit" (Hussain, 2019, p. 171). The change was significant in places such as Gilgit, Skardu, Shigar, and Ghanche.

> Urban centres/towns are expanding, and farmland is being used to construct houses, markets and plazas. A similar trend is seen in rural areas also, as the population grows, more houses are being built, and the tradition of land division among siblings has reduced the total amount of landholding per person. Even though barren land is being developed with the help of machinery, population growth and immigration have increased dramatically and [the] land developed for agriculture is under increasing pressure.
>
> (Hussain, 2019, p. 171)

Because of the available livelihood opportunities, people move between different quadrants as shown in Figure 10.3. The livelihood choices were analysed in reference to tourism activities in relation to improvement in access, both by road and by air at the macro and micro levels. At macro (Gilgit-Baltistan) levels, people earlier affiliated with subsistence farming diversify toward tourism, non-tourism and non-farm-related activities. At the macro levels, people moved from Q3 (traditional subsistence farming) to Q2 (specialist services), Q1 (frequent and specialised tourism) and Q4 (infrequent involvement with tourism), as shown in Figure 10.3. "Because of growth in the tourism market in Gilgit-Baltistan, the number of people involved only occasionally with the tourism industry has increased (Q4). Likely, these people will either move to Q1 or Q2" (Hussain, 2019, p. 171). In relation to farming, at the macro levels, farming activities are diminishing because of lower return on and there is a trend toward cash crops. Similar trends were also seen at the micro levels, with heavy reliance on tourism. At micro levels, the movement of people between the quadrants was different as opposed to macro levels. People either moved from Q3 (traditional subsistence farming) to Q2 (specialist services) or directly from Q3 to Q1 (frequent and specialised tourism).

Nature-based tourism development and the changing visitor profile in Gilgit-Baltistan

Roading infrastructure has improved access to Gilgit-Baltistan and resulted in tourism development (P10, P82, P49, P30, P3, P20, P81, P51, P58. P9, P1, P3, P43). "It was reported that eighty-seven percent of the tourists accessed Gilgit-Baltistan via the Karakoram Highway" (Hussain, 2019, p. 138). Despite an increase in tourism in Gilgit-Baltistan, there has not been proper planning to promote and manage tourism and the investment in tourism has been risky because of both manmade and natural disasters. To date, whatever tourism development has happened in Gilgit-Baltistan is because of the efforts of the private sector (P5, P50, P73, P20, P45, P14, P20, P56, P21). Tourism investment is sceptical as in the past numerous tourism investors have gone bankrupt and there is no security in tourism investment (P7, P30, P5, P69). The other factor which influenced investors is seasonality (P5, P7, P30, P45, P48, P8). Participants argue that tourism is not on the priority list of the government and there is no tourism policy (P28, P30, P29, P48, P45, P20, P12, P73, P50, P45, P20, P44, P81). However, when asked, participants argue that a tourism policy was important (P5, P20, P50, P21, P45).

It was noticed vividly, both through observations and expressions of the participants that there has been a significant change in the tourist profile from international to domestic tourist. "Participants reasoned that it was primarily due to the security situation in Pakistan after 9/11 tightening visa restrictions for Pakistani's international travel creating increased inward visitation of overseas Pakistani's (P50, P37)" (Hussain, 2019, p. 145). While others claimed that it was because of the political reforms of Gilgit-Baltistan in 2009 and reassigning of the name 'Gilgit-Baltistan' instead of 'Northern Areas', which was often affiliated and misunderstood with the troublesome tribal areas. Additionally, promotional coverage of Gilgit-Baltistan on national media (print and electronic) (P24, P50, P5) increased visibility. It was argued that for the first time in the history of Gilgit-Baltistan the disputed political status of Gilgit-Baltistan was highlighted on national media, and a political representative from Gilgit-Baltistan shared views on live TV talk shows and encouraged people to visit Gilgit-Baltistan (P77, P70, P24, P50, P5).

The popularity of Gilgit-Baltistan in the national media of Pakistan can also be credited to several mountaineers from Gilgit-Baltistan. For instance,

> in 2011 Hassan Sadpara a high-altitude porter and mountaineer from Gilgit-Baltistan became the first [Pakistani] to climb six 8,000-meter-high peaks including Mount Everest, without supplementary oxygen. Later in 2013, Samina Khayal Baig a mountaineer from Gilgit-Baltistan became the first and the only [Pakistani] female to climb Mount Everest and following the scaling of the seven highest mountains on seven continents for the cause of gender equality and women empowerment (P50, P3, P5). These successes were cherished at the national and international level (in print, electronic and social media) followed by Baig's appearance in many commercial ads on national TV. Tourism enterprises made documentaries to promote Gilgit-Baltistan with the help of social media to create awareness at the national level (P24, P50, P5). A private tour operator (P50) argued that the increase in domestic tourism was because of the increase in awareness about Gilgit-Baltistan as a tourism destination at the national level. She reasoned: I think there was no awareness and people did not know that our country has such beautiful places as well. People used to go overseas on holiday, e.g. America, London, and France. When we started mountaineering,

> we started making small documentaries and used to post on social media which created awareness, and they could see that they too can-do mountaineering and hiking.
> (Hussain, 2019, p. 145)

Discussion

Mobilities impact on livelihoods

The Karakorum Highway has resulted in profound changes in Gilgit-Baltistan. Formerly isolated and insular communities have now become part of the outside world. Results show that there have been profound impacts, both positive and negative, on livelihoods. People feel socially connected to the rest of the world due to connectivity (air and road) and advances in technology. In this regard, the strategic location of Gilgit-Baltistan has been the driving force in attracting numerous national and international stakeholders to influence the access, mobility, and institutional arrangement in the region. For this reason, sustainable livelihood needed to be understood from a geographical context by including location as a meta-capital. In this respect, the resilience factor has become critical, as livelihood diversification strategies are often called into question by natural and manmade disasters. In the context of nature-based tourism, it is concluded that livelihood outcomes must integrate community capacity and preserve cultural memories and traditions to make these communities long-term sustainable and resilient. Abandoning traditional practices will translate into a significant loss of knowledge that has allowed these communities to survive for thousands of years.

Livelihood diversification activities in Gilgit-Baltistan are risk-minimising strategies to gain desired outcomes. People at micro levels rely heavily on tourism-related activities. The intense diverse livelihood activities since roading infrastructure development have resulted in a significant change in resource availability at local and regional levels. Improvement in access to Gilgit-Baltistan has resulted in a significant change to livelihood capitals. However, because of ongoing mega transport infrastructure development, livelihood capitals, and statuses, are still in transition. The disputed status of Gilgit-Baltistan and mega-infrastructure development results in access to resources, thereby influenced by institutional arrangements at regional levels. The complex and volatile political structure suggests that the change is influenced from both outside and within Gilgit-Baltistan affecting livelihood activities. People affiliation with tourism, and seeing tourism as a significant livelihood diversification strategy, is a prime example. However, the tourism industry itself and the access to Gilgit-Baltistan via a singular roading infrastructure, the Karakoram Highway, are both vulnerable to natural and manmade disasters.

The results of tourism development

The improvement in access to Gilgit-Baltistan has acted as a vector of change in terms of resources access, status, and availability. The introduction of a monetary economy (tourism), as opposed to barter trade, has made people vulnerable in terms of food security as people are moving away from traditional subsistence farming. People now rely heavily on supplies from outside Gilgit-Baltistan. Improvement in access to markets in Pakistan and China has resulted in people having deviated from cereal to cash crops to earn money which can then be used to buy cheaper goods and services as required. The exogenous shocks alter livelihood capitals frequently and the notion of sustainability and security of livelihood strategies remain in limbo. Livelihood vulnerabilities keep changing in response to changing institutional arrangements, at both macro and micro levels. This is mainly because of the disputed status of Gilgit-Baltistan and the sense of security is heavily reliant on the political status of Gilgit-Baltistan and the regional security issues and conflicts. This is why the inclusion of location as a "Meta Capital" is vital to the access and availability of livelihood capitals.

Roading infrastructure is the subject of change from household (micro) to national and international (macro) levels. Transport infrastructure has connected macro with micro and vice versa and influenced the forms and statuses of livelihood capitals. The infrastructure has changed locals' lifestyles. Locals look at the benefits of tourism in general and nature-based tourism in particular. The development of a transit route between China and the Arab Sea across Pakistan opened up the region and led people to the development of new economic activities. In this context, nature-based travel experiences in the mountains are increasingly becoming a 'must-visit' attraction. The traditional livelihood activities provided resilience to indigenous mountain communities for centuries; however, the fast-changing transport infrastructure development has resulted in a swift change in livelihood strategies. The current tourism-related livelihood strategies are neither sustainable nor resilient as tourism has proven to be a fragile industry that is impacted and influenced by changing institutional arrangements. In a disputed territory, tourism cannot be seen as a sustainable source of livelihood. Of numerous changes, the dependency on a monetary economy is the major challenge to sustainable and resilient livelihood status, in both the short run and the long run. While physical mobility has been possible because of the formation of roading infrastructure, however, the very road infrastructure is acting as an oxymoron. This singular roading infrastructure acts as a lifeline for the people of Gilgit-Baltistan and is the major impediment to sustainability and resilience due to volatile manmade and natural disasters.

Tourist area life cycle

The stages of tourist area life cycle are exploration, involvement, development, consolidation, stagnation, and post-stagnation. In this study, before road infrastructure development a few adventurous tourists visited Gilgit-Baltistan as the access was limited. The main purpose of tourist attraction was mountaineering. At this stage, a very limited number of people were involved with tourism-related businesses. However, the improvement in transport infrastructure resulted in a progressive increase in the number of tourist visitation, resulting in the increase (limited) in tourist-related livelihood activities (as shown in Figure 10.3). The road infrastructure development provided a constant inflow of tourists, which resulted in noticeable tourism development where tourist destinations experienced an increase in foreign investment. Gilgit-Baltistan, at this stage, is in the development stage in reference to tourist area life cycle.

Although, it is not yet apparent, there will be some antagonism toward tourists in certain popular tourism destinations and then it will spread across the whole of Gilgit-Baltistan. This has been noticed in some of the top tourist destinations in Pakistan, such as Murree (Dawn, 2018; Yousaf et al., 2021). Once the tourist number reaches its peak, which often leads to a situation in which tourists outnumber local residents, this would result in a decrease in tourist numbers and the destination will enter the consolidation stage, where tourism becomes the major component of the local economy. The tourist life cycle suggests that, at this stage, the destination will try to extend the tourist season which will lead to the stagnation stage. At the stagnation stage a destination will have become well established as a tourist destination; however, it will no longer have been popular. The post-stagnation period would depend on how tourism stakeholders manage a destination and position themselves to explore new markets, ranging from rejuvenation to decline. The outcome will depend on destination planning and management, including the exploration of new experiences.

Conclusion

This chapter has revealed the conflictive relationships between transport infrastructure development and nature-based tourism. While transport infrastructure development has opened up the region of Gilgit-Baltistan to the world, and has also brought tourists to the region, it has had a substantial impact on local communities. In this sense, locals have shifted their business operations to tourism, and tourism, in turn, has challenged traditional livelihoods. A responsible management and marketing of tourism must also protect and promote local heritage and traditions. The main limitation of this research is its qualitative design and also that it relies on a single location. In this sense, Gigit-Baltistan is a patriarchal society. Of the 98 participants in this study there were only three female participants. However, very limited research has been conducted in the rural areas of Gilgit-Baltistan, which results in a unique contribution that sheds light on the understanding of the development of transport infrastructure and its impact on people, places, and practices in the context of nature-based tourism in isolated communities.

Note

1 A term used by the indigenous people to describe the first settlers who developed the barren land and settled permanently in a particular region.

References

AKRSP. (2007). *An assessment of socio-economic trends and impact in the northern areas and Chitral, Pakistan*. Aga Khan Rural Support Programme.

Ali, J., & Benjaminsen, T. A. (2004). Fuelwood, timber and deforestation in the himalayas: The case of basho valley, Baltistan. *Mountain Research and Development, 24*(4), 312–318.

Arif, M. (2001). *Study of petroglyphs of Buddhist period along the Silk Road between Shatial and Khunjerab Pass, Northern Areas, Pakistan*. PhD Dissertation, University of Karachi, Pakistan.

Arora, R. (1940). *In the Land of Kashmir, Ladakh & Gilgit*. Unique Literature Publishing House.

Ashley, C. (2000). *The impacts of tourism on rural livelihoods: Namibia's experience*. Overseas Development Institute.

Béné, C., Newsham, A., Davies, M., Ulrichs, M., & Godfrey-Wood, R. (2014). Resilience, poverty and development. *Journal of International Development, 26*(5), 598–623.

Benson, D., & Whitehead, G. (1985). *Transport and distribution*. Longman.

Bloch, C. (1996). Emotions and discourse. *Text-Interdisciplinary Journal for the Study of Discourse, 16*(3), 323–342.

Bourguignon, F., & Pleskovic, B. (2008). *Rethinking infrastructure for development*. World Bank Publications.

Bradtmöller, M., Grimm, S., & Riel-Salvatore, J. (2017). Resilience theory in archaeological practice–An annotated review. *Quaternary International, 446*, 3–16.

Burki, S. J. (2015). *Historical dictionary of Pakistan*. Rowman and Littlefield.

Butler, R. W. (1980). The concept of tourism area cycle of evolution: Implications for management of resources. *Canadian Geographer, 24*(1), 5–12.

Canning, D., & Bennathan, E. (2000). The social rate of return on infrastructure investments. *World Bank Policy Research Working Paper, 2390.*

Chatzky, A., & McBride, J. (2019). China's massive belt and road initiative. Retrieved from https://www.cfr.org/backgrounder/chinas-massive-belt-and-road-initiative

Chowdhury, D. R. (2013). Pakistan happy to aid in China's quest for land route to the west; India, not so much: Overland access to Middle East oil and markets for western provinces sought. Retrieved from http://www.scmp.com/business/commodities/article/1359761/pakistan-happy-aid-chinas-quest-land-route-west-india-not-so

Church, A., Frost, M., & Sullivan, K. (2000). Transport and social exclusion in London. *Transport Policy, 7*(3), 195–205.

Cohen, E., & Cohen, S. A. (2012). Current sociological theories and issues in tourism. *Annals of Tourism Research, 39*(4), 2177–2202.

Cooper, D. R., & Schindler, P. S. (2008). *Business research methods*. McGraw-Hill.
CPEC-LTP. (2017). *Long Term Plan for China-Pakistan Economic Corridor (2017–2030)*. Retrieved from http://cpec.gov.pk/brain/public/uploads/documents/CPEC-LTP.pdf
Currie, G., Stanley, J., & Stanley, J. (2007). *No way to go: Transport and social disadvantage in Australian Communities Melbourne*. Monash University Press.
Dani, A. H. (2001). *History of northern areas of Pakistan upto 2000 A.D.* Sang-e-meel Publications.
Dawn (2018). 'Boycott Murree' drive launched on social media. Retrieved from https://www.dawn.com/news/1405056
Derissen, S., Quaas, M. F., & Baumgärtner, S. (2011). The relationship between resilience and sustainability of ecological-economic systems. *Ecological Economics*, *70*(6), 1121–1128.
DFID. (1999). *Sustainable livelihood guidance sheet*. Department for International Development, United Kingdom.
Doorn, J. W. V. (1986). Scenario writing: A method for long-term tourism forecasting? *Tourism Management*, 7(1), 33–49.
DoTGB. (2014). Tourism. *Visit Gilgit-Baltistan, Jewel of Pakistan*. Retrieved from http://www.visitgilgitbaltistan.gov.pk/index.php?option=com_content&view=article&id=305&Itemid=271#
DSA. (2017). Understanding the China-Pakistan economic corridor. Retrieved from https://www.silkroadbriefing.com/news/2017/04/03/understanding-the-china-pakistan-economic-corridor
Ehlers, E., & Kreutzmann, H. (2000). *High mountain pastoralism in Northern Pakistan*. Franz Steiner Verlag.
Ellis, F. (1998). Household strategies and rural livelihood diversification. *Journal of Development Studies*, *35*(1), 1–38.
Ellis, F. (2000). *Rural livelihoods and diversity in developing countries*. Oxford University Press.
ESCAP. (1993). *Environmental impact of tourism in the Northern Areas of Pakistan* (ST/ESCAP/1152). United Nations.
Espiner, S., Orchiston, C., & Higham, J. (2017). Resilience and sustainability: A complementary relationship? Towards a practical conceptual model for the sustainability–resilience nexus in tourism. *Journal of Sustainable Tourism*, *25*(10), 1385–1400.
Estache, A. (2008). Infrastructure and development: A survey of recent and upcoming issues. In F. Bourguignon & B. Pleskovic (Eds.), *Rethinking infrastructure for development* (Vol. 2, pp. 47–88). World Bank Publications.
Estache, A., Briceno, C., & Shafik, N. (2004). Infrastructure services in developing countries: Access, quality, costs, and policy reform. *Policy Research Working Paper, 3468.*
FMPRC. (2013). President Xi Jinping delivers important speech and proposes to build a silk road economic belt with Central Asian countries. Retrieved from https://www.fmprc.gov.cn/mfa_eng/topics_665678/xjpfwzysiesgjtfhshzzfh_665686/t1076334.shtml
Force, T. T. (2003). *Down the track; better ways to deliver tourism's land transport infrastructure*. Tourism Task Force Australia.
Fusté-Forné, F., & Michael, N. (2021). Limited tourism: Travel bubbles for a sustainable future. *Journal of Sustainable Tourism*, 1–18.
Ghosh, B., & De, P. (1998). Role of infrastructure in regional development: A study over the plan period. *Economic and Political Weekly*, *33*(47/48), 3039–3048.
Gill, B. (2001). The big man: Surveying sir george everest. *Professional Surveyor Magazine, 21.*
Glaser, B. G. (1992). *Emergence vs forcing: Basics of grounded theory analysis*. Sociology Press.
Glaser, B. G., & Strauss, A. L. (1967). *The discovery of grounded theory: Strategies for qualitative research*. Aldine Transaction.
GoP. (2018). *Gilgit-Baltistan*. Retrieved from: http://www.citypopulation.de/GilgitBaltistan.html
Gössling, S., Scott, D., & Hall, C. M. (2020). Pandemics, tourism and global change: A rapid assessment of COVID-19. *Journal of Sustainable Tourism*, *29*(1), 1–20.
Heron, J. (1981). Philosophical basis for a new paradigm. In Peter Reason & John Rowan (Eds.), *Human inquiry: A sourcebook of new paradigm research* (pp. 19–36).
Hussain, A. (2017). *Transport infrastructural development and livelihood: A case of adaptation and diversification strategies in the face of tourism*. Paper presented at the CAUTHE 2017: Time For Big Ideas? Re-thinking The Field For Tomorrow.
Hussain, A. (2019). *Transport infrastructure development, tourism and livelihood strategies: An analysis of isolated communities of Gilgit-Baltistan, Pakistan*. Doctor of Philosophy, Lincoln University New Zealand.
Hussain, A., Fisher, D., & Espiner, S. (2017). Transport infrastructure and social inclusion: A case study of tourism in the region of Gilgit-Baltistan. *Social Inclusion*, *5*(4), 196–208.

Hussain, A., Fusté-Forné, F., & Simmons, D. (2021). Fear of pandemics or fear of tourism: The challenges for human mobility. *Tourism and Hospitality Management, 27*(1), 223–228.

Hussain, K. (2016). *Experience and status of tourism in Gilgit-Baltistan*. Tourism, Youth Affairs, Sports and Culture, Department Government of Gilgit Baltistan. Regional office Aga Khan Rural Support Program Skardu, Baltistan Division.

IFAD. (2016). *Economic Transformation Initiative Gilgit-Baltistan* (Project No. 2000000836, Report No: 3862-PK). Investing in rural people (IFAD), Islamic Republic of Pakistan, Asia and the Pacific Division, Programme Management Department.

Jennings, G. R. (2005). Interviewing: A focus on qualitative techniques. In B. W. Ritchie, P. M. Burns, & C. A. Palmer (Eds.), *Tourism research methods: Integrating theory with practice* (pp. 99–118). CAB International.

Jensen, O. (2007). *Beyond mountains: The impact of Pakistani territorialisation on Balti livelihoods and migration practices*. PhD Dissertation, Roskilde Universitet.

Kenyon, S., Lyons, G., & Rafferty, J. (2002). Transport and social exclusion: Investigating the possibility of promoting inclusion through virtual mobility. *Journal of Transport Geography, 10*(3), 207–219.

Khadaroo, J., & Seetanah, B. (2008). The role of transport infrastructure in international tourism development: A gravity model approach. *Tourism Management, 29*(5), 831–840.

Khan, S. A. (2013). Geo-economic imperatives of Gwadar Sea Port and Kashgar economic zone for Pakistan and China. *IPRI Journal, 13*(2), 87–100.

Khan, S. R. (2012). *Linking conservation with sustainable mountain livelihoods: A case study of Northern Pakistan*. University of Manitoba.

Kreutzmann, H. (1991). The Karakoram Highway: The impact of road construction on mountain societies. *Modern Asian Studies, 25*(04), 711–736.

Kreutzmann, H. (2004). Accessibility for High Asia: Comparative perspectives on northern Pakistan's traffic infrastructure and linkages with its neighbours in the Hindukush-Karakoram-Himalaya. *Journal of Mountain Science, 1*(3), 193–210.

Kreutzmann, H. (2006). Settlement history of the Hunza Valley and linguistic variegations in space and time. *Karakorum in transition: Culture, development, and ecology in the Hunza Valley* (pp. 251–272). Oxford University Press.

Kreutzmann, H. (2013). *Preservation of built environment and its impact on community development in Gilgit-Baltistan, 42*. Freie Universität Berlin.

Kumaraswamy, M., & Zhang, X. (2001). Governmental role in BOT-led infrastructure development. *International Journal of Project Management, 19*(4), 195–205.

Lapointe, D. (2020). Reconnecting tourism after COVID-19: The paradox of alterity in tourism areas. *Tourism Geographies, 22*(3), 633–638.

Leiper, N. (1992). *Whole tourism systems: Interdisciplinary perspectives on structures, functions, environmental issues and management*. Doctor of Philosophy, Massey University.

Leiper, N. (2004). *Tourism management*. Pearson Education Australia in Frenchs Forest.

Lew, A. A. (2014). Scale, change and resilience in community tourism planning. *Tourism Geographies, 16*(1), 14–22.

MacDonald, K. I. (1996). Population change in the Upper Braldu Valley, Baltistan, 1900–1990: All is not as it seems. *Mountain Research and Development, 16*(4), 351–366.

MacDonald, K. I. (1998). Push and shove: Spatial history and the construction of a portering economy in Northern Pakistan. *Society for Comparative Study of Society and History, 40*(2), 287–317.

Mbaiwa, J. E., & Sakuze, L. K. (2009). Cultural tourism and livelihood diversification: The case of Gcwihaba Caves and XaiXai village in the Okavango Delta, Botswana. *Journal of Tourism and Cultural Change*, 7(1), 61–75.

Murphy, D. (2011). *Where the Indus is young: a winter in Baltistan*. Eland Publishing.

Olsson, J. (2009). Improved road accessibility and indirect development effects: Evidence from rural Philippines. *Journal of Transport Geography, 17*(6), 476–483.

Owen-Edmunds, L. (2013). *Gilgit-Baltistan and Chitral (GBC) Domestic Tourism Value Chain Analysis*. Enhancing Employability and Leadership for Youth, Aga Khan Rural Support Program.

Peters, D. (2000). Old myths and new realities of transport infrastructure assessment: Implication for EU interventions in Central Europe. In A. Pearman, P. Mackie, & J. Nellthorp (Eds.), *Transport projects, Programmes and policies: Evaluation needs and capabilities* (pp. 43–72). Ashgate Publishing.

Post, F. (2017). China-Pakistan economic corridor: The dragon's master plan is making India nervous, and with good reason. Retrieved from http://www.firstpost.com/world/china-pakistan-economic-corridor-the-dragons-master-plan-is-making-india-nervous-and-with-good-reason-3448944.html

Prideaux, B. (1993). *Possible effects of new transport technologies on the tourism industry in the 21st Century*. Paper presented at the Australasian Transport Research Forum (ATRF).

Prideaux, B. (2000). The role of the transport system in destination development. *Tourism Management, 21*(1), 53–63.

PwC. (2018). Belt and road initiative: A strategic pathway. Retrieved from https://nzchinacouncil.org.nz/beltandroad

Reuters. (2017). Pressure on as Xi's 'Belt and Road' enshrined in Chinese party charter. Retrieved from https://www.channelnewsasia.com/news/world/pressure-on-as-xi-s--belt-and-road--enshrined-in-chinese-party-charter-9340138

Riley, R. W. (1996). Revealing socially constructed knowledge through quasi-structured interviews and grounded theory analysis. *Journal of Travel and Tourism Marketing, 5*(1–2), 21–40.

Rogers, C. (1980). *A way of being*. Houghton Mifflin.

Rosen, T., Hussain, S., Mohammad, G., Jackson, R., Janecka, J. E., & Michel, S. (2012). Reconciling sustainable development of mountain communities with large carnivore conservation. *Mountain Research and Development, 32*(3), 286–293.

Scoones, I. (2009). Livelihoods perspectives and rural development. *Journal of Peasant Studies, 36*(1), 171–196.

Seong, Y. B., Bishop, M. P., Bush, A., Clendon, P., Copland, L., Finkel, R. C., Kamp, U., Owen, L. A., & Shroder, J. F. (2009). Landforms and landscape evolution in the Skardu, Shigar and Braldu Valleys, Central Karakoram. *Geomorphology, 103*(2), 251–267.

Shen, F. (2009). *Tourism and the sustainable livelihoods approach: Application within the Chinese context*. Doctor of Philosophy, Lincoln University.

Sigala, M. (2020). Tourism and COVID-19: Impacts and implications for advancing and resetting industry and research. *Journal of Business Research, 117*, 312–321.

Sökefeld, M. (1997). Migration and society in Gilgit, northern areas of Pakistan. *Anthropos*, 83–90.

Strickland-Munro, J. K., Allison, H. E., & Moore, S. A. (2010). Using resilience concepts to investigate the impacts of protected area tourism on communities. *Annals of Tourism Research, 37*(2), 499–519.

Tang, J. C., & Rochananond, N. (1990). Attractiveness as a tourist destination: A comparative study of Thailand and selected countries. *Socio-Economic Planning Sciences, 24*(3), 229–236.

Tao, T. C. H., & Wall, G. (2009). Tourism as a sustainable livelihood strategy. *Tourism Management, 30*(1), 137–152.

Tarar, M. H. (1991). *Nanga Parbat: Baltisan dastan*. Sang-e-meel Publications.

Tashakkori, A., & Teddlie, C. (1998). *Mixed methodology: Combining qualitative and quantitative approaches* (Vol. 46). SAGE Publications.

UNWTO. (2020). *Impact assessment of the COVID-19 outbreak on international tourism*. Retrieved from https://www.unwto.org/impact-assessment-of-the-covid-19-outbreak-on-international-tourism

Wachs, M., & Kumagai, T. G. (1973). Physical accessibility as a social indicator. *Socio-Economic Planning Sciences*, 7(5), 437–456.

Wallace, P. (1996). *A history of Western Himalayas*. Penguin Books.

Wang, L., Zhang, J. L., & Liu, L. M. (2010). Diversification of rural livelihood strategies and its effect on local landscape restoration in the semiarid hilly area of the Loess Plateau, China. *Land Degradation & Development, 21*(5), 433–445.

World Bank. (2011). *Pakistan - Gilgit-Baltistan economic report: Broadening the transformation* (55998-PK). Retrieved from: https://openknowledge.worldbank.org/handle/10986/2984

Yousaf, S., Razzaq, A., & Fan, X. (2021). Understanding tourists' motivations to launch a boycott on social media: A case study of the#BoycottMurree campaign in Pakistan. *Journal of Vacation Marketing, 27*(4), 479–495.

Zheng, D., Luo, Q., & Ritchie, B. W. (2021). Afraid to travel after COVID-19? Self-protection, coping and resilience against pandemic 'travel fear'. *Tourism Management, 83*, 104261.

11

DIGITAL TECHNOLOGIES IN TRAIL-RELATED RECREATION

Future challenges for trail development

Goda Lukoseviciute, Luís Nobre Perreira and Thomas Panagopoulos

Introduction

In recent decades, there has been a substantial increase in the demand for nature-based recreation. Many authors ascribe this to the modern urban resident's desire to escape the pressures of everyday life and get back in touch with nature due to increasing urbanization and the rise of sedentary, indoor pastimes and growing concerns about an emerging disconnect between people and their natural environments (Balmford et al., 2009; Fossgard and Fredman, 2019; Fredman and Tyrvainen, 2010). Lack of exposure to natural environments is well known to have serious consequences for human wellbeing and knowledge and understanding of environmental issues (Korpela et al., 2014). The latter concerns have significantly encouraged many people to explore their nearest natural environments (Hambira et al., 2021; Randler et al., 2020) and especially within the global COVID-19 pandemic context and widespread restrictions on public gatherings and community mobility. As a result, societal changes stemming from the pandemic have reshaped tourists' behaviour and have stimulated increasing interest in domestic tourism, including visiting local natural areas (Fredman and Margaryan, 2020). In such circumstances, in recent years, nature-based activities have become the most popular leisure choice across the globe (Bell et al., 2007; Hardiman and Burgin, 2016; Tyrvainen et al., 2014). With regards to the increasing popularity of natural areas (Balmford et al., 2015), trail-related recreation is expected to continue to increase in the coming decades (United Nations, 2020).

In nature-based recreation, terrestrial and water trails provide access to various types of activities such as walking, jogging, hiking, trekking, kayaking, rafting, horseback riding, biking, and birdwatching (Hall et al., 2017). Today, there are many types of recreational trails worldwide with a wide variety of trail characteristics scattered across various natural environments such as wild forests, mountains, rainforests, deserts, wetlands, rivers, and other aquatic zones (Moore and Ross, 1998; Timothy and Boyd, 2015) which may be within natural parks or other forms of protected zones. Accordingly, many scholars acknowledge that recreational trails play a fundamental role in a system of nature-based tourism attractions (Davies et al., 2012; Kling et al., 2017; Svobodova et al., 2019). Furthermore, many studies of nature-based recreation and its management have concluded that the trail is the most essential attribute in nature-based recreation (Fredman and Tyrvainen, 2010; Kim et al., 2015; Teles da Mota and

 DOI: 10.4324/9781003230748-14

Pickering, 2020) and thus prompt us to conclude that trail-related recreation is an inseparable concept from nature-based recreation management.

Recreational trails utilize natural areas; therefore, according to the literature, trail management is primarily focused on constant monitoring of nature degradation patterns and increasing efficiency of environmental protection methods (Symmonds et al., 2000; Tomczyk et al., 2017). Despite that, trail management needs to also be directed toward the adaptation of the most innovative technologies, which, according to Sturgeon (2021), creates and adds value to the customer experience as well as facilitating site monitoring and challenges for augmented reality (Taylor et al., 2000). Besides, the adoption of new technologies is crucial in enabling tourism organizations to address the impact of the recent COVID-19 pandemic and continue offering tourism value (Sigala, 2020). Moore and Shafer (2001) recognize that even though trails and trail-related topics have been studied within different fields of research, there are still important gaps in the literature related to digitalization and its integration in trail-related recreation that need to be addressed. Concerning the current digital era, which has dramatically accelerated the dramatic impact of technology on the entire tourism industry, including nature-based recreation, it is essential to look at the role which digitalization will play in future trail-related recreation management (Hadjielias et al., 2021). Notwithstanding the discussions that digital technology use may lead to environmental degradation (Carlson et al., 2016) as well as placing greater stress on search and rescue teams (Martin and Blackwell, 2016; Shultis, 2012), many scholars acknowledge that it can increase access to outdoor spaces (Carlson et al., 2016; Wick, 2016) and empower individuals in their nature-based experiences (Weatherby and Vidon, 2018). The applications of digital technologies in nature-based settings have been extensively adopted by nature-based tourism organizations to ease their operations, interact better with their markets, become more competitive, and increase their revenues (Capriello and Riboldazzi, 2020) and by young and active hikers to share experiences (Stanley, 2019). Consequently, nature-based recreation planners and managers have been highly focused on Millennials and the Post-Millennial generations, who tend to be more interested in using technologies than members of prior generations (Bernardi, 2018) and characterized by growing ecocentrism, dominantly related to the growing anthropocentric influences in tourism development contexts (Mandić and Vuković, 2021). Hence, it is crucial to examine what role digitalization will play in trail-related recreation (Martin, 2017). The objective of this chapter is to review the application of digital technologies in nature-based recreation and trail-related recreation, extract the main dimensions of digital technology applications, and discuss the principal future challenges for recreational trail management and development.

The chapter in this volume contains the following sections: 1) theoretical background of sustainable recreational trail development; 2) methodological approach; 3) results (dimensions of digital technologies used in nature-based recreation with the following subsections: information and communication, comfort, safety, environmental management); 4) future challenges in recreational trail management; 5) conclusions.

Theoretical background of sustainable recreational trail development

There are several types of recreational trails in both terrestrial and aquatic settings. Trails are diverse in terms of infrastructure, geographic location, nature and purpose and therefore various types can be distinguished: horseback riding, countryside walking and hiking, cycling, rollerblading, running, wheelchair, wine route, cultural route, rafting, canoeing, themed route, certified hiking tours or off-road recreational motorized activity trail (Al-hagla, 2010; Newman, 2010; Reis and Jellum, 2012; Symmonds et al., 2000).

Sustainable trails are designed and constructed to attract trail users, to make them acceptable to all landowners, funders, and other stakeholders and to be manageable in the long term (Marion and Wimpey, 2017). According to Lukoseviciute et al. (2021), when trails are well managed they are likely to attract consistent use, good publicity and repeat visitors. The pioneering trail management framework, advanced by academic and Forest Service researchers in the United States of America (Clark and Stankey, 1979) proposed three main trail management pillars: 1) physical (access, remoteness, naturalness); 2) social (user density); and 3) managerial (visitor management). More precisely, the National Trails Office (2012) have drafted the guidelines of sustainable trail management with the inclusion of following standards: 1) information and communication, 2) vegetation, 3) physical trail management, 4) route, 5) litter, 6) services, 7) waymarking, 8) trail furniture, 9) trail surface, and 10) insurance. Consequently, trail management standards raise the following questions related to: 1) trail design and how should a resource manager design trails to maximize user enjoyment while minimizing environmental damage; 2) trail impacts and when should a resource manager control trail erosion; 3) management tools and how should a resource manager choose techniques that reduce hazardous impact and protect the environment; and 4) trail visitor management strategies (US Forest Service, 2011).

For a long time, recreational trail management was focused primarily on environmental conservation (Botsch et al., 2018; Miller and Hobbs, 2000) and physical trail maintenance (Kubo et al., 2018). However, the evolution of trail management has led to a more holistic approach with more focus on trail-related experience enhancement (Kelley et al., 2016; Kim et al., 2015) and trail destination management (Curtin, 2013), where one of the management components involves the inclusion of digital technologies and digitalization. Digitalization refers to influencing visitors' experiences, increasing the competitiveness of destination and enabling/ improving the tourism management transformation process by leveraging digital technologies and digitized data (Buhalis and Amaranggana, 2013). The digitalization process incorporates several tools and techniques, such as social media platforms, AI, mobile applications, outdoor sensor technologies such as thermal cameras, footfall counters, digital cameras, drones, and Geographic Information Systems (GIS) (Buhalis, 2020).

The perception of digital technologies may differ among trail users and trail managers (Clark et al., 2021). Looking from a trail user perspective, previous studies argue that digital technologies are perceived differently, and the perception strongly depends on the type of visitor (Clark et al., 2021; Stinson, 2017). Nature-based recreation scholars discuss recreational trail users seeking to disconnect from digitalized life and reconnect with the land (Farkic et al., 2021) may negatively perceive the use of digital technologies as it is supported by previous findings that non-digital trail attributes such as access to the trail, signage, visitor management and constructed trail features are the most critical trail design factors of sustainable recreation provision and natural environment protection (Lukoseviciute et al., 2021). Moreover, regardless of geographical trail context, other non-digital trail attributes, such as parking, toilet access, and trail site cleanliness, are no less important attributes for sustainable trail design and development (Kelley et al., 2016; Ribet and Brander, 2020; Sever and Verbic, 2018). Besides, for both first-time international and frequent national trail users, safety and security, as well as the condition of the path, are the most important trail attributes (Keith et al., 2018; Lukoseviciute et al., 2021). Meanwhile, among mountain bikers, a technically challenging single track is the most desirable trail attribute (Koemle and Morawetz, 2016).

On the contrary, for active hikers aiming to promote and share on-site experiences in outdoor media, the presence of digital technologies may be the most important attribute, in particular for social media influencers and Post-Millennials, known as 'digital natives' (Corbisiero and Ruspini, 2018; Stanley, 2019). Moreover, according to Stinson (2017), digital technologies are positively perceived by a new class of cyborg-recreationists performing wilderness-based recreation and tethered to Internet-connected digital technologies such as mobile digital devices, locative media, and Web 2.0. Looking from a trail manager perspective, the use of digital technologies may be significantly beneficial and facilitative in trail site management as well as the provision of increased trail user value through higher-quality services, personalized interaction and communication, effective and personalized information dissemination or integrated services toward the user (Hadjielias et al., 2021).

Following new tendencies and predictions, it is expected, and already foreseen, that new technologies and digitalisation will be an integral part of various types of trail visitors' and trail managers' activities (Hadjielias et al., 2021; Santos et al., 2021). New technological innovations provide many possibilities for outdoor recreation, for instance, sports connections for mutual interaction through various types of apps, augmented reality and social media platforms on-site (Jepson and Ladle, 2015) or virtually (Stinson, 2017). Digital technologies inform outdoor recreation maintenance schedules and assist in achieving management objectives through outdoor footfall sensor technologies (Zajchowski et al., 2021), support data analysis and creation of tourists' profiles through the power of artificial intelligence (AI) (Samara et al., 2020).

Methodological approach

This study followed the qualitative literature review approach, which is a more suitable approach for gathering a detailed understanding of existing knowledge in a specific topic (Silverman, 2016). A qualitative literature review approach was used in nature-based tourism research to investigate topics such as the significance of natural environments to active outdoor recreation (Andkjær and Arvidsen, 2015), trails as tools for outdoor recreation (Kling et al., 2017), impacts of the private sector and parastatal enterprises in nature-based tourism (Spenceley and Goodwin, 2007). Qualitative research design enables the extensive rigorous empirical investigation of a phenomenon (Strauss and Corbin, 1990). In this kind of research, data analysis is usually performed through coding (identifying categories and dimensions).

This study applied a two-step qualitative literature review analysis approach. The first step was to identify, select, and evaluate relevant literature on digitalization in the nature-based recreation arena. The second step was to code the information in the selected articles and through the application of an interpretative approach extract the main dimensions of digital technologies used in trail-related recreation as this is a key pillar of nature-based recreation and further to discuss challenges and opportunities related to digital technologies use in trail-related recreation (Figure 11.1).

Search criteria

To identify the relevant papers addressing the objective, the research was conducted using specific search terms which were present in the title, abstract, or body of articles authored in English in the Scopus and Web of Science bibliographic databases, which are the most commonly used citation databases for field delineation (Strozzi et al., 2017). These databases search API supporting a Boolean syntax, which is a type of search allowing the combination of keywords with operators such as "AND", "OR" and "NOT" to produce more relevant results. The following search terms were applied to the selected databases: "smart nature-based tourism" OR "smart destinations" OR "digitalisation" OR "digital technologies" AND "nature-based tourism" OR "ecotourism" OR "hiking" OR "recreational trail" OR "greenway" OR "protected area" OR "national park" OR "backpacking" OR "trekking" OR "wilderness area".

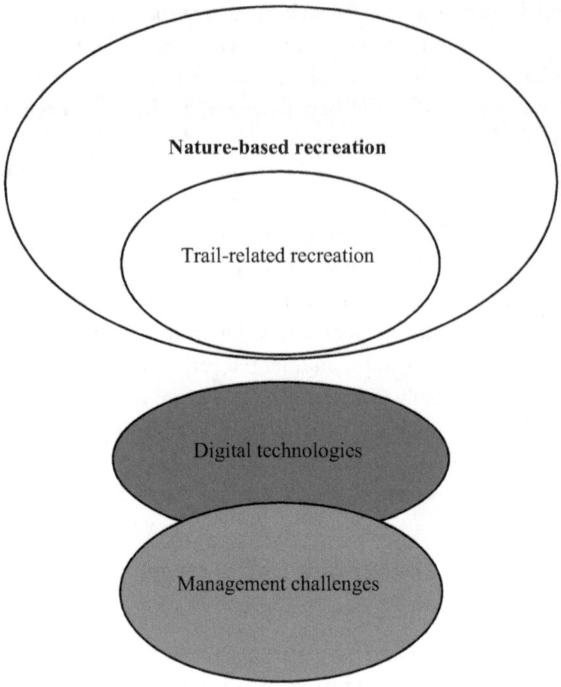

Figure 11.1 The theoretical framework of trail-related recreation and digital technologies.

The research was focused on the papers published from 2000 to 2021. In the Scopus and Web of Science database document type tab, all possible document types were selected. Non-English articles were excluded because the authors are unable to read and analyse publications written in other languages. The cross-checking was performed to find more papers relevant to the study topic. Additionally, a search of grey literature was carried out in Google Scholar about digital technologies used in nature-based tourism and their influence on consumer behaviour and destination management.

Data analysis

Qualitative data analysis and information coding were performed by using the NVivo11 (QSR International) software package to explore the related concepts and the main dimensions of digitalisation in nature-based recreation. NVivo software is a facilitative tool in qualitative data analysis (Sotiriadou et al., 2014) and was widely applied in the field of tourism to investigate topics such as smart tourism destinations (Shafiee et al., 2019), cultural heritage and tourism (Loulanski and Loulanski, 2011), and transitional domestic tourism (Samarathunga et al., 2020).

In the data coding phase, the information in the selected articles was extracted and evaluated, and a coding table for documenting information was created. The information was coded in the table according to the type of nature-based recreation, the type of digital technologies applied and the purpose of application. The interpretative approach was used in the coding of information (Dixon-Woods et al., 2004).

Results

Dimensions of digital technologies use in nature-based recreation

The final yield of 45 papers selected was included in the qualitative analysis and four categories were discovered and labelled as the dimensions of the digital technologies application in nature-based recreation: 1) information and communication; 2) comfort; 3) experience design; and 4) environmental management.

Information and communication

Regarding information and communication, one such important factor in the changing face of the nature-based tourism industry is social media. There has been significant business growth through digital media as well as the overall growth of nature-based recreation. Social media has been recognized as an important source of information to complement traditional nature-based recreation monitoring (Ghermandi and Sinclair, 2019). With the emergence of social media, it has become very easy to share information around the globe. Every recent trend happening in the nature-based tourism industry can be disseminated to millions of people through different tools of social media. It became common to share nature-based experiences through social media channels (Hausmann et al., 2018). Through social media shares, researchers and nature-based destination managers have access to a huge data-rich set in terms of visual and textual content, which is used to analyse visitor satisfaction with destination attributes, motivations, and values (Becken et al., 2017). Teles da Mota and Pickering (2020) have found that data from social media can facilitate and assist in the management of nature-based tourism, for instance identify visitation rates and the relative popularity of sites and parks, indicate both on and off-trail use, and provide data on daily, weekly, or seasonal visitation patterns. Moreover, Toivonen et al. (2019) examined a rapidly growing body of literature on social media data use in nature-based recreation and conservation science. The authors have extracted five key elements complementary to the data set of social media: user information, content (text, image, sound, video), timestamp, geotag and comments and likes. According to Tenkanen et al. (2017), the majority of researchers have used social media data as a proxy of visitation in natural areas, to understand people's behaviour in nature and human dimensions of environmental issues, which are essential to understand for successful nature-based recreation management (Calcagni et al., 2019). Hausmann et al. (2018) have stressed the significance of social media data for studying why visitors prefer certain locations and times.

With regards to 'big data' coming from user shares in social media, a crucial role is played by AI or, in other words, machine learning, with the ability to use all types of data that humans are not using in analytics today. AI can be referred to as the development of computer systems that can perform tasks and activities which require human intelligence (Russel and Norvig, 2016). AI tools offer opportunities to capture the tourist preferences, identify market trends, forecast visitor flows and demand, forecast loyalty and revenue, plan site maintenance as well as management, provide personalized recommendations, and use smart technologies or autonomous transportation. A recent study by Samara et al. (2020) has found that studies incorporating AI can be clustered into three main topics related to 1) forecasting flow and demand, 2) visitor loyalty, and 3) revenue. Regarding forecasting visitor flow and demand, AI-based techniques have been widely applied in nature-based recreation, such as national parks (Clark et al., 2019; Rice et al., 2019), marine protected areas (Rezapouraghdam et al., 2021) or recreational trails (Staab et al., 2021). With regard to examining tourist preferences, service robots, equipped with

AI, have been tested in order to test the effect of crowding on tourists' service (Hou et al., 2021). Finally, examining destination loyalty Mirzaalian and Halpenny (2021) applied machine learning algorithms to provide insight into tourists' loyalty behaviour to support destination marketing organizations with tourism development, management, and planning.

Experience design

With respect to experience design, the main focus here comes to online tools and AI technology tools such as augmented reality, chatbots, robots, language translators, audio-visual tours, and smartphones (Samala et al. (2020). All of the tools mentioned can increase access to outdoor spaces (Carlson et al., 2016) and empower personal experiences in nature (Stanley, 2019). A recent study by Martínez-Graña et al. (2017) has presented a rural geotrail across a high mountain village with the implementation of several georesources such as educational and interpretive panels, videos, QR codes, geoapps and games, and concluded that the provision of digital technologies favour nature-based experience whilst explaining the natural environment.

In nature-based recreation augmented reality is becoming increasingly popular due to its ability to keep visitors connected with their operational site (Hofman et al., 2021). Augmented reality uses the augmented reality headset to create a simulated environment and thus creates an experience of virtual reality. Nature-based tourism scholars have examined how augmented reality contribute to park visitors' emotional experiences revealed that tourists' perceived experiential quality of the augmented reality theatrical performance provided them with a transcendent experience that aroused emotions such as love, romance, and excitement (Hu et al., 2021). It is a very useful tool to provide the virtual journey to the customers with nature-based recreation spots such as zoo parks, recreational trails, or national parks (Jung et al., 2017). In addition, pandemic restrictions have encouraged nature-based recreation managers to provide facilities for visitors safely, simultaneously providing a memorable experience. Consequently, some national parks in Germany and Canada have developed online ways of sharing nature and put up an online format as a substitute for physical guided tours (Spenceley et al., 2021). Through a shift toward digital offerings to create virtual experiences such as 'Safari Talks' or virtual ecological trails, managers are enabled to provide an environmentally-safe form of ecotourism, while visitors are enabled to educate themselves over long distances (Korbut and Tsekina, 2014; Skinner, 2020).

Concerning chatbots, there are no studies conducted in the field of nature-based tourism and it seems that the application of this tool is very likely to occur soon (Calvaresi et al., 2021). Chatbots may provide all the information related to tourism infrastructure and are of interest as a tool that offers new ways of interaction for visitors to learn about the natural environment such as flora, fauna, cultural or natural heritage (Ivanov and Webster, 2017). In relation to robots, Frost et al. (2014) have indicated that they will soon be utilized in leisure attractions, including nature-based recreation for entertainment and as part of the staging of experiences. In particular, the authors emphasized the use of robots in themed parks to recreate experiences, including those related to natural settings.

Regarding audio-visual tours, previous studies found that they may be very attractive to visitors as they are novel, multi-sensory and require the least amount of effort for consuming interpretive information (Moscardo, 1996). Nature-based scholars have tested if recreational trail visitors prefer traditional rather than modern multimedia, which includes audio MP3 tours and found that modern multimedia appealed to more senses than traditional media thus holding visitor attention for longer periods (Wolf et al., 2013).

Safety

Considering the aspects of safety in nature-based recreation sites, digital technologies play an important role for both managers of recreational areas and visitors. From a manager perspective, computation using GIS images and linear models obtaining information from drones is widely used to evaluate the suitability of forest trails to be constructed as hiking trails (Lee et al., 2020). The authors have concluded that a drone-based technique for topographic analysis provides better accuracy when compared to satellite-based and aircraft-based remote sensing techniques. Digital tools such as drones are crucial in monitoring the trail condition potential hazards, such as falling trees or the presence of aggressive wildlife animals, in order to create a safe environment for the trail user.

From a tourist perspective, the list of digital technologies used ranges from smartphones with navigation to digital maps and personal locators. According to nature-based scholars, beacons became more readily available and popular choices to request rescues and may develop unrealistic expectations of rescue (Amerson et al., 2020; Martin and Pope, 2012). Recently, Gonçalves et al. (2020) proposed a digital platform for tourism guidance tracking and safety based on beacons using Bluetooth Low Energy (BLE), a mobile device App and a central cloud server and applied to share trips and advice on natural trail hikes. However, orientation tools working with an Internet connection may face problems with signals in remote and sheltered areas. Accordingly, personal locator beacons are particularly popular among long-distance trail users since long-distance hiking requires experience in navigation; therefore, inexperienced hikers might use personal locators simply to confirm one's location (Rogers and Leung, 2020). In addition, a study in Germany has found that traditional hiking maps were the dominant choice among hikers when compared to digital maps (Moritz et al., 2018). However, the use of digital technologies become inevitable as, according to Amerson et al. (2020), 97% of hikers in the USA carried a smartphone with them for navigation, listening to music, and taking photos, which is supported by Smukavic et al. (2014) that more and more hikers use interactive maps and smart-devices, which makes much easier orientation in the field.

When referring to safety, it is inevitable to discuss the context of the COVID-19 pandemic and its effect on destination management. Pandemic restrictions have significantly impacted nature-based recreation visitation due to newly imposed health and safety standards such as limiting the total capacity of any outdoor area, conducting visitor and staff temperature screening and questionnaires upon entrance to facilities, to anticipating lines forming for attractions, to managing entrance and exit points, to implementing social distancing, and to avoiding the offering of interactive attractions (New Jersey Department of Health, 2020). Many high-density destinations remain closed, but simultaneously many have cautiously reopened (Spenceley et al., 2021). Nevertheless, the new health and safety requirements have spurred managers to be creative and think of innovative as well as effective measures to provide safe recreation for visitors. Accordingly, several digital technologies have been applied in nature-based recreation, including: 1) online scheduling of entrance time windows to distribute visitors and ensure socially distanced outdoor recreation (National Park Service, 2021); 2) mobility records sampled from mobile devices to monitor spatio-temporal patterns and social distancing (Kupfer et al., 2021); 3) site-visitation online using online platforms to ensure socially safe recreation (Alberta Parks Nature Source, 2020); 4) drone or terrestrial video surveillance of visitor and wildlife interactions to avoid disease transmission (Thomsen et al., 2021); 5) obligatory use of COVID-19 contact tracking mobile application within outdoor recreation facilities (Kondylakis et al., 2020); and 6) Global Positioning System (GPS) tracking of animals to monitor their interactions with site visitors (Life View GPS, 2018).

Environmental management

In nature-based recreation where natural settings are utilized as resources to develop a recreational activity, environmental management is a crucial component to manage existing recreational resources efficiently or create new nature-based recreation areas. For this purpose, one of the main commonly applied digital tools in environmental management is GIS, which creates, manages, analyses, and maps all types of environmental variables (Tomlinson, 1987). In nature-based recreation areas, GIS has been used to model the impacts of the degradation of forest roads on recreation opportunities (Queen et al., 1997) or environmental resiliency (Arrowsmith and Inbakaran, 2002). With regards to recreational trails, Whinam et al. (2003) have examined how sustainable carrying capacities might be calculated for walking areas, based on walking-track monitoring, walker-impact trampling trails and sensitivity. Moreover, Tomzcyk (2011) developed a model to assess environmental sensitivity to the impact of recreational trails and forest roads, based on data about topography, soil properties and vegetation obtained from GIS. The spatial recreational area analysis approach facilitates recreation planning by providing information on current recreation opportunities and potential recreation opportunities driven by changes to the road or trail systems in an area. Consequently, the GIS tool has been widely applied to design and plan recreational trails (Olafsson and Skov-Petersen, 2014; Snyder et al., 2008; Tomczyk and Ewertowski, 2013) with an emphasis on the importance of GIS in collaborative and participatory recreational trail planning, characterized by the best possible solution for recreational and conservation function.

Recently, a more sophisticated AI-based random forest-GIS method with different machine learning algorithms was applied to prepare trail susceptibility mapping (Sahani and Ghosh, 2021). The results have shown that GIS and AI-based analyses are a useful source of information to the park and trail managers since predicting areas that are susceptible to trail degradation is a useful means for implementing protective measures and reduction of environmental damage.

A combination of technologies such as pressure-sensor-triggered cameras and deployment of convolutional neural networks were used to assess the performance of a computer vision approach for visitor monitoring (Staab et al., 2021). The authors concluded that camera-based approaches have the potential to be both accurate and deliver comprehensive data about visitor numbers, types, and activities, which is essential not only in monitoring visitors in high-visitation locations but also in sustainable management and valuation tasks. Furthermore, sensory digital technologies are very useful to assess the temperature and air quality in outdoor recreation, which are the most important determinants of the visitor experience and satisfaction with outdoor activity. Using a 24-hour trail camera, Zajchowski et al. (2021) measured trail use, ambient temperature, particulate matter, and ozone in order to assess the impact of air quality on outdoor recreation behaviours. Temperature and air quality monitoring significantly facilitate forecasting the intensity of recreational areas used to aid decision-making.

Referring to monitoring visitors, several footfall sensor technologies have been implemented in nature-based recreation management. According to Kyle et al. (2021), such examples are as follows: 3D-based visual sensors, which utilize cameras to analyse their surroundings and identify if a visitor has entered or exited a certain invisible threshold; Wi-Fi/Bluetooth device monitor 'Sniffer' attempting to quantify Wi-Fi- and/or Bluetooth-equipped devices within the proximity of the sensor; radar sensors, emitting microwave frequencies and utilizing the resultant frequency modulation effect to track an object; pyro sensors, detecting and differentiating objects within the field of view based on their radiant heat 'signature'; passive infra-red sensors, using in a 'break beam' configuration for footfall detection; and a pressure slab, converting the pressure signal into a voltage.

Future challenges in recreational trail management

Recent changes in nature-based recreation development bring new challenges related to the capacities to employ, use and benefit of digital technologies found in each of the four mentioned dimensions of trail-related recreation management.

Concerning data availability from digital technologies, it is important to highlight that social media and other platforms represent only part of trail users, their profiles, activities, and other trail-related recreation data. Nature-based scholars have concluded that datasets related to nature-based recreation are still very limited and usually requires manual collection, which is time-consuming and expensive (Wood et al., 2013). According to Smith and Anderson (2018), the most popular social media platforms are dominated by wealthier, and highly educated users, meaning that a number of trail-related recreationists may be less likely to feature in social media and share experiences. Consequently, it will become challenging to identify whose data are represented in social media and which part of the visitor population it represents due to potential biases in who is uploading the content. Another challenge related to accessing data is restricted access in some platforms such as Panoramio (Becken et al., 2017). In addition, the use of big data requires consideration of the General Data Protection Regulation (GDPR) to guarantee privacy and security. Regarding the data availability from sensory technologies, it might be challenging to correctly interpret the dataset since vision-based systems do not differentiate between different categories of objects. For instance, people's step counting sensors cannot differentiate between people, cars, buses, and bicycles and it creates a biased source of information. Moreover, sensors are not always functional; readings can be erratic, and they cannot be corrected in the database due to recreational trail environmental surrounding changes, as the data are stored by another company (Madden et al., 2021). Finally, facial recognition technologies are challenging to be used in recreational trail development and management due to data privacy and security issues (Leong, 2019). Looking from an environmental management perspective, the main challenge occurring is the accuracy and adequacy of data obtained from digital technologies. Concerning the use of GIS for environmental and visitor management, GIS is not a very useful tool for recreational trail management without accurate datasets (Sipe and Dale, 2003). To obtain accurate data on vegetation, land use, topography, weather factors, and historical data using sensory technologies is challenging mainly due to vandalism, technical issues, weather conditions, and the ability to differentiate between different categories of objects. For instance, camera installation takes time and effort, regular maintenance (batteries, storage cards), and the automated monitoring requires specific software know-how. Moreover, it is costly due to time-consuming manual image evaluation; therefore, lower-budget recreational trail managers might not choose these sensory technologies which are also associated with data security issues, which needs to be considered when setting up the devices. According to Madden et al. (2021), the use of cameras in recreational trail settings is particularly associated with a lack of privacy, and visible installation of these sensors may not complement the atmosphere of an outdoor exploration trail. Moreover, Staab et al. (2021) concluded that sensory digital technologies are relatively expensive to buy, not easily portable and their installation requires time and experience. Another non-trivial challenge of sensory technologies is the short lifespan of batteries, especially in the winter season, where potential snow cover leads to impairment of functions. Further, legality issues must be clarified whenever the use of cameras is planned.

Referring to capacities to employ digital technologies, there are significant challenges with application costs, Internet signal availability and user perception toward digital technologies. For instance, orientation and rescue request devices and other tools working with an Internet connection may be challenged in catching the signal in remote and sheltered recreational trail areas. Regarding augmented reality applications in trail-related recreation related to hardware, software, content, use cases, and regulations include bulky hardware which might be expensive for the user and therefore any attempt at widespread use of the app will not succeed. Nevertheless, current mobile phones with cameras, GPS and inertial sensors, high-resolution screens and graphics processors are the most common way in which people experience augmented reality (Billinghurst, 2021). To provide richer content, for instance, allowing to experience recreational trails through smart glasses might be expensive and complex. In general, the employment of digital technologies based on recreational trail user needs is challenging since very little is currently known about the new features, capabilities, and behaviours afforded by the new 'tourists' that are physically and socially empowered, cognitively augmented and psychologically affected by technologies (Sigala, 2018). However, the provision and use of digital technologies in recreational trails may be negatively perceived by visitors since recent studies revealed that nature-based experiences are often associated with exclusion from everyday routines, digital detox, physical and mental wellbeing (Farkic et al., 2021; Floros et al., 2021) among all types of visitors such as first-time, repeat and prospective travellers (Divinagracia et al., 2012). Technology attachment has been found to lead to new forms of technostress and anxiety affecting satisfaction and wellbeing (Sigala, 2018) and lack of serendipity that is often essential to meaningful nature-based recreation experiences (Gretzel, 2010). Finally, there might be a public scepticism about the use of digital technologies since recreational trails are mainly used by visitors for exercising, self-renewal, relaxation, wildlife viewing, and enjoying the natural environment (Kling et al., 2017).

Conclusions

The purpose of this chapter was to conduct a qualitative literature review of the application of digital technologies in trail-related recreation and discuss the main future challenges for recreational trail management and development. In doing so, four dimensions of digital technology use were extracted: information and communication, experience design, safety, and environmental management. The findings of this research indicate that digital technologies have been applied in the field of trail-related recreation for various purposes, ranging from analysing visitor satisfaction with destination attributes, motivations, and values to requesting rescues, providing and monitoring safe trail-related recreation, and monitoring environmental trail conditions.

This chapter attempted to fill the gap by proposing a qualitative analysis of the application of digital technologies in a theoretical framework of trail-related recreation and identifying the main future challenges. This study revealed that there are a number of new challenges mainly associated with data availability, the adequacy of gathered data, capacities to employ digital technologies, and trail visitor perception toward the provision and use of digital technologies at the trail site. Nature-based recreation is classified as providing opportunities for enjoyment in natural areas and increasing people's holistic wellbeing. Accordingly, several fundamental questions remain, such as: What kind of impact may digital technologies have upon overall trail visitor satisfaction and trail-related experiences?; What are the most necessary digital technologies for sustainable trail-related recreation management?; What kind of digital technologies trail users prefer if such, and, finally, what are the motivations of trail visitors and their behaviour? The results of this study might be used as a basis for further theoretical and empirical research development deepening an understanding of trail user and 'provider' perceptions toward digital technologies, their use, value, and impact on trail-related experiences. In addition, future research should focus on the theoretical framework of digital tools that must recognize the mix of sources that nature-based recreationists user and provide the content that addresses travellers' motives.

Regarding managerial implications, this study provides a classified and structured review of digital technologies that trail managers may use as a guide of local trail destination management and further development considering positive and negative aspects of each digital tool applied.

However, this study is not without limitations. The research data were limited to published information only in the English language. Future research should consider analysing publications written in a non-English language from digital technologies in nature-based recreation articles from academic journals. A second limitation is that the extracted publications were obtained through a qualitative systematic literature review. Future studies should apply for a quantitative systematic literature review first and the results should be quantitatively confirmed. Moreover, future researchers working in this area are advised to explore other nature-based recreation forms and the use of digital technology.

References

Alberta Parks Nature Source, 2020. Alberta Parks Nature Source. Alberta Parks, Edmonton, AB https://www.youtube.com/channel/UCGQKe-Dy4EZs4GaZxJMA4nA (Accessed on 19 December 2021).

Al-hagla, K.S., 2010. Sustainable urban development in historical areas using the tourist trail approach: a case study of the Cultural Heritage and Urban Development (CHUD) project in Saida, Lebanon. *Cities* 27(4): 234–248.

Amerson, K., Rose, J., Lepp, A., and Dustin, D., 2020. Time on the trail, smartphone use, and place attachment among Pacific Crest Trail thru-hikers. *Journal of Leisure Research* 51(3): 308–324.

Andkjær, S., and Arvidsen, J., 2015. Places for active outdoor recreation–a scoping review. *Journal of Outdoor Recreation and Tourism* 12: 25–46.

Arrowsmith, C., and Inbakaran, R., 2002. Estimating environmental resiliency for the Grampians national park, Victoria, Australia: a quantitative approach. *Tourism Management* 23(3): 295–309.

Balmford, A., Beresford, J., Green, J., Naidoo, R., Walpole, M., and Manica, A., 2009. A global perspective on trends in nature-based tourism. *PLoS Biology* 7(6): e1000144.

Balmford, A., Green, J.M.H., Anderson, M., Beresford, J., Huang, C., Naidoo, R., Walpole, M., and Manica, A., 2015. Walk on the wild side: estimating the global magnitude of visits to protected areas. *PLoS Biology* 13(2): e1002074.

Becken, S., Stantic, B., Chen, J., Alaei, A.R., and Connolly, R.M., 2017. Monitoring the environment and human sentiment on the Great Barrier Reef: assessing the potential of collective sensing. *Journal of Environmental Management* 203: 87–97.

Bell, S., Tyrvainen, L., Sievanen, T., Probstl, U., and Simpson, M., 2007. Outdoor recreation and nature tourism: a European perspective. *Living Reviews in Landscape Research* 1(2): 1–46.

Bernardi, M., 2018. Millennials, sharing economy and tourism: the case of Seoul. Journal of Tourism Futures 4(1): 43–56.

Billinghurst, M., 2021. Grand challenges for augmented reality. *Frontiers in Virtual Reality* 2: 578080.

Botsch, Y., Tablado, Z., Scherl, D., Kery, M., Graf, R.F., and Jenni, L., 2018.Effect of recreational trails on forest birds: human presence matters. *Frontiers in Ecology and Evolution* 6: 175.

Buhalis, D., 2020. Technology in tourism-from information communication technologies to eTourism and smart tourism towards ambient intelligence tourism: a perspective article. *Tourism Review* 75(1): 267–272.

Buhalis, D., and Amaranggana, A., 2013. Smart tourism destinations. In Z. Xiang and I. Tussyadiah (Eds.), *Information and communication technologies in tourism 2014*. Cham: Springer.

Calcagni, F., Maia, A.T.A., Connolly, J.J.T., and Lange Meyer, J., 2019. Digital construction of relational values: understanding the role of social media for sustainability. *Sustainability Science* 14(5): 1309–1321.

Calvaresi, D., Ibrahim, A., Calbimonte, J.P., Schegg, R., Fragniere, E., and Schumacher, M., 2021. The evolution of chatbots in tourism: a systematic literature review. *Information and Communication Technologies in Tourism* 2021: 3–16.

Capriello, A., and Riboldazzi, S., 2020. How can a travel agency network survive in the wake of digitalization? Evidence from the Robintur case study. *Current Issues in Tourism* 23(9): 1049–1052.

Carlson, T., Shultis, J., and Van Horn, J., 2016. Technology in the wilderness: emerging issues and directions for research, policy, and management. *International Journal of Wilderness* 22(3): 11–17.

Clark, C., Nyaupane, G.P., and Lichterman, A., 2021. Comparison between millennials' and providers' perceptions of technology use in a nature-based tourism context. *Current Issues in Tourism* 25(13): 2086–2089.

Clark, M., Wilkins, E.J., Dagan, D.T., Powell, R., Sharp, R.L., and Hillis, V., 2019. Bringing forecasting into the future: using Google to predict visitation in U.S. national parks. *Journal of Environmental Management* 243: 88–94.

Clark, R.N., and Stankey, G.H., 1979. The recreation opportunity spectrum: a framework for planning, management, ad research. USDA Forest Service General Technical Report PNW-98.

Corbisiero, F., and Ruspini, E., 2018. Millennials and generation Z: challenges and future perspectives for international tourism. *Journal of Tourism Futures* 4(1): 3–104.

Curtin, S., 2013. Lessons from Scotland: British wildlife tourism demand, product development and destination management. *Journal of Destination Marketing & Management* 2(3): 196–211.

Davies, N.J., Lumsdon, L.M., and Weston, R., 2012. Developing recreational trails: motivations for recreational walking. *Tourism Planning & Development* 9(1): 77–88.

Divinagracia, L.A., Divinagracia, M.R.G., and Divinagracia, D.G., 2012. Digital media-induced tourism: the case of nature-based tourism (NBT) at East Java, Indonesia. *Procedia - Social and Behavioral Sciences* 57: 85–94.

Dixon-Woods, M., Agarwal, S., Jones, D., Young, B., and Sutton, A., 2004. Integrative approaches to qualitative and quantitative evidence. Health Development Agency. Retrieved from www.hda.nhs.uk (Accessed on 12 December, 2021).

Farkic, J., Isailovic, G., and Taylor, S. (2021). Forest bathing as a mindful tourism practice. *Annals of Tourism Research Empirical Insights* 2(2): 100028.

Floros, C., Cai, W., McKenna, B., and Ajeeb, D., 2021. Imagine being off-the-grid: millennials' perceptions of digital-free travel. *Journal of Sustainable Tourism* 29(5): 751–766.

Fossgard, K., and Fredman, P., 2019. Dimensions in the nature-based tourism experience scape: an explorative analysis. *Journal of Outdoor Recreation and Tourism* 28: 100219.

Fredman, P., and Margaryan, L., 2020. 20 years of Nordic nature-based tourism research: a review and future research agenda. *Scandinavian Journal of Hospitality and Tourism* 21(1): 14–25.

Fredman, P., and Tyrvainen, L., 2010. Frontiers in nature-based tourism. *Scandinavian Journal of Hospitality and Tourism* 10(3): 177–189.

Frost, W., Laing, J., and Beeton, S., 2014. The future of nature-based tourism in the Asia-Pacific region. *Journal of Travel Research* 53(6): 721–732.

Ghermandi, A., and Sinclair, M., 2019. Passive crowdsourcing of social media data in environmental research: a systematic review. *Global Environmental Change* 55: 36–47.

Gonçalves, F., Martins, A.L., Ferreira, J.C., Marques, E., Andrade, M., and Mota, L., 2020. Tourism guidance tracking and safety platform. In *International Conference on Intelligent Transport Systems* (pp. 162–171). Cham: Springer.

Gretzel, U., 2010. Travel in the network: redirected gazes, ubiquitous connections and new frontiers. In M. Levina, G. Kien (Eds.), *Post-global network and everyday life* (pp. 41–58). New York: Peter Lang.

Hadjielias, E., Christofi, M., Christou, P., and Drotarova, M.H., 2021. Digitalization, agility, and customer value in tourism. *Technological Forecasting and Social Change* 175: 121334.

Hall, C.M., Ram, Y., and Shoval, N., 2017. *The Routledge international handbook of walking*. Abingdom, UK: Routledge.

Hambira, W.L., Stone, L.S., and Pagiwa, V., 2021. Botswana nature-based tourism and COVID-19: transformational implications for the future. *Development Southern Africa* 39(1): 51–67.

Hardiman, N., and Burgin, S., 2016. Nature tourism trends in Australia with reference to the Greater Blue Mountains World Heritage Area. *Journal of Sustainable Tourism* 25(6): 732–745.

Hausmann, A., Toivonen, T., Slotow, R., Tenkanen, H., Moilanen, A., Heikinheimo, V., and Minin, E.D., 2018. Social media data can be used to understand tourists' preferences for nature-based experiences in protected areas. *Conservation Letters* 11: e12343.

Hofman, K., Walters, G., and Hughes, K., 2021. The effectiveness of virtual vs real-life marine tourism experiences in encouraging conservation behaviour. *Journal of Sustainable Tourism* 30(4): 742–766.

Hou, Y., Zhang, K., and Li, G., 2021. Service robots or human staff: how social crowding shapes tourist preferences. *Tourism Management* 83: 104242.

Hu, R., Wang, C., Zhang, T., Nguyen, T., Shapoval, V., and Zhai, L., 2021. Applying augmented reality (AR) technologies in theatrical performances in theme parks: a transcendent experience perspective. *Tourism Management Perspectives* 40: 100889.

Ivanov, S.H., and Webster, C., 2017. Adoption of robots, artificial intelligence and service automation by travel, tourism and hospitality companies–a cost-benefit analysis. Artificial Intelligence and Service Automation by Travel. Tourism and Hospitality Companies–A Cost-Benefit Analysis.

Jepson, P., and Ladle, R.J., 2015. Nature apps: waiting for the revolution. *Ambio* 44(8): 827–832.

Jung, T., Tom Dieck, M.C., Moorhouse, N., and Tom Dieck, D., 2017. Tourists' experience of virtual reality applications. In *IEEE International Conference on Consumer Electronics (ICCE), IEEE*. Las Vegas, NV, USA.

Keith, S.J., Larson, L.R., Shafer, C.S., Hallo, J.C., and Fernandez, M., 2018. Greenway use and preferences in diverse urban communities: implications for trail design and management. *Landscape and Urban Planning* 172: 47–59.

Kelley, H., Rensburg, T.M., and Jeserich, N., 2016. Determinants of demand for recreational walking trails in Ireland. *Tourism Management* 52: 173–186.

Kim, H., Lee, S., Uysal, M., Kim, J., and Ahn, K., 2015. Nature-based tourism: motivation and subjective well-being. *Journal of Travel & Tourism Marketing* 32: S76–S96.

Kling, K.G., Fredman, P., and Wall-Reinius, S., 2017. Trails for tourism and outdoor recreation: a systematic literature review. *Tourism: An International Interdisciplinary Journal* 65(4): 488–508.

Koemle, D.B., and Morawetz, U.B., 2016. Improving mountain bike trails in Austria: an assessment of trail preferences and benefits from trail features using choice experiments. *Journal of Outdoor Recreation and Tourism* 15: 55–65.

Kondylakis, H., Katehakis, D.G., Kouroubali, A., Logothetidis, F., Triantafyllidis, A., Kalamaras, I., and Tzovaras, D., 2020. COVID-19 mobile apps: a systematic review of the literature. *Journal of Medical Internet Research* 22(12):e23170.

Korbut, V.V., and Tsekina, M.V., 2014. Virtual tours in the environmental education. Vestnik Moskovskogo Universiteta, Seriya 5: Geografiya (3): 10–15.

Korpela, K., Borodulin, K., Neuvonen, M., Paronen, O., and Tyrväinen, L., 2014. Analyzing the mediators between nature-based outdoor recreation and emotional well-being. *Journal of environmental psychology* 37: 1–7.

Kubo, T., Shoji, Y., Tsuge, T., and Kuriyama, K., 2018. Voluntary contributions to hiking trail maintenance: evidence from a field experiment in a national park, Japan. *Ecological Economics* 144: 124–128.

Kupfer, J.A., Li, Z., Ning, H., and Huang, X., 2021. Using mobile device data to track the effects of the COVID-19 pandemic on spatiotemporal patterns of national park visitation. *Sustainability* 13(16): 9366.

Lee, S.Y., Chengju, D., Zhihui, C., Hao, W., Kailang, G., Yirong, L., Yongjie, C., Wenyan, L., Qiang, F., and Wenbo, L., 2020. Assessing safety and suitability of old trails for hiking using ground and drone surveys. *ISPRS International Journal of Geo-Information* 9(4): 221.

Leong, B., 2019. Facial recognition and the future of privacy: I always feel like... somebody's watching me. *Bulletin of the Atomic Scientists* 75(3): 109–115.

Life View GPS, 2018. Elk Being Tracked with GPS in Grand Canyon to Understand their Interaction with Visitors. https://www.liveviewgps.com/blog/elk-tracked-gps-grand-canyon-understand-interaction-visitors/ (Accessed on 19 December 2021).

Loulanski, T., and Loulanski, V., 2011. The sustainable integration of cultural heritage and tourism: a meta-study. *Journal of Sustainable Tourism* 19(7): 837–862.

Lukoseviciute, G., Pereira, L.N., and Panagopoulos, T., 2021. Sustainable recreational trail design from the recreational opportunity spectrum and trail user perception: a case study of the Seven Hanging Valleys. *Journal of Ecotourism*: 1–22.

Madden, K., Ramsey, E., Loane, S., and Condell, J., 2021. Trailgazers: a scoping study of footfall sensors to aid tourist trail management in Ireland and other Atlantic areas of Europe. *Sensors* 21(6): 2038.

Mandić, A., and Vuković, M., 2021. Millennials attitudes, choices and behaviour - integrative analysis. *Journal of Ecotourism* 21(1): 87–97.

Marion, J.L., and Wimpey, J., 2017. Assessing the influence of sustainable trail design and maintenance on soil loss. *Journal of Environmental Management* 189: 46–57.

Martin, S., 2017. Real and potential influences of information technology on outdoor recreation and wilderness experiences and management. *Journal of Park and Recreation Administration* 35(1): 98–101.

Martin, S.R., and Blackwell, J.L., 2016. Personal locator beacons: influences on wilderness visitor behavior. *International Journal of Wilderness* 22(1): 25–31.

Martin, S.R., and Pope, K., 2012. The influence of hand-held information and communication technology on visitor perceptions of risk and risk-related behaviour. USDA Forest Service Proceedings RMRS-P-66:119-126.

Martínez-Graña, A.M., Serrano, L., González-Delgado, J.A., Dabrio, C.J., and Legoinha, P., 2017. Sustainable geotourism using digital technologies along a rural georoute in Monsagro (Salamanca, Spain). *International Journal of Digital Earth* 10(2): 121–138.

Miller, J.R., and Hobbs, N.T., 2000. Recreational trails, human activity, and nest predation in lowland riparian areas. *Landscape and Urban Planning* 50(4): 227–236.

Mirzaalian, F., and Halpenny, E., 2021. Exploring destination loyalty: application of social media analytics in a nature-based tourism setting. *Journal of Destination Marketing & Management* 20: 100598.

Moore, R.L., and Ross, D.T., 1998. Trails and recreational greenways. *Parks & Recreation* 33(1): 68.

Moore, R.L., and Shafer, C.S., 2001. Introduction to special issue trails and greenways: opportunities for planners, managers, and scholars. *Journal of Park and Recreation Administration* 19(3): 1–16.

Moritz, J., Novak, M., and Schwan, S., 2018. Digitalization and mountaineering - Do digital alternatives replace traditional hiking maps? *Kartographische Nachrichten Issue* 3: 134–140.

Moscardo, G., 1996. Mindful visitors: heritage and tourism. *Annals of Tourism Research* 23: 376–396.

National Pak Service, 2021. Plan your visit. https://www.nps.gov/planyourvisit/passes.htm (Accessed on 19 December 2021).

New Jersey Department of Health, 2020. Health and safety standards for outdoor and indoor high-touch amusement and recreation activities pursuant to executive orders nos. 157 and 181 https://nj.gov/health/cd/documents/topics/NCOV/COVID-19-SafetyStandards_HighTouchAmusements.pdf (Accessed on 20 December 2021).

Newman, C., 2010. A feasibility study for a recreational water trail on the river Taff. Doctoral dissertation, Cardiff Metropolitan University.

Olafsson, A.S., and Skov-Petersen, H., 2014. The use of GIS-based support of recreational trail planning by local governments. *Applied Spatial Analysis and Policy* 7(2): 149–168.

Queen, L.P., Vlaming, J.C., Arthaud, G.J., and Lime, D.W., 1997. Modelling impacts of forest roads on recreation opportunities. *Northern Journal of Applied Forestry* 14(4): 194–201.

Randler, C., Tryjanowski, P., Jokimaki, J., Kaisanlahti-Jokimaki, M.L., and Staller, N., 2020. SARS-CoV2 (COVID-19) pandemic lockdown influences nature-based recreational activity: the case of birders. *International Journal of Environmental Research and Public Health* 17(19): 7310.

Reis, A.C., and Jellum, C., 2012. Rail trail development: a conceptual model for sustainable tourism. *Tourism Planning and Development* 9(2): 133–147.

Rezapouraghdam, H., Akhshik, A., and Ramkissoon, H., 2021. Application of machine learning to predict visitors' green behavior in marine protected areas: evidence from Cyprus. *Journal of Sustainable Tourism* 1–25.

Ribet, S., and Brander, L.M., 2020. Willingness to pay of trail runners for sustainable country park use in Hong Kong. *Journal of Outdoor Recreation and Tourism* 31: 100320.

Rice, W.L., Park, S.Y., Pan, B., and Newman, P., 2019. Forecasting campground demand in US national parks. *Annals of Tourism Research* 75: 424–438.

Rogers, A.G., and Leung, Y.F., 2020. Smarter long-distance hike: how smartphones shape information use and spatial decisions on the Appalachian Trail. *Journal of Wilderness* 26(2): 88–103.

Russel, S., and Norvig, P., 2016. Impact of AI and robotics in the tourism sector: a critical insight. *Journal of Tourism Futures* 8(1): 73–87.

Sahani, N., and Ghosh, T., 2021. GIS-based spatial prediction of recreational trail susceptibility in protected area of Sikkim Himalaya using logistic regression, decision tree and random forest model. *Ecological Informatics* 64: 101352.

Samala, N., Katkam, B.S., Bellamkonda, R.S., and Rodriguez, R.V., 2020. Impact of AI and robotics in the tourism sector: a critical insight. *Journal of Tourism Futures* 64: 101352.

Samara, D., Magnisalis, I., and Peristeras, V., 2020. Artificial intelligence and big data in tourism: a systematic literature review. *Journal of Hospitality and Tourism Technology* 11(2): 343–367.

Samarathunga, W.H.M.S., Cheng, L., and Weerathunga, P.R., 2020. Transitional domestic tourist gaze in a post-war destination: a case study of Jaffna, Sri Lanka. *Tourism Management Perspectives* 35: 100693.

Santos, M., Edwards, J., and Laranja, M., 2021. From digital innovation to "smart tourism destination": stakeholders' reflections in times of a pandemic. European Commission, JRC123390.

Sever, I., and Verbic, M., 2018. Providing information to respondents in complex choice studies: a survey on recreational trail preferences in an urban nature park. *Landscape and Urban Planning* 169: 160–177.

Shafiee, S., Ghatari, A.R., Hasanzadeh, A., and Jahanyan, S., 2019. Developing a model for sustainable smart tourism destinations: a systematic review. *Tourism Management Perspectives* 31: 287–300.

Shultis, J., 2012. The impact of technology on the wilderness experience: a review of common themes and approaches in three bodies of literature. In Wilderness Visitor Experiences: progress in research and management (pp. 110–118). *Proceedings RMRS-P-66: USDA Forest Service*, Rocky Mountain Research Station, Fort Collins, CO.

Sigala, M., 2018. New technologies in tourism: from multi-disciplinary to anti-disciplinary advances and trajectories. *Tourism Management Perspectives* 25: 151–155.

Sigala, M., 2020. Tourism and Covid-19: impacts and implications for advancing and resetting industry and research. *Journal of Business Research* 117: 312–321.

Silverman, D., 2016. *Qualitative research*. London: SAGE Publications Inc.

Sipe, N.G., and Dale, P., 2003. Challenges in using geographic information systems (GIS) to understand and control malaria in Indonesia. *Malaria Journal* 2(1): 1–8.

Skinner, T., 2020. Virtual worlds: can we travel without travelling? http://www.bbc.com/travel/story/20200330-covid-19-virtual-travel-during-coronavirus (Accessed on 14 December 2021).

Smith, A., and Anderson, M., 2018. *Social media use in 2018*. USA: Pew Research Centre https://www.pewresearch.org/internet/2018/03/01/social-media-use-in-2018/ (Accessed on 5 October 2021).

Smukavic, M., Posloncec-Petric, V., and Franges, S., 2014. Interactive hiking map - example of the national park Paklenica. *Geonauka* 2(4): 15–22.

Snyder, S.A., Whitmore, J.H., Schneider, I.E., and Becker, D.R., 2008. Ecological criteria, participant preferences and location models: a GIS approach toward ATV trail planning. *Applied Geography* 28(4): 248–258.

Sotiriadou, P., Brouwers, J., and Le, T.A., 2014. Choosing a qualitative data analysis tool: a comparison of NVivo and Leximancer. *Annals of Leisure Research* 17(2): 218–234.

Spenceley, A., and Goodwin, H., 2007. Nature-based tourism and poverty alleviation: impacts of private sector and parastatal enterprises in and around Kruger National Park, South Africa. *Current Issues in Tourism* 10(2–3): 255–277.

Spenceley, A., McCool, S., Newsome, D., Báez, A., Barborak, J.R., Blye, C.J., Bricker, K., Sigit Cahyadi, H., Corrigan, K., Halpenny, E., Hvenegaard, G., Malleret King, D., Leung, Y-F., Mandić, A., Naidoo, R., Rüede, D., Sano, J., Sarhan, M., Santamaria, V., Beraldo Sousa, T., and Zschiegner, A.K., 2021. Tourism in protected and conserved areas amid the COVID-19 pandemic. *Parks* (27): 103–118.

Staab, J., Udas, E., Mayer, M., Taubenbock, H., and Job, H., 2021. Comparing established visitor monitoring approaches with triggered trail camera images and machine learning based computer vision. *Journal of Outdoor Recreation and Tourism* 35: 100387.

Stanley, P., 2019. Unlikely hikers? Activism, Instagram, and the queer mobilities of fat hikers, women hiking alone, and hikers of colour. *Mobilities* 15: 241–256.

Stinson, J., 2017. Geoforum re-creating wilderness 2 .0: or getting back to work in a virtual nature. *Geoforum* 79: 174–187.

Strauss, A.L., and Corbin, J., 1990. *Basics of qualitative research 15*. Newbury Park, CA: Sage.

Strozzi, F., Colicchia, C., Creazza, A., and Noe, C., 2017. Literature review on the 'Smart Factory' concept using bibliometric tools. *International Journal of Production Research* 55(22): 6572–6591.

Sturgeon, T.J., 2021. Upgrading strategies for the digital economy. *Global Strategy Journal* 11(1): 34–57.

Svobodova, K., Monteiro, L., Vojar, J., and Gdulova, K., 2019. Can trail characteristics influence visitor numbers in natural protected areas? A quantitative approach to trail choice assessment. *Environmental & Socio-economic Studies* 7(2): 10–20.

Symmonds, M.C., Hammitt, W.E., and Quisenberry V.L., 2000. Managing recreational trail environments for mountain bike user preferences. *Environmental Management* 25(5): 549–564.

Taylor, J.C., Brewer, T.R., and Bird, A.C., 2000. Monitoring landscape change in the National Parks of England and Wales using aerial photo interpretation and GIS. *International Journal of Remote Sensing* 21(13–14): 2737–2752.

Teles da Mota, V., and Pickering, C., 2020. Using social media to assess nature-based tourism: current research and future trends. *Journal of Outdoor Recreation and Tourism* 30: 100295.

Tenkanen, H., DiMinin, E., Heikinheimo, V., Hausmann, A., Herbst, M., Kajala, L., and Toivonen, T., 2017. Instagram, Flickr, or Twitter: assessingtheusabilityofsocialmediadataforvisitormonitoringinprotectedareas. *Scientific Reports* 7: 17615.

The National Trails Office, 2012. *A guide to planning and developing recreational trails in Ireland*. Dublin: The Irish Sports Council.

Thomsen, J.M., Fowler, J., and Lang, T., 2021. A proposed research agenda on professional and recreational drone use in national forests and national parks. *Journal of Park & Recreation Administration* 39(2): 119–128.

Timothy, D.J., and Boyd, S.W., 2015. *Tourism and trails: cultural, ecological and management issues*. Bristol: Channel View Publications.

Toivonen, T., Heikinheimo, V., Fink, C., Hausmann, A., Hippala, T., Jarv, O., Tenkanen, H., and Minin, E., 2019. Social media data for conservation science: a methodological overview. *Biological Conservation* 233: 298–315.

Tomczyk, A.M., and Ewertowski, M., 2013. Planning of recreational trails in protected areas: application of regression tree analysis and geographic information systems. *Applied Geography* 40: 129–139.

Tomczyk, A.M., Ewertowski, M.W., White, P.C., and Kasprzak, L., 2017. A new framework for prioritising decisions on recreational trail management. *Landscape and Urban Planning* 167: 1–13.

Tomlinson, R.F., 1987. Current and potential uses of geographical information systems The North American experience. *International Journal of Geographical Information System* 1(3): 203–218.

Tomzcyk, A.M., 2011. A GIS assessment and modelling of environmental sensitivity of recreational trails: the case of Gorce National Park, Poland. *Applied Geography* 31(1): 339–351.

Tyrvainen, L., Uusitalo, M., Silvennoinen, H., and Hasu, E., 2014. Towards sustainable growth in nature-based tourism destinations: clients' views of land use options in Finnish Lapland. *Landscape and Urban Planning* 122: 1–15.

U.S. Forest Service, 2011. Trail Assessment and Condition Surveys. Available at: http://bit.do/fg57m (Accessed on 13 January 2020).

United Nations, 2020. Policy Brief: COVID-19 and Transforming Tourism.

Weatherby, T.G., and Vidon, E.S., 2018. Delegitimizing wilderness as the man cave: the role of social media in female wilderness empowerment. *Tourist Studies* 18(3): 332–352.

Whinam, J., Chilcott, N., Ling, R., and Wyatt, P., 2003. A method for calculating environmental sensitivity to walker trampling in the Tasmanian Wilderness World Heritage Area. In R. Buckley, C. Pickering, D.B. Weaver (Eds.), *Nature-based tourism, environment and land management* (p. 151–165). Canberra: Australian Academy of Science.

Wick, R., 2016. Technology brings new challenges to wilderness managers: an example from the Bureau of Land Management-managed Lost Coast of California. *Journal of Forestry* 114(3): 415–416.

Wolf, I.D., Stricker, H.K., and Hagenloh, G., 2013. Interpretive media that attract park visitors and enhance their experiences: a comparison of modern and traditional tools using GPS tracking and GIS technology. *Tourism Management Perspectives* 7: 59–72.

Wood, S. A., Guerry, A.D., Silver, J.M., and Lacayo, M., 2013. Using social media to quantify nature-based tourism and recreation. *Scientific Reports* 3(1): 1–7.

Zajchowski, C.A.B., South, F., Rose, J., and Crofford, E., 2021. The role of temperature and air quality in outdoor recreation behaviour: a social-ecological systems approach. *Geographical Review* 112(4): 512–531.

12

FOSTERING STAKEHOLDER ENGAGEMENT IN SUSTAINABLE CULTURAL TOURISM DEVELOPMENT IN NATURE-BASED SITES

A case study on using methodological layering of art-based methods

Hong Li, Monika Lüthje, Ella Björn and Satu Miettinen

Challenges and opportunities of tourism development in Finnish Lapland

Finnish Lapland is a popular nature-based tourism destination known for its vast wilderness areas and many nature reserves. This image does not come, however, without its challenges. Tourism in the region is highly seasonal, with winter being the peak season. This seasonality brings its own problems, which are mainly socioeconomic. For many inhabitants, tourism offers only seasonal business and job opportunities. Climate change has shortened the peak season, which is also insecure due to fluctuating snow conditions. The regional tourism strategy aims at year-round tourism (Regional Council of Lapland, 2019).

Promoting the region as a wilderness may give tourists a false impression of an 'empty' space, a 'no-man's land', where they are free to do whatever they want because there are no humans living there. This 'wilderness' is, in fact, a cultured landscape for the local people (Olsen et al., 2019, p. 13; Sámediggi, n.d., p. 20), full of meaningful places that play important roles in their lives. This is especially so for the Indigenous Sámi who have lived in the area for thousands of years, but it is also a cultured landscape for the local Finns. Reindeer herding is a traditional nature-based livelihood for both the Sámi and the Finns. Regardless of ethnic background, many inhabitants spend a significant amount of time engaging in nature-based activities, such as picking wild berries, fishing, hunting, skiing, or snowmobiling. Nature is, thus, an important part of the local cultures. Culture is understood here as a way of life, including all material and non-material elements connected to it. These elements make up a shared system of meanings through which the members of the culture make sense of their lives and the world (see Hall, 1995, 2000; Hannerz, 1999; Macnaghten & Urry, 2001). Many tourists do not recognise the cultured nature of the 'wilderness' and behave irresponsibly with

DOI: 10.4324/9781003230748-15

negative effects on local wellbeing (see Alariesto, 2021; Kugapi, Höckert, Lüthje, Mazzullo, & Saari, 2020; Sámediggi, n.d.).

Dividing the world into nature and culture is a western custom, whereas in the Sámi worldview there is no distinction between them (Valkonen & Valkonen, 2014, pp. 30–31, 38; Valkonen & Valkonen, 2018, p. 22). A close, sustainable relationship with nature is commonly seen to characterize Sámi and other Indigenous cultures, and this special nature relationship plays an important role in Indigenous Sámi identity constructions and Sámi politics. The Sámi's relationship with nature is concrete, practical, and anchored to certain places and local environments with special meanings to them in their individual daily lives and practices (Valkonen & Valkonen, 2014).

Although nature remains the chief attraction in Finnish Lapland, tourists are increasingly interested in local ways of life and authentic cultural experiences. The interrelatedness of nature and local cultures could be developed into more culture-based tourism products (Kugapi et al., 2020, pp. 17–18; see also Brattland et al., 2020, pp. 15, 26). Such a shift would make the cultured nature of the 'wilderness' more visible to the tourists and offer an opportunity to educate them about local cultures and nature, which might lead to more respectful behaviour toward both (see De Bernardi, 2021). In Norway, Sámi guides have noticed that explaining the principles behind their nature use as part of the Sámi worldview gives an added value to the tourists' experience (Brattland et al., 2020, p. 23). Kramvig and Førde (2020, p. 5) have described storytelling as a way to translate traditional Sámi knowledge to tourists and a tool to 'encourage tourists to develop a sense of responsibility for the conservation of the destination'.

It can be difficult for local people to see the value of their everyday life as a tourism product or service. Sensitivity and open discussion within the local community are needed to decide which parts of local life can be shared with tourists and which not. From the point of view of social and cultural sustainability, it is essential that the local community can itself decide how its culture(s), lands, and waters are used in tourism (see Brattland et al., 2020, pp. 8–9, 18–19, 27–29; Hurst, Grimwood, Lemelin, & Stinson, 2020, pp. 10–11; Kugapi et al., 2020, pp. 8, 24).

Art-based methods (ABMs) have proved powerful in various realms of research for offering abundant forms and approaches (Coemans & Hannes, 2017) to engage with communities, particularly certain marginalized and vulnerable groups (Redwood, Gale, & Greenfield, 2012); to support participatory development (Seppälä, Sarantou, & Miettinen, 2021); and to empower individuals. Traditional research methods are less effective in engaging such groups as they are often rigid because they are not culturally neutral, and are, therefore, likely to encounter obstacles during the research process. Art-based methods allow more creativity and flexibility for creating suitable approaches that are culturally tuned and sensitive to participants' values and beliefs (Redwood et al., 2012). For instance, Miettinen, Erkkilä-Hill, Koistinen, Jokela, and Hiltunen (2019) have introduced how the creative tourism landscape in Lapland was constructed using creative and cultural activities. The activities are tools to enable creative participation amongst tourists and local communities to increase art, creativity, and innovation in tourism activities in the Arctic. Relatively limited work has used ABMs to engage local stakeholders in tourism development. This chapter introduces how methodological layering of ABMs was used in Utsjoki – a nature-based, remote, Sámi-populated site – to engage the diverse local stakeholders in sustainable cultural tourism planning and development. We offer methodological reflection regarding the role ABMs played in fostering stakeholder engagement and the lessons learned from the use of ABMs.

Participatory methodologies for tourism development

Prior work of relevance to our area of research falls into participatory methodologies for tourism development and studies that have utilized art-based and service design methods for stakeholder engagement and community work.

Participatory tourism planning methods

Much has been written about participatory tourism planning and development. The core aim of the participatory approaches is to 'guarantee local communities' active involvement in their own development' (Höckert, Kugapi, & Lüthje, 2021, p. 15). As cultural tourism development in Utsjokiis in its planning phase, we wanted to gain an understanding of the latest methods used to engage local stakeholders in planning tourism development. For this purpose, we conducted a systematic review (see Kim, Bai, Kim, & Chon, 2018) of participatory tourism planning cases recently reported in peer-reviewed publications in Scopus. As the use of the terms planning, design, and development overlaps in tourism research literature (see e.g., Ferraro, Schilling, Baeza, Oms, & Sá, 2020, p. 500), who speak of designing a strategic development plan for tourism), we used the following search words: 'participatory' and 'tourism' and 'planning' or 'development' or 'design' and 'methodology' or 'method'. We did not focus solely on Indigenous cases, as Utsjoki is multi-ethnic. We searched for both cultural tourism and nature-based cases, as Utsjoki has a mixture of both.

The search resulted in 338 publications. We reviewed the abstracts of 160 recent articles (published between 2022 and 2018) and chose for further review all publications reporting empirical cases where participatory methods were used for engaging local stakeholders in tourism planning. Four potentially relevant cases had to be excluded because no full text was available and two because the full text was not in English. We had 21 relevant cases for full text review. After reading, we excluded three because the methods used were not described in sufficient detail. We, therefore, had 18 cases representing different tourism settings (e.g., rural and urban, cultural and nature-based, Indigenous and non-Indigenous). To categorize the methods used in the articles, we adapted the taxonomy of bottom-up participation tools created by Sève, Redondo Domínguez, Millán Gómez, and Sega (2020) (see Table 12.1). The taxonomy was originally created for urban planning and consists of four categories – tool, duration, place, and purpose – which can be combined in various ways, depending on the planning task at hand. Different combinations of tools result in different methods (Sève et al., 2020).

As can be seen from Table 12.1, in the reviewed planning cases, several participation tools and participatory methods were used per case, and similar tools and methods in different settings. Since planning processes normally have several phases with different aims, this is understandable. In the reviewed cases, the most popular tools were different images and meetings, mainly workshops, which brought stakeholders to discuss, work, and learn together. Images were mainly visualizations and often connected to mappings that were also popular. Images were frequently various kinds of maps. Most mappings were geographical, but sometimes social or systems mappings were used (Bertella, Lupini, Romanelli, & Font, 2021; Boyle, Gallachóir, & Mullally, 2021; Tourais & Videira, 2021). Surveys, interviews, and digital technologies were also commonly used tools. For example, a geographical information system (GIS) was utilized in several mappings, and several surveys were realized online. Writing may have been employed in more cases than indicated in Table 12.1, as it was not always explicitly stated if it was used, for instance, in the workshops or surveys described in the reviewed publications. Various kinds of models and exploration routes were also used, but less frequently. Service design was used in one case (Bertella et al., 2021).

Table 12.1 Participation tools and principal methods used in a set of recently reported participatory tourism planning cases.

Authors	*Participation tools* *Digital technologies*	*Mapping*	*Models*	*Image*	*Participative game*	*Exploration route*	*Arts and crafts*	*Meetings and assemblies*	*Surveys and polls*	*Interviews*	*Writing*	*Main method(s) used*
Arbogast, Butler, Faulkes, Eades and Deng (2020)	x	x		x				x	x	x	x	Social design methods, e.g., asset mapping (PPGIS), visualisations
Bertelli, Lupini, Romanelli and Font (2019)		x		x				x			x	Workshop methodology, service design
Bettelli, Orso, Pluchino and Gamberini (2019)				x				x	x		x	Co-design methods: affinity diagram, survey
Bojorquez-Vargas, Pavon and Quintal (2020)		x		x				x			x	Participatory mapping of tourism resources, workshop
Boyle, Gallachóir and Mullally (2022)		x		x				x			x	Participatory network mapping, workshop
Cattaneo, Giorgi and Ni (2018)		x		x						x		landscape mapping
Đukanovic, Živkovic, Radosavljevic, Lalovic and Jovanovic (2021)	x	x		x		x		x	x	x	x	Participatory urban design methods: survey, interview, guided tour
Ferraro, Schilling, Baeza, Oms and Sá (2020)	x					x		x	x			Workshops
Ferretti and Gandino (2018)		x		x					x	x		Choice experiments based on interviews and survey
Fuldauer, Ives, Adshead, Thacker and Hall (2019)	x		x					x		x		Participatory back-casting
Garcia-Ayllon (2018)	x	x		x				x		x	x	Workshops, GIS participatory mapping
Gkoltsiou and Moukiagou (2021)	x	x	x	x		x		x	x	x		Participatory spatial SWOT (GIS)
Gonçalves et al. (2022)			x	x				x	x			Meetings, workshops, survey, modelling
Herrera-Franco, Alvarado-Macancela, Gavín-Quinchuela and Carrión-Mero, (2018)			x					x	x	x		Surveys, workshops, model
Koens et al. (2022)	x		x	x	x			x		x		Serious game
Plieninger et al. (2018)	x	x		x					x		x	Participatory mapping of landscape values and development preferences (PPGIS)
Romão, Seal, Hansen, Joseph and Piramanayagam, (2021)	x								x			Online survey
Sugimoto, Sugino and Hori (2022)	x			x				x	x	x	x	Workshop, survey
Tourais and Videira (2021)	x	x	x	x				x	x	x		Participatory systems mapping

Source: Adapted from Sève et al. (2020).

No participative games, arts, or crafts were used in the 18 reviewed cases. We, therefore, searched separately in Scopus for publications about them. With the search words 'participatory' or 'serious' and 'game' and 'tourism' and 'planning' or 'design' or 'development', we found 54 publications. Only one of these publications, Koens et al. (2020), was an empirical tourism planning case with local stakeholders engaged in the planning process. (It is included in Table 12.1) Finding tourism planning cases with ABMs was even less successful. In Scopus, we found three publications with the search words 'art-based' and 'method' and 'tourism', but none of them was about engaging local stakeholders in tourism planning (Rydzik, Pritchard, Morgan, & Sedgley, 2012, 2013; Pipere, Mārtinsone, Regzdina-Pelēke, & Griškeviča, 2020). We also tried two other databases, EBSCO Hospitality and Tourism Complete and Leisure Tourism (CABI), but neither gave us any results. All searches were carried out with publication titles, abstracts, and keywords. Based on these searches, ABMs have been rarely used in tourism research.

Art-based and service design methods for the development of sustainable tourism

Art-based methods utilize creativity and creative thinking as resources that are available and can be used by all of us (Leavy, 2015). These methods can create an opportunity for equal collaboration where all participants can contribute with their existing creative skills and capacities (Shapiro, 2020). Artistic practice can play a valuable role in addressing societal challenges such as inequalities (Sonn & Baker, 2016) through inclusion, social justice through response (Arlington, 2018), and climate change through bringing people together (O'Neill & Smith, 2014) for action. The arts enable civic engagement and social change, mobilizing a variety of actors around a common agenda. Thus, ABMs work well in different development contexts, for example, in developing tourism services. Art-based methods can enable connection, empathy, and self-reflection (Leavy, 2015, 2018). Small local communities at the margins can be vulnerable as livelihoods may be scarce or one-sided and services may be far away. Art-based methods can be useful for working with such communities as they are effective tools in bringing together different identity groups through sharing common cultural experiences and facilitate using various cultural traditions when engaging communities in development work (Huhmarniemi & Jokela, 2020).

We integrated ABMs with service design into the tourism development process. Service design has been utilized as a means of tourism development (Zehrer, 2009). Similar to ABMs, service design methods and tools can be used to promote equality and inclusion (Miettinen & Vuontisjärvi, 2016). Service design is a well-suited approach for tourism development projects due to its user-centredness. This approach can contribute to innovation development and create competitive advantages at both strategic and tactical levels (Wetter-Edman et al., 2014), and create transformational change and, yet, keep in mind the sustainable and human-centred agenda.

Using art-based methods in the Utsjoki living lab

Utsjoki – the northernmost municipality in Finland that belongs to the homeland of the Indigenous Sámi people – was chosen as one of the living labs (Bergvall-Kåreborn, Eriksson, Ståhlbröst, & Svensson, 2009) in the SmartCulTour project, to support the development of sustainable cultural tourism with other living labs across European regions. As a nature-based site, Utsjoki provides many opportunities for tourists to seek authentic, nature-related, and unique experiences. The quietness, Arctic nature, and cultural activities of this region are all significant assets for tourism development. The municipality of Utsjoki has attempted to encourage local tourism companies to develop year-round products to pursue sustainable tourism development (Municipality of Utsjoki, 2020). Developing cultural tourism could offer a solution since Utsjoki is rich in Sámi culture. There are several cultural attractions, such as the historic church huts and the old Sámi farm Välimaa, which present information on Sámi culture and the way of life in the past. Modern Sámi culture is also visible in everyday life in Utsjoki, and some local tourism companies offer activities where tourists can familiarize themselves with the local Sámi culture. Cultural tourism development has been and still is a sensitive issue in the municipality, as Sámi cultures have been misrepresented and appropriated in tourism development elsewhere in Finnish Lapland for a long time (Kugapi et al., 2020; Saari, Höckert, Lüthje, Kugapi, & Mazzullo, 2020). Cultural sensitivity (Kugapi et al., 2020; Saari et al., 2020; Viken, Höckert, & Grimwood, 2021) – a concept for recognizing and respecting local cultural heritage and cultural differences – should be taken into consideration when developing sustainable tourism, especially when collaborating with multi-ethnic communities and dealing with different cultures. Sámi culture, cultural sensitivity, authenticity, locality, and sustainability are the foundations for every act of development in Utsjoki. Developing tourism in an ecologically, economically, and socially responsible way is part of Utsjoki's tourism development and land use planning. The aim of developing tourism in Utsjoki is to enhance the vitality of the area by bringing new services and working possibilities. The wellbeing of local people can be enhanced through developing tourism in ecologically, economically, and socially sustainable ways (Municipality of Utsjoki, 2020). Local cultures and nature can be assets for tourism product development, but are also crucial resources for the wellbeing of the local community, towards which tourism products should be respectful.

To develop sustainable cultural tourism in the Utsjoki Lab through a bottom-up approach, we conducted two participatory workshops (see the overview in Table 12.2) with a wide range of stakeholders, including those from local authorities, policymakers, land-use planners, representatives from resident associations, local entrepreneurs from the tourism industry, practitioners

Table 12.2 The overview of the workshops carried out in the Utsjoki Lab.

	Workshop I	*Workshop II*
Number of participants	On-site: 8 participants in two groups. Online: 4 participants in one group.	On-site: 5 participants in two groups. Online: 4 participants in one group.
Number of facilitators	On-site: 1facilitator in each group, and 1 researcher observed and took notes. Online: 1 main facilitator, and 1 researcher observed and took notes.	On-site: 1facilitator in each group, and 1 researcher observed and took notes. Online: 1 main facilitator, and 1 researcher observed and took notes.
Objectives	Engaging the Lab stakeholders in identifying and clarifying the problems and needs regarding (sustainable cultural) tourism development in the municipality Utsjoki.	Collaboratively exploring the potential of cultural tourism in Utsjoki and generating ideas for developing sustainable cultural tourism based on the identified problems and needs.
Stage of the process	Discover → Define	Define → Develop
ABMs used	Opportunity tree	Pictures as probes Multimethod process flow
Service design methods integrated	Sticky notes sorting game	Serious play

from cultural and creative industries, scholars from research centres, and reindeer herders (Li, Miettinen, Luthje, Honkanen, & Björn, 2021). The workshops were carried out in hybrid settings to cope with social distancing during the COVID-19 pandemic. Table 12.2 outlines the ABMs and service design methods implemented as a participatory approach to engage with the local stakeholders. The methods to be implemented in each stage of the cultural tourism development process were chosen based on the well-established double-diamond design model (Design Council, 2005), which consists of four stages – discover, define, develop, and define. As shown in Figure 12.1, a development process of tourism strategies, policies, products, and services was visualized in the Double Diamond Design model for the Utsjoki Lab context. The first workshop in the Lab was positioned at the transition stage between *discover* and *define*, diving deep into identifying problems and needs, and then clarifying the problem definition based on the insights gathered. The second workshop proceeded from *define* to *develop*, where potential solutions to the problem were created collaboratively.

The next section introduces how we implemented each method in the Lab and highlights key findings.

Opportunity tree

The opportunity tree is a collaboration tool initially designed in the AMASS project to enable stakeholders to identify their needs, opportunities, and best practices within their contexts in digital environments (Sarantou, Remotti, Alhonsuo, & Gutierrez Novoa, 2021). We adapted the tool to engage our participants in identifying the problems, opportunities, and good practices around developing sustainable cultural tourism for the Utsjoki Lab. A sticky notes sorting game (Li & Smit, 2021) was used to help identify key needs and problems. First, the participants contributed their insights together on sticky notes, then similar items were grouped together by repositioning the sticky notes and giving a title to each group based on the category they represented. The results created the root of the tree, which represents the primary needs and problems for cultural tourism development in Utsjoki from different viewpoints. Next, the participants were asked to share existing good practices regarding sustainable cultural tourism development. Preparing related practices was assigned to the participants as a preworkshop task. During the workshop, the participants worked together to create a best practice they believed could help address the major needs and problems identified in the root. The best practice served as the trunk of the tree. Building on the insights came from the root and the trunk. The participants identified the opportunities (i.e., branches of the tree) – for example, actions, resources, and people needed to implement the best practice. Lastly, each group presented the results on their tree (see Figure 12.2).

The opportunity tree served as an effective tool for visualization and collaboration. First, it helped to visualize comprehensive outcomes that provided participants with immediate conclusions and results from the group discussions. The results of the roots uncovered the main needs and problems in the Lab area. For example, tourists might not know of every person's rights, such as the free right to roam in the natural land and collect natural products (e.g., mushrooms and berries), provided that they respect nature by not leaving any trash behind and keeping enough distance from local peoples' homes. Trashing and trespassing were revealed as problems for local tourism development. Thanks to the opportunity tree, we gathered useful insights from the Lab stakeholders. Enhancing tourists' knowledge about Sámi culture and giving essential information on how to behave respectfully in the destination were deemed useful for developing sustainable cultural tourism. Informing tourism entrepreneurs on how to handle Sámi culture in tourism was also considered important for protecting the local cultural heritage.

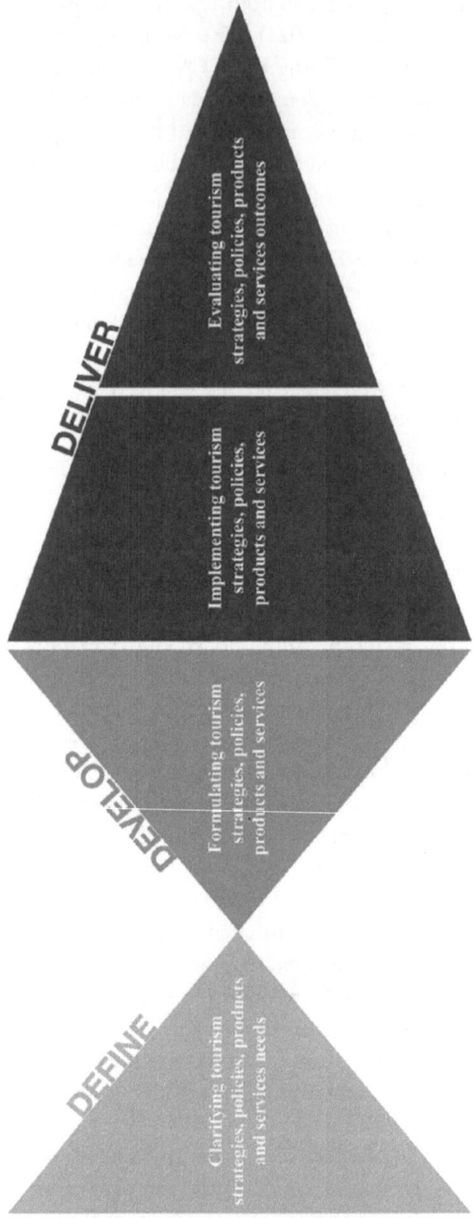

Figure 12.1 The double-diamond design model adapted to the Utsjoki Lab context.

Source: Li & Smit, 2021.

Figure 12.2 A group presenting the results of their opportunity tree in the workshop.

Second, the opportunity tree helped to enhance stakeholder engagement by providing a visual tool with which the participants collaborated and communicated effectively despite representing different interests. There has been a lack of collaboration amongst different actors, especially with land-use plans within tourism and reindeer herding livelihoods (Municipality of Utsjoki, 2020). Therefore, the participants could find it difficult to reach common ground and facilitate coordination amongst themselves. For instance, reconciling new tourism construction and service activities with reindeer husbandry can be challenging. The opportunity tree offered a strategic structure with which they worked together to discuss the needs, problems, existing good practices, and opportunities from different viewpoints. After the workshop, the participants who represented different stakeholders gained a shared understanding and found common interests. Developing sustainable cultural tourism in Utsjoki could bring fresh blood into the municipality's tourism and would support the development of year-round tourism, since cultural tourism is not as seasonal as nature-based tourism. They identified a solution, a mobile application that can guide tourists with all the essential information about Utsjoki and related services in the area.

Pictures as probes

Pictures as probes is a storytelling tool that uses meaningful pictures provided by participants as story props to encourage them to tell stories behind their own pictures (Li & Smit, 2021). Before the workshop, we asked our participants to bring pictures they had taken in the local area that could represent cultural practice, ritual, or place that are educative, creative, aesthetic, emotional, or entertaining. The purpose was to have the participants explore the potential of cultural tourism in Utsjoki based on its definition. This drew on Matteucci and Von Zumbusch

Figure 12.3 A participant showing their picture on the phone.

(2020, p. 19) definition: 'A form of tourism that in which tourists can engage with heritage, local cultural and creative activities and the everyday cultural practices of host communities for the purpose of gaining mutual experiences of an educational, aesthetic, creative, emotional and/or entertaining nature'. The pictures provided by our participants showed a strong link between nature and the local culture as their pictures and the stories behind them were linked to the local way of life, the unique places in the local area, the local art pieces, and the native animals in nature (see Figure 12.3). The stories told by our participants included the history of the Arctic Ocean, salmon migration, unique vegetation, local cultures, and how the Ice Age has shaped the sceneries in history. It was believed that the stories, culture, and history of Utsjoki could be brought alive through tourism, for instance, by associating the stories with the natural surroundings and enriching the existing tourism products and services.

Pictures as probes was an effective ABM for triggering and eliciting stories from participants. It evoked discussion on how the local landscape could be utilized as a natural platform for learning various stories of the local culture and the knowledge of plants, as those would be educational for tourists who come to experience and learn about Utsjoki and its culture. All our participants related to the uniquely beautiful nature of Utsjoki and the interesting stories shared, which inspired them to use visuals as a storytelling approach to spread information about the existing unique places in Utsjoki to potential visitors. On the downside, facilitating pictures as probes took longer than we initially planned, as several participants ran out of time to tell the stories behind their pictures. One participant did not follow the instructions and showed pictures they wanted to criticize rather than telling stories, which created slight tension in the group. The discussion of the criticized topic was continued in the later workshops and an agreement was reached among the Lab stakeholders.

Multi-method process flow

The multi-method process flow (see Figure 12.4) is a four-step collaboration tool for designing sustainable cultural tourism products and services collaboratively (Li & Smit, 2021). Multiple ABMs can be integrated into this tool to help create innovative ideas for further developing culturally interesting touristic assets and enriching the experience for cultural tourists. It was adapted from the 3S principle of community-based design – storytelling, senses, and

Figure 12.4 The Multimethod Process Flow (Li & Smit, 2021).

sophistication – proposed by Richards, Wisansing, and Paschinger (2018), where stories and senses are employed to design a sophisticated experience for tourists.

The multi-method process flow was employed after the pictures as probes exercise. We asked each group to select a local place with tangible or intangible assets based on which they generated ideas for developing sustainable cultural tourism. We provided a template that included several guiding questions to help them in the idea generation process. As shown in Figure 12.3, the template consisted of a four-step process. The process began with identifying local touristic assets, and then bringing them to life by creating narratives behind the assets based on the existing local culture. Next, ideas were generated to provide good sensorial experiences at the places for tourists, engaging any or all the senses. As the last step, the template invited participants to brainstorm on the possible learning experience that could be provided by local individuals, groups, or communities who create and transmit living heritage.

The multimethod process flow served as a helpful tool for developing ideas. It helped to invite creative inputs from the participants who worked collaboratively to explore and discover the potential local assets for supporting sustainable cultural tourism development in Utsjoki. For example, Ailigas Fells – a nature-based site – was identified by our participants as a local place with cultural significance to the local inhabitants, where its cultural importance could be emphasized for tourists. The multimethod process flow also elicited interesting discussions on how to develop tourism in Ailigas Fells in a sustainable way that would lessen the harmful environmental impacts. Our participants pointed out that the basic infrastructure services at Ailigas Fells were still missing and suggested that building guide signs and hiking trails would keep visitors in the guided area. This was believed to be a sustainable way to use the natural resources respectfully. The supporting questions in the template helped our participants to better understand each step and encouraged them to brainstorm in greater depth by providing more open-ended support. By following the steps, they came up with interesting ideas where narratives of the local assets were formulated, along with good sensorial experience and possible learning experience to support the narratives. These ideas had the potential to present a sophisticated experience for tourists. The multimethod process flow evoked discussions regarding the conflicts on land use between the tourism industry and local reindeer husbandry. Reindeer husbandry is an important livelihood in Utsjoki, which involves a lot of know-how regarding moving around in the hinterlands. Finding a solution to combine reindeer herding culture with nature-based tourism in a sustainable way was brought up in the discussion. Storytelling to enable knowledge of nature conservation and showcase how climate change affects nature, especially in reindeer husbandry, was believed to be one possible learning experience for tourists.

As shown in Figure 12.5, we provided our participants with a set of playful objects (e.g., finger puppets) in the workshop, and hoped to use role play – one form of serious play – as a playful way to encourage them to present their ideas created by using the multimethod process flow. Role play helped to create a relaxed atmosphere with much laughter in small groups. By performing the hypothetical concepts created by the participants themselves, using the provided toys, they could envisage what might work well and what could go wrong with their ideas. Engaging different stakeholders who came from the Lab area in serious play allowed for more realistic and holistic views on the social, environmental, and cultural dimensions of the destinations. However, several participants felt shy or embarrassed when asked to role play in a bigger group. Additionally, engaging online participants in role playing without proper tools was problematic.

Figure 12.5 The Lab participants engaging in serious play using finger puppets.

Methodological reflection

In this section, we offer methodological reflections on the roles ABMs played, and the lessons learned from implementing ABMs in the Utsjoki Lab, which may be of use to researchers and practitioners keen to develop ABMs for stakeholder engagement in different realms.

The two-fold role of ABMs

The ABMs generated and collected data with more inclusive and approachable engagement. Using ABMs provided relatively accessible, innovative, and interesting approaches, which engaged diverse stakeholders in tourism planning, including those with no expertise in tourism development. The role ABMs played in the Utsjoki Lab was two-fold. First, ABMs helped to engage stakeholders who came from a multiethnic community to discuss culturally sensitive topics in a collaborative and respectful manner. By using the ABMs, we elicited interesting discussions and insights on, for example, how to combine reindeer herding with tourism and how to represent Sámi culture in tourism in sustainable and respectful ways. In recent academic discussion on cultural sensitivity in tourism, reciprocity and mutuality have been emphasized. Cultural sensitivity can lead to mutual understanding and cultural exchange, which 'promote equity within partnerships/relationships, as well as create shared values and benefits to all parties' (Hurst et al., 2020, p. 9; Kugapi et al., 2020, p. 8; Saari et al., 2020, p. 105; Viken et al., 2021). Our contribution was to suggest methods by which this outcome can be attained in practice. Second, ABMs served as a bottom-up agent, through which we shifted our roles as designers toward roles as facilitators of design. In line with the changing roles in design highlighted by Sanders and Stappers (2008), we, as researchers, supported the Lab stakeholders as 'experts on their destination' by providing ABMs combined with service design tools and methods to help them in idea generation and expression. We encouraged our participants to take the lead in the design process, working collaboratively on identifying the needs and problems regarding sustainable cultural tourism development in the Lab area, and co-designing solutions.

The ABMs used in Utsjoki are a good addition to the toolkit of participatory tourism planning tools and methods reviewed in our chapter. For example, pictures as probes could be an alternative for GIS-based tourism asset mapping. From the perspective of participation, one can ask who should select the methods used and facilitate their use. Participatory tourism development projects have been criticized for being hosted by outside researchers and developers, leaving local people in the role of guests in their own development (Höckert et al., 2021). In the case of Utsjoki, we tried to follow the guidelines of Koster, Baccar, and Lemelin (2012) and employ a local person in our team to plan and run the Lab workshops, but, unfortunately, we could not find anyone for the work. Instead, we planned the aims and contents of the Lab workshops with the project manager of the local tourism development project, but the specific tools and methods used were selected by us, outsider researchers. On the other hand, our outsider presence may have contributed to a respectful discussion between the local stakeholders in the Lab.

There is a vast array of participatory tourism planning methods that can increase inclusivity in tourism development. As Höckert et al. (2021) have suggested, the roles of the 'host and guest' or 'teachers and learners' could constantly change during the development process to improve cooperation and team spirit and create reciprocity between the participants. In the Utsjoki Lab, helped by different participatory methods, the participants were encouraged to talk about local development needs and cultures, which gave them the role of 'teacher' and allowed discussion and decision-making about different development measures. As can be seen from Table 12.1, creative methods such as visualizations and images are commonly used in participatory tourism planning and also evoked discussions in the Utsjoki Lab. Without testing, it is impossible to identify which methods are best suited in which local context. The ABMs used in the Utsjoki context worked well, and, thus, we recommend also testing them elsewhere for discussing local development needs and strategies and for developing new tourism products in culturally sensitive ways. According to recent studies (Brattland et al., 2020; Hurst et al., 2020; Kugapi et al., 2020; Olsen et al., 2019; Saari et al., 2020; Viken et al., 2021), local participation and cultural sensitivity are needed in sustainable tourism planning and development in the Arctic, especially when dealing with tourism in Indigenous lands and waters or drawing on Indigenous cultures. However, this need for local participation and cultural sensitivity is also true for other local people, places, and cultures. When properly facilitated, the ABMs we suggest here can also help in tourism planning and development work beyond the Arctic.

Lessons learned

Art-based methods can offer benefits in the Indigenous agenda on participant engagement, relationship building, Indigenous knowledge creation, capacity building, and community action (Hammond et al., 2018). In the Utsjoki Lab, bringing different stakeholders to the same table to discuss the development of local tourism was already a significant starting point as it was not always easy. It is imperative that the study is relevant and interesting to the local community per se (Smith, 2013). Using ABMs to give voice to different stakeholders to express their needs and include their view of implementation was at the heart of our study. One promising opportunity we saw in our practice was the layering of different ABMs with other data collection methods to facilitate deeper understanding for the research participants. Having acknowledged that ABMs often generate 'interesting types of data' that can be difficult to interpret (Coemans & Hannes, 2017, p. 41), we combined ABMs with service design methods in the project to create dialogue, which enabled alternative ways of expressing needs while also embracing differences. The methods also enhanced engagement by building shared understanding among the participants and between the participants and researchers. When multi-methods are applied collaboratively and practiced with care, a multimethod process flow can emerge, integrating ABMs with established methods—the 3S principles of community-based design, for example (Richards et al., 2018).

The ABMs we created were not rigid, but rather flexible and adaptable. We, therefore, encourage the creative use of ABMs. For example, the multi-method process flow (see Figure 12.4), a stepped approach to incorporate multi-methods including ABMs, is a practical tool for workshop facilitation, but care is needed so this approach does not become rigid and compartmentalized. Strong connections between the four phases should be maintained to ensure flow, offering participants flexibility to move between the phases. Just as steps can lead down, they can also lead up, and so forth. The interconnections between multi-method approaches are important as they connect different processes within the research. It is, however, important that participants clearly understand the purpose of the different tools and methods used to engage in the co-design process.

Although the findings of ABMs in our study were positive, our work was limited by the small sample size. However, every participant in our study represented a key stakeholder in the case area. We believe they were the experts on their destination and could contribute valuable empirical insights and expertise on culturally sustainable tourism development in Utsjoki. The diversity of the sample participants gave solid indications for deepening our knowledge of the

Lab area from wider perspectives. The short-term duration of the study may not have been sufficient to assess the ABMs comprehensively. ABMs are not suitable for all and are not culturally neutral, as arts are perceived differently by individuals. We encountered several challenges while implementing ABMs in the workshops. For instance, some participants could not relate to the metaphor of arts and found some tools and methods too naïve. One reason for this is that the workshop facilitators who had no strong artistic background were not methodologically equipped to guide these arts-based processes (Coemans & Hannes, 2017). Arts offer a myriad of powerful ways to express and understand oneself. For example, using local and meaningful pictures collected by the participants as probes for storytelling helped to elicit stories from the participants. However, without an experienced facilitator, there is a risk of causing a conflict as various issues are expressed from different perspectives. Another challenge was the hybrid facilitation of the workshop due to COVID-19 restrictions. We lacked experience in remote facilitation and encountered technical challenges in facilitating the virtual workshops. The online participants could not use some methods in the same way as the on-site participants. In future work, creating ABMs for the social distancing era would be a timely topic to pursue.

Conclusion

We have illustrated how ABMs were used in Utsjoki – a nature-based living lab – for fostering stakeholder engagement in sustainable cultural tourism development. The value of ABMs for stakeholder engagement in a multi-ethnic community was clear in our study. We provided methodological reflection while implementing the ABMs, highlighting the roles they played and the lessons we learned. Although our findings on the ABMs created within the project were encouraging in this study, ABMs are not a ready-made panacea for all colonized ills, and they still need to be validated and expanded through more thorough investigations. Through the methodological layering of ABMs, we hope to provide different means for communities to express needs, embrace differences, and enhance engagement. We suggest tailoring these ABMs as a starting point for further research in needs-specific contexts.

Acknowledgments

We wish to acknowledge our Lab participants who took part in the project. We really appreciate the precious time, valuable insights, and interesting stories they contributed to the research.

SmartCulTour project has received funding from the European Union's Horizon 2020 research and innovation programme under grant agreement No. 870708.

References

Alariesto, E. (2021). The conflict of sacred and contaminant: The impurifying effects of tourism in Sámi sacred sites. *Finnish Journal of Tourism Research, 17*(1), 64–70.

Arbogast, D., Butler, P., Faulkes, E., Eades, D., & Deng, J. (2020). Using social design to visualize outcomes of sustainable tourism planning: A multiphase, transdisciplinary approach. *International Journal of Contemporary Hospitality Management, 32*(4), 1413–1448.

Arlington, A. K. (2018). Power and control: Responding to social injustice with photographic memes. *Art Education, 71*(6), 51–58.

Bergvall-Kåreborn, B., Eriksson, C. I., Ståhlbröst, A., & Svensson, J. (2009). *A milieu for innovation: Defining living labs.* Retrieved from https://www.diva-portal.org/smash/get/diva2:1004774/FULLTEXT01.pdf

Bertella, G., Lupini, S., Romanelli, C. R. & Font, X. (2021). Workshop methodology design: Innovation-oriented participatory processes for sustainability. *Annals of Tourism Research, 89*(1), 103251.

Bettelli, A., Orso, V., Pluchino, P., & Gamberini, L. (2019). An enriched visit to the botanical garden: Co-designing tools and contents. In *Proceedings of CHItaly'19* (pp. 1–5). Padova, Italy: ACM.

Bojorquez-Vargas, A. R., Pavon, R. G. S., & Quintal, M. D. D. P. (2020). Participatory rural planning and local capabilities toward offering nature tourism. *WIT Transactions on Ecology and the Environment, 248*(1), 215–224.

Boyle, E., Gallachóir, B.Ó., & Mullally, G. (2022). Participatory network mapping of an emergent social network for a regional transition to a low-carbon and just society on the Dingle Peninsula. *Local Environment, 27*(12), 1431–1445.

Brattland, C., Jæger, K., Olsen, K., Dunfjell, O., Elle, M., & Viken, A. (2020). *Cultural sensitivity and tourism. Report from Northern Norway.* Retrieved from https://lauda.ulapland.fi/bitstream/handle/10024/64259/Arctisen_Norway_report_FINAL_13.5.2020.pdf?sequence=1&isAllowed=y

Cattaneo, T., Giorgi, E., & Ni, M. (2018). Landscape, architecture and environmental regeneration: A research by design approach for inclusive tourism in a rural village in China. *Sustainability, 11*(1), 128.

Coemans, S., & Hannes, K. (2017). Researchers under the spell of the arts: Two decades of using arts-based methods in community-based inquiry with vulnerable populations. *Educational Research Review, 22*(1), 34–49.

De Bernardi, C. (2021). The connection between nature and Sámi identity: The role of ecotourism. In D.A. Fennell (Ed.), *The Routledge handbook of ecotourism* (pp. 144–156). Abingdon, England: Routledge.

Design Council. (2005). *Eleven lessons. A study of the design process.* Retrieved from https://www.designcouncil.org.uk/sites/default/files/asset/document/ElevenLessons_Design_Council%20(2).pdf

Đukanovic, Z., Živkovic, J., Radosavljevic, U., Lalovic, K., & Jovanovic, P. (2021). Participatory urban design for touristic presentation of cultural heritage sites: The case of Negotinske Pivnice (wine cellars) in Serbia. *Sustainability, 13*(18), 10039.

Ferraro, F.X., Schilling, M.E., Baeza, S., Oms, O., & Sá, A.A., (2020). Bottom-up strategy for the use of geological heritage by local communities: Approach in the "Litoral del Biobío" Mining Geopark project (Chile). *Proceedings of the Geologists' Association, 131*(5), 500–510.

Ferretti, V., & Gandino, E. (2018). Co-designing the solution space for rural regeneration in a new world heritage site: A choice experiments approach. *European Journal of Operational Research, 268*(3), 1077–1091.

Fuldauer, L. I., Ives, M. C., Adshead, D., Thacker, S., & Hall, J. W. (2019). Participatory planning of the future of waste management in small island developing states to deliver on the sustainable development goals. *Journal of Cleaner Production, 223*(1), 147–162.

Garcia-Ayllon, S. (2018). The Integrated Territorial Investment (ITI) of the Mar Menor as a model for the future in the comprehensive management of enclosed coastal seas. *Ocean and Coastal Management, 166*(1), 82–97.

Gkoltsiou, A., & Moukiagou, S. (2021). The use of islandscape character assessment and participatory spatial SWOT analysis to the strategic planning and sustainable development of small islands. The case of Gavdos. *Land Use Policy, 103*(1), 105277.

Gonçalves. C., Honrado, J. P., Cerejeira, J., Sousa, R., Fernandes, P. M., Vaz, A.S., … Santos, J. A. (2022). On the development of a regional climate change adaptation plan: Integrating model-assisted projections and stakeholders' perceptions. *Science of the Total Environment, 805*(1), 150320.

Hall, S. (1995). New cultures for the old. In D. Massey & P. Jess (Eds.), *A place in the world? Places, cultures and globalization* (pp. 175–213). New York, NY: Oxford University Press.

Hall, S. (2000). Conclusion: The multi-cultural question. In B. Hesse (Ed.), *Un/settled multiculturalisms: Diasporas, entanglements, transruptions* (pp. 209–241). London, England: Zed Books.

Hammond, C., Gifford, W., Thomas, R., Rabaa, S., Thomas, O., & Domecq, M. C. (2018). Arts-based research methods with indigenous peoples: an international scoping review. *Alter Native: An International Journal of Indigenous Peoples, 14*(3), 260–276.

Hannerz, U. (1999). Reflections on varieties of culturespeak. *European Journal of Cultural Studies, 2*(3), 393–407.

Herrera-Franco, G., Alvarado-Macancela, N. Gavín-Quinchuela, T., & Carrión-Mero, P. (2018). Participatory socio-ecological system: Manglaralto-Santa Elena, Ecuador. *Geology, Ecology, and Landscapes*, *2*(4), 303–310.

Höckert, E., Kugapi, O., & Lüthje, M. (2021). Hosts and guests in participatory development. In A. Harju-Myllyaho& S. Jutila (Eds.), *The future of inclusive tourism* (pp. 15–32). Bristol, England: Channel View Publications.

Huhmarniemi, M., & Jokela, T. (2020). Arctic art and material culture: Northern knowledge and cultural resilience in the Northernmost Europe. In L. Heininen, H. Exner-Pirot & J. Barnes (Eds.), *Arctic Yearbook 2020* (pp. 242–259). Akureyri, Iceland: Arctic Portal.

Hurst, C. E., Grimwood, B. S. R., Lemelin, R. H., & Stinson, M. J. (2020). Conceptualizing cultural sensitivity in tourism: A systematic literature review. *Tourism Recreation Research*, *46*(4), 1–16.

Kim, C. S., Bai, B. H., Kim, B. P., & Chon, K. (2018). Review of reviews: A systematic analysis of review papers in hospitality and tourism literature. *International Journal of Hospitality Management*, *70*(1), 49–58.

Koens, K., Klijs, J., Weber-Sabil, J., Melissen, F., Lalicic, L., Mayer, I., Önder, I., & Aall, C. (2022). Serious gaming to stimulate participatory urban tourism planning. *Journal of Sustainable Tourism*, *30*(9), 2167–2186.

Koster, R., Baccar, K., & Lemelin, R.H. (2012). Moving from research ON, to research WITH and FOR Indigenous communities: A critical reflection on community-based participatory research. *The Canadian Geographer*, *56*(2), 195–210.

Kramvig, B., & Førde, A. (2020). Stories of reconciliation enacted in the everyday lives of Sámi tourism entrepreneurs. *Acta Borealia*, *37*(1/2), 27–42.

Kugapi, O., Höckert, E., Lüthje, M., Mazzullo, N., & Saari, R. (2020). *Toward culturally sensitive tourism: Report from Finnish Lapland*. Retrieved from https://lauda.ulapland.fi/bitstream/handle/10024/64276/Arctisen%20Finland%20english%20report%20FINAL.pdf?sequence=1&isAllowed=y

Leavy, P. (2015). *Method meets art: Arts-based research practice*. New York, NY: Guilford Press.

Leavy, P. (2018). *Handbook of arts-based research*. New York, NY: Guilford Press.

Li, H., Miettinen, S., Luthje, M., Honkanen, A., & Björn, E. (2021). In C. Fidelbo & M. Rosati (Eds.), *Specific Terms of Reference for Smart Cul Tour Living Laboratories. Deliverable D6.1 of the Horizon 2020 project Smart Cul Tour (GA number 870708)*. SmartCulTour internal project report: unpublished.

Li, H., & Smit, B. (2021). *Set of service design and art-based methods for co-design and stakeholder work in cultural tourism*. Retrieved from http://www.smartcultour.eu/wp-content/uploads/2021/09/D7.1-Set-of-Service-Design-and-Art-Based-Methods-for-Co-Design-and-Stakeholder-Work-in-Cultural-Tourism.pdf

Macnaghten, P., & Urry, J. (2001). Bodies of nature: Introduction. In P. Macnaghten & J. Urry (Eds.), *Bodies of nature* (pp. 1–11). Thousand Oaks, CA: Sage.

Matteucci, X., & Von Zumbusch, J. (2020). *Theoretical framework for cultural tourism in urban and regional destinations*. Retrieved from http://www.smartcultour.eu/wp-content/uploads/2020/07/D2.1-Theoretical-framework-for-cultural-tourism-in-urban-and-regional-destinations.pdf

Miettinen, S., Erkkilä-Hill, J., Koistinen, S. M., Jokela, T., & Hiltunen, M. (2019). Stories of design, snow, and silence: Creative tourism landscape in Lapland. In N. Duxbury & G. Richards (Eds.), *A Research Agenda for Creative Tourism* (pp. 69–83). Cheltenham, England: Edward Elgar Publishing.

Miettinen, S., & Vuontisjärvi, H. (2016). Service design for participatory development. In P. Rytilahti & S. Miettinen (Eds.) *For Profit, for Good. Developing Organizations through Service Design* (pp. 68–73). Rovaniemi, Finland: University of Lapland.

Municipality of Utsjoki. (2020). Utsjoenmatkailunmaankäyttösuunnitelma - hanke. Retrieved from https://www.utsjoki.fi/project-article/utsjoen-matkailun-maankayttosuunnitelma-hanke/

O'Neill, S. J., & Smith, N. (2014). Climate change and visual imagery. *Wiley Interdisciplinary Reviews: Climate Change*, *5*(1), 73–87.

Olsen, K. O., Abildgaard, M. S., Brattland, C., Chimirri, D., de Bernardi, C., Edmonds, J., s… Viken, A. (2019). *Looking at Arctic tourism through the lens of cultural sensitivity. ARCTISEN – a transnational baseline report*. Retrieved from https://lauda.ulapland.fi/bitstream/handle/10024/64069/Arctisen%20-%20a%20transnational%20baseline%20report.pdf?sequence=1&isAllowed=y

Pipere, A., Mãrtinsone, K., Regzdina-Pelēke, L., & Griškeviča I. (2020). Sailing across the Atlantic: An exploration of the psychological experience using arts-based research. *Frontiers in Psychology*, *11*(1), 572028.

Plieninger, T., Av Rana, H. Á., Fagerholm, N., Ellingsgaard, G. F., Magnussen, E., Raymond, C. M., … Verbrugge, L. N. H. (2018). Identifying and assessing the potential for conflict between landscape values and development preferences on the Faroe Islands. *Global Environmental Change*, *52*(1), 162–180.

Redwood, S., Gale, N. K., & Greenfield, S. (2012). 'You give us rangoli, we give you talk': Using an art-based activity to elicit data from a seldom heard group. *BMC Medical Research Methodology*, *12*(7), 1–13.

Regional Council of Lapland. (2019). *Lapin matkailustrategia 2020–2023*. Retrieved from https://arcticsmartness.eu/wp-content/uploads/Matkailu_tilannekuvaraportti_web.pdf

Richards, G., Wisansing, J., & Paschinger, E. (2018). *Creating creative tourism toolkit*. Retrieved from https://perfectlink.co.th/wp-content/uploads/2019/01/Creating-Creative-Tourism-Toolkit_Version-Eng.pdf

Romão, J., Seal, P. P., Hansen, P., Joseph, S., & Piramanayagam, S. (2021). Stakeholder-based conjoint analysis for branding wellness tourism in Kerala, India. *Asia-Pacific Journal of Regional Science*, *6*, 91–111.

Rydzik, A., Pritchard, A., Morgan, N., & Sedgley, D. (2012). Mobility, migration and hospitality employment: Voices of Central and Eastern European women. *Hospitality & Society*, *2*(2), 137–157.

Rydzik, A., Pritchard, A., Morgan, N., & Sedgley, D. (2013). The potential of art-based transformative research. *Annals of Tourism Research*, *40*, 283–305.

Saari, R., Höckert, E., Lüthje, M., Kugapi, O., & Mazzullo, N. (2020). Cultural sensitivity in Sámi tourism: A systematic literature review in the Finnish context. *Matkailututkimus/Finnish Journal of Tourism Research*, *16*(1), 93–110.

Sámediggi. (n.d.). Principles for responsible and ethically sustainable Sámi tourism. Retrieved from https://www.samediggi.fi/ethical-guidelines-for-sami-tourism/?lang=en

Sanders, E. B. N., & Stappers, P. J. (2008). Co-creation and the new landscapes of design. *Co-design*, *4*(1), 5–18.

Sarantou, M., Remotti, S., Alhonsuo, M., & Gutierrez Novoa, C. (2021). Generating Stakeholder Workshops for Policy Making in Digital Environments through Participatory Service Design. *Malta Review of Educational Research 15*(2), 119–136.

Seppälä, T., Sarantou, M., & Miettinen, S. (2021). *Arts-Based Methods for Decolonising Participatory Research*. Abingdon, England: Routledge.

Sève, B., Redondo Domínguez, E., Millán Gómez, A., & Sega, R. (2020). The urban transformation as a collective creation: Bottom-up and participative tool taxonomy for urbanists and architects. In I. Lombillo, H. Blanco, & Y. Boffill (Eds.), *Rehabend 2020: Construction pathology, rehabilitation technology and heritage management* (pp. 470–477). Santander, Spain: University of Cantabria.

Shapiro, E. R. (2020). Liberation psychology, creativity, and arts-based activism and artivism: Culturally meaningful methods connecting personal development and social change. In L. Comas-Díaz & E. Torres Rivera (Eds.), Liberation psychology: *Theory, method, practice, and social justice* (pp. 247–264). Washington, DC: American Psychological Association.

Smith, L. T. (2013). *Decolonizing methodologies: Research and indigenous peoples*. New York, NY: Zed Books.

Sonn, C., & Baker, A. (2016). Creating inclusive knowledges: Exploring the transformative potential of arts and cultural practice. *International Journal of Inclusive Education*, *20*(3), 215–228.

Sugimoto, A, Sugino, H., & Hori, J. (2022). How bountiful is the ocean? Participatory valuation of human–nature relationships in Yaeyama Islands, Okinawa, Japan. *Sustainability Science*, *17*, 879–898.

Tourais, P., & Videira, N. (2021). A participatory systems mapping approach for sustainability transitions: Insights from an experience in the tourism sector in Portugal. *Environmental Innovation and Societal Transitions*, *38*(1), 153–168.

Valkonen, J., & Valkonen, S. (2014). Contesting the nature relations of Sámi culture. *Acta Borealia*, *31*(1), 25–40.

Valkonen, J., & Valkonen, S. (2018). On local knowledge. In T. H. Eriksen, S. Valkonen, & J. Valkonen (Eds.), *Knowing from the Indigenous North: Sámi approaches to history, politics and belonging* (pp. 12–26). Abingdon, England: Routledge.

Viken, A., Höckert, E., & Grimwood, B. S. R. (2021). Cultural sensitivity: Engaging difference in tourism. *Annals of Tourism Research*, *89*(1), 103223.

Wetter-Edman, K., Sangiorgi, D., Edvardsson, B., Holmlid, S., Grönroos, C., & Mattelmäki, T. (2014). Design for value co-creation: Exploring synergies between design for service and service logic. *Service Science*, *6*(2), 106–121.

Zehrer, A. (2009). Service experience and service design: concepts and application in tourism SMEs. *Journal of Service Theory and Practice*, *19*(3), 332–349.

13

THE PROVISION OF SUSTAINABLE WILDLIFE EXPERIENCES

The role of the wildlife watching tourism guides

Hilde Nikoline Hambro Dybsand and Peter Fredman

Introduction

The role of high-quality guiding and interpretation has been acknowledged in several studies on tourism, and particularly tourism based on experiences of nature and wildlife (Hansen & Mossberg, 2016; Mossberg et al., 2018; Randall & Rollins, 2009; Rokenes et al., 2015). There are several elements at play in high-quality guiding (Curtin, 2010; Rokenes et al., 2015). Some of these elements can be applied to all forms of tourism (Cohen, 1985), while others are specific to the setting or to the particular activity (Randall & Rollins, 2009; Weiler & Davis, 1993). One form of tourism that can benefit greatly from high-quality guiding and interpretation is wildlife tourism: a niche of nature-based tourism activities that are based on interactions with non-domesticated animals (Borges de Lima & Green, 2017). Previous studies on wildlife tourism found that high-quality guiding is key to participants' overall satisfaction (Dybsand & Fredman, 2020), and that wildlife guides need several different skills and attributes to be successful (Curtin, 2010). Since wild animals are unpredictable as main attractions, guides may also find themselves in situations where their main attraction is not found (Margaryan & Wall-Reinius, 2017). In these cases, the guides' ability to adapt to the situation and to highlight other aspects of the experience is the key to provide positive experiences for participants (Dybsand, 2020). Well-executed guiding and interpretation may also improve wildlife tourism activities' potential to foster pro-environmental attitudes and behaviours among participants (Jacobs & Harms, 2014). While these studies all highlight the importance of providing high-quality guiding at wildlife tourism activities, there are fewer studies investigating which exact elements are needed in order to do so successfully (Curtin, 2010).

In this chapter we analyze which elements are needed to provide high-quality guiding at non-consumptive, non-captive wildlife tourism activities. Also referred to as wildlife watching tourism, these activities do not purposefully harvest or remove wildlife from their habitats, and typically involve observations and photography (Hassan & Sharma, 2017). We use Cohen's (1985) conceptualization of the guide's role and Weiler and Davis' (1993) suggested expansion for guiding in natural surroundings as starting points. To investigate and expand these

DOI: 10.4324/9781003230748-16

frameworks we use musk ox safaris in Dovrefjell (Norway), and polar bear tours in Svalbard (Norway) and Churchill (Canada) as case activities. With this approach we compare an activity based on a highly predictable target species (the musk ox safari), an activity based on a highly unpredictable target species (polar bear tours in Svalbard) and an activity somewhere in between (polar bear tours in Churchill). Previous research from these areas indicate that there are more elements involved in guiding non-consumptive, non-captive wildlife tourism activities than for guiding in general (Dybsand, 2020; Dybsand & Fredman, 2020). These elements are related to wildlife encounters and the unpredictability when observing wild animals in nature, both of which are elaborated in this chapter to suggest an expanded model of the guides' role.

Wildlife tourism and wildlife watching tourism

Wildlife tourism can be defined as a niche of nature-based tourism activities, that are based on interactions with non-domesticated animals (Borges de Lima & Green, 2017). The interactions occur in animals' natural environments, semi-captivity or captivity, and include both non-consumptive activities such as safaris or bird watching and consumptive activities such as hunting or fishing (Higginbottom, 2004). Therefore, wildlife tourism is often divided into three main types: captive wildlife tourism (Frost, 2011), consumptive wildlife tourism (Lovelock, 2007) and wildlife watching tourism (Hassan & Sharma, 2017). Wildlife watching tourism has been defined as "tourism organized and undertaken to watch wild animals in natural settings" (Tapper, 2006, p. 7). It has been argued that wildlife watching tourism is the type of wildlife tourism that has grown most in recent years (Hassan & Sharma, 2017). Research interest is also increasing, with studies on topics such as motivations to participate (Mutanga et al., 2017), negative impacts on wildlife (Penteriani et al., 2017), psychological benefits for participants (Curtin & Kragh, 2014), and ethical implications (Green, 2017). While there are also a handful of studies on guiding and interpretation (Ballantyne et al., 2018; Curtin, 2010), more research on which components are involved in high-quality guiding in the wildlife watching tourism context is needed. More information on this topic is especially important in the context of wildlife watching tourism, as the impacts of wildlife watching tourism vary depending on how activities are executed (Burns, 2017). Wildlife watching tourism involves activities that do not purposefully harvest or remove wildlife from their habitats (Burns et al., 2011). However, the activities may still have serious negative impacts on wildlife and the surrounding natural environments (Green, 2017). Negative impacts include alterations of natural behaviors, habituation, food conditioning, crowding, stress, contamination, relocation or displacement, habitat degradation and, in some cases, even local extinction (Green & Giese, 2004). These negative impacts are often the result of incorrect behaviours, such as allowing participants to get too close to the wildlife, damage to the habitat, overcrowding, or poor interpretation (Curtin, 2013), many of which can be managed through a guide. Therefore, we argue that high-quality guiding is especially important at wildlife watching tourism activities, and that knowledge of which elements are involved in guiding such activities is needed.

The guide's role

The role of the tourism guide was first conceptualized by Cohen (1985). He argues that the modern tourist guide mainly stems from two lines of origin: the pathfinder and the mentor. These are the antecedents of the leadership and mediatory spheres in the guides' role. Moreover, as these two spheres each have both an inner- and an outer-directed aspect, he suggests four major components of the guide's role. The leadership sphere consists of the instrumental and the social component. Cohen (1985) argues that the distinction between these two forms of leadership is well founded in general sociological theory and that there is often a tension between the two. To a tourism guide, the instrumental component relates to her or his responsibility for completing the tour smoothly. To accomplish this goal, the tourism guide must provide direction (to find or choose the way), access (sometimes to areas not available to the general public), and control (responsibility of the safe and efficient conduct of the party). The social component relates to the guide's responsibility for cohesion and morale among participants during the activity. It involves tension management (avoiding tensions between participants), integration (social integration within the group), morale (keeping the group in good humor and high morale during the activity), and animation (convincing participants to undertake various activities offered by touristic facilities encountered during the activity).

The mediatory sphere is divided into the interactional and the communicative components. Cohen (1985) argues that this distinction is well founded in the more general distinction between social and cultural mediation – or, in the context of tourism, the difference between the guide as a social middleman and the guide as a culture broker. The interactional component relates to his or her function as a middleman between the travel party and the local population, sites, and institutions. It involves representation (representing the setting to the party, and the party to the setting) and organization (providing services and amenities to the party during the tour). The communicative component relates to the guide's ability to interpose her/himself between the tourist and the site as a culture broker. It includes selection (pointing out objects of interests to show the travel party), information (providing correct and precise information), interpretation (mediating encounters between different cultures), and fabrication (invention or deception such as presenting fake antiques sold in shops as if they were genuine to receive a commission from the shop).

According to Cohen (1985), virtually all of the tangible activities of the tourist guide can be subsumed under one of these principal components of the guide's role. However, this was questioned by Weiler and Davis (1993), who found that the framework does not fully incorporate guides' responsibilities towards the surroundings. They also argue that the tour leader has a special role to play in environmentally responsible tourism, when delivering quality tourist experiences without causing intolerable ecological and social damage. Therefore, they suggest a third sphere which applies to guiding in natural surroundings: the resource management sphere. This sphere requires the guide to focus on the environment, and includes the two components motivation (the modification of participant behaviours and impacts on-site) and environmental interpretation (enhancing understanding and appreciation of environmental issues to facilitate responsible behaviour from participants in the long term).

In a study on kayak paddling in Pacific Rim National Park, Randall and Rollins (2009) investigated participants' support for both Cohen's (1985) original components and Weiler and Davis' (1993) suggested expansion of the guide's role when activities are organized in natural surroundings. They found strong support for five of the six components: the instrumental, social, interactional, motivational and environmental interpretation components, and moderate support for the communicative component. More recently, it has been argued that the role of the tour guide is evolving and changing, due to an increased focus on experiences in the research literature and the guides' need to adapt to new demands and special settings (Rokenes et al., 2015; Weiler & Black, 2015). One such special setting is wildlife watching tourism, in which Curtin (2010) argues that tour guides' or tour leaders are the focal points, who need both social skills and field interpretation skills to be successful. While her suggested social skills are related to Cohen's (1985) framework, Curtin's (2010) suggested field interpretation skills are related to the guides' abilities to locate and interact with wildlife. These skills include habitat awareness, hearing, spotting and identification of wildlife, in addition to providing information on habitats, behaviour and how to identify wildlife to the travel party. These studies suggest that there are more components included in the guide's role in the wildlife watching tourism setting than in the more general nature experience setting. However, to the best of our knowledge there are no previous studies that adapts Cohen's (1985) original framework to the role of the wildlife watching tourism guide.

Study sites and case activities

As shown in Figure 13.1, two case activities at three study sites were used to investigate the guide's role in the wildlife watching tourism setting: musk ox safaris in Dovrefjell-Sunndalsfjella National Park (Norway) and polar bear tourism in Svalbard (Norway) and Churchill (Canada). These areas represent three different environments with different approaches to wildlife watching tourism. Dovrefjell-Sunndalsfjella is one of the largest protected areas on the Norwegian mainland, featuring a large, mainly untouched mountain area (Miljødirektoratet, 2013). The national park is located in the centre of Norway, and is accessible by car and train. Therefore it is popular both as one of more stops for tourists on road trips in Norway and as a nature-based tourism destination. The Svalbard islands in the far north are under jurisdiction of the Kingdom of Norway, and feature arctic nature attractions such as glaciers, the midnight sun and northern lights. Due to the islands' remote location, tourists mainly arrive by air. Churchill in Canada is located on the west coast of Hudson Bay, at the juncture of three eco regions: the boreal forest to the south, the arctic tundra to the northwest and the Hudson Bay to the north. Although Churchill is a popular place for wildlife tourism, there is no road access and tourists arrive by either air or train. Each of the study activities and sites are briefly described in the following sections.

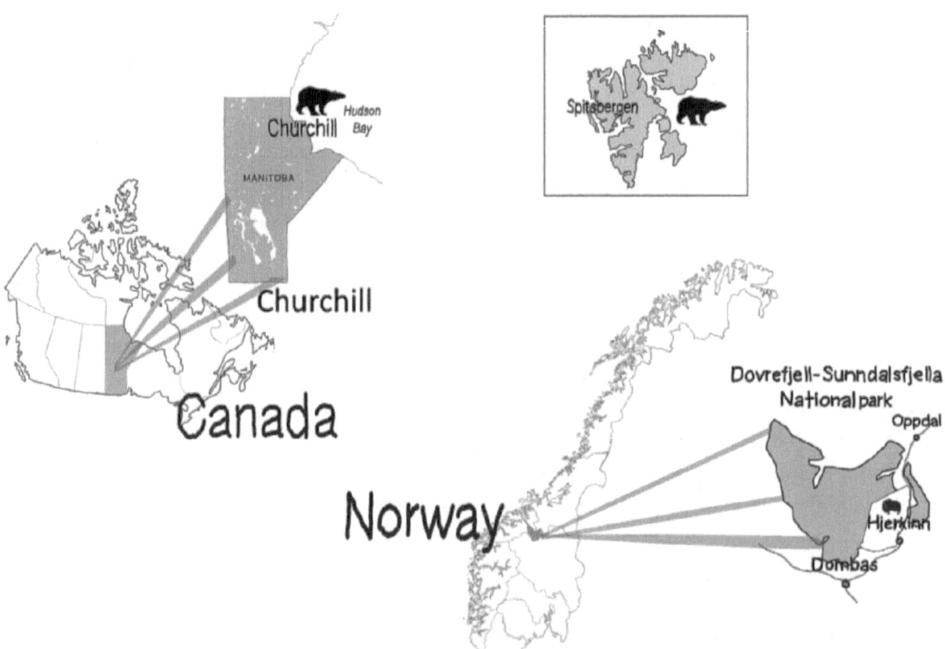

Figure 13.1 The location of our case activities and study sites.

Musk ox safaris in Dovrefjell-Sunndalsfjella National Park, Norway

Dovrefjell-Sunndalsfjella National Park was first established in 1974 to conserve a large, mainly untouched mountain area with an intact ecosystem, and to preserve the habitat of wild reindeer herds in the area (Dovrefjell nasjonalparkstyre, 2017). The national park is also the home of Europe's only viable herd of musk oxen (*Ovibos moschatus*). This iconic species was native to the area in pre-historic times, but became locally extinct in Europe after the last ice age. However, musk ox remains were found in Dovrefjell in 1932, inspiring a reintroduction of the species from 1932 to 1953 (Miljødirektoratet, 2019). The species is now formally considered reintroduced in Norway. However, the individuals found in the national park are not considered a threat to other wildlife or ecosystems in the area and are allowed to develop as naturally as possible in designated area of 340 km^2, marked in white on the map shown in Figure 13.2. Since its reintroduction, the musk ox has become an important tourist attraction, and there are five operators organizing walking musk ox safaris in the area with approximately 3000–3500 participants per year in total (Rangbru & Seljevoll, 2017). Chances of encountering musk oxen are considered high, and the respondents of a survey among musk ox safari participants during the summer of 2019 had all encountered at least one musk ox during the safari they had joined (Dybsand & Stensland, 2021).

	Outer-directed (resourced from outside the tour group)	Inner-directed (resourced from inside the tour group)
Tour Management (focus on group, based on Cohen, 1985)	The instrumental component	The social component
Experience Management (focus on individual, based on Cohen, 1985	The interactional component	The communicative component
Resource management (focus on environment, based on Weiler & Davis, 1993)	The motivational component	The environmental Interpretation component
Wildlife management (focus on wildlife, based on our findings)	*The encounter component*	*The uncertainty component*

Figure 13.2 The main components of guiding wildlife watching tourism activities.

(Expanded from Cohen, 1985 & Weiler and Davis, 1993).

Polar bear tourism in Svalbard, Norway

The Arctic island group Svalbard is located approximately 74°–81° north, and is included in the Kingdom of Norway (Thuesen & Barr, 2018). The islands are famous for their wildlife, including the polar bear (*Ursus maritimus*), which is the most popular species among visitors. While many visitors visit Svalbard hoping to encounter a polar bear, Norwegian law states that it is forbidden to lure, pursue or in any other active act seek polar bears out to interfere with them or endanger humans or polar bears (The Svalbard Environmental Protection Act, 2001). Furthermore, the local destination company warns potential visitors that there are no polar bear safaris (Visit Svalbard, 2018). However, it is possible to book trips and boat cruises that enter polar bear territory (Better Moments AS, 2017; Visit Svalbard, 2018), and visitors often join such trips hoping to encounter a polar bear. As companies offering such trips are formally not allowed to actively pursue Polar Bears, the chances of encountering the target species are considered relatively low, and when polar bears are found they are mainly seen from a distance.

Polar bear tourism in Churchill, Canada

Churchill is the northernmost seaport of Canada, located on the west coast of Hudson Bay. The city is famously referred to as the "*polar bear capital of the world*", due to a large number of polar bears who aggregate along the shores of Hudson Bay once a year in early to mid-November waiting for the formation of sea ice (Lemelin, 2006). The number of commercial tour operators and vehicles in this area are limited to protect polar bears, and measures taken to restrict travels to existing trails (Manitoba.ca, 2018). However, polar bear tourism is possible on a larger scale than in Svalbard as providers are allowed to approach the polar bears, and the polar bears are considered relatively predictable during the peak season while they await the formation of sea ice. Meanwhile, activities organized outside of the peak season are less predictable and if the sea ice forms earlier than normal this affects activities organized during the peak season. Chances of encounters with polar bears are therefore considered higher than in Svalbard, but lower than encounters with musk oxen in Dovrefjell.

Data collections and analyses

Data for this study are from a project about participant experiences in wildlife watching tourism (Dybsand, 2021). A mixed-methods research approach was applied, and data collection was based on the convergent research design, in which complementary data are utilized to investigate a research topic (Morse, 1991). The findings presented in this chapter are all based on qualitative methods; participant observation, semi-structured interviews and content analysis of online reviews.

Participant observations and semi-structured interviews were conducted at musk ox safaris in Dovrefjell during the summer season (June–September) of 2018. The participant observations were conducted at 14 randomly selected safaris, and followed the entire safari experience including pre-safari briefings, hiking with the group to find musk oxen, navigating challenging terrain, short stops on the way with guiding and interpretation, lunch breaks with the groups, encounters with the musk oxen, and, on a few occasions, chance encounters with other wildlife. Guidelines from DeWalt and DeWalt (2010), Guest et al. (2013) and Thaagard (2009) were followed. Participants were informed about the presence of the researcher and short jot notes were taken during the safaris. More detailed field diaries were written after the safaris, recording each day's events, participants' responses to the experience as well as the author's own reflections. These observations were combined with short on-site interviews with 33 travel parties. The travel parties were selected at random, but were all participants at the 14 musk ox safaris selected for participant observation. They ranged from one to four participants per travel party, and were 49 in total. The interviews were semi-structured, using an interview guide as a starting point, with the options of changing the order or phrasing of the questions and asking follow-up questions (Kvale et al., 2009). Interviews lasted from 5 to 15 minutes and were either conducted during the safaris or directly after the safaris. Interview transcripts and field diaries were analyzed applying thematic analysis (Clarke, 2006). A combination of predefined and open coding was applied, and predefined codes were quality controlled by both authors.

Data on polar bear tourism from Churchill, Canada and Svalbard, Norway was collected through content analysis of online reviews written between 2021 to 2017 (obtained in January 2018). The online platform chosen for the analysis was TripAdvisor.com, which is one of the largest online travel forums in the world (Ayazlar, 2017). Since the characteristics of the polar bear watching activities varied depending on the provider and the study site, reviews were collected and analyzed manually rather than applying a scraping tool to ensure that reviews using different wording to explain similar elements of the reviewers' experiences were all included in the analysis. When analyzing reviews, a combination of content analysis (Joffe & Yardley, 2004) and thematic analysis (Clarke, 2006) was applied. First, all reviews mentioning polar bear tourism or polar bear safaris were collected, resulting in 154 reviews from Svalbard and 697 reviews from Churchill. A content analysis with three pre-defined categories was then undertaken; 1) Polar bears not found; 2) Fewer polar bears found than expected; and 3) statements related to the unpredictability of wildlife encounters. 152 reviews belonged to at least one of the pre-defined categories: 64 from Svalbard and 87 from Churchill. These reviews were included in a second phase, in which reviews were analyzed more thoroughly, applying thematic analysis. No pre-defined codes were applied in this phase, but re-occurring themes were identified and recorded applying open coding. Examples of re-occurring themes found in this phase were "*Staff effort and skill locating wildlife*", "*Staff dedication*" and "*inaccurate information provided on chances of polar bear sightings*".

Main findings

Our findings support previous studies that underline the importance of high-quality guiding – both in nature-based tourism and, more specifically, in wildlife watching tourism (Curtin, 2010; Dybsand & Fredman, 2020; Mossberg et al., 2018; Rokenes et al., 2015). Having a dedicated and capable guide was important, both to participants who encountered their target species as well as to those who did not get that experience.

During the interviews, participants at the musk ox safari at Dovrefjell mentioned that they had chosen to join an organized safari rather than search for wildlife on their own because they wanted to get as close as possible to the musk oxen, and learn about wildlife, plants and history. Additional motivations to go with a guide was to make sure that they did not disturb wildlife and for safety reasons. Safety was especially important to participants in this area, as male musk oxen can sometimes become aggressive and attack people who get too close in order to protect their family group. However, participants also mentioned that it was important to them that their visit to the area did not have negative consequences for the musk oxen or other wildlife in the area. When travel parties were asked which part of the safari they enjoyed most, besides encountering the musk oxen, the element mentioned most often was high quality guiding and interpretation. Two of the travel parties that were interviewed even mentioned that they had been on a safari with the same company previously, and were so happy with the guiding that they decided to come back and join a safari with the same company again hoping the tour would be led by the same guide:

> Ah… because we liked it, three years ago.. we met a young guide –and we had much contact during the last three years, and we wanted to meet him again and to do again the safari
>
> (German couple, 60 and 59 years old)

Additionally, participants mentioned that guides knew where to find musk oxen, took care of the group and made wildlife encounters better because they helped the group approach musk oxen correctly, were knowledgeable, told interesting stories and facts the participants would not have thought about researching on their own, and helped the rest of the travel party form a social bond with each other:

> Even if it's a little related to the musk oxen, it was great to get so many facts and good stories from the guide – that was a really positive experience. It's easy to overlook it, but the stories were a very important part of my experience and were even better because we were such a small group and the guide and his stories helped us bond socially and talk amongst ourselves about our experiences. It is a lot… you gain a lot of knowledge you wouldn't have gained otherwise and that is incredibly exciting, at least for me
>
> (Norwegian man, 27 years old)

While most participants had positive experiences, there were two travel parties interviewed that were unhappy with the safaris they had joined. Both parties mentioned the guiding, and in particular the role of the guides, at their safaris' lack of ability to handle a large group. Thus, the findings from our interviews indicate that the quality of the guiding provided had a substantial impact on participants' experiences, both for the better and sometimes also for worse. The interview findings were supported by the main findings of the participant observation. At the 14 safaris we joined, guides actively attempted to improve participants' experiences by providing expected services such as locating the musk oxen, navigating natural surroundings, and providing information about the area to the best of their abilities. Moreover, they often went beyond these expected services and surprised participants by, for example bringing coffee or cinnamon buns from the local bakery for the entire travel party, bringing telescopes the participants could use to take better pictures of the musk oxen or finding musk ox wool for participants to bring home with them after the safari. Guides also provided nature interpretation, focusing on musk oxen, other wildlife in the area such as reindeer, lemming and the arctic fox, plants in the area, the national park and threats to the local ecosystem and/or musk oxen. In other words, guides made experiences better both by adding elements to the experience, and by highlighting and connecting other elements such as the natural surroundings, plants, and wildlife.

TripAdvisor reviews from participants at polar bear tourism activities in Svalbard and Churchill show that the guides' role is just as important in cases when the target species is not found, or in cases when participants do not get to see as many animals as they had expected. Reviews from participants who found themselves in these situations were often positive, and generally mentioned that other elements of the experience than the lack of wildlife was the main reasons why they were happy or unhappy. The element mentioned most frequently was the quality of the guiding provided during the polar bear tourism activity. Negative reviewers mentioned guides who were not engaged, who did not take the participants' wish to see polar bears seriously, who spoke in a language they did not understand, who did not actively search for polar bears the entire time, or who gave up the search for polar bears early. In many cases, reviewers also mentioned that they had not received accurate information about how high the chances of encountering a polar bear actually were. This finding is partially linked to information provided before the experience, for example on provider websites, but also to the guides' abilities to manage expectations.

> Wildlife is unpredictable, so it's not (the name of the tour company)'s responsibility that we had no good polar bear sightings on this tour at the peak of bear season. What is within their control is advertising a day tour for $479 and then giving up the search for bears and heading back to base at 2pm (the tour was over by 3pm).
>
> (Reviewer 1, Churchill)

Positive reviewers wrote about guides who were highly engaged in the search for polar bears, showed the travel party other interesting attractions in the area such as glaciers, rock formations or other wildlife species, were entertaining, provided relevant information, showed participants polar bear tracks or other signs that polar bears were in the area even if they were not found during the safari and provided understandable reasons why they were unable to find polar bears during the safari.

> Unfortunately we only saw some polar bear tracks, but we did see plenty of seals, reindeer and bird life. (Guide's name) went out of his way to try and find us some bears, but despite his best efforts, the sea ice is so big, and the bears are so few! But his effort really showed how dedicated the staff are to ensuring you achieve the best Arctic experience.
>
> (Reviewer 1, Svalbard)

These comments indicate that guides who were able to divert participants' attention away from the lack of polar bear sightings towards other elements of the experience were able to provide positive experiences for the participants, even in cases when the target species could not be found.

An expanded model of the guides' role when guiding wildlife watching tourism activities

Our findings support previous studies on guided wildlife watching tourism, that acknowledge the importance of high-quality guiding and interpretation and suggest that guides need a variety of skills to be successful (Ballantyne et al., 2009; Curtin, 2010; Ham & Weiler, 2002). Having a dedicated and capable guide was important to participants at all of our case activities, and guides were expected to play several roles to deliver good experiences. Our study supports all the components in Cohen's (1985) original framework of the tourism guide's role. The activities were all organized in natural areas with limited infrastructure, and the guides were responsible both for locating wildlife and navigating relatively challenging surroundings. Thus, the instrumental component which involves leading the way, providing access, safety and efficiency played a key role. Another effect of the activities being organized in natural areas was that providers could not alter their surroundings to strengthen the theme of their activities. Therefore, the communicative component of the guide's role, which entails providing relevant information and selecting points of interest, became especially important to making sure that the activities entertained the participants. So did the interactional component, which involves acting as a link between the area and the tourist party. Not all of the participants encountered the animal they wanted to see during their activity and even participants who did encounter their target species sometimes had to wait for a long time before the encounter took place. Therefore, the social component which involves tension management, social integration, group morale and cohesion was also important during all of our activities. Our findings also support Weiler and Davis' (1993) suggested expansion of the guide's role when activities take place in natural surroundings. All of our activities took place in vulnerable natural areas, and managing participants' behaviour on-site was important. Therefore, the motivational component of the guide's role which has to do with managing tourists' behaviour and impacts during the experiences was important at all of our case activities. As wildlife watching tourism has been justified by its potential to foster pro-environmental attitudes and behaviours among participants (Hughes, 2013), guides should also try to influence participants' environmental behaviors after the experience. This supports the environmental interpretation component, which involves improving tourists' environmental behaviour in the long term through a relevant, organized and entertaining experience with a strong theme (Ham, 2013).

Although we find support for all the six components suggested by Cohen (1985) and Weiler and Davis (1993) our findings suggest that there are additional sub-elements of guiding wildlife watching tourism activities that are not fully covered by these frameworks. We refer to them as "The encounter component" and "The uncertainty component", both presented in Figure 13.2 (besides the six components mentioned above) and further elaborated below.

During musk ox safaris in Dovrefjell, one of the most important aspects of the participants' experiences was the quality of the encounter they had with the musk oxen. The guides improved these encounters in several respects by making sure the group approached the musk oxen correctly: walking close to each other rather than spreading out, taking short breaks to avoid scaring the musk oxen away, and staying quiet until the musk ox had accepted the presence of the travel party. Since male musk oxen can attack people if they feel threatened, the guides continuously studied the musk oxen's body language to ensure a safe encounter. To make encounters even better for participants, two of the guides carried telescopes with them and one guide brought binoculars for his entire travel party. They did this so that participants could get a closer look at the musk oxen and take pictures through the telescopes with their cell phones. Other guides improved the encounter by arranging for participants to have their lunch at a spot overlooking the musk oxen. Delivering high-quality wildlife encounters is partly covered from the components suggested by Cohen (1985), Weiler and Davis (1993), such as the instrumental component (which includes access and safety), the motivational component (which includes managing participants' behaviours on site), and the communicative component (which includes providing information). However, the guides in Dovrefjell took this a step further by bringing extra equipment for their participants, taking the best possible path through the terrain and organizing lunches at strategic spots with good views of the target species. These ways of improving the wildlife encounter are not fully covered by the other components of guiding in natural surroundings, and because there are only certain aspects of the existing components that are applicable to the wildlife encounter, we suggest an additional component linked to the improvement of the wildlife encounter: *The encounter component*. This involves managing encounters with wildlife through securing correct behaviour from participants (distance to wildlife, correct ways of approaching the wildlife etc.) and finding ways to enhance these encounters through staging, interpretation and equipment. This component supports

Curtin's (2010) suggested field interpretation skills that are related to the guides' abilities to locate and interact with wildlife.

Findings from Svalbard and Churchill clearly show that another key aspect of guiding wildlife watching tours is to manage the uncertainty surrounding whether the target species will be encountered or not. This aspect is also only partially covered by the Cohen (1985), Weiler and Davis (1993) components, such as the communicative component (which includes tension management and securing group morale) as well as the communicative component (which includes providing information and selecting points of interest to show the party). However, our findings suggest that guides also manage the uncertainty elements of their activities in ways not fully covered by these existing components. Examples from the polar bear watching guides include keeping participants' expectations at realistic levels, offering secondary but more guaranteed side experiences and shifting participants' attention towards other elements of the natural environment that can (partially) substitute the direct polar bear experience. The uncertainty element was also challenging at some of the musk ox safaris in Dovrefjell. Although guides had very high success rates, they were not in control of the distance travel parties would have to hike in order to reach the musk oxen, or how challenging the terrain would be. This affected the length of the safaris, as the guides' intention was to hike until the group encountered musk oxen. The guides were also not able to control whether the encounter happened at the beginning, middle or end of the safari. To reduce the uncertainty element, the guides in the area all worked together to locate the musk oxen and informed each other about the location of the animals they encountered. They also managed participants' expectations by explaining how they worked to find musk oxen, where musk oxen were last seen, how long of a hike they estimated was necessary to find the musk oxen and what type of equipment participants should bring. Since these approaches to managing uncertainty are not fully covered by the existing components of guiding in natural surroundings, we argue there is an additional component in the wildlife watching context: *The uncertainty component.* This component involves managing participants' expectations, shifting participants' attention to other elements of the natural surroundings, providing secondary experiences and reducing uncertainty through securing as much information as possible on the target species' last-known location. Support for this component is found in previous studies on strategies used by guides at wildlife watching tourism experiences as well as northern lights experiences which also rely on a spatially and temporally discontinuous phenomena (Heimtun & Lovelock, 2017; Margaryan & Wall-Reinius, 2017).

Conclusions

Our study contributes to the tourism literature by exploring the role of the guide in the context of wildlife watching tourism. Providing high-quality guiding is particularly important in the context of wildlife watching tourism, as wild animals are both unpredictable as main attractions, and vulnerable to inappropriate behaviours from tourists who would like to experience them (Curtin, 2013; Dybsand, 2020). Our findings indicate that high-quality guiding can contribute to more sustainable, enjoyable wildlife watching tourism experiences, both by managing participants' interactions with wildlife and by creating positive experiences in the absence of the participants' target species. Support was found for all four components of the guide's role proposed by Cohen (1985), and for the two additional components of the guide's role at nature-based experiences proposed by Weiler and Davis (1993). However, due to the unpredictable and vulnerable nature of wild animals as main attractions, we suggest two additional components which apply when guiding wildlife watching tourism activities: The uncertainty component and the encounter component.

The findings are based on qualitative case studies at three study sites. This should be considered if generalized to other wildlife watching tourism activities and contexts. Moreover, the activities we investigated at these sites all focused on charismatic mega fauna species, which often attract tourists who do not have much experience with wildlife watching (Stensland et al., 2021). Future studies on the guide's role at more specialized activities such as, for example, birdwatching may provide further insight on whether our suggested components also apply at other forms of wildlife watching tourism. Although our findings may not be generalized to all forms of wildlife watching tourism, they may provide useful insights and suggestions for similar activities.

Our findings are also helpful to wildlife watching tourism providers, as they provide insight on which components are necessary to provide high-quality guiding in these settings. They highlight the differences between guiding more general experiences of nature and guiding wildlife watching experiences. Additionally, these findings are important to managers of areas rich in wildlife, as they indicate that guides can play vital roles when it comes to managing how visitors interact with vulnerable wild animals. In cases when restrictions on interactions with wildlife are needed, high-quality guiding is likely to both enforce such restrictions and provide positive experiences for participants who are unable to interact with wildlife as closely as they want.

To conclude, we argue that our extended model of key components in guiding wildlife watching tourism adds new knowledge to this field of research that is also applicable in other parts of the broader nature-based tourism sector when the main attraction involves featured encounters and/or uncertainty. An important feature of nature-based tourism is that the provider seldom has complete control of the service and experience delivery (Fredman & Tyrväinen, 2010), which makes this sector special compared with other forms of tourism. This also calls for more attention for a sustainable delivery of nature-based tourism in general, and wildlife-related tourism in particular. In our case, improved wildlife encounters and management of uncertainty will reduce negative impact on the animals and provide better experiences, to the benefit of wildlife, tourists, tourism companies and destinations.

References

Ayazlar, R. A. (2017). Major wildlife attractions and choices in Turkey and the tourist's reports about their experiences and learning in this tourism niche. In I. Borges de Lima & R. J. Green (Eds.), *Wildlife tourism, environmental learning and ethical encounters* (pp. 195–209). Cham, Switzerland: Springer International Publishing AG.

Ballantyne, R., Hughes, K., Lee, J., Packer, J., & Sneddon, J. (2018). Visitors' values and environmental learning outcomes at wildlife attractions: Implications for interpretive practice. *Tourism Management, 64*, 190–201. https://doi.org/10.1016/j.tourman.2017.07.015

Ballantyne, R., Packer, J., & Hughes, K. (2009). Tourists' support for conservation messages and sustainable management practices in wildlife tourism experiences. *Tourism Management, 30*(5), 658–664. https://doi.org/10.1016/j.tourman.2008.11.003

Better Moments AS. (2017, 10. september). *Snowmobile Trips*. https://www.bettermoments.no/our-trips/

Borges de Lima, I., & Green, R. (2017). *Wildlife tourism, environmental learning and ethical encounters* Cham, Switzerland: Springer International Publishing AG.

Burns, G. L. (2017). Ethics and responsibility in wildlife tourism: Lessons from compassionate conservation in the anthropocene. In I. Borges de Lima & R. J. Green (Eds.), *Wildlife tourism, environmental learning and ethical encounters* (pp. 213–220). Cham, Switzerland: Springer International Publishing AG.

Burns, G. L., MacBeth, J., & Moore, S. (2011). Should dingoes die? Principles for engaging ecocentric ethics in wildlife tourism management. *Journal of Ecotourism, 10*(3), 179–196. https://doi.org/10.1080/14724049.2011.617450

Clarke, V. (2006). Using thematic analysis in psychology. *Qualitative Research in Psychology*, *3*(2), 77–101. https://doi.org/10.1191/1478088706qp063oa

Cohen, E. (1985). The tourist guide: The origins, structure and dynamics of a role. *Annals of Tourism Research*, *12*(1), 5–29. https://doi.org/10.1016/0160-7383(85)90037-4

Curtin, S. (2010). Managing the wildlife tourism experience: The importance of tour leaders. *International Journal of Tourism Research*, *12*(3), 219–236. https://doi.org/10.1002/jtr.747

Curtin, S. (2013). The intrinsic motivations and psychological benefits of eco and wildlife tourism experiences. In R. P. Ballantyne (Ed.), *International handbook on ecotourism* (pp. 203–216). Cheltenham, UK: Edward Elgar.

Curtin, S., & Kragh, G. (2014). Wildlife tourism: Reconnecting people with nature. *Human Dimensions of Wildlife*, *19*(6), 545–554. https://doi.org/10.1080/10871209.2014.921957

DeWalt, K. M., & DeWalt, B. R. (2010). *Participant observation: A guide for fieldworkers* (2nd ed.). Lanham, New York, Toronto, Plymouth: Altamira Press.

Dovrefjell nasjonalparkstyre. (2017, May 7). *Verneområder*. http://www.nasjonalparkstyre.no/Dovrefjell/Verneomrade/

Dybsand, H. N. H. (2020). In the absence of a main attraction – perspectives from polar bear watching tourism participants. *Tourism Management*, *79*, 104097. https://doi.org/10.1016/j.tourman.2020.104097

Dybsand, H. N. H. (2021). *Participant Experiences in Wildlife Watching Tourism* [PhD Thesis, The Norwegian University of Life Sciences]. Ås.

Dybsand, H. N. H., & Fredman, P. (2020). The wildlife watching experiencescape: The case of musk ox safaris at Dovrefjell-Sunndalsfjella national park, Norway. *Scandinavian Journal of Hospitality and Tourism*, 1–21. https://doi.org/10.1080/15022250.2020.1850347

Dybsand, H. N. H., & Stensland, S. (2021). Centrality to life and the theory of planned behavior: The case of musk ox safaris in Dovrefjell-Sunndalsfjella National Park Norway, *Human Dimensions of Wildlife*, 1–19. https://doi.org/10.1080/10871209.2021.1876187

Fredman, P., & Tyrväinen, L. (2010). Frontiers in Nature-Based Tourism. *Scandinavian Journal of Hospitality and Tourism*, *10*(3), 177–189. https://doi.org/10.1080/15022250.2010.502365

Frost, W. (2011). Rethinking zoos and tourism. In W. Frost (Ed.), *Zoos and tourism: Conservation, education, entertainment?* (pp. 1–8). Bristol, Buffalo, Toronto: Channel View Publications.

Green, R. (2017). Disturbing skippy on tour: Does it really matter? Ecological and ethical implications of disturbing wildlife. In I. Borges de Lima & R. J. Green (Eds.), *Wildlife tourism, environmental learning and ethical encounters* (pp. 221–245). Cham, Switzerland: Springer International Publishing AG.

Green, R., & Giese, M. (2004). Negative effects of wildlife tourism on wildlife. In K. Higginbottom (Ed.), *Wildlife tourism: Impacts, management and planning*. (pp. 81–97). Altona, Australia: Common Ground Publishing.

Guest, G., Namey, E. N., & Mitchell, M. L. (2013). Collecting qualitative data: A field manual for applied research. In *Collecting qualitative data* (pp. 75–112). London, UK: SAGE Publications, Ltd. https://www.doi.org/10.4135/9781506374680

Ham, S. (2013). *Interpretation: Making a difference on purpose* (2nd ed.). Golden, USA: Fulcrum Publishing.

Ham, S. H., & Weiler, B. (2002). Interpretation as the centerpiece of sustainable wildlife tourism. In R. Harris, T. Griffin, & P. Williams (Eds.), *Sustainable tourism* (2nd ed., pp. 35–44). New York, US & London, UK: Routledge.

Hansen, A. H., & Mossberg, L. (2016). Tour guides' performance and tourists' immersion: facilitating consumer immersion by performing a guide plus role. *Scandinavian Journal of Hospitality and Tourism*, *17*(3), 1–20. https://doi.org/10.1080/15022250.2016.1162347

Hassan, A., & Sharma, A. (2017). Wildlife tourism: Technology adoption for marketing and conservation. In J. K. Fatima (Ed.), *Wilderness of wildlife tourism* (pp. 61–85). Oakville, Canada & Waretown, USA: Apple Academic Press Inc.

Heimtun, B., & Lovelock, B. (2017). Communicating paradox: Uncertainty and the northern lights. *Tourism Management*, *61*, 63–69. https://doi.org/10.1016/j.tourman.2017.01.017

Higginbottom, K. (2004). Wildlife tourism: An introduction. In K. Higginbottom (Ed.), *Wildlife tourism: impacts, management and planning* (pp. 2–7). Altona, Australia: Common Ground Publishing Ltd.

Hughes, K. (2013). Measuring the impact of viewing wildlife: do positive intentions equate to long-term changes in conservation behaviour? *Journal of Sustainable Tourism*, *21*(1), 42–59. https://doi.org/10.1080/09669582.2012.681788

Jacobs, M. H., & Harms, M. (2014). Influence of interpretation on conservation intentions of whale tourists. *Tourism Management*, *42*, 123–131. https://doi.org/10.1016/j.tourman.2013.11.009

Joffe, H., & Yardley, L. (2004). Content and thematic analysis. In D. F. Marks; & L. Yardley (Eds.), *Research methods for clinical and health psychology* (pp. 56–68). London, UK: Sage Publications.

Kvale, S., Brinkmann, S., Anderssen, T. M., & Rygge, J. (2009). *Det kvalitative forskningsintervju*. Oslo, Norway: Gyldendal akademisk.

Lemelin, R. H. (2006). The gawk, the glance, and the gaze: Ocular consumption and polar bear tourism in Churchill, Manitoba, Canada. *Current Issues in Tourism*, *9*(6), 516–534. https://doi.org/10.2167/cit294.0

Lovelock, B. (2007). An introduction to consumptive wildlife tourism. In B. Lovelock (Ed.), *Tourism and the consumption of wildlife: Hunting, shooting and sport fishing* (pp. 3–11). New York, USA: Routledge.

Manitoba.ca. (2018, November 03). *Watchable Wildlife Manitoba*. https://www.gov.mb.ca/sd/wildlife/viewing/churchill.html

Margaryan, L., & Wall-Reinius, S. (2017). Commercializing the unpredictable: Perspectives from wildlife watching tourism entrepreneurs in Sweden. *Human Dimensions of Wildlife*, *22*(5), 406–421. https://doi.org/10.1080/10871209.2017.1334842

Miljødirektoratet. (2019, June 06). *Moskus*. https://miljostatus.miljodirektoratet.no/tema/arter/fremmede-arter/moskus/

Miljødirektoratet. (2013). *Dovrefjell-Sundalsfjella nasjonalpark [Brosjyre]* Trondheim: Miljødirektoratet.

Morse, J. M. (1991). Approaches to qualitative-quantitative methodological triangulation. *Nursing Research*, *40*(2), 120–123. https://journals.lww.com/nursingresearchonline/Fulltext/1991/03000/Approaches_to_Qualitative_Quantitative.14.aspx

Mossberg, L., Hanefors, M., & Hansen, A. H. (2018). Guide performance: Co-created experiences for tourist immersion. In N. K. Prebensen, J. S. Chen, & M. Uysal (Eds.), *Creating experience value in tourism* (2nd ed., pp. 203–214). Wallingford, UK: Cabi.

Mutanga, C. N., Vengesayi, S., Chikuta, O., Muboko, N., & Gandiwa, E. (2017). Travel motivation and tourist satisfaction with wildlife tourism experiences in Gonarezhou and Matusadona National Parks, Zimbabwe. *Journal of Outdoor Recreation and Tourism*, *20*, 1–18. https://doi.org/10.1016/j.jort.2017.08.001

Penteriani, V., López-Bao, J. V., Bettega, C., Dalerum, F., Delgado, M. D. M., Jerina, K., Kojola, I., Krofel, M., & Ordiz, A. (2017). Consequences of brown bear viewing tourism: A review. *Biological Conservation*, *206*, 169–180. https://doi.org/10.1016/j.biocon.2016.12.035

Randall, C., & Rollins, R. B. (2009). Visitor perceptions of the role of tour guides in natural areas. *Journal of Sustainable Tourism*, *17*(3), 357–374. https://doi.org/10.1080/09669580802159727

Rangbru, B., & Seljevoll, K. V. (2017). *Forvaltningsplan for moskusbestanden på Dovrefjell* (2017: 4). (Miljøvernavdelingen, Issue. https://www.oppdal.kommune.no/globalassets/pdfdokumenter/plan-miljo-og-landbruk/miljo/forvaltningsplan-for-moskus-12.12.2017-endelig-versjon.pdf

Rokenes, A., Schumann, S., & Rose, J. (2015). The art of guiding in nature-based adventure tourism: How guides can create client value and positive experiences on mountain bike and backcountry ski tours. *Scandinavian Journal of Hospitality and Tourism*, *15*(sup1), 62–82. https://doi.org/10.1080/15022250.2015.1061733

Stensland, S., Aas, Ø., Dybsand, H. N. H., & Haugen, T. (2021). A Nordic perspective on wildlife tourism experiences. In P. Fredman & J. V. Haukeland (Eds.), *Nordic perspectives on nature-based tourism. From place-based resources to value-added experience* (pp. 263–280). Cheltenham, UK: Edward Elgar Publishing. https://doi.org/10.4337/9781789904031

The Svalbard Environmental Protection Act. IV. Fauna (2001). https://lovdata.no/dokument/NL/lov/2001-06-15-79/KAPITTEL_4#KAPITTEL_4

Tapper, R. (2006). *Wildlife watching and tourism – a study on the benefits and risks of a fast growing tourism activity and its impacts on species*. Bonn, Germany: UNEP/CMS Secretariat.

Thaagard, T. (2009). *Systematikk og innlevelse: En innføring i kvalitativ metode* (3nd ed.). Bergen, Norway: Fagbokforlaget Vigmostad & Bjørke AS.

Thuesen, N. P., & Barr, S. (2018, 26. september). *Svalbard*. https://snl.no/Svalbard

Visit Svalbard. (2018, October 03). *Polar Bear, The King of the Arctic*. https://en.visitsvalbard.com/visitor-information/polar-bears

Weiler, B., & Black, R. (2015). The changing face of the tour guide: One-way communicator to choreographer to co-creator of the tourist experience. *Tourism Recreation Research*, *40*(3), 364–378. https://doi.org/10.1080/02508281.2015.1083742

Weiler, B., & Davis, D. (1993). An exploratory investigation into the roles of the nature-based tour leader. *Tourism Management*, *14*(2), 91–98. https://doi.org/10.1016/0261-5177(93)90041-I

14
THE ROLES OF TOUR GUIDES IN NATURE-BASED TOURS

Gül Erkol Bayram

Introduction

Some important developments in the world have enabled the movement of the masses. For example, the increase in production and their resources has increased the free time of the employees and created their rights to paid leave. However, the need to keep up with the developing age has always led to increasing on expectations. Although the common belief for holidays as a way of life or ritual coincides by the 20th century; history shows that famous royal families or some wealthy families traveled with urges such as entertainment, curiosity, pleasure and seeking different pleasures centuries ago. It is very difficult to travel around the world even it has become a necessity today.

Role is defined as "the behaviour expected of a person in a certain position or assigned with a certain function". Guides are people who guide, lead and set examples in areas and environments that visitors have never known or experienced before. Accordingly, an examination of the roles of tourist guides in nature-based tourism activities is important for sustainable tourism. There has been a significant increase in nature-based tourism activities and these activities are increasing – every year (Gartner, 2004).

In recent years, there has been an increase in the number of people participating in nature-based tours. This development has both positive and negative consequences for both nature and local people. The positive results can be seen in terms of making significant contributions to both the local people and the country's tourism infrastructure. However, these developments also bring about adverse situations with regard to nature, the environment, the local flora and fauna and local people. An awareness of the positive and, most importantly, the negative situations that tourism can create on nature and the environment is of great importance in terms of spreading the positive results over large areas and limiting the negative results. Limiting or minimizing the damage that may occur to nature is a social duty that should be emphasized across the whole of society, including all civil society organizations, all sectors of the economy, the authorities and all people. In terms of achieving the purpose of nature-based tourism activities, it is also very important that the tour guides to be assigned in these activities are expert and well-trained guides (Köroğlu, Köroğlu and Alper, 2012).

DOI: 10.4324/9781003230748-17

Exploring tour guides

The tour guide is a person who represents his country and region to tourists, enabling visitors to experience the historical, natural and cultural accumulation of this region, helps them in many matters during their trips, takes care of them, and gives accurate and complete information about the destinations. In Turkey, guides are given their authority through an official route since they are licensed by the Ministry of Culture and Tourism (Köroğlu, 2013). In tours organized by travel agencies, tour guides are with the guests throughout. They take care of them, not only helping with translation issues, but also helping the tourists to define and experience the culture in the region, and to be informed about natural, historical and cultural aspects. It is clear that the tour guide is one of the most significant stakeholders in the tourism industry. According to the definition of the World Federation of Tourist Guides Associations (WFTGA), "tourist guides can visit monuments, museums, guide the natural and cultural environment and historical places and convey them to visitors with an entertaining interpretation."

According to Erkol Bayram (2017), a tour guide is the professional who helps the tourists with many issues related to the historical, natural and cultural issues of their country and region. Tour Guides give information about natural and cultural resources with accurate and effective way, makes a great contribution to the development and strengthening of a positive image related destinations and instills the necessary sensitivity. Furthermore, the tourist guide often takes part in the tours on behalf of the travel agency, staying with the tourists throughout the tour, dealing with them on many issues, and successfully undertaking many roles, such as translator, culture broker, teacher, mediator, and leader.

Visiting destinations that pique the public's interest while accompanied by experts Visiting the destinations that attract the masses with curiosity, accompanied by experts such as tourist guides, is a concept accepted by law today. Tour guides, with their numerous qualifications, play a unique role in the tourism sector since they direct, lead, accompany both domestic and foreign tourists. The importance and function of tourist guides are so great that if a tourist is satisfied with his tour, enjoys his vacation, approaches the country he visits with a warm, friendly, smiling face and conveys it, a large part of this success belongs to the tour guide, and this will continue to be the case (Şahin and Balta, 2007).

Ahipaşaoğlu (2001) emphasizes their versatility in the definition of a professional tour guide and their roles for realizing the expectation of tourists and needs including transhumance, mountain and nature walks, water sports, hunting, skiing, caving, ornithology, parachuting, botany, cycling and horse tours. In another study, Zhang and Chow (2004: 82), emphasizing the versatility of guides, state that tour guides take on roles such as actors, mediators, cultural ambassadors, commentators, teachers, translators, balance elements, and salespeople.

Guide, in its most primitive form, is a trustful person due to cultural knowledge, road language, geography informations Considering that people have been traveling for millions of years for different reasons, there have been millions of people who have done this job, even though they are not called guides. Considering that everything is born out of necessity, making an unknown place known with reliable information has revealed the profession of guidance with its most up-to-date name, even though it is called by names such as leader, chief, or local (Rabotić, 2010).

Tour guides are mixed with employees from other sectors, such as tour managers or group leaders in charge of the realization of the tour program such as field guides, who accompany a group during the trip and do or help them to do legal or routine transactions on their behalf (Türkmen, 2017). On the other hand, Köroğlu (2013) discussed those who are not tour guides to avoid confusing tour guides with other tourism industry officials.She states that people in the following positions are not tourist guides:

- Those who communicate with tourists in front of a museum on the pretext of providing information about the museum, for purposes such as improving their foreign language or taking tourists to a nearby store.
- Group transfers to a hotel or holiday village.
- Those who work in a travel agency but do not have a license and accompany the group for a tour or have an "information" officer certificate.
- Foreign nationals who come to Turkey as an employee of a foreign tour operator, provide information to visitors on behalf of their agency, and mostly work illegally in our country without a work permit.

The quality of tour guiding is mainly affected by the interdisciplinary nature of the profession. In this context, first of all, a tour guide should take on some roles. In this context, a tourist guide should be (Ap and Wong, 2001):

- A leader who can take responsibility;
- A tool that helps tourists understand the places they visit;
- An educator;
- A hospitable ambassador;
- A host who can create a comfortable environment for the guest

A tour guide's field of interest spans a growing number of disciplines as a result of its interdisciplinary structure, and in order to be an effective guide, the guide must constantly advance his knowledge in those fields. Tourist guides are very important for the tourism industry as they perform a range of different functions. For example; A tour guide may be the reason why a tourist who is not satisfied during his stay leaves that city happily, or he can solve the problems of a tourist who cannot find the service he expects in the restaurant they visit during the tour. Şenel (2011) drew attention to the importance of tour guides in the tourism sector and highlighted the following points:

- The tour guide has a great role in the promotion of the country.
- One of the factors that contribute to tourists' satisfaction when they depart a region or nation is the guide's accurate and thorough information dissemination. One of the reasons why tourists leave a region or country satisfied is; It is the correct and complete information transfer carried out by the guide.
- In addition to explaining the natural, cultural and historical features of the region, the tour guide should also provide accurate and clear information about the region, the country's politics, social structure, ethnic and religious characteristics.
- Intermediary between local people and tourists and their cultures are binding.
- The guide is in one-to-one contact with tourists and therefore plays a very important role in representing their country.
- Tourist feedback affects a large portion of tourism sales. Positive feedback from tourists about the tour will have a significant impact on the tour guide.The feedback of tourists visiting that region affects a large part of sales in tourism. Tourists' positive feedback about the tour will greatly affect the tour guide.

- Tourist guides shape the perception of tourists and this affects the marketing activities of all stakeholders in the tourism system.
- The tour guide is the transmitter, interpreter and summarizer of the regions visited during the tour.
- The tour guide provides information on important or interesting details about the region and encourages tourists to revisit the region at some future time.
- Tour Guide provides information to travel agencies about the tour, conveys the issues that it considers important about tourists, and allows the agency to carry out the tour with firmer steps in future tours.

The tour guide requires many areas of expertise. They should be a psychologist when they see a worried tourist, an anthropologist when they come across a tourist who is curious about cultures, and an archaeologist when they see a tourist who has little understanding of ancient sites. In addition, the tourist guide, in general terms, must take responsibility for the tour during the tour and a good educator (when conversing, for example, with a visitor who has never before visited a museum). Tourist guides must also be friendly, hospitable, tolerant, and unique hosts when the tour party visit them in their own home. Above all, they must know and recognize where and when to use these skills (Ap and Wong, 2001: 552 cited. Ak, Kargiglioğlu and Bayram, 2019: 24).

Contributions of tour guides to the tourism industry

Tourists participate in guided tours or benefit from the guidance service for a variety of reasons. Wishing to experience a new culture or avoiding the difficulties of being in a foreign country is central to 'tourists' reasons for using guide services. As many researchers agree, the performance of the tour guide will affect the tourists' experience. For example, the guide has a great influence on where tourists will go, what they will see, and what aspects of the local culture they will relate to. Tour guides play a direct role in providing cultural mediation between local culture and visitors, in addition to the important role they will play in improving the experience of tourists (Yu, Weiler and Ham, 2001). In this respect, tour guides undertake important duties as cultural mediators and provide an important service in terms of communication between tourists and local people (Leclerc and Martin, 2004).

Tour guides also connect travelers with historical and cultural heritage, helping them to understand natural and social life, past and present. Guides represent the most relevant part of the tourism sector with human, environmental, social and cultural needs. The relations and interactions of the guides with the tourists also have a directly affect on the relations of the tourists with the local people and tourist satisfaction. In addition, tour guides play an important role in the long-term use of resources and in directing tourist behaviour, and they have an active role in the adoption of a sustainable tourism approach (Avcıkurt, 2009). When viewed from this viewpoint, it is noteworthy that tour guides play an important role both in ensuring that tourists participate in activities that aim to be intertwined with nature without harming the environment, and in the realization of tourism activities that support the people of the region, both economically and socially, and also that these roles impose important responsibilities on tour guides in terms of sustainable tourism (Taylı, 2003).

The tourist, who constitutes the building block of the tourism sector, goes beyond his normal life to an environment where he is foreign, whether he is visiting another region in his own country or traveling to another country. In this unfamiliar environment, he needs help in everything from information to shopping, from attitudes and behaviours to entertainment.

Today, most tourists browse internet sites relating to the region or country to which they will travel, from guidebooks, magazines, etc. They can obtain an opinion on almost any subject from available publications. Depending on the performance and quality service delivery, the tour guide affects the success of the tourism sector, influencing whether or not the tourists will continue their travels. They play an extremely important role for the image of the country and destination (Zhang and Chow, 2004: 82). In addition to knowing the mother tongue very well, religion, culture, and the history and geography of the country, the tour guide should also have a good command of subjects such as archeology, mythology, art history, history of religions and iconography. They should also possess strong communication skills and have sufficient foreign language knowledge. Most of the tourists prefer to benefit from the knowledge of tour guides; that is, people who have chosen guiding as a profession (Değirmencioğlu and Ahipaşaoğlu, 2003). It is possible to argue that this tourist attitude, as well as any concerns about the future of tour guiding, are unfounded, but that tour guides must provide services by adapting to the innovations that are required.

As a result of his behaviour, the tourist guide is in a situation to inform tourists about the society and the country. Although they may bring static guides such as books, maps and brochures with them, these are not usually sufficient. In the end, reasons such being unfamiliar with the language, customs and traditions of the destination country and feeling fearful and uneasy about this provide a reason for tourists to acquire the assistance of a tour guide. In addition to meeting this need, the tour guide is the person who provides the opportunity to better introduce the tourists to the country and society, and to correct any possible wrong or erroneous information which the tourists may have. Therefore, it is possible to regard a tour guide as being a volunteer tourism ambassador representing his country and society. Naturally, the reason why countries, and our country in particular, have been very meticulous and disciplined in the selection and training of tour guides in recent years is explained by the fact that the profession directs international movements and creates interaction (Değirmencioğlu and Ahipaşaoğlu, 2003).

Tourists, who are among the basic components of the tourism sector, travel from their neighborhood to a different region, even if they have visited their own country or a different destination. He will come to an environment that he has never experienced before, and of which they perhaps have no knowledge, to carry out a touristic activity and will participate in some tourism activities such as eating and drinking, entertainment, accommodation. Meanwhile, the tour guide wants to ensure that many of the steps will be reliable. Furthermore, there is always a need for an expert's knowledge and experience in almost every subject. Travel guides are one of the professions that fill this need. Tour guides will not only provide satisfaction in their tour with their services, verbal and non-verbal communication skills, experience and experience; they will also lead to positive feedback within the tourism sector. It is possible to state that tour guides are at a key point in terms of both strengthening the image of the region and the country and the development of the branding. For this reason, tour guides know the language indicated on their cockade and their own language, as well as art history, archeology, history of religions, mythology, iconography process (Zhang and Chow, 2004).

Tourism is a service-based industry and tourist guides play a very important part of the system. In general, although he is on the tour representing the travel agency, it is possible to reflect that the guide also represents both the country and many stakeholders in the tourism system. In recent years, tourist guides have been one of the most important promoters of word-of-mouth marketing, which has becoming increasingly important. In international touristic circulation, where new destinations are added continuously, the tourist guide is one of the main reasons for the next trip. Such guides have a significant importance in national and regional development. The stakeholders will also be content if the tour guide has good communication skills and makes the tourists and their environment happy. A tour guide who can communicate well and makes his environment and himself happy makes both the tourist, the stakeholders and himself happy. It is known that tour guides make a positive contribution to the development of destinations, the development of the region, and the positive development of the 'country's image. Hughes (1991) focuses on the role played by tourist guides in making tourists feel at home in an environment they are unfamiliar with, and states that tour guides have a great responsibility in creating a positive impression about the local people, destination and region.

Eker and Zengin (2016) also emphasized the ambassador role of tourist guides. They stated that they contribute to issues such as tourists leaving their trips happy, enhancing the quality of the holiday, having a positive perspective towards the region and country they visit, and, in this context, increasing the income obtained from tourism and providing competitive advantage. Huang, Hsu, and Chan (2010) stated in their studies that tourist guides have an active role in regional development and tour experiences can positively affect the regional or national image. Tour guides are an indispensable resource for tourists; they have a great role in the international arena, and they can make their country competitive with many destinations with the service they provide (Yıldız, Kuşluvan and Şenyurt, 1997).

Roles of tour guides in nature-based tours

Gülnar (2009) states that one of the first important studies on the concept of role was carried out by introducing Biddle's role theory in 1979; role theory was structured in such a way that behaviour in the context of a role is associated with a shared identity (positions) that people share and an awareness of their roles. According to the theory, roles are continuous because they are the result of a larger social system and these roles have to be taught to people. The integration of roles, positions and expectations is all considered within role theory. The concept of role, which is one of the basic concepts of theater art, offers an effective perspective in terms of interpreting human behaviour in the real world. Shakespeare famously remarked that 'All the world's a stage'; in other words, social life works like a theater in terms of its decor, mask and atmosphere. In fact, while playing our role in real life, there is the use of a special language required by the role. The concepts of role and status were brought into the social sciences terminology with Linton's work titled "Human Study" published in 1936 (Kasapoğlu, 1994). Linton argues that the concepts of role and status cannot be considered independent of one another; they are interrelated, with every role depending on a status and every status depending on a role. Linton also explains the concept of status as the sum of the powers and responsibilities that come with having a certain status, independent of the individual. The individual demonstrates his/her role by fulfilling the status, authority and responsibilities that he/she has obtained by being included in a certain social system. While roles and status are the models that regulate the attitudes and behaviours of individuals, they enable other individuals to adapt to the behavioural models that are expressed by these roles and status The individual assumes multiple statuses and various roles depending on their participation in social systems (Linton, 1936).

Based on these definitions, it is possible to talk about three elements that make up a role: the expectations of the environment, the perceptions and behaviours of the person. The demands of the environment are the pressures that the environment expects from and transfers to the person regarding the role of the person. The second element that constitutes a role is the definition about oneself. A role constitutes the definitions of the person himself as well as the expectations of the environment. When a person combines the expectations of his or her environment with their own (Heken, 2006). Individuals in a particular society fulfill certain roles they assume. This is just like the role played by a theater actor on stage. A role is a game in which people in a group or society play with defined boundaries. In real life, every society expects its members to fulfill their role in the best way possible. For example, when taking a taxi, the driver is expected to take it to the desired address, a mother is expected to take good care of her children, a teacher is expected to take care of her students, and teach them the best manners. Each role has some expectations which are required from the individual and also comes with some privileges. This is because every role is related to a status. People occupy a certain status and play a role appropriate to that status. Role and status are like two sides of the same coin. Although each individual's status is unique, the roles he/she fulfills are many. In other words, there are many different roles played by a person of a certain status. A status requires or reveals many role behaviours (Özkalp, 2004).

It is not possible to think of organized tours without the service of a guide. This is provided by both the tour leader accompanying the group throughout the tour and the guide meeting the group at the destination (Rabotić, 2010). Tour guides play a very important function in the tourism sector and tourist guides have many roles that they are obliged to fulfill in business life. In the literature on tourist guiding, the role and functions of the tourist guide have been emphasized, and different definitions have been given about the roles of the guides (Zhang and Chow, 2004). In order to become a professional tourist guide, it is necessary to have many qualifications. After starting the profession, the guide is expected to fulfill some requirements. These include the professional roles of tourist guides. The roles of tour guides are quite complex and subordinate. The origin of the tourist guide's roles were explained by Eric Cohen who deal with tour guidance scientificall. He is one of the people (Rabotić, 2010). The roles of tourist guides is a subject that has been the subject of a substantial amount of research (Black and Weiler, 2005). A number of different researchers have made contributions to clarify the subject. First of all, in Cohen (1985) the roles of guides were examined under four different headings. Tourist guides' outward-oriented role (intermediary roles) and interactive roles that enable them to meet local people are related to organization and management. By contrast, intrinsic roles imply that tourist guides focus on leadership, which is a feature of social interaction (social roles) and the dissemination of knowledge (communicative role). It is possible to summarize the four roles of Cohen (1985) as follows;

- Agent Role: This refers to orientation, directions, reaching the area.
- Interactive Role: This refers to the promotion of the region to the visitors with a non-threatening and harmonious approachSocial Role: Through the use of humor, social interaction focuses on stress management, harmony, and group psychology.
- Communicative Role: This entails identifying the topics that the group is interested in and conveying the (Cohen, 1985 as cited in Randall and Rollins, 2009).

According to Cohen (1985), tourist guides are not only individuals who guide, set up camp tents and read road signs; they also teach and lead, helping tourists to interpret what they are seeing. Tourist guides have to be professional and well-trained not only in terms of the geography and history of the regions they guide, but also in sociological and psychological areas such as group characteristics, motivation, cultural and ethnic characteristics. The functional leadership and social leadership tasks of guides should be complementary. The leadership role of guides in nature-based tourism activities is divided into two distinct parts: instrumental leadership and social leadership. As a leader, the tour guide's instrumental leadership role is to organize (provide direction, transport and control) and provide safety and security for visitors. The guide must also give a thorough presentation on the subject in order to serve as an instrumental leader and provide orientation to the visitors about the tour and the destination before the tour even begins, as well as the tools that visitors should bring with them, appropriate clothing and footwear, and medications to be taken. Warning visitors where necessary and making sure that the visitors understand the rules are among the most essential parts of this role. Taking these precautions is important in terms of preventing possible accidents (Howard et al., 2001).

The leadership role of tourist guides contributes to the experiences of tourists in three distinct ways. They provide detailed information on the options available so that tourists can make good choices about what to do or where to go; they provide detailed information that will create a safe and suitable environment for tourists to understand the messages given on the warning signs, to manage and cope with the problems they encounter in the best way; and they create and live real experiences for tourists to participate in various activities such as guided walking, ecotours, visiting art galleries, and visiting botanical gardens or zoos (Moscardo, 1998). Moscardo (1998) defines tourist guides as people who transform tourist visits from tour to experience through their knowledge, cultural understanding, communication skills and leadership qualities and states that the leadership role of tourist guides plays an important role in increasing the experience of tourists.

It is seen that the leadership roles of tour guides in nature-based tourism activities exhibit a great similarity with the leadership in leisure activities. Leadership in recreational activities is defined as the process of helping individuals and groups define and achieve goals by their leaders during the recreational experience. Leaders in recreational activities do not want participants to just complete their activities and complete their recreational activities, and they often expect the recreational activity to make a change in the lives of the participants. Leading and facilitating leisure experiences on the past of leaders help develop understanding, knowledge, emotions and skills that participants can apply in other areas of their lives. Leaders are also expected to be people who have the ability to plan, organize, implement, direct and supervise different programs. In this context, all the responsibilities of the leaders in recreational activities should also be fulfilled by the guides. Randall and Rollins (2009) emphasize, in addition, that the leadership role of guides has a positive effect on the tour experience. In the literature, In nature-focused tourism activities, it can be seen that the mediator role comes up (Köroğlu, 2013).

The role of intermediary is defined in terms of providing information, providing access, creating group interaction, and mediating communication between tourists and their guests (Holloway, 1981). The mediating role of mentors is defined as eliminating communication difficulties between the tourist and the social environment, affecting the places to see, transportation and communication with the local people (Schmidt, 1979). The mediating role of guides should focus on selection (what tourists will and will not visit) and mediate the cohesion of groups (Cohen, 1985). In these roles, guides should have a range of duties including communicating with the group, creating positive perceptions on tourand having a mediating position between tourists and local people. In addition, the tour guide should act as an interactive tool and as an educator. The success of guides in interactive mediation is based primarily on defining themselves as representatives of the local environment and local people, and creating a peaceful environment between the local people and the tourist (Köroğlu et al., 2012).

According to Yu, Weiler, and Ham (2001: 79), tourist guides have to have extensive knowledge about nature and what it offers. In addition to all kinds of knowledge about nature and the environment, guides should have cultural knowledge, including cultural values, language, communication styles, traditions and artistic values. In addition, the guide should provide tourists with a deep view of the destination, enabling them to evaluate what is real and reliable about the destination. However, a guide's foreign language knowledge is also important in cultural mediation. A fluency in speaking the languages of both cultures by the guide provides the opportunity to eliminate the communication problem between the two cultures and the conflicts that may arise due to this.

Tourist guides must have extensive knowledge about nature and what it offers. In addition to all kinds of knowledge about nature and the environment, guides should have cultural knowledge, including cultural values, language, communication styles, traditions and artistic values. In addition, the guide should provide tourists with a deep view of the destination, enabling them to evaluate what is real and reliable about the destination. However, a guide's knowledge of foreign languages is also important in cultural mediation. Fluently speaking the languages of both cultures by the guide provides the opportunity to eliminate the communication problem between the two cultures and the conflicts that may arise due to this (Yu et al., 2001). The guides do not have sufficient knowledge, especially in the field of flora and fauna; at the same time, they either express or mispronounce many plant names in their own language. This may lead to misunderstandings. Thus, one of the most important deficiencies in the activities carried out in nature is the lack of expert guides; the other is that the guides do not have sufficient foreign language knowledge. In addition, a nature guide should be able to speak fluently in a variety of languages, have access to remote, wild, and previously uncharted areas, possess scientific knowledge, be able to build relationships with both locals and visitors, and contribute to the preservation of nature. This emphasizes the need for them to have experience. They also emphasize that the appropriate material to be used in the promotion of nature should be used by the guides (Fennell and Eagles, 1989).

Guides to interpret local culture in ecotourism and nature-based tourism and to motivate tourists to change their behaviour in a way that reduces the negative effects on natural resources, etc. indicate that they should have roles at the level of expertise (Black and Weiler, 2005). Many researcher defines tourist guides more as a teacher, interactive tool, and person responsible for long-term nature conservation goals and behaviour towards the area. It has been concluded that fulfilling the responsibilities affects customer satisfaction and tour experience (Ham and Weiler, 2002). The intermediary role of guides plays a key role in tourists' unforgettable experiences and in increasing the quality of the nature tour experience (Rabotić, 2010). The educational role of tour guides in developing tourists' awareness of nature conservation. The authors state that the educative role of the guides is in the direction of developing environmentally friendly behaviour on the part of the tourists and thus has a significant impact on the tour experience (Randall and Rollins, 2009 cited by Köroğlu et al., 2012).

Conclusions

Nature-based tourism is one of the fastest-growing sectors of the travel industry (Ewert and Shultis, 1997). Ecotourism emerges as a form of tourism that gains traction in the tourism industry and assigns some responsibilities to certain stakeholders.One of these stakeholders is the tourist guide, who have many responsibilities, roles and duties. The protection of nature by tourist guides is one of the important parts of ecotourism principles. Tourist guides are used to reduce the negative effects of tourism, and it aims to contribute to the conservation process of ecotourism, based on the educational and interpretation functions of tourist guides. Tour guiding plays a great role in the realization of nature-based tours with its roles such as leadership, mediation, and resource management. First of all, guides have to be aware of their roles and their responsibilities and they have to put this into practice. In this regard, the first thing to be done is to determine and define the roles of the guides by determining the appropriate strategies for nature throughout the country and for each region. Thus, the roles and responsibilities of guides in nature-oriented tourism activities will be revealed more clearly and guides will be informed about them.

Tourists can easily cross the borders with the comfort of being on holiday in the regions they visit. For this reason, guides have to set limits for visitors in nature-oriented activities. Visitors sometimes do not like the limits set for them and may be disappointed. Sometimes, despite showing respect, these boundaries can be ignored and they do not follow the rulesVisitors will be happy to listen to the guides and follow the rules.

If the guide explains these rules clearly and gives a good explanation of how and why the rules were created, visitors will be happy to listen to the guides and follow the rules. Taylı (2003) states that the majority of the damage to the environment is not caused by the excessive number of visitors, but by the poor management of the visitors. For this reason, the guide has to be a good manager by fulfilling the responsibility given to him by the leadership task at this point and to manage the visitors in accordance with the environment and environment. Being aware of the leadership and responsibilities of the guides will enable them to easily manage their visitors. In terms of achieving the purpose of nature-based tourism activities, it is also of great importance that the guides to be assigned in these activities are expert, well-trained and well-trained guides (Köroğlu et al., 2012).

References

Ahipaşaoğlu, H. S. (2001). *Turizmde Rehberlik*, Detay Yayıncılık, Ankara.

Ak, S., Kargiglioğlu, Ş., & Bayram, G.E. (2019). Turist Rehberliği Mesleği, B. Zengin, G. Erkol Bayram ve O. Batman (Ed.) *Turist Rehberliği Mesleği (Dünü-Bugünü-Yarını)* Detay Yayıncılık, Ankara.

Ap, J., & Wong, K. K. F. (2001). Case Study on Tour Guiding: Professionalism, Issues and Problems. *Tourism Management*, 22(5), 551–563.

Avcıkurt, Cevdet. (2009). *Turizm Sosyolojisi: Genel ve Yapısal Yaklaşım*. 3. Baskı: Ankara Detay Yayıncılık.

Black, R., & Weiler, B. (2005). Quality Assurance and Regulatory Mechanisms in the Tour Guiding Industry: A Systematic Review. *The Journal of Tourism Studies*, 16(1), 24–37.

Cohen, E. (1985). The Tourist Guide: The Origins, Structure and Dynamics of a Role. *Annals of Tourism Research*, 12, 5–29.

Değirmencioğlu, Ö., & Ahipaşaoğlu, S. (2003). *Anadolu'da Turizm Rehberliği Temel Bilgileri*. 4. Baskı. Ankara: Gazi Kitabevi (Fersa Maatbacılık).

Eker, A. G. N., & Zengin, B. (2016). 6326 Sayılı Turist Rehberliği Meslek Kanunun Profesyonel Turist Rehberlerince Değerlendirilmesi: İstanbul Örneği. *3rd International Congress of Tourism & Management Researches*, 1692–1710, 20–22 May 2016, Antalya.

Erkol Bayram, G. (2017). Motivasyonun Demografik Ve Mesleki Özelliklere Göre Farklılaşması: Turist Rehberleri Üzerine Bir Araştırma. *Yaşar Üniversitesi E-Dergisi*, 12 (48), 257–271. Retrieved from https://dergipark.org.tr/en/pub/jyasar/issue/31740/347869

Ewert, A. W., & Shultis, J. (1997). Resource Based Tourism: An Emerging Trend in Tourism Experiences. *Parks and Recreation*, 32(9), 95–103.

Fennell, D. A., & Eagles, P. F. J. (1989). Ecotourism in Costa Rica: A Conceptual Framework. *Journal of Parks and Recreation Administration*, 8(1), 23–34.

Gartner, William C. (2004). Rural Tourism Development in the USA. *International Journal of Tourism Research*, 6, 151–164.

Gülnar, B. (2009). İletişim Doyumu Boyutları İle Örgütlenme Yapısı İlişkisi: Selçuk Üniversitesi Akademisyenleri Örneği. *Selçuk İletişim*, 5(4), 62–82.

Ham, S. H., & Weiler, B. (2002). Interpretation as the Centrepiece of Sustainable Wildlife Tourism. Chapter 3 in Harris, R., Griffin, T. & Williams, P (eds.). *Sustainable Tourism: A Global Perspective*. London: Butterworth-Heinemann, 35–44.

Heken, H. (2006). Toplumsal Cinsiyet Olgusu Temelinde Mesleğe İlişkin Rol İle Aile İçi Rol Etkileşimi: Türk Silahlı Kuvvetlerindeki Kadın Subaylar. *Selçuk Üniversitesi Sosyal Bilimler Enstitüsü Dergisi*, 15, 247–281.

Holloway, J. C. (1981). The Guided Tour: A Sociological Approach. *Annals of Tourism Research*, 8(3), 377–401.

Howard, J., Rik, T., & Brenda, S. (2001). Investigating the Roles of the Indigenous Tour Guide. *The Journal of Tourism Studies*, 12(2), 32–39.

Huang, S. S., Hsu, C. H., & Chan, A. (2010). Tour Guide Performance and Tourist Satisfaction: A Study of the Package Tours in Shanghai. *Journal of Hospitality & Tourism Research*, 34(1), 3–33.

Hughes, K. (1991). Tourist Satisfaction: A Guided "Cultural" Tour in North Queensland. *Australian Psychologist*, 26, 166–172.

Kasapoğlu, A. M. (1994). Aile ve Kadın Araştırmaları İçin Yedi Temel Rol ve Statü. *Ankara Üniversitesi Dil ve Tarih-Coğrafya Fakültesi Felsefe Bölümü Dergisi*, 15, 217–233.

Köroğlu, Ö. (2013). Turist Rehberlerinin İş Yaşamındaki Rolleri Üzerine Kavramsal Bir Değerlendirme. *Pamukkale Üniversitesi Sosyal Bilimler Enstitüsü Dergisi*, (16), 91–112.

Köroğlu, Ö., Köroğlu, A., & Alper, B. (2012). A Study on the Roles of The Tour Guides in Nature Based Tourism Activities. *KMÜ Sosyal ve Ekonomik Araştırmalar Dergisi* 14(22), 131–139.

Leclerc, D., & Martin, J. N. (2004). Tour Guide Communication Competence: French, German and American Tourists' Perceptions. *International Journal of Intercultural Relations*, 28(3–4), 181–200.

Linton, R. (1936). *The Study of Man*. New York: D. Appleton-Century Camp, Inc.

Moscardo, G. (1998). Interpretation and Sustainable Tourism: Functions, Examples and Principles. *The Journal of Tourism Studies*, 9(1), 2–13.

Özkalp, E. (2004). *Örgütsel Davranış* (2nd edn.). Eskişehir: Anadolu Üniversitesi Web-Ofset.

Rabotić, B. (2010). *Tourist Guides in Contemporary. Tourism International Conference on Tourism and Environment*, Philip Noel-Baker University, Sarajevo, Bosnia & Herzegovina, 4–5 March, 353–364.

Randall, C., & Rollins, R. (2009). Visitor Perceptions of The Role of Tour Guides in Natural Areas. *Journal of Sustainable Tourism*, 17(3), 357–374.

Şahin, Y. G., & Balta, S. (2007). Distance Education Techniques to Assist Skills of Tourist Guides. *Educational Technology & Society*, 10(2), 213–224.

Schmidt, C. J. (1979). The Guided Tour: Insulated Adventure. *Urban Life*, 7(4), 441–467.

Şenel, G. (2011). *Stres Faktörlerinin İş Motivasyonuna Etkisi: Turist Rehberleri Üzerine Bir Uygulama Yayınlanmamış Yüksek Lisans Tezi. T.C.* Balıkesir: Balıkesir Üniversitesi Sosyal Bilimler Enstitüsü.

Taylı, Ü. (2003). *Turizm Rehberleri Sürdürülebilir Turizm İçin Ne Yapabilir? Rehber Dünyası, Ağustos, Sayı: 33*, İstanbul: TUREB Yayınları, 16–17.

Türkmen, M. T. (2017). A Research on Determination of Leadership Orientations of Tourist Guides. *Gaziantep University Journal of Social Sciences*, 16(3), 913–927. DOI: 10.21547/jss.314860.

Yıldız, R., Kuşluvan, S., & Şenyurt, Y. (1997). Turist Rehberliği Öğretiminde Yeni Bir Model: Nevşehir Turist Rehberliği Bölümü Programı ve Değerlendirilmesi. *Erciyes Üniversitesi Nevşehir Turizm İşletmeciliği ve Otelcilik Yüksekokulu Hafta Sonu Semineri IV*, 12, 9–14.

Yu, X., Weiler, B., & Ham, S. (2001). Intercultural Communication and Mediation: A Framework for Analysis the Intercultural Competence of Chinese Tour Guides. *Journal of Vacation Marketing*, 8(1), 75–87.

Zhang, H., & Chow, Q.I. (2004). Application of Importance-Performance Model in Tour Guides' Performance: Evidence From Mainland Chinese Outbound Visitors in Hong Kong. *Tourism Management*, 25(1), 81–91.

15

BUSINESS NETWORKING IN THE FIELD OF NATURE-BASED TOURISM

Key factors for successful management

Marios Sotiriadis and Ziyi Yan

Introduction

Tourism enterprises – especially those that ar small and medium-sized (SMEs) – are exposed to strategic risks related to increased uncertainty in their business and market environments, mainly due to globalization and the digital revolution (Okumus, Altinay & Chathoth, 2010). Literature has extensively discussed the relevance of collaboration, the related issues and aspects, as well as its contribution to the tourism business environment. Academic research and business intelligence in tourism have documented the role and benefits of a strategic approach to business partnerships and suggested suitable forms and strategies (Sotiriadis, 2018). The methods for business venturing applied in tourism were analysed by different studies, indicating that collaborative methods/strategies can: (i) make a significant contribution to designing, managing, and marketing experiences; (ii) contribute towards the small business ventures' development and success; and (iii) ultimately be beneficial and positive to local communities (Gursoy, Saayman & Sotiriadis, 2015). There are various collaborative forms at the macro/destination level (the most common being partnership, cluster, and product club) and collaborative strategies at the micro/business level between enterprises for business venturing (such as business alliance, network, consortium). The focus of this chapter is on networking.

It is generally accepted that a business network constitutes a valuable alternative to internal growth strategies and enhances SMEs, allowing them to address the challenges in a highly competitive business environment (Asian Development Bank & Asian Development Bank Institute, 2015). Therefore, networking represents the best organizational model for promoting the growth of SMEs that lack the skills, resources, and knowledge needed for their development and successful operation (Valeri, 2016). Business networks face a series of challenges and issues throughout the process, from the initial stage of planning and designing to the implementation phase. That is why they require adequate planning and management (Morrison, 2019). This is imperative for tourism suppliers and destinations whose main aim is to provide their visitors with unique experiences. In this regard, over the recent decades, many regions and businesses have discovered the benefits of combining local and natural resources with tourism offerings, a pairing that resulted in various forms of experiential tourism such as rural, nature-based, culinary/gastronomic tourism, to name a few (Morrison, 2019; Jolliffe, 2007).

DOI: 10.4324/9781003230748-18

It is believed that local natural resources and attractions have the potential for development and business opportunities because they offer a spectrum of experiences, including all services and activities – such as processing, dining, tasting, entertainment, education – related to tourism experiences (Merkesdal, 2013; Sotiriadis, 2017). This kind of experience opportunities is beneficial to all involved stakeholders (Dangi & Jamal, 2016), that is, local communities (business and job opportunities, farmers/producers, retail), natural and cultural resources (preservation and improvement), and visitors (getting better knowledge and understanding of places visited). This chapter aims to explore networking from a business perspective within the management discipline by examining the prerequisites and key factors for successfully managing business networking within the context of nature-based tourism. A review of the extant literature reveals that the research on nature-based tourism has grown rapidly over the last three decades; see, for instance, Hall and Boyd (2005), Vespestad and Lindberg (2011), Coghlan and Buckley (2012). Previous research focuses mainly on various aspects, such as tourists' motivations and experiences, economic contribution, stakeholders' collaboration, sustainable development, etc. The literature review also indicates that studies exploring the topic of management and collaborative forms are rare. The effective management of business networking is under-researched. To our best knowledge, there are currently no studies exploring this topic from a management perspective based on the blended theoretical foundations of community-based development and theming marketing strategy. Our study attempts to address this knowledge gap.

This chapter argues that the successful and sustainable management of a business network can be achieved by elaborating a master plan/framework that is based on the needs and requirements of the local communities (community-based development) and is adapted to the needs and expectations of the targeted market segments (World Tourism Organization, 2017; Morrison, 2019). The chapter is structured as follows. Thus, the study's research problem is: 'What are the main prerequisites and key elements for successful and effective management business networking in nature-based tourism?" The study aims to address this question. The business environment and the drivers for and benefits of collaboration between tourism enterprises are outlined in the 'Literature review' section. This is followed by a review of literature focusing on the theoretical backgrounds adopted by this study and then the suggested management framework for business networking. An empirical study within the Chinese context tested and validated this framework. Two are the main contributions of this study, namely: (i) elaborate on an integrated framework for the effective and sustainable management of business networking; and (ii) suggest guidance for industry practitioners and involved stakeholders alike to enhance them effectively plan, design, and implement this collaborative strategy in this field. The component elements and the main findings are presented in the 'Empirical study' section. The chapter is completed by a discussion of the conclusion, management implications of the study's findings, and limitations.

Literature review

The issues of collaboration and the theoretical backgrounds adopted by this study are briefly discussed.

Collaboration and business networking

Academic research indicates that tourism experiences are multidimensional and complex in nature, and tourism providers must address the resulting challenges by adopting adequate approaches. Studies indicate that the global tourism market no longer involves single businesses but rather consists of spatial or thematic destinations composed of a cluster or network of tourism-related operations (Gursoy, Saayman & Sotiriadis, 2015; Lazzeretti & Petrillo, 2006). Because most tourism-related industries operate in a highly competitive business environment and deal with rapidly changing market conditions, working closely with all actors involved in producing and delivering a tourism experience is vital for survival and success (Morrison, 2019). It is believed that: (i) collaboration is necessary rather than a luxury in a competitive tourism business environment; and (ii) partnering could significantly contribute to designing, managing, and marketing services and experiences.

The various models of business collaboration constitute the efficient framework for management and marketing purposes, especially in the field of niche tourism markets, such as experiential tourism, gastronomic/culinary, nature-based, and creative tourism, because of the nature of tourists' experience and the variety of involved actors/businesses. Investment in business collaborative ventures and alliances is a good investment because it constitutes a potential source of sustainable competitive advantage.

Tourism providers can also provide appealing experience opportunities and extra customer value by entering into a business venture. Nevertheless, many key issues need to be considered and resolved in order to realize the above-mentioned potential benefits; the main factors are as follows (Eisingerich & Bell, 2008; Weidenfeld, Butler & Williams, 2011; Gursoy, Saayman & Sotiriadis, 2015). The collaboration and combination of forces and resources is not a magic recipe for business success. It must be well planned; only a collaborative platform wisely designed creates a series of business benefits. The main aims of a collaborative strategy are threefold: creating an appealing and attractive offering; achieving a competitive advantage; and generating business and market diversification. Members/partners should share consistent values and focus on the same market segments; therefore, it is imperative to select the appropriate partners. The appropriate collaboration structure is equally important (Sotiriadis, 2018).

Collaborative strategies in the business, such as alliances and networks, are seen as a framework providing small and medium-sized tourism enterprises with managerial and marketing opportunities to operate locally and in a globalized and competitive business environment (Fyall & Garrod, 2005; Gursoy, Saayman & Sotiriadis, 2015; Sotiriadis, 2018). The most common collaborative strategies/methods for business venturing (at the business/micro-level) are business alliance, network, and consortium; the chapter focuses on networking. A business network is a social network for businesses (Kokkonen & Tuohino, 2007). Networking is a socio-economic activity performed by groups of likeminded business people to identify, create, or act upon business opportunities (Osterle, Fleisch & Alt, 2001). Many managers contend that business networking is a more cost-effective method of generating a new business than marketing plans. This is because business networking is a low-cost activity that involves more personal commitment than company resources.

Several studies examined the formation and contributions of collaborative approaches in tourism, for instance, spa and health tourism (Novelli, Schmitz & Spencer, 2006) and creative tourism (Couret, 2015). They all reported that partnerships are vital strategies for an enterprise's survival and growth in a saturated and highly competitive environment and address tourists' requirements. Networking enables the partnering members to exploit their synergies and complementarities. This collaborative framework has been extensively used in tourism industries, with niche markets gaining from network building when their businesses compete and collaborate. They provide a series of business benefits for partners (Novelli, Schmitz & Spencer, 2006; Sotiriadis, 2018). A successful example is the Creative Tourism Network (Couret, 2015).

Theoretical foundations

The key issues in the studied field are sustainable community-based tourism development and strategic marketing (theming) from a management perspective. The careful consideration of these foundations and extant literature should allow us to determine the prerequisites and critical success factors (CSF) for the efficient and successful management of a business networking in management of experiential tourism, especially in rural areas, with fragile and volatile environment and limited resources base (WTO, 2017; Morrison, 2019).

Community-based tourism

It is believed that the main idea of community-based tourism (CBT) is the involvement of the local population in planning and managing tourism development in order to create a more sustainable and successful – financially viable, socially and culturally beneficial – tourism (Hall, 1998). The main principles and key suggestions of the CBT paradigm are as follows. CBT is usually owned, planned, and managed by and serving the community (APEC Tourism Working Group, 2010). Community participatory development and tourism have a mutually synergetic impact (Morrison, 2019). CBT is often regarded as an alternative pathway to mass tourism and a more sustainable approach (Dodds, Ali & Galaski, 2018). The determining or critical success factors for monitoring and evaluating the CBT development are local ownership, management and leadership, community participation, economic benefits fairly distributed, preservation of tourism resources/assets, small-scale projects and infrastructure, partnerships within and outside the community, and inter-stakeholder relationships (Vajirakachorn, 2011). Engagement, communication and interaction, job opportunities, and tourists' satisfaction are key issues and challenges to economic prosperity and social welfare (Dangi & Jamal, 2016). The CBT projects are usually small in scale and involve the interaction between the host community and visitors. This model is particularly suitable for rural and nature-based spatial zones.

Strategic marketing planning: theming strategy

For companies, theming is a marketing tool that can help them stand out and distinguish themselves from competitors (Astrøm, 2017). Likewise, in the tourism industry, the theming strategy is mainly used to distinguish places and their offering, increasing their attractiveness (Astrøm, 2020). Usually, theming strategy in the tourism field is used for differentiation and branding purposes (Sotiriadis & Shen, 2017; Morrison, 2019). A tourist destination usually has fascinating features, such as nature, culture, industry, local produce, or a combination of these elements. Tourists' visit motivation is based on one of these elements (or a combination), which is the foundation of thematic offering/strategy (Morrison, 2019; Tsai & Chen, 2019). Themed routes and circuits constitute an alliance/partnership evolving touring itineraries for visitors under the umbrella/logo of a specific theme. A good example of themed partnerships is the product clubs – such as the 'Wine Routes' of Spain – created by tourism stakeholders at the local/destination level (Del Campo Gomis et al., 2010). A review of the extant literature indicates that there have been some studies on the adoption and implementation of theming strategy in tourism at the destination level, predominantly for marketing purposes (Morrison, 2019). Scholars began to be interested in the key issues of theming strategy, such as the impact of theming on the tourist experience (Astrøm, 2017), its use for branding and promotional purposes (Sotiriadis & Shen, 2017), and the foundations of the formation of tourism themes (Astrøm, 2020). This strategy could be very effective in the field of nature-based tourism.

Nature-based tourism and tea tourism

Nature-based tourism (NBT) is a broad concept. Its activities represent a large spectrum of passive to active behaviours. Examples include passive enjoyment of the scenery, landscapes, geology, flora, and fauna; outdoor recreation and adventure; and consumptive uses such as hunting and fishing (Coghlan, 2016). NBT includes a diverse range of products with different conceptualizations of the importance of nature within the tourism experience (Buckley, 2009). Nature-based tourism is now firmly embedded in the experience economy, where different tourism market segments are expected to pick and mix components and activities to form customized holidays (Vespestad & Lindberg, 2011). Coghlan (2016) highlights the management of tourism experience and indicates the importance of the partnership approaches among local stakeholders, tour operators, and individual tourists to deliver high-quality and low-impact nature-based tourism experiences. This is particularly valid in the case of tea tourism, which is a form combining nature and culture.

Jolliffe (2007) defines tea tourism as "tourism motivated by the interest in the history, tradition, and consumption of tea" (Jolliffe, 2007, p. 26). Sun (2005) argues that cultural tourism integrates sightseeing, business, learning, tea picking/harvesting, and shopping (Sun, 2005). This form of tourism is predominantly based on natural resources. Most of the studies focus on the practical significance of tea tourism, principally analysing the development stage of tea tourism in different contexts (Yi & Li, 2017). Tea resources/assets also include tea cultural activities, such as visiting tea gardens, tea factories, and houses, experiencing tea-drinking customs, and purchasing tea (Jolliffe, 2007; Aslam & Jolliffe, 2015). These social and cultural activities are very attractive to tourists due to their experiential nature (Gunasekera & Momsen, 2007). The study by Yan et al. (2021) analysed the tea offering as a tourism asset and suggested adequate strategies to achieve an effective pairing/partnership between the tea industry and tourism/leisure activities. Their study indicated the key elements for the successful integration, effective pairing, and sustainable operation of tea tourism offering.

The above discussion indicates that collaborative strategies and adequate management in this field are under-researched. There is a need for more conceptual suggestions (based on robust theoretical foundations) and empirical investigations in specific contexts and settings to better understand the conditions and prerequisites for designing and managing a business network. This is the aim of our study by suggesting a management framework.

Suggesting a framework

The suggested management framework – prerequisites to and key elements of successful operation – for business networking is based on the above outlined theoretical foundations and the suggestion of main prerequisites and key factors (CSF) for the successful implementation, management, and operation of this collaborative form. The suggested framework encompasses two core strategies and some support strategies based on the value chain concept suggested by Porter (1985).

There are two core strategies. The first is the choice of collaborative form between local businesses providing tea tourism services. This should be a business network including businesses and other tea tourism organizations which have a common/shared interest. The business network should be able to offer visitors and tourists opportunities for knowledge and taste and cherish all aspects of tea assets and resources through all stages of the process, from farm to consumer. Production (breeding, planting, farming); processing, harvesting, packaging; tasting and sales; tea culture and traditions (rituals). The second strategy is theming.

The efficient management would be designed and achieved through themed products/packages. Strategic marketing planning (SMP) suggests the following operational strategies: a local alliance (network), design and offering unique sales value propositions of themed products, segmenting, positioning and targeting through theming, and integrated marketing communications. Therefore, the local tea tourism offering should be designed by employing three to four themed experiences (visits or packages) addressed to different market segments. These themed products are designed/crafted based on specific criteria: targeted market segments, duration, and included activities. Support strategies should be implemented to enhance and improve the above-outlined offering and ensure the sustainable and effective management of local resources. It is believed that four fields are critical in supporting the management of the LT Tea Tourism Network, namely: (i) human resources: developing skills and know-how; (ii) product development: innovation and creativity; (iii) quality management: customer care and continuous quality improvement; and (iv) integrated marketing communications: promotional activities, based on the 7Ps model (Product, Price, Place, Promotion, People, Process/Physical evidence and Partnership).

Several studies have concentrated on analysing the dynamics of network development and which factors are responsible for their success and financial sustainability (Soda, Usai & Zaheer, 2004; Valeri, 2016). In the tourism context, networks are created to: (i) design and promote tourist projects (Braun 2013; Scott, Cooper & Baggio, 2008; Lemmetyinen & Go, 2009); and (ii) market tourism services and experiences (Novelli, Schmitz & Spencer, 2006; Sfandla & Bjork, 2013). Based on the literature review, this study identifies and suggests several prerequisites/conditions and key factors (CSFs) for this partnership's successful management and operation. This study, therefore, posits the following elements and factors. First, the prerequisites suggested are creating a platform of common interests, long-term commitment, full support, achievement of mutually shared benefits, and the fair distribution of results. In terms of interrelationships between network members, a basic prerequisite for creating and developing networks is mutual trust between the participants. Trust is essential in establishing relationships between partners (Gulati & Singh, 1998; Ruggieri & Iannolino, 2012; Gabor, 2015). This also stands for the involvement, engagement, and interaction of local stakeholders. These prerequisites become much easier when the trust between participants is based on the previous common experience of collaboration (Gulati & Singh, 1998).

The second is the CSF. It is worth pointing out that CSFs are those key elements required for an organization or a project to either accomplish or exceed its desired goals (Johnson & Friesen, 1995). These factors must be given proper attention and attain the desired objective. A CSF is a critical factor or activity required for ensuring the success of an organization or a project. Therefore, a CSF is exactly what it sounds like: something that is critical to the success of a project or organization. CSFs are limited (usually five to eight) key characteristics or conditions that directly impact a project's viability and must be performed at the highest possible level in order to achieve success, the intended overall objectives, or expected outputs (Businessdictionary.com, 2021). The CSFs proposed by this study are: (1) development of complementary competencies, capabilities, and skills; (2) knowledge and intelligence sharing and dissemination; (3) market orientation and customer-minded; (4) coordination, control, monitoring, and assessment; (5) continuous improvement in all domains; (6) investments (devote sufficient human and financial resources); and (7) appropriate managerial structure. Figure 15.1 depicts these suggestions.

The value and utility of the above suggestions (framework, prerequisites, and CSFs) were investigated and tested through an empirical study detailed in the following section.

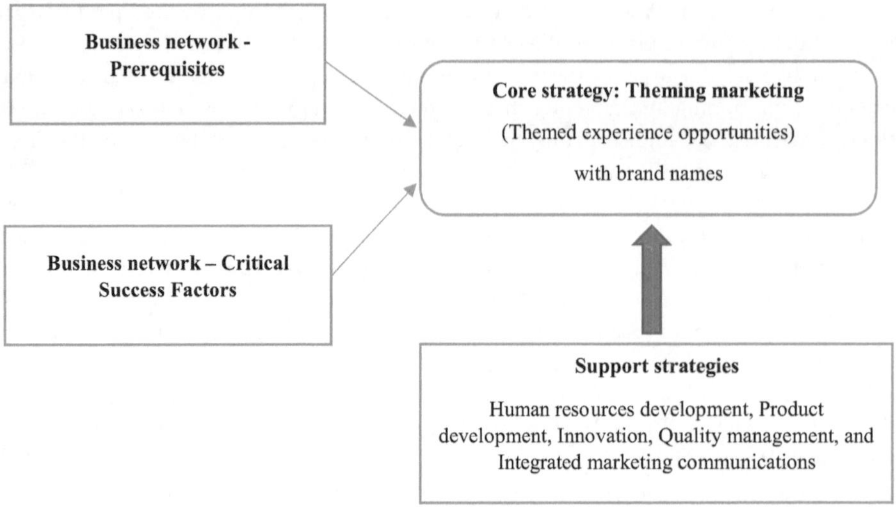

Figure 15.1 A suggested framework for designing and managing the Tea Tourism Network.

Empirical study

Study area

Longwu Tea Town is located on the southwest side of Hangzhou, Zhejiang, SE China, and grows Longjing tea, the most famous tea in China. It is the largest producing area in West Lake Longjing. In addition, Longwu Town has brought together more than 150 key specialty tea companies and institutions, including Nongfu Spring Headquarters, China Tea Industry Alliance, and the Technology Transformation Centre of the Zhejiang University Tea Research Institute (Figure 15.2).

Today, Longwu is the most popular tourist town in the Yangtze River Delta with a unique natural environment and historical heritage, and is rated as a 4A national tourist attraction. 'Longwu Tea Town Tour' was also launched in 2018. Longwu Tea Town currently has 102 characteristic homestays. In 2018, Longwu produced 187 tons of spring tea, hosted 2.38 million Chinese tourists, and produced a tourism revenue of 170 million yuan. Longwu Tea Town is also a hub for official tea events across the country, such as the 'West Lake Longjing Tea Festival', the 'China International Tea Expo', and the 'Chinese Tea Olympics'.

Additionally, the West Lake Longjing tea-making skills have been successfully included in the 'National Intangible Cultural Heritage' list for preservation and promotion purposes. All of the above-outlined features and elements of the spatial zone formed a solid foundation for selecting Longwu Tea Town as the study area for our research. To sum up, our choice was based on the following reasons: (i) outstanding geographical location and resources; (ii) the main local resource is tea, which has high quality and an outstanding reputation; and (iii) focus on community-based development and the interlinkages of local industries and activities.

Research instrument and technique

In order to test and validate the suggested framework, an empirical study was performed which focused on analysing the perceptions of local stakeholders about its value and utility. Therefore, the implemented research method is in the form of qualitative, in-depth interviews.

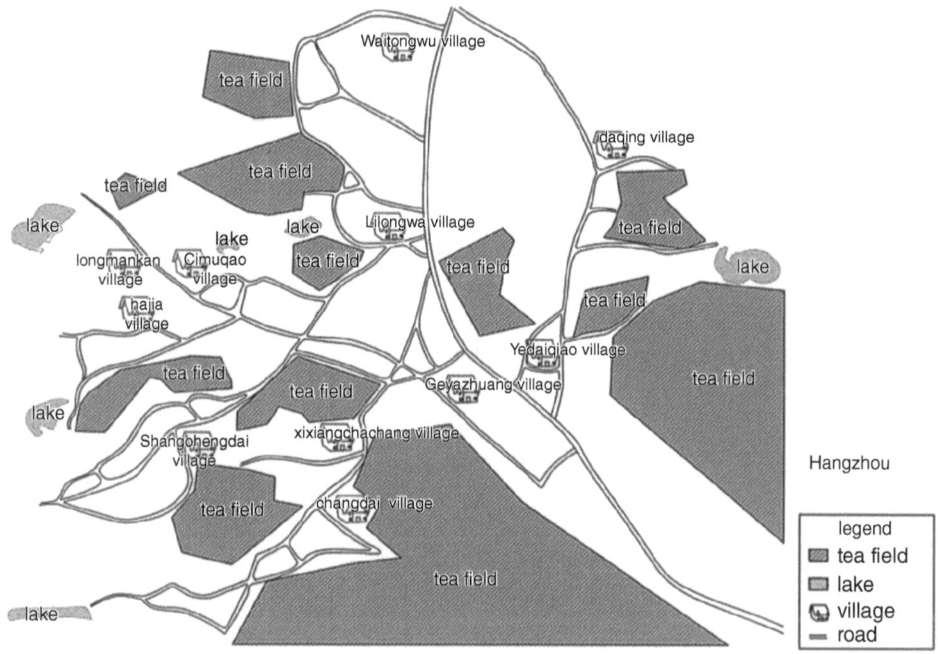

Map of Longwu Tea Town

Figure 15.2 Presents a map and plan of Longwu Tea Town.

These individual interviews were conducted in Chinese with selected stakeholders. Our study opted for this technique due to the nature and aim of the research project, i.e., to explore the 'stakeholders' views on a specific topic (tea tourism) and its key aspects (Finn, Elliot-White & Walton, 2000; Jordan & Gibson, 2004). The participants were asked to express their opinion and rate the level of importance or agreement about the various items measured using a five-point Likert scale, except the demographics.

Sample and data collection

The sample population included 25 businesspersons and key local stakeholders classified into two groups. The first group included 19 local businesspersons. More specifically, this group consisted of 6 tea growers (Code: TG) and 13 tea-related businesspersons, distributed as follows: Two from homestay businesses (Code: HS), two from catering businesses (Code: CB), two from processing businesses (Code: PB), two from retail business (Code: RB), one from cultural activity business (CAB), and four from leisure and tourism/accommodation (Code LT). The second group included three public sector managers (Code: PSM) and three industry practitioners (Code: IP). Table 15.1 depicts the profile of participant stakeholders with a code for the interviewed persons to keep their anonymity.

The personal interviews were conducted by a research team member, following an appointment, between April and May 2021. Each interview lasted 45 minutes and all were fully transcribed for data analysis.

Table 15.1 Profile of sample: Participant stakeholders

No	*Interviewee's Code*	*Capacity/Main occupation*	*Educational level*	*Gender*	*Age group*
1	TG1	Tea grower	Junior High School	Male	46 to 55
2	TG2	Tea grower	High school	Female	36 to 45
3	TG3	Tea grower	Primary school	F	46 to 55
4	TG4	Tea grower	Primary school	F	56 to 65
5	TG5	Tea grower	Primary school	M	56 to 65
6	TG6	Tea grower	Primary school	M	Over 65
7	HB1	Businessperson	Junior high school	M	46 to 55
8	HB2	Service business	Junior high school	F	26 to 30
9	CB1	Businessperson	Junior high school	M	46 to 55
10	CB2	Businessperson	University degree	F	31 to 35
11	PB1	Artisan	Primary school	F	56 to 65
12	PB2	Businessperson	Junior high school	M	46 to 55
13	RT1	Businessperson	Junior high school	F	31 to 35
14	RT2	Businessperson	Junior high school	F	56 to 65
15	CAB1	Office employee	University degree	F	31 to 35
16	LT1	Services	High school	M	36 to 45
17	LT2	Businessperson	University degree	F	31 to 35
18	LT3	Businessperson	University degree	M	36 to 45
19	LT4	Services	University degree	M	36 to 45
20	PSM1	Civil servant	University degree	F	36 to 45
21	PSM2	Civil servant	Postgraduate degree	M	36 to 45
22	PSM3	Civil servant	University degree	M	46 to 55
23	IP1	Technician	Postgraduate degree	F	36 to 45
24	IP2	Businessperson	High school	M	46 to 55
25	IP3	Services	University degree	M	56 to 65

Results

This study adopted the content analysis conducted at two-levels data analysis, encoding, and processing. At the first level, the pre-processing of data was conducted, encoding all interviews to process the collected data. Data were then processed at the second level using keywords for classification to discuss the related issues in a context. The main elements of this analysis are discussed in the following points per issue.

Strategies

Most of the participant stakeholders agreed on the high utility of the suggested framework (Table 15.2). According to the local stakeholders' opinion, the 'Thematic strategy' and 'Support strategies' were considered very important and valuable. It was also pointed out that risk management should be added to the support strategies. Some comments: Participant LT2 argued that "providing tea travel experience can enhance tourists' sense of participation, increase tourism enjoyment, make tourists impressed, and then increase the rate of tourists revisit and recommendation." Most of the interviewees agreed with stakeholder PSM3 that "the future will be bright and promising if we design and develop some more creative tea tourism activities based on this asset." PSM1 stressed that "based on excavating and consolidating the existing characteristics, the government also attaches importance to the product development.

Table 15.2 Perception about strategies

Strategies	*Description*	*Perception/ opinion of stakeholders*
Theming	Experience opportunities to know, taste, and value all aspects of tea assets and resources	Very important
	Design and creation of theme products	Very important
Support	Development of human resources (skills and know-how)	Important
	Product development (Creativity and innovation)	Important
	Quality management (Customer care and continuous improvement)	Important
	Integrated marketing communications	Important

Participant TG3 indicated that "quality management is the priority for Longjing. From tea harvesting to packaging and making, the business competition is on quality."IP1 pointed out that" marketing is conducive to opening up the market."

Prerequisites and critical success factors

The prerequisites construct included seven factors/conditions for the successful and effective management of the network. A summary of perceptions about the suggested prerequisites is presented in Table 15.3.

Some comments: "Building alliances is the basis for reducing costs, increasing benefits and making common progress" (PSM1). "Stakeholders are not only the direct beneficiaries of the tea tourism but also the direct participants. We have the right and obligation to express our opinion and act." "I think the involvement of local stakeholders is a sign of the full support of all stakeholders." "Mutual benefits are the basis/premises to establish and operate a business alliance. Without profitability and financial viability, the tea tourism project cannot be carried out" (PSM2).

This study also considered and investigated seven factors as key elements for efficient business network management (Table 15.4).

Some comments: "Developing complementary capabilities is a way to integrate resources, which can save costs and closely fit the business relationships between actors/players" (PB2). "This will make the team more skilled, effective and efficient, enhancing the tourism appeal, attractiveness and productivity of Longwu compared to competing areas" (IP1). "Knowledge and technology skills are component elements of alliance competitiveness and should not be spread." "No matter how the market changes, we must keep up with the market pace and even

Table 15.3 Perception about prerequisites

Prerequisites	*Perception/opinion of stakeholders*
Creation of a platform of common interests	Important
Long-term commitment	Not important; it cannot be assumed
Full support	Important
Achievement of mutually shared benefits	Very important
Fair and equitable distribution of results	Important
Trust	Very important
Active engagement and interaction	very important

Table 15.4 Perception about CSFs

CSFs	*Perception/opinion of stakeholders*
Development of complementary competencies, capabilities, and skills	Very important
Knowledge and intelligence sharing and dissemination	Very significant
Market orientation and customer-minded	Very important
Coordination, monitoring, and assessment	Very important
Continuous improvement	Very important
Investments	Very significant/Key factor
Appropriate managerial structure	Fairly important/essential

anticipate by innovating" (HB1). "The premise of improvement and sustainability is to continuously evaluate and improve in all fields, from inside and outside, and continuous learning is also very important, capitalize on gained experience" (PSM3) "the management structure could be a type of public and private partnership" (IP2).

Discussion, conclusion, and management implications

Discussion

The study's main purpose was to suggest and validate a framework for effectively managing a business network. The study's aim was achieved by crafting a framework drawn on the blended theoretical foundations of the CBT model and strategic marketing planning. The elaborated framework was then empirically tested, employing an explorative study within the Chinese context. The suggested framework was seen as valuable because it allows for achieving a comprehensive and implementable approach to the studied issue, as suggested by literature (Novelli, Schmitz & Spencer, 2006; Valeri, 2016; Sotiriadis, 2017). Participant stakeholders made an interesting suggestion that was the management approach should be adaptive and flexible.

The study's findings indicated that 'theming strategy' was seen as very important regarding the core and support strategies. As for the support strategies, it was found that were important or very important the following strategies: development of human resources, product development, quality management, and integrated marketing communications. All four strategies were considered as having high value and utility to attain effective management. Concerning the prerequisites and CSFs for the effective management of a business network, the following list could be established based on the fact that six prerequisites and seven CSFs are important/very important., according to opinions expressed by key stakeholders. The study's findings confirm those indicated by previous studies. More specifically, the suggestions and recommendations by Gunasekera and Momsen (2007), Jolliffe (2007), Merkesdal (2013), Aslam and Jolliffe (2015), and Star, Rolfe and Brown (2020).

The endeavours should be planned, designed, and offered in an integrated fashion (WTO, 2017). The suitable strategies are similar to those recommended by previous studies; see, for instance (Jolliffe, 2007; Star, Rolfe & Brown, 2020). The validated prerequisites and CSFs confirm the findings and suggestions made by other studies, such as (Sotiriadis, 2017; Yi & Li, 2017). It is believed that clear and specific aims and the necessary conditions lead to determining the appropriate means/ways, i.e., the suitable strategies and adequate actions. Local stakeholders should consider a series of key elements for the successful integration, effective pairing, and sustainable operation of tea tourism offering. The study concludes that the suggested framework for managing a business network could contribute to the proper consideration, planning, and implementation of this kind of endeavor in experiential tourism.

Conclusion and management implications

Planning, building, and managing a business network are not easy tasks. The challenge is to manage this partnership sustainably and effectively. That is why the management requires the adoption of suitable approaches and the elaboration and implementation of suitable frameworks enhancing the achievement of related benefits. This study aimed to identify the prerequisites and CSFs and the adequate strategies for a sustainable and financially viable business network within the context of nature-based tourism. The suggested management framework encompasses the main prerequisites, strategies, operational actions, and critical factors to effectively manage a business network. Clear and specific development aims and the necessary conditions lead to determining suitable strategies and adequate actions. The study's findings contribute to the topic of collaboration in the tourism field. From a theoretical perspective, the study adopted a blended approach to comprehensively consider the topic, drawn on the CBT model and the strategic marketing planning. An integrated and comprehensive consideration and approach enhancing to capitalize on extant literature and elaborate on an implementable framework. It constitutes the first research endeavour adopting such an approach in this field. It has a theoretical contribution by extending the knowledge in this field as a practicable framework, useful to industry practitioners and local stakeholders.

This study contributes to the knowledge body and practice alike. This study argues that the proposed management framework comprehensively considers this field's key elements and factors. The framework encompasses the prerequisites and the ways and means, that is, a set of adequate strategies to achieve the desirable outcomes, and the key elements local stakeholders should take into account and use to monitor the progress and assess the results. That is why it is believed that the above-outlined findings have several implications for industry practitioners, i.e., managers and local stakeholders. It is believed that such an integrated framework could contribute to the sustainable and effective management of the alliance at a local level. The rational management approach provides potential and opportunities to focus efforts on and dedicate limited resources to specific strategies and factors critical in the efficient and sustainable management of a business network at a local level.

Although the study does make a contribution to nature-based tourism, it does have some limitations. First, the exploratory nature of the qualitative research method has a number of issues. Future endeavours could implement other qualitative methods (for instance, multiple case studies) to provide more solid and robust findings. The second limitation is the context of empirical investigation. The suggested framework was tested and validated within the Chinese context. The study's findings cannot be generalized and automatically extended. Future research could probably contribute to the knowledge body by implementing the same framework in different geographical zones, such as India and Sri Lanka. These projects could collaborate and expand the study's framework. Another interesting pathway is to explore possible improvements to the suggested framework by adding some elements.

References

APEC Tourism Working Group (2010). *Effective community-based tourism: A best practice manual*, Sydney: Asia-Pacific Economic Cooperation.

Asian Development Bank and Asian Development Bank Institute (2015). *Integrating SMEs into global value chains: Challenges and policy actions in Asia*, Tokyo, Japan: ADB & ADBI.

Aslam, M.S.M., & Jolliffe, L. (2015). Repurposing colonial tea heritage through historic lodging. *Journal of Heritage Tourism*, *10*(2), 111–128. doi:10.1080/1743873X.2014.985226

Astrøm, J.K. (2017). Theme factors that drive the tourist customer experience. *International Journal of Culture, Tourism, and Hospitality Research*, *11*(2), 125–141. doi:10.1108/IJCTHR-07-2015-0070

Astrøm, J.K. (2020). Why theming? Identifying the purposes of theming in tourism. *Journal of Quality Assurance in Hospitality and Tourism, 21*(3), 245–266. DOI:10.1080/1528008X.2019.1658147

Braun, P. (2013). Regional tourism networks: The nexus between ICT diffusion and change in Australia. *Information Technology & Tourism, 6*(1), 1–13.

Buckley, R. (2009). *Ecotourism: Principles and practices*, London: CABI.

Businessdictionary.com. (2021). Definitions. Retrieved from http://www.businessdictionary.com/definition/critical-success-factors-CSF.html

Coghlan, A. (2016). Nature tourism. In J. Jafari & H. Xiao (Eds.), *Encyclopaedia of tourism* (pp. 653–655). Wien & New York: Springer International Publishing.

Coghlan, A. and Buckley, R.C. (2012). Nature-based tourism. In A. Holden & D. Fennell (Eds.), *A handbook of tourism and the environment* (pp. 334–344). Oxon, UK: Routledge.

Couret, C. (2015). Collaboration and partnerships in practice: The creative tourism network. In D. Gursoy, M. Saayman & M. Sotiriadis (Eds.), *Collaboration in tourism businesses and destinations: A handbook* (pp. 191–203). Bingley, UK: Emerald.

Dangi, T.B. & Jamal, T. (2016). An integrated approach to "sustainable community-based tourism". *Sustainability, 8*, 475. DOI:10.3390/su8050475

Del Campo Gomis, F.J., López Lluch, D., Sales Civera, J.M., Agulló Torres, A.M., Mollá-Bauzá, M.B., Martínez Poveda, A., Camacho de los Ríos, F., & Nogués Pedregal, A.M. (2010). Wine tourism product clubs as a way to increase wine added value: The case of Spain. *International Journal of Wine Research, 2*, 27–34. DOI:10.2147/IJWR.S7317

Dodds, R., Ali, A., & Galaski, K. (2018). Mobilizing knowledge: Determining key elements for success and pitfalls in developing community-based tourism. *Current Issues in Tourism, 21*(13), 1547–1568. DOI:10.1080/13683500.2016.1150257

Eisingerich, A., & Bell, S. (2008). Managing networks of inter-organizational linkages and sustainable firm performance in business-to-business service contexts. *Journal of Services Marketing, 22*(3), 494–504. DOI:10.1108/08876040810909631

Finn, M., Elliot-White, M., & Walton, M. (2000). *Tourism and leisure research methods*. Harlow, UK: Pearson.

Fyall, A., & Garrod, B. (2005). *Tourism marketing: A collaborative approach*. Clevedon: Channel View.

Gabor, M. (2015). A content analysis of rural tourism research. *Journal of Tourism, Heritage & Services Marketing, 1*(1), 25–29.

Gulati, R., & Singh, H. (1998). The architecture of cooperation: Managing coordination costs and appropriation concerns in strategic alliances. *Administrative Science Quarterly, 43*, 781–814. DOI:10.2307/2393616

Gunasekera, R.B., & Momsen, J.H. (2007). Amidst the misty mountains: The role of tea tourism in Sri Lanka's turbulent tourist industry. In L. Jolliffe (Ed.), *Tea and tourism: Tourists, traditions and transformations* (pp. 84–97). Clevedon: Channel View.

Gursoy, D., Saayman, M., & Sotiriadis, M. (2015). Introduction. In D. Gursoy, M. Saayman, & M. Sotiriadis (Eds.), *Collaboration in tourism businesses and destinations: A handbook* (pp. xv–xxvi). Bingley: Emerald Publishing.

Hall, C.M. (1998). *Introduction to tourism: Development, dimensions and issues*. South Melbourne, Australia: Addison, Wesley, and Longman.

Hall, C.M., & Boyd, S. (Eds) (2005). *Nature-based tourism in peripheral Areas: Development or disaster?* Clevedon: Channel View.

Johnson, J.A., & Friesen, M. (1995). *The success paradigm: Creating organizational effectiveness through quality and strategy*. New York: Quorum Books.

Jolliffe, L. (2007). *Tea and tourism: Tourists, traditions and transformations*. Clevedon: Channel View.

Jordan, F., & Gibson, H. (2004). Let your data do the talking. In J. Phillimore & L. Goodson (Eds.), *Qualitative research in tourism: Epistemologies, ontologies, and methodologies* (pp. 215–235). London: Routledge.

Kokkonen, P., & Tuohino, A. (2007). The challenge of networking: Analysis of innovation potential in small and medium-sized tourism enterprises. *The International Journal of Entrepreneurship and Innovation, 8*(1), 44–52.

Lazzeretti, L., & Petrillo, C. (Eds.) (2006). *Tourism local systems and networking*. Oxford, UK: Elsevier.

Lemmetyinen, A. & Go, F. (2009). The key capabilities required for managing tourism business networks. *Tourism Management, 30*(1), 31–40. DOI:10.1016/j.tourman.2008.04.005

Merkesdal, C. (2013). Green experience value: An investigation whether thoroughly green service providers enhance the value for the customer experience. Unpublished thesis, Frederiksberg, Denmark. Retrieved from https://hdl.handle.net/10417/4009

Morrison, A.M. (2019). *Marketing and managing tourism destinations* (2nd ed). New York, NY: Routledge.

Novelli, M., Schmitz, B., & Spencer, T. (2006). Networks, clusters, and innovation in tourism: A UK experience. *Tourism Management, 27*(6), 1141–1152. DOI:10.1016/j.tourman.2005.11.011

Okumus, F., Altinay, L., & Chathoth, P. (2010). *Strategic management for hospitality and tourism*. Oxford, UK: Elsevier.

Osterle, H., Fleisch, E., & Alt, R. (2001). *Business networking: Shaping collaboration between enterprises*. New York, NY: Springer.

Porter, M.E. (1985). *Competitive advantage: Creating and sustaining superior performance*. New York, NY: The Free Press.

Ruggieri, G., & Iannolino, S. (2012). Tourism destination and the role of trust. In A. Morvillo (Ed.), *Advances in tourism studies* (pp. 209–229). New York: McGraw-Hill.

Scott, N., Cooper, C., & Baggio, R. (2008). Destination networks: Four Australian cases. *Annals of Tourism Research, 35*(1), 169–188. DOI:10.1016/j.annals.2007.07.004

Sfandla, C., & Bjork, P. (2013). Tourism experience network: Co-creation of experiences in interactive processes. *International Journal of Tourism Research, 15*(5), 495–506. DOI:10.1002/jtr.1892

Soda, G., Usai, A., & Zaheer, A. (2004). Network memory: The influence of past and current networks on performance. *Academy of Management Journal, 47*, 893–906. DOI:10.2307/20159629

Sotiriadis, M. (2017). Pairing intangible cultural heritage with tourism: The case of the Mediterranean diet. *Euro Med Journal of Business, 12*(2), 269–284. DOI:10.1108/EMJB-07-2016-0019

Sotiriadis, M. (2018). Collaborative forms and strategies for business venturing in tourism industries. In M. Sotiriadis (Ed.), *The handbook of entrepreneurship in tourism, travel, and hospitality: Skills for successful ventures* (pp. 131–150). Bingley, UK: Emerald Publishing.

Sotiriadis, M., & Shen, S.W. (2017). The contribution of partnership and branding to destination management in a globalized context: The case of the UNWTO Silk Road programme. *Journal of Tourism, Heritage & Services Marketing, 3*(2), 8–16. DOI:10.5281/zenodo.1209121

Star, M., Rolfe, J., Brown, J. (2020). From farm to fork: Is food tourism a sustainable form of economic development? *Economic Analysis and Policy, 66*, 325–334. DOI:10.1016/j.eap.2020.04.009

Sun, Y. (2005). On the analytic hierarchy of tea culture and the tourism development of tea cultural resources. *Market Modernization, 2*, 63–64.

Tsai, T., & Chen, C. (2019). Evaluating tourists' preferences for attributes of thematic itineraries: Holy folklore statue in Kinmen. *Tourism Management Perspectives, 30*, 208–219. DOI:10.1016/j.tmp.2019.02.010

Vajirakachorn, T. (2011). Determinants of success for community-based tourism: The case of floating markets in Thailand. Ph.D. Thesis, Texas, USA: Texas A&M University, College Station.

Valeri, M. (2016). Networking and cooperation practices in the Italian tourism business. *Journal of Tourism, Heritage & Services* Marketing, *2*(1), 30–35. DOI:10.5281/zenodo.376333

Vespestad, M. K., & Lindberg, F. (2011). Understanding nature-based tourist experiences: An ontological analysis. *Current Issues in Tourism, 14*(6), 563–580. DOI:10.1080/13683500.2010.513730

Weidenfeld, A., Butler, R., & Williams, A. (2011). The role of clustering, cooperation, and complementarities in the visitor attraction sector. *Current Issues in Tourism, 14*(7), 595–629. DOI:10.1080/13683500.2010.517312

World Tourism Organization (2017). *Tourism and the sustainable development goals – Journey to 2030*. Madrid: UNWTO.

Yan, Z., Sotiriadis, M., & Shen, S. (2021). Integrating a local asset/resource into tourism and leisure offering: The case of tea resources in Longwu Town, Zhejiang Province, China. *Sustainability, 13*(4), 1920. DOI:10.3390/su13041920

Yi, K., & Li, J. (2017). Integration and interactive development of tea tourism: Model construction and effect measurement based on the empirical study in Zhejiang Province. *Journal of Tea Science, 37*(5), 532–540.

PART III

Visitor experience design and management

16

EXPERIENCE DESIGN AND VALUE CO-CREATION IN NATURE-BASED TOURISM

The role of customer interaction types

Dorthe Eide and Frank Lindberg

Introduction

> **Box 16.1 Excerpt from Facebook about case K2**
>
> "Here we are received with an approach far beyond most of what I have experienced before. The warmth and care you met us with was totally unexpected and made unforgettable imprints. Thank you very much to all at Lånan"
>
> (informant K1-2-female-4)

Nature-based tourist experiences have been subject to considerable discussion in tourism research, and the concept of value is attracting increased attention among practitioners. However, empirical research on how consumer value is created in nature-based tourism (NBT), where undisturbed natural surroundings often represent the primary attraction or setting, remains scarce (Vespestad et al., 2019). Compared to other forms of tourist experiences, it is expected that value creation relates to both consumptive and adventurous activities as well as non-consumptive contemplative activities (Buckley & Coghlan, 2012).

Many consumers seek adventures in nature to escape from their everyday lives and because they want excitement and stimulation (Beedie & Hudson, 2003) or simply want to experience meaningful, memorable and personal experiences (Sundbo & Sørensen, 2013). Kayaking is a nature-based adventure tourism activity in which people are offered 'authenticity and simplicity, an opportunity to connect with nature and themselves, and an outlet to test physical limits' (Adventure Travel Trade Association/ATTA, 2016, p. 2). The duration, potential challenge, commitment, skills, and active consumers imply that value co-creation during a multiday kayaking experience is a complex topic. Furthermore, kayaking that is offered commercially requires guiding and the co-presence of fellow tourists. It is expected that various interaction types would influence value co-creation in addition to consumer preconditions (e.g., skills, physical condition), transportation, sleeping and meals during longer trips. Customer interactions that matter for value co-creation are the point of departure for this chapter.

DOI: 10.4324/9781003230748-20

Researchers agree that value is not an objective phenomenon (Peñaloza & Venkatesh, 2006) and that it does not exist in the object consumed but rather is mainly co-created during the experience itself (Prebensen et al., 2014). This means that value as a function of the exchange, i.e. based on the perceptions of what is received and what is given (Zeithaml, 1988), is less relevant for NBT experiences. Rather, meaning and value creation can be understood as multi-relational phenomena (Eide & Mossberg, 2013; Lindberg et al., 2014). Three main interaction types have been documented as being central in NBT, i.e. communion with nature, communitas or connecting with other tourists and personnel (Arnould & Price, 1993). However, outdoor interactions are influenced by less controllable elements, e.g. when the weather turns bad, or animals do not show up. Since consumers' motives vary across segments and value propositions vary across types of experiences, the design of nature-based experiences can rely on different types of sensations, stories, and customer engagements. Vespestad et al. (2019) argue that value co-creation depends on the experiential styles, i.e. the manner of experiencing nature genuinely, entertaining, as a state of being, or through community affiliations. They argue that designing experiential styles would influence how and to what extent interactions matter for customer value. Thus, there is a need for research that focuses on the tourists from a web of interaction perspective (Campos et al., 2016).

Whereas much research on NBT has focused on the role of guides, i.e. in facilitating safety, skills and tourists' interaction with natural surroundings, the role of interactions with local culture, residents and group atmosphere appears to be lacking. This chapter aims to explore the web of customer interaction types involved, and we pose the question: *How do different customer interaction types enable or constrain consumer value co-creation, and what are the consequences for experience design?* The context is four cases of multi-day guided sea kayaking packages offered by tourist providers in Northern Norway. First, inspired by consumer culture research, we elaborate the theoretical framework by focusing on the concept of value and value co-creation. Second, experience research aid our conceptualizations of experience design and particularly six main customer interaction types (with tour personnel, other humans, other customers, self, objects and animals). Third, we account for the cases and method approach before we present the empirical findings. Finally, we discuss the implications of the findings for theory and practice.

Theoretical framework

Consumer value as socio-cultural and interactive

Consumer value has attracted considerable attention in consumer research (Karababa & Kjeldgaard, 2014) and in tourism (Prebensen et al., 2013), but remains an elusive concept. Traditionally, value is understood in terms of the economic or functional benefits received or as some sort of exchange value. Karababa and Kjeldgaard (2014) argue that three abstract value types dominate the literature. Economic value and exchange value refer to the "value for money" logic, social value refers to the "goodness" for individuals, and semiotic value refers to the sign value that embodies persons and is embedded in processes and things in culture. Their review shows that the notion of value is subjective, context-dependent, complex and interrelated and that various value types are separable analytically, but not in practice.

Recent decades are marked by an increased awareness of consumers' active role during the consumption of experiences (Carù & Cova, 2007). Several contributions suggest a value-in-use approach (Holbrook, 2005; Prahalad & Ramaswamy, 2004), and central to this is the service-dominant logic (SDL) approach (Vargo & Lusch, 2008), which calls attention to how the consumer integrates resources proposed by the company with their resources during an experience. Whereas the SDL approach has been criticized for being preoccupied with the customer-firm interaction, we suggest viewing *value as a socio-cultural and interactive phenomenon* where consumers mainly integrate physical resources (e.g., objects, nature, preconditions), social resources (e.g., fellow tourists, community) and cultural resources (e.g., conventions, traditions, rules, imagery) during experiences (Arnould et al., 2006). The view on value has thus shifted from a firm-centric view towards one within which consumers not only interact based on firm-enabling resources, such as a kayak, food and guiding in our case, but where invisible intangible resources (e.g., cultural resources) also mark their entrance into the value-creation processes.

The classical river rafting study of Arnould and Price (1993) shows how various intangible resources are at stake during NBT. In western culture, hedonically motivated tourism experiences are often referred to as 'intense, positive and intrinsically enjoyable' (Arnould & Price, 1993, p. 25). The rafting tourists argue that the value is co-created during the experience due to resources such as close connections with the other members of the trip (social resources), self-renewal (physical resource), and romantic unity with nature (cultural resource). Consumers often seek nature-based experiences out of a desire for adventure and escape (Varley, 2011), and consumers are actively involved throughout such experiences (Tumbat & Belk, 2011). Relevant for value-creation processes are consumer-dependent resources that consumers bring to the experience. Lindberg and Østergaard (2015) show how a strenuous job situation and complicated projects suddenly become symbolic resources during consumers' multi-day canoeing trips in the wilderness. Furthermore, previous research shows that NBT experiences such as horse riding, adventure travel, and dog sledging can be marked by challenges due to unrealistic expectations, trouble with learning and a lack of previous outdoor life practices (Lindberg & Eide, 2016).

To sum up, we have theorized consumer value co-creation as a socio-cultural and interactive phenomenon that dynamically emerges during experiences, nurtured by physical, social, cultural, and consumer-dependent resources that are integrated and after which the value outcome might appear. Such outcomes might be extrinsic (utilitarian) and/or intrinsic (end-in-itself, autotelic) (Holbrook, 2005). For the purpose of the present research, examples of value types are such as efficiency (e.g. kayaking as a mean for reaching a destination) and play (e.g. kayaking as a fun activity). It is also reasonable to refer to potential outcomes as social, semiotic, emotional, epistemic and functional (Karababa & Kjeldgaard, 2014; Prebensen, 2014). While some experiences contribute to hedonic values, others might contribute to eudaimonic values – yet sometimes they may co-exist (Hetland et al., 2018). Finally, contextualization is important for understanding consumer value co-creation, and it is argued that "what may be valued societally is articulated semiotically and hence is also valorized in exchange terms" (Karababa & Kjeldgaard, 2014, pp. 123–124).

Experience design through different customer interactions

Experience providers develop and offer NBT experience and value propositions through experience design and then try to facilitate co-creation during marketing and production. Different disciplines contribute to a conceptual understanding of experience design (Jantzen et al., 2011), and different focuses and tools are provided. For example, Jantzen (2013, p. 164) provides a psychology approach, addressing how sensing and perception change emotions that may challenge habitual patterns and develop into a new identity. Pine and Gilmore (1999) describe four main experience designs (realms) using two main dimensions, i.e. the degree of active–passive and absorption–immersion participation. Their main focus is on 'staging', assuming that the firm has the control with employees as role-players on the 'scene' and that customers are the (passive) audience. Boswijk et al. (2012) label this conception 'generation one' of the experience economy and design theory, whereas 'generation two' focuses on facilitating value co-creation that reflects a view of the customer as active and the firm with less control of the experience.

Different experience design tools address partly different dimensions. For example, Sundbo and Hagedorn-Rasmussen (2008) suggest a model in which sensing and storytelling are argued as being most important for value creation. The Finnish LEO-model includes six elements to work with when designing and facilitating customer involvement and transformation: sensing, contrast (to everydayness), individualization, authenticity, interactions and storytelling (Tarsannen & Kylänen, 2009). Similarly, Eide and Mossberg (2019) suggest facilitating experience value and quality through a framework focused on five main dimensions: customer journey (before–during–after), experiencescapes, customer interactions, storytelling and dramaturgy (low/high intensity with peak moments). The environment where the customer interactions occur has been referred to as 'scapes', i.e. in the context of experiences as *experiencescapes* (O'Dell, 2005). Such environments we see as physical, social and symbolic resources that influence the interactions (Eide & Mossberg, 2013). Some experience offerings, not least within NBT, can involve several experiencescapes and even experience areas. For example, what route is selected and where to stop for rests can shape how the weather and landscape influence the activities and what interaction types become relevant.

According to Demirkan (in Ostrom et al., 2010), service innovations and design change the roles of the involved actors and the patterns of customer interactions. It can be similar within experience innovations and design. The web of customer interactions can be the main place where consumers get motivated and involved in value co-creation in the customer journey (before–during–after). Carù and Cova (2007) claim that enclavement, thematizing, and safety are important for immersion and thus value creation. Hansen and Mossberg (2016) claim that *tour guides* therefore should facilitate for these three, i.e. they are basic dimensions. If not caring for the basic, then guiding has too low quality. They further argue that comprehensive guiding ('guide plus') also depends on relevant and engaging storytelling, clear instructions and facilitation of social interactions.

Research on NBT suggests focusing on interactions with *other customers* as co-producers of atmosphere, emotions, belonging/bonding, and teamwork when exploring tourist experiences (Hallin & Mykletun, 2006). Ladwein (2007) argues that tourists might create social fear when lacking skills or fitness because they are afraid of constraining value for the other group members, but they may also fear that others hamper their own value creation. In a study of safari experiences, Sørensen and Sundbo (2014) find that tourists often report that interactions with family/friends are the most important value driver, whereas one might have expected that interactions with animals are the explicit reason to go. Such knowledge has implications for innovation and the facilitation of customer interactions in experience designs.

Eide and Mossberg (2013) suggests conceptualizing the web of interactions and included *six main types of customer interactions*, i.e. with personnel (tour guides), other customers, other humans (e.g., residents), animals (wild or tame), objects and self (feelings, reflections) (see Figure 16.1).

Some of these can be planned, while others are more ad hoc. Which of these customer interactions that are most important in experience-based offerings might vary, not only across NBT types, but also within a subsector such as sea kayaking or birdwatching, since there are different segments within both. In particular, interactions with objects can be divided into different subtypes. For a sea kayak tour, relevant objects are kayak equipment (for example, kayak,

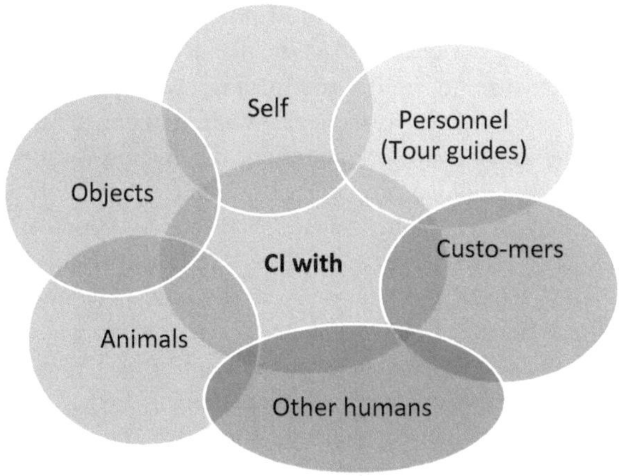

Figure 16.1 Six main types of customer interactions (inspired by Eide & Mossberg, 2013).

paddle, drysuit, west), food-related equipment (cutlery, pans, cooking equipment), accommodation equipment (tent, sleeping bag, toilet organizing), natural elements such as the weather and waves, botanical arts, geology, terrains, and cultural landscape.

Different combinations of customer interaction types can be considered when involved with experience innovations (Eide & Mossberg, 2013). For example, a high degree of interactions with other customers and a low degree of interactions with guides can be a good match for segments with a high level of skills, good physical condition, and being motivated by a social gathering and bonding with other customers. However, it can be less relevant for segments with low skills and who need to learn central aspects of the activity (see Varley, 2011). Therefore, innovation, marketing, and the production of successful experience design depend largely on segmentation, with a good match between the designed and facilitated value propositions and the consumers' motivation and resources. For example, a three-day tour of birdwatching can focus only on the birdwatching and being transported safely to relevant spots by a guide. An alternative design would be where the provider also includes meetings with locals and storytelling about the birdlife throughout seasons. Then both the web of interactions and value propositions would become more multi-dimentional. The change might be relevant for tourist segments being more generalists, i.e. interested in varied experiences and not only bird watching as such.

Methods

The research strategy chosen is a hermeneutical-phenomenological approach (Alvesson & Sköldberg, 2009), for which understanding the phenomena, context, and meaning creation are central during dialogues and reflections (Kvale & Brinkmann, 2009). A qualitative multi-case design was chosen for in-depth explorations within and between cases (Flyvbjerg, 2001) without ambitions of generalizing. Four multi-day guided sea-kayaking packages were chosen that focused on nature and culture.

Research setting

The fieldwork took place in Northern Norway in the region of Helgeland, just below the Arctic circle. This setting was chosen because the Helgeland coast is particularly attractive for kayaking due to its shallow waters, thousands of islands and skerries, rich birdlife and the small inhabited communities along the coastline. Many islands have been populated for hundreds of years, based on a combination of fishing and small farms. The first archaeological evidence is from the Stone Age (about 10.000 years ago). While traditional fishing and farming have declined, fish farming and tourism are both increasing. The region consists of 19 municipalities, and south in this region is the municipality of Vega, which received UNESCO World Heritage (WH) status in 2004 based on its unique coastal culture and traditions of producing luxurious down from Eider Ducks (verdensarvvega.no). The Vega archipelago consists of about 6500 islands and skerries. The Helgeland coastline covers about 200 km, and contains over 12,000 islands. The largest islands host the municipality and service centres and include mountains and varied terrain (see Figure 16.2). The smaller islands tend to be rather flat (see Figure 16.3). This region is also branded as being ideal for 'island jumping' by kayak and bikes.

The tour packages are designed for tourists, and three of them, in particular, involve advanced kayaking. Only K4 had international guests (two). Participants were a mix of couples, friends, families and individual travellers, equal across gender and age, mainly between the ages of 30 and 60. The four cases (K1–K4) varied regarding duration, from 2.5 to 6 days. K2 and 4 were situated in the southern part of Helgeland coastlands (Vega World Heritage – Herøy). K1 was situated just below the middle part of the region, while K3 was in the northern part (Table 16.1).

K2 was the most demanding due to the length of paddling, the time constraints on the last day, and heavy headwinds all day. K2, K4 and parts of K3 were also demanding due to waves, streams and many longer open crossings since they mostly took place far out on the ocean and seldom in smooth fjords. K2–4 requested paddling skills (including a 'wet-card' certificate) and good physical condition. However, such demands are relative and become a problem (particularly in the case of K3). K2 and K3 were new experience products, and the study showed they were too little tested and fine-tuned. For example, one trip involved an overly long paddling route within too short a time frame (K2), the other involved challenges

Figure 16.2 Paddling toward the largest island in Vega municipality. Photograph by the first author.

Figure 16.3 Paddling close to one of the many smaller islands, which now are secondary homes. Photograph by the first author.

Table 16.1 Case descriptions

Cases	*Days*	*Guests*	*Paddle route and trip descriptions*
K1	2,5	6 (N)	Starting with a focus on rescue practising and some techniques, then going on tour. First-day they pass through area that was important during the World War 2 (Tjøtta). First night in old boarding school used for children of small islands. Second night accommodation in tourism firm, visit to the gallery and handcraft firm day three. Stops and guided hiking on small islands and paddling in labyrinths. Tour guides do almost all the storytelling, also on land, about culture, botanic, history. Except from one shorter (500–1000 m) crossing of open water in waves and wind, most paddling was close to islands/skerries with a rather easy paddling. Ferry back from Herøy.
K2	4	12 (N)	Starting in Vega WH area with gathering and night at tourism firm. Second night at Gåkka, a semi-professional firm with accommodation and restaurant at old fishing village Hysvær. Third night at Lånan with local families being voluntary/semi-professionals providing accommodation and the living museum. The museum focus on the Eider-duck culture, and other topics related to living on the island in the past. Extensive storytelling by locals at Hysvær and Lånan, by the guides other places. The paddling started along the main island before crossing a longer area of open water in rather heavy wind (group spread, and some got afraid). Many crossings between smaller islands, heavy headwind all days, but sunny. Last day, too short of time, hard work. Closing with a short visit at café and ferry back.

(*Continued*)

Table 16.1 (Continued)

Cases	*Days*	*Guests*	*Paddle route and trip descriptions*
K3	6	6 (N)	Starting with the ferry to Solværøyan, an area of many small islands, two nights camping (guests in tip formed tent) on a small island without houses. Paddling two days in the area, with experiences such as watching hubris and a planned visit to locals with a farm on a small island telling stories about living in the place as a child and adult. Day three, crossing in heavy wind and stream over larger open stretch to another community (Lovund), passing fish-farm plants. Hard struggling, but mainly perceived as safe. Night in hotel, shower, good bed and gourmet meal enjoyed after two nights in camp. Day four, visiting bird mountain, famous for puffins, museum and fish-farming plant, with a local guide. Catamaran to new archipelago and municipality (Træna) at the end of civilization. Due to bad weather, one night in budget accommodation instead of camping. Paddling to the second largest island with old houses being second homes, with a unique raw nature. Two more camping nights. Different smaller guided tours, including hiking in the mountain and fishing from kayaks for dinner. No visits to locals planned, but random interaction. The guides do most of the storytelling. One guide and three guests had physical body problems in periods, and one guest complained most of the trip. Ferry back.
K4	4	12 (N)	Opposite direction as K2, but shorter, starting with catamaran to local village. Visiting Lånan (first night) this time accommodation in tents, Hysvær (second night, in house) and mainland Vega (last night). Last day (day four) in Vega WH, paddling in labyrinths and a stream. As in K2, there're rich storytelling at Lånan and Hysvær, in Norwegian language and heavy dialect. The international tourists and even some Norwegian could not follow it.

of carrying heavy kayaks and scarce food stores (K3). However, after the feedback about K2 challenges, the provider changed the design into what became K4 and they got ideas also to develop K3. The same chief guide participated in all cases, in addition to a part-time guide (which varied across cases). A focus on the unique nature and socio-cultural landscapes was communicated online. Drysuits were used most of the time. The cases can be argued as being hybrid in nature, containing periods of soft and slow adventure, and also periods of hard and fast adventures involving open sea crossings and/or taking place in challenging weather.

Data gathering and analysis

Data were gathered through two main methods: participative observations and semi-structured interviews with participants and guides. Pictures and documents (e.g. tour descriptions, package list) supplement. The first author participated in all cases and carried out the interviews. Field logs were written each evening and at the end of the tour. Interview guides were developed and adjusted, focused on a number of factors such as: previous experience with paddling and other outdoor leisure; expectations and motivations for choosing this experience offering; in-depth descriptions and reflections about each day regarding what enabled or constrained meaning and value; the natures and roles of different customer interaction types; suggestions on how the product could be improved; and implications for own future choices of holidays and leisure activities. Some participants were returning guests who were already familiar with the setting, the chief guide and some of the other guests. In addition to the first author, four participants in K1 were also in K2, and four (one different than in K2) also in K3. Some have been interviewed two times, opening for comparing and critical reflections across cases. Those participating in K1 and K2 were first selected, and these interviews were about both cases. In addition, others in K2 and K4 were strategically selected to get variation in gender, age and how they tended to group into sub-groups during the journey.

A total of 16 participants were interviewed. In addition, three interviews were conducted with the main guide and one with one of the semi-professionals at a small island. The interviews lasted between 1.5 and 2.5 hours, most of them through Skype or on the phone. All interviews were recorded and transcribed. The data analyses were conducted through qualitative content analyses informed by the theoretical framework, i.e. primarily focusing on customer interaction types and value co-creation and value outputs. The analyses relied on meaning patterns and condensations (Kvale & Brinkman, 2009). The triangulation of data and methods provided a comprehensive understanding which was less likely if only using interviews. The comparing of trips and critical reflections in interviews also increased the understanding. In sum, the triangulation strengthened reliability and validity since we were able to assess evidence across interviews and cases.

Findings

Customer interactions as dynamic, intertwined, and multi-dimensional

The first illustrations of how customer interactions are essential in value co-creation can be introduced by the quote at the start of the chapter and the two quotes below describing the peaks:

> I was very proud that I did such a hard *physical* tour. I experienced lots of great *nature*, and met many beautiful *people* at Hysvær and Lånan, good food. Their hospitality and great stories - amazing people.
>
> (K1–2, female-1, about K2)

> For instance, Lånan was great, as mentioned with those guys, how they lived there, talking with those *people* was great. And the *kayaking on the ocean* was really lovely. At first, I was a bit concerned about the rough open water toward Vega. It was new to us, then I got used to it, and I enjoyed it really… Also, it was a great experience to really have a totally unknown *group* that didn't exist before; we came along very well with everybody in the group.
>
> (K4–man-1)

The italics address the main customer interactions (CI) involved. The analysis shows that different CI types came into play in varying degrees throughout the journey and even within a single day, i.e. the CIs are *dynamic*. Some of them are pre-designed by the provider as planned visits and then facilitated by tour guides, while interactions with animals happen mostly ad hoc. Further, the CIs are *intertwined* in different combinations in different periods and not separate when enabling and constraining value co-creation. These two claims are further illustrated in the quote below:

> The *landscape* was great and the evening *light*. It was not commercial, it was *authentic places*, and the *people* we met were very engaged and gave lots of *stories*. Varied

> experiences increase memories and not only physical activities. It is exciting *to hear about the place, to understand it*… At the start of the paddling, I was a bit insecure but became more confident through learning. In between, I was very tired, but really I am glad to be pushed a bit, and *to cope* in *difficult water*, it was great. Also, the parts when you are not thinking, just *feeling one with the water*, just glide through it, it feels good… But *the wind* was hard, and the stress with *time pressure* the last day, it drained energy….
> (K2–female-7)

According to these claims, interactions with other humans, storytelling on islands, and interaction with objects (water in particular) and self while paddling are essential for value creation. However, informants, particularly females, in case-K2 argued that value co-creation related to CIs with other customers, animals and landscapes were hampered due to lack of time or the fatigue of the participants. This can be argued as being, in part, a design problem (not enough prepared before the journey), a segmentation problem (too low access barriers regarding physics and pre-experiences, age did not influence), and partly a leadership problem (not seeing or doing enough to ease the challenges).

Furthermore, informants also describe how each main type of CI have a *multi-dimensional* nature, meaning that each CI can both enable and constrain value co-creation, although some CI types are more important for value co-creation than others. For analytical purposes, it is helpful to inspect each CI separately to reveal patterns and subcategories within each type. The core findings are shown in Table 16.2.

Table 16.2 How the six main CI's enable or constrain value co-creation

CI with	*Enabling value*	*Constraining value*
Tour guides	*Care and leadership during:* • Safe and skillful guiding during open stretches and demanding wind waves, • Instructions (packing, paddling), • See individuals and facilitate teamwork *Knowledge sharing and storytelling:* about the place, people and animals. Offer short trips/extra activity (botanic, extra paddling, hiking). *(Re-)bonding with guides.*	*Care and leadership during:* • Did not feel enough safe on open crossing, group to spread, too pushy on tempo. • Too little information, supervising and communication, particularly at the start and each day (e.g. lack use of map, where to go). • Camping and toileting: Two nights in the same place without organized toileting influenced the landscape negatively; guides could have done more (K3). • One guide bad-tempered (had physical problems K3) *Design and marketing before* (K2 and K3): • Timing & length problematic (K2). • Limited pre-research about the route, too much carrying, food store, (K3). • Too low access barrier.

(*Continued*)

Table 16.2 (Continued)

CI with	*Enabling value*	*Constraining value*
Other humans (local residents)	*Hospitality and care:* Amazing planned meetings on small islands all cases. K2 even accommodation in people's houses. *Knowledge sharing and storytelling:* Voluntary and semi-professional locals about life and culture in past and present, birds, fishing, boarder schools, war, etc. Contribute to fun, learning, connection to place and people. *Random meetings:* mostly residents, on road, in shop, ferry, sharing ad hoc stories.	*Too long storytelling:* Could hamper informal interactions. *Language:* Heavy dialect and not English (K3) in storytelling hampered the understanding and involvement of some participants in periods.
Other customers	*Teamwork and care:* Different activities depend on teamwork and are necessary for security (functional and emotions). *Social needs and emotions:* Being together for days can bring social bonding and fun. *Share experiences:* Experiencing something together can enrich the experiences and bonding during, but also after when talking about it and sharing pictures.	*Teamwork and negative communication:* • Complaining and bad temper related to bodily problems and tiredness (cf. access barrier). • Continuing delaying the group (attitude). • Continuing need of help and never helping others. *Socializing hampered:* Lack of time and too tired for social interactions (last day K2). *Group match and size:* Too small group (K1 and particularly K3), too little variation to provide energy. Complaining person drain can overshadow the total experiences and create fear of group tours.
Self	*Sensing and emotions* open for reflections (positive approach) and learning, including more long term transformations of identity and practice.	*Body problems*: pain, tired, low energy, bad temper Reflections (negative thoughts, irritation of self).
Animals	*Sensing*: Watching and hearing wild and tame animals, talking about, photographing, wow emotions. *Practicing wildlife* (kayaking, fishing) with animals nearby, wow and learning.	*Feeling unsafe*: too close meeting-with running livestock (K1)
Objects	*Watching and practicing in nature:* • varied nature, interesting to watch and enter into • The power of nature: Light, mountains, bays, sandy beaches, labyrinths, waves *Weather:* creates different interactions, conditions, and experiences *Watching and visiting places:* cultural landscapes and communities	*Weather:* can make the paddling too challenging and reduce the feeling of safety *Kayak equipment:* This provider had high-quality equipment so seldom mentioned, e.g. having to use double kayak, use of the largest single kaya (harder to paddle), leaking dry suit, too heavy loaded kayak (luggage), repeating repackaging of kayaks.

(*Continued*)

Table 16.2 (Continued)

CI with	*Enabling value*	*Constraining value*
	Practicing wildlife/kayaking: • Interact and become one with the kayak equipment and water/elements • Physical exercise, cope, learn, proud/identity Accommodation and food: • In privet homes amazing hospitality, • in the hotel with a shower, good bed and gourmet meal after camping.	Accommodation and food: • Food boring (K3), • Sharing tent (4 nights K3) too little privet life, • Camping two nights in the same place without organized toileting influenced the landscape negatively.

Customer interactions enabling value co-creation

Two main types of CI were argued as being very important for value co-creation by all informants, i.e. *interactions with other humans* (local residents, including the semi-professional visited) and the *tour guides*, particularly through the subcategories of 'knowledge sharing and storytelling', and 'care and hospitality'. Knowledge sharing and storytelling are not only about nature and animals but also culture and people, which point at sea-kayaking packages as not only nature-based tourism but also potentially community-based tourism, with elements of cultural tourism. The places visited on land or paddle by or within (e.g. labyrinths) *can* be perceived only as physical environments with physical objects (e.g. stones, sand beaches, plants) as resources around the activity. In addition, the storytelling can be focused on the physical and natural. However, in these cases, the experience designs involve interactions and stories so that experiencescape also becomes cultural/symbolic and social environments. Such multi-dimensional interactions facilitate meaning and value co-creation:

> One thing is the experience of *nature*, but the meeting *with locals* made it more into a *cultural* experience. Therefore, the tour became multi-dimensional. We moved from being observers, although being very close to the landscape and nature. The meeting with them made us part of something and helped us understand how it was to live in the area.
>
> (K4–female-1)

Informants also argue that CI with residents and their stories makes it easier to remember the experience. It is what they tell stories about when returning home to friends and family. The stories are more possible to communicate than the embodied feeling of paddling.

Interactions with other customers' are seen as important. Some are motivated to participate in packages because they want to meet new people or want to do something new with people (e.g. partner, family, friend) they arrive with. However, interactions with other customers were seldom argued the most important enabler of value co-creation, i.e. not the 'wow' during the tour or contributing most to memory. For some, it is more like a hygiene factor, important for safety: "When so far out in the ocean, and long open stretches, it is preferable to be more people together" (K1–2, man-2). Some also point to doing active adventure experiences together and having entertaining talks both when paddling and when not. Multiday tours also mean doing more ordinary things together (eat, sleep in a tent) and cooperating for functionality (keep together, help each other's carrying kayaks). Even more for those in double kayaks. Facilitation of positive social interactions and understanding of teamwork in this setting is, therefore, an important task for tour guides from the start and during the journey, as customers co-create this as well (Figure 16.4).

Interactions with objects can involve relations with different types of objects. The most valuable happened during paddling when an embodied person–kayak–water relational[1] practice takes place. Other object interactions enabling value are when appreciating, for example, the landscape, lights, plants (e.g. local orchids) or geology.

Interactions with 'self' (e.g. sensing, dialogic reflections) help informants to focus on the positive (i.e. take a positive approach), dwell in activities (be immersed) and create meaning. For some, it starts a transformation of more long term consequences.

Taken that most time is spent in nature, it is surprising that *interactions with animals and objects in nature* seldom is argued peaks. Although there are some examples:

> The ocean was flat, and there were porpoise [small whales] playing up front and a sea eagle. It was fantastic. We stopped on the ocean, kayaks together.
>
> (K1–female-5)

> It was fun during a hike to a small height, to see the four sea eagles surfing on the wind.
>
> (K4–female-1)

CI with animals involved close or distanced interactions with tame (e.g. sheep and cows) and wild animals (sea eagles, hubris, eider ducks, other sea birds, otter and small whales). Case K3 had the most varied interactions with animals in the form of sea birds, as it also involved puffins and hubris. The two spices are rare and where attractions addressed in the marketing of this package. Case K3 also included fishing for dinner, a new activity from a kayak for some.

Figure 16.4 Making a float for relaxing and social interactions.

Photographed by the first author.

Customer interactions constraining value co-creation

The main constraining CI's are with 'tour guides', 'other customers' and 'objects'. One described it as such:

> Too much carrying of kayaks, too little information, *tour leadership and planning ahead*... Too long trip with such a small group, some of the other participants became increasingly *annoying due* to constant complaining, need of help and unable to cope. I am not sure I will take the risk again on group tours when not controlling who the others are... One *guide started yelling* and seemed fed up, and with a bad temper, it made the atmosphere low.
>
> (K3–female-1)

Some of the challenges can be argued due to too little work with the experience design during innovation by the provider, as it was new kayak products that were still in the making. Particularly case K3 used a too low access barrier when selling the product. There were also argued too limited or 'bad' leadership (not keeping the group together, not enough safety, lack of communication, lack of facilitating teamwork). Being on a kayaking tour for—between two and six days means interacting rather closely with other customers and the tour guides. It becomes irritating if some constantly paddle too fast or slow down the group when paddling or packing the kayak. Similarly, if always needing help and not offering help. Even more, if constantly complaining, it can start draining energy (case K3). The quote even expresses that it is risky to go on new group tours without controlling who the others are. In addition, kayaking equipment not working as it should (e.g. leaking or too small/large dry suit, heavy kayak, having to be in a double kayak when wanting a single) are constraining interactions with objects.

Weather and waves create hard work and narrow focus when struggling, constraining some CI and value co-creation as elaborated above when combined with long periods and time pressure. On the other hand, after the journey, several argued they had felt coping and were proud, i.e. it had spurred skills, identity change and narrating, and the more long lasting eudaimonian emotion. Further, this points at the dynamic nature of CI for value co-creation.

Multi-dimensional, intertwined and dynamic consumer value as output

The main types of value co-created are presented summarized in Table 16.3.

All informants argue that a central motive for going on the selected tour is getting more balance in life by doing and learning something not part of everyday life. Active physical holidays creates energy and value in life. However, there is often more than one reason to go, and, not least, more types of value co-created; therefore, the types of value co-created are intertwined, multi-dimensional, *and dynamic*. The multi-dimensional nature of the values co-created was shown in Table 16.3. The quotes below also point at the interwinedness of experience value created:

> I got the experience of *nature*, the *safety* and the *physical exercise*. To me the struggling is part of the experience. The landscape was more monotone this time [K2 than last time K1], while the group larger made it better since it was more *social*. The meeting with locals was great this time, *amazing stories and hospitality*, people with personality expressing themselves, that is unforgettable and is what you tell about when getting home.
>
> (K1 and 2, man-1)

Table 16.3 Seven main categories of value with subcategories

Value types	*Subcategories*
Social	1. Social needs like laughter, talking, opening up, connecting, togetherness, bonding; 2. Becoming and behaving like a pair/group (attentive, interactions within, bonding, identity, community); 3. Team working like carrying kayaks, camping tasks; 4. Caring for each other's: help, sharing food/drinks, looking after
Learning/epistemic	5. Do something new, go to a new place, meet new people, develop oneself 6. Learn about self, others, nature, culture, objects, techniques. 7. Become more experienced within the wildlife practice, increase or change meaning/understanding, identity, skills, etc., through doing. 8. Existential, transformational (e.g. realize that need more exercise, adventure in nature good alternative to city holidays).
Emotional	9. During: Mix of calm, no stress, enjoyment, feeling whole and alive, bonding with/connect with others/nature, pressure, struggle, unsafe, anger, pain; two different main immersions (calm peaceful; attentive active), coping with waves. 10. After: Coping and pride (the more struggles, the more pride)
Physical/bodily activity	11. Engage in physical activity, sensing and involving larger/other parts of the body 12. Open up for other types of value 13. Exercise muscles, fitness, health 14. Feel alive, wellness
Identity	15. Meaning, self-development, life quality 16. Testing oneself, 17. Coping with challenge-> proudness, confidence, self-image 18. Personal branding, showing off after 19. Existential, transformational
Nostalgic	20. Stories to tell; pictures to show; objects to show (shell, stone) 21. Share with others after (friends/family not participating, with others in the group), 22. Help to remember, relive, keep the value of the tour alive and long
Functional	23. Transportation from A to B; 24. Cover basic physical needs (eat to get energy, sleep, get to the toilet, get warm) 25. Safety: cannot go alone; the package tour gives access to safety (local guide, rescue each other) and find the way, provides access to the motivating experiences

> I wanted to *see nature* from another angle than from fast boats, you see more details with a kayak. And I wanted to be active, to get the *physical exercise*. Also, I wanted to *meet new people*, and some I knew from before. I wanted to develop myself, *learn* things like planning, packing and orienting on tours like that.
>
> (K3, female-2)

The functional value is not the main value of the experience, but it is necessary to co-create other values, such as epistemic and emotional value. The customers need to eat, sleep and have an available restroom, but such experiences are not the main motivations for participating nor the type of value for which meaningful memories are created. The customer journey involves periods of different CI mixes as well as different value outputs. The dynamic nature of value has been addressed in the sections above, not least related to how time pressure and strong wind created anger and hard work during, but after the coping brought proudness.

Discussion and conclusions

This chapter aims to explore how customer interaction (CI) types enable or constrain consumers' value co-creation and its implications for experience design. Below, we discuss four propositions (P) based on the main findings before suggesting implications and further research.

> P1: Customer interactions are multi-dimensional, and each type can enable and constrain customers' value co-creation.

The study shows that multiday nature-based tourist experiences, like sea kayaking, can enable and constrain value co-creation through six main customer interaction types. Research on adventure tourism often focuses on the intensity of experiences and the role of immersion (Varley, 2006). However, the valuation of experiences also depends on the importance of CI that varies across a person, product and subsector (Holbrook, 2005). The present study confirms the claim of Eide and Mossberg (2013), who argue that six main CI's potentially can enable and constrain value co-creation. Our study expands their research by contributing to knowledge about how *all* the six CI types can influence value co-creation. In particular, the study provides insight on the nature and role of CI with other humans as residents, tour guides and other customers, i.e. socio-cultural interactions. It contributes to knowledge about how customers actively and dynamically contribute to co-creating experiences, which goes beyond an individual approach focusing on a rather passive consumer who senses and perceives nature during experiences. The study confirms that tour guides are important facilitators of immersion, not least for safety and meaningful storytelling (e.g. Sundbo & Hagedorn-Rasmussen, 2008; Caru & Cova, 2007). Hansen and Mossberg's (2016) guide theory suggests that guides can help connect customers with contexts through storytelling. Guides can be seen as mediators that stage experiences and facilitate customer interactions (Arnould & Price, 1993). Our study extends this knowledge by showing how also residents and local semi-professionals can do so. Further, the study points at tour guides' role in involving customers as co-creators in the tourist community as these packages depend on teamwork.

> P2: Customer interactions are dynamic and intertwined during customers' value co-creation.

The study confirms and extends the claim of Lindberg et al. (2014) that meaning and value creation are dynamic and multi-relational. First, the study shows how different CI's are involved in varying degrees and importance across time during the customer journey. Much can be designed for and facilitated during a journey, but some CI's happen more ad hoc due to, for example, weather conditions and confronting animals. Second, we argue for expanding the literature on NBT from mainly focusing on interactions with tour guides and other customers to including four other interaction types (i.e., objects, animals, self, other humans). An increased interactional focus would provide an elaborated understanding of value co-creation as a dynamic and intertwined phenomenon and how different combinations can inspire varied experiential designs and facilitation during the customer journey.

P3: Customer interactions with residents and socio-cultural places can increase value co-creation for relevant segments, providers and local communities.

Pearce (2013) argues that what is regarded as ordinary life and culture for locals would often be experienced as extraordinary for tourists. Tourism firms should thus design NBT trips where tourists could meet locals and learn about their everyday life and culture. Authentic meetings and being touched and involved in personal ways can enable value co-creation that is less standardized offerings (see Wang, 1999). Interactions with local residents (communities) can give access to relevant and attractive stories and objects as resources that might impact customers' value co-creation (Arnould et al., 2006). The findings elaborate on the possible role of residents, local communities, and hence socio-cultural experiences, potentially also local meal experiences, in NBT. First, the study shows how and why socio-cultural/symbolic resources connect tourists more intensively to the places, relevant for tourists' and providers' value creation. Second, such connection, we argue, contributes to a broader partnership in NBT that can increase local sustainability (Høegh-Guldberg et al., 2022) in line with regenerative tourism and community-based sustainability in tourism (e.g. Saarinen, 2006, 2014; Saarinen & Gill, 2019). NBT can then not only 'use' the nature for free, but give something back by involving locals and local communities. Third, the extensions of CI types in value co-creation processes in NBT means involvement of 'nested' experiencescapes and broader value networks (-chains).

P4: Consumer value as output are multi-dimensional, intertwined and dynamic.

First, the study suggests that value is multi-dimensional, and a matrix of seven main types of consumer value outcomes are suggested (cf. Table 16.3). The value types are not new and have been reported previously (Holbrook, 2005; Karababa & Kjeldgaard, 2014), but identity value and nostalgic value are less elaborated in previous studies. Our findings show that functional value is not very relevant for NBT experiences. Prebensen (2014) includes functional, emotional, social, and epistemic value, arguing that the three latter are most central for the experience economy. Multi-day trips open for bonding with other customers, tour guides and even with animals, which can contribute to social value (Arnould & Price, 1993). NBT trips can also result in challenges related to social incompatibles, scary nature, and difficult practices that need to be attended to (Tumbat & Belk, 2011). The findings show that NBT and adventure tourists do not only appreciate value related to the physical and skilful activity of paddling, nor only related to sensing and learning about the natural environments and animals. Culture and socializing can be just as important. However, valuing social and symbolic/nostalgic experiences would not match all kayaking segments because some would prefer kayaking packages that focus more on technical practices such as Inuit roll and wave surfing. Consequently, matching tourists motivations and packages value propositions are vital for co-creating relevant value output, which depends on consumer research, segmentations and tailored design. Varley (2011) argues that most tourists are motivated to learn the skills of paddling for being able to enter an "alien, marginal, and liminoid world" (p. 85); however, he studied paddling courses, not multi-day kayaking packages. In our multi-day guided cases, increasing paddling skills were just one among mixed motives. Some participants accounted for a pull motive where the escape from ordinary life mattered most. Consequently, the results show a mix of push-and-pull motives that have consequences for the multi-dimensional matrix of value outcomes during multi-day nature-based experiences. We find overlap with soft adventure within which tourists are driven by a push drive towards "feeling, sensing and investing in place, community, belonging, sociality, and tradition over time and in nature" (Varley & Semple, 2015, p. 78).

Second, the findings contribute to knowledge about the dynamics of value outcomes. In particular, this was evident in case K2, day three, where strong winds and time pressure created struggles and even anger among the customers. Then the activity turned functional (transportation), which constrained relevant experience-based value co-creation. However, after the strenuous paddling period, the experience was re-created into proudness of achievement (symbolic/nostalgic value) that boosted eudaimonia emotion value important for memorable, meaningful experiences (e.g. Hetland et al., 2018). Finally, our results show that value outcomes are *intertwined*, as argued by Karababa and Kjeldgaard (2014). The just mentioned example from case K2 shows this well.

Management implications and further research

The results of this study have several implications for management and providers of multi-day nature-based experiences. First, the study shows that the view of nature-based experiencescapes should be developed from being regarded as mainly a physical environment, within which one can offer activities that allow tourists to do physical exercises and/or conduct gazing activities, towards being viewed as contexts within which value co-creation takes place through customer interactions (CI) that rely on resources that are symbolic/cultural and social just as much as they are natural and physical. One implication for innovative experience design in NBT, and in the pre-phase of customer journeys (plan, information, access barriers), is to prepare for all of the six CIs carefully and facilitate for combinations throughout the dynamic customer journey. Providers should focus on CI types that capacitate to enable customers' value co-creation and avoid those that might constrain value. Second, the challenges and constraining of value elaborated in cases K2 and 3 show the importance of segmentation in experience design and who the experiential products should be sold to. Too low access barriers (skills, physical condition), or if the customer does not match the product regarding motivation and interests (cf. e.g. personas segmentation, not mainly the traditional of age, gender, income, nationality), would create challenges during the trip. Third, some of the challenges in cases K2 and 3 can be traced back to a rather ad hoc innovation process (also described as 'Doing, Using, Innovation (DUI)' process, e.g. Jensen et al., 2007). Both experience offerings were new, and they were not enough concretized, tested, and 'polished' before being offered to paying tourists. This created quality challenges that could have been avoided with more knowledge-based systematic work with innovation and experience design, such as planning and testing the customer journey, route to use, access barriers.

The findings can have relevance for other geographical contexts that offer sea kayaking and NBT. The propositions and other implications should be studied in other multi-day adventures and NBT subsectors, such as horseback riding, dog sledging, biking and hiking. We encourage more studies on multi-day sea-kayaking packages and other subsectors to investigate how different experience designs of CI combinations matches different segments. Finally, and not least, there is a need for more knowledge about how community-based approaches in NBT may influence sustainability and how such knowledge has consequences for CI, value co-creation and experiential design in vulnerable rural areas.

Acknowledgement

This research could not have been done without the great participation of informants sharing their experiences and opinions. The research was done as part of the research program 'Northern Insight', therefore partly financed by the Research Council of Norway.

Note

1 Has overlaps with bodily phenomenology by Merleau-Ponty (1962) and as tacit knowing (e.g. Strati, 2007).

References

Adventure Travel Trade Association (2016). *Adventure Tourism Development Index: An Adventure Travel Scorecard*. 6th edn. Washington, DC: International Institute of Tourism Studies, The George Washington University and Adventure Travel Trade Association. Available at: ATDI16-web.pdf (adventureindex.travel) (accessed 1 February 2018).

Alvesson, M., & Sköldberg, K. (2009). *Reflexive Methodology: New Vistas for Qualitative Research*. London: SAGE Publications.

Arnould, E. J., & Price, L. L. (1993). River magic: Extraordinary experiences and the extended service encounter. *Journal of Consumer Research*, 20, 24–45.

Arnould, E. J., Price, LL, & Malshe, A. (2006). Toward a cultural resource-based theory of the customer, pp. 91–104 in R. F. Lusch & S. L. Vargo (Eds.), *The Service-Dominant Logic of Marketing: Dialog, Debate and Directions*. Armonk, NY: Sharpe.

Beedie, P., & S. Hudson (2003). Emergence of mountain-based adventure tourism, *Annals of Tourism Research* 30(3): 625–643.

Boswijk, A., Peelen, E., & Olthof, S. (2012). *The Economy of Experiences*. Amsterdam: Pearson Education Benelux.

Buckley, R., & A. Coghlan (2012). Nature-based tourism in breadth and depth, pp. 304–306, in T. Singh (Ed.), *Critical Debates in Tourism*. Bristol: Channel View Publications.

Campos, A.C., Mendes, M., Valle, P. O., & Scott, N. (2016). Co-creation experiences: Attention and memorability, *Journal of Travel & Tourism Marketing*, 33(9), 1309–1336, DOI: 10.1080/10548408.2015.1118424

Carù, A., & Cova, B. (2007). Consumer immersion in an experiential context, pp. 34–47, in A. Carù & B. Cova (Eds.), *Consuming Experience*. London: Routledge.

Eide, D., & Mossberg, L. (2013). Towards more intertwined innovation types: Innovation through experience design focusing on customer interactions, pp. 248–268, in J. Sundbo & F. Sørensen (Eds.), *Handbook on the Experience Economy*. Cheltenham, UK: Edward Elgar Publishing.

Eide, D. & Mossberg, L. (2019). Toward a framework of experience quality assessment: Illustrated by cultural tourism, pp. 101–120, in D. Jelinčić & Y. Mansfeld (Eds.), *Creating and Managing Experiences in Cultural Tourism*. Singapore: World Scientific.

Flyvbjerg, B. (2001). *Making Social Science Matter: Why Social Inquiry Fails and how It Can Count Again*. New York: Cambridge University Press.

Hallin, C. A., & Mykletun, R. J. (2006). Space and place for BASE: On the evolution of a BASE-jumping attraction image. *Scandinavian Journal of Hospitality and Tourism*, 6(2), 95–117.

Hansen, A. H., & Mossberg, L. (2016). Tour guides' performance and tourists' immersion: Facilitating consumer immersion by performing a guide plus role. *Scandinavian Journal of Hospitality and Tourism*, 17(3), 259–278.

Hetland, A. Vittersø, J.O., Bø, S.O.B., Kjelstrup, E., Mittner, M. & Dahl, T.I. (2018). Skiing and thinking about it: Moment-to-moment and retrospective analysis of emotions in an extreme sport, *Frontiers in Psychology*, 9(979), 1–16.

Høegh-Guldberg O., Maziliauske, E., Eide, D. & Ryan, A.W.F. (2022), Innovation for sustainability in world heritage destinations: Opportunities and challenges of the idealistic paradigm, pp. 56–83, in C.A. Almeida, C. Martins, A.R. Gonçalves, S. Quinteiro & M.L. Gasparini (Eds). *Handbook of Research on Cultural Tourism and Sustainability*. Hershey: IGI Global.

Holbrook, M. B. (2005). Customer value and autoethnography: Subjective personal introspection and the meanings of a photograph collection. *Journal of Business Research*, 58, 45–61.

Jantzen, C. (2013). Experiencing and experiences: A psychological framework, pp. 146–170, in J. Sundbo & F. Sørensen (Eds), *Handbook on the Experience Economy*. Cheltenham: Edward Elgar Publishing.

Jantzen, C., Vetner, M., & Bouchet, J. (2011). *Oplevelsesdesign [Experience design] DK*. Fredriksberg: Samfundslitteratur.

Jensen, M. B., Johnson, B., Lorenz, E. & Lundvall, B. A. 2007. Forms of knowledge and modes of innovation. *Research Policy*, 36(5), 680–693.

Karababa, E., & Kjeldgaard, D. (2014). Value in marketing: Toward socio-cultural perspectives. *Marketing Theory*, 14(1), 119–127.

Kvale, S., & Brinkmann, S. (2009). *Interviews: Learning the Craft of Qualitative Research Interviewing* (2nd ed.). Los Angeles, CA: Sage.

Ladwein, R. (2007). Consumption experience, self-narrative, and self-identity: The example of trekking, pp. 95–108, in A. Carù & B. Cova (Eds.), *Consuming Experience*. London: Routledge.

Lindberg, F., & Eide, D. (2016). Challenges of extraordinary experiences in the arctic. *Journal of Consumer Behaviour*, 15(1), 15–27.

Lindberg, F., Hansen, A-H., & Eide, D. (2014). A multi-relational approach for understanding consumer experiences within tourism. *Journal of Hospitality Marketing & Management*, 23(5), 487–512.

Lindberg, F., & Østergaard, P. (2015). Extraordinary consumer experiences: Why immersion and transformation cause trouble. *Journal of Consumer Behaviour*, 14, 248–260.

Merleau-Ponty, M. (1962). *Phenomenology of Perception*. London: Routledge & Kegan Paul Ltd.

O'Dell, T. (2005). Experiencescapes. Blurring borders and testing connections, 11–33, in T. O'Dell & P. Billing (Eds.), *Experiencescapes: Tourism, culture, and economy*. Køge: Copenhagen Business School Press.

Ostrom, A.L., Bitner, M.J., Brown, S.W., Burkhard, K.A., Gould, M., Smith-Daniels, V., Demirkan, H., & Rabinovich, E. (2010). Moving forward and making a difference: Research priorities for the science of service. *Journal of Service Research*, 3(1), 4–36.

Pearce, P. L. (2013). *The Social Psychology of Tourist Behaviour*. International Series in Experimental Social Psychology (Vol. 3). Butterworth-Heinemann, Amsterdam: Elsevier.

Peñaloza, L., & Venkatesh, A. (2006). Further evolving the new dominant logic of marketing: From services to the social construction of markets. *Marketing Theory*, 6(3), 299–316.

Pine, B. J., & Gilmore, J. H. (1999). *The experience economy: Work is theatre & every business a stage*. Boston, MA: Harvard Business School Press.

Prahalad, C. K., & Ramaswamy, V. (2004). Co-creation experiences: The next practice in value creation. *Journal of Interactive Marketing*, 18(3), 5–14.

Prebensen, N. K. (2014). Facilitating for enhanced experience value, pp. 154–177, in G. A. Alsos, D. Eide & E. L. Madsen (Eds.), *Handbook of Research on Innovation in Tourism Industries*. Cheltenham: Edward Elgar.

Prebensen, N. K., Vittersø, J., & Dahl, T. I. (2013). Value co-creation significance of tourist resources. *Annals of Tourism Research*, 42, 240–261.

Prebensen, N. K., Woo, E. & Uysal, M. (2014). Experience value: Antecedents and consequences. *Current Issues in Tourism*, 17(10), 910–928.

Saarinen, J. (2006). Traditions of sustainability in tourism studies. *Annals of Tourism Research*, 33(4), 1121–1140.

Saarinen, J. (2014). Critical sustainability: Setting the limits to growth and responsibility in tourism. *Sustainability*, 6(1), 1–17.

Saarinen, J., & Gill, A. M. (2019). *Resilient Destinations and Tourism. Governance Strategies in the Transition Towards Sustainability in Tourism*. London: Routledge.

Sørensen, F., & Sundbo, J. (2014). Potentials for user-based innovation in tourism: the example of GPS tracking of attraction visitors, pp. 132–154, in G. A. Alsos, D. Eide, & E. L. Madsen (Eds.), *Handbook of Research on Innovation in Tourism Industries*. Cheltenham: Edward Elgar Publishing.

Strati, A. (2007), Sensible knowledge and practice-based learning. *Management Learning*, 38(1), 61–77.

Sundbo, J., & Hagedorn-Rasmussen, P. (2008). The backstaging of experience production, pp. 83–110, in J. Sundbo & P. Darmer (Eds.), *Creating Experiences in the Experience Economy*. Northampton, MA: Edward Elgar.

Sundbo, J., Sørensen, F., & Fuglsang, L. (2013). Innovation in the experience sector, pp. 228–247, in J. Sundbo & F. Sørensen (Eds.), *Handbook on the Experience Economy*. Cheltenham: Edward Elgar Publishing Ltd.

Sundbo, J., & Sørensen, F. (Eds.). (2013). *Handbook on the Experience Economy*. Cheltenham: Edward Elgar Publishing.

Tarssanen, S. & Kyläinen, M. (2009). *Handbook for Experience Stagers*. Rovanemi.

Tumbat, G., & Belk, R. W. (2011). Marketplace tensions in extraordinary experiences. *Journal of Consumer Research*, 38(1), 42–61.

Vargo, S. L., & Lusch, R. F. (2008). Service-dominant logic: Continuing the evolution. *Journal of the Academy of Marketing Science*, 36(1), 1–10.

Varley, P. (2006) Confecting adventure and playing with meaning: The adventure commodification continuum. *Journal of Sport & Tourism*, 11(2), 173–194.

Varley, P. & Semple, T. (2015). Nordic slow adventure: Explorations in time and nature, *Scandinavian Journal of Hospitality and Tourism*, 15(1–2), 73–90.

Varley, P. J. (2011). Sea kayakers at the margins: The liminoid character of contemporary adventures. *Leisure Studies*, 30(1), 85–98.

Vespestad, M. K., Lindberg, F., & Mossberg, L. (2019). Value in tourist experiences: How nature-based experiential styles influence value in climbing. *Tourist Studies*, 19(4), 453–474.

Wang, N. (1999). Rethinking authenticity in tourism experience, *Annals of Tourism Research*, 26(2), 349–370.

Zeithaml, V. A. (1988). Consumer perceptions of price, quality and value: A means-end model and synthesis of evidence. *Journal of Marketing*, 52, 2–22.

17

TRANSFORMATIVE BIRDWATCHING TOURISM EXPERIENCES IN THE PERUVIAN AMAZON

Sandra Sotomayor, Pablo Romo and Karen Guillén

Introduction

Nature provides resources that can be used for the generation of a vast diversity of tourism products worldwide. It is believed that when these products have a component of wildlife observation and, more specifically, birdwatching – one of its most popular and rapidly expanding forms (Connell, 2009; Kronenberg, 2014), they have a great potential to foster a process of personal development that awakes their consciousness, better known as a transformative tourism experience. Given that this kind of process promotes long-lasting personal and social benefits and therefore contributes to the sustainable tourism agenda by an improvement in tourists' attitudes and consciousness towards social and environmental issues (Wolf et al., 2017; Mkhize & Ivanovic, 2020; Teoh et al., 2021), the topic has started to receive academic attention, especially in the last five years. However, there is still room for a deeper analysis regarding the characteristics that visitor experiences should have in natural areas to attain these results. Further, there is a lack of nature-based tourism products which have been designed (i.e., in the case of the new ones) or redesigned (i.e., in the case of previously existing ones) with the intention to promote a transformative outcome.

Therefore, biodiversity-rich countries, like Peru (i.e., the second country with the highest number of bird species), can take advantage of this gap as a great opportunity. The present case study examines the potential of a sector of the Northern Peru Birding Route, which includes ten birdwatching sites in total over an area that covers around 350 km, to generate an emotional bond between the tourist, the species, their habitat, and the people involved in their conservation (Pung & Chiappa, 2020), leading to a Transformative Birdwatching Experience. For this purpose, the authors described the components of such experience based on the theoretical model suggested by Kirillova et al. (2017), which is built upon a "peak," extraordinary or transcendent experience, starting with aspects related with the circumstantial tourism environment that can lead to the co-creation of a triggering episode which meaning favors a transformation.

DOI: 10.4324/9781003230748-21

Birdwatching is an activity which is highly in demand by tourists that visit natural settings with abundant bird species, for which it is considered part of nature-based tourism (Connell, 2009). Moreover, when birdwatching provides tourists with information and opportunities to learn with and relate to the species and their habitat in a personal and transformative level, considering not only nature but also the local culture, it becomes an immersive, valuable, and significant nature-based experience with transformative potential. If managed responsibly by promoting conservation and respect for the environment as well as the local economy and culture, transformative birdwatching would be considered a sustainable nature-based experience (Lean, 2009).

Nature-based tourism and wildlife observation activities

Nature-based tourism – defined as the collection of activities that people engage in to seek connection and new experiences with nature (Silvennoinen & Tyrväinen, 2001) – has developed and continues to grow rapidly across the globe (Madin & Fenton, 2004; Moore & Polley, 2007; Winter et al., 2020). It has been estimated that over 8 billion people every year visit terrestrial protected areas around the world (Balmford et al., 2015). Interacting with nature benefits visitors on several levels, including psychological, physiological, and social (Kyle, Mowen & Tarrant, 2004; Kyle et al., 2006; Sotomayor et al., 2014). These benefits are even more evident today because of the COVID-19 pandemic (Spenceley et al., 2021).

Even though there has been a concern in the tourism industry for several years about sustainability issues – particularly among natural settings – some argue that there is still a significant gap between what academics publish and the implementation of projects that contribute to this cause on the destinations (Winter et al., 2020). It is not easy to achieve the desired results, but for these good intentions to be reflected in the growing diffusion of sustainable practices in natural areas among visitors and practitioners, the team in charge of managing natural areas must have adequate skills to find a balance between natural environment conservation and the visitor satisfaction level (Moore & Polley, 2007). Thus, the visitor's enjoyment and experience with nature needs to continue generating resources that are used for the preservation and conservation of natural areas for the continuing benefit of future visitors.

Nature-based tourism can also play a significant role in shaping people's attitudes toward the environment in addition to generating funds for conservation (Balmford et al., 2009). On the one hand, visitors' interactions with nature leads to increased awareness in their behaviour (Priskin, 2003; Ballantyne, Packer & Falk, 2011), resulting in greater regard for it during their visits. Local communities, on the other hand, might profit from this sort of tourism by increasing environmental awareness and education. Indeed, nature-based tourism is an economic activity where several entrepreneurs participate, the majority of whom represent small entrepreneurships from local rural regions, and balance working in tourism with other industries such as agriculture and fishing (Weaver & Lawton, 2007). This is because they must face the seasonality that defines tourist demand as well as challenges and confrontations with other stakeholders (e.g., landowners and protected area managers) due to the various possible uses of natural resources. Thus, nature-based tourism should involve and benefit all stakeholders, as it encourages the creation of non-extractive economic activities that can support local economies (Wall-Reinius & Fredman, 2007).

Wildlife tourism provides an ideal platform for scientific study (Dobson, 2008) in addition to from having the ability to support conservation by increasing environmental awareness and education among visitors and local populations (Apps, Dimmock & Huveneers, 2016). Wildlife observation activities, which are a popular aspect of global nature-based tourism (Orams, 2000), are associated with an animal species or group of species, among which the UNWTO (2015) highlights the following: the "big 5" (i.e., buffaloes, elephants, leopards, lions, and rhinos); gorillas; lemurs; birds; whales and dolphins. These activities connect to tourism in several ways as they can overlap with different travel modalities. For example, within a specialized tourism offer where these are the major focus or attraction (e.g., birdwatching); they can be part of a product with less specialization where they are complementary offers; or they can even be part of a mass tourism product (Tapper & Waedt, 2006).

In any case, interactions between wildlife and human systems in the context of tourism are complex, and both parties can be either positively or negatively influenced. Thus, sustainability aspects (e.g., maintaining ecological processes, respecting the socio-cultural authenticity of local communities, and ensuring the project's long-term economic sustainability and the equitable distribution of benefits) should be considered to ensure the non-affectation of the species and contribute to their conservation and care (Tapper & Waedt, 2006). Challenges must not be overlooked, even though there are numerous potentials for sustainable development (Dou & Day, 2020). For instance, safaris, which were originally associated with hunting and mobilization through four-wheel-drive vehicles, are becoming increasingly associated with the observation and photography of wildlife species, mostly in protected areas (Apps & Dimmock, 2018); furthermore, new types of safaris (e.g., trekking, kayaking, or camel safaris) are emerging (UNWTO, 2015).

Birdwatching

Birdwatching, often known as avitourism, is the practice of observing and identifying birds in their natural habitats by sight and sound (Sekercioglu, 2002). Some may prefer to observe with their own eyes, whereas others employ visual enhancing gadgets and listen to birds sing or collect data on various species (Istomina et al. 2016). Birdwatching is one of the most rapidly expanding forms of wildlife observation tourism (Connell, 2009). Its popularity has grown since the early 20th century (Kronenberg, 2014), owing to the realization that birdwatching has a great potential to offer economic advantages and financial wellbeing to local communities located near well-known birdwatching sites (Callaghan et al., 2018). For example, birdwatching is the most popular wildlife observation activity in the United States, with over 45 million enthusiasts (US Fish and Wildlife Service, 2016). However, it has recently started to acquire popularity in other areas, in addition to being one of the most popular outdoor recreation activities in Western countries (Ma et al., 2013), particularly in remote areas where more rare species can be found (Connell, 2009).

It is generally known that varying levels of commitment and experience are shown by birdwatchers (Kim et al., 1997; Scott et al., 2005; Callaghan et al., 2018). Furthermore, Lee et al. (2015) proposed that while males and females demonstrate equal levels of dedication and seriousness toward birdwatching, they exhibit distinct types of engagement, with females being more emotionally devoted to this activity. Recent years have also seen the emergence of some specialized kinds of birdwatching activities. Connell (2009) stated that "birders" are those people who are interested in scientific classification and the detailed study of birds, whereas "twitchers" are birders whose special interests are rare or new birds and who are willing to spend significant time and financial resources and travel long distances to see them.

Despite the diversity observed by researchers within the birdwatching community, birdwatchers may be broadly defined as well-educated, rich, and devoted individuals (Sekercioglu, 2002) who are more conscious of environmental issues related to habitat protection than typical visitors. Consequently, for decades birdwatchers have been positively connected with environmental protection (Kellert, 1985). They are likely to support community-based conservation efforts and help build and disseminate environmental knowledge and awareness by educating the public about bird habitats and their conservation, which can be accomplished by funding new projects or donating and/or paying membership fees to be part of active birdwatching NGOs (Kellert, 1985; Ma et al., 2013; Kronenberg, 2014).

However, birdwatching can have negative consequences and become unsustainable if they are not effectively managed, as with any other type of tourism. Indeed, increased ecosystem penetration may affect not just avian communities but also the habitat on which they rely (Kronenberg, 2014). Jones and Buckley (2001) stated that birdwatchers may disturb species during specific times of the year (e.g., breeding, migration, and wintering seasons); further, they may get too close to vulnerable sites (e.g., nesting, roosting, and migration lay-over sites), they may disrupt bird social organization by overusing taped vocalizations, and habitats may be damaged in the attempt to provide better services for birdwatchers (e.g., viewing sites, parking lots, and camping areas). Therefore, tourism planning that leads to increasing the benefits for birds, their environment, and the surrounding communities is especially important when operating in ecologically sensitive areas that are becoming birdwatching hotspots (Puhakka et al., 2011). It is also recommended to incorporate resource-interpretative services and other tools that may add value to the product while playing a role in managing tourism resources (Lee et al., 2009) to provide visitors with long-term, transformative experiences.

Transformative tourism experiences

It has long been noted that tourism has the potential to facilitate a positive human transformation when developed properly (Reisinger, 2013; Wolf et al., 2017; Teoh, Wang & Kwek, 2021). The concept of transformative tourism, which has received significant attention in the last five years, refers to the practice of organized tourism experiences where tourists are pushed out of their comfort zone to go through a process of growth and development that leads to the reevaluation of their attitudes and values (Christie & Mason, 2003), reflected in the change of behaviour and the way they relate with the world (i.e., natural, socio-cultural, economic, political, and technological environments; Reisinger, 2013). Sheldon (2020) stated that this awakening of consciousness or enlightenment allows tourists to experience several personal inner benefits (e.g., a sense of peace and freedom) that boosts a desire to contribute to the greater good. It has also been suggested that tourists become agents of change who embrace inclusive worldviews, cross-cultural understanding, and social empowerment in the long term (Soulard et al., 2021). Thus, transformative tourism has powerful personal and societal implications (Mkhize & Ivanovic, 2020; Teoh et al., 2021) and has also been linked to sustainable tourism (Wolf et al., 2017).

Christie and Mason (2003) stated that even though any educative experience – including tourism experiences – inevitably originates changes, these may be slight (e.g., acquisition of new knowledge). Not everyone, therefore, experiences real transformation and even those who do experience it in different ways. Although some tourists may have in situ awareness, others may require post-experience awareness to identify the transformation (Teoh et al., 2021). Although some may anticipate this transformation, others may experience it unintentionally (Wolf et al., 2017). Pung et al. (2020) also found generational differences where the female's transformation process is characterized by bodily feelings and self-consciousness, whereas flow, adaptation, and a sense of community with other travelers prevail in the male's process. This process is considered individualized and co-created because various elements, such as visitors' emotions, personalities, prior experiences, and the staged experience play a vital part in mediating change (Kirillova et al., 2017; Teoh et al., 2021).

This complex network of factors makes it difficult to have a precise understanding of the transformation process (Lean, 2009). For example, it has been claimed that for a "peak" (or extraordinary and transcendent) tourism experience to be transformative, triggering moments – typically occurring near the conclusion of the trip – must be personally significant to the tourist (Kirillova et al., 2017). Cavender et al. (2020) encouraged visitors to focus their energy on themselves, their experiences, and the people and places they encounter along the way. Some transformation factors that have been identified relate to the setting (e.g., awe-inspiring scenery and experiencing the sense of the place), the experience (e.g., long stays, immersion, challenges, novelty, and post-travel reflection), and human dimension (e.g., interacting and connecting with locals and travelers and access to unfamiliar cultural authenticity) (Kirillova et al., 2017; Pung & Chiappa, 2020; Magrizos et al., 2021). Regarding the latter, Soulard et al. (2021) highlighted that the major source of transformation may be deep regard and respect for destination residents. Thus, tourism practitioners must encourage participation and recognize the perspectives of local people to create tourist experiences that might generate meaningful transformation (Soulard et al., 2019).

While the transformative tourism concept has been studied in several forms of tourism such as volunteer (Coghlan & Gooch, 2011; Magrizos et al., 2021), educational (i.e., study abroad) (Cavender et al., 2020) and cultural/rural tourism (Vidickien et al., 2020), the design of visitor experiences intended specifically for natural areas to trigger this transformation is only beginning now. However, it is generally understood that these environments are excellent for fostering a deeper knowledge and appreciation of the natural environment (e.g., through the interpretation of attractions and ecotourism programs) and eliciting overall emotions of well-being (Lean, 2009). Wolf et al. (2017) also stated that being surrounded by different stimuli in natural settings may make tourists feel humble, encouraging self-exploration, critical reflection, and, thus, intended or unintended transformation, while increasing participants' satisfaction and loyalty to these settings.

Study approach

In the following case study, we identified key characteristics or elements linked to the concept of transformative tourism experiences present in a birdwatching tourism cluster composed of observation sites that are mostly managed by small rural entrepreneurs and local organizations.

Methodologically, a qualitative research approach was followed, as one of the authors participated in a long immersion process, given that he is a partner of an ecotourism company that operates in the area. He first visited the study site in August 2018. Since then, he had revisited the location at least eight times. He lived in the study site from March 2020 to June 2021 and, to date, he continues to visit and work with projects in the area. Therefore, through his observations, he acquired vast knowledge regarding the tourism operation, the key birding sites, the managers or owners of the birding sites, and the conservation and management of each site, among other topics.

Study site: the Northern Peru Birding Route

Peru is mainly known for its cultural heritage. Thus, international tourism is concentrated in southern regions like Cusco, the location of Machu Picchu and other Inca archaeological sites (MINCETUR, 2021). However, there are many lesser-known destinations where wildlife observation activities take place. In fact, Peru is the country with the second-highest number of bird species (1,876, of which 110 are endemic) (Plenge, 2021). Specifically, the Northern

Peru Birding Route, located in one of the 35 biodiversity hotspots identified around the world, concentrates endemic species that are under a great threat of extinction (Young et al., 2015). It runs over 1,200 km of land routes and rivers, from 0 to 3,500 m.a.s.l. within eight regions (i.e., La Libertad, Lambayeque, Cajamarca, Amazonas, San Martín, Loreto, Piura, and Tumbes). There are 55 birdwatching sites identified throughout this route (Promperu, 2019).

This case focuses in an area along 350 km of the Interoceanic North Highway, which includes ten birding sites (see Figure 17.1) in Amazonas and San Martin regions. This area is a transition zone between the Andes mountains and the Amazon. It includes diverse ecosystems and rare endemic bird species (Young, 2007). The birdwatching sites are in the vicinity of numerous natural protected areas, the most notable of which are:

- Huembo Ecological Easement[1]
- Abra Patricia Private Conservation Area[2]
- Alto Mayo Protection Forest[3]
- Ecosystem Conservation and Recovery Zone (ECRZ), Humedal del Alto Mayo[4]
- Ecosystem Conservation and Recovery Zone (ECRZ), Morro de Calzada[5]
- Ecosystem Conservation and Recovery Zone (ECRZ), Rumiyacu, Misquiyacu, and Almendra
- Cordillera Escalera Regional Conservation Area[5]

Currently, most of these sites receive few visitors interested in birdwatching or other wildlife species. They are mainly international as domestic tourists rather visit nearby areas for family recreation, relaxation, and fun (Promperu, 2020); this is especially the case with regard to the cities of Tarapoto and Moyobamba, where there is a concentration of lodging, food, and other services. In contrast, in the birdwatching sites and surrounding rural areas, offer limited services. Abra Patricia stands out for its lodging capacity and services, while other sites offer basic services and lodging for small groups of up to 8 passengers.

Figure 17.1 Long-whiskered Owlet in Abra Patricia.

The growth of economic activities that are incompatible with conservation has resulted in the loss of tropical forests despite attempts to prioritize the conservation of the area's biodiversity. The proximity of the birdwatching sites to one of the main national highways has assisted access to these sites, but has also accelerated the deforestation process because of economic activities carried out by migrants (Conservación Internacional & SERNANP, 2017). The Peruvian Ministry of the Environment – MINAM (2021) – stated that the San Martín and Amazonas regions occupied the second and seventh places between 2001 and 2019, respectively, in terms of accumulated deforestation levels of Amazonian forests in Peru. Given this scenario, the Northern Peru Birding Route emerges as a sustainable cluster of enterprises that promote the Amazonian forests and their wild species.

The Abra Patricia Private Conservation Area was established in 2007 under the direction of the Andean Ecosystems Association (ECOAN, 2021). Among the most common species in this area are the Long-whiskered Owlet (*Xenoglaux loweryi*) (see Figure 17.2) and Johnson's Tody-Flycatcher (*Poecilotriccus luluae*). ECOAN also maintains the Huembo Ecological Easement, where the Marvelous Spatuletail (*Loddigesia mirabilis*) is a common sight. Later, the Alto Nieva Farm – an area teeming with fascinating species, and an observatory for the Ochre-fronted Antpitta (*Grallaricula ochraceifrons*) and Rusty-tinged Antpitta (*Grallaria przewalskii*) – was established (see Figure 17.3). Similarly, locals were assisted in the construction of bird observatories in the buffer zone and inside the area of the Alto Mayo Protection Forest as part of a collaborative initiative of the Alto Mayo Protection Forest Headquarters and the NGO Conservation International (Conservación Internacional & SERNANP, 2017). These activities resulted in the establishment of the following three places dedicated to bird conservation and the provision of services for observing and photographing birds:

- Royal Path: Royal Sunangel (*Heliangelus regalis*), Long-tailed Sylph (*Aglaiocercus kingii*), and Botted Racket-tail (*Ocreatus underwoodii*).
- Parador La Llanteria: Andean Cock-of-the-rock (*Rupicola peruvianus*), Ecuadorian Piedtail (*Phlogophilus hemileucurus*), and White-tipped Sickbill (*Eutoxeres aquila*).

Figure 17.2 Ochre-fronted Antpitta.

Figure 17.3 Wildlife monitoring in the Arena Blanca Reserve.

- Arena Blanca Reserve: Rufous-breasted Wood-Quail (*Odontophorus speciosus*), Little Tinamou (*Crypturellus soui*), Wire-crested Thorntail (*Discosura popelairii*), and Rufous-crested Coquette (*Lophornis delattrei*).

The ECRZ Humedal del Alto Mayo, the ECRZ Morro de Calzada, and the ECRZ Rumiyacu, Misquiyacu, and Almendra, three conservation areas with interesting bird species, are located between the cities of Rioja and Moyobamba. Tourism services are provided to allow visitors to explore and wander around the wetland in the case of the ECRZ Humedal del Alto Mayo. A concession has been granted in this space to two local groups that must follow a site management plan and offer treks where you can observe a broad range of wildlife, including numerous bird species associated with rivers and Amazon lakes. In the case of the ECRZ Morro de Calzada and the ECRZ Rumiyacu, Misquiyacu, and Almendra, the Painted Manakin (*Machaeropterus eckelberryi*), the Black-bellied Tanager (*Ramphocelus melanogaster*), and the Mishana Tyrannulet (*Zimmerius villarejoi*) stand out. In these two areas, there are also nearby private enterprises that offer bird observatory services (i.e., Morro de Calzada Amazon Center of the Ikam Expeditions and Waqanki company). Further, in the Cordillera Escalera Regional Conservation Area, the Association for Bird Conservation and Biodiversity Koepcke's Hermit (ACONABIKH) manages a birdwatching site where the Koepcke's Hermit (*Phaethornis koepckeae*) is the most noteworthy species.

Study analysis: a transformative experience in the Peruvian Amazon

This birding route shows the following characteristics:

- There is a heterogeneity of habitats because of the geographical and altitudinal variety, and each site specializes in different species, contributing to the complementarity between sites.
- At least four sites monitor the main bird species they have in their space (see Figure 17.4).
- Four sites present a high success of species observation: the Marvelous Spatuletail in Huembo (see Figure 17.5), the Ochre-fronted Antpitta in the Alto Nieva Fundo, the

Figure 17.4 Marvelous Spatuletail in Huembo.

Figure 17.5 Hummingbird Garden and Observatory at the Arena Blanca Reserve.

Rufous-breasted Wood-Quail and the Wire-crested Thorntail in the Arena Blanca Reserve, and Koepcke's Hermit in ACONABIKH.

- Some species are national icons and have great stories, such as the Marvelous Hummingbird, the most beautiful hummingbird in the world that only exists in this area. Observers recognize it as the favorite species and of great interest for observation and the most promoted species in Peru (Puhakka, Salo & Saaksjarvi, 2011). Another iconic species is the Long-Whiskered Owlet, the smallest owl in the world, which is only to be found in this area. There are also other endemic, rare, and threatened species and those that stand out for their beauty, color, and behaviour.
- At least eight sites contain facilities to assist in birdwatching. These facilities are a complement to the conservation and monitoring actions. They include hummingbirds' observatories (including a gardening facility that attracts species and hummingbird drinkers), Wood-Quail and Tinamou observatories (hiding places from where the species that are attracted to corn can be observed), and a Antpitta's observatory (a 3-km trail that has a natural hiding place where the species is attracted by the supply of worms) (see Figures 17.6–17.8).
- New birdwatching sites are emerging, such as some lodgings or restaurants and tourist sites.

Therefore, this birding route possesses all the components of the "mechanism of transformative tourism experience", as suggested by Kirillova et al. (2017) (see Figure 17.9). Within the *circumstantial tourism environment*, this route offers: (1) the natural beauty of its various landscapes and ecosystems, due to its location between the Andes and the Amazon for which it was recognized as a hot spot of global biodiversity (Critical Ecosystem Partnership Fund, 2015); (2) encounters with unique wildlife, among which their endemic birds' species (e.g., Marvelous Spatuletail, Long-whiskered Owlet, Ochre-fronted Antpitta) stand out; (3) cultural exchange with the local residents involved in nature conservation (i.e., mostly migrants from the Andes);

Figure 17.6 Alto Mayo Protected Forest.

Figure 17.7 Huembo Lodge, Marvelous Spatuletail Home.

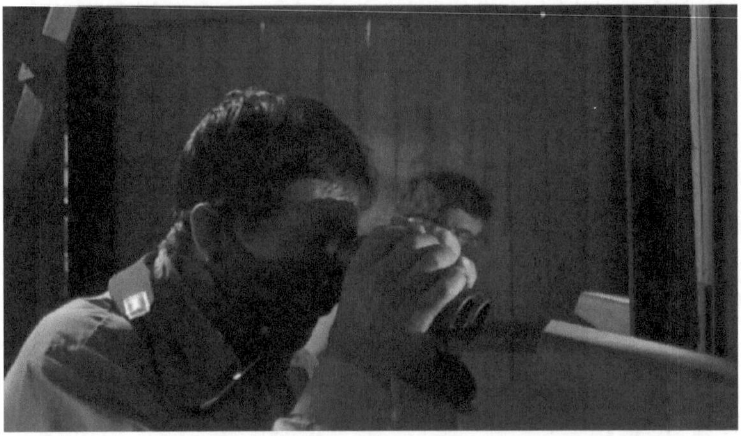

Figure 17.8 Quail and partridge observatory in the Arena Blanca Reserve.

and (4) social interactions with tourism practitioners and other individuals committed to the environmental conservation of the route.

According to Kirillova et al. (2017), these aspects can lead to the *co-creation* of a *triggering episode* whose *meaning* favors a *transformation*. Furthermore, in this case the following can be identified as the main facilitators for the peak experience to take place:

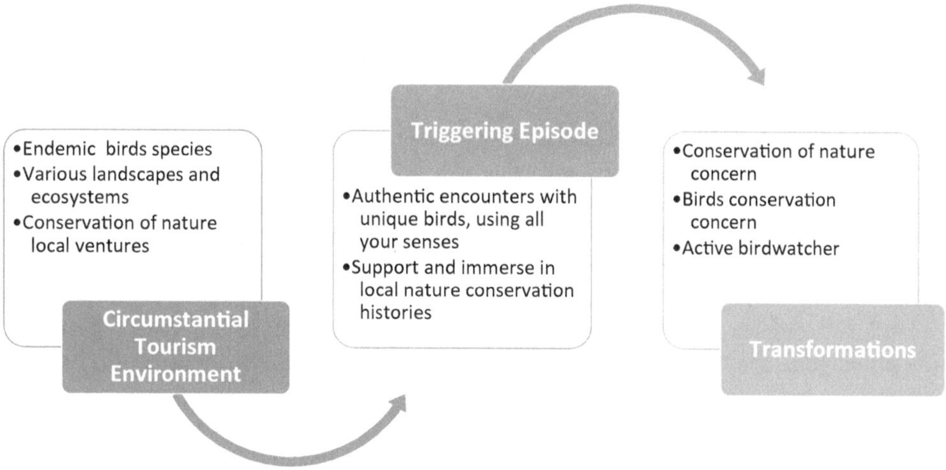

Figure 17.9 Transformative Birdwatching Experience in The Northern Peru Birding Route.

- Birds are inputs for reflection. As Zarankin (2021) mentions, birds "… *serve as powerful reminders that we are part of something larger than just the world inside our own heads*"; therefore, birdwatchers experience mindfulness by feeling fully immersed in the present moment.
- Through the challenge of observing unique species, an awareness of limits and personal strengths is promoted among the tourists. In fact, according to Lund (2017), for birders there is no greater excitement and dread than that which comes from possibly seeing a rare species.
- An emotional bond emerges between the tourists and the bird species, natural spaces, and people that conserve them. For example, the Marvelous Spatuletail is an endemic and an endangered hummingbird that lives only in this area (American Bird Conservancy, n.d.). So, people may develop a special bond with this species as well as with the people who facilitated this visit, with the other places they had to stop by before encountering it, and even with the other species seen while anticipating finding the Spatuletail (Sekercioglu, 2002).
- Constant learning and questioning are generated by the conservation stories that tourists get to hear from people and organizations that fight on a daily basis to reach the conservation goals of the forests and the sustainable living standard against regular living standard from where the tourists normally come from.

Discussion

Conservation stories and observation success elicit powerful and diverse emotions in visitors, increasing cognition and connection with something greater, with the ability to promote personal development. Therefore, it appears clear that this birding route has strong potential to become a transformative tourism experience. However, it is crucial to highlight that currently the sites that constitute the birding route (i.e., the tourism product) focus mostly on a specialized sector offering exclusively birdwatching activities. Thus, three of the components of the *circumstantial tourism environment* (i.e., the natural beauty of landscapes and ecosystems; encounters with unique wildlife; and interactions with tourism practitioners) (Kirillova et al., 2017) seem to be fulfilled. Meanwhile, it is suggested that one last component should be strengthened by incorporating visits and activities that encourage further cultural exchange, both with the migrants who currently live in the area and with nearby native communities such as "Bosque de las Nuwas" or "Women's Forest", from the native Awajún people.

In addition, it is necessary to continue supporting the significant efforts that are currently being made by the local community and other stakeholders dedicated to conservation so as to develop alternative birdwatching products (e.g., shorter itineraries complementing conventional itineraries with visits to some of the sites in the birding route, mainly those closest to cities or tourism products close to cities) directed to visitors who, compared with "avitourists" whose travel motivations are focused on birdwatching (Steven et al., 2015), would like to enjoy this activity as a complement of their trips. Thus, these products aim to attract less specialized travelers looking for bird conservation-oriented tourism projects (Puhakka et al., 2011), a segment on the rise in post-pandemic times and other alternative markets to offer them transformative tourism experiences. In all cases, it would be important to plan on enhancing the triggering episodes by the end of the trip, as suggested by Kirillova et al. (2017), which can only be attained with the collaboration of tour guides who play a key role in the experience and therefore need to be trained and supported as needed (Christie & Mason, 2003).

The present case study aims to contribute to the tourism academic literature and expand the existing knowledge regarding overall transformative tourism experiences through the application of such concept in a natural setting where a specific wildlife observation activity (i.e., birdwatching) has the potential of triggering transformative experiences. The attention that transformative tourism has received as a research topic has grown exponentially in recent years. To date, most authors have focused on unveiling the aspects and circumstances that facilitate the transformative process and those which prevent it, as well as the main characteristics and outcomes of these transformations, both in general (Kirillova et al., 2017; Robledo & Batle, 2017; Pung et al., 2020; Pung & Chiappa, 2020; Soulard et al., 2021), and in the case of specific types of tourism and/or destinations (Wolf et al., 2017; Cavender et al., 2020; Mkhize & Ivanovic, 2020; Vidickienė et al., 2020; Magrizos et al., 2021). The tourism product proposed for this case study was designed on the basis of all these contributions. In particular, it builds upon the work of Kirillova et al. (2017), who proposed a theoretical model for the "mechanism of transformative tourism experience" where the circumstantial tourism environment can lead to the co-creation of a triggering episode which meaning favors a transformation.

Managerial implications

From a practical point of view, by presenting valuable insights with regard to each of the product components, tourism practitioners from different birdwatching, or overall nature-based tourism sites around the world where wildlife observation activities are promoted, may adopt a similar approach. Consequently, transformative tourism practitioners may envision the right balance at meeting the desires of both travelers and residents (Soulard et al., 2019). The importance of these managerial implications lies in the fact that it has been claimed not only that the tourism sector must focus on developing this kind of long-lasting behaviour change to achieve sustainability (Lean, 2009), but also that the practice of offering transformative tourism experiences is already becoming a popular sustainable marketing tool (Wolf et al., 2017), and a tourism form that appears ideally suited in a time when the sector and the world both need optimism (Pung & Chiappa, 2020).

Limitations and future research directions

Despite the contributions of the present work, the main limitation is that while it has been recognized that peak transformative experiences can be planned and designed (Sheldon, 2020), it is also well known that the entire transformative process is subjective and, in addition to the actual experience, its co-creation depends on factors related to the tourists' minds, and their past experiences, among other aspects (Teoh et al., 2021) that cannot be easily anticipated. Thus, several opportunities for further research emerge. First, for the case study presented and other settings where different products are offered to tourists based on their level of interest or previous experience in a particular activity, a comparative study would be recommended to further understand if there are significant differences among groups regarding the level of the transformations they experience or their duration, which may require a longitudinal approach. Further, it would be recommended to compare transformative outcomes in places where staged experiences have been put in place, as compared to those where experiences have not been intentionally designed, to measure the impacts of these interventions.

Notes

1 Private legal agreement of commitment for the conservation of biodiversity.
2 Natural Protected Area of private administration.
3 Natural Protected Area of national administration (protected by the National Service of Natural Protected Areas by the State- SERNANP).
4 Unregistered rustic properties administered by the Regional Government of San Martin.
5 Natural Protected Area of regional administration, by the Regional Government of San Martin.

References

American Bird Conservancy (n.d.). Marvelous Spatuletail. Retrieved from https://abcbirds.org/bird/marvelous-spatuletail/

Andean Ecosystems Association (2021). Retrieved from https://www.ecoanperu.org/amaz.html. Amazonas

Apps, K., Dimmock, K., & Huveneers, C. (2016). Turning wildlife experiences into conservation action: Can White Shark cage diving tourism influence conservation behaviour?. *Marine Policy Journal*, 88, 108–115.

Ballantyne, R., Packer, J., & Falk, J. (2011). Visitors' learning for environmental sustainability: Testing short- and long-term impacts of wildlife tourism experiences using structural equation modelling. *Tourism Management*, 32(6), 1243–1252.

Balmford, A., Beresford, J., Green, J., Naidoo, R., Walpole, M., & Manica, A. (2009). A global perspective on trends in nature-based tourism. *PLOS Biology*, 7(6), e1000144.

Balmford, A., Green, J. M., Anderson, M., Beresford, J., Huang, C., Naidoo, R., Walpole, M., & Manica, A. (2015). Walk on the wild side: Estimating the global magnitude of visits to protected areas. *PLOS Biology*, 13(2), e1002074.

Callaghan, C. T., Slater, M., Major, R. E., Morrison, M., Martin, J. M., & Kingsford, R. T. (2018). Travelling birds generate eco-travellers: The economic potential of vagrant birdwatching. *Human Dimensions of Wildlife*, 23(1), 71–82.

Cavender, R., Swanson, J. R., & Wright, K. (2020). Transformative travel: Transformative learning through education abroad in a niche tourism destination. *Journal of Hospitality, Leisure, Sport and Tourism Education*, 27, 100245.

Christie, M. F., & Mason, P. A. (2003). Transformative tour guiding: Training tour guides to be critically reflective practitioners. *Journal of Ecotourism*, 2(1), 1–16.

Coghlan, A., & Gooch, M. (2011). Applying a transformative learning framework to volunteer tourism. *Journal of Sustainable Tourism*, 19(6), 713–728.

Connell, J. (2009). Birdwatching, twitching and tourism: Towards an Australian perspective. *Australian Geographer*, 40(2), 203–217.

Conservación Internacional & SERNANP (2017). Voces del Alto Mayo. Los Acuerdos de Conservación cambian vidas en el Bosque de Protección Alto Mayo. Lima, Perú. Retrieved from https://www.conservation.org/docs/default-source/peru/voces-del-alto-mayo_.pdf?Status=Master&sfvrsn=684059d0_3.

Critical Ecosystem Partnership Fund (2015). Resumen Técnico del Perfil del Ecosistema. Hotspot de Biodiversidad de Los Andes Tropicales. Retrieved from https://www.cepf.net/sites/default/files/tropicalandes_techsummary_sp.pdf

Dobson, J. (2008). *Shark! a new frontier in tourist demand for marine wildlife. Marine Wildlife and tourism Management: Insights from the natural and social Sciences.* CABI, Wallingford, 49–65.

Dou, X., & Day, J. (2020). Human-wildlife interactions for tourism: A systematic review. *Journal of Hospitality and Tourism Insights*, 3(5), 529–547.

Istomina, E. A., Luzhkova, N. M., & Khidekel', V. V. (2016). Birdwatching tourism infrastructure planning in the Ria Formosa natural park (Portugal). *Geography and Natural Resources*, 37(4), 371–378.

Jones, D. N., & Buckley, R. (2001). *Bird-watching tourism in Australia.* CRC for sustainable Tourism, Gold Coast, QLD.

Kellert, S. R. (1985). Birdwatching in American society. *Leisure Sciences*, 7(3), 343–360.

Kim, S. S., Scott, D., & Crompton, J. L. (1997). An exploration of the relationships among social psychological involvement, behavioral involvement, commitment, and future intentions in the context of birdwatching. *Journal of Leisure Research*, 29(3), 320–341.

Kirillova, K., Lehto, X., & Cai, L. (2017). What triggers transformative tourism experiences? *Tourism Recreation Research*, 42(4), 498–511.

Kronenberg, J. (2014). Environmental impacts of the use of ecosystem services: Case study of birdwatching. *Environmental Management*, 54(3), 617–630.

Kyle, G. T., Absher, J. D., Hammitt, W. E., & Cavin, J. (2006). An examination of the motivation–involvement relationship. *Leisure Sciences*, 28(5), 467–485.

Kyle, G. T., Mowen, A. J., & Tarrant, M. (2004). Linking place preferences with place meaning: An examination of the relationship between place motivation and place attachment. *Journal of Environmental Psychology*, 24(4), 439–454.

Lean, G. L. (2009). Transformative travel: Inspiring sustainability. In Wellness and tourism. In *Mind Body. Spirit, Place. Cognizant*, New York edited by Robyn Bushell and Pauline J. Sheldon, 191–205.

Lee, C. K., Lee, J. H., Mjelde, J. W., Scott, D., & Kim, T. K. (2009). Assessing the economic value of a public birdwatching interpretative service using a contingent valuation method. *International Journal of Tourism Research*, 11(6), 583–593.

Lee, S., McMahan, K., & Scott, D. (2015). The gendered nature of serious birdwatching. *Human Dimensions of Wildlife*, 20(1), 47–64.

Lund, N. (2017). *Birdist Rule #42: Discover the Thrill of Chasing a Rare Bird.* Audubon. Retrieved from https://www.audubon.org/news/birdist-rule-42-discover-thrill-chasing-rare-bird

Ma, Z., Cheng, Y., Wang, J., & Fu, X. (2013). The rapid development of birdwatching in mainland China: A new force for bird study and conservation. *Bird Conservation International*, 23(2), 259–269.

Madin, E. M. P., & Fenton, D. M. (2004). Environmental interpretation in the Great Barrier Reef Marine Park: An assessment of programme effectiveness. *Journal of Sustainable Tourism*, 12(2), 121–137.

Magrizos, S., Kostopoulos, I., & Powers, L. (2021). Volunteer tourism as a transformative experience: A mixed methods empirical study. *Journal of Travel Research*, 60(4), 878–895.

MINAM (2021). Apuntes del bosque 3. Monitoreo de la pérdida de Bosques Húmedos Amazónicos en el año 2019. Retrieved from http://www.bosques.gob.pe/archivo/Apuntes-del-bosque-3.pdf. Programa Nacional de Conservación de Bosques para la Mitigación del Cambio Climático (PNCBMCC).

MINCETUR (2021). Llegada de visitantes a sitios turísticos, museos y áreas naturales protegidas por el Estado. Retrieved from http://datosturismo.mincetur.gob.pe/appdatosTurismo/Content2.html

Mkhize, S. L., & Ivanovic, M. (2020). Building the case for transformative tourism in South Africa. *African Journal of Hospitality, Tourism and Leisure*, 9(4), 717–731.

Moore, S. A., & Polley, A. (2007). Defining indicators and standards for tourism impacts in protected areas: Cape Range National Park, Australia. *Environmental Management*, 39(3), 291–300.

Orams, M. B. (2002). Feeding wildlife as a tourism attraction: A review of issues and impacts. *Tourism Management*, 23(3), 281–293.

Plenge, M. (2021). List of the birds of Peru. Retrieved from https://sites.google.com/site/boletinunop/checklist

Priskin, J. (2003). Tourist perceptions of degradation caused by coastal nature-based recreation. *Environmental Management*, 32(2), 189–204.

PromPeru (2019). The Northern Peru birding route. Retrieved from https://www.peru.travel/birdwatching/Complemento/brochures/ruta_aves_norte.pdf

Promperu (2020). Perfil del Vacacionista Nacional que visita San Martín 2019.

Puhakka, L., Salo, M., & Sääksjärvi, I. E. (2011). Bird diversity, birdwatching tourism and conservation in Peru: A geographic analysis. *PLOS One*, 6(11), e26786.

Pung, J., & Chiappa, G. (2020). An exploratory and qualitative study on the meaning of transformative tourism and its facilitators and inhibitors. *European Journal of Tourism Research*, 24, 2404–2404.

Pung, J. M., Yung, R., Khoo-Lattimore, C., & Del Chiappa, G. (2020). Transformative travel experiences and gender: A double duoethnography approach. *Current Issues in Tourism*, 23(5), 538–558.

Reisinger, Y. (2013). *Transformational tourism: Tourist perspectives*. CABI, Wallingford.

Robledo, M. A., & Batle, J. (2017). Transformational tourism as a hero's journey. *Current Issues in Tourism*, 20(16), 1736–1748.

Scott, D., Ditton, R. B., Stoll, J. R., & Eubanks, T. L. (2005). Measuring specialization among birders: Utility of a self-classification measure. *Human Dimensions of Wildlife*, 10(1), 53–74. doi:10.1080/10871200590904888

Sekercioglu, C. H. (2002). Impacts of birdwatching on human and avian communities. *Environmental Conservation*, 29(3), 282–289.

Sheldon, P. J. (2020). Designing tourism experiences for inner transformation. *Annals of Tourism Research*, 83, 112–123.

Silvennoinen, & Tyrväinen (2001). Demand for nature tourism in Finland and customers' environmental wishes. *Communications from the Finnish Forest Research Institute*, 802, 112–127.

Sotomayor, S., Barbieri, C., Wilhelm Stanis, S., Aguilar, F. X., & Smith, J. W. (2014). Motivations for recreating on farmlands, private forests, and state or national parks. *Environmental Management*, 54(1), 138–150.

Soulard, J., McGehee, N. G., & Stern, M. (2019). Transformative tourism organizations and glocalization. *Annals of Tourism Research*, 76, 91–104.

Soulard, J., McGehee, N. G., Stern, M. J., & Lamoureux, K. M. (2021). Transformative tourism: Tourists' drawings, symbols, and narratives of change. *Annals of Tourism Research*, 87. 103141.

Spenceley, A., McCool, S., Newsome, D., Báez, A., Barborak, J. R., & Blye, C. J., ... & Zschiegner, A. K. (2021). Tourism in protected and conserved areas amid the COVID-19 pandemic. *Parks*, 27, 103–118.

Steven, R., Morrison, C., & Castley, J. G. (2015). Birdwatching and avitourism: A global review of research into its participant markets, distribution and impacts, highlighting future research priorities to inform sustainable avitourism management. *Journal of Sustainable Tourism*, 23(8–9), 1257–1276.

Tapper, R., & Waedt, D. K. (2006). Wildlife watching and tourism. In a study on the benefits and risks of a fast growing tourism activity and its impacts on species. *United Nations Environment Program/Convention on Migratory Species Secretariat*. Bonn, Germany.

Teoh, M. W., Wang, Y., & Kwek, A. (2021). Conceptualising co-created transformative tourism experiences: A systematic narrative review. *Journal of Hospitality and Tourism Management*, 47, 176–189. doi:10.1016/j.jhtm.2021.03.013

US Fish and Wildlife Service (2016). National survey of fishing, hunting, and wildlife-associated recreation. Retrieved from https://www.fws.gov/wsfrprograms/subpages/nationalsurvey/nat_survey2016.pdf. 2016

United Nations World Tourism Organization (2015). *UNWTO. Towards Measuring the Economic Value of Wildlife Watching Tourism in Africa*.

Vidickienė, D., Vilkė, R., & Gedminaitė-Raudonė, Ž. (2020). Transformative tourism as an innovative tool for rural development. *European Countryside*, 12(3), 277–291.

Wall-Reinius, S. W., & Fredman, P. (2007). Protected areas as attractions. *Annals of Tourism Research*, 34(4), 839–854.

Weaver, D. B., & Lawton, L. J. (2007). Twenty years on: The state of contemporary ecotourism research. *Tourism Management*, 28(5), 1168–1179.

Winter, P. L., Selin, S., Cerveny, L., & Bricker, K. (2020). Outdoor recreation, nature-based tourism, and sustainability. *Sustainability*, 12(1), 81.

Wolf, I. D., Ainsworth, G. B., & Crowley, J. (2017). Transformative travel as a sustainable market niche for protected areas: A new development, marketing and conservation model. *Journal of Sustainable Tourism*, 25(11), 1650–1673.

Young, B. (2007). *Distribución de las especies endémicas en la vertiente oriental de los Andes en Perú y Bolivia*. NatureServe, Arlington, Virginia, EE UU.

Young, B., Josse, C., Stern, M., Vasconez, S., Olander, J., Sanchez de Lozada, A., Zador, M., Smyth, R., Comer, P., Moull, K., Echavarría, M., & Hak, J. (2015). *Ecosystem profile tropical Andes Biodiversity Hotspot*. Critical Ecosystem Partnership Fund. Retrieved from https://www.cepf.net/sites/default/files/tropical_andes_profile_final_4_2015.pdf

Zarankin, J. (2021). *More Birds Bring More Happiness, According to Science*. Audubon. Retrieved from https://www.audubon.org/news/more-birds-bring-more-happiness-according-science

18

RE-DESIGNING SUSTAINABLE EXPERIENCES FOR MOUNT FUJI CLIMBERS

Turning peak-baggers back into pilgrims?

Thomas E. Jones, Huong T. Bui and Michal Apollo

Introduction

Nature-based tourism (NBT) ranks among the industry's rapid-growth markets and mysterious mountain peaks make up a major sub-segment. Regardless of secular or sacred motivation, mountain tourists face exposure to various types of risk that range from inclement weather, altitude sickness to a variety of environmental hazards (landslides and avalanche to slips, risk of trips and falls) (Ziegler, et al. 2021; Apollo & Wengel, 2022). Nonetheless, mountains have long since attracted pilgrims who seek spiritual enlightenment by scaling the physical and psychological peaks. Historically limited by extreme topography, climate or socio-cultural barriers, mountainous NBT expanded in the twentieth century from its pilgrim roots to cater to increasingly heterogenous demand spectrums (Nyaupane et al. 2015; Bui et al. 2021; Gawlik et al. 2022). If tourism is a post-modern extension of pilgrimage, self-stated pilgrims nonetheless acknowledge secular comforts such as "comfortable amenities, satisfying food, reliable information by guides and interpretive markers" (Di Giovine, 2011). Experience design is thus needed to proactively plan pilgrimage events and itineraries that can ensure today's climbers acquire relevant experiences at a physiological, emotional, and cognitive level (Jernsand et al., 2015). Designing authentic experiences and memories can promote sustainability as when participation in traditional festivals foster a sense of stewardship (Hung Lee, Jan & Lin, 2021).

Demand drivers push visitors to seek new and unique NBT experiences (Swarbrooke, Beard, Leckie, & Pomfret, 2003), resulting in the commodification of the mountain stage on which it unfolds (Beedie & Hudson, 2003). Mountains and NBT have long been economic bedfellows and even during the coronavirus period, many such destinations have continued to see seasonal influxes of visitation, albeit now driven by domestic instead of international demand. Highland destinations become 'honeypots' whose geophysical and temporal barriers channel climbers into "swarms to the summit" during peak places at high season (Pickering & Buckley, 2003). The improved access enabled by technological innovations and infrastructure development (Apollo 2017a) thus threatens mountains by exacerbating congestion during peak periods, or spilling over into impacts such as trail widening, dumped trash, or overflowing toilets (Apollo, 2017b; Apollo, 2021). However, the overpowering popularity of certain peaks has had a range of consequences that includes impacts on the environment (e.g. those related to trails, trash and toilets) and impacts on the visitor experience (e.g. congestion: see Manning,

DOI: 10.4324/9781003230748-22

2009). The latter research arena has attracted less attention despite the serious psychological and spatial congestion that can undermine the sustainability of visitor experience at crowded mountainous sites (Pickering & Buckley, 2003). It is worth noting that the fragile ecosystems of the highlands and alpine meadows – where pilgrims spots are located – have largely been affected by the pilgrims and tourists as their numbers have increased dramatically, and the policy towards promoting pilgrimages has turned into earn money; this together has resulted in exploitation and natural and cultural resources (Gawlik et al., 2022; Apollo et al., 2021).

To that end, the principal aim of this research is to investigate whether re-designing sustainable experiences can support site managers' efforts to infuse some historical aspects of pilgrimage into the current trend toward peak-bagging (Blake, 2002), which suggests that mountains are "collectibles" to be ticked off a list (Kedrowski, 2009). The case study of Fuji climbers past and present provides a platform to compare and contrast the 'pilgrims' of yesteryear with today's 'peak-baggers' in the context of contemporary NBT motivations and behaviours. Mount Fuji (3,776 m), as a case in point, is one of the most recognizable, most visited, and most climbed peaks in Japan and the whole world. Following the heightened media presence linked to listing as a UNESCO World Heritage Site in 2013, the aggregate footfall on Fuji's four trails exceeded 300,000 per season. This included over 10,000 climbers converging on the summit to see sunrise on a typical Sunday morning in July–August (MoE, 2016). But few of the contemporary climbers during a typical 10-week season are aware of the worship-ascent heritage, even though it represents one of the two criteria that underpinned the UNESCO listing (Jones & Yamamoto, 2016). Religion ranks amongst the oldest travel motivators (O'Gorman, 2010) and sacred peaks such as Fuji's have long attracted pilgrims. But few of the current crop of climbers start from the traditional trailhead at the foot of the mountain or visit the shrines that dot the way. Instead, most opt to drive to the 5th station (2,300 m) before setting out on an overnight hike to see sunrise from the summit.

Mountain climbing: peak-baggers or pilgrims?

This section sets the scene for our investigation by reviewing motivations for mountain climbing, along with the symbolic role of iconic peaks and factors behind their popularization.

On a spectrum of motivation-based visit objectives, two extremes of mountainous NBT position secular adventure at one end of the scale and pilgrim aspiration at the other. Secular goals, such as adventure and adrenaline, can be distinguished from sacred aims such as self-enlightenment to be achieved via sacrosanct beliefs or performing religious or cultural rites (Apollo & Wengel, 2022). However, borders have been blurred by progressive advances in access and technology, along with the erosion of fragile ecosystems and changes in climatic and socio-economic conditions in the upland areas (Apollo & Andreychouk, 2022). In the twentieth century, the upper slopes and more challenging technical climbing sections of the world's mountains have evolved from the domain of elite mountaineers to become more easily accessible playgrounds for new waves of 'peak-baggers' (Blake, 2002), which Apollo and Wengel (2022) refer to as "novice mountaineers". "Peak-bagging implies that the mountains are collectibles, with climbers and hikers devoted to completing all of the peaks (technical or non-technical) in a meticulous, scripted fashion" (Kedrowski, 2009). Like a baseball-card collection, the "allure of conquering all of the highest mountains in a region, such as Colorado's 14,000 foot (4267 m) summits" is a timeless attempt to impose order on chaos by checking the names of peaks off a list (Shelton, 1998 cited in Kuby et al., 2001). The original 'Munro-bagging' phenomenon dates back to the mid-eighteenth century. Colorado's Fourteeners followed in the footsteps of "Sir Hugh Thomas Munro (1856–1919) who identified and climbed the 284 summits in Scotland over 3,000 feet (914 m)" (Kuby et al., 2001, p. 456). Promoted by climber clubs and guidebooks, and more recently by websites such as TripAdvisor that filter and 'rank' the target sites, other prominent examples include the Adirondacks' 46ers from the US and Japan's 100 Peaks that became ubiquitous in an era of mass mobility.

Yet mountains do not only form the backdrop for lists of landscape superlatives, but also contain sacred sites for religion and culture. This is evident in Asia's mountainous peaks, many of which have long been revered as holy sites. Volcanoes such as Mt. Bromo (2,329 m) in East Java have featured prominently for pacification purposes, with at least 50 eruptions recorded since 1804 (Roscoe, 2013). Bromo is located within a caldera that forms the ancestral home of the Tengger, people who make a pilgrimage to the top every May in the climax of a month-long celebration called *Yadnya Kasada* or *Kosodo* festival (Choe & Hitchcock, 2018). Crossing the desert sands, pilgrims pause at the temple at the foot of the volcano, before climbing up to throw offerings into the crater. Distinguishing between Bromo pilgrims and peak-baggers is problematic partly due to the integral role of the local Tengger as guides and tour operators, and also because of the influx of revellers to join the celebrations – a crowd of 20–25,000 visitors that accounts for over a quarter of annual national park visits (Cochrane, 2009). Visitation at other Southeast Asian peaks such as Mt. Ramelau in Timor-Leste is also clustered around certain dates due to religious ceremonies (Silva & Bui, 2021). These festivals further blur the lines between demand segments, attracting quasi-'locals' from regional or national hubs that often climb for a combination of secular and sacred purposes. As with the Tengerrese tour guide taxiing groups of Jakartan 'pilgrims' up to the lip of Mt. Bromo in a 4x4 jeep, drawing a clear distinction between 'guests' and 'hosts' can also be problematic given the deep ties of economic dependency. Neither do national borders necessarily provide clear demographic distinctions, since domestic visitors are often outnumbered by pilgrims to sacred sites in the same or neighbouring mountain range. For example, various peaks in the Himalayas are visited by hundreds of thousands of inter- and intra-regional pilgrims each year, sourced from a range of different countries (Bui, Jones & Apollo, 2021). Numbers are substantial – in 2011, the Badrinath Temple (3,133 m) attracted around one million visits and Amarnath Holy Cave (3,888 m) 600,000 pilgrims (Gawlik et al., 2022). Such visits include Buddhist or Hindu practices or cultural rites, although sacrosanct beliefs do not prevent pollution (Timothy & Nyaupane, 2009), with air and water contaminated by open dumping of waste (Sati, 2015).

Despite the lack of an effective management system, this religious symbolism ensures that the iconic peaks of many mountains are designated as conservation areas. But because the philosophy of Asia's mountainous national parks and protected areas often aspire to the 'Yellowstone model', spectacular scenery and sightseeing tend to be prioritized over science-based conservation (Jones, Apollo & Bui, 2021). This approach differs from the Swiss National Park, for example, where tourist access is permitted but numbers tightly regulated to conserve the environment for scientific research (Kupper, 2009). This strict Swiss stance stands in contrast to the *laissez-faire* approach adopted across premier Asian mountains such as Huangshan (1,864 m) in China or Hallasan (1,947 m) on Jeju Island, South Korea – both estimated to attract in excess of 3 million annual visits (Bui, Jones, & Apollo, 2021).

Why do certain iconic peaks attract so many mountain climbers? Prior to the pandemic, the rapid rise in climber numbers ensures that this question has renewed relevance for site managers. Peak-bagging by alpine climbers, or environmental education for 'harder' and 'softer' segments of ecotourists have been frequently discussed in the literature (Ewert, 1994; Weaver, 2013). Overall, tourists that participate in high-altitude mountaineering for adventure purposes can be further subdivided into two groups, self-guided or pre-organized expeditions, with the latter reliant on the logistical support of service companies and guides. The motives of these two groups broadly reflect Maslow's (1954) hierarchy of needs, with self-guided climbers seeking self-actualisation while status drives climbers using paid guides (Apollo & Wengel, 2022). A comparison of climber motivations at Mt. Fuji in Japan and Mt. Kinabalu in Malaysia found the 'peak-bagging' pull of conquering a world-famous icon to be the most important factor, followed by the social bonding of climbing together with friends (Jones & Bui, 2018).

However, the pre-pandemic era of mass mobility has encouraged considerable cross-over of mountain climbers' secular goals with pilgrims' sacred aims. This hybridization reflects a similar transition a century earlier from Romantic trailblazers to a generation of New Mountaineers who were more "disciplined, spartan, professional, concerned with the tangible, and somewhat lacking in aesthetic appreciation" (McNee, 2016). Graburn (1983) argued that tourism itself can concurrently represent a 'sacred journey' and a 'secular ritual' and mountain climbing contains elements of both. Yet modern mass mobility has de-aggregated much of the sacred journey into disjointed, secular stages. The sustainability of visitor experience is threatened requiring proactive intervention to manage peak-baggers at crowded mountainous sites as will now be examined from the case of Fuji.

Mount Fuji: legacy of the past, modernization and mass tourism

A heritage of pilgrims past

Mt. Fuji's worship ascent heritage is not only a legacy of pilgrims' past but also provides a platform to reconsider current climber experience along a more sustainable trajectory. In many of the world's mountains or upland areas, historical access was restricted or denied altogether. Mt. Kailash (6638 m) is one example of an administrative ban that was initiated in 2001 due to its sacred role in four religions: Bön, Buddhism, Hinduism and Jainism. Other restrictions such as bans based on gender (females were not allowed to climb Fuji until 1872 a gender ban that continues at other sites e.g. Mt. Omine), are primarily grounded in customs or religious beliefs that can severely restrict mountaineering activities (Apollo, 2017a).

Most mountains were either 'off-limits' or the exclusive domain of elite mountaineers. Despite the ban on females, Fuji has a long history of pilgrimage and 'worship ascent', and by the sixteenth century lay persons and urbanites were already climbing *en masse.* Understanding this historical context is imperative to finding Fuji's future design and has ramifications for other mountains that have more recently been rendered easily accessible playgrounds for new generations of 'peak-baggers.' Fuji's symmetrical cone has long attracted the attention of pilgrims. Externally, it appears a 'solitary' volcano but the slopes that rise steeply from the shores of the Pacific Ocean are, in fact, a composite of three older volcanoes. The summit crater has remained inactive since its last eruption about 2000 years ago, but there have been numerous destructive outbursts from the flanks, including the Jogan Eruption of 864, and the Hoei Eruption of 1707 (Miyaji, 2007; Chakraborty & Jones, 2018). Interspersed with these periodic eruptions, Fuji became a target for 'worship ascent,' with pilgrims seeking out the summit of the sacred stratovolcano considered to be the spiritual home of the mountain goddess Konohana-Sakuyahime (Hirano, 1987).

In the seventh century, the ascetic monk En no Gyōja began leading small bands of disciples on mountain-climbing expeditions that combined physical and spiritual training. The roots of Fuji's faith are thus inextricably intertwined with Shugendō. This evolved from an eclectic mix of beliefs, philosophies, and doctrines based on religions such as Shinto, Buddhism and Taoism that in turn merged with elements of local folklore and their associated rites (Kornicki & McMullen, 1996). Fuji offered a full range of sites suitable for Shugendō's blend of esoteric retreats and ascetic training conducted over steep mountain terrain, so regular pilgrimages to the region began based on a concept of gaining bodily and spiritual well-being through physical exertion. By the fourteenth century, Shugendō practitioners had established a trail that led groups of pilgrims up to Fuji's summit and institutionalized the stop-off points *en route* to conduct ascetic practises (Polidor, 2007). Over time Fuji's fame spread as one of Japan's three holy mountains, along with Tateyama and Hakusan, which were also climbed by large numbers of pilgrims (Koizumi & Chakraborty, 2016).

After the period of civil war, the Edo era from the seventeenth century onwards ushered in a period of peace and stability that facilitated domestic travel. Fuji's summit soon became the goal for guided groups of pilgrims known as *kō*. In trailhead towns dotted around Fuji's foothills, clusters of *oshi* houses developed to cater for the pilgrims' spiritual and logistical needs. Working closely with local shrines and temples, *oshi* presided over the Fuji-*ko* style of worship-ascent, and also provided in-house accommodation for the pilgrims. With the help of *sendatsu* guides, the *oshi* organized logistics such as Sherpa-like porters and provided spiritual services, including blessings, rites and ablutions (Matsui & Uda, 2015). Pilgrims paid a package fee which included meals prepared in accordance with the devotional diet, purification and banners, as well as interpretation of local customs, thus ensuring a cost-recovery mechanism for local communities. The role of the *oshi* as cultural mediators, combined with their substantial input for regional economies, has even drawn parallels with contemporary ecotourism goals (Ito, 2009).

Fuji *kō* came from confraternities located all across Japan, pooling contributions from its members to send representatives on a summer pilgrimage (Bernstein, 2008). After the Hoei Eruption in 1707 climber numbers temporarily tapered off, but the *kō* began to rise again, especially in the Kantō and Tōkai regions from 1733 onwards (Murakushi, 2003). By the early nineteenth century there were 808 Fuji- *kō* groups in Tokyo (Bernstein, 2008). Admission criteria broadened over time to include ever-increasing swathes of the general population so that even lay persons were permitted to climb. Most were organized into groups of 20–30 climbers of whom a handful were selected to climb periodically as representatives of their sect. The scale of the sect membership can be inferred from the fact that in 1892 the Murayama *kō* alone had 56 branches with an estimated 1.4 million members (Murakushi, 2003). Pre-modern pilgrim numbers are contested, but Murakushi (2006) suggests over 10,000 climbed Fuji in 1877, whereas Ito (2009) estimates 20,000 per season in their heyday. It was this history of 'worship-ascent,' along with Fuji's perennial appeal as subject matter for artists and writers, that inspired the volcano's listing as a UNESCO *cultural* heritage site in 2013. Nonetheless, pre-modern Fuji climbing excluded certain sections of the population, notably females[1] and foreigners (Fumiko, 2005). Such restrictions were swept aside by a modernization process that also opened up Fuji's summit ascent to a new era of mass mobility introduced next.

A modern era of national parks and mass mobility

Mount Fuji forms the centrepiece of the Fuji-Hakone-Izu National Park. Designated in 1936, it covers an area of 1,227 square kilometres, including Fuji and the surrounding Lake District, Hakone, the Izu Peninsula and island chain that stretches south as far as Hachijō-jima. Fuji-Hakone-Izu is Japan's busiest national park and ranks among the most crowded parks in the world with over 132 million estimated annual visits in 2018 (MoE, 2021). How did such a sacred site become a mecca of mass tourism? This next section will review Fuji's pre- and post-World War II access development to deepen the debate over a sustainable visitor experience design.

One of Japan's earliest efforts to pass protected area legislation was a 1911 petition to parliament to designate Fuji as an 'imperial park' (Koizumi, 2001). However, it was not until 1931 that the National Parks Act was passed. Subsequently the first 12 national parks were designated from 1934, culminating with Fuji-Hakone in1936 (Kato, 2008).[2] In 1937, three further parks were gazetted in the mountains of Taiwan, then a Japanese colony (Kanda, 2012). Fujisan lay at the heart of the Shinto emperor-worship cult, with a conservation narrative adapted to accommodate the growing drumbeats of militarization. Yet Fuji's foothills were also subject to commercial pressure. Fujisanroku, the forerunner to today's Fujikyuko Corporation, was established in 1926 and in 1929 the Fuji Sanroku Electric Railway began operation, connecting trains with the Chuo line to Tokyo (Tsuchiya, 1981) (Figure 18.1).

Local transport conglomerates developed holiday resorts and second homes in tandem with an extensive transport infrastructure that paved the way for the post-war boom in tourism. Fuji's lakes and foothills were the frontline for development, but in 1935, the first of several blueprints for a funicular railway to ascend the flanks of Fuji was proposed. Although rejected by the Department of the Interior on conservation grounds, the economic potential of access improvements was evident. *Oshi* houses and trailhead towns were increasingly being bypassed by day-travellers from Tokyo and other urban hubs. Private profit, together with 'distance demolishing technologies', dragged Fuji's mountain communities into the mix of modernization in a precursor to the 'Zomia Highlands' described by Scott (2009) across upland Southeast Asia a half-century later.

(a) (b) (c) (d)

Figure 18.1 Images at Mount Fuji.

Mass tourism: blurring the boundary of peak-bagging and pilgrims

During the post-war period of extended economic growth in Japan, numbers of national park visitors – and Fuji climbers – began to accelerate in earnest. Increased car ownership and demand for improved access infrastructure saw construction of many toll roads inside protected areas (Oyadomari, 1989). On Fuji's north flank, a bus service was extended up to the mid-section, reaching the 5th Station (2,300 meters above sea level) on the Yamanashi side in 1952 (Sayama & Nishida, 2001). On the occasion of the Tokyo Olympics in 1964, the Subaru line toll road opened up a 30-km paved dual carriageway to transport vehicles up to the Yoshida 5th station. Automobile tourism and consecutive feats of civil engineering radically reduced the travel time from urban hubs. One example was the Misaka tunnel toll road, which opened in 1967. By 1969, the opening of the Chuo Highway linked Chofu in West Tokyo to Kawaguchiko at Fuji's northern foot, further shortening the journey (Sato & Shibuya, 2015). Meanwhile completion of the Fuji Skyline road in Shizuoka in 1970 reduced access time for the Chubu and Kinki regions to the southwest.

The convenience of this road network, which extends almost two-thirds of the way up Fuji's flanks, and the ease of access from urban hubs led to a progressive rise in visitation. There were around 100,000 summer climbers on the Yoshida trail in 1981.[3] The figure subsequently doubled, exceeding 200,000 in 2008,[4] and since then the four trails have maintained a combined total[5] of approximately 300,000 climbers per season, although this figure is under-reported by up to 30% (Jones, Yang & Yamamoto, 2018).

Each of the four main summit trails now has a 5th station trailhead that ranges in altitude from Gotemba (approx. 1400m ASL) to Fujinomiya (2400m). Yoshida has the second highest trailhead (2305m ASL), from where the ascent is just 1450 vertical meters to the summit. Due to its trailhead elevation and proximity to Tokyo, Yoshida is the most popular of the four routes, accounting for 60% of all summit ascents in 2015. This included many young, first-time climbers without guides, as well as an increasing number of foreigners.

The exact proportion of international climbers is not known. According to sampling by visual observation on the Yoshida Route, foreigners accounted for just 5% to 7% of climbers in August 2009 (Jones et al., 2016). But by the 2015 season, the proportion had risen to 20% (weekends) and 30% (weekdays) based on appearance and language. Multiplying this estimated 20–30% market share by the total climber population during the summer season suggests that some 60–90,000 foreigners were summiting Fuji each summer prior to the pandemic (Jones & Yamamoto, 2016). However, this figure is likely to be an underestimate, due to difficulties in distinguishing between different nationalities for logistical reasons, such as the lack of an official registration or permit system, and ethical ones, such as avoiding potential invasions of privacy by directly interrogating climbers on their nationality.

In an echo of pilgrims past, climbers depart huts before dawn or ascend directly overnight in time to reach the summit for sunrise. But the peak of modern Mt. Fuji is a much more congested and commercial environment, especially on weekends and public holidays. Saturday nights are the most crowded with a high of 6831 climbers recorded on the Yoshida trail on 25 August 2012. Most descending climbers return promptly to Tokyo without staying another night to explore other heritage sites or contribute to the regional economy (Yamamoto, Jones & Aramaki, 2014). However, the climbers do leave behind significant environmental impacts that places pressure on visitor facilities and flashpoints such as trails and toilets (Aoki, 2008; Apollo, 2017b).

In short, modern access arrangements have transformed the logistics of summiting Fuji from a week-long pilgrimage to an overnight adventure. Today, many climbers eschew a stay in a mountain hut altogether, opting to set out from the 5th station trailhead and complete the 'bullet climb' to the summit as fast as possible. Most climbers aim to reach the summit via round-trips that last less than two days in total, including the return journey from Tokyo and a median 16–19 hours spent on the climb itself (Jones & Yamamoto 2016). Three in four climbers do not undertake any special training prior to climbing, in terms of either physical or spiritual preparation. Meanwhile the ascetic practises that once formed the core of Shugendō belief are passed down as traditional rites, if at all. The trailheads have moved higher up the mountain and less than 5% of climbers visit the shrine at the 5th station before beginning their hike up to the summit (Figures 18.2 and 18.3).

Study methods

Our mixed-method research approach combined environmental history, ethnography and on-site visitor surveys to intercept climbers. The ethnographic study used techniques of participant observation and interviews that concluded with the recording of a focus group later translated into English. The re-enacted pilgrimage began at Nihonbashi bridge near the centre of Tokyo, proceeding due west from the metropolis via suburban post-towns to Fuji's foothills in rural Yamanashi Prefecture. The ethnographic notes, based on five days of observations and periodic interviews over the duration of the walk, were transcribed and translated from Japanese to further contextualise the behaviour and the experience of pilgrims and contrast it with the secular equivalent of contemporary Fuji climbers.

Our visitor survey has monitored NBT at Mt. Fuji over ten years from 2008 to 2017. A report published the results of intercept surveys conducted near the 5th station to collect responses from climbers descending from Fuji's summit (Yamamoto, Jones & Aramaki, 2014). In this research, we reported demographic profile, climbing behaviour and experience were captured using a multilingual questionnaire and compared with those of participants on a

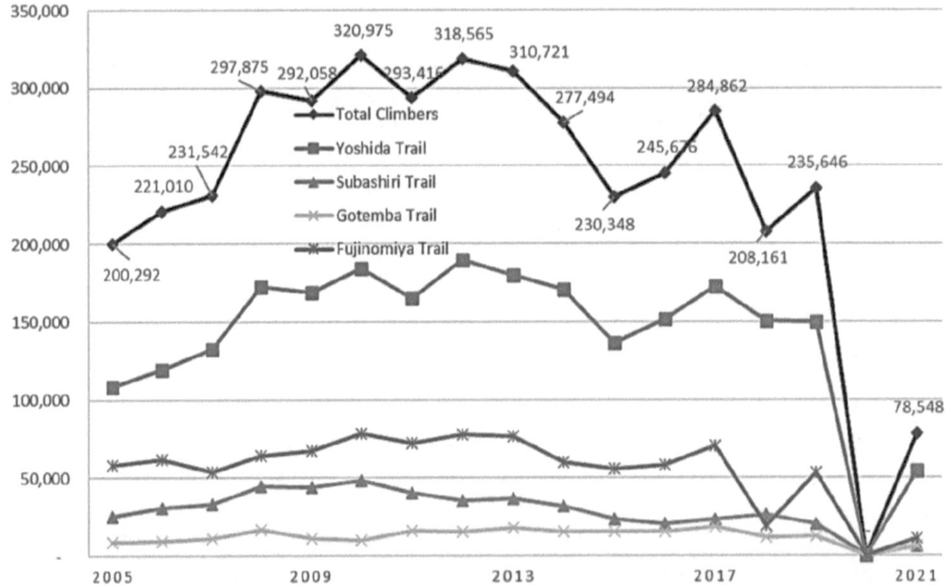

Figure 18.2 Number of climbers year wise.

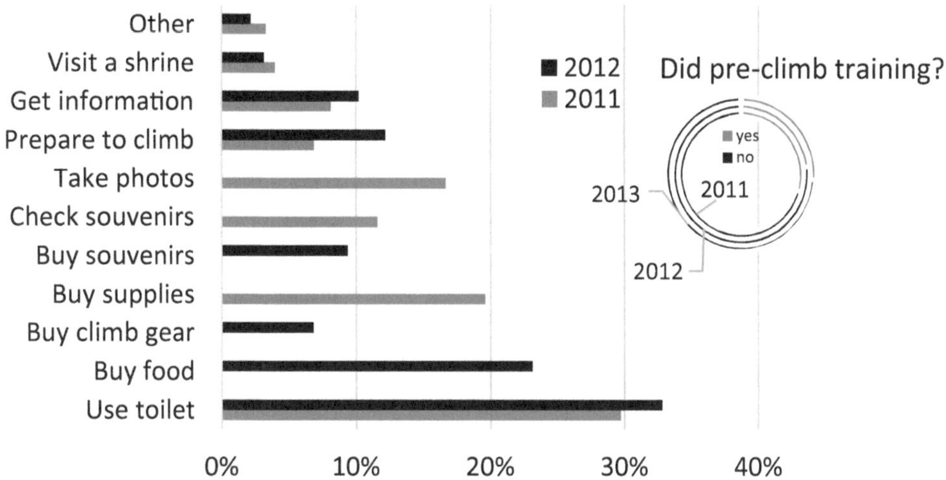

Figure 18.3 5th station activities (L) & per cent that did pre-climb training (R).

heritage tour in June 2015 that re-enacted an original pilgrimage from Edo (modern Tokyo) to the Sengen Shrine at Fuji's northern foot. The 120-km trek was conducted over four nights and five days, and timed to coincide with the official start of Fuji's summer season.

Ethnography study: re-enacting the pilgrim's trail

The first author undertook an ethnographic study in 2015 to better understand the pilgrim trail from Tokyo to Mt. Fuji, rediscover the heritage that has been better conserved and reimagine which resources could be used to revitalize the contemporary climber experience. The re-enacted 120-km hike was timed to coincide with the official 'mountain-opening' ceremony held at the Fuji Sengen shrine on June 30th, the start of the summer climbing season. Our group had 24 participants, of whom over half were local residents from Yamanashi Prefecture. Most were aged in their 60s or 70s, with one gentleman aged 80, and the majority were repeaters who had joined the heritage tour before. The four-night, five-day pilgrimage was first organized in 1999 by the Chamber of Trade and Commerce of the aforementioned Fuji-Yoshida City which originally developed as an *oshi* base camp. Other partners included local financial institutions, government agencies and conservations NGOs. We followed in the pilgrims' footsteps by walking along one of the original *go-kaidō* – the five principal trade routes of pre-modern Japan that radiate outward from the Nihonbashi terminus in central Tokyo like spokes on a bicycle wheel. In order to cement their control over the local feudal lords via a system called *sankin kotai*,[6] the Edo Bakufu government established post towns at strategic locations along these five main roads. Our itinerary followed the authentic pilgrims' selection of overnight rest points in *shuku* post towns. These were the pre-modern equivalent of highway service stations, offering services that once ranged from food, drink and accommodation, through to entertainment and red-light districts.

Our pilgrimage began at Teppozu inari Shrine in Tokyo, where a 'meet and greet' session was conducted to familiarize walking companions for the next five days. After a brief ceremony we set off for Nihonbashi, the official starting point of the *kaidō* and the pilgrimage. The route passed the Imperial Palace before traversing Shinjuku, a busy commercial centre and transport hub (Figure 18.4). Our pilgrimage garb, including conical bamboo hat, white robe

Figure 18.4 Pilgrimage tour from Tokyo to Mt. Fuji. Photographs by the author.

and wooden staff, contrasted with the dark-suited morning commuters on their way to work. The pilgrimage continued westward along the National Route 20, tracing the historical path of Koshu Kaidō. We observed how *shuku* post towns such as Fuchu in the Western Tokyo suburbs still existed in name, but had been completely rebuilt. Others further out from the urban centre retained more of their ambience and characteristic wooden architecture, albeit with some modern additions.

Accommodation for the first night was a business hotel in Fuchu. The second day began with a warm-up at a local shrine in Fuchu. The calisthenic exercises were directed by the *sendatsu* guide who led our group throughout the five days accompanied by a logistical support team that forwarded our luggage and provided periodic food and drink stops during the daily hikes. Of the 45 post towns that remain along the Koshu Kaidō, the inn at Hino is one of only three Edo-era buildings that are still standing.[7] As we reached the western extremes of Tokyo, the path skirted the base of Mt. Takao, another former Shugendō site that is now known for recreation and trail running. As the route relaxed into a more rural landscape, local residents began to appear at their doorways to watch the band of modern pilgrims and wish us well on our journey. This opportunity to mingle with locals and meet community members was an unexpected benefit of the slower pace of travel. We ascended a mountain trail and traversed over the Kobotoke Pass, chanting a traditional mountain song to lift the spirits as the ascent grew steeper. This chant became a familiar over the course of the five days, and was sung whenever the group morale was ebbing, including sections of steep climbs or adverse terrain. The lyrics loosely translate to remind climbers that the power of mind and spirit can overcome the temporal discomforts of the flesh, while also including a prayer for fine weather on the mountain.

The fourth day included the final long-distance section of the journey (25km). After the morning warm-up, the route led us to Katsura-gawa, a river that pilgrims used to purify themselves in with a mixture of ablutions and prayer. The further out from Tokyo that we travelled, the more well-preserved period architecture we found. Saru-hashi Bridge in Ōtsuki, for example, is a distinctive timber-framed structure held together without the use of nails. Its design is said to have been inspired by watching a troop of monkeys (*saru*) holding onto each other's legs to form an animal ladder for the troop's leader to cross the river. Along with local legends and architectural heritage, we also encountered folklore such as a traditional picture story show (*kamishibai*), enjoyed with some *tsukemono* pickles apparently homemade by the elderly storyteller herself. Shortly after passing Otsuki City Hall, the pilgrim trail turned off the Route 20, the modern equivalent of the Koshu Kaidō which we had followed faithfully all the way from Tokyo. By now we were treated to ever more frequent glimpses of Mt. Fuji's peak in the distance, heightening the anticipation. Our path passed under an experimental stretch of linear maglev *shinkansen* track and Maiko, one of our enthusiastic team of local supporters, excitedly reminisced about a test ride she had once taken on the next-generation bullet train. At a rest stop soon after in a small park overlooking Tahara-no-taki waterfalls, we were met by a life-sized stone effigy of Matsuo Bashō (1644–1694), a renowned poet who had walked the length and breadth of the Japanese archipelago, composing *haiku* poems. We were reminded of Fuji's links with cultural heritage and literature, and a juxtaposition between ancient and modern aspects of heritage.

We ended day four of the walk in Nishikatsura, where we were greeted by officers from the local town hall and alumni members from the 1999 pilgrimage. We swapped stories over a meal of Yamanashi *hōtō* noodles and rice wine, and had our white robes (*happi*) stamped with an authentic wood block by one of the descents of the same *oshi* innkeepers who had welcomed groups of pilgrims from Edo in the pre-modern era. On the last day, our group set out for Sengen shrine, the pilgrimage's terminus. For this final leg of the marathon we were greeted by many supportive local residents who lined the way, cheering us on with cries of encouragement and refreshments including hand-rolled *udon* noodles. With Mt. Fuji's peak now in full view, we passed many stone route markers that confirmed we were following in the footsteps of the Fuji-*ko*. Finally, a throng of priests, revelers and local residents gathered at Kanatorii Park for a speech from the mayor of Fujiyoshida, before making a procession up Kanaidorii. After five days we found ourselves in the dense, cedar groves that surround the Fuji Sengen Shrine (850m above sea level.[8] After being ceremoniously congratulated for successfully completing our journey, we bid farewell to our group. Some of the members, including the venerable 80-year-old Tomio, planned to carry on the climb up to Fuji's peak the next day, thereby completing the authentic pilgrimage route. Other visual evidence of the worship-ascent heritage was evident at the shrine, where a costumed 'deity' cut the rope to open the Kanatorii Gateway. This was the opening ceremony to pray for a safe and prosperous climbing season in the summer months to follow.

Turning peak-baggers back into pilgrims?

The next section compares and contrasts the re-enacted pilgrimage tour toward a more sustainable re-design of current climbing experience at Mt. Fuji. In pre-modern Japan, pilgrims approached on foot from everyday locales, especially Tokyo, conducting ascetic practices to seek spiritual enlightenment. This religious heritage was one of the twin criteria that underpinned Fujisan's 2013 listing as a UNESCO world cultural heritage site. However, modern access arrangements have transformed the logistics of summiting Fuji from a week-long pilgrimage to an overnight expedition.[9] The Fuji-*ko* ideology was already in decline due to modernization, and the sacred dimension has disappeared almost entirely except for certain festival and religious rites carried on as occasional events. For example, few of today's 300,000 annual climbers start out from the trailhead shrine at the foot of the mountain, nearly all drive up the Subaru Line by car or bus to the 5th station trailhead (2300m). Less than 5% of climbers visit the shrine at the 5th station (Figure 18.3), and even fewer undertake any *harae*[10] purification rituals. According to Yamamoto, Jones and Aramaki (2014), less than half of all contemporary climbers perform a *hachi-meguri* summit loop of the crater.[11]

The Fuji-*ko* of yesteryear selected a smaller number of climbers who had ample opportunities to acclimatize during their pilgrimage, which usually lasted for a week or more. After the journey to Fuji, they stayed overnight at the foot before beginning an ascent from the base punctuated by numerous breaks en route to pray and perform ablutions. This search for speed is one of the most notable changes in the climbing style. Almost all contemporary climbers depart from the 5th station trailhead (approx. 2300m ASL) at the terminus of one of four toll roads, predominantly the Subaru Line.[12] A considerable proportion of climbers ascend overnight from the 5th station without staying in a mountain hut. In addition to the four main ascent trails, the Ochudō Loop trail girdles the mountain's midriff. This 19km path was once used for ascetic training, and access was only allowed to pilgrims that had reached Fuji's summit on at least three previous occasions. Today, several sections of the trail are impassable for topographical reasons as it skirts the Hoei crater and crosses the Osawa ravine.[13] Nonetheless, a surviving section that traverses horizontally from the parking lot at the 5th station could be reinvigorated to provide better interpretation of natural and cultural heritage for casual tourists. UNESCO World Heritage Center documents acknowledge these threats to visitor experience, noting that the pressure brought by numbers and alteration of the landscape to cater for the demands of tourism affects both the integrity and the authenticity of the property (UNESCO, 2021).

Table 18.1 Date, itinerary & approximate distance walked per day on tour

Date	*Itinerary & approximate distance walked per day*	
26-Jun-15	Teppobashi Nihonbashi Shinjuku Takaido Chofu Fuchu (o/night)	30 km
27-Jun-15	Fuchu Hino Hachioji Takao Kobotoke Gohonmatsu (o/night)	30 km
28-Jun-15	Gohonmatsu Yose Uenohara Tsurugawa Ohno Inume (o/night)	25 km
29-Jun-15	Inume Sarubashi Otsuki Nodajuku Tanimura Nishikatsura (o/night)	25 km
30-Jun-15	Nishikatsura Shimoyoshida Kanaidorii Fuji Sengen shrine	10 km

Nonetheless, UNESCO and other international organizations could try harder to 'nudge' site managers toward more sustainable climber experiences, for example by encouraging them to climb from the foot of the mountain.

In order to do so, targeted marketing is necessary to better understand current climbers. Table 18.2 reports the findings from 773 questionnaires collected at Fuji's 5th station in 2016 (Jones & Bui, 2018). International respondents accounted for 68%, and male climbers for 62% of Fuji climbers surveyed during their descent from the summit. The median age was relatively young, with only 13% aged over 41 years old. 83% were climbing Fuji for the first time, reflecting the volcano's attractive status among Japan's mountains for younger, inexperienced and first-time climbers (Jones & Yamamoto, 2016). Such demographics are distinct from other peaks and the contemporary Fuji climbing style also bears scant resemblance to the 'worship-ascent' of yesteryear. Today, hut reservations tend to fill up from the top down, in an attempt to beat the crowds and be the first to see sunrise from the summit. Moreover, the lack of a guide entails safety and environmental management issues.

Apart from climber etiquette, Fuji's 5th station on the Yamanashi side attracts an estimated three million annual visitors. One historical legacy is that Fuji, unlike other sacred sites and mountain peaks that were off-limits to all except social elites or religious elders, has a long history of pilgrimage and 'worship ascent' by lay persons and citizens. The majority of visits remain clustered in the peak summer season that runs from July to mid-September. However, the climber profile has also diversified in a positive manner, with female and foreign climbers permitted and gradually on the increase. Reflecting on Fuji's legacy of worship ascent offers not only a record of past pilgrimages but also a means to reimagine current climber experience along a more sustainable trajectory. This chapter outlined the pre- and post-war infrastructure overhaul that made Fuji one of the world's most accessible and arguably one of the most 'overused' peaks. Survey results show that few of the contemporary climbers recognize the sacred rites of their pilgrim forefathers, seeking to summit as fast as possible without following Shugendō customs such as pre-hike ablutions or a shrine visit. One potential counter-strategy is proposed via observation and interviews with participants that re-enacted one of the original pilgrimage route from Tokyo to Fuji's northern foot.

Table 18.2 Climbers' demographic profile (N = 773)

Sample	
International climbers	495
Domestic climbers	238
Gender	
Male	61.8%
Female	38.2%
Age group	
<20	8.2%
20 – 30	54.6%
31–40	24.4%
41–50	12.6%
>51	0.3%
Previous experience	
First-time climbers	83.0%
Repeat climbers	17.0%

Adapted from Jones and Bui (2018)

Local municipalities, including Fuji-Yoshida city, are now seeking to restore some of the routes used to climb Mt. Fuji. Since 2011, a movement has been underway to encourage more climbers to ascend gradually from the foot of the mountain without driving up to the 5th station. The number of hikers is still small compared to the overall climber population.[14] However, the initiative has the potential to diffuse some of the pressure from spatial and temporal congestion at the 5th station during the summer peak. In addition, climbing from the foot of the mountain enables climbers to enjoy cultural heritage sites such as shrines and temples like the ones experienced on our Edo-Fuji walking tour. Climbing up through the lower levels of the mountain also provides opportunities to connect with some old growth woodland which is missed by most climbers that start from the 5th station that is already located on or above the timber line. Other natural heritage includes over 100 lava caves dotted around the base of the mountain, providing habitat for local wildlife, such as meadow voles, foxes, serows and black bears. For Fuji to reclaim its legacy as an NBT destination, better interpretation of such natural and cultural heritage is required to modify the mass tourism market and steer the current 'peak-baggers' back towards the authentic pilgrim roots.

Conclusion

This chapter investigated the pilgrimage spectrum from an NBT experience perspective that contains vestiges of both a 'sacred' and a 'secular' journey (Graburn, 1983). Fuji's pre-modern style of worship-ascent involved a lengthy approach hike and gradual ascent. Conducting ascetic practices symbolized the search for spiritual enlightenment, thereby 're-creating' mind and body, and purifying the pilgrims' soul for everyday life. Climbers comprised 'confraternities' known as Fuji-*ko*, divided into a national network of sects centred on Edo. However, thanks to 'distance-demolishing technologies,' a rapid reduction in climbing time from the 5th station trailhead has turned Fuji's summit into an overnight trip from Tokyo. Today's NBT experience consists of self-guided 'bullet-climbs' undertaken in round-trips lasting less than 48 hours, but a pilot project is underway to encourage climbers to start from the base of the mountain, via an ancient Shinto shrine. Setting out from this lower trailhead, instead of the 5th station, would enable more gradual acclimatization to altitude gain and filter out less-committed climbers, promoting a more sustainable and authentic experience.

As a cultural World Heritage Site, Fuji is in urgent need of such holistic planning and conservation measures to mitigate the footprint of mass tourism and cultivate experiences more meaningful than mere 'peak-bagging.' Commodification of mountains turns destinations into 'honeypots' (Pickering & Buckley, 2003), yet the more susceptible such sensitive environments are to visitor impacts, the harder they are to restore (Eagles, 2002). Via active involvement in the re-created pilgrimage, along with interviews with climbers, guides and pilgrims, this chapter proposed a more proactive approach to visitor management that could help re-design a more sustainable climbing experience for Fuji's future. A better understanding of the demographic profile of emerging segments is indispensable along with a reciprocal need for targeted management to modify the experience of 'peak-baggers' that select this summit as a readily accessible, nontechnical 'trophy.'

Fuji could be a precursor for other major mountain destinations, not least since its proximity to the mega-cities results in visitation levels among the planet's highest. Beyond its accessibility from Tokyo, Fuji's iconic shape and cultural legacy ensure it stands out as a 'must-see' NBT destination on any itinerary to Japan. Few who have visited during the peak periods would doubt that Fuji-Hakone-Izu is among the world's most visited national parks. Designing a more sustainable visitor experience here would thus have implications for other sacred sites that suffer from 'peak bagging.' Congestion at Fuji is usually exacerbated by the clustering of summer holidaymakers, especially during the official climbing season from July to mid-September. However, the 2020 summer season was closed entirely, and 2021 regulated to mitigate the risk of a climber-related coronavirus cluster. Albeit a desperate decision for local businesses, the pandemic could thus provide an unprecedented opportunity to re-design the Fuji climbing experience along a more sustainable trajectory. Local communities that once thrived as base camps for *oshi* (guides) are now seeking to restore more trails and encourage more climbers to start from lower down the mountain for a more authentic experience that follows more closely in the footsteps of the pilgrim predecessors.

Notes

1 Although the first recorded female climb was in 1832, women were not officially allowed to climb until 1872 (Fumiko, 2005).
2 The Izu island chain were added on to the national park in 1950.
3 A manual counting system at the 6th station was introduced in 1981 along with a 'Safety Centre' after 12 climbers were killed and 29 injured by a large-scale rock slide on 14 August 1980.
4 The year in which Fuji was placed on the tentative list for inscription as UNESCO cultural heritage.
5 As there is no system of climber registration, this data is estimated from visual monitoring by police (6th station) and infrared trail counters (8th station). It includes some double counts of staff, guides and repeat climbers etc.
6 Each lord was required to rotate alternate years between his home fief and Edo, while his family were held hostage in the capital (www.nakasendoway.com/sankin-kotai-and-the-hostage-system/).
7 Although Hino's Honjin (first-class inn) was reconstructed in 1864 after a fire in 1849.
8 Also known as the *Kitaguchi Hongu Sengen Jinja*, this was the trailhead of the Yoshida climbing route in the Edo era where the pilgrims came to pray before starting their final ascent.
9 The round-trip climbing distance (approx. 14 km from the Yoshida 5th station), graded trails and switchbacks through soft volcanic scoria makes for an easily accessible, non-technical climb.
10 One of four essential elements involved in a Shinto ceremony to purge 'sins' including bad luck, disease or guilt.
11 700 m in diameter and 250 m deep, at its highest point it rises 3776 m above sea level.
12 A 30-km-long, twin-lane, paved toll road that provides access from the north.
13 Osawa failure is one of the largest scale collapse areas in Japan. It exists on the west slope of Mt. Fuji between the summit and 2,200 meters, 2.1 km in length, 500 m maximum width, 150 m maximum depth, 1 km^2 collapsed area, and about 75,000,000 km^3 of collapsed soil. www.cbr.mlit.go.jp/fujisabo/en/bosai/bosaikatudo/bosairekisi-genjyou.html
14 For example, 2011 had an estimated 10,171 and 2012 had 12,706 climbers.

References

Aoki, N. (2008). Tackling illegal dumping and human waste problems on Mount Fuji: Situation before and after World Heritage Site designation. *Material Cycles and Waste Management Research 26*(3), 207–214.

Apollo, M. (2017a). The true accessibility of mountaineering: The case of the High Himalaya. *Journal of Outdoor Recreation and Tourism, 17*, 29–43.

Apollo, M. (2017b). The good, the bad and the ugly–three approaches to management of human waste in a high-mountain environment. *International Journal of Environmental Studies, 74*(1), 129–158.

Apollo, M. (2021). *Environmental Impacts of Mountaineering: A Conceptual Framework*. Cham: Springer.

Apollo, M., Andreychouk, V. (2022). *Mountaineering Adventure Tourism and Local Communities: Social, Environmental and Economic Interactions*. Cheltenham: Edward Elgar.

Apollo, M., Andreychouk, V., Mostowska, J., Jin, Z., Maciuk, K., Rettinger, R., & Jones, T. E. (2021). Himalayan nature-based tourism. Potential, regional diversity, nature conservation and touristic load. *International Journal of Conservation Science, 12*(4), 1527–1546.

Apollo, M., & Wengel, Y. (2022). *Mountaineering Tourism: A Critical Perspective*. New York: Routledge.

Beedie, P., & Hudson, S. (2003). Emergence of mountain-based adventure tourism. *Annals of Tourism Research, 30*(3), 625–643.

Bernstein, A. (2008). Whose Fuji?: Religion, region, and state in the fight for a national symbol. *Monumenta Nipponica, 63*(1), 51–99.

Blake, K. S. (2002). Colorado fourteeners and the nature of place identity. *Geographical Review, 92*(2), 155–179.

Bui, H. T., Jones, T. E., & Apollo, M. (2021). Reflections for trans-regional mountain tourism. In T.E. Jones, M. Apollo, & H. T. Bui (Eds.), *Nature-Based Tourism in Asia's Mountainous Protected Areas* (pp. 293–316). Cham: Springer.

Chakraborty, A., & Jones, T. E. (2018). Mount Fuji: The volcano, the heritage, and the mountain. In Chakraborty, A., Mokudai, K., Cooper, M., M. Watanabe, & S. Chakraborty (Eds.), *Natural Heritage of Japan* (pp. 167–175). Cham: Springer.

Choe, J., & Hitchcock, M. (2018). 15 Pilgrimage to Mount Bromo. *Religious Pilgrimage Routes and Trails: Sustainable Development and Management, 180*, 180–195.

Cochrane, J. (2009). Spirits, nature and pilgrimage: The "other" dimension in Javanese domestic tourism. *Journal of Management, Spirituality and Religion, 6*(2), 107–119.

Di Giovine, M. A. (2011). Pilgrimage: Communitas and contestation, unity and difference-An introduction. *Turizam: međunarodni znanstveno-stručni časopis, 59*(3), 247–269.

Eagles, P. F. J. (2002). Trends in park tourism: Economics, finance & management. *Sustainable Tourism 10*(2), 132–153.

Ewert, A. W. (1994). Playing the edge: Motivation and risk taking in a high-altitude wilderness like environment. *Environment and Behavior, 26*(1), 3–24.

Fumiko, M. (2005). Female pilgrims and Mt. Fuji: Changing perspectives on the exclusion of women. *Monumenta Nipponica, 60*(3), 339–391.

Gawlik, A., Apollo, M., Andreychouk, V., & Wengel, Y. (2022). Pilgrimage tourism to sacred places of the high Himalaya. In Nyaupane, G. & Timothy, D. (Eds.), *Tourism and Sustainable Development in the Himalayas: Social, Environmental, and Economic Encounters*. New York: Routledge.

Graburn, N. H. H. (1983). The Anthropology of Tourism. *Annals of Tourism Research, 10*(1), 9–33.

Hirano, E. (1987). *Fujisan Sengen Shinko* (Fujisan Sengen Faith). Tokyo: Yuzankaku.

Hung Lee, T., Jan, F. H., & Lin, Y. H. (2021). How authentic experience affects traditional religious tourism development: Evidence from the Dajia Mazu Pilgrimage, Taiwan. *Journal of Travel Research, 60*(5), 1140–1157.

Ito, T. (2009). Trail and climber management and its cost recovery from the climbers at Mt. Fuji in Edo Period. *Journal of the Japanese Forest Society, 91*(2), 125–135.

Jernsand, E.M., Kraff, H. & Mossberg, L. (2015) Tourism experience innovation through design. *Scandinavian Journal of Hospitality and Tourism, 15*(sup1), 98–119.

Jones, T. E., Apollo, M., & Bui, H. T. (2021). Mountainous protected areas & nature-based tourism in Asia. In T. E. Jones, M. Apollo, & H. T. Bui (Eds.), *Nature-based tourism in Asia's mountainous protected areas* (pp. 3–25). Cham: Springer. https://www.amazon.com/Nature-Based-Tourism-Asias-Mountainous-Protected/dp/3030768325

Jones, T. E., & Bui, H. T. (2018). Comparing international and domestic climber profiles, motivation and the influence of world heritage site status at mount Fuji and mount Kinabalu. *Journal of Environmental Information Science, 2018*(1), 67–72.

Jones, T.E. & Yamamoto, K. (2016). Visitor perspectives of UNESCO World Heritage Site branding: A comparative study of domestic and international climbers at Mount Fuji. *Proceedings of JITR Annual Conference, 31*, 77–80, 2016–12.

Jones, T. E., Yang, Y., & Yamamoto, K. (2018). Comparing automated and manual visitor monitoring methods: Integrating parallel datasets on mount Fuji's North Face. *Journal of Park & Recreation Administration, 36*(1), 22–38.

Kanda, K. (2012). The selection process of National Park landscape areas and the imaginative geographies in Taiwan during the Japanese colonial period. *Academic World of Tourism Studies, 1*, 77–87.

Kato, M. (2008). *National park system of Japan, national park & protected area management series* (Vol. III). Tokyo: Kokon Shoin.

Kedrowski, J. J. (2009). Determining the relative annual mountain climbing frequency on Colorado's 14,000-foot peaks. *Mountain Research and Development, 29*(1), 82–92.

Koizumi, T. (2001). *The Birth of Climbing: Why Man Climbs Mountains*. Tokyo: Chuokoron Press.

Koizumi, T., & Chakraborty, A. (2016). Geoecotourism and environmental conservation education: Insights from Japan. *GeoJournal, 81*(5), 737–750.

Kornicki, P.F. & McMullen, I. J. (1996). *Religion in Japan: Arrows to Heaven and Earth*. Cambridge: Cambridge University Press.

Kupper, P. (2009). A commonwealth of alpine nature: The Swiss National Park. In *4th Symposium of the Hohe Tauern National Park for Research in Protected Areas*. Nationalpark Hohe Tauern; ALPARC.

Kuby, M.J., Wentz, E.A., Vogt, B.J. & Virden, R. (2001). Experiences in developing a tourism web site for hiking Arizona's highest summits and deepest canyons, *Tourism Geographies*, *3*(4), 454–473.

Manning, R. E. (Ed.). (2009). *Parks and people: Managing outdoor recreation at Acadia National Park*. USA: UPNE.

Maslow, A. H. (1954). *Motivation and Personality*. New York: Harper & Row Publishers Inc.

Matsui, K., & Uda, T. (2015). Tourism and religion in the Mount Fuji area in the pre-modern era. *Journal of Geography (Chigaku Zasshi)*, *124*, 895–915.

McNee, A. (2016). Resisting the new mountaineer. In A. McNee (Ed.), *The New Mountaineer in Late Victorian Britain* (pp. 73–107). Cham: Palgrave Macmillan.

Miyaji, N. (1988). History of younger Fuji volcano. *The Journal of the Geological Society of Japan*, *94*, 433–452.

MoE (Japanese Ministry of Environment) (2016). Report on Fuji climbers. Retrieved 15.2.18 at http://kanto.env.go.jp/to_2016/28_3.html

MoE (Japanese Ministry of Environment) (2021). *Shizenkoen riyou shasuu* (Numbers of visits to nature parks). Retrieved 27.11.2021 at http://www.env.go.jp/park/doc/data.html

Murakushi, N. (2003). The Establishment of Fujisan National Park (1), p. 28.

Murakushi, N. (2006). *Kokuritsu kouen no seiritsu-shi (The History of the National Parks' Establishment)*. Tokyo: Housei University.

Nyaupane, G. P., Timothy, D. J., & Poudel, S. (2015). Understanding tourists in religious destinations: A social distance perspective. *Tourism Management*, *48*, 343–353.

O'Gorman, K. D. (2010). *Introduction to the Origins of Hospitality and Tourism* (pp. vii–xii). Oxford: Goodfellow.

Oyadomari, M. (1989). The rise and fall of the nature conservation movement in Japan in relation to some cultural values. *Environmental Management*, *13*(1), 23–33.

Pickering, C. M., & Buckley, R. C. (2003). Swarming to the summit. *Mountain Research and Development*, *23*(3), 230–233.

Polidor, A. (2007). Sacred Site Reports. http://www.sacredland.org/mountfuji. Accessed 11 July 2014.

Roscoe, R. (2013). Volcanic cultures of risk: Photographing sites of memory. *Global Environment*, *6*(11), 216–227.

Sati, V. P. (2015). Pilgrimage tourism in mountain regions: Socio–economic and environmental implications in the Garhwal Himalaya. *South Asian Journal of Tourism and Heritage*, *8*(2), 164–182.

Sato, D., & Shibuya, K. (2015). Development and usage patterns of second-home areas at the foot of Mt. Fuji. *Journal of Geography Chigaku Zasshi*, *124*(6), 965–977.

Sayama, H. & Nishida, M. (2001). Transition of the beautification and cleaning activities in Mt. Fuji area after World War II. *Journal of the Japanese Institute of Landscape Architecture*, *64*(5) 485–488.

Scott, J. C. (2009). *The Art of Not Being Governed*. New Haven, Connecticut: Yale University Press.

Silva, A. D., & Bui, H. T. (2021). Collaborative management of protected areas in timor-leste: Stakeholder participation in community-based tourism in mount ramelau. In T. E. Jones, M. Apollo, & H. T. Bui (Eds.), *Nature-Based Tourism in Asia's Mountainous Protected Areas* (pp. 133–151). Cham: Springer.

Swarbrooke, J., Beard, C., Leckie, S., & Pomfret, G. (2003). *Adventure tourism: The new frontier*. London: Butter-Worth.

Timothy, D. J., & Nyaupane, G. P. (Eds.). (2009). *Cultural Heritage and Tourism in the Developing World: A Regional Perspective*. London: Routledge.

Tsuchiya, T. (1981) Tourism development through transport capital in the pre-war period; a case study of Fujikyu Company, *Journal of Forest Economics*, *100*, 30–35.

UNESCO (2021). Fujisan, sacred place and source of artistic inspiration. Retrieved 25.12.2021 at https://whc.unesco.org/en/list/1418/

Weaver, D. B. (2013). Protected area visitor willingness to participate in site enhancement activities. *Journal of Travel Research*, *52*(3), 377–391.

Yamamoto, K., Jones, T. E., & Aramaki, S. (Eds.). (2014). *Fuji tozan ni kansuru ankeeto chōsa hōkoku*. Yamanashi: Yamanashi-ken kankyō kagaku kenkyūjo.

Ziegler, A. D., Wasson, R. J., Sundriyal, Y., Srivastava, P., Sasges, G., Ramchunder, S. J., … & Apollo, M. (2021). A call for reducing tourism risk to environmental hazards in the Himalaya. *Environmental Hazards*, 1–28. https://doi.org/10.1080/17477891.2021.1984196

19

TOWARDS EXCLUSIVELY INCLUSIVE VISITOR EXPERIENCE

The case for equity in sustainable tourism

Agnes M. Nowaczek and Carolin Seiferth

Introduction

Immersive and transformative visitor experience is typically promoted by the sustainable tourism industry, yet true equity and application of related frameworks, such as the 17 United Nations Sustainable Development Goals (UNSDG), is still lacking in most travel companies (Gillovic and McIntosh, 2020). In an effort to expand the sustainability discussion and ground it in practical experience, the chapter authors present two detailed examples of Costa Rican sustainable industry leaders in the areas of transformative visitor experience (Senda Monteverde) and inclusion (Il Viaggio Travel). Anchored in the theoretical underpinnings of the Guests' Site-Adapted Sustainability Tour Experience Model (GSASTEM) (Seiferth, 2020), the authors further adapt this conceptual framework to serve as a potential management tool for considerations around issues of sustainable and equitable visitor experience. Based on the modified conceptual framework (see Figure 19.1), the authors apply relevant co-creation components based on Senda Monteverde's current sustainability tour, and physical aspects based on Il Viaggio Travel's leading Accessible Beach project as part of the Costa Rican Network of Accessible Tourism. Each case study is applied to a different portion of the framework, the main contribution being its conceptual extension and practical applicability to the tourism industry, while the main limitation is the narrow but detailed focus of these two examples. Moreover, the geographical proximity and philosophical ideology of these two companies offer an excellent opportunity for symbiotic co-learning and the exchange of ideas in an effort to deliver truly holistic sustainable and accessible experiences to all potential visitors.

Literature review

Important interdisciplinary intersections

Recognizing the interconnectedness and interdependence of various thematic areas – as is the nature with true sustainable philosophy – several authors (Fennell and Garrod, 2021; Scheyvens and Biddulph, 2018; Morris and Kazi, 2016; Portales, 2015) point to important areas that were historically absent from the debate. Beyond ecotourism and the crucial elements of programming and education, new areas come to the fore: accessibility, ethics and law,

 DOI: 10.4324/9781003230748-23

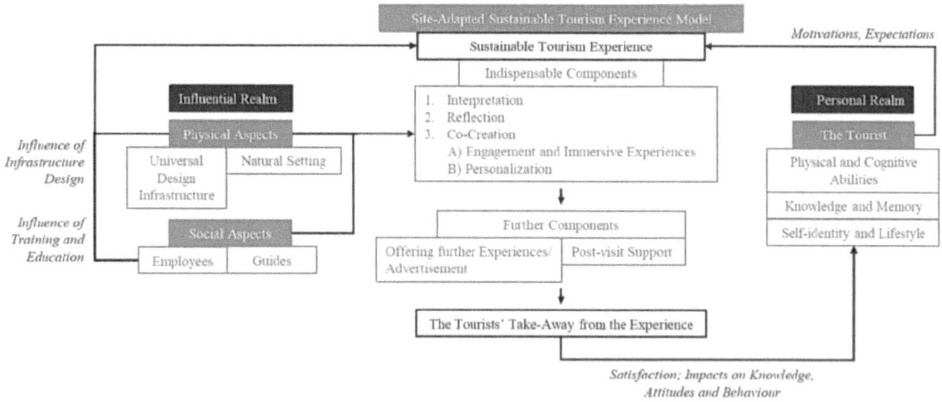

Figure 19.1 Site-Adapted Sustainable Tourism Experience Model (SASTEM).

(adapted from Seiferth, 2020).

universal design (i.e., whole-of-life approach), physical infrastructure, and even the extension to digital ecosystems. At the highest level, support for accessible tourism comes from the global directive, including the Convention on the Rights of Persons with Disabilities (CRPD) (United Nations, 2006) and the UNSDG (United Nations, 2015) – the International Accessible Tourism Destination Distinction being the most recent award introduced by the UNWTO. These initiatives have been promoted throughout Europe, visible in the robust European Network of Accessible Tourism (ENAT), whereas developing countries like Costa Rica are beginning to make strides in their own right. Currently, the Accessible Tourism Network has only two companies in this important category: Il Viaggio Travel and Red Costarricense de Turismo Accesible (see ENAT).

Consistent with these global directives, Fennell and Garrod (2021) emphasize important intersections between sustainability (and ecotourism as the pinnacle of sustainable tourism), accessibility, and ethics. The authors view responsibility as a consideration of moral duty at the intrinsic level, and as part of strategic business practice to remain competitive. Their recommendation for tourism providers is to adopt ways of emphasizing duty and justice instead of following accepted business practice. Similar to these authors, Portales (2015) employs ethics in the examination and evaluation of accessible tourism stakeholders (including the differently abled and aging populations), the necessary cooperation among them, and the issue of prejudice that permeates all other circumstances. Consequently, whether businesses are fearful of losses or certain of benefits, empathy (i.e., ethical approach) can be the basis of a profitable accessible tourism based on intergenerational standardization and government's commitment to cooperation between stakeholders (Portales, 2015). Moreover, in their definition of inclusive tourism, Scheyvens and Biddulph (2018) emphasize all forms of marginality and intersectionality grounded in the application of ethics, as well as opportunities for participation among marginalized groups.

Extending the debate to policy and law, Morris and Kazi (2016) propose an ethical and legal framework as the minimum regulatory standard of certification to address accessible tourism in Dubai. This is particularly important for countries which may rely on international legislation (i.e., the Americans with Disabilities Act) and self-directed ethical models which may be plagued with inconsistencies across various sectors, especially with regard to hospitality and tourism. Michopoulou et al. (2015) see the future of accessible tourism closely related to major policy initiatives on both accessibility and sustainability, with unique lessons on the diversity of tourism markets. Similar to other studies, the authors identify stakeholder collaboration as the key theme and crucial factor for the successful development of accessible tourism, where the market itself needs to play a central role.

Pertaining to physical infrastructure, Darcy et al. (2020) advance similar recommendations for integrating universal design principles, information systems, and assistive technology (e.g., smart cities), all based on wider definition and the participation of stakeholders. In the interpretation of physical spaces and the digital ecosystem (see Cassia et al., 2021; Darcy et al., 2020), Nowak Da Costa and Bielski (2018) discuss content, the availability of accessibility information, and improved communication with users. They frame mobility on a continuum from temporary to permanent constraints or conditions, where users can take advantage of smart data such as universal accessibility maps. Applicable to digital ecosystems, Gretzel et al. (2015) emphasize an increasing reliance of tourism destinations and tourists on the emerging forms of information and communication technologies (ICTs) integrated into physical infrastructure. This trend may present an opportunity for differently abled travelers to interact with smart destinations, smart business ecosystems, and smart experiences as the three basic components supported by layers of data creation, processing, and exchange.

Chosen accessibility terminology and background

To move away from 'disability' as a marginalizing and stigmatizing social construct, the authors draw on Leshota and Sefotho (2020), who proposed the term 'differently abled', thereby assigning transformative, anti-discriminatory, and empowering language properties. This holistic approach is in line with Portales (2015), who sees differently abled and aging sectors of the population as stakeholders of accessible tourism. Darcy and Dickson (2009) use the term 'whole-of-life approach' to advocate similar ideas encompassing access, disability, and ageing, especially as they estimate that 30 percent of the population will have access requirements at some stage during their life. These more recent approaches have been built upon a variety of historically and culturally grounded disability models (DMs). Exemplary in this case is the comprehensive research and review of DMs by Zajadacz (2015), which aims at outlining the most suitable course of action to direct future planning and the development of accessible tourism. Examining the evolution of various DMs (i.e., medical, social, geographical, economic, ethical), the author develops a 'biopsychosocial' DM that combines all approaches and extends this contribution to universal design (UD). Zajadacz (2015) sees the value of UD as lying in the elimination of physical barriers and social limitations, as well as respecting the needs of society at large (i.e., the whole-of-life approach) without differentiating individuals with special needs as a distinct market.

Similar to various DMs presented by Zajadacz (2015), Pereira (2009) offers critical and historically grounded analysis of the social image of people whose bodies are differently abled. In the investigation of social, educational, political, and therapeutic realms from ancient to current times, the author demonstrates the current trend in terminology that avoids placing an emphasis on the negative aspects. More recently, a study by Leshota and Sefotho (2020) explores the connotations of the term 'disability' as a marginalizing and stigmatizing social construct based on the anthropological, linguistic, and legal considerations. In the context of travel and tourism, Darcy and Dickson (2009) expand the definition of accessible tourism to include all dimensions of accessibility, including: an overview of the market, the built environment, accessible destination experiences, destination management processes, and contemporary approaches to accessible destinations. Likewise, Fennell and Garrod (2021) present an expanded definition of ecotourism based on a sample of Ecotourism Australia-certified company websites to examine inclusivity barriers based on the social model of disability (i.e., physical, attitudinal, and informational). Consequently, more studies point to the importance of participation by differently abled individuals not only as tourism consumers, but also as producers and participants in decision-making (Gillovic and McIntosh, 2020).

Site-adapted sustainable tourism experience model

In 2020, Seiferth developed the GSASTEM as a conceptual framework for designing and delivering customized sustainable visitor experiences on location. The model is based on an extensive literature review with a focus on the 'experience economy' and experience realms such as entertainment and education (Pine and Gilmore, 1999). Today's tourists show an increased demand for immersive experiences and connections to nature, places, and people (see Binkhorst and Dekker, 2009; Seeler and Schänzel, 2019), thereby leading to participation in transformative tourism offerings (Jantzen, 2013; Pine and Gilmore, 2011). To allow for the design of such transformative experiences, the model draws on Cutler and Carmichael's (2010) study into the dimensions of tourist experiences. Further, important components of experiences, such as quality interpretation (Hughes and Ballantyne, 2013), reflection (Boud, Keogh, and Walker, 1985), co-created experiences (Prahalad and Ramaswamy, 2004), and the personalization of experiences (Caldeira et al., n.d.) have been studied. To ensure the model's content validity, Seiferth (2020) adapted it to Senda Monteverde, Costa Rica to further improve the visitor experience during the sustainability tour. During the tour, local guides take guests around the property to showcase ethical practices and convey an image of a sustainable operator. Cayuga Collection, which manages the hotel (hereafter Cayuga), used the model to conceptualize similar experiences at other properties across their portfolio, thereby allowing their guests to experience, learn, connect, and relax (Cayuga Collection, 2021). Extending Seiferth's (2020) model to accessible tourism, the authors modified and renamed it as the Site-Adapted Sustainable Tourism Experience Model (SASTEM) (see Figure 19.1).

Connection to thematic dimensions in the model

In an effort to apply the SASTEM to accessible tourism, the authors sought thematic conceptualizations in literature, especially in the co-creation component, with a focus on the physical and social aspects of the influential realm (see Figure 19.1). Presenting an interesting overlap in terminology with immersive and sustainable tourism experiences in our model, Scheyvens and Biddulph (2018) define inclusive tourism as "transformative tourism in which marginalized groups are engaged in ethical production or consumption of tourism and the sharing of its benefits" (p. 592). Their definition encompasses all forms of marginality and intersectionality (see, for example, the framework by Kimberlé W. Crenshaw) and further reaffirms the component of co-creation in our model. Indeed, Zajadacz (2015) points to the need for a variety of DMs in the creation of accessible tourism supply due to the high degree of diversity among the differently abled visitors and the complexity of delivering optimal solutions. This view is supportive of the personalization component of co-creation in our model, where tourism experiences are approached and developed with clients and their specific needs.

In addition to the acknowledged barriers (i.e., cultural, architectural, and relational) typically found in the literature, Cassia et al. (2021) emphasize the issue of lacking coordination amongst the supply chain. This is especially true if operators wish to co-create tourism experiences for and with differently abled visitors – both with and without disabilities – who are the focal point and the active participants in the process. With focus on the physical aspects in the influential realm of our model, and drawing on the principles of UD, Cassia et al. (2021) advocate for tourism locations to be accessible to differently abled visitors. According to the authors, digital ecosystems (the result of a major shift from tourism products to tourism experiences) could contribute to making tourism locations more accessible by enabling information-sharing and coordination amongst the supply chain in the co-creation of tourism experiences. Using the tourism customer journey, all actors in the supply chain would work together to assure the tourist's safety and care via a digital tourism service network (e.g., the Costa Rican Network of Accessible Tourism). The authors see the digital engagement platforms as the first step in removing informational and relational barriers, such as the difficulty to verify information and being perceived as 'less-than'. They also see a contribution to the lessening of cultural barriers; social indifference on one level and entrepreneurial myopia at the business level, where only a few successful operators recognize the differently abled as a robust market (Cassia et al., 2021).

The discussion of digital ecosystems further supports the co-creation component in our model, particularly when it comes to understanding the necessary personalization of experiences and products, as well as the need for engagement amongst the supply chain. The ideas of co-creation and collaboration evident in our model are further supported by Darcy et al. (2020). Accordingly, "accessible tourism is a form of tourism that involves collaborative processes between stakeholders that enables people with access requirements, including mobility, vision, hearing, and cognitive dimensions of access, to function independently and with equity and dignity through the delivery of universally designed tourism products, services, and environments" (Darcy et al., 2020). This definition nicely captures concepts in the personal realm of our model, particularly when visitors consider their abilities and needs in relevance to their self-identity and lifestyle.

Applicable to the physical aspects and the co-creation component in our model, Gretzel et al. (2015) discuss smart business in the complex business ecosystem as an important element in creating and supporting the exchange of tourism resources and the co-creation of tourism experience. The authors define smart tourism as "tourism supported by integrated efforts at a destination to collect and aggregate/harness data derived from physical infrastructure, social connections, government/organizational sources, and human bodies/minds in combination with the use of advanced technologies to transform that data into on-site experiences and business value-propositions with a clear focus on efficiency, sustainability and experience enrichment" (p. 181).

Congruent with the physical aspects in the influential realm of our framework, Michopoulou et al. (2015) make a good case for UD to play a major role in enhancing the competitiveness of tourism destinations that would transform the future of accessible tourism for all. Unfortunately, there are still major gaps in accessibility standards across tourism products and services globally, particularly in developing and third-world countries (see Meskele et al., 2018 in the context of Ethiopia), which presents tremendous challenges for these destinations to remain competitive. In addition to the importance of developing accessible infrastructure (i.e., physical aspects in our model), Naniopoulos et al. (2016) also emphasize the applicable training of stakeholders (e.g., employees and guides) which is consistent with the social aspects found in the influential realm of our conceptual model.

Sustainable tourism experience

Visitors' sustainable tourism experience presents the central part of the SASTEM. Accordingly, the tourism experience consists of three indispensable components which contain both educational and entertainment dimensions. Guides often deliver quality interpretation – the first indispensable component – with the goal of impacting tourists' behaviours and attitudes by increasing their knowledge. They must ensure accessibility to information, which refers to tourists making sense of the experience, regardless of their abilities (Bindu and Devi, 2016). To illustrate, audio-descriptive tours which describe the visual content of the experience cater to visually impaired tourists while captions provided on tablets enable tourists suffering from hearing loss to participate equitably in the experience. During the sustainable tourism experience, guides can use examples of existing environmental problems at the destination to draw connections to global environmental challenges. As an example, guides stimulate emotions when communicating empathy for plants and animals to promote more responsible environmental attitudes (Berenguer, 2007). They further apply storytelling to increase tourists' awareness of holistic sustainability, hold their interest, and encourage reflection both throughout and after the experience.

Reflection, as the second indispensable component, relates to mindful learning which allows tourists to better appreciate and understand the content (Langer, 2000; Moscardo, 2009; Packer and Ballantyne, 2013). In the context of experiential and transformative learning, engaging tourists in critical reflection where one questions taken-for-granted beliefs and assumptions presents a powerful tool to identify reasons for changing one's behaviour (Mezirow, 2002). To exemplify, when guides provide time for critical self-reflection after they introduce tourists to environmental and social problems through storytelling, tourists might acknowledge their compliance in constituting these problems. The learning process could result in a desire to change current behaviour, thereby creating lasting and life-changing impacts of sustainable tourism experiences.

Co-creation as the third indispensable component entails (A) engagement and immersive experiences and (B) personalization. Engagement and immersive experiences refer to tourists' active involvement in experiential education, such as hands-on approaches to interact with their environment. Opportunities to observe, taste, smell, hear, and touch for sensory engagement offer possibilities to personalize the experience based on tourists' physical abilities. Visitors learn, discover, and explore the destination in an authentic manner, with the addition of touchpoints for social interaction within these 'sensory-scapes' catering to all senses (Efstathiou and Zippelius, 2019). Pine and Gilmore (1999) emphasize the importance of personalized experiences in the proposed 'experience economy'. Personalization allows for the creation of relationships through a process of social interaction. To create these touchpoints for differently abled tourists, experience providers need to understand tourists' personal realm to cater to physical abilities and deliver suitable content. Nyanjom, Boxall and Slaven (2018) suggest collaborations with differently abled tourists since they argue one cannot develop accessible travel without involving its main target market. While hospitality and tourism providers contribute service expertise, differently abled tourists can share experiential knowledge when designing new experiences. When delivering the experience, guides must consider individual needs to ensure inclusiveness, thereby further personalizing the experience.

Other components proposed in the model allow for flexibility since their incorporation depends on both the guide and the group dynamics. When group dynamics reflect a high degree of tourists' interest, the guide can offer additional experiences or post-visit support to extend relationship-building. When tourism providers fulfill visitors' expectations of such experience, tourists show a high degree of satisfaction and loyalty. Quality interpretation and reflection alongside tailored and personalized immersive experiences increase visitors' knowledge, allow for personal growth, and impact upon their attitudes towards holistic sustainability and pro-environmental behaviour. Consequently, tourists' take-aways from the experience will impact their personal realm, thus shaping their motivations and expectations of taking part in future sustainable tourism experiences.

Personal realm

Following Cutler and Carmichael's (2010) dimensions of the tourist experience, this model reflects the influence of personal and influential realms on tourists' experience. While many tourists see sustainable tourism experiences as opportunities to better understand themselves and the world (Kelly and Freysinger, 2000), operators need to understand tourists' key motivations and their expectations of specific experience. Since tourists' physical and cognitive abilities often tend to determine the experience in which they take part, experience providers who wish to provide sustainable and inclusive experiences need to ensure accessibility which might require them to personalize the tour for some individuals or the entire group. Consequently, such operators allow differently abled tourists to participate in the experience (Sisto et al., 2021), and, more importantly, to travel with independence, equity, and dignity (Darcy and Dickson, 2009). Furthermore, operators need to consider tourists' prior knowledge to deliver suitable educational content, for instance, CR's leadership in reforestation. Being familiar with participants' self-identity and lifestyle allows the guide to better understand how tourists might use the experience (e.g., tree planting) to enhance or change their perception and sense of identity with the goal to assist them throughout this process (Black, 2005; De Bernardi, Pashkewich and Wagner, n.d).

Influential realm

Physical aspects, such as the natural setting, shape the content of quality interpretation (Seiferth, 2020). Storytelling on environmental issues and specific local ecosystems stimulate tourists' reflection process (Barrett, 2006; McDrury and Alterio, 2002, 2003). UD infrastructure allows for inclusive co-creation and ultimately determines which activities guides can offer and easily incorporate into the sustainable tourism experience. UD opens the availability of products, services, and information with a low degree of physical effort in flexible, simple, and intuitive ways (Center for Universal Design, 2009). Consequently, there is no need for further infrastructure adaptation since UD caters to all tourists (Darcy and Dickson, 2009).

Social aspects refer to employees, and especially guides, who deliver and shape the experience. Since guides take on various roles, such as interpreters, cultural brokers (Black, 2005), or facilitators, well-designed training and education programs need to foster their self-reflective skills and cultural sensitivity, among others (Ap and Wong, 2001; Black and Weiler, 2013; Weiler and Ham, 2002). Polat and Hermans (2016) note the importance of training experience providers to cater to the needs of differently abled tourists to ensure the provision of truly sustainable and inclusive experiences.

Senda Monteverde and Il Viaggio Travel

The two Costa Rican travel companies Senda Monteverde and Il Viaggio Travel are the leaders in the sustainable tourism sector, and, as such, they were used to exemplify in detail the applicability of the modified conceptual framework and its further evolution. These companies serve as case studies applicable to different portions of the conceptual framework, based on the profile and focus of each company. Consequently, the relevant elements of the Sustainability Tour at Senda Monteverde (see also experiences at Cayuga) are applied to the personal and influential realm of the model, whereas the Accessible Beach project and accessible tours by Il Viaggio Travel are predominantly applied to the physical aspects in the influential realm of the model. Although these case studies are different, they complement each other in the conceptual areas of the model, which is another reason for their inclusion.

Cayuga Collection

As a management company founded by Andrea Bonilla and Hans Pfister in 1999, Cayuga operates eight independently owned ecolodges in Central America. Six of these lodges are located in Costa Rica, with the other two properties in Nicaragua and Panama. As part of Cayuga's corporate philosophy to create a symbiosis between luxury and sustainability (Cayuga Collection, 2021), all ecolodges implement a sound triple bottom line in their operations to account for environmental and social impacts next to traditional economic profitability (Stoddard, Pollard and Evans, 2012).

Senda Monteverde

Located nearby Monteverde Cloud Forest Reserve, Senda Monteverde became part of Cayuga in 2018. Offering 28 guest rooms, a bar and restaurant area, internal walking trails, and access to the neighbouring private Aguti Reserve, Senda Monteverde caters predominantly to travelers who want to immerse themselves in nature-based activities. Translating Cayuga's corporate philosophy into practice, Senda supports wildlife conservation (e.g., by planting over 5500 native plants in their gardens to attract local wildlife), exclusively employs local staff and guarantees social benefits and opportunities for professional development.

Senda Monteverde sustainability tour

Senda Monteverde's sustainability tour takes guests through gardens surrounding the hotel and back of the house. The tour combines experiential education with entertainment. An experienced guide usually leads the tour and introduces guests to Cayuga's sustainability philosophy by presenting the triple bottom line. Consequently, connected stories on planet, people, and prosperity allow guests to understand sustainability holistically. During different stops, guests learn more about the Monteverde Cloud Forest as a unique ecosystem, Senda's actions to protect the sensitive environment, various projects supporting surrounding communities, and collaborations with Monteverde's entrepreneurs to positively impact the local economy. During the tour, guides deliver environmental education through quality interpretation, for instance, explaining the effects of climate change on sensitive Cloud Forest ecosystems. To exemplify environmental problems at the destination, the guide uses an example of the extinct Golden Toad and communicates empathy when talking about various animals on the property as 'members of the Senda family'. Quality interpretation combined with storytelling and the use of emotions hold much potential to impact behaviours and attitudes of participants.

Tourists can engage in reflection and mindful learning throughout the tour. Connecting new knowledge to personal experience helps participants understand new content. To exemplify, the guide invites visitors to think about where the effects of climate change are visible in their home countries or how they perceive forests at home in comparison to the Cloud Forest. When visiting kitchen staff, participants reflect on where their food comes from and the guide shares how to decrease one's own environmental footprint by changing dietary habits. At the end of the tour, participants summarize new ideas about sustainability and share them with one another. The guide delivers experiential education through immersive, hands-on experiences. In diverse sensory-scapes, participants can observe plants and animals, they can plant seedlings, smell and taste herbs while collecting them for lunch, listen to birds, and sample homemade lemonade. The variety of social interactions between participants on the tour and various touchpoints with employees, such as the gardener or the kitchen staff, contribute to important relationship building and future loyalty.

At the beginning of each tour, it is important for the guide to assess the tourists' prior knowledge on sustainability, participation in similar experience, topics of interest, and expectations of the experience so that the content can be tailored according to the tourists' personal realm. Further, there are opportunities for one-on-one personalization, such as referring to participants by their first name and asking direct questions. During the tour, visitors can choose activities of interest to participate in, thereby further customizing their personal experience. Throughout the tour, the guide can offer additional experiences ranging from night walks in Aguti Reserve, to cooking and mixology classes at Senda. Post-visit support presents an opportunity to continue the discussion on sustainability issues. Guides can also initiate online information sharing, sending email reminders and building a virtual community on social media where tourists can share their sustainability initiatives at home to prolong relationship building (Hughes and Moscardo, 2019).

With the goal of turning the sustainability tour into a truly inclusive activity, Cayuga's experience managers should involve differently abled people to better design and deliver the experience in an inclusive way. The first step in reducing inequality is to increase access on each property (Nyanjom, Boxall and Slaven, 2018). Second, Cayuga can involve differently abled tourists, both prior and potential visitors, whose lived experience and knowledge can inform the design and delivery process of tourism products and services (Nyanjom, Boxall and Slaven, 2018). Third, Cayuga could also collaborate with accessible organisations, such as Il Viaggio Travel, to explore new perspectives about differently abled visitors by challenging negative perceptions and deficit-based attitudes (Beresford, 2016). By allowing all members of society to participate in sustainable tourism activities, these experiences hold much potential to initiate more meaningful change by connecting equity and inclusiveness with more holistic delivery of sustainability.

Personal realm of Senda's guests

In an interview conducted with Hans Pfister in 2020 (personal communication, May 16, 2020), the owner of Cayuga describes Senda Monteverde's guests as nature-based tourists with a slightly higher interest in sustainability compared to other target markets of Cayuga hotels. Guests tend to be more knowledgeable on environmental issues since their main activities include visiting preserves and participating in wildlife tours or birdwatching. While curiosity presents one motivation to take part in the sustainability tour, the short duration allows guests to easily incorporate this experience into their stay.

Influential realm of Senda

Located in proximity to Monteverde Cloud Forest Reserve, Senda's native plant gardens attract local wildlife. When remodeling and refurbishing the abandoned hotel on property into the current Senda Monteverde, visual pollution was minimized by using the natural landscape colour scheme. To guard against light pollution, big windows allow for the maximum use of natural light while ensuring cross-ventilation. On a professional level, Senda's employees participate in cross-training and career planning, whereas Cayuga further offers training sessions on service excellence, leadership, and emotional intelligence. Corporate managers share the business philosophy with employees who then bring the brand image to life.

Il Viaggio Travel

Il Viaggio Travel is a Costa Rican ground operator founded in 2005 by Stephanie Sheehy and Emilio Zúñiga, and specializing in custom-made travel experiences. They are pioneers in accessible travel and the first local travel operator to secure all government certifications at an elite level based on the principle of accessible tourism for all. The first contact with the operator's website with a wide range of accessibility tools (e.g., text size, grayscale, contrast, readable font) is a good indication of what clients can expect on their tours (Il Viaggio Travel, 2021b). Their 2020 Catalogue lists crucial partnerships in support of their market, including divisions specializing in family travel and accessible travel, as well as certified local guides and drivers. Further, for every trip booked by Il Viaggio Travel, the company finances inclusion and accessibility projects throughout Costa Rica.

Among the 14 different travel profiles and associated packages, Il Viaggio Travel lists: multi-generational bonding, special needs (with 'permanent' or 'temporary disability'), golden relax, blended family, single parent journey, and first encounter with children under four years of age (Il Viaggio Travel, 2021c). The broad range of demographic characteristics, life stages, and temporary health issues captures more realistically the gamut of visitors and their needs. In addition to the operator's 'classic' itineraries designed for specific ages and abilities, Il Viaggio also offers: (1) 'personalized' itineraries where visitors can modify activities to their particular needs and interests; and (2) 'custom' itineraries where visitors can work with local experts to craft unique experiences. In the 'accessible' package the focus is on ability: "Our goal is to create experiences that every family member on the trip – regardless of ability – will enjoy" (Il Viaggio Travel, 2021c). The operator offers a wide range of equipment available on all trips in this package: transportation with lifts and anchors, amphibian chairs, Joelette wheelchairs, bathroom chairs, shower chairs, room lifts, walkers, and manual wheelchairs.

As pioneers in accessible travel and family-friendly holidays, the operator has been featured in the media and has received numerous awards, e.g., Wheel the World, Travel Xperience, Responsible Travel, Turismo Social con Integridad, CATA award 2019, Costa Rica Incluye 2017 (Il Viaggio Travel, 2021c). In July 2019, Il Viaggio Travel was part of a select group of innovative organizations from around the world and their Accessible Beach project received funding from Booking.com in the category of sustainable tourism (Travel2latam, 2021). Most telling of the operator's success and effort is the fact that Costa Rica's government itself refers differently abled visitors to Il Viaggio Travel, which is the only operator in the country to provide accessible travel Lonely Planet, 2020). Costa Rica's steps towards accessible tourism were further recognized by the Lonely Planet travel guide with the award to the country of the Best in Travel 2021 Accessible Destination title (Lonely Planet, 2020). In addition to the Accessible Beach program, the country boasts accessible pathways in seven different micro-climates, including national parks with accessible trails, and numerous towns with developed accessibility programs and infrastructure.

Costa Rican Network of Accessible Tourism

The Costa Rican Network of Accessible Tourism (Red Costarricense de Turismo Accesible – CRNAT) is a non-profit organization founded in 2017 by ten partners, including Emilio

Zúñiga Cubillo as the president and Stephanie Sheehy as the vice-president (Il Viaggio Travel, 2017). The organization is built on three main pillars: to inform visitors, to verify accessible businesses, and to educate tourism service providers. Philosophically, the organization's work is based on the UNSDG and the Global Code of Ethics with a focus on human rights. Translated to practice, CRNAT's focus is on domestic and international visitors seeking accessible tourism products which have high levels of service and accurate information. The key aim of the organization is to position Costa Rica as an accessible tourism destination for all (Il Viaggio Travel, 2017). Currently, the country's accessible beaches include three beaches in Guanacaste, two in Puntarenas, and one each in Dominical and Limón (Zúñiga, 2021).

One of the best examples of CRNAT's accomplishments is the Accessible Beach project and Beach Club which incorporates sustainable design, and which showcases Playa Hermosa in Carrillo, Guanacaste as the global example. The project was selected as one of the six best proposals worldwide by Booking.com Cares Fund, with an important initial economic contribution for the construction of the first phase (Il Viaggio Travel, 2019). Based on this important support, CRNAT signed an agreement with the Municipality of Carrillo to develop an accessible and sustainable model of basic services which will be provided free of charge to differently abled individuals. This space will include facilities such as parking, showers, toilets, changing rooms, lockers, picnic areas, bicycle racks, elevated paths, retractable walkways, and opportunities to rent specialized equipment such as amphibian wheelchairs. What makes this project special is that many partners came together to realize the vision of accessible tourism. CRNAT, Booking.com, the Costa Rican Tourism Institute (ICT), and the Municipality of Carrillo have joined together to provide universally accessible space for those visiting Playa Hermosa in Carrillo, Guanacaste through the innovative Beach Club (Il Viaggio Travel, 2019).

Dona Tapa [Donate Lid] is one of the leading projects of the organization and focuses on social and ecological responsibility. It was developed by CRNAT in March 2018, seeking to make the beaches of Costa Rica accessible to individuals with reduced mobility, and, by doing so, creating a model for accessible beaches (Il Viaggio Travel, 2017). Costa Rican beaches are among the greatest natural tourist attractions, yet they present two problems: increasing levels of plastic pollution, and inaccessibility for differently abled visitors. Through Dona Tapa, the goal is to repurpose and downcycle plastic pollutants found along the shoreline into products with environmental and social purpose (Il Viaggio Travel, 2017). These products are then marketed to promote the consumption of recycled products; simultaneously, they help finance accessible beach projects for people with reduced mobility, and environmental education projects among tourism operators. Three types of recycled plastic – type 2 HDPE (high-density polyethylene), type 4 LDPE (low-density polyethylene), and type 5 PP (polypropylene) – are used to create retractable walkways on beaches which facilitate wheelchair access, as well as beach chairs and amphibian chairs that can enter the ocean and safely float the user for an immersive experience. Launched in May 2019 at the Costa Rican Tourism Trade Show, Expotur, Dona Tapa has proven very successful. In seven months, the project secured 50 plastic donors and opened 10 plastic collector centres (Il Viaggio Travel, 2017). The project further contributed to the inauguration of Jaco beach as the first accessible beach in Central America, with 21 pathways manufactured from 521 kg of recycled plastic (Family Travel Association, 2021), and also the development of a second prototype accessible beach chair which was tested by differently abled visitors.

Another important dimension of CRNAT's success is the trained and accessible value chain. Thanks to the support of various municipalities, the Instituto Costarricense de Turismo, and the Costa Rican private sector, the Accessible Beach project was able to continue even during COVID-19 (ISTO, 2021). The Accessible Beach complex rotates every six months to different beaches in Costa Rica where CRNAT provides training to interested tourism operators to have lasting impact in each part of the country (ISTO, 2021). In view of the challenges posed by COVID, an Accessible Tourism E-Learning Program was developed as a virtual option in the continuation of accessibility training, provided in partnership with the State Distance University (UNED) and available on the Noun educational platform (Il Viaggio Travel, 2021a). Topics included in the training and development across tourism careers emphasize social, psychological, and physical barriers in the implementation of universal tourism. Structured into three modules, each of which contained three units, this certification is aimed at anyone interested in the accessible tourism industry, including students and graduates with appropriate degrees, and professionals who are part of the value chain (Il Viaggio Travel, 2021a).

Necessary cooperation in accessible tourism outlined by Naniopoulos et al. (2016) and Portales (2015) is showcased in a multitude of initiatives and projects carried out by Il Viaggio Travel and its partners. In particular, the signed agreement between CRNAT and the Municipality of Carillo pertaining to an accessible model of basic services is a serious and legally binding commitment which reflects the necessary policy and legislation to guide the development of services for differently abled individuals (Morris and Kazi, 2016). At the highest level of legislation, Costa Rica is the only Central American country to include tourism requirements. This is enshrined in its accessibility law no. 7600, which requires provisions to accommodate differently abled individuals (Lonely Planet, 2020). This equal opportunity law for differently abled persons defines 'disability' as "any physical, mental or sensory impairment which substantially limits one or more of an individual's major activities" (DREDF, 2021). Article 3b) of the law aims to "Ensure equal opportunities for the Costa Rican population in areas that include health, education, work, family life, recreation, sports, culture and other dimensions" (i.e., tourism), especially in Chapter 7, Access to Culture, Sports and Recreation (DREDF, 2021). Beyond equitable access to education, work, and health services, this law aims to ensure equitable access to physical space and transportation, meaning that Il Viaggio Travel's focus on the Accessible Beach program and the Beach Club are most appropriate for the direction of travel of tourism in Costa Rica.

Personal realm of Il Viaggio Travel's clients

The operator's consideration of wide-ranging physical and cognitive abilities among their clients is evident in the equally wide-ranging travel profiles and packages offered with additional flexibility for customization and personalization. Further distinction between 'permanent' and 'temporary disability', various demographic characteristics, life stages, and temporary or permanent health issues demonstrates the company's intimate understanding of the broad range of requirements and needs among their market (Il Viaggio Travel, 2021c). In consideration of the wide-ranging physical and cognitive abilities among travellers, as well as their varied self-identities and lifestyles, Il Viaggio Travel offers the option of personalized and custom itineraries. Visitors are welcome to modify activities to their particular needs or work with local experts to design unique experiences of interest. To assist further with these experiences, the operator offers a wide range of equipment available on all trips, along with certified guides. As an affiliated organization of Il Viaggio Travel, CRNAT further ensures accessible tourism service of high-quality, accurate information provided to differently abled domestic and international visitors.

Another important contribution of CRNAT, and particularly of the Accessible Beach project (and Dona Tapa), is the support of self-identity and lifestyle of differently abled visitors – for instance, that of an adventurous traveler or one immersed in nature. By making Costa

Rican beaches accessible to individuals with reduced mobility, CRNAT created a model of accessibility that could be applied to other outdoor spaces in order to attract adventure-seeking visitors of all abilities. Unsurprisingly, Costa Rican beaches are one of the greatest natural tourist attractions, which, until recently, were inaccessible to differently abled visitors.

Influential realm of Il Viaggio Travel

In the context of tropical destinations like Costa Rica, teaming with opportunities for nature and adventure, physical aspects can be addressed with UD infrastructure incorporated in various natural settings, in this case accessible beaches and oceanfront. One of the best examples is the Accessible Beach project and Beach Club, which incorporate UD into the most popular beaches throughout the country. This space includes the necessary infrastructure (e.g., retractable walkways on the beach), facilities (e.g., showers and changing rooms), specialized equipment (e.g., amphibian chairs), as well as certified guides and trained industry partners. Furthermore, Dona Tapa contributes recycled plastic, which is used to create retractable walkways on these beaches to allow wheelchair access. The operator's commitment to sustaining these projects is seen in financing inclusion and accessibility projects throughout the country.

Among the social aspects in the influential realm are Il Viaggio Travel's certified employees, guides, drivers, and local experts. One of the company's reps in the Accessible Division is Natalia Vindas, a local expert who, as a wheelchair user who has been a single traveler to many countries, is well equipped to design travel experiences for differently abled visitors. In order to have a lasting impact on tourism operators in each part of the country, the Accessible Beach complex rotates every six months to different beaches where training is provided to interested industry partners (ISTO, 2021). Even in the face of the real challenges presented by COVID, Il Viaggio Travel shifted to the provision of the Accessible Tourism E-Learning Program developed as a virtual option in the continuation of accessibility training (Il Viaggio Travel, 2021a).

Management implications and final considerations

The two case studies outlined in this chapter present a range of bottom-up initiatives, such as Cayuga's creation of transformational experiences and Il Viaggio Travel's provision of accessible tourism offerings. Costa Rica's accessibility law and continuous efforts regarding policy development as top-down approaches contribute to systemic change towards more the provision of more inclusive and accessible tourism. Furthermore, organizational co-learning and the creation of networks contribute to sharing knowledge and leveraging new insights which can be applied in future decision-making (Reed et al., 2010). These broad ideas of co-learning and co-creation in the SASTEM can be successfully applied to organizational collaborations beyond tourism operators and visitors, as well as various interdisciplinary approaches – from social psychology to architecture and design, program planning, law, digital technologies, and so on. Multiple actors collaborating across different governance levels not only contribute to reducing inequalities in the tourism industry, but potentially support broader systemic transformation towards more equitable and just futures.

The issue of inclusive experience

Although Costa Rica is only just starting on the path to a fully accessible tourism destination, its holistic approach is very much on par with some of the leading destinations in this category. Examples in the Australian context reveal similarities with Costa Rican experiences which focus on 'sense of place' within broader destination management approaches, applicable legislation, proactive planning, and wide segmentation approach (i.e., whole-of-life) encompassing access, disability, and aging (Darcy and Dickson, 2009). Consistent with the expanded definition of accessible tourism by Darcy and Dickson (2009), Il Viaggio Travel targets many dimensions of accessibility in its provision of tourism services, including a wide overview of the market, accessible destination experiences, aspects of the built environment (i.e., Accessible Beach Club), and destination management processes, including CRNAT and various local governments. Cayuga's innovative ideas on design and delivery of ecotourism experiences could inspire Il Viaggio Travel to incorporate further educational and entertaining components into their accessible tourism portfolio. As part of every experience offered to visitors, operators could offer accessibility education contributing to a holistic understanding of sustainability and the importance of accessible travel. This approach would create space for interaction and co-learning among differently abled visitors, acknowledge their dignity, and provide equity in the country's tourism offerings.

However, since Costa Rica is a developing country, it lacks large-scale UD, particularly in combination with contemporary approaches such as digital ecosystems. Pertaining to UD on the scale of smart cities, De La Fuente-Robles et al. (2020) emphasize that successful market opportunity requires technological systems (e.g., big data) and greater education on the part of providers to facilitate the co-creation of tourism experiences based on personalized profiles with specific needs of the traveler. In this scenario, the needs of the entire visitor population would be incorporated in the design of tourism experiences from the start, not only as tourism consumers but also as producers and participants in the decision-making based on continued consultations incorporated into all processes (Gillovic and McIntosh, 2020). Il Viaggio Travel seems well versed in co-creation with their market and incorporates at least some technological systems discussed above, such as its E-Learning Program, which offers disability awareness training to interested partners in the industry. Cayuga could learn from this experience through a potential partnership with Il Viaggio Travel to help in the implementation of UD principles and the application of technological systems to better understand the needs of differently abled guests.

In consideration of the entire travel experience, Naniopoulos et al. (2016) emphasize that all tourism sectors need to be connected. This connectivity and interdependence across the destination is something that Il Viaggio Travel still struggles with, having to personally inspect hotel rooms and facilities to approve specific requirements of their diverse market. Although Costa Rica's accessibility law is exemplary in research informing policy, the country needs more work in developing accessible infrastructure alongside applicable training of stakeholders led by operators like Il Viaggio Travel. As an established and well-connected brand, Cayuga could use its influence to promote and support the development of accessible infrastructure and inclusive visitor experiences throughout Costa Rica. Naniopoulos et al. (2016) speak of accessible tourism leadership as one method of gaining advantage in this highly competitive industry. Accordingly, Il Viaggio Travel is on a very good footing as the first operator in the country to attract the accessible tourism market, and it can use its experience to inspire other well-established brands like Cayuga to add their expertise in transformative ecotourism experiences.

Conclusion

The main contribution of this study is the adapted SASTEM being applied to sustainable tourism operators with a focus on transformative visitor experience (e.g., Senda Monteverde), and those focusing on accessible and inclusive tourism experience (e.g., Il Viaggio Travel).

The conceptual model serves as a potential management tool for considerations around both sustainable and equitable visitor experience. There is much opportunity to use the model for symbiotic co-learning and exchange among various operators and the development of collaborative networks across various disciplines, organizations, and levels of government. The main limitation of this work is the narrow – however detailed – focus of the two case studies applied to the model as practical industry examples. The similarity in the geographical location of the two companies (i.e., Costa Rica) and the shared business philosophy (i.e., sustainable tourism) may present unforeseen challenges in the application of this model elsewhere. However, the main contribution to the area of sustainable tourism is the conceptual extension of the SASTEM and its practical applicability to the tourism industry, thereby encouraging the further sharing and testing of this model among other destinations and business models.

References

Ap, J., & Wong, K. (2001). Case study on tour guiding: Professionalism, issues and problems. *Tourism Management, 22*(5), 551–563.

Barrett, H. (2006). Researching and evaluating digital storytelling as a deep learning tool. In C. Crawford, et al. (Eds.), *Proceedings of Society for Information Technology and Teacher Education International Conference 2006* (pp. 647–654). Orlando, FL: AACE.

Berenguer, J. (2007). The effect of empathy in pro-environmental attitudes and behaviors. *Environment and Behavior, 39*(2), 269–283.

Beresford, P. (2016). *All our welfare: Towards participatory social policy*. Bristol: Policy Press.

Bindu, V. T., & Devi, E. (2016). Accessible tourism: Determinants and constraints. A demand side perspective. *Journal of Business and Management, 18*(9), 1–8.

Binkhorst, E., & Dekker, T. (2009). Agenda for co-creation tourism experience research. *Journal of Hospitality Marketing & Management, 18*(2), 311–327.

Black, G. (2005). *The Engaging Museum: Developing Museums for Visitor Involvement*. London/New York: Routledge.

Black, R., & Weiler, B. (2013). Current themes and issues in ecotour guiding. In J. Packer & R. Ballantyne (Eds.), *International Handbook on Ecotourism* (pp. 336–350). Cheltenham, UK: Edward Elgar Publishing.

Boud, D., Keogh, R., & Walker, D. (1985). Promoting reflection in learning: A model. In D. Boud, R. Keogh, & D. Walker (Eds.), *Reflection: Turning Experience Into Learning* (pp. 18–40). London: Kogan Page.

Caldeira, A., Carneiro, M. J., Vasconcelos, A., Mesquita, S., & Kastenholz, E. (n.d.). Personalising cultural heritage co-creative experiences. In W. Gronau (Ed.), *E-Cul-Tours* (pp. 206–218). Open Source Textbook. https://www.kultopia.dk/CustomerData/Files/Folders/7-pdf/22_e-cultourstext-book-final.pdf

Cassia, F., Castellani, P., Rossato, C., & Baccarani, C. (2021). Finding a way towards high-quality, accessible tourism: The role of digital ecosystems. *The TQM Journal, 33*(1), 205–221.

Cayuga Collection. (2021). *Why Cayuga*. Retrieved from https://www.cayugacollection.com/why-cayuga/

Center for Universal Design. (2009). *Universal Design Principles*. Retrieved from http://www.ncsu.edu/ncsu/design/cud/about_ud/docs/use_guidelines.pdf

Cutler, S. Q., & Carmichael, B. A. (2010). The dimensions of the tourist experience. In M. Morgan, P. Lugosi, & J. R. Ritchie (Eds.), *The Tourism and Leisure Experience: Consumer and Managerial Perspectives* (pp. 3–26). Bristol/Tonawanda/New York: Channel View Publications.

Darcy, S., & Dickson, T. J. (2009). A whole-of-life approach to tourism: The case for accessible tourism experiences. *Journal of Hospitality and Tourism Management, 16*(1), 32–44.

Darcy, S., McKercher, B., & Schweinsberg, S. (2020). From tourism and disability to accessible tourism: A perspective article. *Tourism Review (Association Internationale d'Experts Scientifiques Du Tourisme), 75*(1), 140–144.

De Bernardi, C., Pashkevich, A., & Wagner, J. (n.d.). Improving the performance of cultural tourism attractions: Tour guides experiences at the world heritage site in Falun, Sweden. In W. Gronau (Ed.), *E-Cul-Tours* (pp. 161–170). Open Source Textbook. https://www.kultopia.dk/CustomerData/Files/Folders/7-pdf/22_e-cultourstext-book-final.pdf

De La Fuente-Robles, Y. M., Muñoz-de-Dios, M. D., Mudarra-Fernández, A. B., & Ricoy-Cano, A. J. (2020). Understanding stakeholder attitudes, needs and trends in accessible tourism: A systematic review of qualitative studies. *Sustainability, 12*, 1–23.

Disability Rights Education & Defense Fund (DREDF). (2021). *Costa Rica – Law no. 7600*. Retrieved from https://dredf.org/legal-advocacy/international-disability-rights/international-laws/costa-rica-law-no-7600-en/

Efstathiou, A. & Zippelius, E. (2019). *Cultural tourism, destination branding and spatial experience design: Contemporary hospitality design in Cyprus*. Paper presented at *the Strategic Innovative Marketing and Tourism*, Cham.

Family Travel Association. (2021). *Il Viaggio Travel makes Costa Rican beaches more accessible*. Retrieved from https://familytravel.org/il-viaggio-travel-makes-costa-rican-beaches-more-accessible/

Fennell, D. A. & Garrod, B. (2021). Seeking a deeper level responsibility for inclusive (eco)tourism duty and the pinnacle of practice. *Journal of Sustainable Tourism, 30*(6), 1–20.

Gillovic, B., & McIntosh A. (2020). Accessibility and inclusive tourism development: Current state and future agenda. *Sustainability, 12*(22), 1–15.

Gretzel, U., Sigala, M., Xiang, Z., & Koo, C. (2015). Smart tourism: Foundations and developments. *Electron Markets, 25*(3), 179–188.

Hughes, K., & Ballantyne, R. (2013). Winning hearts and minds through interpretation: Walking the talk. In J. Packer, & R. Ballantyne (Eds.), *International Handbook on Ecotourism* (pp. 322–335). Cheltenham, UK: Edward Elgar Publishing.

Hughes, K., & Moscardo, G. (2019). Fun with fungi in the forest: Making tourist experiences more sustainable with quality interpretation. In J. Pearce (Ed.), *Proceedings of the BEST Education Network Think Tank & BEST EN Think Tank XIX* (pp. 58–74). Townsville: James Cook University Press.

Il Viaggio Travel. (2017). *Business Plan 2017–2020: Costa Rica Network of Accessible Tourism*. [Unpublished Report].

Il Viaggio Travel. (2019). *Press Kit for Accessible Beach Club*. [Unpublished Report].

Il Viaggio Travel. (2021a). *Accessible Tourism E-Learning Program*. Retrieved from https://costaricaturismoaccesible.com/servicios/formacion/?fbclid=IwAR0bBGlbX0BnDATsiHIijKbNv62ayp4z5Mw_ZRV4mM9P-cVhI_qEs1Bq2Hw

Il Viaggio Travel. (2021b). *Accessible Tours*. Retrieved from https://accessible.ilviaggiocr.com/

Il Viaggio Travel. (2021c). *How a Costa Rica family holiday should be*. Packages. Retrieved from https://costaricafamilyholidays.com/tours-and-packages/

International Social Tourism Organization (ISTO). (January 27, 2021). *Costa Rica develops a project to make its beaches accessible for all*. Retrieved from https://isto.international/cp_news/costa-rica-develops-a-project-to-make-its-beaches-accessible-for-all/

Jantzen, C. (2013). Experiencing and experiences: A psychological framework. In J. Sundbo & F. Sørensen (Eds.), *Handbook on the Experience Economy* (pp. 146–165). Cheltenham, UK: Edward Elgar.

Kelly, J. R., & Freysinger, V. J. (2000). *21st Century Leisure: Current Issues*. State College, PA: Venture Publishing.

Langer, E. J. (2000). Mindful Learning. *Current Directions in Psychological Science, 9*(6), 220–223.

Leshota, P. L. & Sefotho, M. M., (2020). Being differently abled: Disability through the lens of hierarchy of binaries and Bitso-lebe-ke Seromo. *African Journal of Disability 9*(0), a643.

Lonely Planet. (November 19, 2020). *Costa Rica: Best in Travel 2021 Accessible Destination*. Retrieved from https://www.youtube.com/watch?v=c5E13RmVH50&t= 7s

McDrury, J. & Alterio, M. (2002). *Learning through Storytelling: Using Reflection and Experience to Improve Learning*. London, Sterling: Routledge.

McDrury, J. & Alterio, M. (2003). *Learning through Storytelling in Higher Education:Using Reflection and Experience to Improve Learning*. London: Kogan Page.

Meskele, A. T., Woreta, S. L., Weldesenbet, E. G. (2018). Accessible tourism challenges and development issues in tourist facilities and attraction sites of the Amhara Region World Heritage Sites, Ethiopia. *International Journal of Hospitality & Tourism Systems, 11*(1), 26–37.

Mezirow, J. (2002). How critical reflection triggers transformative learning., In J. Mezirow (Ed.), *Fostering Critical Reflection in Adulthood: A Guide to Transformative and Emancipatory Learning* (pp. 1–20). San Francisco: Jossey-Bass Publishers.

Michopoulou, E., Darcy, S., Ambrose, I., & Buhalis, D. (2015). Accessible tourism futures: The world we dream to live in and the opportunities we hope to have. *Journal of Tourism Futures, 1*(3), 179–188.

Morris, S., & Kazi, S. (2016). Planning an accessible expo 2020 within Dubai's 5 star hotel industry from legal and ethical perspectives. *Journal of Tourism Futures, 2*(1), 88–94.

Moscardo, G. (2009). Understanding tourist experience through mindfulness theory. In M. Kozak & A. Decrop (Eds.), *Handbook of Tourist Behaviour: Theory and Practice* (pp. 99–115). New York: Routledge.

Naniopoulos, A., Tsalis, P., & Nalmpantis, D. (2016). An effort to develop accessible tourism in Greece and Turkey: The MEDRA project approach. *Journal of Tourism Futures, 2*(1), 56–70.

Nowak Da Costa, J., & Bielski, C. (2018). Towards "tourism for all" - Improving maps for persons with reduced mobility. *International Archives of the Photogrammetry, Remote Sensing and Spatial Information Sciences*. Delft, The Netherlands.

Nyanjom, J., Boxall, K. & Slaven, J. (2018). Towards inclusive tourism? Stakeholder collaboration in the development of accessible tourism. *Tourism Geographies, 20*(4), 675–697.

Packer, J., & Ballantyne, R. (2013). Encouraging reflective visitor experiences in ecotourism. In J. Packer & R. Ballantyne (Eds.), *International Handbook on Ecotourism* (pp. 169–177). Cheltenham, UK: Edward Elgar Publishing.

Pereira, R. (2009). Differently abled: Difference and the historical model of the standard man. *História, Ciências, Saúde-Manguinhos, 16*(3), 715–728.

Pine, B. J. & Gilmore, J. H. (1999). *The Experience Economy: Work is Theater & Every Business a Stage*. Boston, MA: Harvard Business School Press.

Pine, B. J., & Gilmore, J. H. (2011). *The Experience Economy*. Updated Edition. Boston, MA: Harvard Business Review Press.

Polat, N. & Hermans, E. (2016). A model proposed for sustainable accessible tourism (SAT). *Tékhne, 14*(2), 125–133.

Portales, R. C. (2015). Removing "invisible" barriers: Opening paths towards the future of accessible tourism. *Journal of Tourism Futures, 1*(3), 269–284.

Prahalad, C., & Ramaswamy, V. (2004). Co-creation experiences: The next practice in value creation. *Journal of Interactive Marketing, 18*(3), 5–14.

Reed, M. S., Evely, A. C., Cundill, G., Fazey, I., Glass, J., Laing, A., Newig, J., Parrish, B., Prell, C., Raymond, C. & Stringer, L. C. (2010). What is social learning? *Ecology and Society, 15*(4).

Scheyvens, R., & Biddulph, R. (2018). Inclusive tourism development. *Tourism Geographies, 20*(4), 589–609.

Seeler, S., & Schänzel, H. A. (2019). From the fluidity of the experience economy to transformative experiences as a catalyst for positive change: The future of experiential travel. *Journal of Tourism Futures, 5*(2), 111–113.

Seiferth, C. (2020). *It is all about traveler experience: Towards a site-applicable conceptual model exemplified with Sustainability Tours by Cayuga Collection, Costa Rica*. BA Thesis, Leisure and Tourism Management, University of Applied Sciences, Stralsund.

Sisto, R., Cappelletti, G. M., Bianchi, P., & Sica, E. (2021). Sustainable and accessible tourism in natural areas: A participatory approach. *Current Issues in Tourism, 25*(8), 1–18.

Stoddard, J. E., Pollard, C. E., & Evans, M. R. (2012). The Triple Bottom Line: A Framework for Sustainable Tourism Development, *International Journal of Hospitality & Tourism Administration, 13*(3), 233–258.

Travel2latam. (January 25, 2021). *Guanacaste, the first beach with 100% accessible facilities in Costa Rica*. Retrieved from https://en.travel2latam.com/nota/65195-guanacaste-the-first-beach-with-100-accessible-facilities-in-costa-rica

United Nations. (2006). *Convention On the Rights of Persons with Disabilities*. United Nations, New York, NY. United Nations General Assembly A/61/611.

United Nations. (2015). *The 17 Goals*. Department of Economic & Social Affairs, Sustainable Development. Retrieved from https://sdgs.un.org/goals

Weiler, B., & Ham, S. H. (2002). Tour guide training: A model for sustainable capacity building in developing countries. *Journal of Sustainable Tourism, 10*(1), 52–69.

Zajadacz, A. (2015). Evolution of models of disability as a basis for further policy changes in accessible tourism. *Journal of Tourism Futures, 1*(3), 189–202.

Zúñiga, A. (July 16, 2021). Playa Hermosa (Osa) inaugurates accessible beach project. *The Tico Times*. Retrieved from https://ticotimes.net/2021/07/16/playa-hermosa-osa-inaugurates-accessible-beach-project

20

DESIGNING ECOTOURISM EXPERIENCES THROUGH CO-CREATION

The case of small central Mediterranean islands

Karl Agius and Samantha Chaperon

Introduction

Tourism provides significant economic gains for small Mediterranean islands, but its activity concentrated on coastlines in peak summer months also means that islands in the region suffer environmental pressures and other negative consequences of seasonality (Said, 2017). To tackle challenges arising from tourism trends before the COVID-19 pandemic, and to respond to an increasing awareness and demand for sustainable tourism (UNWTO, 2019), small islands in the Mediterranean region are beginning to develop nature-based tourism products. This development is likely to be consolidated by the COVID-19 pandemic, with early market research suggesting that post-COVID-19 people will actively seek out natural spaces and quality experiences (GEF, 2020). This growing interest is corroborated by TripAdvisor, which recently added 'parks' as a category of 'travellers' choice' that had previously included beaches and resorts (TripAdvisor, 2021).

One form of nature-based tourism is ecotourism, a type of tourism that takes place in natural settings, involves interpretation and embraces sustainability (Weaver & Lawton, 2007). Until two decades ago, ecotourism in Mediterranean islands was still in its 'infancy stage' (Diamantis, 2000). Recent studies have shown that ecotourism development on small central Mediterranean islands has great potential, particularly in the more peripheral islands and archipelagos which have the most pristine environments, even if the development of such experiences is still lacking (Agius et al., 2019; Agius et al., 2021a). Darmawan et al. (2018) argue that there is a significant lack of policy direction and guidance to help destination stakeholders, to design effective ecotourism experiences and to optimize their nature-based tourism potential. The development, or consolidation, of nature-based tourism is challenging in terms of being able to continue to offer a sustainable and high-quality visitor experience which meets, or ideally exceeds, visitors' expectations. To achieve this, it is important to take into account core conservation objectives, the unique characteristics of a destination, and the capacity of local operators and communities in its design (McCool, 2006).

The notion of co-creation has started to gain traction in relation to tourism experience design. (Campos et al., 2018; Phi & Dredge, 2019). Co-creation involves interaction and collaboration of organizations, groups and individuals in order to jointly generate solutions and

DOI: 10.4324/9781003230748-24

to create value (Russo-Spena & Mele, 2012). This approach has been increasingly employed to engage tourists in the design process and to complement the expertise of local communities. Customers have a key role to play in co-creating tourism experiences as they can offer vital information to tourism operators (Hoarau, 2014) helping them to understand travel motivations, tourists' interests and preferences, based on both their past experiences and their latent desires, all contributing to new innovations (Hoarau, 2016; Yachin, 2018). Studies have found that, for small tourism operators in particular, involving customers in the experience design can be hugely beneficial (Komppula & Lassila, 2015). For example, the testing of trips in co-creation studies have provided small operators with useful data in relation to the appeal and functionality of their services and how they can be improved. The operators can learn which activities were enjoyed the most, what the ideal group size is, about the quality of the interpretation provided, and how itineraries can be amended, all with the aim of improving the experience, meeting and exceeding tourists' expectations, and increasing the commercialization of the tourism offer (Konu, 2015). Despite the obvious benefits, this approach is not much used by small-scale tourism firms, such as ecotourism operators (Yachin, 2017) as they lack time, money and/or expertise to employ these methods (Komppula & Lassila, 2015; Yachin, 2018). Further research into the co-creation of tourism experiences, with the cooperation of external organizations, is vital for small tourism operators (Konu & Komppula, 2016).

Designing ecotourism products in a way that meets both the generic and more destination-specific motivations and expectations is vital for success (Scholtz et al., 2015). 'Motivation', defined as 'a state of need, a condition that exerts a push on the individual towards certain types of action that are seen as likely to bring satisfaction' (Decrop, 2006: 9). Satisfaction in tourism is generally referred to as an evaluation of pre-travel expectations and post-travel experiences (Chen & Chen, 2010). If expectations are largely fulfilled, tourists will consider their experience as satisfactory (Higham et al., 2008). For this study, satisfaction is evaluated in terms of overall satisfaction (Chen & Tsai, 2007) and as compared with expectations (Kim & Park, 2017). If visitor satisfaction is high, the likelihood of revisiting the destination, commenting positively about it and recommending it to others is also high (Higham & Lück, 2007; Shakoori & Hosseini, 2019). Satisfaction is considered as a significant antecedent for destination loyalty among tourists (Kim & Park, 2017). Loyal tourists are particularly desirable because they require fewer targeted marketing efforts to encourage them to revisit the destination, they are more interested in a destination's characteristics, and they tend to be more sensitive to the values of the place (Rivera & Croes, 2010; Agius, 2021a). The tourism industry strives to design quality experiences to achieve competitive advantage, and to build relationships with consumers to gain their loyalty (Rickly & McCabe, 2017). Tourism operators are therefore increasingly engaging in co-creation to encourage revisit and repurchase intentions (Assiouras et al., 2019). Co-creation and its potential to develop tourism in a way that meets visitors' expectations can become an important tool for small central Mediterranean islands seeking to offer high quality and sustainable experiences that lead to visitor loyalty.

In recent years, researchers have examined different methods of co-creation (Konu, 2015). While a number of studies have looked at the application of co-creation by small tourism operators, there is still limited research when it comes to testing different types of customer involvement in tourism service development (Komppula & Lassila, 2015), and how this can lead to satisfaction and customer loyalty. Konu (2015) also calls for further research into ethnographic approaches, multi-data sources, and studies involving more than one test group.

Using three cases – two archipelagos and an island destination in the Mediterranean – the aim of this empirical study is to examine how consumers can be effectively engaged in the co-creation of ecotourism experiences. In particular, it aims to explore motivations for participating in ecotours, examine participants' views on tour design, and assess whether participation can lead to satisfaction and loyalty. In this case, loyalty is examined with respect to both the specific ecotour experienced and also the intention to consume similar experiences offered in other islands in the central Mediterranean region. Recommendations are made which will support stakeholders (including ecotourism operators, site managers, Destination Management Organizations, and tourism policy makers) in making decisions about the design and management of sustainable ecotourism products.

Area of study

The study area consists of three tourism destinations which collectively form part of what is known as the Sicilian Archipelago (see Figure 20.1). The two archipelagos are the Pelagian Islands (comprising Lampedusa, Linosa and the uninhabited islet of Lampione) and the Aegadian Islands (comprising Favignana, Levanzo and Marettimo). The third site is the island of Pantelleria (see Table 20.1). The islands' economies are heavily dependent on tourism which is mainly sun, sand, and sea in nature and most popular during the summer months. Lampedusa and Pantelleria have their own airports, but the other islands can only be reached via ferry boat or hydrofoil service (Agius et al., 2021b). The islands have various terrestrial and marine protected areas (MPAs) which have been designated under regional, national or EU legislation (EUR-Lex, 2015) (see Table 20.2). Such protected areas play an important role in the ecotourism experience (Newsome, 2013). Furthermore, these islands are home to several endemic, rare and charismatic species that have conservation value and serve as key attractions for ecotourists (Agius et al., 2019).

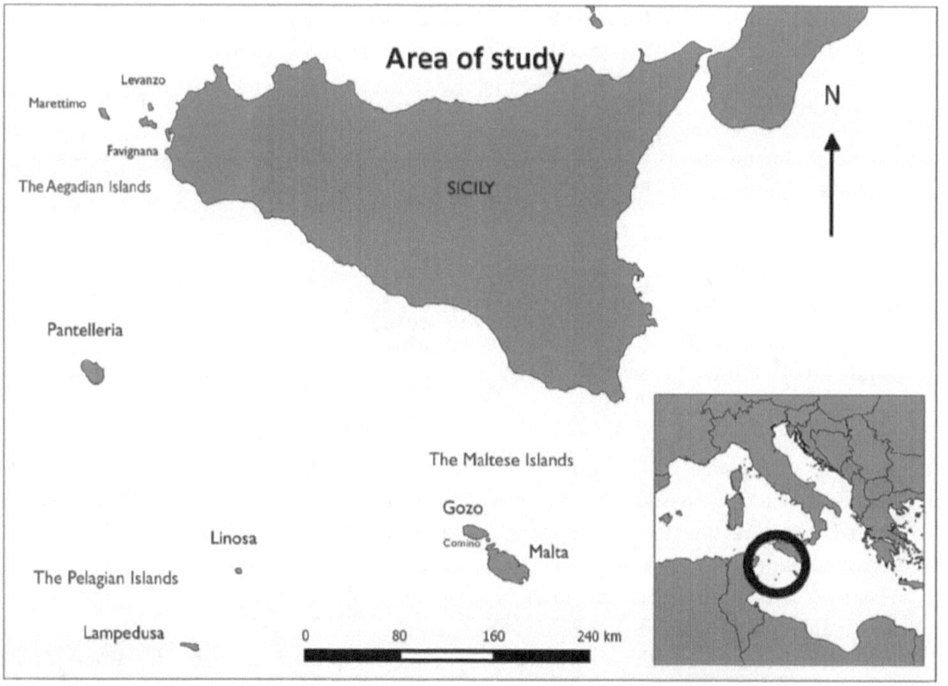

Figure 20.1 Map showing islands under study. Drawn by Andrea Pace for the authors.

Table 20.1 Characteristics of the islands in the area of study

Factor	*Aegadian Islands*	*Pelagian Islands*	*Pantelleria*
Population	5,540	6,141	7,759
Land area (Km^2)	38	44.3	51.5
Inbound Tourists per year	207,843 (2019)	253,710 (2017)	151,917 (2017)

Sources: ENAC, 2018; Gallia, 2012; Himes, 2007; ISTAT, February 6, 2017; Libero Consorzio Comunale di Trapani, 2019; Nicolosi et al., 2018; Peronaci & Luciani, 2015; Serio et al., 2006; Tudisca et al., 2013.

Table 20.2 Protected areas in the area of study

Island/Archipelago	*Protected area category*	*Hectares (ha)*
Pantelleria	Natura 2000: Isola di Pantelleria: Montagna Grande e Monte Gibele	3099
Pantelleria	Natura 2000: Isola di Pantelleria — Area Costiera, Falesie e Bagno dell'Acqua	3402
Pantelleria	National Park	6560
Aegadian Islands	Natura 2000: Fondali dell'Arcipelago delle Isole Egadi	54281
Aegadian Islands	MPA	53992
Favignana	Natura 2000: Isola di Favignana	1832
Levanzo	Natura 2000: Isola di Levanzo	552
Marettimo	Natura 2000: Isola di Marettimo	1111
Lampedusa and Lampione	Natura 2000: Isola di Lampedusa e Lampione	1406
Lampedusa	Reserve: Isola di Lampedusa - Isola dei Conigli	370
Linosa	Natura 2000: Isola di Linosa	435
Linosa and Lampione	Reserve: Isola di Linosa e Lampione	267
Pelagian Islands	MPA	4136

Source: EUR-Lex, 2015; Federparchi, 2021.

Primary research

This study involved three consecutive phases of research, comprising ethnography, surveys, and focus groups (see Figure 20.2). The use of different methods in different phases of research allowed for data to be collected before, during, and after the ecotourism experience, in what Ingram et al. (2017) refer to as the prospective, active and reflective phases. This combination of methods to study the design of nature-based tourism experiences has also been adopted by Konu (2015) and Yachin (2018). Applying multiple methods in a way that closely connects to the actual experience is considered ideal for obtaining information about tourists' needs (Komppula & Lassila, 2015) and in gaining a more comprehensive picture of the phenomenon under examination (Konu, 2015).

The first data collection phase involved participant observation, allowing the researcher to get to know the islands as tourism destinations, and, more specifically, as ecotourism destinations. As part of this phase, informal meetings and conversations were held with ecotourism

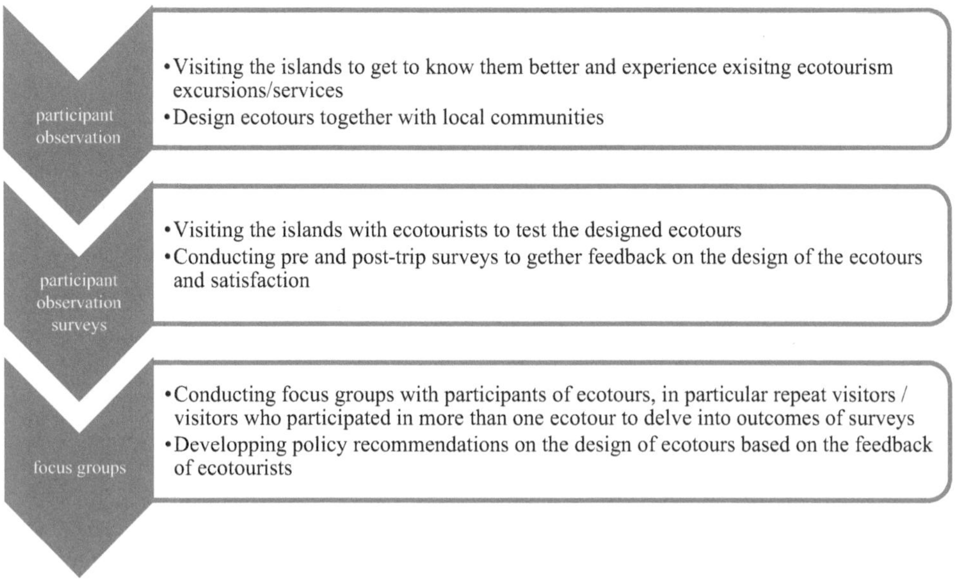

Figure 20.2 Method adopted to design the ecotourism experience together with tourists.

operators, those managing the MPAs, relevant nongovernmental organizations (NGOs), and members of the local communities to hear their views on tourism in the islands and, more specifically, to identify distinctive ecotourism sites and activities. The ethnographic phase also served as an opportunity to experience existing nature-based tours first-hand and to discuss with fellow ecotourists. This phase concluded in the design of three ecotours, one in each study site, which were then used as the basis for the following data collection phases. Similar to Komppula and Lassila (2015) and Konu (2015), the tours included accommodation, transportation, meals and activities. Each itinerary included a variety of nature-based activities such as wildlife watching, trekking, boat trips with local fishermen, snorkelling, horse-riding, and cycling. The tours offered locally owned, traditional accommodation, and there were opportunities to eat local dishes, buy locally produced goods, and visit cultural sites. Educational activities included guided tours on the botany and geology of the island, and visits to nature-related interpretation centres and sea turtle rescue centres. Priority was given to suppliers explicitly offering sustainable ecotourism products and services and those actively working towards conservation objectives.

The second phase of data collection involved joining each of the ecotours organized to carry out further participant observation. The ecotour participants were from Malta and recruited through the support of an eNGO. They all paid fully for the tour, thus avoiding preferential bias (Konu, 2015). All participants were aware that research was taking place. It is acknowledged that this may have influenced participants' behaviours and responses and could be considered a limitation of the study. There were 53 participants in total, with 23 'loyal' ecotourists who participated in two out of the three ecotours, and five who participated in all three, generating a total of 81 ecotourism experiences. The first ecotour was from Malta to the Aegadian islands, the second was to Pantelleria, and the third was to the Pelagian islands.

During this phase of research participants were invited to anonymously complete pre- and post-tour questionnaires to note their expectations, levels of satisfaction with the ecotour experience(s), and loyalty. This method of conducting surveys before and after a trip has been

Table 20.3 Actual number of attendees and respondents for the ecotours organized

Ecodestination	Actual number of attendees		Actual number of respondents pre-trip		Actual number of respondents post-trip	
	Count	Percentage	Count	Percentage	Count	Percentage
Aegadian Islands	43	53	35	48	37	49
Pantelleria	27	33	27	37	27	36
Pelagian Islands	11	14	11	15	11	15
Total	81	100	73	88	75	90

regularly used in the field of tourism (Papadimitriou & Gibson, 2008), ecotourism (Jacobs & Harms, 2014) and co-creation research to test experiences (Konu, 2015). The questionnaire was group-administered, also known as the captive group survey (Veal, 2006), and achieved a high response rate (see Table 20.3).

The questionnaire consisted of a combination of pre-coded (closed-ended) and open-ended questions for richer data. A range of techniques were used, including single and multiple response checklists, ranking, scaling and 4- and 5-point Likert scales. The pre-tour questionnaire was used to primarily investigate (Agius, 2021a) the level of environmental awareness and ethical concerns and (Agius, 2021b) their motivations for participating in the ecotour as well as their expectations of it. The post-tour questionnaire examined (Agius, 2021a) the organization and programming of the ecotour (Agius, 2021b), levels of enjoyment, fulfilment and satisfaction with the experience and (Agius et al., 2021a) willingness to participate in future, similarly designed ecotours in other Mediterranean island destinations. The scales used were adapted from previous studies on ecotourism satisfaction and loyalty (e.g. see Nepal, 2007; Rivera & Croes, 2010). The questionnaire was piloted with six participants in an earlier ecotour organized on the Maltese Islands and some minor changes were made to increase the clarity of the questions being asked. The closed questions in the questionnaire were pre-coded and the data from the open-ended questions were coded into categories using quantitative content analysis techniques (Veal, 2006).

The third phase of data collection involved three focus groups. An approach widely adopted in ecotourism research (Backman & Morais, 2001). The aim was to further examine answers provided in the pre- and post-tour questionnaires (Veal, 2006), to learn more about participants' ecotour motivations, expectations, experiences, and levels of satisfaction, and to explore the reasons for destination and ecotour loyalties. Three focus groups were held with 12 participants in total, with 11 out of the 12 participants having taken part in either two or three of the ecotours organized. During the focus groups, the researcher took field notes, rather than using a recording device, allowing for a more informal approach and encouraging participants to talk openly (Gorman & Clayton, 2005). Data from the focus groups were analysed manually following the approach adopted by Stoffelen (2019).

Study findings

This section integrates the findings from all three phases of research to present a comprehensive understanding of ecotourists' motivations, expectations, and their levels of satisfaction with the co-created ecotour experiences, all of which can provide useful insights and guidance to ecotourism operators.

Ecotourist motivations

A number of motivations for joining the ecotours were identified. More than half of the respondents indicated that the presence of protected areas, nature reserves, and conservation initiatives in ecotour destinations were the main factors that motivated them to join the ecotours. More broadly, the ecotourists' motivations included the opportunity to immerse themselves in nature, to experience pristine environments, to be in more remote and less crowded islands, to learn about wildlife, and to contribute to conservation efforts. A curiosity to visit new destinations and to experience something different to their usual holidays – 'to have an adventure' – were also cited as motivations, though proximity of the ecotour destination to the point of departure was considered important. These motivations identified in the survey were confirmed in the focus groups, but the focus groups revealed an additional motivation to do 'island hopping', particularly in the case of the Pelagian and Aegadian archipelagos.

35 respondents (94.6 percent) in the first organized ecotour, in the Aegadian Islands, felt that their expectations had been met. This clearly motivated their further participation, with 23 ecotourists (53.5 percent) joining the second ecotour, in Pantelleria, and 5 (18.5 percent) of those who participated in the second ecotour also participated in the third ecotour, in the Pelagian Islands. It was revealed in the focus groups that word-of-mouth recommendations from friends, who had taken part in earlier ecotours, had also influenced participation in later tours. A particularly interesting finding was that there was a strong preference amongst respondents to join future ecotours in smaller groups.

In addition to the ecotours organized as part of this research, some respondents revisited the Aegadian Islands on their own initiative, explaining that they were motivated by the sense of tranquillity and fresh air. They were very keen on the almost complete absence of cars and boats, and even aircraft in the off-season. A safe environment, a simple lifestyle, and a sense of local community were other key aspects that motivated their return. These motivations are summarized in Figure 20.3.

Ecotourists' satisfaction with ecotour experiences

After the tours, all respondents claimed high satisfaction rates. Immersing in nature was a key motivation for them and they felt that the ecotour provided them with the opportunity to do this. Conservation efforts were also deemed an important factor in choosing the ecotour destination and these expectations were met, with participants making donations to eNGOs managing reserves and turtle rescue centres, and also wanting to learn about how their participation in the ecotour would support conservation initiatives. It is worth highlighting here that most participants (77 percent) considered themselves either 'fairly' or 'very' environmentally conscious, and 40 percent claimed to be affiliated to an eNGO. The importance of environmental sustainability was discussed by participants and they impressed the importance for them that ecotours are organized and delivered in an environmentally conscious manner. They were pleased with the efforts made to, for instance, recycle waste and to stick to designated pathways to avoid trampling of sensitive ground.

The preferred activities undertaken as part of the ecotours were varied, but particular favourites were trekking, excursions relating to geology, and boat tours. The participants also enjoyed horse-riding, visits to cultural sites and cycling activities. In the Aegadian Islands, trekking and boat trips were most enjoyed. Trekking was similarly enjoyed in Levanzo, as well as a cultural excursion to a natural feature, La Grotta del Genovese. In Favignana, it was cycling, and in Marettimo it was a boat trip with local fishermen to the coastal caves which

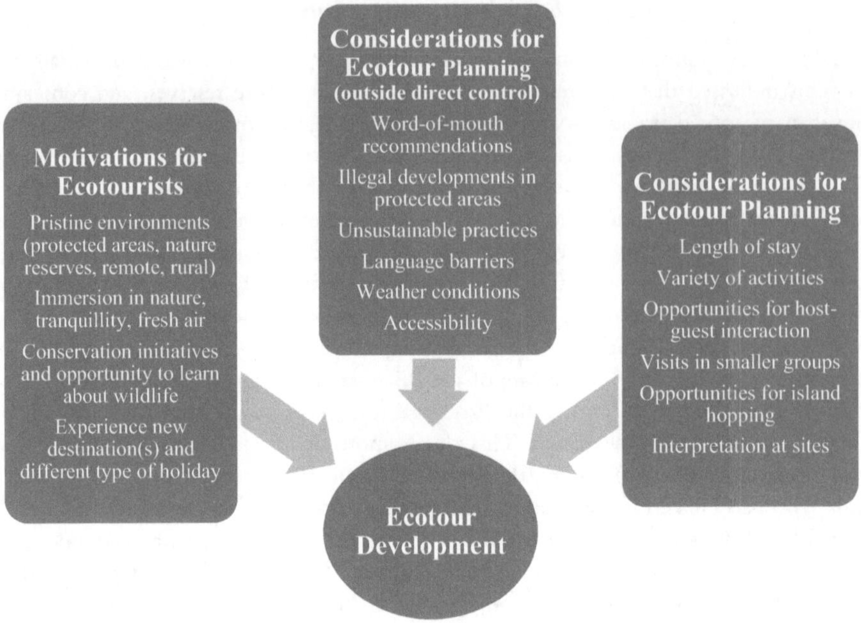

Figure 20.3 Key considerations for ecotour development.

were the activities enjoyed the most. For Pantelleria, in addition to trekking, horse-rising, and an excursion to visit volcanic phenomena, participants cited their stay in a remote *dammuso* (traditional dwelling) as a particularly enjoyable and memorable element of the ecotour. A one-day excursion on the island of Linosa (which involved cycling, snorkelling and buying local and organic products) along with a boat tour off Lampedusa (which involved snorkelling) were the activities most enjoyed by respondents participating in the ecotour taking place in the Pelagian Islands. A visit to the turtle rehabilitation centres on both Lampedusa and Linosa was mentioned as a particular highlight of the tour. This wide range of ecotourism activities which were identified as contributing most significantly to the participants' satisfaction with the trip were enjoyed within and across the island archipelagos visited. This reflects the notion of 'island hopping' as a main motivator and as a key contributor to trip satisfaction; being able to take part in a variety of activities in several different locations was something that significantly and positively impacted on the ecotourists' experiences.

Some ecotourism activities were considered less enjoyable. For instance, there were a lot of jellyfish off the coast of Pantelleria which negatively impacted the snorkelling activities, and the very warm temperatures experienced during the trip made the trekking activities fairly challenging. Some ecotourists complained of a lack of adequate interpretation in the MPA centre on Lampedusa, and at the heritage sites situated within the nature reserve on Pantelleria.

Operational issues also negatively impacted levels of satisfaction with the ecotours. Some commented that the duration of the ecotours were not long enough to fully experience the islands' natural environment. During the development phase, local operators had recommended extending the length of the tours, but a decision was made to keep most of the trips to 3 or 4 days to make them more convenient and to fit in with standard holiday periods. However, the highest level of satisfaction with the length of trip was expressed by the participants of the ecotour to Pantelleria which was 7 days long, and most other participants considered 7 days to be the ideal length for an ecotour.

For future trips, in addition to extending the length of trip to 7 days, participants explained that they would prefer to travel in smaller groups. Furthermore, they called for more interaction with members of the local communities at the ecotour destinations, through specifically designed activities which would support this. In fact, the ecotourists that chose to return to the Aegadian Islands outside of the scope of this research project, said that they made a point of increasing their interactions with the locals throughout their stay. For the trips carried out as part of the research, most interaction was limited to that with locals involved in the hospitality sector and those providing ecotourism excursions. It is true that the ecotour schedules were very busy with several planned activities which left very little free time for the participants to engage with other services and to interact with the local community. Where interactions were possible, for instance with local boat trip operators, language was a barrier as the ecotourists were Maltese who also spoke English, but most operators did not.

Some elements of the tours which were reported to have negatively impacted the overall experience were very much out of the control of the organizers. Participants raised concerns about what were evidently illegal developments on Favignana and also within the nature reserve on Lampedusa. They also witnessed the use of trawlers and sponge fishing taking place on Lampedusa, which caused disappointment. These sentiments are unsurprising given the nature of many of participants, who declared themselves to be concerned about environmental sustainability. Another element of the trip mentioned, which is again out of the control of the trip organizer, is the issues associated with accessibility to the islands; ferry trips between islands can be unpleasant (or cancelled) in bad weather conditions.

Taking into consideration the overall experience, almost all survey respondents rated their tour as either positive (50.6 percent) or very positive (42.7 percent). They indicated that their expectations of the trip had been successfully met, with 100 percent satisfaction rates for tour elements for the Pelagian islands, 95 percent for the Aegadian Islands, and 70 percent for the island of Pantelleria. Following the ecotours organized as part of this research project, all but one respondent expressed a willingness to participate in a future ecotour and said that they were very likely to recommend the tours to their friends and family. In fact, word-of-mouth recommendations had already been made for the second and third ecotours planned as part of the research, and some participants had even collected the contact details of tourism operators, such as those offering accommodation, due to their interest to return with family members. Participants that had taken part in all three tours expressed interest in taking part in future ecotours, similar to those experienced as part of the research project, in neighbouring islands and archipelagos in the region, such as the Aeolian islands, north of Sicily.

Recommendations for ecotourism development in Mediterranean islands

Results show that the 'nature' dimension is the most significant predictor of satisfaction and loyalty in island ecotourism destinations, supporting studies by Carvache-Franco et al. (2021). The islands used in this study still have pristine environments owing to their peripheral location, they have experienced only limited tourism development, and they benefit from extensive nature reserves and MPAs, all of which contributes to their attractiveness to, and satisfaction of, visiting ecotourists. These islands have an opportunity to develop or consolidate their nature-based tourism offer, but this will only be successful if illegal development and environmentally unsustainable practices in protected areas, which can impact their potential, are addressed by governmental authorities.

The fact that ecotourists enjoyed different activities on different islands reflects Baldacchino's (2015) 'nature of differentiation' as a key characteristic of islands in archipelagos. Distinct island characteristics were highlighted by local community members when designing the ecotour itineraries and supports the argument for needing to include all islands, including peripheral ones, in ecotour programmes. For islands in archipelagos, each island's distinctiveness can serve as a unique selling point. This differentiation promotes island hopping, which was considered as a major motivator for participating in the tour and as an activity that contributed greatly to satisfaction. This supports Weaver's (2017) assertion that the experience of island hopping is proving to be a favourite activity for ecotourism on islands and Baldacchino's (2015) argument that archipelago tourism requires further attention. Tourism is a major economic activity for many islands (Butler, 2017) and tourism operators from different islands of the same archipelago are often in fierce competition with each other to attract tourists, with those at the main, more easily accessible, islands usually benefiting most (Chaperon & Bramwell, 2013). Planned visits to all islands in an archipelago ensures that the benefits of ecotourism reach the more peripheral islands too.

Less crowded and more remote destinations were identified as motivating factors for participating in the ecotours. While peripherality is normally considered a limitation in destination accessibility terms, the associated remoteness has been identified as an opportunity for creating destination distinctiveness (Chaperon & Theuma, 2015). As a result, the negative connotations associated with the periphery are challenged and the periphery becomes an experiential core and a centre for tourism based on ecological and cultural distinctiveness (Weaver, 2017).

The proximity of the ecotour destination, identified as a key motivation, could be influenced by a desire to be environmentally friendly since the carbon footprint associated with air transport is considered to have a hugely significant negative impact (Lenzen et al., 2018). Furthermore, the reduced travelling time allows for more time at the destination to immerse in nature. In addition, 'proximity tourism' is playing a key role in revitalizing the tourism industry post COVID-19, and this includes sustainable travel such as ecotourism, as visitors prefer to travel closer to home in times of crisis and in their aftermath (Romagosa, 2020). To take advantage of this trend, Mediterranean islands should focus on attracting ecotourists from neighbouring countries and islands, and those with the most environmentally sustainable travel options.

Ecotour itineraries, co-created together with local operators and community members, have proved successful in terms of achieving high levels of satisfaction and loyalty. However, there are some key considerations for ecotour planners which could lead to increased satisfaction amongst ecotourists (see Figure 20.3). Operators should take into consideration the time of year that the tour is taking place and alter the programme of activities to best suit the weather conditions. For example, intense trekking in very hot weather could be replaced with more marine-based activities. The length of the tour also needs to be considered. There should be more leisure time in the schedule to allow for ecotourists to interact freely with the local community, outside of the ecotour context. Furthermore, programmes that involve island hopping need to take into account the extra time needed for travel, inter-island travel and allocate time for interaction with locals. The group size of the ecotour needs to be considered and large groups should be avoided. A group of less than 25 people is considered optimal, as was indicated by the ecotourists in the Pelagian Islands and in Pantelleria. Interpretation at sites visited also needs attention. One of the expectations for ecotourists was the learning experience, and this supports the findings of Balantine and Eagles (1994) who identified the educational element as one of the pillars of ecotourism. Without appropriate interpretation of heritage and natural sites, this is not possible to achieve. This issue is often related to the lack of expertise of the eco-guides on small islands, and is further compounded by a lack of language competencies needed to cater for international visitors (Agius, 2021b). This needs to be addressed through education and training opportunities for the local tourism stakeholders.

Ecotourists visiting the islands in this study commented very positively about their experiences, and for most this led to loyalty to the ecotourism experience in the region. This was confirmed by visitors who participated in more than one ecotour, who revisited a destination on their own initiative with friends, who expressed willingness to make or made a positive word of mouth or who asked for contacts/information related to the designed ecotours. Aside from destination loyalty, ecotours like those designed in this study, with high levels of community involvement in their design, can generate their own product-based loyalty. The development of similarly designed tours in neighbouring destinations has the potential to attract loyal tourists, and lead a loyalty to the wider region. At the very least, these ecotours can complement the existing 3S provision in Mediterranean islands, and overcome some of the issues associated with seasonality if encouraged in the off-peak months. This calls for better transport connections all year round and regional promotional efforts which focuses on the distinctive nature-based elements of the smaller islands.

Ecotourism taking place on small islands is predominantly run by small-scale operators (often referred to as micro-enterprises) (Romagosa, 2020) with limited resources for research and development of tourism products. Hence, co-creation with the support of a research institution has offered such operators key information that can help to add further value to their services facilitating development/consolidation of ecotourism experience on central Mediterranean islands that meet expectations of a new segment of tourists that the islands have potential to target. Lessons learnt can be applied by regional tourism organizations, especially in the light of initiatives aiming to market central Mediterranean islands and an ecotourism hub (Agius, 2021a). This research approach has led to the identification of a series of services based on the local service system run by micro-operators. The range of services can then be used as part of a tourism package, but they can also be purchased individually for interested customers willing to organize their travel independently. In line with Komppula and Lassila (2015), this study points out that that co-creation of new ecotourism experiences with the involvement of customers, but also other stakeholders including research institutes and small operators, can lead to the development of packages with high value and innovation leading to satisfaction and loyalty towards island destinations in the region.

Conclusion

Results show that small islands in the central Mediterranean region can serve as ecotourism destinations. Nature-based experiences designed with local stakeholders and taking into account the rich natural and cultural heritage within protected areas can give rise to satisfactory visitor experiences on small islands and generate loyalty towards a destination or ecotourism experience in a region among visitors.

Feedback from tourists shows that such experiences must ensure maximum contact with nature but at the same time facilitate interaction with local communities. Duration of the tour needs to take into account the number of islands and inter-island travel to ensure that ecotourists have a comprehensive experience of the island or archipelago in question. Excursions in the programme must be well spaced and take into account weather conditions as well as the different characteristics of the islands. Adequate interpretation in English should be made available to increase satisfaction of ecotourists. This further confirms the need to involve stakeholders from all islands in the design process.

Group size must not be greater than 25 and sustainability in the wider sense needs to be given more attention. Visitor management is key to keep both the protected areas' system and tourism development sustainable (Petrić & Mandić, 2014). Taking this into account, managers and policy makers need to take visitor management initiatives to keep the islands attractive to ecotourists who prefer to be away from the crowds and consider the pristine state of the environment as vital elements of their experience. Therefore, enforcement of existing regulations in MPAs, limiting activities and operations need to be seriously adhered too.

Findings also confirm an opportunity to use nature-based tourism to relaunch tourism - an important economic sector on small islands – following the pandemic as visitors seek nature-based experiences with limited risk. In this regard, proximity tourism and close markets must be targeted by small ecotourism operators, further contributing the to the sustainability dimension. Additionally, this study shows that these islands can attract another form of loyal tourists, apart from those who revisit islands annually for 3S, which can help to address seasonality, a major challenge for operators on small islands.

This study contributes to nature-based tourism management literature by providing an empirical example of how both consumers and service providers can get involved in designing ecotourism experiences. The study which adopted an ethnographic approach generating multiple-data sets in 3 cases run in parallel confirms that that co-creation can generate key considerations (see Figure 20.3) for the design process of nature-based tourism experiences in novel destinations.

References

Agius, K. (2021a). The Ecotourism Hub: A joint cross-border marketing strategy for peripheral islands. *Shima*, https://doi.org/10.21463/shima.124

Agius, K. (2021b). Ecoguides: Legal challenges and impacts on visitor experiences in central Mediterranean. In J. N. Albrecht (Ed.), *Managing visitor experiences in nature-based tourism* (pp. 67–79). CABI Publishing, Wallingford, UK.

Agius, K. Sindico, F., Sajeva, G., & Baldacchino G. (2021a). 'Splendid isolation': Embracing islandness in a global pandemic. *Island Studies Journal*. https://doi.org/10.24043/isj.163

Agius, K., Theuma, N., & Deidun, A., (2021b). So close yet so far: Island connectivity and ecotourism development in central Mediterranean islands. *Case Studies on Transport Policy*, *9*(1), 149–160.

Agius, K., Theuma, N., Deidun, A., & Camilleri, L. (2019). Small islands as ecotourism destinations: A central Mediterranean perspective. *Island Studies Journal*, *14*(2), 115–136.

Assiouras, I., Skourtis, G., Giannopoulos, A., Buhalis, D., & Koniordos, M. (2019). Value co-creation and customer citizenship behavior. *Annals of Tourism Research*, *78*, 102742.

Backman, K. F., & Morais, D. B. (2001). Methodological approaches used in the literature. In D. B. Weaver (Ed.), *The encyclopedia of ecotourism* (pp. 597–609). CABI Publishing, Wallingford, UK.

Balantine, J., & Eagles, P. J. (1994). Defining the Canadian ecotourist. *Journal of Sustainable Tourism*, *2*(4), 210–214.

Baldacchino, G. (2015). More than Island tourism: Branding, marketing and logistics. In G. Baldacchino (Ed.), *Archipelago tourism: Policies and practices* (pp. 1–18). Ashgate Publishing Ltd., Surrey, England.

Butler, R. (2017) Thoughts on core-periphery and small island tourism. *Tourism Recreation Research*, *42*(4), 537–539.

Campos, A. C., Mendes, J., Valle, P. O. D., & Scott, N. (2018). Co-creation of tourist experiences: A literature review. *Current Issues in Tourism*, *21*(4), 369–400.

Carvache-Franco, M., Pérez-Orozco, A., Carvache-Franco, W., Víquez-Paniagua, A. G., & Carvache-Franco, O. (2021). Motivations and their influence on satisfaction and loyalty in eco-tourism: A study of the foreign tourist in Costa Rica. *Anatolia*, https://doi.org/10.1080/13032917.2021.1933115

Chaperon, S. & Bramwell, B. (2013) Dependency and Agency in Peripheral Tourism Development. *Annals of Tourism Research*, 40, 132–154.

Chaperon, S., & Theuma, N. (2015). The Malta-Gozo-Comino story: Implication of a Malta-Gozo fixed link on tourism activity. In G. Baldacchino (Ed.), *Archipelago tourism: Policies and practices* (pp. 51–66). Ashgate Publishing Ltd., Surrey, England.

Chen, C.-F., & Chen, F.-S. (2010). Experience quality, perceived value, satisfaction and behavioral intentions for heritage tourists. *Tourism Management*, *31*(1), 29–35.

Chen, C. F., & Tsai, D. (2007). How destination image and evaluative factors affect behavioral intentions? *Tourism Management*, *28*, 1115–1122.

Darmawan, E., Arfa, M., & Purbawati, D. L. (2018). The development of nature based-tourism in Tembarak District, Temanggung regency. *Asia Pacific Journal of Tourism Research*, *23*(9), 924–934.

Decrop, A. (2006). *Vacation decision-making*. CABI Publishing, Wallingford, UK.

Diamantis, D. (2000). Ecotourism and sustainability in Mediterranean islands. *Thunderbird International Business Review*, *42*(4), 427–443.

ENAC. (2018). *Traffic data 2017*. Retrieved on June 9, 2021 from https://www.enac.gov.it/sites/default/files/allegati/2018-Ago/ENAC_Traffic_data_2017_en.pdf

EUR-Lex (2015). Commission Implementing Decision (EU) 2015/74 of 3 December 2014 adopting an eighth update of the list of sites of Community importance for the Mediterranean biogeographical region (notified under document C(2014) 9098) L18/696. *Official Journal of the European Union*. Retrieved on June 9, 2021 from https://eur-lex.europa.eu/legal-content/EN/TXT/PDF/?uri=CELEX:32015D0074&from=EN

Federparchi. (2021). Parks.it. Retrieved on July 1, 2021 from http://www.parks.it/

Gallia, A. (2012). La valorizzazione dei beni culturali e ambientali per lo sviluppo delle isole minori italiane. *Rivista giuridica del Mezzogiorno*, *26*(4), 929–962.

Global Environment Facility. (GEF). (2020). White paper on a GEF COVID-19 response strategy: The complexities and imperatives of building back better. Retrieved on June 9, 2021 from https://www.thegef.org/sites/default/files/council-meeting-documents/EN_GEF_C.59_Inf.14_White%20Paper%20on%20a%20GEF%20COVID-19%20Response%20Strategy_.pdf

Gorman, G. E., & Clayton, P. (2005). *Qualitative research for the information professional*. Facet Publishing, London.

Higham, J. & Lück, M. (2007). Marine wildlife and tourism management: In search of scientific approaches for sustainability. In J. Higham and M. Lück (Eds.), *Marine wildlife and tourism management: Insights from the natural and social sciences* (pp. 1–18). CABI Publishing, Wallingford, UK.

Higham, J. E. S., Lusseau, D., & Hendry, W. (2008). Wildlife viewing: The significance of the viewing platforms. *Journal of Ecotourism*, 7(2–3), 137–146.

Himes, A. H. (2007). Fishermen's opinions of MPA performance in the Egadi Islands marine reserve. *Mast*, *5*(2), 55–76.

Hoarau, H. (2014). Knowledge acquisition and assimilation in tourism-innovation processes. *Scandinavian Journal of Hospitality and Tourism*, *14* (2), 135–151.

Hoarau, H. (2016). *Open innovation in the tourism experience sector: The role of practice based knowledge explored*. Open tourism, Springer, Berlin, Heidelberg, pp. 137–152.

Ingram, C., Caruana, R., & McCabe, S. (2017). Participative inquiry for tourist experience. *Annals of Tourism Research*, *65*, 13–24.

ISTAT. (2017, February, 6). Personal communication – request for data.

Jacobs, M. H., & Harms, M. (2014). Influence of interpretation on conservation intentions of whale tourists. *Tourism Management*, *42*, 123–131.

Kim, K. H., & Park, D. B. (2017). Relationships among perceived value, satisfaction, and loyalty: Community-based ecotourism in Korea. *Journal of Travel & Tourism Marketing*, *34*(2), 171–191.

Komppula, R., & Lassila, H. (2015). Co-creating tourism services–a multiple case study of methods of customer involvement in tourism. In H. Pechlaner & E. Smeral (Eds.), *Tourism and leisure* (pp. 287–303). Springer Gabler, Wiesbaden.

Konu, H. (2015). Developing a forest-based wellbeing tourism product together with customers – An ethnographic approach. *Tourism Management*, *49*, 1–16.

Konu, H., & Komppula, R. (2016). Customer involvement in a new service development process: The case of "Feelgood in Lapland". *Anatolia*, *27*(4), 421–432.

Lenzen, M., Sun, Y. Y., Faturay, F., Ting, Y. P., Geschke, A. & Malik, A. (2018). The carbon footprint of global tourism. *Nature Climate Change*, *8*, 522–528.

Libero Consorzio Comunale di Trapani. (2019). TAV.4 Favignana: movimento turisito ricettivo mensile anno 2019 – dati provvisori. Retrieved on June 9, 2021 from http://www.consorziocomunale.trapani.it/provinciatp/images/FAVIGNANA_agg2019.pdf

McCool, S. F. (2006). Managing for visitor experiences in protected areas: Promising opportunities and fundamental challenges. *Parks, 16*(2), 3–9.

Nepal, S. K. (2007). Ecotourists' importance and satisfaction ratings of accommodation-related amenities. *Anatolia, 18*(2), 255–276.

Newsome, D. (2013). An 'ecotourist's recent experience in Sri Lanka. *Journal of Ecotourism, 12*(3), 210–220.

Nicolosi, A., Laganà, V. R., Cortese, L., & Privitera, D. (2018). Using the network and MCA on tourist attractions the case of Aeolian Islands, Italy. *Sustainability, 10*(11), 4169.

Papadimitriou, D. & Gibson, H. (2008). Benefits sought and realized by active mountain sport tourists in Epirus, Greece: Pre-and post-trip analysis. *Journal of Sport & Tourism, 13*(1), 37–60.

Peronaci, M., & Luciani, R. (2015). Anthropic pressures on Egadi Islands. *Energia Ambiente e Innovazione, 61*(4), 9–11.

Petrić, L., & Mandić, A. (2014). Visitor management tools for protected areas focused on sustainable tourism development: The Croatian experience. *Environmental Engineering and Management Journal, 13*(6), 1483–1495.

Phi, G. T., & Dredge, D. (2019). Collaborative tourism-making: An interdisciplinary review of co-creation and a future research agenda. *Tourism Recreation Research, 44*(3), 284–299.

Rickly, J. M., & McCabe, S. (2017). Authenticity for tourism design and experience. In D. R. Fesenmaier & Z. Xiang (Eds.), *Design science in tourism* (pp. 55–68). Springer, Cham.

Rivera, M. A., & Croes, R. (2010). Ecotourists' loyalty: Will they tell about the destination or will they return? *Journal of Ecotourism, 9*(2), 85–103.

Romagosa, F. (2020). The COVID-19 crisis: Opportunities for sustainable and proximity tourism. *Tourism Geographies, 22*(3), 690–694.

Russo-Spena, T. & Mele, C. (2012). "Five Co-s" in innovating: A practice-based view. *Journal of Service Management, 23* (4), 527–553.

Said, N. (2017). From mass tourism to ecotourism: The case of the Balearic Islands, the Maltese Islands and the republic of Cyprus. *Occasional papers on islands and small states*, 03/2017.

Scholtz, M., Kruger, M., & Saayman, M. (2015). Determinants of visitor length of stay at three coastal national parks in South Africa. *Journal of Ecotourism, 14*(1), 21–47.

Serio, D., Alongi, G., Catra, M., Cormaci, M., & Furnari, G. (2006). Changes in the benthic algal flora of Linosa Island (Straits of Sicily, Mediterranean Sea). *Botanica Marina, 49*(2), 135–144.

Shakoori, A., & Hosseini, M. (2019). An examination of the effects of motivation on visitors' loyalty: Case study of the Golestan Palace, Tehran. *Tourism Management Perspectives, 32*, 100554.

Stoffelen, A. (2019). Disentangling the tourism sector's fragmentation: A hands-on coding/post-coding guide for interview and policy document analysis in tourism. *Current Issues in Tourism, 22*(18), 2197–2210.

TripAdviser. (2021). Travellers' Choice 2021. https://www.tripadvisor.com/TravelersChoice

Tudisca, S., Di Trapani, A. M., Sgroi, F., & Testa, R. (2013). Marketing strategies for Mediterranean wineries competitiveness. The case of Pantelleria. *Calitatea, 14*(137), 101–106.

UNWTO. (2019). International Tourism Highlights. Retrieved on June 9, 2021 from https://www.e-unwto.org/doi/pdf/10.18111/9789284421152

Veal, A. J. (2006). *Research methods for leisure and tourism: A practical guide*. Pearson Education, Harlow, UK.

Weaver, D. B. (2017). Core–periphery relationships and the sustainability paradox of small island tourism. *Tourism Recreation Research, 42*(1), 11–21.

Weaver, D. B., & Lawton, L. J. (2007). Twenty years on: The state of contemporary ecotourism research. *Tourism Management, 28*(5), 1168–1179.

Yachin, J. M. (2017). The entrepreneur–opportunity nexus: Discovering the forces that promote product innovations in rural micro-tourism firms. *Scandinavian Journal of Hospitality and Tourism, 19*(1), 47–65.

Yachin, J. M. (2018). The 'customer journey': Learning from customers in tourism experience encounters. *Tourism Management Perspectives, 28*, 201–210.

21
A HIDDEN PLAYGROUND
Airsoft-related experience design in nature-based tourism

Ivana Damnjanović

Introduction

Nature-based tourism (NBT) and outdoor recreation have been a growing trend worldwide (Fredman & Margaryan, 2020). The reason for this phenomenon lies in the rising need to counterbalance contemporary stress-related health threats and a meaningful connection with nature. The unbreakable bond between healthy nature and human health and wellbeing (Hockings et al., 2020, pp. 13–14) ignites needs and motives that create new travel trends. They include travel for transformation through authentic experiences and the pursuit of healthy lifestyles, including sports tourism and a growing awareness of sustainability (UNWTO, 2019, p. 5). More specifically, sustainable, experiential, mentally restorative travel and physical adventure in natural surroundings are on the rise (Joyner et al., 2018). Its design is crucial for a travel experience to fulfil its described functions. Nature-based and adventure tourism consumers value the experiences based on a particular sequencing of activities that correspond to their specific interests and preferences.

There is a spectrum of existing NBT and adventure tourism activities with transformational potential and the proliferation of emerging ones. Such activity is Airsoft – an adventure sport based around simulated combat games, usually placed in an isolated natural environment. Developing under the radar, Airsoft can potentially cause environmental and socio-economic issues. However, by carefully observing every aspect of the complex system of experience design and delivery, particularly in NBT (Powell et al., 2009), Airsoft could realize its potential for positive impacts towards becoming a sustainable tourism product.

Extensive research on the concepts such as NBT (Buckley, 2020; Fuchs et al., 2021; Winter et al., 2019), adventure tourism (Giddy & Webb, 2017; Houge Mackenzie et al., 2021), transformative role of tourism (Wolf et al., 2017), and their interrelations is available. In addition, tourism experience design has also been the focus of numerous studies (Breiby et al., 2020; Gardiner & Scott, 2018; Rickly & McCabe, 2017) regarding NBT (Fossgard & Fredman, 2019) and transformative tourism (Neuhofer et al., 2020).

The following proposed research avenues underpin this study: (i) the crossroads of NBT and human–nature relationships (Fredman & Margaryan, 2020); (ii) transformational outdoor recreation and nature-based tourism experiences (Winter et al., 2019); (iii) reconceptualizing adventure nature sports to reflect participant experiences better (Houge Mackenzie & Brymer,

DOI: 10.4324/9781003230748-25

2018); (iv) gameplay motivation factors in different market segments (Xu et al., 2015); and (v) the demands and opportunities for public land management (McCool & Kline, 2020).

Airsoft-related research exists in terms of its medical-related aspects (injuries and health benefits associated with playing Airsoft) (Khalaily et al., 2018; Tornero-Aguilera et al., 2020), technical features of Airsoft equipment (Horák et al., 2014), and equipment-manufacturing market characteristics (Menet & Szarucki, 2020). However, the research on Airsoft as an adventure sport is sparse. It has been limited to single-country studies of a general consumer profile and players' perceptions of the benefits of active involvement in Airsoft (Nowacki & Wiśniewska, 2015; Roşu, 2015). However, no travel-related research on Airsoft is available.

Quality tourism experience and tourists' value perception have been correlated in various research (Kim & Thapa, 2018; Wu & Li, 2017) and tourists' subjective perceptions and motivation (Kim et al., 2015). Therefore, this study will take the perspective of those aspects of Airsoft players' profiles focused on their motives and experience-related organizational preferences. Through this prism, this initial study of Airsoft in the tourism context aims to find the answers to: (1) wherein the tourism typology spectrum Airsoft belongs; (2) how to design such experiences so their participants value them as high quality; and (3) how such experiences fit into the sustainability paradigm.

Finally, the study could point towards potential implications of certain Airsoft experience design elements on natural land and resource management and its entrepreneurial potential.

Following the introduction, the study comprises four main sections. The first section (Literature review) examines the elements of interconnected concepts evident in Airsoft. These concepts are crucial for formulating the study survey and answering the research questions. They include fundamental features of Airsoft, NBT with elements of adventure and transformational travel, their experience design, and managerial and sustainability implications. A detailed description of applied research methods follows. The following section outlines the results from the qualitative and quantitative research phases. The conclusion section frames the answers to research questions by combining the literature review and research results. Finally, it concludes with the study's limitations and suggestions for future research.

Literature review

To set the scene for the research and understand if and how Airsoft fits into nature-based, adventure, transformational and sustainable travel, this literature review contains two sections:

(i) an explanation of Airsoft; and (ii) relating Airsoft to these travel concepts based on their shared characteristics.

Fundamental features of Airsoft

According to the European Airsoft Federation (EAF) (2017), Airsoft is one of the fastest-growing outdoor adventure sports activities. They describe it as a simulated combat game "enjoyed by men and women of all ages at thousands of playing sites throughout Europe." Its main motives are releasing pent-up energy, escaping everyday pressures, gaining a sense of community, spending time with friends, strong impressions and experiences, and physical exercise (EAF, 2017; Nowacki & Wiśniewska, 2015). While Airsoft players are aware of its psycho-motor and psycho-intellectual benefits (Roşu, 2015), Airsoft-related physical activity often exceeds recommended levels (Tornero-Aguilera et al., 2020).

Airsoft originated in Japan (Martinez, 2008) in the 1980s, from where it first spread to neighbouring Asian countries, reaching America and Europe in the early 1990s (Airsoft Dome, 2021). Today, there are numerous gaming clubs worldwide (Grand View Research, 2019) with no actual estimated number of Airsoft players worldwide (Popular Airsoft, 2008). Airsoft gun market size may indicate the number of players as the replica guns are the main piece of the sport's equipment. Airsoft gun market reports state that the global market was valued at US$1.5 billion in 2018 (Grand View Research, 2019), with an annual growth rate of around 6 percent from 2015 to 2020 and projected future growth (Imarc, 2020). Some Airsoft leagues gather about 200,000 players from 105 countries at their events annually (Grand View Research, 2019).

Being a simulated combat game, Airsoft relies heavily on military tradition, primarily in emulating actual tactical equipment and accessories (loadout) and its team nature and organization. It is played with so-called replicas of real weapons (real-size guns that fire small plastic projectiles). Airsoft players often wear uniforms that belong to a particular military and era (e.g., World War I French soldiers). Due to its nature, the game includes military-related skills, e.g., tactics, orienteering, adequate movement, specific communication forms and language (Larsen, 2008), discipline, and scenarios. Each game scenario (i.e., gameplay, a game story, or plot) has missions, objectives, and tasks and is either imagined or based on a real-life event. It includes action shooting, live-action role-playing, close-quarters battle skirmishes, military simulations (MilSim), or historical re-enactments. Players are divided into units and assume roles assigned by the scenario. Each scenario defines the number of players, units, roles, game timeframe, the size of the used terrain, end goals, and rules of play (e.g., re-spawn and medic rules for resuscitating and healing hit players, respectively). There is an unlimited number of scenarios: free a hostage, take over/defend a territory, map a terrain, relocate/secure civilians.

Games are played under stringent general rules (Airsoft Dome, 2021; Roger, 2021) overseen by designated supervisors. The most important are the honour system of hit recognition (a player reports being shot/hit and "plays dead") and safety rules (e.g., obligatory wearing protective goggles/glasses). Depending on their purpose, Airsoft gaming events can include a diverse number of players – from a few participants (team training) to official events ranging from the typical approximate of one hundred up to 2300 recorded at famous traditional Border Wars in the Czech Republic ("Borderwar," 2021).

Certain Airsoft features relate to tourism: playgrounds as travel destinations, a focus on experience, experience design complexity, and the diversity of the Airsoft-specific stakeholders involved. Airsoft equipment market reports recognize a close connection between rising demand for such products and growth trends in travel and tourism, outdoor team sports, and growing fondness for real shooting simulation sports (Fact MR, 2019; Imarc, 2020). Carlingford, Ireland (2020) is an example of a destination that perceives the potential of Airsoft for tourism development and places it as a niche tourism product into its marketing features.

Airsoft experience design in the context of nature-based, transformational, adventure tourism

Understanding Airsoft as an outdoor sports activity first requires focusing on the natural environment. Namely, the natural environment provides the resource essential for many forms of tourism, particularly nature-based and adventure tourism (Leung et al., 2008). Furthermore, NBT and outdoor recreation experiences have been increasingly recognized as beneficial for, among other aspects, enhancing social connections, connecting people with local nature and culture, and contributing to local economies and human health (Winter et al., 2019) particularly pertinent to tech-connected and nature-disconnected urban dwellers (Global Wellness Summit, 2019, pp. 57–58).

High-value quality experience in NBT relates to its tourists' perception. In this context, outdoor nature activities improve and maintain mental health and happiness with short-term emotional benefits, mid-term stress-recovery, and long-term changes in worldview (Buckley, 2020). More precisely, specific outdoor tourism experiences, such as Airsoft, can provide: (1) positive emotions and social interactions; (2) clarity, insight, and decisiveness; and (3) an improved sense of the purpose and meaning of life (Buckley, 2019). In addition, if there are vigorous physical activity levels involved, there are greater benefits to tourists' wellbeing (Hansmann et al., 2007). We can conclude that Airsoft, with its recognized physical, social, experiential, and mental health benefits, belongs to the NBT activities spectrum.

Since certain Airsoft features need a more specific consideration than that offered solely within the NBT paradigm, an examination of two other rising trends that intersect with the NBT narrative follows – adventure and transformative travel.

Over recent years, adventure travel has grown into a global phenomenon in the tourism sector (Giddy, 2016). It includes at least two of the three essential components: participation in physical activity, a visit to a natural environment, and a culturally immersive experience (Adventure Travel Trade Association [ATTA] & [IITS], 2020, p. 8). Since the main motives for playing Airsoft match two components (physical and experiential), it can be referred to as an adventure travel activity. Understanding distinct adventurous nature sports' motivation requires examining their context: the adventure sports activity (e.g., event structure, team or individual activity, need-supportive motivational climate) and the tourism destination (e.g., the natural and social environment) (Houge Mackenzie et al., 2021), both of which are essential elements of Airsoft. Adventure travellers' key motives revolve around their mental and physical wellness, unique experiences, challenges, and transformation accompanied by the need to positively impact the destination (ATTA & IITS, 2020, p. 8).

From the other perspective, transformative travel enables a positive change in an individual through becoming "somebody else," which is a focal point of a transformational process (Pine & Gilmore, 1999) and adds value to experience portfolios through improved conditions and satisfaction (Wolf et al., 2017). A person needs to assume a role within a set gameplay scenario to play Airsoft. Therefore, such a feature builds transformational potential into Airsoft.

In some instances, an emerging gamification trend supports a transformative role of travel, based on consumers' desires being seamlessly blended into memorable experiences (Xu et al., 2015). Laing et al. (n.d., p. 151) further examine the relationship between experiential and transformative travel with fantasy enactment and narrative-based role-play in natural settings. As a simulated, scenario-based game often placed in nature, Airsoft also fits into this transformative travel paradigm.

Experience creation is a repeating theme in all discussed types of tourism. Careful sequencing of activities is a crucial element of experience design. That includes paying attention to the uniqueness and complexity of such activities, as in Airsoft. Therefore, in Airsoft experience design, it is vital to be attentive to dimensions focused on transformational, nature-based, adventure tourism activities. In NBT experiencescapes, combining resource diversity with guests' shifting needs, skills development, and activity performance is crucial (Fossgard & Fredman, 2019). Experience design extends to building in material and symbolic (simulated) notions of authenticity into the staging for subjective, meaningful performances and psychological engagement. The process of co-creation through immersion, tourists' imagination, a sense of achievement, camaraderie, and learning (Rickly & McCabe, 2017) are the qualities fundamental to Airsoft as a scenario-based, mission-oriented team game. In his discussion of the design of transformative experiences, Neuhofer et al. (2020) propose including dimensions such as positive emotions, engagement, relationships, meaning, and accomplishment. The same authors emphasize theming, ambience, storytelling, and liminality, particularly in events design and social-gathering spaces, community formation, and surprise effects. The very principles and organizational features of Airsoft as a transformational NBT adventure activity require applying these experience design dimensions as its core elements.

New experiences design is a source of innovation for destinations expanding their appeal to new visitor markets (Gardiner & Scott, 2018), such as Airsoft. Including tourists in identifying experiences that enhance perceived value and sustainability is essential for destination managers and marketers (Breiby et al., 2020) and a basis of this research.

Growing demand for NBT, suddenly boosted by the COVID-19 pandemic, is expected to become a strong future trend (Fredman & Margaryan, 2020; Spenceley, 2021). This calls for creative and innovative NBT entrepreneurship practices (Fuchs et al., 2021), Airsoft being such an opportunity. However, that will require the transformation and adaptation of some natural areas to this tourism activity. Along with positive impacts on the environment, there will also be negative ones, such as activity-related soil and vegetation damage and wildlife harassment (Leung et al., 2008). Due to its nature, Airsoft has a responsibility in this matter. Management interventions can significantly influence and mitigate such adverse impacts. However, public land managers have lacked detailed data on recreation use levels and the specific types of experiences visitors desire (McCool & Kline, 2020), particularly in the case of often-unperceived NBT activities, such as Airsoft. In that context, it is crucial to know how willing such tourists are to comply with the rules and to contribute to their implementation, particularly in NBT niches, Airsoft included. In addition, high-quality nature-based recreation experiences with incorporated educational opportunities can potentially lead to developing pro-environmental attitudes (Kil et al., 2014). Therefore, it is important to investigate the willingness of Airsoft players to be actively involved and positively contribute to the sustainability of this NBT activity.

Methodology

A mixed-sequential qualitative-quantitative method was applied to answer the research questions. Qualitative research served as a foundation upon which further quantitative research was designed and conducted, and, finally, results were interpreted with regard to the literature review. This method was chosen as it provides sustainable tourism academics with more opportunities for pragmatic transformative research and increasing reliability regarding transdisciplinarity (Molina-Azorín & Font, 2015), noted in the literature review of Airsoft and its part in the transformative adventure NBT paradigm. Qualitative and quantitative research phases are described in detail.

The qualitative research phase included fieldwork (ethnographic observation) (as in the work of Buckley, 2015) at over 30 Airsoft events in the period 2016–2019. This defined the events' participants as the sample population.

Description of three phases of qualitative research

First, *shadowing* as a research technique allowed observation for detailed-grained data collection (Meunier & Vasquez, 2008) and an in-depth systematic, direct *in situ* observation of behaviours within this particular social setting (Bartkowiak-Theron & Robyn Sappey, 2012). In accordance with that, the participants were followed during three Airsoft gaming events. The researcher assumed roles suitable for shadowing in such a tightly closed community: a game photographer and an organizational assistant. Therefore, it was possible to follow the players "invisibly" (without disrupting the game) from multiple perspectives:

- in-game: (i) in a group of Airsoft players (a fraction) on their game missions, and (ii) Airsoft game organizing team.
- off-game: with organization teams before and after the games.

This phase produced a list of the sport's main features (Figure 21.1). All observed aspects were subject to the researcher's interpretation, leading to a decreased objectivity level.

Next, an extended period of *immersion into Airsoft activities* helped further understanding of the aspects perceived and noted in the shadowing phase. The researcher's participation in studied activities has been used in adventure tourism research on a previous occasion (Buckley, 2018). *The immersion* phase aimed at further understanding Airsoft players' preferences regarding the experience design elements. Based on the variations in Airsoft features observed through shadowing, the researcher played: (i) as a member of two teams at gaming events open only to selected Airsoft community members (local and international); and (ii) independently at free access games. The results from this research phase are available in Figure 21.1. The question arose

research phase	results	categories
	QUALITATIVE RESEARCH	
shadowing	main airsoft features	in-game: • *community nature* • *teams' culture* • *specific loadout* • *game size* • *scenario* • *role-play* off-game: • *transport* • *accommodation* • *free time activities*
↓ **immersion**	experience design elements	motives and drivers to play criteria for rating gaming experience quality
↓ **semi-structured interwievs**	multiple-choice questions	comprehensive lists of offered answers
↓	***questionnaire design***	
↓	**QUANTITATIVE RESEARCH**	
survey deployment	292 respondents/ 33 countries	excell data
↓ **data analysis**	statistical analysis	airsoft-related tourism experience design elements

Figure 21.1 Research structure and results according to each phase.

whether the categorized results were comprehensive enough to encompass all relevant elements of Airsoft experience design.

Therefore, *semi-structured interviews*, underpinned by a review of pertinent literature, made it possible to determine various aspects of Airsoft players' motives and experience design needs. The interviews structured around three research questions were now possible. The questions contained lists of offered answers derived from the previous research phases and literature review. The interviewees were selected based on their rich Airsoft experience (five team leaders) and asked to revise the given lists of answers (refine, add, or delete items).

A survey was constructed based on the described three-phase qualitative research and relevant literature review, including nature-based, adventure, and transformative travel elements in Airsoft experience design.

The survey contained 13 questions closely related to and aligned with the three research questions. They are grouped into five themes: (i) general respondents' info (gender, age, country, education, and Airsoft experience); (ii) motivation; (iii) Airsoft event preferred location type; (iv) experience quality (rating criteria; organizational features, scenario, complementary services, opportunities for independently organized experience); (v) sustainability (preferences and behaviour). The survey was available from April 2019 to October 2020. It was distributed with the support of the European Airsoft Federation and numerous national associations. Since such associations network and cooperate almost entirely through social media (predominantly Facebook), the survey was published on the associations' official Facebook pages. The survey was also available to various Airsoft teams around the globe, Airsoft equipment manufacturers and vendors, and Airsoft celebrities/influencers via email. Respondents were asked to forward the survey to further respondents. Such targeted network sampling allowed gathering relevant responses from a relatively specific, small, and hard-to-reach population scattered globally (Dusek, Yurova, & Ruppel, 2015; Onwuegbuzie & Leech, 2005), such as the Airsoft community. The same method has was applied in previously published Airsoft-related research (Nowacki & Wiśniewska, 2015) and Airsoft gun market reports (Popular Airsoft, 2008). Finally, it reached 292 respondents from 33 countries.

IBM SPSS Statistics 21 and JASP 0.14.1 were used for descriptive statistical analysis and hypothesis testing.

Results

Quantitative research results are presented in alignment with five question groups (themes), including correlations between certain data sets where appropriate

Sample size considerations: It was not possible to determine the validity of the gathered data through the comparison to the total Airsoft population as there is no actual estimated number of Airsoft players worldwide (Popular Airsoft, 2008). Therefore, it is determined in comparison to sample size, geographical scope, and in alignment with the findings from previously conducted Airsoft-related research (in terms of country of origin, age, gender, motivation, and Airsoft experience). One such scientific research collected 100 responses in Poland (Nowacki & Wiśniewska, 2015) and another 88 in Romania (Roşu, 2015), the scope of each limited to one country. An Airsoft publisher's global research project reached 4,861 respondents worldwide (Popular Airsoft, 2008). Based on these, the sample size of this individual scientific global scope research allows drawing conclusions regarding travel experience design and sustainability.

i) *General respondents' info Geographical origin*: A total of 97 percent of the respondents came from Europe, which was a result of national and regional Airsoft associations' support of the research. Approximately 31 percent of all responses came from the Balkans region. The author's origin and active Airsoft participation in the region was potential reason. The responses came from 33 countries, most of which came from Serbia (15.8 percent), France (14.7 percent), Greece (9.9 percent), and Croatia (9.6 percent).

Gender, age, education level, and Airsoft experience:
A total of 97.6 percent of the respondents were male. Respondents belong to five age categories: 51 percent in 26–40 age group; 33.6 percent in 19–25 age group, 13.7 percent in 41–59 age group; 1 percent of those younger than 18, and 0.3 percent older than 60.

In terms of education level, 46 percent of the respondents have secondary education, and 42% have a university or master's degree.

Regarding Airsoft experience, 42.5 percent of respondents play up to 5 years, 34 percent for 5–10 years, 16.8 percent for more than ten years, and 6.2 percent for under a year.

The general profile of Airsoft players, who predominantly make the market for Airsoft-related travel, comprises male players, mostly between 25 and 40 years of age, about half of them with a university degree.

ii) *Motivation* for playing Airsoft is presented in Figure 21.2.

iii) *Airsoft event preferred location* types are presented in Figure 21.3.
Location is highly rated as an Airsoft event quality criterion (Figure 21.1). Preferred location types are provided in Figure 21.3. There is no statistically significant difference between types of location and (1) age groups, (2) the respondents' countries of origin (Table 21.1).

iv) *Experience quality*
Airsoft players' criteria for rating the *quality of an Airsoft gaming event* are provided in Figure 21.4.

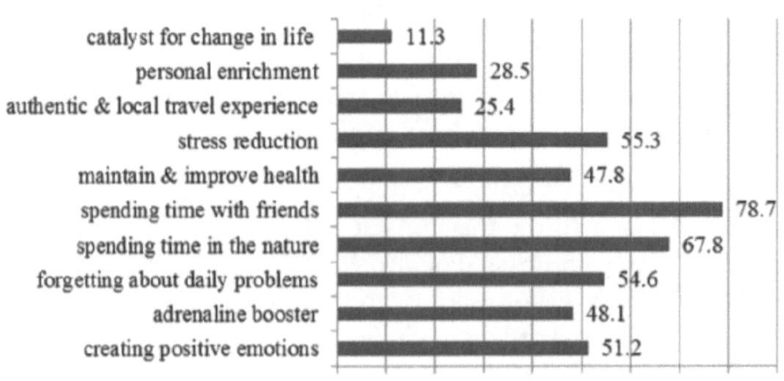

Figure 21.2 Motives for playing Airsoft (%).

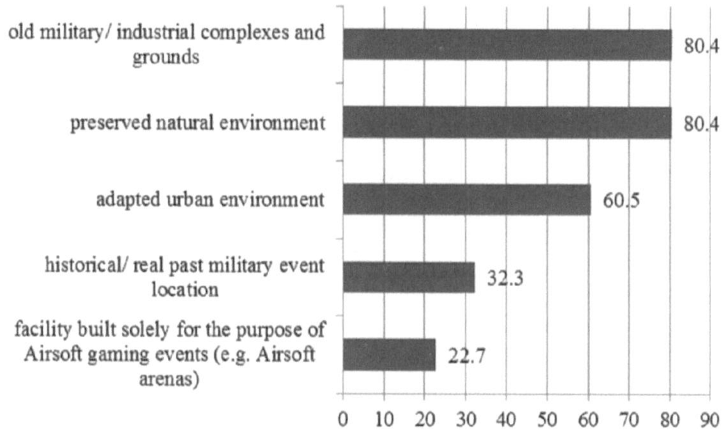

Figure 21.3 Preferred locations for playing Airsoft gaming events (%).

Table 21.1 Correlation between chosen variables

	Chi-Square Tests			
Choice of gaming **locations vs. age groups**	Pearson Chi-Square	356,908[a]	110	,000
	Likelihood Ratio	67,101	110	1,000
	N of Valid Cases	292		
	a. 124 cells (89.9%) have an expected count of less than 5. The minimum expected count is .00.			
Choice of gaming **locations vs. country** of origin	Pearson Chi-Square	1192,077[a]	726	,000
	Likelihood Ratio	326,095	726	1,000
	N of Valid Cases	292		
	a. 771 cells (98.6%) have an expected count of less than 5. The minimum expected count is .00.			
Criteria for rating Airsoft gaming event **quality vs. age groups**	Pearson Chi-Square	709,809[a]	435	,000
	Likelihood Ratio	230,313	435	1,000
	N of Valid Cases	292		
	a. 518 cells (98.1%) have an expected count of less than 5. The minimum expected count is ,00.			
Criteria for rating Airsoft gaming event **quality vs. length of playing**	Pearson Chi-Square	1356,613[a]	1235	,009
	Likelihood Ratio	530,958	1235	1,000
	N of Valid Cases	292		
	a. 1487 cells (99.9%) have an expected count of less than 5. The minimum expected count is .00.			

Source: Author.

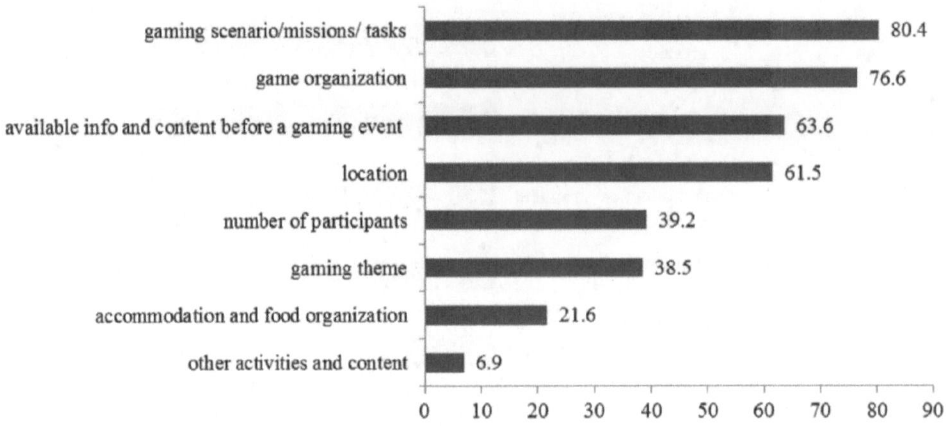

Figure 21.4 Criteria for rating an Airsoft gaming event quality (%).

There is a statistically significant difference between age groups, Airsoft experience, and criteria that players use to rate the quality of an Airsoft event (Table 21.1). Respondents who play for 1–5 years and belong to the age groups 19–25 and 26–40 use the criteria "available information and content before a gaming event" and "location." Players that belong to the same age group but with less than a year of playing experience rate the events according to the "number of participants," "game organization," and "accommodation and food organization."

Available info and content before an event are also strong determinants of quality experience, followed by the *number of participants*. However, when designing experiences for Airsoft players, organizers need to pay attention to various aspects that matter to different groups. So-called "newbies" to the sport (playing for less than a year) pay attention to game and accommodation/food organization and the number of participants. We can presume that for more experienced, "hardcore" players, such aspects are either implicit or have been refined over the years of playing. Therefore, they now focus on the determinants of the game quality itself – location and pre-event available content. The latter allows game immersion far before the event starts and allows adequate physical, technical, mental, and emotional preparation.

Organizational features of an Airsoft gaming event that increase the quality of players' experience are available in Figure 21.5.

Game-related features of an Airsoft event are among its most prominent event quality (Figure 21.4) and experience-rating criteria (Figure 21.5). Nature and built areas' attributes, game theme, and vehicle use for creating realness are on the top of the list. Realism can also be achieved through the participation of, e.g., military, parachuters, or free-climbers and the inclusion of extras/supers in the scenario. Finally, a reasonable participation fee contributes to the experience.

Certain aspects of off-game zone organization contribute moderately towards quality experience design. Providing camping areas is a valid cue for Airsoft event designers, emphasizing sanitary facilities, a favourable camping area equipped with a bonfire, barbecue, tables, and benches, followed by Airsoft retailers' stands and catering outlets as experience boosters.

Criteria	Frequency	%
access – concrete/ paved road to the game field	50	17.2
access – not too far away from the main road	59	20.3
access – sufficient number of parking lots	105	36.1
off-game zone – locker rooms	30	10.3
off-game zone – interesting surrounding area for complementary (non-*airsoft*) activities	37	12.7
off-game zone – present *Airsoft* and other equipment retailers	80	27.5
off-game zone – catering outlets (places to buy/ consume food)	84	28.9
off-game zone – recreation facilities (bonfire, barbecue, tables, benches, etc.)	112	38.5
off-game zone – favourable camping area	117	40.2
off-game zone – sanitary facilities	120	41.2
game – participation of various stakeholders (e.g. military, parachuters, free-climbers, divers)	101	34.7
game – acceptable registration/participation fee	123	42.2
game – use of equipment such as boats, helicopters, road vehicles, tanks, etc.	174	59.8
game – interesting theme and gaming scenario	180	61.9
game field – existing buildings and built obstacles	184	63.2
game field – existing natural obstacles (hills, rivers, etc.)	186	63.9

Figure 21.5 Experience-rating criteria – Organizational features of Airsoft events (%).

Scenario type (gaming theme) is the most prominent Airsoft event quality rating criterion (Figure 21.4). However, the background story it relies on is irrelevant for 52.2 percent of respondents. A completely (1) made-up scenario can increase experience quality for 33.3 percent of respondents; a scenario based on (2) an actual historical military event for 28.9 percent, (3) a popular film or series for 8.9 percent, and (4) a popular video game for 5.8 percent of respondents. The most frequent combination of preferred scenario types is "completely made-up" and "based on a real historical military event" (15.1 percent).

Complementary services that may increase an Airsoft gaming event experience quality are available in Figure 21.6.

In addition to Figure 21.6, for 15.1 percent of players, "participation of extras/supers for a better scenario" is the only complementary service that can increase their experience quality. Extras' participation and "other specialized training" represent the most frequent combination of Airsoft gaming complementary services (11.3 percent) with experience-increasing potential.

Opportunities for *experiences that are organized independently* from an Airsoft gaming event that can increase players' travel experience are provided in Figure 21.7.

Over a third of Airsoft players focus only on the game aspects when rating an Airsoft gaming event quality. However, for the same number of players, independently organized visits to a natural attraction or a local restaurant/café would increase the overall Airsoft-related experience. The same applies to about a fifth of them through an extension of stay in a nearby area and visiting a cultural attraction.

v) Sustainability

Players' attitudes towards sustainability are moderately high (see Table 21.2).

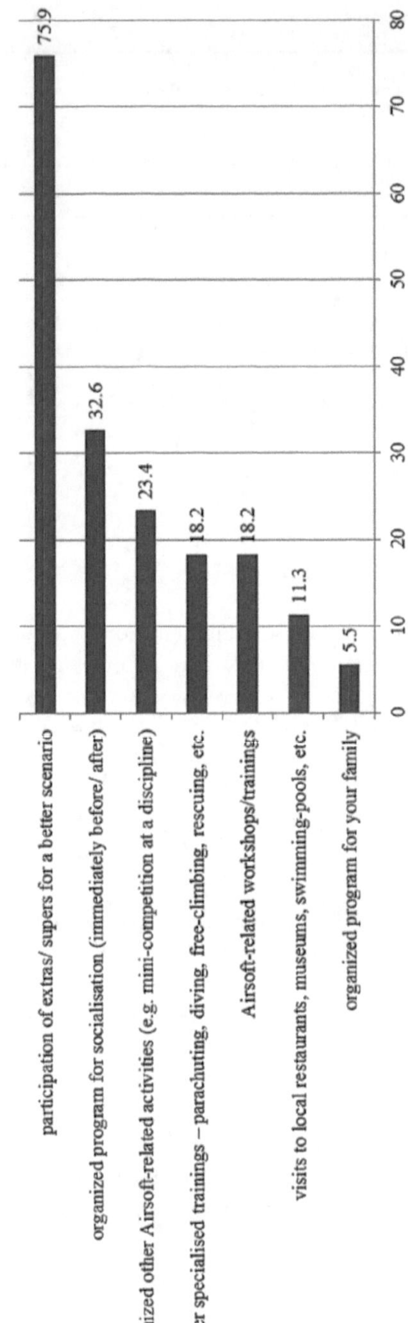

Figure 21.6 Experience-rating criteria – Complementary services (%).

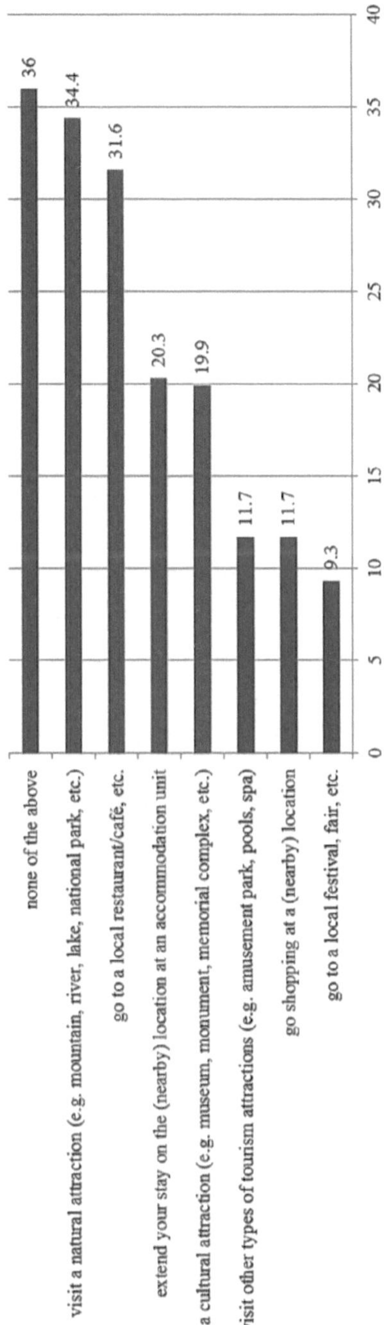

Figure 21.7 Experience-rating criteria – Opportunities separate from Airsoft event (%).

Table 21.2 Players' attitudes towards sustainability aspects

13. How important is it to you that during an Airsoft gaming event you:

a. learn about nature & culture of the location				b. do not disturb environment preservation				c. do not disturb daily routine and local communities' lifestyle				d. spend money that goes to local communities (e.g., renting location, food, accommodation)			
		Frequency	Valid Percent			Frequency	Valid Percent			Frequency	Valid Percent			Frequency	Valid Percent
Valid	1	56	19,2	Valid	1	13	4,5	Valid	1	14	4,8	Valid	1	30	10,3
	2	65	22,3		2	12	4,1		2	18	6,2		2	48	16,5
	3	78	26,8		3	47	16,2		3	57	19,6		3	74	25,4
	4	55	18,9		4	77	26,5		4	73	25,1		4	85	29,2
	5	37	12,7		5	142	48,8		5	129	44,3		5	54	18,6
	Total	291	100,0		Total	291	100,0		Total	291	100,0		Total	291	100,0
Missing	System	1		Missing	System	1		Missing	System	1		Missing	System	1	
Total		292			Total	292			Total	292			Total	292	

Source: Author

Discussion and conclusions

This section is organized according to the three research questions. A multidisciplinary literature review was necessary, followed by complex sequential qualitative-quantitative research to answer them. First, a three-phase researcher-immersive qualitative research phase was designed to construct a meaningful, Airsoft-specific questionnaire underpinned by the three research questions. The researcher spent time with Airsoft players following and observing them, playing with them, organizing games for them, and including them in the questionnaire design.

On that basis, 13 questions grouped into five themes corresponded to the three research questions. After the questionnaire deployment to a "secluded" Airsoft community worldwide, an analysis of the results was possible.

(1) *Place of Airsoft in tourism typology spectrum*
Airsoft's fast-growing nature, players' play-related motivations, play-environment, game-specific, and general organizational and experience design preferences place it into nature-based, adventure, and transformational tourism frameworks. This conclusion was firstly based on a literature review that compared the fundamental features of Airsoft and the determinants of these types of tourism. Further research results confirmed these conclusions. For example, among a multitude of motives that fuel Airsoft, the most prominent ones are spending time with friends and in nature, followed by stress reduction, forgetting daily problems, and creating positive emotions. This agrees with Buckley's (2020) and Winter et al.'s (2019) findings and defines Airsoft as a nature-based activity. Also, its physical, nature-associated, immersive experiential character and specific activity-related features place Airsoft into an adventure tourism framework. In addition, as a simulated, scenario-based, role-play game, Airsoft belongs to the transformative travel paradigm.

(2) *Airsoft-related experience design valued as high quality by participants/players*
An increasing interest in Airsoft requires attention to its experience design. Due to its nature (events design and organization), Airsoft-related travel sits at the crossroads between typically small-scale adventure nature sports and large-scale festival activities. Furthermore, simulated notions of authenticity (e.g., number of participants, game field, and use of real vehicles), players' immersion and co-creation (mission-oriented nature), and a sense of achievement and camaraderie are fundamental to Airsoft, which is in agreement with Rickly and McCabe's findings (2017). Therefore, for a quality transformative experience (as recognized by Neuhofer et al., 2020), theming (scenario), storytelling (in-game organization and related available content), and ambience (location) are essential aspects, as viewed by Airsoft event participants, and in accordance with their motives.

As it is expected that NBT and adventure tourism will be the first to recover after the COVID19 pandemic, it is crucial to be aware of the characteristics of Airsoft as a nature-based adventure activity. This knowledge has multifaceted benefits. First, it provides opportunities for innovation and creativity in NBT entrepreneurship for experience designers, as Airsoft has organizational features for creating a high-value experience environment, ambience, and atmosphere (supporting the conclusions by Neuhofer et al., 2020).

Airsoft players' interest in visiting or extending their stay in the vicinity of the gaming event harbours the enormous potential for underdeveloped regions. As a part of its integrated experience, Airsoft can be a hub for a complex system of tourism suppliers and various stakeholders in the vicinity of Airsoft playing ground, particularly for off-game zone organization (e.g., organized camping areas and socialization programme). Preferred gaming locations (preserved natural environment combined with old military and industrial complexes) hold possibilities for sustainable re-use of secluded and deserted resources with no particular need for building or refurbishing investment.

Based on this knowledge, authorities of the land used for playing Airsoft can not only prevent harm and conflicts but set up synergetic relationships. With the diversification of the demand for recreation, supplying recreation opportunities cannot remain fixed (McCool & Kline, 2020). Although the number of Airsoft players worldwide yet needs to be determined, at destinations unprepared for visitors, even one is excessive (Damnjanović, 2021), particularly true for those with unique needs. In Airsoft, this can be achieved through adequate concession permissions and public-private partnerships, as destination managers, especially in undeveloped regions, rely on the private sector to develop and maintain recreational facilities and sites (Mandić, Mrnjavac, & Kordić, 2018).

The findings lead to the conclusion that there is no need for massive investments in creating a high-quality Airsoft gaming experience (e.g., facilities, access), as is the case in many tourism products. Instead, investments could help preserve nature and facilitate the use of abandoned built complexes and in-game aspects (e.g., props and vehicles for creating realness). Developing partnerships with various stakeholder groups can create a high-quality experience and a competitive product. These can include industries and sports associated with Airsoft (e.g., parachuting, free climbing, rescuing) or developers of films and video games on which Airsoft game scenarios are based.

(3) *Sustainability aspects of Airsoft-based experiences*

Giddy (2016) recognized that adventure tourism operators have a limited sense of environmental awareness. However, this needs to change, particularly in niche adventure nature-based activities, to correspond to the participants' preferences. Airsoft players' attitudes towards sustainability aspects are moderately high. It is noticeable that they find it highly important not to disturb the environment and the daily routine of local communities at Airsoft event locations. They also demonstrate a moderately high attitude towards learning about the locale and economic contribution to local communities.

Academic and practical implications

These research results introduce Airsoft into academia in the context of nature-based, adventure, and transformational tourism. They open new avenues for research that can contribute to its overall, deeper understanding and a balanced recognition of Airsoft through its benefits for individual and societal wellbeing.

In a practical sense, this research could improve the quality of experience design in Airsoft-related travel, bring in innovation and creativity in NBT and adventure entrepreneurship, and present an opportunity for local/regional development by providing associated services. If Airsoft equipment manufacturers recognize a close connection between their sector and tourism, not paying attention to it from the perspective of tourism experience design would be an untapped opportunity. Instead, Airsoft can help destinations position their offerings towards emerging or unperceived niche segments. Simultaneously, those with a strong position in the nature-based adventure market can provide a more diverse and competitive, increasingly sought-after transformative experience. On a general level, this research indicates specific managerial, policy, and marketing aspects of NBT and opens a door for experience designers and land managers.

Limitations and recommendations for future research

The research represents a snapshot view and includes limited numbers of representatives from countries, mainly from Europe. The respondents' participation is unequally geographically distributed. Respondents from countries known to have numerous and developed Airsoft communities (e.g., the USA, Russia) are missing. The sample size cannot be compared to the Airsoft players' population size as there is no available estimate.

Differences in Airsoft players' profiles and preferences between countries need to be determined and compared, leading to a better experience design at international Airsoft events. For the same purpose, determining the differences in subtypes of Airsoft gaming is necessary. More needs to be examined regarding the sport market size and its economic potential, travel organizational aspects (e.g., transport, accommodation, and supply chain), promotion, obstacles to its recognition and development, and the like. Finally, research on Airsoft in the context of NBT, adventure, and transformational travel can lead to a better understanding of them all.

Acknowledgments

The author thanks the European Airsoft Federation and other national associations that supported the survey distribution (e.g., France, the Netherlands, Greece, Serbia, Croatia, Bosnia and Herzegovina) and Airsoft team BIG1 from Serbia for their continuous support.

Declaration of interest statement

The author reported no potential conflict of interest.

References

Adventure Travel Trade Association (ATTA), & The George Washington University International Institute of Tourism Studies (GW). (2020). *Adventure Tourism Development Index (ATDI) 2020*. https://www.adventuretravel.biz/research/2020-adventure-tourism-development-index

Bartkowiak-Theron, I., & Robyn Sappey, J. (2012). The methodological identity of shadowing in social science research. *Qualitative Research Journal*, *12*(1), 7–16. https://doi.org/10.1108/14439881211222697

Borderwar. (2021). https://www.borderwar.cz/sites/index.php

Breiby, M. A., Duedahl, E., Øian, H., & Ericsson, B. (2020). Exploring sustainable experiences in tourism. *Scandinavian Journal of Hospitality and Tourism*, *20*(4), 335–351. https://doi.org/10.1080/15022250.2020.1748706

Buckley, R. (2019). Therapeutic mental health effects perceived by outdoor tourists: A large-scale, multi-decade, qualitative analysis. *Annals of Tourism Research*, 77, 164–167. https://doi.org/10.1016/j.annals.2018.12.017

Buckley, R. (2020). Nature tourism and mental health: Parks, happiness, and causation. *Journal of Sustainable Tourism*, *28*(9), 1409–1424. https://doi.org/10.1080/09669582.2020.1742725

Buckley, R. C. (2015). Adventure thrills are addictive. *Frontiers in Psychology*, *6*. https://doi.org/10.3389/fpsyg.2015.01915

Buckley, R. C. (2018). To analyze thrill, define extreme sports. *Frontiers in Psychology*, *9*. https://doi.org/10.3389/fpsyg.2018.01216

Carlingford Tourist Office. (2020). *Bellurgan Park Airsoft*. Retrieved May 23, 2021, from https://carlingfordandcooleypeninsula.ie/carlingford-activities/carlingford-adventure-activities/bellurgan-park-Airsoft

Damnjanović, I. (2021). Overtourism and the local community well-being. In A. Mandić & L. Petrić (Eds.), *Mediterranean protected areas in the era of overtourism: Challenges and solutions* (pp. 93–115). Springer Nature. https://link.springer.com/book/10.1007/978-3-030-69193-6

Dusek, G. A., Yurova, Y. V., & Ruppel, C. P. (2015). Using social media and targeted snowball sampling to survey a hard-to-reach population: A case study. *International Journal of Doctoral Studies, 10*, 279–299. Retrieved from http://ijds.org/Volume10/IJDSv10p279-299Dusek0717.pdf

European Airsoft Federation. (2017, July 12). Retrieved January 14, 2021, from https://www.euroAirsoft.org/what-is-Airsoft/

Fact.MR. (2019). *Airsoft guns market forecast, trend analysis & competition tracking - Global market insights 2019 to 2029* (FACT1429MR). https://www.factmr.com/report/1429/airsoft-guns-market

Fossgard, K., & Fredman, P. (2019). Dimensions in the nature-based tourism experiencescape: An explorative analysis. *Journal of Outdoor Recreation and Tourism, 28*, 100219. https://doi.org/10.1016/j.jort.2019.04.001

Fredman, P., & Margaryan, L. (2020). 20 years of Nordic nature-based tourism research: A review and future research agenda. *Scandinavian Journal of Hospitality and Tourism*, 1–12. https://doi.org/10.1080/15022250.2020.1823247

Fuchs, M., Fossgard, K., Stensland, S., & Chekalina, T. (2021). Chapter 12: Creativity and innovation in nature-based tourism: A critical reflection and empirical assessment. In P. Fredman & J. V. Haukeland (Eds.), *Nordic perspectives on nature-based tourism: From place-based resources to value-added experiences*. Edward Elgar Publishing, Cheltenham, UK, https://doi.org/10.4337/9781789904031.00008

Gardiner, S., & Scott, N. (2018). Destination innovation matrix: A framework for new tourism experience and market development. *Journal of Destination Marketing & Management, 10*, 122–131. https://doi.org/10.1016/j.jdmm.2018.07.002

Giddy, J. K. (2016). Environmental values and behaviours of adventure tourism operators: The case of the Tsitsikamma, South Africa. *African Journal of Hospitality, Tourism and Leisure, 5*(4), Article 10.

Giddy, J. K., & Webb, N. L. (2017). Environmental attitudes and adventure tourism motivations. *Geo Journal, 83*(2), 275–287. https://doi.org/10.1007/s10708-017-9768-9

Global Wellness Summit. (2019). *2019 Global Wellness Trends Report*. https://www.globalwellnesssummit.com/2019-global-wellness-trends/download-report/

Grand View Research. (2019). *Airsoft Guns Market Size, Share & Trends Analysis Report 2019 - 2025* (GVR-3-68038-531-1). https://www.grandviewresearch.com/industry-analysis/airsoft-guns-market

Hansmann, R., Hug, S., & Seeland, K. (2007). Restoration and stress relief through physical activities in forests and parks. *Urban Forestry & Urban Greening, 6*(4), 213–225. https://doi.org/10.1016/j.ufug.2007.08.004

Hockings, M., Dudley, N., Elliott, W., Napolitano Ferreira, M., MacKinnon, K., Pasha, M. K., Phillips, A., Stolton, S., Woodley, S., Appleton, M., Chassot, O., Fitzsimons, J., Galleries, C., Kroner, R. G., Goodrich, J., Hopkins, J., Jackson, W., Jonas, H., Long, B., … Yang, A. (2020). Editorial essay: COVID-19 and protected and conserved areas. *Parks, 26*(1), 7–24. https://parksjournal.com/wp-content/uploads/2020/06/Hockings-et-al-10.2305-IUCN.CH_.2020.PARKS-26-1MH.en_-1.pdf

Horák, Vladimir, Linh, Duc, Roman Vítek, S. Beer, & Quang Huy, Mai. 2014. Prediction of the air gun performance. *Advances in Military Technology* 9, 31–44.

Houge Mackenzie, S., & Brymer, E. (2018). Conceptualizing adventurous nature sport: A positive psychology perspective. *Annals of Leisure Research, 23*(1), 79–91. https://doi.org/10.1080/11745398.2018.1483733

Houge Mackenzie, S., Hodge, K., & Filep, S. (2021). How does adventure sport tourism enhance well-being? A conceptual model. *Tourism Recreation Research*, 1–14. https://doi.org/10.1080/02508281.2021.1894043

Imarc. (2020). *Airsoft guns market share, size, growth, opportunity and forecast 2021-2026*. https://www.imarcgroup.com/airsoft-guns-market

Joyner, L., Lackey, Q., & Bricker, K. (2018). *Ecotourism Outlook 2018*. Outdoor Recreation, Education, and Tourism Lab, University of Utah. https://ttra.com/wp-content/uploads/2018/11/Ecotourism-Outlook-2018.pdf

Khalaily, S., Tsumi, E., Lifshitz, T., Kratz, A., & Levy, J. (2018). Airsoft gun-related ocular injuries: Long-term follow-up. *Journal of American Association for Pediatric Ophthalmology and Strabismus, 22*(2), 107–109. https://doi.org/10.1016/j.jaapos.2017.10.019

Kil, N., Holland, S. M., & Stein, T. V. (2014). Structural relationships between environmental attitudes, recreation motivations, and environmentally responsible behaviors. *Journal of Outdoor Recreation and Tourism, 7–8*, 16–25. https://doi.org/10.1016/j.jort.2014.09.010

Kim, H., Lee, S., Uysal, M., Kim, J., & Ahn, K. (2015). Nature-based tourism: Motivation and subjective well-being. *Journal of Travel & Tourism Marketing, 32*(sup1), S76–S96. https://doi.org/10.1080/10548408.2014.997958

Kim, M., & Thapa, B. (2018). Perceived value and flow experience: Application in a nature-based tourism context. *Journal of Destination Marketing & Management, 8*, 373–384. https://doi.org/10.1016/j.jdmm.2017.08.002

Laing, J., Voigt, C., & Frost, W. (n.d.). Fantasy, authenticity, and the spa tourism experience. In *health & wellness tourism: A focus on the global spa experience* (pp. 147–164). Routledge. https://www.routledge.com/rsc/downloads/Health_Wellness_Tourism_FB_final.pdf

Larsen, C. (2008). *Paintball and Airsoft battle tactics*. Voyageur Press.

Leung, Y. F., Marion, J. L., & Farrell, T. A. (2008). Recreation Ecology in Sustainable Tourism and Ecotourism: a Strengthening Role. In S. F. McCool & R. N. Moisey (Eds.), *Tourism, recreation, and sustainability: Linking culture and the environment* (2nd ed.) (pp. 19–37). CABI.

Mandić, A., Mrnjavac, Ž. I., Kordić, L. (2018). Tourism infrastructure, recreational facilities and tourism development. *Tourism and Hospitality Management, 24* (1), 41–62. https://doi.org/10.20867/thm.24.1.12

Martinez, L. E. (2008). *Force-on-force police training using Airsoft 2008: A manual for the law enforcement trainer on the use of Airsoft non-lethal technology*. Outskirts Press.

McCool, S.F.; Kline, J.D. 2020. Chapter 11: A systems thinking approach for thinking and reflecting on sustainable recreation on public lands in an era of complexity, uncertainty, and change. In: Selin, S.; Cerveny, L.K.; Blahna, D.J.; Miller, A.B., eds. *Igniting research for outdoor recreation: linking science, policy, and action*. Gen. Tech. Rep. PNW-GTR-987. Portland, OR: U.S. Department of Agriculture, Forest Service, Pacific Northwest Research Station: 161–172.

Menet, G., & Szarucki, M. (2020). Impact of value Co-creation on international customer satisfaction in the Airsoft industry: Does country of origin matter? *Journal of Risk and Financial Management, 13*(10), 223. https://doi.org/10.3390/jrfm13100223

Meunier, D., & Vasquez, C. (2008). On shadowing the hybrid character of actions: A communicational approach. *Communication Methods and Measures, 2*(3), 167–192. https://doi.org/10.1080/19312450802310482

Molina-Azorín, J. F., & Font, X. (2015). Mixed methods in sustainable tourism research: An analysis of prevalence, designs and application in JOST (2005–2014). *Journal of Sustainable Tourism, 24*(4), 549–573. https://doi.org/10.1080/09669582.2015.1073739

Neuhofer, B., Celuch, K., & To, T. L. (2020). Experience design and the dimensions of transformative festival experiences. *International Journal of Contemporary Hospitality Management, 32*(9), 2881–2901. https://doi.org/10.1108/ijchm-01-2020-0008

Nowacki, M., & Wiśniewska, I. (2015). Determinants of recreational activity – through the example of ASG players (Airsoft gun). *Studia Periegetica, 14*(2), 69–81.

Onwuegbuzie, A. J., & Leech, N. L. (2005). Taking the "Q" out of research: Teaching research methodology courses without the divide between quantitative and qualitative paradigms. *Quality & Quantity, 39*(3), 267–295. https://doi.org/10.1007/s11135-004-1670-0

Pine, B. J., & Gilmore, J. H. (1999). The experience economy: Work is theatre & every business a stage. *Choice Reviews Online, 37*(04), 22–54. https://doi.org/10.5860/choice.37-2254

Popular Airsoft. (2008). *Global Airsoft Survey Project*. Billion Tribes Limited. http://kam-airsoft.ru/pub/GASP_2008_Results_FULL.pdf

Powell, R. B., Kellert, S. R., & Ham, S. H. (2009). Interactional theory and the sustainable nature-based tourism experience. *Society & Natural Resources, 22*(8), 761–776. https://doi.org/10.1080/08941920802017560

Rickly, J. M., & McCabe, S. (2017). Authenticity for Tourism Design and Experience. In D. R. Fesenmaier & Z. Xiang (Eds.), *Design science in tourism. Tourism on the Verge* (pp. 55–68). Springer, Cham. https://doi.org/10.1007/978-3-319-42773-7_5

Roger, R. (2021, January 6). *Airsoft rules 2021: Do's and don'ts [A complete guide]*. Pick Airsoft. Retrieved May 23, 2021, from https://pickAirsoft.com/Airsoft-rules/

Roşu, D. (2015). Participants profile in Airsoft sport. *Procedia - Social and Behavioral Sciences, 180*, 1316–1321. https://doi.org/10.1016/j.sbspro.2015.02.270

Spenceley, A. (2021). *The future of nature-based tourism: Impacts of COVID-19 and paths to sustainability*. Luc Hoffmann Institute. https://www.ltandc.org/wp-content/uploads/2021/04/luchoffmanninstitute-future-nature-based-tourism-report-2021.pdf

Tornero-Aguilera, J. F., Sanchez-Molina, J., & Clemente-Suárez, V. J. (2020). Airsoft: An efficient and motivating cardiovascular training choice. *The Journal of Sports Medicine and Physical Fitness, 61*(1). https://doi.org/10.23736/s0022-4707.20.11167-8

Winter, P. L., Selin, S., Cerveny, L., & Bricker, K. (2019). Outdoor recreation, nature-based tourism, and sustainability. *Sustainability*, *12*(1), 81. https://doi.org/10.3390/su12010081

Wolf, I. D., Ainsworth, G. B., & Crowley, J. (2017). Transformative travel as a sustainable market niche for protected areas: A new development, marketing and conservation model. *Journal of Sustainable Tourism*, *25*(11), 1650–1673. https://doi.org/10.1080/09669582.2017.1302454

World Tourism Organization (UNWTO) (2019), *International Tourism Highlights, 2019 Edition*, UNWTO, Madrid. https://doi.org/10.18111/9789284421152

Wu, H., & Li, T. (2017). A study of experiential quality, perceived value, heritage image, experiential satisfaction, and behavioral intentions for heritage tourists. *Journal of Hospitality & Tourism Research*, *41*(8), 904–944. https://doi.org/10.1177/1096348014525638

Xu, F., Tian, F., Buhalis, D., Weber, J., & Zhang, H. (2015). Tourists as mobile gamers: Gamification for tourism marketing. *Journal of Travel & Tourism Marketing*, *33*(8), 1124–1142. https://doi.org/10.1080/10548408.2015.1093999

22

LINKING TOURISTS' VISITATION SATISFACTION, DEMOGRAPHIC, AND TRAVEL CHARACTERISTICS

John T. Mgonja

Introduction

The Travel and Tourism (T&T) industry is one of the most dynamic and important industries in Tanzania and it plays an important role in the country's economy. The T&T industry not only generates foreign revenues, but is also one of the most important engines for growth and development. For instance, for the year ending 2018, the T&T industry in Tanzania accounted for nearly 18 percent of the country's GDP. By the year 2018, foreign exchange receipts from tourism had grown from the figure of US$1,288.70 million in 2008 to US$2,412.30 million, while the corresponding number of international tourist arrivals for these two years grew from 770,376 to 1,505,702 (MNRT, 2019). Compared to other sectors of the economy in the country, the T&T industry is one of the fastest-growing, accounting for more than 13 percent of the country's labour force (1,389,000 jobs) in 2018.

Tanzania's rapidly growing tourism industry depends mainly on the six protected areas (PAs) located within the northern circuit (Serengeti, Lake Manyara, Tarangire, Kilimanjaro, Arusha national parks and the Ngorongoro conservation area). Nearly 38 percent of Tanzania's land area is set aside in protected areas for conservation (Tegegn, 2014). Currently, there are 22 national parks, which cover an area of approximately 99,306.5 sq.km. (TANAPA, 2020). Part of the revenues derived from park-entrance fees is used to support management and operations of all protected areas in Tanzania. Thus, tourism is not only important for economic growth and development but also an instrument for conserving PAs (Balmford et al., 2009; Munanura et al., 2016).

However, tourism is one of the world's most highly fragile and challenging industries. Its success is dependent upon many factors that are both directly and indirectly related to tourism such as the host country's political situation, the quality of attractions and services, infrastructure development, climate change, energy, health & safety and the occurrence of disasters (Roe & Urquhart, 2001) such as the ongoing COVID-19 pandemic.

From that perspective, tourist destinations have to be managed in a more strategic manner in order to be competitive and sustainable in both the short and the long run. From the standpoint of strategic management, destinations need to carry out continuous planning, monitoring, analysis and assessment of all the essentials which a destination requires to meet the demands of tourism. This is particularly important because in tourism terms some sites are sometimes

DOI: 10.4324/9781003230748-26

considered "once in a life time experience" where the traditional rules related to return visits do not apply. These might include, for example, hiking up Mount Kilimanjaro. One way of understanding tourist needs, motivations, behaviour and preferences is through the assessment of the demographic and travel characteristics of tourists. Studies suggest that the demographic characteristics of tourists are good indicators of their motivations, preferences and behaviour (Adam, Adongo, & Amuquandoh, 2019; Kim et al., 2008) and to a large extent reflect the nature of activities that tourists like to do and the places they like to visit. Studies suggest further that demographic characteristics of tourists are important in evaluating tourists' level of satisfaction and future travel behaviour (Lee, 2009; Ozdemir et al., 2012; Tsiotsou & Vasioti, 2006).

Despite the significance of demographic characteristics of tourists to PAs, very little information on tourist characteristics and preferences exist in relation to Tanzania (see e.g. Jani, 2018; Kara, 2016; Kara & Mkwizu, 2020; Kazuzuru, 2014; Okello & Yerian, 2009). Nevertheless, most of these studies focused on tourist values around issues of sustainability (Jani, 2018), links between demographic factors and tourist travel activity preferences (Kara, 2016), demographic factors and travel motivations (Kara & Mkwizu, 2020), the length of stay and trip-related characteristics (Kazuzuru, 2014) and tourist satisfaction (Okello & Yerian, 2009). None of these studies focused on the link between tourist demographic profiles and preferences within PA settings.

Further insights into the demographic characteristics of tourists to PAs can benefit Destination Marketing Organizations (DMOs), specifically with regard to market segmentation, product development, service quality evaluation, image development and promotional activities (Fodness, 1994: 555; Yoon & Uysal, 2005: 45; Kozak, 2001: 222).

Thus, this chapter describes a study that seeks to understand the relationship between the demographic and travel characteristics of tourists, tourists' preferences for watching diverse wildlife species in PAs and tourists' willingness to revisit the northern circuit in Tanzania and how these demographic profiles might be used to address the diversity of visitor needs and, consequently, promote transformative and memorable visitor experience.

This chapter will first discuss previous studies which focused on the correlations between demographic characteristics, tourist satisfaction and preferences. We will then introduce and explain the methods used to carry out this study. This section will be followed by study findings. The chapter will then end up with a discussion and an assessment of the implications of the findings.

Literature review

The relationship between tourist demographic characteristics, preferences, and satisfaction has been widely studied in tourism by various researchers (e.g. Martinez-Garcia & Raya, 2008; Shi, Liu, & Li, 2018; Veisten et al., 2015). Studies on demographic characteristics of tourists provide fundamental information for the management of tourists, marketing strategy, product development, and service provision (Cheung & Fok, 2014; Mehmetoglu & Normann, 2013).

Demographic characteristics and tourist segmentation

In understanding tourist behaviour, a large body of tourism literature has used demographic characteristics, activities, travel expenditure, benefits, and motivation to segment tourists (Birdir, 2015; Shi, Liu, & Li, 2018). The term demographics is derived from the Greek words for people (*demos*) and study/picture (*graphy*) and it generally refers to the particular characteristics of a population (Salkind, 2010). The demographic characteristics most commonly used in tourism studies include: age, race, gender, ethnicity, nationality, household income, education, marital status, family size, occupation, and the frequency of travel (Salkind, 2010; Weaver & Oppermann, 2000).

Several studies have examined the role of tourist demographic characteristics in understanding tourist behaviours in various settings. Many of these studies have established that there is a positive correlation between tourist demographic characteristics and tourist participation in various activities (Collin & Tisdell, 2002; Demir & Oral, 2007; Kim et al., 2008). A number of studies on tourism have also used demographic characteristics in assessing a range of issues: tourist travel demand (e.g. Collins & Tisdell, 2002); perceived image of tourist destinations (e.g. Beerli & Martín, 2004); tourist decision-making (e.g. Yusuf & Naseri, 2005); destination loyalty formation (e.g. Chi, 2011); travel motivation and satisfaction (e.g. Ma et al., 2018); and tourist activity choices (e.g. Slak-Valek, Shaw, & Bednarik, 2014). Other studies on tourism have also used demographic characteristics in understanding tourists' profile and preference of tourists' destination choice (e.g. Rahman et al., 2014) as well as predicting travelers' vacation type (e.g. Williams, Deslanders, & Crawford, 2007). However, some scholars contend that demographic characteristics are less effective than psychographic factors in predicting tourist behaviour (Fennell et al., 2003; Reisinger & Mavondo, 2004). Psychographic characteristics refer to the psychological attributes of people and include aspects such as personality, lifestyle, social status, AIO (Activities, Interests & Opinions) and attitudes (Senise, 2007). Nevertheless, studies that compared demographic and psychographic characteristics have concluded that the multi-combination variables of demographic segmentation exhibit market-segmenting capabilities equivalent to those of psychographic segmentation (Lin, 2002). The demographic segmentation of consumers has been widely employed by many researchers because it is admittedly more straightforward than psychographic segmentation.

Demographic characteristics and tourist preferences

Tourism literature shows that without any analysis of tourists' characteristics according to segments, Destination Marketing Organizations (DMO) as well as park managers cannot correctly select target markets to cater for their specialized needs, wants, and preferences (Otoo, Kim, & Choi, 2020). Thus, assessment of tourists' preferences is fundamental in the tourism industry because such assessments provide clues in developing tourism products, programmes, and services that are likely to attract more tourists in the future (Otoo, Kim, & Choi, 2020).

Tourist preference refers to the act of selecting from a set of choices as influenced by one's motivations (Tran & Ralston, 2006), and normally include aspects such as choices of transport, accommodation, services, food, travel companion, or destination. Motivation is a process of starting, directing, and maintaining behaviour, while preference serves as an intermediary step between motives and behaviour. Thus, tourist preference is the act of selecting from among a set of choices as influenced by one's motivations (Tran & Ralston, 2006).

Many tourism researchers have explored a range of tourist preferences and associated them with aspects of the tourists' demographic profile, such as age, gender, nationality, household income, education and marital status. For instance, a study conducted by Correia, Pimpao and Crouch (2008) found that younger tourists are more likely than older tourists to seek novelty. Similarly, an earlier study by Iso-Ahola et al. (1994) found that older tourists may be limited from taking part in some leisure activities because of health problems. In a similar study, Tran and Ralston (2006) makes the case that older tourists are unlikely to prefer adventure tourism activities because such activities take place in unusual, exotic, remote, or wilderness destinations and tend to be associated with high levels of involvement and activity by the participants. Several studies have also concluded that factors such as education, income, occupation (Biernat & Lubowiecki-Vikuk, 2012), gender and marital status (Kattiyapornpong & Miller, 2008) have an impact on the decision of a tourist to participate in particular leisure activities when travelling (Tran & Ralston, 2006).

Tourist preferences reflect the perceptions and comments of tourists regarding the destination following their actual holiday. The awareness of the preferences of segments of tourists can help PA managers to understand what aspects tourists like most and what they don't like most. Such awareness can also be useful in utilizing available resources effectively while meeting the needs and wants of the tourists. Similarly, the knowledge of activity preferences of tourists can benefit park managers by identifying what is perceived as favorable and desirable among potential tourists (Prebensen & Tkaczynski, 2012).

Based on the literature review, it can be seen that there is relatively little empirical research about the influence of demographic characteristics on tourist preferences in Tanzania. Thus, focusing on the preceding discussion, the following hypotheses are proposed.

Hypothesis [H1]: Demographic characteristics of tourists significantly influence tourist willingness to revisit nature-based destinations

Hypothesis [H2]: Demographic characteristics of tourists significantly influence tourist preferences for watching the diversity of wildlife species in parks visited.

Research method

Data collection for the study was done at the Kilimanjaro International Airport (KIA) in Tanzania. KIA was selected as a data collection point because many international tourists who visit the key PAs in the northern part of the country i.e. Kilimanjaro, Arusha, Serengeti, Lake Manyara, Tarangire and Ngorongoro Conservation Area use that airport. Data were collected for a period of three weeks in August. This period was selected for data collection because the best wildlife viewing took place during this period (Mgonja, 2015). A self-administered questionnaire was developed based on a comprehensive literature search. The first draft of the questionnaire was pretested using a sample of tourism professionals in the Department of Tourism and Recreation at the University to make sure that the questionnaire items were accurate and precise.

Each dependent variable from a section appropriate to this chapter in the questionnaire was measured using a 7-point Likert scale. With respect to preferences for wildlife species diversity in various parks visited, respondents were asked to rate the level of preferences, which ranged from 1 = not highly preferred to 7 = highly preferred. With regards to willingness to revisit the destination, respondents were asked to rate the level of agreement on a range from 1 = I will definitely not come back to 7 = I will definitely come back. In terms of how the questionnaire was administered, once the group made it through passport control and got a seat in the departure terminal, the researcher approached a group of tourists and requested one person in each group to fill out a survey (convenient sampling). The survey was written in the English language. A sample size of 520 was used in the survey, and there was a response rate of 88%. The collected data were then analysed using multiple regression analysis from SPSS version 20 software.

Study findings

This section presents the findings of the study. The first section presents demographic profiles of the respondents followed by testing of hypotheses.

The final data analysis involved 520 respondents (Table 22.1). Of these, 206 (39.6%) were above 60 years (older adults) and 204 (39.2%) were between 40 and 60 years (middle-aged adults). 275 (52.9%) of the respondents were male and 245 (47.1%) were female. About 386 (74.8%) of the respondents reported an annual household income of US$40,000–US$140,000 while about 87 (16.9%) reported a yearly income of above US$140,000. About 269 (52.4%) of the respondents attended some college or had a Bachelor's degree, while 194 (37.8%) of the respondents were postgraduates (with a Master's degree or PhD). 314 (60.4%) respondents originated from Europe while 127 (24.4%) originated from North America. Similarly, the majority of the respondents, 449 (86.3%), were first-time visitors.

Hypothesis testing

To test hypothesis (H1), the researcher used General Linear Model (GLM) univariate. Under this analysis, willingness to revisit the destination was treated as a Dependent Variable (DV)

Table 22.1 Demographic profile of respondents

Demographic	*Variable*	*Number*	*%*	*Mean DV1*	*Mean DV2*
Age	Ages 18–39 Years (Young adults)	110	21.2	6.3	5.6
	Ages 40–60 Years (Middle aged adults)	204	39.2	6.4	5.9
	Ages above 60 years (Older adults)	**206**	**39.6**	6.4	5.8
Gender	**Male**	**275**	**52.9**	6.2	5.9
	Female	245	47.1	6.6	5.8
Annual Income	Less than 40,000 USD (Low income)	43	8.3	6.3	5.4
	40,000–140,000 USD (Middle class)	**386**	**74.8**	6.4	5.9
	More than 140,000 USD (High income)	87	16.9	6.5	5.8
Education	Did no finish high school	7	1.4	5.0	6.0
	Finished high school	43	8.4	6.3	5.4
	Bachelor degree/attended some college	**269**	**52.4**	6.3	5.9
	Master degree or PhD	194	37.8	6.5	5.9
Nationality	North America	127	24.4	6.5	5.8
	South America	31	6.0	6.7	5.9
	Europe	**314**	**60.4**	6.4	5.8
	Australia and New Zealand	48	9.2	5.7	6.0
Are you a repeat visitor?	Yes	71	13.7	6.4	5.8
	No	**449**	**86.3**	6.4	5.9
Would you return to this destination?	Yes	**495**	**95.2**	6.4	6.4
	No	27	4.8	5.5	5.4

DV 1: If time or money is not a limitation, would you return to this destination in the future?

DV 2: Please tell us how important is wildlife species diversity to you in the parks visited.

and demographic characteristics of the respondents were treated as an Independent Variables (IV). The GLM is one of the most important tools in the statistical analysis of data. It allows researchers to include an enormous amount of information and summarize a wide variety of research outcomes. GLM (or the general multivariate regression model) is a compact way of simultaneously writing several multiple linear regression models. One of the assumptions of GLM is that the errors are usually assumed to be uncorrelated across measurements, and follow a multivariate normal distribution.

The researcher selected "post hoc" function for significant variables only to decompose the main effect for all variables with more than two categories (i.e. age, income, education and nationality) as shown in Table 22.1. Gender and revisit variables were not moved in to the "post-hoc" window since they have only two categories (Table 22.1). The researcher selected LSD test in the equal variances assumed window, since it is the most commonly used test among researchers (see e.g. Amuquandoh, 2010; Schaarschmidt & Vaas, 2009; Waterman et al., 2020)

The results in Table 22.2 indicated that only gender (p = 0.000) and education (p = 0.011) had a significant main effect. The results also indicated that interactions, Nationality * Gender (p = 0.007) and Nationality * Repeat visitor (p = 0.012) each had significant effects. The results also show that the coefficient of determination is small ($R^2 = 0.369$), suggesting that the decision to revisit the destination is determined by more factors than the factors presented in the model.

After decomposing the main effect for education, the post-hoc tests result (Table 22.3) indicated that as far as the willingness to revisit the destination is concerned, there was a significant mean difference between respondents who did not finish high school and those who finished high school (p = 0.001), those with Bachelor's degree/attended some college (p = 0.000), and those with Master's degree or PhD (p = 0.000). Similarly, there was a significant mean difference between those who have Bachelor's degree/attended some college and those who have Master's degree or PhD (p = 0.037).

Table 22.2 Test of main effect: DV- Revisiting the destination

Source	*Type III Sum of Squares*	*df*	*Mean Square*	*F*	*Sig.*
Corrected Model	208.883[a]	122	1.712	1.852	.000
Intercept	3001.876	1	3001.876	3246.837	.000
Age	.443	2	.221	.239	.787
Gender	**11.470**	**1**	**11.470**	**12.406**	**.000**
Income	.785	2	.393	.425	.654
Education	**10.400**	**3**	**3.467**	**3.750**	**.011**
Nationality	2.362	3	.787	.852	.466
Repeat visitor	1.453	1	1.453	1.571	.211
Age * Gender	2.130	2	1.065	1.152	.317
Nationality * Gender	**11.459**	**3**	**3.820**	**4.131**	**.007**
Nationality * Repeat visitor	**8.286**	**2**	**4.143**	**4.481**	**.012**
Corrected Total	565.760	508			

a R Squared = .369 (Adjusted R Squared = .170), Only few statistically non-significant variables are also reported for illustrative purposes.

* = Interaction

Table 22.3 Post-hoc results for education (DV- willingness to revisit the destination)

(I) Education	*(J) Education*	*Mean Difference (I-J)*	*Std. Error*	*Sig.*	*95% Confidence Interval*	
					Lower Bound	*Upper Bound*
Did not finish high school	Finished high school	−1.3095*	.39255	**.001**	−2.0813	−.5377
	Bachelor/some college	−1.3097*	.36814	**.000**	−2.0335	−.5859
	Master degree or PhD	−1.5000*	.36999	**.000**	−2.2275	−.7725
Finished high school	Did not finish high school	1.3095*	.39255	.001	.5377	2.0813
	Bachelor/some college	−.0002	.15957	.999	−.3139	.3136
	Master degree or PhD	−.1905	.16379	.246	−.5125	.1316
Bachelor/some college	Did not finish high school	1.3097*	.36814	.000	.5859	2.0335
	Finished high school	.0002	.15957	.999	−.3136	.3139
	Master degree or PhD	−.1903*	.09091	**.037**	−.3690	−.0116
Master degree or PhD	Did not finish high school	1.5000*	.36999	.000	.7725	2.2275
	Finished high school	.1905	.16379	.246	−.1316	.5125
	Bachelor/some college	.1903*	.09091	.037	.0116	.3690

The error term is Mean Square(Error) = .925.
* The mean difference is significant at the 0.05 level.

The simple effect results for Gender * Education showed that the effect of education was significant for both males and females (Table 22.4). Within the group of males, there was a significant mean difference between respondents who finished high school and those who have Bachelor's degree or attended some college ($p = 0.022$). There was also a significant mean difference between those who have Bachelor's degree/attended some college and those who have Master's degree or PhD ($p = 0.023$). The results in Table 22.4 also showed that within the group of males, there were no respondents who did not finish high school.

Within females, there was a significant mean difference between; respondents who finished high school and those with Bachelor's degree or attended some college ($p = 0.000$), those who finished high school and those with Master's degree or PhD ($p = 0.001$), and those who finished high school and those who did not finish high school ($p = 0.003$). There was also a significant mean difference between those who have Bachelor's degree or attended some college and those who did not finish high school ($p = 0.000$). Likewise, there was a significant mean difference between those who have Master's degree or PhD and those who did not finish high school ($p = 0.000$).

The study results (Table 22.2) showed that nationality has no significant main effect. However, after decomposing the main effect for nationality, the post-hoc tests results (Table 22.5) indicated

Table 22.4 Simple effect of Gender *Education interaction (DV- willingness to revisit the destination)

Gender	*(I) Education*	*(J) Education*	*Mean Difference (I–J)*	*Std. Error*	*Sig.*	*95% Confidence Interval*	
						Lower Bound	*Upper Bound*
Male	Finished high school	Bachelor/some college	**.6048***	**.26208**	**.022**	.0882	1.1215
		Master degree or PhD	.2789	.27089	.304	−.2551	.8129
	Bachelor/some college	Finished high school	−.6048*	.26208	.022	−1.1215	−.0882
		Master degree or PhD	**−.3260***	**.14281**	**.023**	−.6075	−.0445
	Master degree or PhD	Finished high school	−.2789	.27089	.304	−.8129	.2551
		Bachelor/some college	.3260*	.14281	.023	.0445	.6075
Female	Finished high school	Bachelor/some college	**−.6726***	**.17629**	**.000**	−1.0205	−.3246
		Master degree or PhD	**−.6316***	**.17899**	**.001**	−.9848	−.2783
		Did not finish high school	**1.0000***	**.32828**	**.003**	.3521	1.6479
	Bachelor/some college	Finished high school	.6726*	.17629	.000	.3246	1.0205
		Master degree or PhD	.0410	.10530	.698	−.1668	.2488
		Did not finish high school	**1.6726***	**.29465**	**.000**	1.0910	2.2541
	Master degree or PhD	Finished high school	.6316*	.17899	.001	.2783	.9848
		Bachelor/some college	−.0410	.10530	.698	−.2488	.1668
		Did not finish high school	**1.6316***	**.29627**	**.000**	1.0469	2.2163
	Did not finish high school	Finished high school	−1.0000*	.32828	.003	−1.6479	−.3521
		Bachelor/some college	−1.6726*	.29465	.000	−2.2541	−1.0910
		Master degree or PhD	−1.6316*	.29627	.000	−2.2163	−1.0469

The error term is Mean Square(Error) = .572.

* The mean difference is significant at the 0.05 level.

Table 22.5 Post-hoc results for education (DV- willingness to revisit the destination)

(I) Nationality	*(J) Nationality*	*Mean Difference (I-J)*	*Std. Error*	*Sig.*	*95% Confidence Interval*	
					Lower Bound	*Upper Bound*
North America	South America	−.1975	.19324	.307	−.5774	.1824
	Europe	.1368	.10251	.183	−.0648	.3383
	Australia and New Zealand	.8600*	.16618	**.000**	.5333	1.1868
South America	North America	.1975	.19324	.307	−.1824	.5774
	Europe	.3343	.18115	.066	−.0219	.6904
	Australia and New Zealand	1.0575*	.22344	**.000**	.6182	1.4968
Europe	North America	−.1368	.10251	.183	−.3383	.0648
	South America	−.3343	.18115	.066	−.6904	.0219
	Australia and New Zealand	.7232*	.15196	**.000**	.4245	1.0220
Australia and New Zealand	North America	−.8600*	.16618	.000	−1.1868	−.5333
	South America	−1.0575*	.22344	.000	−1.4968	−.6182
	Europe	−.7232*	.15196	.000	−1.0220	−.4245

The error term is Mean Square(Error) = .925.
* The mean difference is significant at the 0.05 level.

that as far as the willingness to revisit the destination is concerned, there was a significant mean difference between respondents who originated from Australia and New Zealand and those who originated from North America (p = 0.000), those who originated from South America (p = 0.000) and those who originated from Europe (p = 0.000).

The simple effect results (Table 22.6) for (Nationality * Gender) showed that the effect of nationality was only significant for males but not for females. Within males, there was a significant mean difference between those who originated from Australia and New Zealand and those who originated from North America (p = 0.000), South America (p = 0.000) and Europe (p = 0.000).

The simple effect results for (Nationality * Repeat visitor) showed that the effect of nationality was significant for both first time and repeat visitors (Table 22.7). Within the group of first-time visitors (those who answered 'No'), there was a significant mean difference between those who originated from South America and those who originated from Europe (p = 0.048). Similarly, there was a significant mean difference between those who originated from Australia and New Zealand and those who originated from North America (p = 0.000), from South America (p = 0.000) and from Europe (p = 0.000). Within repeat visitors (those who answered yes), the only significant mean difference was between those who originated from Australia and New Zealand and those who originated from North America (p = 0.000 and from Europe (0.000).

In testing hypothesis (H2), the researcher used SPSS General Linear Model (GLM) univariate, where preferences for wildlife species diversity in the park was treated as a Dependent Variable (DV) and demographic characteristics were treated as Independent Variables (IV). Other procedures were similar to Hypothesis (H1) testing.

The results in Table 22.8 indicated that gender (p = 0.000), education (p = 0.030) and nationality (p = 0.034) all have significant main effects. The results also indicated that interactions

Table 22.6 Simple effect of Gender ⋆Education interaction (DV- willingness to revisit the destination)

Gender	*(I) Nationality*	*(J) Nationality*	*Mean Difference (I–J)*	*Std. Error*	*Sig.*	*95% Confidence Interval*	
						Lower Bound	*Upper Bound*
Male	North America	South America	−.1822	.26607	.494	−.7067	.3423
		Europe	.1878	.16045	.243	−.1285	.5040
		Australia and New Zealand	1.3495*	.25434	.000	.8482	1.8509
	South America	North America	.1822	.26607	.494	−.3423	.7067
		Europe	.3700	.24648	.135	−.1159	.8559
		Australia and New Zealand	1.5318*	.31575	.000	.9093	2.1542
	Europe	North America	−.1878	.16045	.243	−.5040	.1285
		South America	−.3700	.24648	.135	−.8559	.1159
		Australia and New Zealand	1.1618*	.23377	.000	.7010	1.6226
	Australia and New Zealand	North America	**−1.3495***	**.25434**	**.000**	−1.8509	−.8482
		South America	**−1.5318***	**.31575**	**.000**	−2.1542	−.9093
		Europe	**−1.1618***	**.23377**	**.000**	−1.6226	−.7010
Female	North America	South America	−.3818	.28625	.184	−.9468	.1831
		Europe	.1052	.11883	.377	−.1293	.3397
		Australia and New Zealand	.2182	.19753	.271	−.1717	.6080
	South America	North America	.3818	.28625	.184	−.1831	.9468
		Europe	.4870	.27432	.078	−.0544	1.0284
		Australia and New Zealand	.6000	.31646	.060	−.0246	1.2246
	Europe	North America	−.1052	.11883	.377	−.3397	.1293
		South America	−.4870	.27432	.078	−1.0284	.0544
		Australia and New Zealand	.1130	.17980	.531	−.2419	.4679
	Australia and New Zealand	North America	−.2182	.19753	.271	−.6080	.1717
		South America	−.6000	.31646	.060	−1.2246	.0246
		Europe	−.1130	.17980	.531	−.4679	.2419

The error term is Mean Square(Error) = .572.

⋆ The mean difference is significant at the 0.05 level.

Table 22.7 Simple effect of Nationality ★Repeat visitor (DV-willingness to revisit the destination)

Are you a repeat visitor	*(I) Nationality*	*(J) Nationality*	*Mean Difference (I–J)*	*Std. Error*	*Sig.*	*95% Confidence Interval*	
						Lower Bound	*Upper Bound*
No	North America	South America	−.1924	.20348	.345	−.5926	.2078
		Europe	.1865	.11257	.098	−.0349	.4079
		Australia and New Zealand	.8887*	.19409	.000	.5069	1.2704
	South America	North America	.1924	.20348	.345	−.2078	.5926
		Europe	**.3789***	**.19135**	**.048**	.0026	.7553
		Australia and New Zealand	1.0811*	.24822	.000	.5929	1.5693
	Europe	North America	−.1865	.11257	.098	−.4079	.0349
		South America	−.3789*	.19135	.048	−.7553	−.0026
		Australia and New Zealand	.7022*	.18133	.000	.3455	1.0588
	Australia and New Zealand	North America	**−.8887***	**.19409**	**.000**	−1.2704	−.5069
		South America	**−1.0811***	**.24822**	**.000**	−1.5693	−.5929
		Europe	**−.7022***	**.18133**	**.000**	−1.0588	−.3455
Yes	North America	Europe	−.1676	.12180	.177	−.4144	.0792
		Australia and New Zealand	.7013*	.14627	.000	.4049	.9977
	Europe	North America	.1676	.12180	.177	−.0792	.4144
		Australia and New Zealand	.8689*	.10040	.000	.6654	1.0723
	Australia and New Zealand	North America	**−.7013***	**.14627**	**.000**	−.9977	−.4049
		Europe	**−.8689***	**.10040**	**.000**	−1.0723	−.6654

The error term is Mean Square(Error) = .092.
★ The mean difference is significant at the 0.05 level.

Gender ★ Nationality (p = 0.000), Gender ★ Repeat visitor (p = 0.047) and Income ★ Repeat visitor (p = 0.023) all have significant effects.

After decomposing the main effect for education, the post-hoc tests result (Table 22.9) indicated that there was a significant mean difference between those who finished high school and those who have Bachelor's degree or attended some college (p = 0.000). There was also a significant mean difference between those who finished high school and those who

Table 22.8 Test of main effect: DV- Species diversity

Source	*Type III Sum of Squares*	*df*	*Mean Square*	*F*	*Sig.*
Corrected Model	144.671[a]	122	1.186	1.918	.000
Intercept	2320.084	1	2320.084	3752.101	.000
Age	.165	2	.082	.133	.875
Gender	14.588	1	14.588	23.593	**.000**
Income	.899	2	.449	.727	.484
Education	5.610	3	1.870	3.024	**.030**
Nationality	5.414	3	1.805	2.918	**.034**
Repeat visitor	.085	1	.085	.137	.711
Gender * Nationality	28.582	3	9.527	15.408	**.000**
Gender * Repeat visitor	2.454	1	2.454	3.969	**.047**
Income * Repeat visitor	4.725	2	2.362	3.820	**.023**
Corrected Total	383.352	508			

a R Squared = .377 (Adjusted R Squared = .181). Only few statistically non-significant variables are also reported for illustrative purposes.

Table 22.9 Post-hoc results for education (DV- Species diversity)

(I) Education	*(J) Education*	*Mean Difference (I–J)*	*Std. Error*	*Sig.*	*95% Confidence Interval*	
					Lower Bound	*Upper Bound*
Did not finish high school	Finished high school	.5952	.32103	.064	−.0359	1.2264
	Bachelor/some college	.1045	.30107	.729	−.4875	.6964
	Master degree or PhD	.1250	.30258	.680	−.4699	.7199
Finished high school	Did not finish high school	−.5952	.32103	.064	−1.2264	.0359
	Bachelor/some college	−.4908*	.13050	**.000**	−.7473	−.2342
	Master degree or PhD	−.4702*	.13395	**.001**	−.7336	−.2069
Bachelor/some college	Did not finish high school	−.1045	.30107	.729	−.6964	.4875
	Finished high school	.4908*	.13050	.000	.2342	.7473
	Master degree or PhD	.0205	.07435	.783	−.1257	.1667
Master degree or PhD	Did not finish high school	−.1250	.30258	.680	−.7199	.4699
	Finished high school	.4702*	.13395	.001	.2069	.7336
	Bachelor/some college	−.0205	.07435	.783	−.1667	.1257

The error term is Mean Square(Error) = .618.
* The mean difference is significant at the 0.05 level.

have Master's degree or PhD (p = 0.001). The simple effect results for Gender * Education (Figure 22.1) showed that the effect of education was significant for both males and females, with highly educated male showing more preferences for wildlife species diversity than female.

After decomposing the main effect for nationality, the post–hoc tests result (Table 22.10) indicated that there was a significant mean difference between those who originated from North America and those who originated from Australia and New Zealand (p = 0.037).

The simple effect results for Gender * Nationality (Table 22.11) showed that the effect of nationality was significant for both males and females. Within the group of males, there was

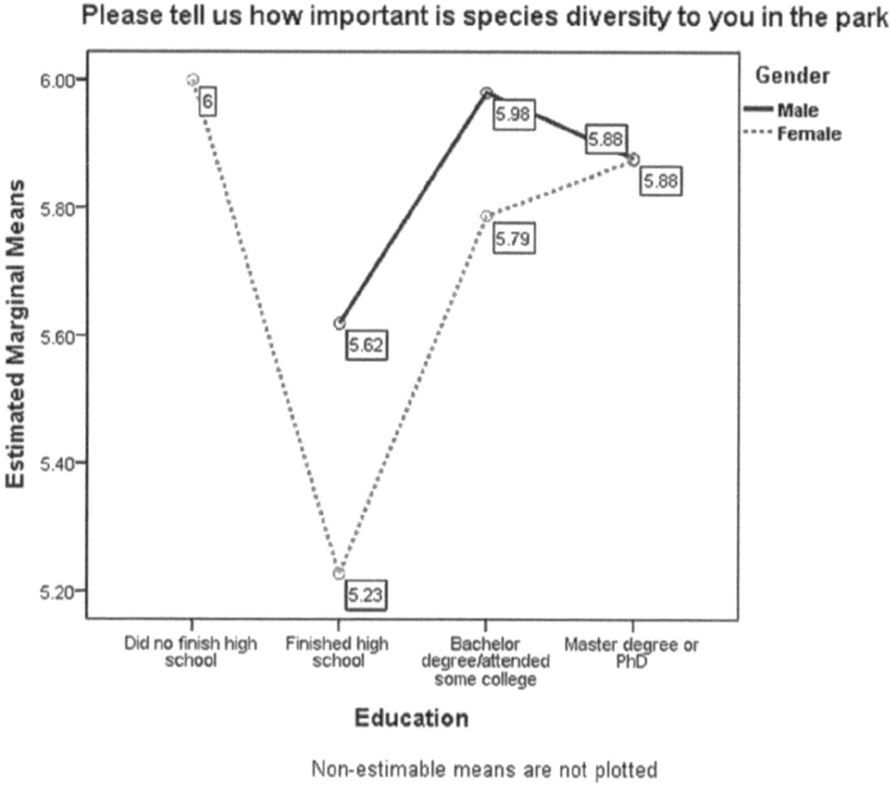

Figure 22.1 Gender * Education for DV – Species diversity.

Table 22.10 Post-hoc results for nationality (DV- Species diversity)

(I) Nationality	*(J) Nationality*	*Mean Difference (I–J)*	*Std. Error*	*Sig.*	*95% Confidence Interval*	
					Lower Bound	*Upper Bound*
North America	South America	−.0905	.15803	.567	−.4012	.2202
	Europe	−.0609	.08383	.468	−.2258	.1039
	Australia and New Zealand	−.2847*	.13590	**.037**	−.5519	−.0175
South America	North America	.0905	.15803	.567	−.2202	.4012
	Europe	.0295	.14815	.842	−.2617	.3208
	Australia and New Zealand	−.1942	.18273	.288	−.5535	.1650
Europe	North America	.0609	.08383	.468	−.1039	.2258
	South America	−.0295	.14815	.842	−.3208	.2617
	Australia and New Zealand	−.2238	.12427	.073	−.4681	.0205
Australia and New Zealand	North America	.2847*	.13590	.037	.0175	.5519
	South America	.1942	.18273	.288	−.1650	.5535
	Europe	.2238	.12427	.073	−.0205	.4681

The error term is Mean Square(Error) = .618.

* The mean difference is significant at the 0.05 level.

Table 22.11 Simple effect of Gender *Nationality (DV− Species diversity)

Gender	*(I) Nationality*	*(J) Nationality*	*Mean Difference (I–J)*	*Std. Error*	*Sig.*	*95% Confidence Interval*	
						Lower Bound	*Upper Bound*
Male	North America	South America	**−.5396***	**.18788**	**.004**	−.9100	−.1693
		Europe	−.0417	.11329	.713	−.2651	.1816
		Australia and New Zealand	**−.8122***	**.17960**	**.000**	−1.1663	−.4582
	South America	North America	.5396*	.18788	.004	.1693	.9100
		Europe	**.4979***	**.17404**	**.005**	.1548	.8410
		Australia and New Zealand	−.2726	.22296	.223	−.7121	.1669
	Europe	North America	.0417	.11329	.713	−.1816	.2651
		South America	−.4979*	.17404	.005	−.8410	−.1548
		Australia and New Zealand	**−.7705***	**.16507**	**.000**	−1.0959	−.4451
	Australia and New Zealand	North America	.8122*	.17960	.000	.4582	1.1663
		South America	.2726	.22296	.223	−.1669	.7121
		Europe	.7705*	.16507	.000	.4451	1.0959
Female	North America	South America	**1.1750***	**.30092**	**.000**	.5811	1.7689
		Europe	−.0766	.12492	.540	−.3232	.1699
		Australia and New Zealand	.4000	.20765	.056	−.0098	.8098
	South America	North America	−1.1750*	.30092	.000	−1.7689	−.5811
		Europe	**−1.2516***	**.28837**	**.000**	−1.8208	−.6825
		Australia and New Zealand	**−.7750***	**.33268**	**.021**	−1.4316	−.1184
	Europe	North America	.0766	.12492	.540	−.1699	.3232
		South America	1.2516*	.28837	.000	.6825	1.8208
		Australia and New Zealand	**.4766***	**.18902**	**.013**	.1036	.8497
	Australia and New Zealand	North America	−.4000	.20765	.056	−.8098	.0098
		South America	.7750*	.33268	.021	.1184	1.4316
		Europe	−.4766*	.18902	.013	−.8497	−.1036

The error term is Mean Square(Error) = .632.

* The mean difference is significant at the 0.05 level.

a significant mean difference between those who originated from North America and those from South America (p = 0.004). There was also a significant mean difference between those who originated from North America and those from Australia and New Zealand (p = 0.000). Similarly, there was a significant mean difference between those who originated from South America and those from Europe (p = 0.005). There was also a significant mean difference between those who originated from Australia and New Zealand and those from Europe (p = 0.000).

Within females, there was a significant mean difference between those who originated from North America and those from South America (p = 0.000). There was also a significant mean difference between those who originated from South America and those from Europe (p = 0.000). Similarly, there was a significant mean difference between those who originated from South America and those from Australia and New Zealand (p = 0.021). There was also significant mean difference between those who originated from Europe and those from Australia and New Zealand (p = 0.013).

The simple effect results for Gender * Repeat visitors (Table 22.12) showed that the effect of repeat visitation was only significant for females but not for males. Within females, there was a significant mean difference between those who visited the destination for the first time and the repeat visitors (p = 0.000).

The simple effect results for Income * Repeat visitation (Table 22.13) showed that the effect of income was significant for both first-time visitors and repeat visitors. Within first-time visitors, there was a significant mean difference between those earning less than US$40,000 per year and those earning US$40,000–140,000 per year (p = 0.000). There was also a significant mean difference between those earning less than US$40,000 per year and those earning more than US$140,000 per year (p = 0.002). Within repeat visitors, there was a significant mean difference between those earning US$40,000–140,000 per year and those earning more than US$140,000 per year (p = 0.002).

Table 22.12 Simple effect of Gender *Repeat visitation (DV- Species diversity)

Gender	*Repeat visitor Yes/No*	*Mean*	*Mean Difference*	*Std. Error*	*Sig.*	*95% Confidence Interval*	
						Lower Bound	*Upper Bound*
Male	Yes	6.1125	.2611	.0592	.494	.20531	1.08552
	No	5.8514	.2611	.0592	.494	.20531	1.08552
Female	Yes	**5.1077**	**.5133**	**.12856**	**.000**	**−.45206**	**.05412**
	No	5.6210	.5133	.12856	.000	−.45206	.05412

Table 22.13 Simple effect of Income *Repeat visitation (DV- Species diversity)

Repeat Visitor (Yes/No)	*(I) Income*	*(J) Income*	*Mean Difference (I–J)*	*Std. Error*	*Sig.*	*95% Confidence Interval*	
						Lower Bound	*Upper Bound*
No	Less than 40,000 USD	40,000–140,000 USD	**−.5785***	**.14335**	**.000**	−.8605	−.2966
		More than 140,000 USD	**−.5180***	**.16594**	**.002**	−.8444	−.1917
	40,000–140,000 USD	Less than 40,000 USD	.5785*	.14335	.000	.2966	.8605
		More than 140,000 USD	.0605	.10506	.565	−.1461	.2671
	More than 140,000 USD	Less than 40,000 USD	.5180*	.16594	.002	.1917	.8444
		40,000–140,000 USD	−.0605	.10506	.565	−.2671	.1461
Yes	Less than 40,000 USD	40,000–140,000 USD	−.0535	.17273	.759	−.4034	.2965
		More than 140,000 USD	.3788	.20352	.071	−.0336	.7912
	40,000–140,000 USD	Less than 40,000 USD	.0535	.17273	.759	−.2965	.4034
		More than 140,000 USD	**.4322***	**.13287**	**.002**	.1630	.7015
	More than 140,000 USD	Less than 40,000 USD	−.3788	.20352	.071	−.7912	.0336
		40,000–140,000 USD	−.4322*	.13287	.002	−.7015	−.1630

The error term is Mean Square(Error) = .161.

* The mean difference is significant at the 0.05 level.

Discussion

The objective of this study was to investigate the influence of the demographic characteristics of tourists on willingness to revisit the destination (DV1) and wildlife species diversity preferences in the parks visited (DV2). This study not only investigated differences within segments of tourists but also focused on differences across segments. Overall, this empirical study partially confirms the research hypotheses and the relevant models from which these hypotheses were derived from.

Demographic characteristics and tourists' willingness to revisit the destination

The study findings (Table 22.2) show that gender, education, and the interactions between gender and nationality, nationality and repeat visitation significantly influence tourists' willingness to revisit selected protected areas. Simultaneously, age and income of respondents are not significant predictors. The findings of this study contradict previous findings (e.g. Correia, Zins, & Silva, 2015; Garcia, Juaneda, Raya, & Sastre, 2015; Ozturk & Gogtas, 2016) which showed that certainly age and income are significant predictors for repeat visitation. For instance, a study by Correia, Zins and Silva (2015) in Portugal concluded that individuals with advanced age and higher incomes (over €5,000 monthly) exhibited a greater willingness to repeat visits than younger and lower-income visitors due to a lack of interest in finding cheaper alternatives and scarcity of time to plan holidays elsewhere, with some trust being conferred on an option already well known. The findings from Correia, Zins and Silva (2015) are supported by an earlier study by Teaff and Turpin (1996) who found that older Americans travel more frequently and longer distances, stay away longer, and rely more on travel agents than other segments of the population. Although age and income are not significant predictors in this study, it is pertinent to suggest that tourism managers particularly those dealing with tourists in the parks should provide more conducive environment for older visitors since this is the segment of the population with more discretionary income and time to travel. Older people have less recreation opportunities compered to younger people due to many reasons, including their healthy status, and therefore park managers should design activities that encourage the participation of older people. A good example of this is the recent use of hot air balloons as an alternative of game drive using cars at Serengeti National Park.

With respect to gender, the study findings (Table 22.2) indicate that there was a significant difference between males and females on their willingness to revisit the destination, with female showing more interest than males in revisiting the destination (Table 22.1). As pointed out by Chi, Gursoy and Qu (2009), gender-based research has inspired growing interests in tourism literature, as females become an increasingly important segment in the tourism industry. Thus, destinations should critically address the constraints, needs, and preferences of female tourists. For instance, studies show that female tourists are highly involved in shopping and are more affected by intrapersonal or structural constraints compared to males (Josiam, Kinley, & Kim, 2005).

In relation to education, the study findings (Table 22.3) showed that there was a significant difference between tourists who were highly educated and those who were less educated, with highly educated tourists showing more willingness to revisit the destination than less-educated ones (Table 22.1). With regards to gender and education interaction, the study findings showed that the effect of education was significant for both males and females (Table 22.4). However, further analysis showed that for a low level of education (high school level) male tourists were more likely to revisit the destination than female tourists, but as education increased (e.g. at Bachelor's level), female tourists were more likely to revisit the destination than males. At the highest level of education (Master's degree or PhD), the willingness of female tourists to revisit the destination appeared to be higher than that of male tourists. Overall, as the level of education increased, the likelihood of female respondents to revisit the destination became higher than that of male respondents. This study corroborates the findings from Yilmaz et al. (2009), who found that more educated female tourists had more positive image perceptions and were likely to return to the destination than did male travelers. Similarly, a study conducted by Shahrivar (2012) concluded that tourists who are highly educated are more satisfied and likely to revisit the destination compared to others.

With respect to nationality, the study findings (Table 22.5) showed that in general tourists from all nationalities are likely to revisit the destination in the future (mean above 6 for all nations). But, comparatively, tourists from South America and North America (both males and females), revealed that they were more likely to return to this destination than tourists from Europe, Australia, and New Zealand. Similarly, female tourists (regardless of nationality) indicated that they were more likely to revisit the destination than male tourists. Findings from this study corroborate previous research (e.g. Chi, 2011) which found that female tourists held more positive image perceptions than did male travelers.

Demographic characteristics and tourists' preference for wildlife species diversity in the parks

The study findings (Table 22.8) show that gender, education, nationality and the interactions between (gender and nationality, gender and repeat visitation, and between income and repeat visitation) are significant predictors of tourists' preferences for watching diverse wildlife species. The results also demonstrated that age and income of respondents were not significant predictors. The findings of this study differ to some extent from previous findings (e.g. Correia, Pimpao, & Crouch, 2008), which established that older tourists are less adventurers than younger tourists. The findings from Correia, Pimpao and Crouch (2008) are supported by another finding from Kattiyapornpong and Miller (2009), who found that retirees are significantly more likely to be constrained by the perception of age, disability, health conditions, and physical energy.

With respect to gender, the study findings (Table 22.8) demonstrate that there was a significant mean difference between males and females on their preference for watching a high diversity of wildlife species in the parks visited, with male showing more preference for wildlife species diversity than females (Table 22.1), although generally both males and females showed higher preferences for wildlife species diversity (mean above 5). This study corroborates the work by Muratet et al. (2015), who found that both male and female respondents preferred a high variety of natural attractions when visiting parks.

With respect to education (Table 22.9), the study results indicated that there was a significant mean difference between tourists who were highly educated (with Master's degree or PhD) and those who were less educated (finished high school), with highly educated tourists showing more preference for wildlife species diversity than less-educated ones (Table 22.1). Further analysis showed that the effect of education was significant for both male and female respondents. At a low level of education (Figure 22.1), the results showed that male tourists preferred more wildlife species diversity than female tourists. However, at the highest level of education (Master's degree or PhD), both male and female tourists exhibited the same level of preference for wildlife species diversity. Overall, the findings showed that the preference for wildlife species diversity increased with the increase in the level of education. This study corroborates the findings from Randler, Höllwarth, and Schaal (2007), who concluded that more educated tourists have more knowledge and preference for wildlife species diversity than less educated ones.

With respect to nationality, the study findings (Table 22.11) showed that the effect of nationality on wildlife species diversity preference was significant for both male and female respondents. However, wildlife species diversity appeared to be more important for male than female respondents for all nations (Figure 22.1), with the exception of North America and Europe, where the mean for male was equal to the mean for female. The results indicated further that within male respondents, wildlife species diversity was more important for tourists who originated from Australia and New Zealand than tourists who originated from South America, North America and Europe. Similarly, within female respondents, wildlife species diversity was more important for tourists who originated from Europe and North America than tourists who originated from Australia and New Zealand and South America.

With respect to revisiting the destination, the study results (Table 22.12) showed that the effect of repeat visitation on wildlife species diversity preference was significant for female tourists but not for male tourists. Further analysis showed that, in general, male tourists preferred more wildlife species diversity than female tourists for both segments (i.e. first-time and repeat visitors). Among male tourists, repeat visitors preferred more wildlife species diversity than first-time visitors, while among female tourists, first-time visitors preferred more wildlife species diversity than repeat visitors The results of this study contradict the findings of other studies, notably Coghlan and Prideaux (2008), who found that male respondents were less interested in wildlife viewing than female ones. However, a study by Ho et al. (2005) concluded that men and women did not differ significantly in perceived preferences for wildlife viewing (Ho et al. 2005).

In relation to the income of tourists, the study findings (Table 22.13) showed that the income of tourists moderates the effects of repeat visitation on wildlife species diversity preferences among tourists. For low-income tourists (those earning less than US$40,000 per year), preferences for wildlife species diversity was significantly higher for repeat visitors than for first-time visitors. However, for high-income tourists (those earning more than US$140,000 per year), preferences for wildlife species diversity was significantly higher for first-time visitors than for repeat visitors. For the middle class (those earning between US$40,000 and US$140,000 per year), the mean difference was not significant. Further analysis showed that, for repeat visitors, as income increased, the preference for wildlife species diversity decreased. Interestingly, for first-time visitors, as income increased, the preference for wildlife species diversity also increased. This study corroborates with the earlier observations that income of tourists is a significant predictor of tourist preferences (Kim et al., 2008; Torres-Sovero et al., 2012).

Conclusions and recommendations

The findings of this study has demonstrated an influence of tourist demographic characteristics on: (1) intentions to revisit the destination; and (2) wildlife species diversity preference in the parks visited. Thus, this study extends the existing literature on the importance of demographic factors in meeting the needs of tourists.

Preference for wildlife species diversity in the visited parks and willingness to revisit the destination was significantly influenced by gender, income, and education. These findings are comparable with those reported by previous researchers who found that demographic characteristics of tourists, including gender (Collin & Tisdell, 2002; Demir & Oral, 2007), income

(Martinez-Garcia & Raya, 2008; Torres-Sovero et al., 2012), and education (Martinez-Garcia & Raya, 2008; Torres-Sovero et al., 2012) were significant predictors of tourists' motivations and preferences. Surprisingly, age was not a significant predictor in this study as reported in other previous studies, notably Martinez-Garcia and Raya (2008).

Overall, this study has demonstrated that tourists visiting the northern protected areas of Tanzania are likely to come back to this destination in the future to enjoy the high diversity of wildlife. This signifies that, to a large extent, tourists had exciting experience during their trips in this destination but would like to return to acquire either more experience on the same or something different within this destination. Thus, park managers and Destination Marketing Organizations should devise special mechanisms to attract more tourists to come back to Tanzania. Such mechanisms may involve: designing new tourism products to cater for various tourist needs within and outside the parks (e.g. introduction of golf courses); improving fundamental facilities and systems (e.g. hotels, roads, power supplies and communication facilities) to support sustainable tourism particularly in the southern part of the country which has less visited national parks; increasing promotions of less visited parks in the country; improving and diversifying promotion efforts outside the country (e.g. using various promotion channels, such as social media, televisions, websites, etc.); improving tour guide trainings; and improving university and other tourism training institutions (e.g. improving training curriculums) etc. Such improvements will be important for enhancing destination image and competitiveness. The decisions for tourists to revisit the destination should be viewed and used as a core evaluative construct to help the destination develop more robust tourism products. The fact that majority of tourists who participated in this study showed interest in revisiting the destination (95.2%, n = 495) but that only about 10% were repeat visitors is an interesting finding and needs more research in relations to factors deterring tourists from revisiting the destination.

Destination Marketing Organizations and tourism managers, particularly in protected areas in Tanzania, can utilize these findings to minimize the barriers to participate in various recreation activities by particular population segment. For example, advanced age tourists cannot participate in game drive activities comfortably when park roads are too rough. Attractive (less bumpy) roads can be designed around park areas with high diversity of wildlife species or getaway attractions. Specific tour packages can be designed around constraint groups defined by age, income and life stage as an incentive to participate in recreation activities around the parks. Similarly, some tourists are time-constrained. Park management can design less time-consuming recreation activities in the park which enhances visitors' experience but less time-consuming. For example, park management can increase the number of hot air balloons in Serengeti to attract more visitors who are time-constrained and who can't do game drive in rough roads for a variety of reasons (e.g. health problems). Furthermore, this research assumes that repeat visits are related to some sort of risk aversion for some tourists; thus, future studies should assess the effects of perceived risk on the decision to return.

On a related note, the ongoing COVID-19 pandemic calls for a reexamination of tourism architecture to promote an approach to sustainability that significantly increases investment in health, safety and emergency preparedness during service delivery (Connor, 2020). It is clear that the COVID-19 pandemic has generated intense confusion and disorientation among tourism scholars and practitioners regarding the future of the tourism industry in many countries (Crouch, Dolnicar, & Hsu, 2020). Thus, post COVID-19 pandemic, it is likely that there will be significant changes in the tourism sector, causing stiff competition among products, services and tourist destinations, particularly in developing countries. It is also most likely that there will be substantial changes in tourists' expectations, demands and habits. Thus, understanding and paying attention to tourists' preferences post-COVID-19 pandemic will be significantly important for the continued growth of tourism in Tanzania.

With regard to limitations, this study used convenient sampling and, therefore, the results of this study cannot be generalized to the larger population because of the potential bias of the sampling technique due to the possibility of under-representation of subgroups in the sample in comparison to the population of interest. Similarly, data collection was done using a questionnaire written in the English language; thus, some respondents whose English is not their first language might have not understood some questions precisely.

References

Adam, I., Adongo, C. A., & Amuquandoh, F. E. (2019). A structural decompositional analysis of eco-visitors' motivations, satisfaction and post-purchase behaviour. *Journal of Ecotourism, 18*(1), 60–81.

Amuquandoh, F. E. (2010). Residents' perceptions of the environmental impacts of tourism in the Lake Bosomtwe Basin, Ghana. *Journal of sustainable tourism, 18*(2), 223–238.

Balmford, A., Beresford, J., Green, J., Naidoo, R., Walpole, M., & Manica, A. (2009). A global perspective on trends in nature-based tourism. *PLoS Biology*, 7(6), e1000144–e1000144

Beerli, A., & Martín, J. D. (2004). Tourists' characteristics and the perceived image of tourist destinations: A quantitative analysis—a case study of Lanzarote, Spain. *Tourism Management, 25*(5), 623–636.

Biernat, E., & Lubowiecki-Vikuk, A. P. (2012). Tourist activity among urban singles in view of socio-demographic factors. *Studies in Physical Culture and Tourism, 19*(2), 86–93.

Birdir, S. S. (2015). Segmentation of tourist using demographic and travel characteristics: The case of Istanbul. *International Review of Management and Marketing, 5*(4), 221–229.

Cheung, L. T., & Fok, L. (2014). The motivations and environmental attitudes of nature-based visitors to protected areas in Hong Kong. *International Journal of Sustainable Development & World Ecology, 21*(1), 28–38.

Chi, C. G., Gursoy, D., & Qu, H. (2009, August). Demographic variables and loyalty formation: A systematic examination. In *International CHRIE Conference-Refereed Track* (p. 18).

Chi, C. G. Q. (2011). Destination loyalty formation and travelers' demographic characteristics: A multiple group analysis approach. *Journal of Hospitality & Tourism Research, 35*(2), 191–212.

Coghlan, A., & Prideaux, B. (2008). Encounters with wildlife in Cairns, Australia: Where, what, who? *Journal of Ecotourism*, 7(1), 68–76.

Collins, D., & Tisdell, C. (2002). Gender and differences in travel life cycles. *Journal of Travel Research, 41*(2), 133–143.

Connor, C. (2020). A focus on sustainable service delivery. Retrieved on 12 March, 2020 from https://www.devex.com/news/after-the-pandemic-how-will-covid-19-transform-global-health-and-development-96936.

Correia, A., Pimpao, A., & Crouch, G. (2008). Perceived risk and novelty-seeking behavior: The case of tourists on low-cost travel in Algarve (Portugal). *Advances in Culture, Tourism and Hospitality Research, 2*, 1–26.

Correia, A., Zins, A. H., & Silva, F. (2015). Why do tourists persist in visiting the same destination? *Tourism Economics, 21*(1), 205–221.

Crouch, G., Dolnicar, S., Hsu, C., (2020). The future of tourism research in a post-pandemic world. Retrieved on 12 March, 2020 from https://sthm.temple.edu/events/the-future-of-tourism-research-in-a-post-pandemic-world.

Demir, C., & Oral, S. (2007). Relationship among demographic variables and the preferences of water park visitators: An implication from Turkey. *Journal of Yasar University, 2*(6), 537–553.

Fennell, G., Allenby, G. M., Yang, S., & Edwards, Y. (2003). The effectiveness of demographic and psychographic variables for explaining brand and product category use. *Quantitative Marketing and Economics, 1*(2), 223–244.

Fodness, D. (1994). Measuring tourist motivation. *Annals of Tourism Research, 21*(3), 555–581.

Garcia, J., Juaneda, C., Raya, J. M., & Sastre, F. (2015). A study of traveller decision-making determinants: Prioritizing destination or travel mode? *Tourism Economics, 21*(6), 1149–1167.

Ho, C. H., Sasidharan, V., Elmendorf, W., Willits, F. K., Graefe, A., & Godbey, G. (2005). Gender and ethnic variations in urban park preferences, visitation, and perceived benefits. *Journal of Leisure Research, 37*(3), 281–306.

Iso-Ahola, S. E., Jackson, E., & Dunn, E. (1994). Starting, ceasing, and replacing leisure activities over the life-span. *Journal of Leisure Research, 26*(3), 227–249.

Jani, D. (2018). Segmenting tourists using sustainable values: A cluster analysis of Tanzanian inbound tourism. *Journal of Quality Assurance in Hospitality & Tourism, 19*(3), 322–340.

Josiam, B. M., Kinley, T. R., & Kim, Y. K. (2005). Involvement and the tourist shopper: Using the involvement construct to segment the American tourist shopper at the mall. *Journal of Vacation Marketing, 11*(2), 135–154.

Kara, N. (2016). Demographic factors and preference for travel activities among tourists in Tanzania. *International Journal of Science Arts and Commerce, 1*, 9–26.

Kara, N. S., & Mkwizu, K. H. (2020). Demographic factors and travel motivation among leisure tourists in Tanzania. *International Hospitality Review, 34*(1), 81–103.

Kattiyapornpong, U., & Miller, K. E. (2008). A practitioner's report on the interactive effects of socio-demographic barriers to travel. *Journal of Vacation Marketing, 14*(4), 357–371.

Kattiyapornpong, U., & Miller, K. E. (2009). Socio-demographic constraints to travel behavior. *International Journal of Culture, Tourism and Hospitality Research, 3*(1), 81–94.

Kazuzuru, B. (2014). Determinants of tourist length of stay in Tanzania. *International Journal of Business and Social Science, 5*(9), 204–214.

Kim, S. S., Kim, M., Park, J., & Guo, Y. (2008). Cave tourism: Tourists' characteristics, motivations to visit, and the segmentation of their behavior. *Asia Pacific Journal of Tourism Research, 13*(3), 299–318.

Kozak, M. (2001). Repeaters' behavior at two distinct destinations. *Annals of Tourism Research, 28*(3), 784–807.

Lee, T. H. (2009). A structural model to examine how destination image, attitude, and motivation affect the future behavior of tourists. *Leisure Sciences, 31*(3), 215–236.

Lin, C. F. (2002). Segmenting customer brand preference: Demographic or psychographic. *Journal of Product & Brand Management, 11*(4), 249–268.

Ma, A. T., Chow, A. S., Cheung, L. T., Lee, K. M., & Liu, S. (2018). Impacts of tourists' sociodemographic characteristics on the travel motivation and satisfaction: The case of protected areas in South China. *Sustainability, 10*(10), 3388.

Martinez-Garcia, E., & Raya, J. M. (2008). Length of stay for low-cost tourism. *Tourism Management, 29*(6), 1064–1075.

Mehmetoglu, M., & Normann, Ø. (2013). The link between travel motives and activities in nature-based tourism. *Tourism Review, 68*(2), 3–13.

Mgonja, J. T. (2015). *Evaluating local food-tourism linkages as a strategy for promoting sustainable tourism and economic development: A case of Tanzania* (Doctoral dissertation, Clemson University).

MNRT (2019). Ministry of Natural Resources and Tourism (MNRT). The 2018 Tourism Statistical Bulletin. Dodoma, Tanzania.

Munanura, I. E., Backman, K. F., Hallo, J. C., & Powell, R. B. (2016). Perceptions of tourism revenue sharing impacts on Volcanoes National Park, Rwanda: A Sustainable Livelihoods framework. *Journal of Sustainable Tourism, 24*(12), 1709–1726.

Muratet, A., Pellegrini, P., Dufour, A. B., Arrif, T., & Chiron, F. (2015). Perception and knowledge of plant diversity among urban park users. *Landscape and Urban Planning, 137*, 95–106.

Okello, M. M., & Yerian, S. (2009). Tourist satisfaction in relation to attractions and implications for conservation in the protected areas of the northern Circuit, Tanzania. *Journal of Sustainable Tourism, 17*(5), 605–625.

Otoo, F. E., Kim, S., & Choi, Y. (2020). Understanding senior tourists' preferences and characteristics based on their overseas travel motivation clusters. *Journal of Travel & Tourism Marketing, 37*(2), 246–257.

Ozdemir, B., Aksu, A., Ehtiyar, R., Çizel, B., Çizel, R. B., & İçigen, E. T. (2012). Relationships among tourist profile, satisfaction and destination loyalty: Examining empirical evidences in Antalya region of Turkey. *Journal of Hospitality Marketing & Management, 21*(5), 506–540.

Ozturk, U. A., & Gogtas, H. (2016). Destination attributes, satisfaction, and the cruise visitor's intent to revisit and recommend. *Tourism Geographies, 18*(2), 194–212.

Prebensen, N. K., & Tkaczynski, A. (2012). Domestic nature-based tourism: a case study of Norway. *Tourism Analysis, 17*(6), 733–745.

Rahman, S. A. A., Osman, M. M., Bachok, S., & Ibrahim, M. (2014). Understanding tourists' profile and preference of tourists' destination choice: A case study in Melaka World Heritage City. *Planning Malaysia, 12*(3), 81–94.
Randler, C., Höllwarth, A., & Schaal, S. (2007). Urban park visitors and their knowledge of animal species. *Anthrozoös, 20*(1), 65–74.
Reisinger, Y., & Mavondo, F. (2004). Exploring the relationships among psychographic factors in the female and male youth travel market. *Tourism Review International, 8*(2), 69–84.
Roe, D., & Urquhart, P. (2001). Pro-poor tourism: Harnessing the world's largest industry for the world's poor. Opinion Paper for World Summit on Sustainable Development, International Institute for Environment and Development, London.
Salkind, N. J. (2010). *Encyclopedia of research design*. Thousand Oaks, CA: SAGE Publications, Inc. doi: 10.4135/9781412961288.
Schaarschmidt, F., & Vaas, L. (2009). Analysis of trials with complex treatment structure using multiple contrast tests. *HortScience, 44*(1), 188–195.
Senise, J. (2007). Who is your next customer? Strategies for targeting potential consumers in foreign markets. *Strategy and Business, Autumn, 1*(48), 1–5.
Shahrivar, R. B. (2012). Factors that influence tourist satisfaction. *Journal of Travel and Tourism Research (Online), 12*(1), 61.
Shi, T. T., Liu, X. R., & Li, J. J. (2018). Market segmentation by travel motivations under a transforming economy: Evidence from the Monte Carlo of the Orient. *Sustainability, 10*(10), 3395.
Slak-Valek, N., Shaw, M., & Bednarik, J. (2014). Socio-demographic characteristics affecting sport tourism choices: A structural model. *Acta Gymnica, 44*(1), 57–65.
TANAPA (2020). Tanzania National Parks (TANAPA). National Parks. Retrieved on 12 March, 2020 from https://www.tanzaniaparks.go.tz/pages/history
Teaff, J. D. and Turpin, T. (1996), Travel and elderly. *Parks and Recreation, 31*(6), 16–19.
Tegegn, M. (2014). The impact of dominant environment policies on indigenous peoples in Africa. *Indigenous People in Africa. Pretoria, ZAF: Africa Institute of South Africa*, 45–63.
Torres-Sovero, C., González, J. A., Martín-López, B., & Kirkby, C. A. (2012). Social–ecological factors influencing tourist satisfaction in three ecotourism lodges in the southeastern Peruvian Amazon. *Tourism Management, 33*(3), 545–552.
Tran, X., & Ralston, L. (2006). Tourist preferences influence of unconscious needs. *Annals of Tourism Research, 33*(2), 424–441.
Tsiotsou, R., & Vasioti, E. (2006). Using demographics and leisure activities to predict satisfaction with tourism services in Greece. *Journal of Hospitality & Leisure Marketing, 14*(2), 69–82.
Veisten, K., Haukeland, J. V., Baardsen, S., Degnes-Ødemark, H., & Grue, B. (2015). Tourist segments for new facilities in national park areas: Profiling tourists in Norway based on psychographics and demographics. *Journal of Hospitality Marketing & Management, 24*(5), 486–510.
Waterman, J. O., Campbell, L. A., Marechal, L., Pilot, M., & Majolo, B. (2020). Effect of human activity on habitat selection in the endangered Barbary macaque. *Animal Conservation, 23*(4), 373–385.
Weaver, D., & Oppermann, M. (2000). *Tourism management*. John Wiley & Sons, Brisbane, Australia.
Williams, D. A., Deslandes, D. D., & Crawford, D. (2007). Demographic factors that influence tourists' vacation choice. *IDEAZ, 6*, 5–21.
Yilmaz, Y., Yilmaz, Y., İçigen, E. T., Ekin, Y., & Utku, B. D. (2009). Destination image: A comparative study on pre and post trip image variations. *Journal of Hospitality Marketing & Management, 18*(5), 461–479.
Yoon, Y., & Uysal, M. (2005). An examination of the effects of motivation and satisfaction on destination loyalty: a structural model. *Tourism Management, 26*(1), 45–56.
Yusuf, F., & Naseri, M. (2005). A study of domestic and overseas holidays taken by Australian households. In *Australia New Zealand Marketing Academy* (pp. 139–145). ANZMAC, Fremantle, Australia.

23

TRAVEL CHARACTERISTICS, TYPOLOGIES, AND MOTIVATIONS OF NATURE-BASED TOURISTS

A Literature Review

Sıla Karacaoğlu

Introduction

There have been a range of ongoing cultural and social changes since the end of the 1980s and the beginning of the 1990s. In addition, increased environmentalism has emerged with cultural and structural changes that society has been experiencing, including the evolution of post-modern values (Mehmetoglu, 2007a). In this period, alternative development has been the most accepted approach in the world. With these developments, a wide range of products has emerged, which cover a range of alternative and niche tourism types. Apparently, new types of tourism, such as nature-based tourism (NBT), ecotourism, and sustainable tourism are supported as an environmentally safe basis for economic purposes across the world (Shafer & Choi, 2006; Marasinghe, Perera, Simpson, & Newsome, 2021). In particular, it is fair to state that NBT is the fastest-growing sustainable tourism type with a demand that is growing faster compared to the entire global tourism industry (Hall & Boyd, 2005; Mehmetoglu, 2007b; Chikuta, du Plessis, & Saayman, 2017), particularly in those areas which have rich natural resources such as countries in Africa, north Europe and the Asia-Pacific region (Fredman & Tyrväinen, 2010; Tyrväinen, Silvennoinen, & Hallikainen, 2017; Uusitalo, 2017; Kosenius, Juutinen, & Tyrväinen, 2020). Recently, it has been reckoned that some 10–20 percent of all international travel is done for nature experiences despite the fact that there exist no exact figures for the sector since there is no global database (Newsome, More, & Dowling, 2012). The reason behind this growth in demand is the global awareness that is being experienced throughout the world regarding the natural environment (Slabbert & Du Plessis, 2013). This demand can also be related to the phenomenon of growing urbanization, which has alienated people from nature. Therefore, to be able to meet their needs such as mental and physical health, people tend to spend more time in nature, which can be defined as not only protected areas, but also rural and suburban areas (Derek, Woźniak, & Kulczyk, 2019). In addition, it can be stated that this growth in demand can be related to the demand for several types of recreation, relaxation, and entertainment in nature besides prospering marketing campaigns (Olafsdottir, 2013). It is estimated that the growing demand for nature-based activities will

 DOI: 10.4324/9781003230748-27

remain in future which makes the sustainable development a more significant issue in nature-based destinations (Meng & Uysal, 2008). Consequently, in order to better understand nature tourists, it is vital to understand their attitudes and opinions with regard to NBT. In this context, the travel characteristics, segmentation, and motivation of nature-based tourists is the focal point of this chapter.

Nature-Based Tourism and Nature-Based Tourists

Tourism in nature has been defined in a range of terms: "natural area tourism", "nature tourism" or "nature-based tourism" (NBT). Nonetheless, it must be stated that there is little consensus on the meaning and the differences among these terms which results in a confusion in the field of tourism since some use these terms with clear-cut distinctions while others use them interchangeably (Fredman & Tyrväinen, 2010; Tangeland, Aas, & Odden, 2013a; Chen, Wang, & Prebensen, 2016; Sumanapala & Wolf, 2020). Other different terms that are used synonymously with NBT include ecotourism, green tourism, sustainable tourism, responsible tourism, and alternative tourism (Weiler & Hall, 1992; Priskin, 2001; Luo & Deng, 2008; Mehmetoglu, 2007a). It can be stated that all the mentioned terms have their origins in the framework of alternative tourism (Valentine, 1992; Weaver, 2001; Mehmetoglu, 2010).

Laarman and Durst (1987) regarded NBT as a nature trip that combines education, fun and, frequently, adventure. According to Boo (1990), it is an act of travelling to unspoiled natural areas with the aim of studying, enjoying, and adoring its scenery and inhabitants such as animals and plants in addition to any existing cultural manifestations. Valentine (1992: 108) defined NBT as "primarily concerned with the direct enjoyment of some relatively undisturbed phenomenon of nature". The enjoyment that the tourist experience in nature is deeply related to the degree to which their travel experience depends on nature. He states that there exist three kinds of relation between nature and tourist within the framework of his definition and these relations can be stated as activities dependent on nature, activities for which the natural setting is identical, and activities enhanced by nature. Priskin (2001: 638), in short, defined NBT as "tourism that features nature". With regard to Fredman & Tyrvainen (2010: 179), NBT can be defined as "travelling to and staying overnight in locations close to protected areas, forests, lakes or the sea or the countryside and participating in activities compatible with the location's natural qualities". In other words, this type of tourism capitalizes the cultural and natural attractions of the destination. Chikuta, du Plessis and Saayman (2017: 2), based on the definitions in the literature, defined NBT as "tourism primarily concerned with direct enjoyment of some relatively undisturbed phenomenon of nature for at least an overnight stay at the destination". In addition, they stated that there are three recurrent themes in the previous definitions. These themes can be stated as: (i) visiting a natural area; (ii) experiencing the nature; and (iii) sustaining the natural environment. In short, there are various definitions and frame of references about the NBT.

It is possible to define NBT as a travel to natural areas in which we see examples of environmental conversation, social equity, environmental education, and economic practicality regardless of the damaging inflicted on the host environment. Since it is seen as an alternative to traditional, destructive forms of development, NBT has been praised as an environmental education which is an informal, sustainable development tool in undisturbed areas (Powell, Kellert, & Ham, 2009). NBT differs from other tourism types with its emphasis on sustainability and experiencing natural scenery (Line & Costen, 2011). There are various types of NBT such as wildlife tourism, cultural travels in which tourists have a desire to learn about cultures of other communities inhabiting in natural attractions, and travels which support the aims of sustainable ecosystems (Divinagracia, Divinagracia, & Divinagracia, 2012). In spite of the lack of consensus over the definition, various tourism products are based upon outdoor activities such as bicycling, kayaking, canyoning, and hiking, etc. (Tangeland, 2011). The participants of these outdoor recreation activities need a set of skills and also need specialized equipment in order to do the activity (Tangeland et al., 2013a; Kosenius, Juutinen, & Tyrväine, 2020).

Based on the above definitions and studies in the literature, it is possible to summarize the general characteristics of the nature-based tourist. Nature-based tourists are considered to be environmentally conscious (Arnegger, Woltering, & Job, 2010). It is a common notion that nature-based tourists are keen on nature, have a greater tendency to travel long distances, and tend to stay longer at the destination (Priskin, 2003). Moreover, they are well-informed, have high levels of both individual and household income, and are enthusiastic about spending more while of holiday (Mehmetoglu, 2007b). Employed, educated, and highly paid nature-based tourists have motivation to experience the preserved nature and meanwhile to enjoy a cultural and a learning experience with family and friends (Slabbert & Du Plessis, 2013). Nature-based tourists are searching for natural, authentic and beautiful environments and experiences. In addition, they are socially and environmentally conscious (Pennington-Gray & Kerstetter, 2002). In short, the typical characteristics of nature-based tourists can be summarized as having an interest in traveling to uncrowded destinations, experiencing preserved nature, being informed regarding nature and culture, getting in touch with local people and taking part in challenging activities in terms of physicality (Palacio, 1997).

Benefits and Costs of Nature-Based Tourism

NBT can be regarded as an important economic activity, which has grown more quickly than any other tourism types (Mehmetoglu, 2007a). In addition, NBT is commonly employed as a regional development strategy thanks to its possible contribution to income variousness and job creation with its low requirements in training and substructure (Margaryan & Fredman, 2017). On condition that it is managed successfully, NBT can sustain additional income to both public and private landowners and managers besides additional reasons for wielding the resources for sustainable use. In addition to its economic significance, development in terms of tourism can result in the protection of cultural and social values such as the historical sites that might otherwise be lost. This is especially true for nature tourism since it supports cultural and ecological travels (Luzar, Diagne, Gan, & Henning, 1995). For instance, tourists who rather would like the money they spend to go to the parks, local residents, restaurants since they figure that the mentioned process would help preserve the nature (Lindberg, 2001).

Although the development of NBT in a destination provides many benefits in terms of local economy, socio-cultural structure and environmental issues, if not properly planned and managed, it can also cause undesirable costs in the dynamics of that destination and its community. Nevertheless, it must be stated that NBT may result in huge costs to local communities such as damage to farmlands, conflicts which can be named as the conflicts among tourism stakeholders and human-wildlife, stress to local residents which is a direct result of the increase in human population, and cultural degradation (Masuruli, 2014). However, tourism can also have negative effects on the environment. For instance, tourism operations may result in damaging ecological components, decrease ecological integrity, and degrade natural process. Therefore, it can be stated that tourism effects may be several such as air, water, and land pollution and soil and vegetation damage and damage to wildlife (Leung, Marion, & Farrell, 2008).

Segmentations of Nature-Based Tourists

To comprehend the relationship between consumers and products and to identify target segments, market segmentation holds vital importance since it determines groupings of consumers. Therefore, it is commonly used for product development, promotion, and management in travel and tourism (Nyaupane & Graefe, 2008). Segmenting tourists according to their motives for travelling to a destination is helpful and valuable since this type of psychographic segmentation base is useful for marketers and researchers for identifying the particular needs of travelers and how the needs affect their travel behaviours in order to create market strategies accordingly (Kruger, Saayman, & Hull, 2019). It is fair to state that segmenting tourists heterogeneously is the most significant factor in terms of creating influential marketing campaigns to reach the target cohort (Otoo, Kim, & Choi, 2020). In segmentation, there are significant variables, which include accommodation type, the number of travelers, preparation time for the trip, duration of trip, means of transportation, type of travel, the motivation for travel, socio-demographic features of the population, and so on. These help describe the consumer behaviour. Interdependence and their subordination on some situations, to economic, time, or family barriers are the main characteristics of these mentioned variables (Alén, Losada, & Domínguez, 2016).

Recently, the NBT market; with the demand growing rapidly; is becoming increasingly heterogeneous and the studies on ecotourists/nature-based tourists have revealed different segments of tourists according to their demographic, behavioural, and motivational features (Sumanapala, Kotagama, Perera, Galahitiyawe, & Suranga, 2017). Therefore, several different typologies which refer to different kinds of nature-based experiences, activities, and tourists have been advanced (Mehmetoglu, 2005). Lindberg (1991), Laarman and Durst (1987), and Orams (1995) are among examples of scholars who have created ecotourist typologies (Fennell, 2000). In the following years, for different types of nature-based experiences, activities, and tourists, there are various typologies that were developed in literature (Mehmetoglu, 2006). According to Laarman and Durst (1987), the level of interest and the degree of physical competence are significant in distinguishing between hard and soft nature tourists. While soft nature tourists is a term which refers to those who have a general glimpse on the nature activities, hard nature tourists have a deep knowledge regarding them. Lindberg (1991) categorizes nature-based tourists into four: (1) Hard-Core Nature Tourists: This category includes scientific researchers or tour members who have educational motivations. (2) Dedicated Nature Tourists: Tourists who have a specific desire to see the preserved areas and learn about the culture of the locals. (3) Mainstream Nature Tourists: Tourists who prefers unusual trips such as trip to Amazon or Safari. (4) Casual Nature Tourist: Tourists who participate in nature-based activities incidentally as a part of their broader trips. It is essential to note at this point that an individual can be categorized in more than one category at different times. However, the above typology is vital for tourism planners since it reveals the variety in the nature tourism. For instance, Type 1 and Type 2 tourists are expected to be more tolerant to the limitations regarding amenities, while Type 4 tourists will hope to be in uncrowded places.

Hvenegaard (2002) put forward four forms of tourist typologies via a case study regarding a national park in Thailand. According to Hvenegaard, there are four main empirical categories of definitional and typological approaches: "researcher-based", "respondent-based", "motivation-based", and "activity-based". In studies where we employ "researcher-based" typologies, the researcher is attempting to define different types of nature-based tourists. The researcher determines the categories of nature-based tourists before the empirical investigation which means before the contribution of the participants, therefore, it can be stated that these typologies become the supposed etic mode of classifying tourists. On the contrary, in "respondent-based" typologies, the tourists are asked to determine the type of nature-based tourists that they reckon they belong to. Here, it is possible for participants to have a decision on the category that best describes them best. In "motivation-based" studies, the tourists are requested to reveal the reasons behind their trips. In other words, they are asked to state the degree of the nature factor on their decisions. Via the data collected, the researcher makes a categorization of nature-based tourists. In "activity-based" typologies, inputs such as places visited, activities participated, and the knowledge regarding the activities are used by the researcher in order to define the varieties of nature-based tourists (Mehmetoglu, 2007c). In addition, Mehmetoglu (2005) claimed that nature-based tourists can be categorized ranging from specialist to generalist with the data collected from 162 visitors in northern Norway. With regard to their travel motives, it can be stated that nature-based tourists can be grouped into two distinct: "specialists" and "generalists". The "specialists" are the ones who travel mainly for nature, while the "generalists" are the ones that travel both for nature and other reasons. His study also depicts some distinctions in the mentioned two dimensions regarding origin, and duration in the target destination, and also reveals that the specialists tend to travel with historic and cultural motivations. According to Beh and Bruyere (2007), there exist three segments among tourists who visit three national attractions in north central Kenya: "escapists", "learners", and "spiritualists". The "escapist" can be defined as travelers who have a motivation to escape from routines of daily life and to have adventures. For the "spiritualists", self-reflection is vital. On the other hand, the "learners" would like to have a beneficial educational experience from reserve staff.

As a result, segmenting the NBT market is very important in comprehending it in a better way. Providing tourism products that are adapted to segments in this market is a must (Tangeland, 2011). In addition, segmentation is beneficial with regard to tourism management and planning. For those natural areas with high natural value, the stakeholders and managers must know the variety and diversity of experiences which are in demand in order to decide the diversity of setting where these activities are experienced. The diversity of experiences is vital in terms of planning since it can help with the defining typologies of nature-based tourists (Derek et al., 2019).

Travel Motivations of Nature-Based Tourists

Socio-demographics and travel characteristics are commonly used to comprehend several tourist types; however, they are still insufficient to explain the reason behind the travel choices of tourists and it is partly due to the fact that they are descriptive in nature. Conceivably, the mentioned motivational concerns affect all tourist behaviours (Mehmetoglu, 2006). Tourist motivation concept can be defined as "the global integrating network of biological and cultural forces which gives value and direction to travel choices, behaviour and experiences" (Slabbert & Du Plessis, 2013: 642). In addition, the analysis of tourist motivations allows a better understanding of the expectations, needs, and goals of tourists, and is also very important for designing the competitive and marketing strategies of businesses. Therefore, several theories have been put forward to explain and define the travel motives since the 1970s (Plog, 1974; Crompton, 1979; Dann, 1981; Iso-Ahola, 1982; Pearce, 1988). In the mentioned studies, the aim was to clarify the behaviours and actions of people when they have physical, social, and psychological needs and when they desire to reach their utmost physical, social, and mental states (Karn & Swain, 2017).

In his study, Plog (1974) reveals a starting point regarding travel motivation analysis. His work has been generally accepted as the major tourism motivation theory in the literature of travel and tourism. As to Plog, it is possible to represent the travelers based on their position on a single dimension or a scale. They are shown on a single dimension which extends from the categorization of "psychocentric" (introvert, non-adventurous) through to "mid-centric" (neither extrovert nor introvert) through "allocentric" (adventurous regarding looking for new destinations) (Pearce, 2011). In the following years, the "Push and Pull" theory was put forward by Dann (1977) and subsequently developed by Crompton (1979). This model proposes that internal derives (push factors) and external drives (pull factors) have an effect on the traveler in the course of decision process. It can be stated that the model is beneficial in analyzing traveler behaviour and decision-making process (Xu & Chan, 2016). According to Dann, there exist two factors as push travel motives: "Anomie" and "ego-enhancement". "Anomie" refers to the desire to precede the isolation in social life. In other words, the traveler wants to escape from the monotony and routines of daily life. Conversely, "ego-enhancement" refers to the need of approval which is obtained via the status by travel (Kassean & Gassita, 2013). With regard to Dann, the traveler's own push factor is more significant than any other motivation in choosing the target destination (Nikjoo & Ketabi, 2015). These inner motivations can be counted as "escape from personal/social pressures", "social recognition/prestige", "socialization/bonding", "self-esteem", "learning/discovery", "novelty/thrill", and "distancing from the crowds" (Kim & Lee, 2002). In 1979, Crompton put forward another model for push-and-pull factors and expanded on the importance of destination's role in attracting tourists. As to Crompton (1979), the unbalance among traveler's cultural, social, and psychological requirements can be considered a major motivation for travel. He indicated that the life of an individual can be unbalanced at some point since he lives in a social and psychological equilibrium. The routine in everyday doings at home or at work can result in this unbalance. The desire to escape from a repetitive environment, to change, and to relax may cause psychological disequilibrium. There is also a behavioural dimension to this interaction and it includes both the reasons for travel and the attitudes toward destination. There are nine motives in his model, seven of which are defined as push factors (socio-psychological): "escape from a mundane environment", "exploration and evaluation of self", "relaxation", "prestige", "regression", "enhancement of kinship relationships", and "facilitation of social interaction". The other two motives (pull factors) are "novelty" and "education" which emerge via some qualities of a destination (Uysal, Li & Sirakaya-Turk, 2008). As to Iso-Ahola (1982), "escaping" (routine environments) and "seeking" (intrinsic rewards) are the forces that creates leisure and tourism motivation. With Iso-Ahola's theory, it is referred to "escaping" motivation as the desire to move away from everyday life and also to "seeking" motivation as the desire to get an authentic reward for future behaviours. In addition, these motivational factors are two-dimensional: the personal dimension and the interpersonal dimension (Kim, Borges, & Chon, 2006). On the other hand, McIntosh and Goeldner (1986) focus on four need categories and motivators which trigger the desire of travelling in an individual: "physical motivators", "cultural motivators", "interpersonal motivators", and "status and prestige motivators". Another travel motivation theory is a "Travel Career Ladder" (TCL) developed by Pearce (1988, 1993). The conceptual scheme stresses the significance of the change in the motivational needs of travelers with more travel experience. The mentioned needs are depicted by building on the "Maslow's Hierarchy of Needs" approach and this depiction results in the five levels of a Travel Career Ladder. The steps can be counted as "physiological needs", "level of stimulation" control, "relationship needs", "self-esteem needs" and "self-actualization or personal growth needs". It was put forward that the more a traveler get experienced, the more he/she get motivated by the needs of self-esteem and self-actualization (Pearce, 2011).

The "Push and Pull" motivation framework (Kim, Lee, & Klenosky, 2003; Slabbert & Viviers, 2012; Scholtz, Kruger, & Saayman, 2013; Chen & Chen, 2015; Kim, Lee, Uysal, Kim, & Ahn, 2015) and the "Travel Career Ladder" (Kim & Lee, 2000; Hartley & Harrison, 2009; Shieh & Wang, 2012); the motivation theories mentioned above and widely accepted in the tourism literature; have been put to the test in several scenarios which are related to the travel motives of nature-based tourists. Generally, the studies that focus on the motivations in nature-based tourism have defined the following common push factors: "escape from daily routine" (Kim, Lee, & Klenosky, 2003; Kruger & Saayman, 2010; Hermann, Van der Merwe, Coetzee, & Saayman, 2016; Kim et al., 2015), "relaxation" (Holden & Sparrowhawk, 2002; Hartley & Harrison, 2009; Shieh & Wang, 2012; Ma, Chow, Cheung, & Liu, 2018), "social interaction" (Hartley & Harrison, 2009; Saayman & Saayman, 2009; Shieh & Wang, 2012), "education and learning about nature" (Luo & Deng, 2008; Kruger & Saayman, 2010; Chow, Cheng, & Cheung, 2019), "novelty seeking" (Kruger & Saayman, 2010; Mehmetoglu & Normann, 2013; Kim et al., 2015), "excitement/adventure/risk taking" (Holden & Sparrowhawk, 2002; Tangeland, Vennesland, & Nybakk, 2013b; Acquah, Dearden, & Rollins, 2016), "physical health" (Tangeland et al., 2013b; Kim et al., 2015; Jones & Nguyen, 2021), "self-fulfilment" (Hartley & Harrison, 2009; Nguyen & Pham, 2015; Massi & De Nisco, 2018), "self-esteem" (Kim & Lee, 2000; Hartley & Harrison, 2009; Shieh & Wang, 2012); "prestige and status" (Mehmetoglu, 2007a; Iversen, Hem, & Mehmetoglu, 2016; Xu & Chan, 2016). Common pull factors are as follows: "information and convenience of facilities and accessibility" (Kim et al., 2003; Chen & Chen, 2015; Xu & Chan, 2016); "nature scenery, experience and attractions" (Saayman & Saayman, 2009; Meng, Tepanon, & Uysal, 2008; Scholtz, Kruger, & Saayman, 2013), "photography" (Saayman & Saayman, 2009; Kruger & Saayman, 2010; Kruger et al., 2019), "local culture", "traditions and history" (Chan & Baum, 2007; Chen & Chen, 2015; Acquah et al., 2016; Nduna & van Zyl, 2020).

People who are disappointed with the ever-growing urban/suburban growth have either push motivational forces or pull motivational forces are in contact with nature as a part of recreation (Hartley & Harrison, 2009; O'Neill, Riscinto-Kozub, & Hyfte, 2010). Therefore, stating that nature-based tourists are commonly heterogeneous regarding their origins, interests, motivations, and behaviours is a must (Kitouna & Kim, 2017) and, in this case, the need to have a better understanding about tourists' expectations on outdoor recreation environments is becoming more significant (Tyrväinen et al., 2017).

Conclusion

With the awareness of increasing environmental problems in the 1960s and 1970s, environmental movements and approaches began to gain momentum all over the world. As in all sectors, sustainability has also begun to be questioned in tourism. It has been realized that mass tourism exhibited a rapid, unplanned, and uncontrolled development and it has begun to cause irreversible damage to natural and cultural assets. In this process, as a result of the argument that the sustainability of tourism will depend on the protection and development of the resources used for tourism purposes, the concept of sustainable tourism development has emerged. Nature-based tourism can be regarded as a tool for sustainable development.

Recently, parallel to the increasing demand for nature-based tourism, the factors (preferences, motivations, attitudes, behaviours, demographic characteristics, etc.) in tourists' travel decisions have begun to be studied more closely by destination planners, tourism marketers, and other tourism stakeholders. In this context, motivations are important needs that affect the behaviour of nature-based tourists and which are extremely helpful in finding the segmentation that provides the necessary information for the NBT market and facilitating the development of appropriate products. In this context, it is necessary to determine the intrinsic (psychological) motivations of the individuals participating in nature-based activities, as well as their motivation to the attractive aspects of the destination. In addition, demographic characteristics of the segment should also be considered.

In short, via segmentation, the specific characteristics, needs and interests of those participating in nature-based tourism activities can be determined and a higher level of marketing strategies can be developed. At this point, tour operators, travel agencies, accommodation businesses, and other tourism stakeholders interested in nature-based tourism should develop touristic products by taking into account the different characteristics of each group. The tourism products they develop should reach their potential customers by choosing the most effective and appropriate communication method. Thus, while ensuring the satisfaction of nature-based tourists and their re-visiting the destination, sustainable development can be in question in the destination where nature-based tourism takes place.

References

Acquah, E., Dearden, P., & Rollins, R. (2016). Nature-based tourism in Mole National Park, Ghana. *African Geographical Review, 35*(1), 53–69.

Alén, E., Losada, N., & Domínguez, T. (2016). The impact of ageing on the tourism industry: An approach to the senior tourist profile. *Social Indicators Research, 127*(1), 303–322.

Arnegger, J., Woltering, M., & Job, H. (2010). Toward a product-based typology for nature-based tourism: A conceptual framework. *Journal of Sustainable Tourism, 18*(7), 915–928.

Beh, A., & Bruyere, B. L. (2007). Segmentation by visitor motivation in three Kenyan national reserves. *Tourism Management, 28*(6), 1464–1471.

Boo, E. (1990). *Ecotourism: The Potentials and Pitfalls*. Washington, DC: World Wildlife Fund.

Chan, J. K. L., & Baum, T. (2007). Ecotourists' perception of ecotourism experience in lower Kinabatangan, Sabah, Malaysia. *Journal of Sustainable Tourism, 15*(5), 574–590.

Chen, J. S., W. Wang, & N. K. Prebensen (2016). Travel companions and activity preferences of nature-based tourists. *Tourism Review, 71* (1), 45–56.

Chen, L. J., & Chen, W. P. (2015). Push–pull factors in international birders' travel. *Tourism Management, 48*, 416–425.

Chikuta, O., du Plessis, L., & Saayman, M. (2017). Nature-based travel motivations for people with disabilities. *African Journal of Hospitality, Tourism and Leisure, 6*(1), 1–16.

Chow, A. S., Cheng, I. N., & Cheung, L. T. (2019). Self-determined travel motivations and ecologically responsible attitudes of nature-based visitors to the Ramsar wetland in South China. *Annals of Leisure Research, 22*(1), 42–61.

Crompton, J. L. (1979). Motivations for pleasure vacation. *Annals of Tourism Research, 6*(4), 408–424.

Dann, G. M. (1977). Anomie, ego-enhancement and tourism. *Annals of Tourism Research, 4*(4), 184–194.

Dann, G. M. (1981). Tourist motivation an appraisal. *Annals of Tourism Research, 8*(2), 187–219.

Derek, M., Woźniak, E., & Kulczyk, S. (2019). Clustering nature-based tourists by activity: Social, Economic and Spatial dimensions. *Tourism Management*, 75, 509–521.

Divinagracia, L. A., Divinagracia, M. R. G., & Divinagracia, D. G. (2012). Digital media-induced tourism: The case of nature-based tourism (NBT) at East Java, Indonesia. *Procedia-Social and Behavioral Sciences, 57*, 85–94.

Fennell, D. A. (2000). What's in a name? Conceptualizing natural resource-based tourism. *Tourism Recreation Research, 25*(1), 97–100.

Fredman, P., & Tyrväinen, L. (2010). Frontiers in nature-based tourism. *Scandinavian Journal of Hospitality and Tourism, 10*(3), 177–189.

Hall, C. M., & Boyd, S. (2005). *Nature-based tourism in peripheral areas: Development or disaster?*. Clevedon: Channel View Publications.

Hartley, N., & P. Harrison (2009). *An exploration of motives for attending Australian ecotourism locations and their influence on future intentions* (pp. 1–9). ANZMAC, Melbourne, Australia, 30 November–2nd December.

Hermann, U. P., Van der Merwe, P., Coetzee, W. J., & Saayman, M. (2016). A visitor motivational typology at Mapungubwe National Park and World Heritage Site. *Acta Commercii, 16*(1), 1–7.

Holden, A., & Sparrowhawk, J. (2002). Understanding the motivations of ecotourists: The case of trekkers in Annapurna, Nepal. *International Journal of Tourism Research, 4*(6), 435–446.

Hvenegaard, G. T. (2002). Using tourist typologies for ecotourism research. *Journal of Ecotourism, 1*(1), 7–18.

Iso-Ahola, S. (1982). Toward a social psychology theory of tourism motivation. *Annals of Tourism Research, 12*, 256–262.

Iversen, N. M., Hem, L. E., & Mehmetoglu, M. (2016). Lifestyle segmentation of tourists seeking nature-based experiences: The role of cultural values and travel motives. *Journal of Travel and Tourism Marketing, 33*(sup1), 38–66.

Jones, T. E., & Nguyen, M. H. (2021). Nature-based tourism motivations and visit profiles of domestic and international segments to a Japanese National Park. *Quaestiones Geographicae, 40*(2), 77–92.

Karn, S., & Swain, S. (2017). Health consciousness through wellness tourism: A new dimension to new age travelers. *African Journal of Hospitality, Tourism and Leisure, 6*(3), 1–9.

Kassean, H., & Gassita, R. (2013). Exploring tourists push and pull motivations to visit Mauritius as a tourist destination. *African Journal of Hospitality, Tourism and Leisure, 2*(3), 1–13.

Kim, E., & Lee, D. (2000). Japanese tourists' experience of the natural environments in North QLD Region—Great Barrier Reef experience. *Journal of Travel & Tourism Marketing, 9*(1–2), 93–113.

Kim, H., Borges, M. C., & Chon, J. (2006). Impacts of environmental values on tourism motivation: The case of FICA, Brazil. *Tourism Management, 27*(5), 957–967.

Kim, H., Lee, S., Uysal, M., Kim, J., & Ahn, K. (2015). Nature-based tourism: Motivation and subjective well-being. *Journal of Travel & Tourism Marketing, 32*(sup1), S76–S96.

Kim, S. S., & Lee, C. K. (2002). Push and pull relationships. *Annals of Tourism Research, 29*(1), 257–260.

Kim, S. S., Lee, C. K., & Klenosky, D. B. (2003). The influence of push and pull factors at Korean national parks. *Tourism Management, 24*(2), 169–180.

Kitouna, S., & Kim, Y. G. (2017). Tourists' novelty-seeking motivation in nature-based tourism destinations. *International Journal of Tourism and Hospitality Research, 31*(10), 45–58.

Kosenius, A. K., Juutinen, A., & Tyrväinen, L. (2020). The role of state-owned commercial forests and firm features in nature-based tourism business performance. *Silva Fennica, 54*, 1–23.

Kruger, M., & Saayman, M. (2010). Travel motivation of tourists to Kruger and Tsitsikamma National Parks: A comparative study. *South African Journal of Wildlife Research, 40*(1), 93–102.

Kruger, M., Saayman, M., & Hull, J. (2019). A motivation-based typology for natural event attendees. *Journal of Policy Research in Tourism, Leisure and Events, 11*(1), 35–53.

Laarman, J. G., & Durst, P. B. (1987). Nature travel in the tropics. *Journal of Forestry, 85*(5), 43–46.

Leung, Y. F., Marion, J. L., & Farrell, T. A. (2008). Recreation ecology in sustainable tourism and ecotourism: A strengthening role. In Stephen F. McCool & R. Neil Moisey (Eds.), *Tourism, Recreation and Sustainability* (2nd Edition) (pp. 19–37). London, UK: CABI International.

Lindberg, K. (1991). *Policies for Maximizing Nature Tourism's Ecological and Economic Benefits*. Washington, DC: World Resources Institute.

Lindberg, K. (2001). Economic impacts. In D. B. Weaver (Ed.), *The Encyclopaedia of Ecotourism* (pp. 363–378). New York, USA: CABI International.

Line, N., & Costen, W. (2011). Environmental attitudes, motivation, and attachment: Toward a model of nature-based tourism (pp. 1–16), *International CHRIE Conference-Refereed Track*, Denver, Colorado USA, 27 July.

Luo, Y., & Deng, J. (2008). The new environmental paradigm and nature-based tourism motivation. *Journal of Travel Research*, 46(4), 392–402.

Luzar, E. J., Diagne, A., Gan C., & Henning B. R. (1995). Evaluating nature-based tourism using the new environmental paradigm. *Journal of Agriculture and Applied Economics*, 27 (2), 544–552.

Ma, A. T. H., Chow, A. S. Y., Cheung, L. T. O., Lee, K. M. Y., & Liu, S. (2018). Impacts of tourists' sociodemographic characteristics on the travel motivation and satisfaction: The case of protected areas in South China. *Sustainability*, 10,3388, 1–21.

Marasinghe, S., Perera, P., Simpson, G. D., & Newsome, D., (2021). Nature-based tourism development in coastal wetlands of Sri Lanka: An importance–performance analysis at Maduganga Mangrove Estuary. *Journal of Outdoor Recreation and Tourism*, 33, 100345.

Margaryan, L., & Fredman, P. (2017). Natural amenities and the regional distribution of nature-based tourism supply in Sweden. *Scandinavian Journal of Hospitality and Tourism, 17*(2), 145–159.

Massi, M., & De Nisco, A. (2018). The internet-based marketing of ecotourism: Are ecotourists really getting what they want?. In Camilleri, M. A. (Ed.) *Tourism Planning and Destination Marketing*, (pp. 161–182), Bingley, UK: Emerald Publishing Ltd.

Masuruli, M. B. (2014). *Costs and Benefits of Nature-Based Tourism to Conservation and Communities in the Serengeti Ecosystem*. Doctorate dissertation, University of Victoria, Victoria, BC, Canada.

McIntosh, R. W., & Goeldner C. R. (1986). *Tourism: Principles, Practices, Philosophies* (5th ed.). New York: John Wiley.

Mehmetoglu, M. (2005). A case study of nature-based tourists: Specialists versus generalists. *Journal of Vacation Marketing, 11*(4), 357–369.

Mehmetoglu, M. (2006). Segmenting the nature-based tourists based on travel mode choice. *Journal of Hospitality & Leisure Marketing, 14*(4), 47–67.

Mehmetoglu, M. (2007a). Nature-based tourism: A contrast to everyday life. *Journal of Ecotourism, 6*(2), 111–126.

Mehmetoglu, M. (2007b). Nature-based tourists: The relationship between their trip expenditures and activities. *Journal of Sustainable Tourism, 15*(2), 200–215.

Mehmetoglu, M. (2007c). Typologising nature-based tourists by activity—Theoretical and practical implications. *Tourism Management, 28*(3), 651–660.

Mehmetoglu, M. (2010). Accurately identifying and comparing sustainable tourists, nature-based tourists, and ecotourists on the basis of their environmental concerns. *International Journal of Hospitality & Tourism Administration, 11*(2), 171–199.

Mehmetoglu, M. & Normann, Ø. (2013). The link between travel motives and activities in nature-based tourism, *Tourism Review, 68*, 2, 3–13.

Meng, F., Tepanon, Y., & Uysal, M. (2008). Measuring tourist satisfaction by attribute and motivation: The case of a nature-based resort. *Journal of Vacation Marketing, 14*(1), 41–56.

Meng, F., & Uysal, M. (2008). Effects of gender differences on perceptions of destination attributes, motivations, and travel values: An examination of a nature-based resort destination. *Journal of Sustainable Tourism, 16*(4), 445–466.

Nduna, L. T., & van Zyl, C. (2020). A benefit segmentation framework for a nature-based tourism destination: The case of Kruger, Panorama and Lowveld areas in Mpumalanga Province. *International Journal of Tourism Cities, 6*(4), 953–973.

Newsome D., Moore S. A., & Dowling R. K. (2012). *Natural Area Tourism: Ecology, Impacts, and Management*. Bristol, United Kingdom: Channel View Publications.

Nguyen, T. H., & Pham, T. M. L. (2015). The influence of tourism motivation to foreign tourist's intention to engage in Vietnam ecotourism. *International Journal of Culture and Tourism Research, 8*, 60–68.

Nikjoo, A. H., & Ketabi, M. (2015). The role of push and pull factors in the way tourists choose their destination. *Anatolia, 26*(4), 588–597.

Nyaupane, G. P., & Graefe, A. R. (2008). Travel distance: A tool for nature-based tourism market segmentation. *Journal of Travel & Tourism Marketing, 25*(3-4), 355–366.

O'Neill, M. A., Riscinto-Kozub, K. A., & Van Hyfte, M. (2010). Defining visitor satisfaction in the context of camping oriented nature-based tourism—the driving force of quality!. *Journal of Vacation Marketing, 16*(2), 141–156.

Olafsdottir, G. (2013). On nature-based tourism. *Tourist Studies 13* (2), 127–138.

Orams, M. B. (1995). Towards a more desirable form of ecotourism. *Tourism Management, 16*, 3–8.

Otoo, F. E., Kim, S., & Choi, Y. (2020). Understanding senior tourists' preferences and characteristics based on their overseas travel motivation clusters. *Journal of Travel & Tourism Marketing, 37*(2), 246–257.

Palacio, V. (1997). Identifying ecotourists in Belize through benefit segmentation: A preliminary analysis. *Journal of Sustainable Tourism, 5*(3), 234–243.

Pearce, P. L. (1988). *The Ulysses Factor: Evaluating Visitors in Tourist Settings*. New York: Springer-Verlag.

Pearce, P. L. (1993). Fundamentals of tourist motivation. In D. G. Pearce, & R. W. Butler (Eds.), *Tourism Research: Critiques and Challenges* (pp. 113–134). London: Routledge.

Pearce, P. L. (2011). Travel motivation, benefits and constraints to destinations. In Y. Wang & A. Pizam (Eds.), *Destination Marketing and Management: Theories and Applications* (pp. 39–52). Wallingford: CABI.

Pennington-Gray, L. A., & Kerstetter, D. L. (2002). Testing a constraints model within the context of nature-based tourism. *Journal of Travel Research, 40*(4), 416–423.

Plog, S. C. (1974). Why destination areas rise and fall in popularity. *Cornell Hotel and Restaurant Administration Quarterly, 14*(4), 55–58.

Powell, R. B., Kellert, S. R., & Ham, S. H. (2009). Interactional theory and the sustainable nature-based tourism experience. *Society and Natural Resources, 22*(8), 761–776.

Priskin, J. (2001). Assessment of natural resources for nature-based tourism: The case of the Central Coast Region of Western Australia. *Tourism Management, 22*(6), 637–648.

Priskin, J. (2003). Tourist perceptions of degradation caused by coastal nature-based recreation. *Environmental Management, 32*(2), 189–204.

Saayman, M., & Saayman, A. (2009). Why travel motivation and socio-demographics matter in managing a national park. *Koedoe: African Protected Area Conservation and Science, 51*(1), 49–57.

Scholtz, M., Kruger, M., & Saayman, M. (2013). Understanding the reasons why tourists visit the Kruger National Park during a recession. *Acta Commercii, 13*(1), 1–9.

Shafer, E. L., & Choi, Y. (2006). Forging nature-based tourism policy issues: A case study in Pennsylvania. *Tourism Management, 27*(4), 615–628.

Shieh, W. L., & Wang, W. B. (2012). Consumer attitudes and participate intentions in relation to ecotourism in Taiwan. *桃園創新學報, 32*, 413–428.

Slabbert, E., & Du Plessis, E. (2013). Do socio-demographic factors influence the travel behaviour of visitors to nature-based tourism products in South Africa?. *African Journal for Physical Health Education, Recreation and Dance, 19*(3), 639–660.

Slabbert, E., & Viviers, P. (2012). Push and pull factors of national parks in South Africa. *Journal of Contemporary Management, 9*(1), 66–88.

Sumanapala, D., & Wolf, I. D. (2020). Think globally, act locally: Current understanding and future directions for nature-based tourism research in Sri Lanka. *Journal of Hospitality and Tourism Management, 45*, 295–308.

Sumanapala, H. D. P., Kotagama, S. W., Perera, P. K., Galahitiyawe, N. W. K. & Suranga, D. A. (2017). Comparison of characteristics of Asian and nonAsian tourists visiting eco-lodges in Sri Lanka. *Sri Lanka Journal of Social Science, 40*(2), 119–126.

Tangeland, T. (2011). Why do people purchase nature-based tourism activity products? A Norwegian case study of outdoor recreation. *Scandinavian Journal of Hospitality and Tourism, 11*(4), 435–456.

Tangeland, T., Aas, Ø., & Odden, A. (2013a). The socio-demographic influence on participation in outdoor recreation activities–Implications for the Norwegian domestic market for nature-based tourism. *Scandinavian Journal of Hospitality and Tourism, 13*(3), 190–207.

Tangeland, T., Vennesland, B., & Nybakk, E. (2013b). Second-home owners' intention to purchase nature-based tourism activity products–A Norwegian case study. *Tourism Management, 36*, 364–376.

Tyrväinen, L., Silvennoinen, H., & Hallikainen, V. (2017). Effect of the season and forest management on the visual quality of the nature-based tourism environment: A case from Finnish Lapland. *Scandinavian Journal of Forest Research, 32*(4), 349–359.

Uusitalo, M. (2017). *How to Maintain Naturalness in Nature-based Tourism Resorts? Approaches to Assessments of Landscape Quality for Tourism Planning*. Rovaniemi: University of Lapland Printing Centre.

Uysal, M., Li, X., & Sirakaya-Turk, E. (2008). Push-pull dynamics in travel decisions. In Haemoon Oh (Ed.), *Handbook of Hospitality Marketing Management* (pp. 412–439). Oxford: Butterworth-Heinemann.

Valentine, P. (1992). Nature-based tourism. In C. Hall & B. Weiler (Eds.), *Special Interest Tourism* (pp. 105–127). London: Belhaven Press.

Weaver, D. B. (2001). Ecotourism as mass tourism: Contradiction or reality?. *Cornell Hotel and Restaurant Administration Quarterly, 42*(2), 104–112.

Weiler, B. & Hall, C. M. (1992). Special interest tourism: In search of an alternative. In C. Hall & B. Weiler (Eds.), *Special Interest Tourism* (pp. 199–204). London: Belhaven Press.

Xu, J. B., & Chan, S. (2016). A new nature-based tourism motivation model: Testing the moderating effects of the push motivation. *Tourism Management Perspectives, 18*, 107–110.

24

EXPERIENTIAL TOURISM

Nature-based Tourism Trends in India

N. Sneha and G. Nagarjuna

Introduction

Ecotourism most likely has a convergent evolution, where many places and people independently responded to the need for more nature-based travel opportunities in line with society's efforts to become more ecologically minded (Khursheed et al., 2014). Nature-based tourism encompasses, among others, mass tourism, adventure tourism, low-impact tourism and ecotourism, which use natural resources in a wild or undeveloped form, including species, habitat, landscape, scenery and salt and fresh-water features. Nature tourism is travel for the purpose of enjoying undeveloped natural areas or wildlife. Nature tourism involves marketing natural landscapes and wildlife to tourists (Goodwin, 1996). It has the potential to provide developing countries with the finance and motivation required to boost conservation efforts. Nature-based tourist destinations must compete in a competitive environment. The emphasis is on the ability of nature-based destinations to be used as experiential tourism destinations through innovative marketing.

Tourism is a service-oriented industry that aims at offering unique experiences to tourists (Camilleri, 2017). Experiential travel is one form of tourism in which people travel to other places, cities, or countries to experience the culture, history, food, and environment. The concept of experiential travel emerged during the 1980s (Westcott, 2015). The decade witnessed the growth of experience-based mass tourism. Experiential tourism as a term cannot be described in an exact definition. Broadly speaking, experiential tourism means different things to different people (Meacci & Liberatore, 2015). It indicates a type of tourism that is different from more conventional tourism (Stamboulis & Skayannis, 2003). Pine and Gilmore, the American researchers in their 'The Experience Economy', explained how the service industry, hospitality and tourism gains from the experiential activities (Mehmetoglu & Engen, 2011). Tourists initially are motivated by physiological needs, but different experienced travellers may use their travel experience (Cohen et al., 2014). A few researchers believed that the destination is not essential, but the experience matters (Breiby et al., 2020). Every travel is an experience. People visit the monument, natural attractions, and socio-cultural and environmental prominent sights. However, experiential travel focuses on joining any activity that offers hands-on experience on something (Fuggle, 2016). The experiential tourists immerse themselves in discovery, insight and inspiration toward a destination. Uniqueness is the biggest

DOI: 10.4324/9781003230748-28

asset of experiential tourism. Because of this nature, experiential travel has become a new way of defining a location. Anything that deviates from the regular sightseeing to the usual tourist spots may focus on interacting with local people. Thus, experiential tourism may be defined subjectively, but the objective of travellers who are driven to immersive experiential tourism looks forward to an insight, discovery or inspiration (Coca-Stefaniak et al., 2017). To sum up, "Experiential Tourism engages visitors in a series of authentic, memorable travel activities, revealed over time, that engage the senses, are inherently personal and make connections on a physical, emotional, spiritual, intellectual or social level" (Rajan, 2015). Every traveller who takes up the tour indulges in one or another activity (Ogilvie et al., 2012). Even though this travel philosophy may indicate that experiential tourists are independent travellers who manage their trip fully by themselves without the help of tour operators. However, recent trends have shown that tour operators have jumped on the bandwagon to provide various experiential tourism packages for tourists seeking transformative experiences (Ali, 2014) A survey conducted by American Express found that 72 per cent of respondents prefer to spend money on experiences that they earn from tourism activities (Ali, 2014).

Experiential tourism can be observed in different types of tourism. However, experiential tourism in nature tourism can be a game-changer that will shape the future of travel by propelling the tourism industry forward to stimulate experiences that generate extraordinary value by placing those who experience in direct contact with the richness of nature, which is not detached from the local reality. Every nature-based destination has various tourism products to offer unique experiences and can differentiate itself with the help of experiential tourism. It can be said that the ingredients for experiential travel are mainly emotion, hands-on activities and authentic experiences for its tourists (Ali, 2014). The success of these ingredients is when they establish an emotional connection with the nature-based destination where feeling takes priority and just seeing a place takes a backseat. When travellers are attached to some destinations in particular not for their monuments or any tourist hotspots but just by the sheer charming allure of the destination due to personal experiences.

The objective of experiential tourism is to get into the roots of the travel instead of the usual traditional visitor experiences. Keeping this in mind, travel marketers are beginning to present destinations to tourists in an authentic manner by shedding typical touristy stereotypes (Cohen, 1972). Tourists always look for multiple interactive methods to earn a lifetime experience (Odunga, 2005). According to John Steinbeck, 'People do not take trips, trips take people' (Travelsmartwoman, 2018). Modern tourists also look for hands-on experiences that could be novel and memorable. (Larsen et al., 2019) only if natured-based-tourism combines with experiential tourism in perfect harmony. Hence, the marketer in the travel industry should showcase authentic experiential aspects to pull tourists towards the destination. Places, people, and their stories are the heart of every tourist destination (Barcelona, 2019). They bring a positive vibe to make the tourists rejuvenated and recreated. Experiencing nature, resources, and culture are the relevant takeaways that add on tangible experiences. A tourist opts to take up experiential tourism activities that focus on environmental sustainability leading to nature-oriented tourism activities (Stone, 2002).

Today, the majority of tourists prefer experienced-based tourism (Barcelona, 2019). Experiential tourism also leads to personal growth and the enrichment of values; thus, personal enrichment, engagement, and stimulation become motivators (Khazami & Lakner, 2021). The concept of Experiential Tourism became increasingly popular in recent years when the travel service providers started focusing on special interest tourism. However, the concept of experiential tourism in a nature-based tourism context has not received significant attention.

Ideally, experiential tourists would be the reason for the development of nature-based experiential tourism. This is also noticed in larger trends witnessing the social media accounts of several tourists sharing their experiential travel diaries. The Instagram effect is as real as it gets. Experiential travel is currently redefining itself in the social media space. For instance, the Destination Management organization started inviting photographers who have more Instagram followers to press meet. This has helped redefine travel, and people wanted the same experience as shown in social media (Lew, 2017). One organization, known as Outdoor World Sustainable Tourism, is an experience-based alternate tourism service provider in Bengaluru, India, that offers outdoor tourism experiences in exciting and offbeat locations. It offers experiential voluntourism in Banavasi, including nature trails and the opportunity to learn to make *Dia's* (lamp) with mud (*Banavasi Voluntourism*, 2018). This organization posts pictures on Facebook and Instagram. Active promotion in social media may connect like-minded experiential travellers and play a massive role in promoting the destination.

Tourism service providers are keen to promote experiential tourism, leaving no stone unturned, with every follower wanting the same experiences. Some of the finest examples of nature-based experiential tourism can be seen through conserving the ecology and supporting the communities (Kanchwala, 2020). Tourist eco-trails are the ultimate example of experiential tourism, which are very popular among youth travellers. Experiential tourism involves gaining an understanding of the destination visited. People are keen to write about their travel experiences to set them apart from the rest of the social media community. Travel agents and tour operators are breaking the barrier of traditional packages to presenting novel ideas through adventure-based trips (Westcott, 2015). Tour operators are becoming more flexible in their planning, collaborating with the local communities to benefit the local community. Modern tourists show much interest in interacting with the local community and are involved in various activities, which generates income for the local community (Rasoolimanesh & Jaafar, 2016).

The chapter reviews the literature focused on trends in nature-based experience tourism in India and the marketing of experiential tourism as supported by available literature from marketing reports, industry surveys, blogs, news articles, and related reviews. The study focuses on experiential tourism trends in India through the lens of marketing and provides much-needed innovative marketing solutions for nature-based experiential tourism planning and development. This discusses implications for marketers and policymakers and reflects on various marketing strategies adopted to promote and market the trending experience-based nature tourism products with the help of a conceptual framework.

Trends in Experiential Nature-based Tourism in India

Generally, the most demanded form of travel is experiential travel. The main intention of experiential tourism is immersive, local, authentic, adventurous, and active travel (Ali, 2014). According to a cultural mindset study from the cultural trip, 'younger generations are more interested in collecting memorable experiences than material possessions.' (Naudts, 2019). The current generation (Generation Zs) consider travel is fundamental to their lives and not only ready to compromise on their daily expenses but also save money to travel. They do not mind spending more money on an experience or a trip. Nature is one of the most critical factors for them to travel to different destinations and think about how they can reduce their impact on the environment and opt for locally sourced food (Naudts, 2019).

In experiential tourism, the intention of travellers to visit the destination is to get immersed in local culture, interact with the local community, be involved in the traditional activities of the destination, be interested in local cuisine, and stay in eco-friendly cottages or heritage homes. Many travellers choose experiential travel that allows them to live, breathe, and understand the destination they visit. These next generation travellers are shifting their interest from regular packaged tours or traditional sightseeing to so-called 'slow travel' that promotes experiential tourism, as it provides sufficient time to explore the destination and establish a connection between travellers and the destination they visit (Sheth, 2021). The travellers who opt for experiential tourism should be responsible during vacations because their involvement in experiencing the natural and cultural resources of the destination differs from the regular travellers. Experiential tourism benefits the destination's sustainability since it adopts principles of sustainable tourism such as conserving nature, respecting the local community and the economic growth of both industry and the host community. It also benefits travellers to transform their lives and learn more about the places and interact with local people and the local culture. Experiential tourism also helps in taking a break from monotonous life and makes experience something new (Ian, 2021).

The travel tech company in India offers experiential tours by adopting smart and electric bicycles. Their tours offer local experience and the joy of cycling. They have their services in Goa, Himachal, Karnataka, Madhya Pradesh, Rajasthan, Tamil Nadu and Uttarakhand. Most of their tours in nature trails offer a rich experience in meeting local farmers and understanding livestock farming in all these destinations. They also take tourists to local shops and restaurants to experience the authenticity of the destination (BLive, 2021).

Mad about India is a travel company providing experiential travel services and an excellent example to understand experiential tourism in the Indian context. One of their services offered is the 'Wonders of Western Ghats.' The major highlights of this tour are accommodation provided in an eco-cottage, a guided trek to waterfalls, and a jungle and sunset point, traditional folk dances and music performances by locals, a visit to an organic spice farm, the agro-processing of cashews and vanilla, and, finally, an authentic Goan lunch at a spice farm served in earthenware pots by bowls made from coconut shells and eaten in plates made from banana leaves. 'Whispers of Kerala', 'Rustic North', 'Colours of Kutch', 'Shades of Malabar', 'Rural Rajasthan', 'Experiential Leh Ladakh', 'Scenic Kinnaur' are the other experiential nature-based tourism services offered by Mad about India (Mad about India, 2021).

India is a country with a wide destination choice for travel because of its uniqueness in catering to different types of tourists. The destination's uniqueness offers a rich heritage, history, culture, flora and fauna, natural attractions, and people (Rajan, 2015). There is a need to change the promotional activity of tourism in India. The promotion activity should focus on wellness, less crowded tourism, experience and learning-based trips. The Indian tourism industry should create an opportunity to focus on high-margin, low-impact travellers as they give importance to nature and the outdoors, health and wellness, local, regional, seasonal food, and cuisines (Conde Nast Traveller, 2020). There is a paradigm shift in the Indian tourism industry, and the reason for this is that Indian millennials are seeking more experiential travel. They would like to experience local culture, customs, food, festivities, and a lot more. According to the study conducted by Thrillophilia, a travel-based company, the demand for experiential travel has grown by 178 percent between 2017 and 2019 and is estimated to grow at 17.4% from 2017 to 2023. The data also revealed that out of 100%, 38% of the travellers would like to go for wildlife and birding tours (Daga, 2019). The north Guwahati in Assam is known for its diversified natural and cultural landscape. It has vast potential for creating experiential-based natural tourism by increasing and stimulating tourists' engagement with locals in terms of interaction and diversified activities (Saikia & Chakrabarti, 2019). Banni region, Kutch in Gujarat, is known for its rich grassland reserves and cultural landscape. The adoption of experiential tourism in this area has helped to preserve its rich natural and cultural landscape and boost the sustainable development of this region (Jayabharathi & Vedamuthu, 2017). Through its responsible tourism mission, Kerala tourism decided to introduce experiential tourism packages that reflect the rural life experience (Kumar, 2022). India as a country can offer an enriching experience to tourists, with its tranquil beaches, heritage trails, and tea in the lap of nature. Keeping experiential travel in mind, the Ministry of Tourism drafted a policy to promote India as a 'must-see experience' destination (Kaur, 2021). Wayanad, a district in Kerala state, can offer a rich experience to the tourists through its varied natural landscape which makes tourists involved in cycling, trekking, mountaineering, fishing, wildlife safari and heritage walk (Sibi, 2017). Maharashtra tourism promotes nature-based experiential tourism in the name of ecotourism as it not only supports conservation but also helps in improving the economic standards of the region (Maharashtra Tourism, 2021). Rajasthan is the state of India known for its rich cultural and natural heritage. In order to promote experiential tourism in the state, the Department of tourism in Rajasthan started developing guidelines for the recognition of experiential tourism agencies (Rajasthan Tourism, 2020). States such as Sikkim, Punjab and Kerala offer farms stay that enrich tourist experiences as they can be involved in various farming activities along with gaining knowledge of organic farming. Similarly, Sundarbans of West Bengal allows tourists to experience the technique of honey collection inside the dense forest that is known for its Bengal tigers (Kanchwala, 2018). Kerala tourism launched STREET projects explicitly to promote experiential tourism. Green street, experiential tourism street and water street are all parts of this project, and all these street projects will focus on experiential tourism (Press Trust of India, 2021). Through a combination of village walks, trekking and rural ecotourism activities, the government is keen on promoting experiential tourism in Nagaland (Ambrocia, 2022). Staying in an Igloo in Himachal Pradesh, experiencing facial and hand tattoos of Konyak tribe of Nagaland, a waterfall in Maharashtra, a farm stay in the interiors of Leh, and the highest frozen lake, Gurudongmar Lake in North Sikkim, all combine to create a unique nature-based experiential travel in India (Saumiabee, 2021).

Marketing Experiential Tourism

Experiential tourism engages all five senses. Various channels of communication are effectively used to reach out to the general public on travel experiences. Today, tourists prefer social media networks to talk about their travel experiences (Liu et al., 2020). Modern customers prefer to purchase goods and services that give them experiences (Epsilon, 2018). The experiences may be either tangible or intangible. Products give tangible experiences, whereas services offer intangible experiences (Placko, 2021). Products may be anything produced in a factory or manufacturing unit with a physical identity (Naudts, 2019). Experiential marketing should create an experience for the consumers that makes them involved in the services, and their later memories may result in a preference for the services and loyalty. Therefore, experiential marketing should understand the 'consumer as a person, who experiences, lives, gets emotional, and participates in what is happening.' In tourism, elements of experience also add value to the product. So, to market experiential tourism, the product should provide 'a holistic experience, creating an environment that will appeal to the consumer's five senses and overall involvement in experience' (Dieguez & Conceição, 2021). The segment of travellers based on their five senses and emotions is essential while offering experiential tourism services. Apart from segmentation, effective communication strategies such as 'websites, social media, consumer portals, blogs or influencers' to reach out to their target customers may create the image of authentic experiential tourism resources (Garcia, 2018). Visitor segmentation can be used as a marketing technique since it helps match the product as per the desired experience of the visitors (Dodds & Jolliffe, 2016). Experiential tourism should focus on creating a personalized experience, so the traveller talks about their vacation when they come back home. To market this experience, tourism marketing should adopt a strategy that will create a sensory world in the consumers' minds (Giorgio, 2019). Elements of emotional attributes need to be considered while promoting a destination to generate memorable experiences to the visitors (Lagiewski & Zekan, 2006). Experiential marketing helps in influencing tourists' behavior, but it depends on tourists' imagery processing. So, experiential marketing strategies should adopt promotional measures creating imagination, fantasy, and memory recall among the tourists (Le et al., 2019). Customers' tourism engagement in experiential marketing activities affects the customer experience and behavioral intention toward the destination (Rather, 2020). The jungle book, an eco-resort in Goa, used 'Fantasy', 'Fun' and 'Feeling' as experiential tourism marketing tools (D'Souza & Desai, 2020). The nature-based experience product should adopt a co-design strategy since it helps destination management organizations understand tourists' perceived experience value (Breiby et al., 2020).

Marketing Experiential Nature-based Tourism

Reflecting marketing images related to performing arts activities in nature-based servicescapes attracts customers' attention toward nature-based tourism products (Wang et al., 2018). The products and services offered in experiential tourism should focus on enriching activities, the interaction between local community and travellers, enabling emotional and memorable connection with destination, authentic cuisine, volunteer activity (Lew, 2017).

The use of social media channels, storytelling, content marketing, and developing appropriate messages can all be used as a marketing strategy to promote nature-based tourism destinations (Lundgren, 2020). Appropriate pricing and marketing strategy helps identify heterogenous consumer preferences and results in 'triple bottom line' gains for nature-based tourism (Naidoo et al., 2021). Online photography on Instagram creates experiential value in nature-based tourism, and types of experience value depend on photography based user-generated content (Conti & Lexhagen, 2020). So, social media may be among the most effective platforms to market experiential nature-based tourism. While marketing experiential nature-based tourism, one should focus on wildlife experience and outdoor activities, and price factor should base on uniqueness of experience, describing the right market segment, and the use of social media and Google analytics for customized marketing campaigns (Phillips, 2021). Visitors purchasing nature-based experiences tend to be young, high-income earners and risk-taking in nature (Tangeland et al., 2013). Aesthetic experiential qualities such as 'cleanliness', 'scenery', 'harmony', 'architecture' and 'genuineness' should be given importance while developing a nature-based tourism product as it provides a competitive edge (Breiby & Slåtten, 2018). Nature-based tourism products should be classified based on travel motivation regarding whether visitor intentions are nature protection, nature experience, sports and adventure, or hedonistic. It also depends on the service arrangements required by each individual (Arnegger et al., 2010). Preference toward nature-based tourism products or activities depends on the travel companion. So, nature-based tourism service providers should design tailor-made products and services (Chen et al., 2016). Nature-based tourist experience depends on its genuine entertainment, state of being and socio-cultural elements of the region. So one should consider all these elements for marketing and presentation of such tourism products (Vespestad & Lindberg, 2011). Focusing on sustainability and understanding the target market is essential in marketing experiential nature-based tourism (Byron, 2016) (Figure 24.1).

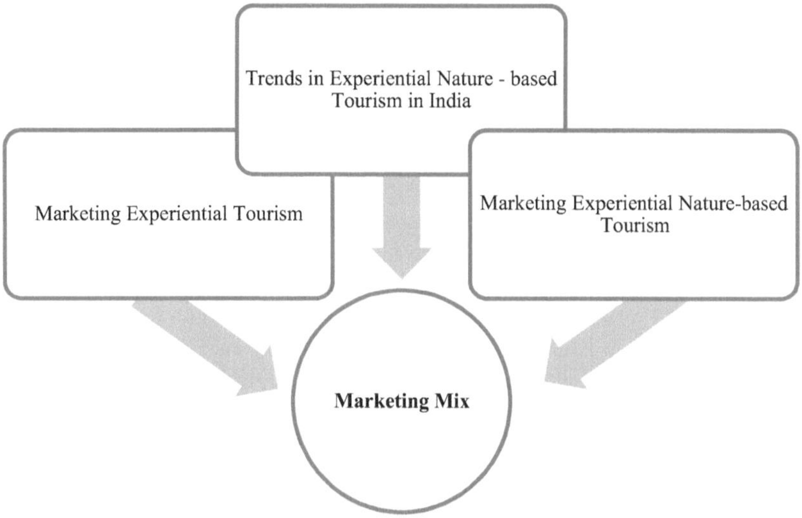

Figure 24.1 Influence of Trends, Marketing Experiential and Experiential Nature-Based Tourism on Marketing Mix.

Source: Developed by Author

There is a change in the trend in nature-based tourism as travellers are looking for experience-based activities. They are more eager for nature-based tourism products that offer something unique and enable immersion in local culture, enjoying seasonal food (Conde Nast Traveller, 2020). They want to be involved in cycling, trekking, mountaineering, fishing, heritage walk (Sibi, 2017), farming activities, knowledge of organic farming, learn the honey collection process in the dense forest (Kanchwala, 2018), staying in unique accommodation, experiencing a different side of the tribal community, and visiting unique natural attractions (Saumiabee, 2021), to participate in tribal music and other art farms (Mad about India, 2021). To market such a complex demand, is it important to have expertise of about experiential marketing tourism. Experiential tourism marketing helps to understand the senses of consumers (Liu et al., 2020). A consumer is a person who experiences, lives, gets emotional, and participates in what is happening' (Dieguez & Conceição, 2021), how to create a sensory world in the minds of the consumers (Giorgio, 2019), and imagination, fantasy, and memory recall among the tourists (Le et al., 2019). Marketing experiential nature-based tourism provides direction in understanding the elements of nature-based tourism required while promoting experiential nature-based tourism products. The elements such as creating nature-based servicescapes while advertising the product (Wang et al., 2018), appropriate pricing and marketing strategy (Naidoo et al., 2021), the use of social media while promoting a product (Conti & Lexhagen, 2020) describing the right market segment and use of technology in a marketing campaign (Phillips, 2021), giving importance to aesthetic experiential qualities, understanding travel motivations and travel companion while developing tourism products (Breiby & Slåtten, 2018; Arnegger et al., 2010; Chen et al., 2016). Thus, it is imperative to understand trends as it influences developing the right marketing mix for nature-based experiential tourism.

Discussion and Conclusion

Below, we discuss different aspects of the marketing mix concerning the challenges associated with experiential nature-based tourism development and visitor experience design in the Indian context.

Product: Authentic nature-based experiential tours are heterogeneous products and capture the visitor's interest due to their uniqueness. The experiential tours as a product help take a break from the usual monotonous life and create a new experience (Ian, 2021). Nature-based experiential tourism products must be marketed so that it caters to the five senses and emotional elements of customers. Innovative experiential products can now satisfy their search for strong emotions, high involvement, and unusual situations, together with their desire to discover new tastes and scents. These products focus on providing sensory inputs, which allow them to be immersed entirely in authentic tourist experiences (Maccannell, 1973). They also must establish a connection between community involvement and sustainability to conserve and promote nature-based experiential tour products.

Place: The location of the destination where the nature-based experiential tours are offered is the focus of the place mix. Nature-based tourists will visit destinations to attain specific experiences which they cannot attain in other areas. For instance, natural areas afford opportunities to learn about the environment (e.g., nature trails in National Parks), relieve stress (e.g., coastal areas to escape crowds and urban life), and bonds with family and friends (e.g., family and group wildlife camping areas) (Bustam & Stein, 2021).

Price: Pricing is vital in meeting the nature-based experiential tourist demands, satisfying the business's profit margins, and increasing the visitor dwell time. Pricing experiential tour products according to demand and value are essential. Service providers need to offer users a valuable and authentic experience for the activities and facilities. Pricing strategies must be paired with factors that influence pricing, such as product quality, competition with other private business owners, and publicly owned natural attractions in the local area offering similar services. Pricing strategies must also consider new opportunities such as price skimming and penetration pricing (Goeldner & Ritchie, 2009).

Promotion: Promotion is used to create demand for a service or product, attract attention, create interest or desire, and generate action to sell that service/product (Goeldner & Ritchie, 2009). Communicating the unique value given through experiential tourism products through innovative marketing campaigns such as a co-design strategy to capture the target market will boost sales. The use of social media for communication is an increasing trend. Standard outlets such as Twitter, Facebook, and Instagram, and informal outlets such as blogs and wikis, provide a platform for marketing ecotourism business on the internet (Bustam & Stein, 2021).

People: Nature-based tourism is a people business; tourists expect a distinctive experience from their participation (Bustam & Stein, 2021). Investing in the right employees and vendors can draw more customers and enhance the overall customer experience before and during the experiential tour. Recognizing that not all nature-based tourists are alike will give service providers the advantage in funneling their marketing efforts toward a specific target market in the delivery of the distinctive experience. The public sector should focus on framing policies and regulations, providing financial and fiscal incentives, land use planning, and promoting experiential tourism along with host communities. It is one of the prime responsibilities of the public sector to conserve natural areas with partnerships with the private sector where experiential tourism activities occur.

Tour operators, transport carriers, travel media, hospitality, entertainment, shopping, and tourist guides play a huge role in creating a memorable experience. Experiential tourism aims to intensify the sensations and emotions lived during holidays through the amplification of not only visual sensations but also gustatory, olfactory, tactile, and auditory stimuli (Ferrari & Adamo, 2008).

Physical Evidence: This is a part of the conscious effort to make the intangible experience more tangible for the visitor while getting exposed to local culture, authentic cuisine, eco-trails and routes and unique local transport. During the service delivery, the physical evidence influences customers' perceptions and levels of satisfaction; in some cases, it could create scenic atmospheres that offers enriching consumption experiences (Shostack, 1977) (Figure 24.2).

One of the most significant and popular trends in global tourism today is the demand for "experiential travel," typically meant to convey the idea of more authentic, local, immersive, and active travel. Travel is profitable because it makes the traveller more satisfied and positively impacts the post-trip behavioral intentions. Nature-based tourism can provide an experiential feel to the tourists when they visit natural destinations. The various activities, such as hiking, bird watching, staying in huts and camps, and experiencing local culture and traditional dishes, make travellers immerse themselves in nature-based tourism.

The intangibility aspect of experiential tourism causes a certain degree of challenge in the marketing of this type of holiday. With travel consumer behaviors varying wildly in this space, it is a matter of degree of the way marketers define and approach the overall trend. The challenges of experiential tourism marketing can be addressed by developing the right marketing mix that helps develop and promote nature-based experiential tourism.

Other limitations of experiential nature-based tourism include pseudo-experiences which are the fabricated version far away from reality. In addition, all stakeholders involved in nature-based tourism may not sync to co-create an experience value and merely limit traditional tourism product offerings. Many nature-based tourist spots could have restrictions resulting

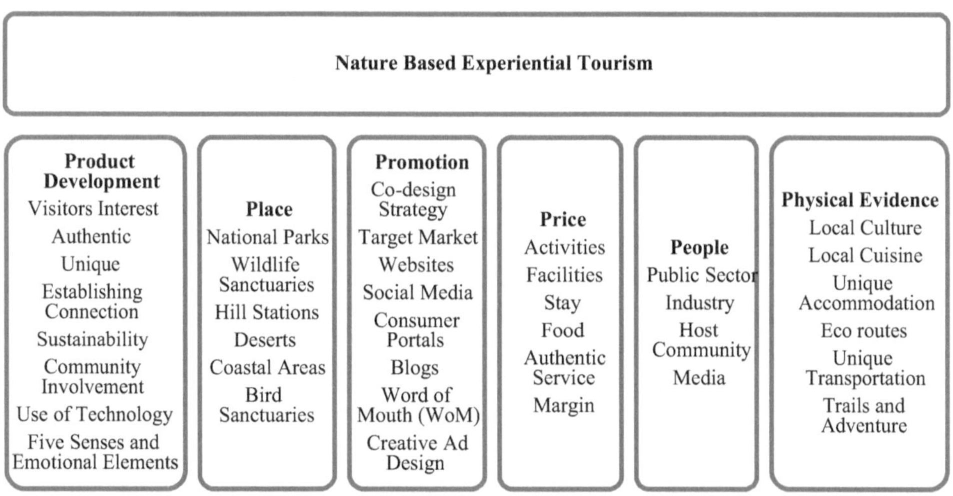

Figure 24.2 Marketing Mix for Nature-Based Experiential Tourism.

Source: Developed by Author

in zero prospects of experiential tourism. Shortcomings of the tourism policies have serval implications on promoting only certain types of tourism and thereby neglecting potential. A sound marketing strategy for experiential nature-based tourism to work on the untapped potential of nature-based tourism trends by the government tourism boards as a part of a tourism policy needs to be the new focus area. Although this work represents a study about nature-based experiential tourism in India, the research could be extended to other countries working with markets as experiential tourism. The future line of research proposed to design an experiential smart tourism model not only for a nature-based destination but also for lifestyle, culture, and so on with the help of a robust Public–Private Partnership Model to amplify the marketing synergies brought together. In addition, it would be interesting to analyze the trends in experiential tourism in both urban and rural areas, which will help formulate the experiential tourist profiles to use the right marketing strategies for each group. These aspects can be analysed in the following works and research in-depth to explore the possibilities of the practical application of the experiential tourism model globally.

References

Ambrocia, M. (2022, January 25). *Nagaland govt working to boost experiential tourism: CM Adviser Abu Metha*. EastMojo. http://www.eastmojo.com/nagaland/2022/01/25/nagaland-govt-working-to-boost-experiential-tourism-cm-adviser-abu-metha/

Ali, R. (2014). *Free Skift Report: The Rise of Experiential Travel*. Skift. https://skift.com/2014/06/03/launching-free-skift-report-the-rise-of-experiential-travel/

Arnegger, J., Woltering, M., & Job, H. (2010). Toward a product-based typology for nature-based tourism: A conceptual framework. *Journal of Sustainable Tourism*, *18*(7), 915–928. https://doi.org/10.1080/09669582.2010.485680

Banavasi Voluntourism by Outdoor World Sustainable Tourism - Banavasi Voluntourism by Outdoor World Sustainable Tourism. (n.d.). Retrieved February 20, 2022, from https://www.facebook.com/Banavasi.in/photos/a.268173676683643/1019681141532889/

BLive. (2021). *E-Bike Tours | EV Tourism | Top-rated Experiential Tourism Trips | B:Live*. Blive. https://www.blive.co.in/

Breiby, M. A., Duedahl, E., Øian, H., & Ericsson, B. (2020). Exploring sustainable experiences in tourism. *Scandinavian Journal of Hospitality and Tourism*, *20*(4), 335–351. https://doi.org/10.1080/15022250.2020.1748706

Breiby, M. A., & Slåtten, T. (2018). The role of aesthetic experiential qualities for tourist satisfaction and loyalty. *International Journal of Culture, Tourism and Hospitality Research*, *12*(1), 1–14. https://doi.org/10.1108/IJCTHR-07-2017-0082

Bustam, T. D., & Stein, T. (2021). *How to Develop a Marketing Plan for Your Ecotourism Business*. UF/IFAS Extension University of Florida. https://edis.ifas.ufl.edu/publication/FR340

Byron, K. L. (2016). *Marketing the magic in a nature-based experience: A case study on the Walt Disney company*. DigitalCommons@CalPoly. Retrieved February 16, 2022, from https://digitalcommons.calpoly.edu/rptasp/73/

Camilleri, M. A. (2017). The Tourism Industry: An Overview. In M. A. Camilleri, *Travel Marketing, Tourism Economics and the Airline Product: An Introduction to Theory and Practice*. Italy: Springer.

Chen, J. S., Wang, W., & Prebensen, N. K. (2016). Travel companions and activity preferences of nature-based tourists. *Tourism Review*, *71*(1), 45–56. https://doi.org/10.1108/TR-06-2015-0024

Coca-Stefaniak, J. A., Powell, R., Morrison, A. M., & Paulauskaite, D. (2017). Living Like a Local: Authentic Tourism Experiences and the Sharing Economy. *International Journal of Tourism Research*, *19*. https://doi.org/10.1002/jtr.2134

Cohen, S. A., Prayag, G., & Moital, M. (2014). Consumer behaviour in tourism: Concepts, influences and opportunities. *Current Issues in Tourism*, *17*(10), 872–909. https://doi.org/10.1080/13683500.2013.850064

Conde Nast Traveller. (2020). '*Experiential travel way forward for tourism…*' Condé Nast Traveller India. https://www.cntraveller.in/story/experiential-travel-way-forward-for-tourism-india/

Conti, E., & Lexhagen, M. (2020). Instagramming nature-based tourism experiences: A netnographic study of online photography and value creation. *Tourism Management Perspectives, 34*, 100650. https://doi.org/s10.1016/j.tmp.2020.100650

D'Souza, L. M., & Desai, V. P. (2020). Jungle book - fantasy, feeling and fun: Experiential marketing in ecotourism. *IUP Journal of Business Strategy, 17*(3), 46–68.

Daga, A. (2019). *Off the beaten path: How millennials are powering India's experiential travel boom.* https://www.cnbctv18.com/views/off-the-beaten-path-how-millennials-are-powering-indias-experiential-travel-boom-3327741.htm

Dieguez, T., & Conceição, O. (2021). *Experiential Tourism and Experiential Marketing: An Innovative Approach* (pp. 550–559). https://doi.org/10.1007/978-981-33-4260-6_47

Dodds, R., & Jolliffe, L. (2016). Experiential Tourism: Creating and Marketing Tourism Attraction Experiences. In M. Sotiriadis & D. Gursoy (Eds.), *The Handbook of Managing and Marketing Tourism Experiences* (pp. 113–129). Emerald Group Publishing Limited. https://doi.org/10.1108/978-1-78635-290-320161005

Epsilon. (2018). *New Epsilon research indicates 80% of consumers are more likely to make a purchase when brands offer personalized experiences.* https://www.epsilon.com/us/about-us/pressroom/new-epsilon-research-indicates-80-of-consumers-are-more-likely-to-make-a-purchase-when-brands-offer-personalized-experiences

Ferrari, S., & Adamo, G. (2008). Experiential and multisensority holidays as a form of creative tourism. In G. Riachard & J. Wilson (Eds.), *Cultural Tourism to Creative Tourism* (pp. 11–24). ATLAS, Arnhem.

Fuggle, L. (2016, August 2). The rise of experiential travel and its impact on tours and activities. Retrieved from www.trekksoft.com: https://www.trekksoft.com/en/blog/rise-of-experiential-travel

Garcia, H. B. (2018). Urban experiential tourism marketing: Use of social media as communication tools by the food markets of Madrid. *Journal of Tourism Analysis: Revista de Análisis Turístico, 25*(1), 2–22. https://doi.org/10.1108/JTA-01-2018-0002

Giorgio, A. (2019). *Tourism Trends According to Industry Experts.* We Are Marketing. https://www.wearemarketing.com/whitepapers/tourism-trends/experiential-tourism.html

Goeldner, C. R., & Ritchie, B. J. R. (2009). *Tourism: Principles, Practices, Philosophies* (Eleventh Edition). New Jersey: Wiley.

Goodwin, H. (1996). In pursuit of ecotourism. *Biodiversity and Conservation, 5*(3), 277–291.

Ian, J. (2021). *Experiential Travel: Manipulative Marketing or Transformative Trend?* https://travelcollecting.com/experiential-travel-manipulative-marketing-or-transformative-trend/

Jayabharathi, P., & Vedamuthu, R. (2017). Experiential tourism as a response to the sustenance of a cultural landscape: The case of Banni, Kutch, Gujarat, India. In C. Silver, L. Marques, H. Hanan, & I. Widiastuti (Eds.), *Proceedings of the 6th International Conference of Arte-Polis* (pp. 149–166). Springer. https://doi.org/10.1007/978-981-10-5481-5_15

Kanchwala, S. (2018). *Take your family for out-of-the-box experiences to enjoy first showers of monsoon.* The Economic Times. https://economictimes.indiatimes.com/magazines/travel/take-your-family-for-out-of-the-box-experiences-to-enjoy-first-showers-of-monsoon/articleshow/64763462.cms

Kanchwala, S. (2020). *Slow travel, ecotourism, micro-cations: The ultimate trend guide for 2020.* The Economic Times. https://economictimes.indiatimes.com/magazines/travel/slow-travel-eco-tourism-micro-cations-the-ultimate-trend-guide-for-2020/articleshow/73165759.cms

Kaur, T. S. G. (2021). *Out-of-the-Box Tourism: India encashing on experiential holidays.* http://travelscapesonline.com/out-of-the-box-tourism-india-encashing-on-experiential-holidays/

Khazami, N., & Lakner, Z. (2021). Influence of experiential consumption and social environment of local tourists on the intention to revisit Tunisian guesthouses: Mediating role of involvement in the experience. *Sustainability, 13*(12), 6584. https://doi.org/10.3390/su13126584

Khursheed, W., Khah, S., & Rao, R. J. (2014). Ecotourism and the impact of the conventional tourism on the fragile ecosystems. *International Journal of Science and Nature, 2*(3), 432.

Kumar, P. K. (2022). *Kerala RT Mission to introduce new experiential tourism packages from next month—ET TravelWorld.* ETTravelWorld.Com. https://travel.economictimes.indiatimes.com/news/destination/states/kerala-rt-mission-to-introduce-new-experiential-tourism-packages-from-next-month/88893844

Lagiewski, R., & Zekan, B. (2006). Experiential marketing of tourism destinations. *RIT Scholar Works, 15*, 163–175. https://scholarworks.rit.edu/other/633

Larsen, S., Wolff, K., Doran, R., & Øgaard, T. (2019). What makes tourist experiences interesting. *Frontiers in Psychology, 10*, 1603. https://doi.org/10.3389/fpsyg.2019.01603

Le, D., Scott, N., & Lohmann, G. (2019). Applying experiential marketing in selling tourism dreams. *Journal of Travel & Tourism Marketing*, *36*(2), 220–235. https://doi.org/10.1080/10548408.2018.1526158

Lew, J. (2017). *What Is Experiential Tourism?* Treehugger. https://www.treehugger.com/what-is-experiential-tourism-4869029

Liu, X., Mehraliyev, F., Liu, C., & Schuckert, M. (2020). The roles of social media in tourists' choices of travel components. *Tourist Studies*, *20*(1), 27–48. https://doi.org/10.1177/1468797619873107

Lundgren, M. (2020). *DMOs Strategies to Attract Tourists to Nature-Based Destinations in Norrbotten County*. DiVA. Retrieved from http://www.diva-portal.org/smash/record.jsf?pid=diva2%3A1439092&dswid=6881

Maccannell, D. (1973). Staged authenticity: Arrangement of social space in tourist settings. *American Journal of Sociology - AMER J SOCIOL*, *79*. https://doi.org/10.1086/225585

Mad about India. (2021). *Travel Experiences*. Experiential Travel in India. http://madaboutindia.co.in/

Maharashtra Tourism. (2021). *Eco Tourism—DOT-Maharashtra Tourism—Maharashtra Tourism*. DOT-Maharashtra Tourism. https://www.maharashtratourism.gov.in

Meacci, L., & Liberatore, G. (2015, October 1). *Towards a senses-based model for experiential tourism: The youtooscany.com case*. IMIC 15 Santorini. https://www.researchgate.net/publication/282877756_Towards_a_senses-based_model_for_experiential_tourism_the_youtoosca

Mehmetoglu, M., & Engen, M. (2011). Pine and Gilmore's Concept of Experience Economy and Its Dimensions: An Empirical Examination in Tourism. *Journal of Quality Assurance in Hospitality & Tourism*, *12*(4), 237–255. https://doi.org/10.1080/1528008X.2011.541847

Naidoo, R., Beytell, P., Malherbe, A., Middleton, A., Perche, J., & Muntifering, J. R. (2021). Heterogeneous consumer preferences for local community involvement in nature-based tourism drive triple-bottom-line gains. *Conservation Science and Practice*, *3*(6), e425. https://doi.org/10.1111/csp2.425

Naudts, K. (2019, June 24). *How curiosity and globalization are driving a new approach to travel*. World Economic Forum. https://www.weforum.org/agenda/2019/06/curiosity-globalization-driving-new-approach-to-travel/

New Epsilon research indicates 80% of consumers are more likely to make a purchase when brands offer personalized experiences. (2018). https://www.epsilon.com/us/about-us/pressroom/new-epsilon-research-indicates-80-of-consumers-are-more-likely-to-make-a-purchase-when-brands-offer-personalized-experiences

Odunga, P. O. (2005). Choice of Attractions, Expenditure and Satisfaction of International Tourists to Kenya. Wageningen University.

Ogilvie, D., Bull, F., Cooper, A., Rutter, H., Adams, E., Brand, C., Ghali, K., Jones, T., Mutrie, N., Powell, J., Preston, J., Sahlqvist, S., & Song, Y. (2012). Evaluating the travel, physical activity and carbon impacts of a 'natural experiment' in the provision of new walking and cycling infrastructure: Methods for the core module of the iConnect study. *BMJ Open*, *2*(1), e000694. https://doi.org/10.1136/bmjopen-2011-000694

Phillips, M. (2021). *Marketing and Nature Tourism*. Texas A&M AgriLife Extension Service. https://agrilifeextension.tamu.edu/library/community-development-parks-recreation-tourism/marketing-and-nature-tourism/

Placko, W. (2021). Making the Intangible Tangible: Selling Services &; Experiences. Retrieved from https://neuger.com: https://neuger.com/news/making-the-intangible-tangible-selling-services-experiences/

Press Trust of India. (2021). *Kerala Tourism launches STREET project for experiential tourism*. Republic World. https://www.republicworld.com/india-news/general-news/kerala-tourism-launches-street-project-for-experiential-tourism.html

Rajan, R. B. (2015). Experiential tourism: Understanding tourism trends today to prepare for tomorrow. *Shanlax International Journal of Commerce*, *3*(1), 121–126.

Rajasthan Tourism. (2020). Guidelines for recognition of Experiential Tourism Agencies in Rajasthan. https://www.tourism.rajasthan.gov.in/content/dam/rajasthan-tourism/english/pdf/acts-and-policy/Recognition-of-Experiential-Tourism-Agencies-2021.pdf

Rasoolimanesh, S. M., & Jaafar, M. (2016). Sustainable tourism development and residents' perceptions in World Heritage Site destinations. *Asia Pacific Journal of Tourism Research*, *22*(1), 34–48. https://doi.org/10.1080/10941665.2016.1175491

Rasoolimanesh, S. M., Roldán, J. L., Jaafar, M., & Ramayah, T. (2017). Factors influencing residents' perceptions toward tourism development: Differences across rural and Urban world heritage sites. *Journal of Travel Research*, *56*(6), 760–775. https://doi.org/10.1177/0047287516662354

Rather, R. A. (2020). Customer experience and engagement in tourism destinations: The experiential marketing perspective. *Journal of Travel & Tourism Marketing*, *37*(1), 15–32. https://doi.org/10.1080/10548408.2019.1686101

Saikia, S. B., & Chakrabarti, D. (2019). Experiential design intervention to motivate tourist local interaction and connect with unexplored landscape: A case study of North Guwahati. In A. Chakrabarti (Ed.), *Research into Design for a Connected World* (pp. 757–767). Springer. https://doi.org/10.1007/978-981-13-5974-3_66

Saumiabee. (2021). *22 Unique Travel Experiences In INDIA That You Need Have In 2022*. Tripoto. https://www.tripoto.com/india/trips/20-unique-travel-experiences-in-india-that-you-need-have-in-2020-5deddec58c592

Sheth, A. (2021). *Experiential Travel will be the next big thing in tourism—ET Hospitality World*. ETHospitalityWorld.Com. https://hospitality.economictimes.indiatimes.com/news/speaking-heads/experiential-travel-will-be-the-next-big-thing-in-tourism/80463174

Shostack, G. L. (1977). Breaking free from product marketing. *Journal of Marketing*, *41*(2), 73–80. https://doi.org/10.1177/002224297704100219

Sibi, P. (2017). The experiential facets of special interest tourism offerings of Wayanad, Kerala. *Atna Journal of Tourism Studies*, *12*(1), 35–48. https://doi.org/10.12727/ajts.17.3

Stamboulis, Y., & Skayannis, P. (2003). Innovation strategies and technology for experience-based tourism. *Tourism Management*, *24*(1), 35–43. https://doi.org/10.1016/S0261-5177(02)00047-X

Stone, M. J. (2002). *Ecotourism & Community Development: Case Studies from Hainan, China*. Ecotourism & Community Development: Case Studies from Hainan, China. Retrieved July 12, 2021, from https://www.collectionscanada.gc.ca/obj/s4/f2/dsk3/OWTU/TC-OWTU-79.pdf?origin=publication_detail

Tangeland, T., Vennesland, B., & Nybakk, E. (2013). Second-home owners' intention to purchase nature-based tourism activity products – A Norwegian case study. *Tourism Management*, *36*, 364–376. https://doi.org/10.1016/j.tourman.2012.10.006

Travelsmartwoman. (2018, February 7). People Don't Take Trips … Trips Take People | TravelSmart Woman. https://travelsmartwomen.com/people-dont-take-trips/

Universitat Pompeu Fabra – Barcelona. (2019, April 8). What most attracts us to a tourist destination? Attractions, culture and gastronomy. *ScienceDaily*. Retrieved November 11, 2022 from www.sciencedaily.com/releases/2019/04/190408114002.htm

Vespestad, M. K., & Lindberg, F. (2011). Understanding nature-based tourist experiences: An ontological analysis. *Current Issues in Tourism*, *14*(6), 563–580. https://doi.org/10.1080/13683500.2010.513730

Wang, T. C., Tsai, C. L., & Tang, T. W. (2018). Exploring advertising effectiveness of tourist hotels' marketing images containing nature and performing arts: An eye-tracking analysis. *Sustainability*, *10*(9), 3038. https://doi.org/10.3390/su10093038

Westcott, M. (2015). *Introduction to Tourism and Hospitality in BC*. BCcampus. https://opentextbc.ca/introtourism

PART IV

The impacts of nature-based tourism

25

REDEFINING SUSTAINABLE TOURISM CHALLENGES AND ASSOCIATED INDICATORS IN PROTECTED AND CONSERVED AREAS

Marković Vukadin Izidora, Zovko Mira and Ante Mandić

Introduction

The human population and its activities have a tremendous impact on the environment. It is enough to look at any landscape to see the changes caused by anthropological influence. Never before in the history of the Planet Earth have such significant landscape changes taken place. Before the increasing urbanization and industrialization took place, the only visible transformations in the landscape were associated with geological agents such as erosion, abrasion, tectonics and deflation. Simultaneously, today, human manipulation, exploitation and destruction of the natural environment are so great that the entire physical and biological system is subordinate to the need for intensive use of planetary resources. Consequently, the establishment of the protected areas (Pas) is critical for a successful global conservation strategy (Leung et al., 2018).

Today, tourism is the most extensive use of Pas (Mandić, 2020), and ensuring sustainability remains a critical challenge of nature-based tourism planning and development (Marković Vukadin, 2017; Mandić, 2022). The impacts associated with nature-based tourism and their severity depend primarily on the form of recreational and sightseeing activities offered, the sensitivity of the environment, and management capacities to deal with them (Newsome & Lacroix, 2011). Monitoring and an explanation of the processes and causal connection in the ecosystem and social environment in which tourism takes place is critical (Hobbs et al., 2010) in order to inform the planning decisions.

The sustainable coexistence of tourism and conservation is of utmost importance worldwide, from the perspective of the environment, economic development and local community well-being. Over recent decades, monitoring emerged as a critical tool to achieve these goals. It has become evident how it is necessary to identify challenges and develop effective approaches to monitor the progress toward the sustainability of tourism development, taking into account the diversity and the differences between protected environments. The appeal of pristine and unique natural environments is to be blamed for both positive and negative impacts of tourism development, as it draws people to PAs. For example, increasing visitor numbers often lead to

DOI: 10.4324/9781003230748-30

environmental pressures, including waste, wastewater, traffic congestion, environmental noise, visual intrusion, urbanization, and overcrowding. Once initiated, tourism can have an impact on the local community's socio-cultural identity and transforms the local economy, usually by replacing traditional economic activities. Having said that, as tourism has an increasingly significant influence on PAs, efficient development policies become critical. However, it is evident how planning and management within PAs often only superficially promotes the concepts of integrated management and sustainable development, while risks, including those related to excessive visitation, remain marginalized. For example, a number of such cases have been reported in the recently published "The Mediterranean Protected Areas in The Era of Overtourism: Challenges and Solutions", (Mandić & Petrić, 2021).

Monitoring the sustainability of tourism development remains a critical challenge for the global tourism industry, with limited ability to inspire the transformation of tourism development policies (Font et al., 2021). When it comes to the PAs, the existing frameworks of indicators often rely, almost exclusively, on visitor perception (Klanjšček et al., 2018; Tverijonaite et al., 2018), neglecting rather important attitudes of the local communities (Ristić et al., 2019) and knowledge of management specialists (Mandić & Marković Vukadin, 2021). Currently, Pas, globally, are witnessing the increasing demand for spending time in nature, the diversity of visitor experiences, challenges associated with financing, but also climate change, and the increasing number of adverse impacts, including poaching, tourism revenue dependence, and the depletion of ecosystems. Consequently, the need to setting up a course toward the establishment of the symbiotic relationship between PAs and visitors (Weaver & Lawton, 2017) is of utmost importance. The critical questions that arise are *how do we monitor these challenges and the associated impacts? and if current frameworks are sufficient to inform managerial decisions?*

This chapter begins with a brief reflection on established indicator systems in the sustainable tourism domain. It continues with an identification of the critical challenges associated with sustainable tourism development in PAs and developing a genuine indicator list that could be considered to monitor the associated impacts. Finally, within the final section we discuss the managerial implications and the avenues for future research.

Basic indicators systems in tourism

While sustainable tourism, indeed, has been recognized as the catalyst for positive change, the measurement of progress toward the achievement of higher levels of sustainability, particularly at the destination scale where the impacts occur (Joppe, 2018), remains a critical challenge (Boluk et al., 2019). In his recent analysis of the global progress toward achieving this goal, Sharpley (2020) concludes that although specific positive shifts have been made, tourism, as it is today, is in a far from sustainable state. With our full support, he further explains how global tourism requires a radical rethinking of the current consumption and production patterns and the shift away from the neoliberal politics in which it is engraved. Higgins-Desbiolles (2018) even claims that "tourism is killing tourism" and calls for acknowledgement and understanding that the industry indeed suffers from the condition, calling for the required joint commitment for a shift toward greater sustainability and resilience. However, it is evident that not all of us share such an outlook. Recently, the pandemic has inspired a debate on Trinnett (an online community of tourism academicians) about the shape of sustainable industry recovery in the post-pandemic period. As a result, academic papers have been published in respectful journals (for more information, see Higgins-Desbiolles, 2020), suggesting the fundamental divide in the opinions and perceptions between those supporting the rebounding in a more fair, just, responsible and equitable way and those advocating a more "laissez-faire" approach toward the development of the tourism industry. Although it has been recognized as a precondition for a positive change (see Mandić & Petrić, 2021), on consensus has yet emerged.

Sustainable tourism, particularly in protected and conserved areas, requires effective governance and adaptive management (Mandić, 2020, 2022; Mandić & Marković Vukadin, 2021; Mandić & Petrić, 2021; Plummer & Fennell, 2009), a critical aspect of which is effective monitoring (Petrić & Mandić, 2014). In essence, monitoring informs the decision-making by facilitating the understanding of the progress toward achieving specific goals or in the context of natural ecosystems, usually the state of the environment or protected features. To be effective, the monitoring must be comprehensive and involve all three tiers of sustainability, including the environment, economy and society (Eagles et al., 2018). Some authors (Pan et al., 2018) call for the fourth tier, referring to cultural sustainability. However, this is one aspect of the more extensive debate on whether culture is considered separately or as an inherent part of all other tiers (for more information, see *Agenda 21 Culture*). Monitoring tourism development's sustainability received significant attention from tourism scholars, with a substantial number of studies being published continuously (Bhuiyan et al., 2016; Blancas et al., 2015, 2016, 2018; Farinha et al., 2019; Franzoni, 2015; Gössling, 2015; Jurigova & Lencsesova, 2015; Lozano-Oyola et al., 2012, 2019; Pérez et al., 2013; Tanguay et al., 2013; Thongdejsri & Nitivattananon, 2019; Torres-Delgado & Palomeque, 2014; Weng et al., 2019; Ziaabadi et al., 2017). The authors rely on established frameworks of indicators such as the European Tourism Indicatory System (ETIS)(European Commission, 2016), or develop their own by either selecting established indicators, or proposing new ones, to cater to specific destination needs (e.g. focusing on particular types of destinations such as wetlands, or niches such as ecotourism), or to retain the analysis on a particular geographical level to facilitate the comparative assessment of destinations.

The selection of useful indicators to monitor particular sites or destinations is affected by a range of factors, including, among others, data availability, costs, focus (e.g. drivers, state, impacts), the complexity of challenges that have to be monitored, characteristics of the site (destination), and the effectiveness of governance and management. The final selection of indicators that are relevant and valid representations of the human–environment systems we wish to sustain is a value-based, political and collaborative process requiring the understanding of the function of indicators in policymaking and implementation (McCool & Stankey, 2004). To establish an effective framework, indicators could be drawn from different sources and established lists, including European Commision (2016): ETIS; Global Sustainable Tourism Council (GSTC) – Destination Criteria (2029); United Nations World Tourism Organization (UNWTO) (2004): *Indicators of Sustainable Development for Tourism Destinations A Guidebook*, United Nations (2007): *Indicators of sustainable development*, and OECD (2008): Key environmental indicators: *Towards Sustainable Development*.

The ETIS system was developed as an extension of the work of the Tourism Sustainability Group. The core of 43 indicators cover the critical information that a destination needs to understand, monitor, and manage to improve its performance and destination management. They also allow for comparison over time and for benchmarking between destinations. The indicators are designed to address: *destination management, economic value, social and cultural impacts, and environmental impacts of tourism development.* Along with core indicators, ETIS provide additional indicators focused on *beach and maritime destinations, urban and cultural destinations, cruising destinations, recreational boating destinations, nature and ecotourism destinations, and several pilot area-specific indicators* that might be used.

The GSTC destination criteria were first developed through a stakeholder consultation process in 2013 and revised in 2019 (Global Sustainable Tourism Council, 2019). The framework's scope is broad and could be applied to a wide range of destinations. The GSTC criteria fit into four sections, each with two to three sub-sections, including *sustainable management, socio-economic sustainability, cultural sustainability, and environmental sustainability*. The indicators are framed to support the achievement of the 2030 Agenda for Sustainable Development Goals.

The UNWTO (United Nations World Tourism Organization, 2004) Guidebook has been produced to inform decision-makers regarding sustainable development for tourism. This document gives attention to the *enterprise-specific, local, regional and global challenges* that could affect the tourism sustainability. UNWTO guidelines involve the indicators related to critical sustainability issues, including the *wellbeing of host communities, sustaining of cultural assets, community participation in tourism, tourist satisfaction, health and safety, capturing economic benefits from tourism, protection of valuable natural assets, managing scarce natural resources, limiting impacts of tourism activity, controlling tourist activities and levels, destination planning and control, designing products and services, and sustainability of tourism operations and services.*

The indicators proposed by OECD involve a core set of about 50 indicators, covering issues that reflect the main environmental concerns in OECD countries. It incorporates major indicators derived from sectoral sets and environmental accounting. Indicators are classified following the pressure–state–response (PSR – Pressure–State–Response) model. with the focus on specific *sectoral trends of environmental significance, their interactions with the environment and related economic and policy considerations*. The environmental indicators cover *climate change, ozone layer depletion, air quality, waste, water quality, water resources, forest resources, fish resources, and biodiversity. Socio-economic indicators cover GDP and population, consumption, energy, transport, agriculture, expenditure.*

The study approach

This chapter aims to determine which existing internationally recognized systems and initiatives meet the needs of monitoring sustainability at the PA level. The establishment of sustainability parameters and indicators is the foundation for informed decision-making, as it facilitates creating a deeper understanding of challenges by forcing us to explicate and "work through" their various dimensions (McCool, 2016).

The current study involved three major stages. The first stage included an in-depth analysis and critical reading of the relevant literature (Dimitrakopoulos & Jones, 2021; He & Cliquet, 2020; Leung et al., 2018; McGinlay et al., 2020; Múnera & Kerkhoff, 2019; Rannow et al., 2014; Squeo et al., 2012) to identify the main challenges associated to sustainable tourism development within PAs. The analysis yielded 27 challenges, which were grouped into three main categories: (1) *climate changes*, (2) *responsible and efficient management*, and (3) *strengthening the local community and their role in PA management* (Table 25.2). It should be noted that we retained 24 challenges for the final analysis, as we found no relevant indicators associated with three challenges, including *Managing ecosystem services, Managing geodiversity*, and *Implementation of smart solutions*.

The second stage of the research involved the identification of the relevant indicators which could be used to monitor the identified challenges within the existing frameworks, including the European tourism indicator system – ETIS (European Commission, 2016); Indicators of Sustainable Development for Tourism Destinations A Guidebook – UNWTO (United Nations World Tourism Organization, 2004); Indicators of sustainable development – UN (United Nations World Tourism Organization (2007); OECD Key environmental indicators (OECD, 2008) and GSTC Destination Criteria (Global Sustainable Tourism Council, 2019). At this stage, we have identified more than 300 potential indicators. This number has been reduced in the preliminary analysis to 168, reflecting the consensus of all three co-authors.

Table 25.1 The members of the expert group

Category	*Number of experts involved*	*Country of origin*
Academic researcher in the field of nature-based tourism development	3	Slovenia, Croatia, USA
PA management expert	1	Croatia,
Non-governmental organizations aimed at indicators development	1	France
Consultant in tourism or PA management	2	Italy, Austria
Academic researcher in the field of indicator development	2	UK, USA

In the final stage of the analysis, all indicators identified and associated with specific challenges were listed in an Excel spreadsheet and sent for evaluation to the nine expert group members, involving researchers and scientists in the field from the USA and Europe (Table 25.1). The experts were invited to evaluate the *feasibility and usefulness* of the existing indicators to *monitor the detected challenge* on a five-point Likert scale. The feasibility reflected their measurability regarding the data availability for PAs in general. Once the responses were collected, the arithmetic mean was determined for both parameters. The co-authors agree to retain Indicators with an arithmetic value in both parameters equal to or higher than 3.5 for further analysis.

Before proceeding with the analysis of critical challenges and indicators, we would like to acknowledge the two fundamental limitations of our approach. Firstly, we relied predominately on five primary sources to identify the potential indicators that could be used to monitor the proposed challenges. Secondly, our expert group involved only nine established academics and nature-based tourism experts, and the final framework reflects the subjective opinions. Although we do understand the limited generalizability of the analysis results, we believe that this attempt to identify some critical, common and underlying indicators to monitor proposed challenges could be the initial step toward establishing more effective monitoring of tourism in PAs globally.

Detecting PAs' sustainability challenges

During the last two decades, many authors agreed that it is crucial to establish a relevant monitoring framework to elucidate the impacts, uncertainties, and issues arising from different crises or disasters (Aliperti et al., 2019; Zhong et al., 2021). Considering the climate and environmental issues, socio-economic trends, and rapid development of technology, it is obvious that we are facing a more comprehensive range of challenges.

Climate change is a trigger to consider the sustainable transition of the tourism industry more seriously. Many studies present evidence of climate change effects, such as sea-level rise, flooding, shorter skiing seasons, outbreaks of fire and altered migration patterns of animals and birds (Agnew, 2001; Aretano et al., 2015; Thonfeld et al., 2020; Tranos et al., 2014). Without appropriate mitigation and adaptation measures, climate change could significantly jeopardize the attractiveness of nature-based tourism around the globe. From the perspective of climate change **mitigation**, tourism should primarily rely on increasing the energy efficiency and the share of renewable energy sources, including implementing tourism activities with zero CO_2 emissions (Gössling & Lund-Durlacher, 2021). Simultaneously, **adaption** depends on climate, spatial, environmental and socio-economic characteristics. The common for all destinations is that risk assessment and forward-looking analysis should be done to determine climate change adaptation measures (European Commission, 2021). For example, when modelling climate impacts on adaptation responses to biodiversity, it is crucial to predict future species' distribution patterns (Sinclair et al., 2010). Accordingly, the place and intensity of tourism in PAs need to be adjusted. Additionally, touristic infrastructure and activities should be designed to provide a "response to actual or expected climatic stimuli or their effects, which moderates harm or exploit beneficial opportunities" (IPCC, 2007).

As ecologically fragile environments, PAs could be placed under various pressures that are strongly connected with recreational tourism activities. Simultaneously, non-recreational activities in PAs, such as **agricultural and fisheries**, are considered part of the landscape heritage that supports wildlife (Eagles et al., 2002). In terms of sustainable yields and genetic resilience, **ecosystem services** (on which agriculture and fisheries depend) are under systemic climate, pollution, and anthropogenic pressures that reflect their productivity (Braat & de Groot, 2012). Systemic pressures on **forests and forest land** are caused by forest fire, transboundary air pollution or extreme climate events (Forzieri et al., 2021). Still, most of the PAs' tourist accommodation and recreational capacities are **watered** directly from natural sources. This situation is challenging to protect the quality and quantity of water flows and develop ecologically sustainable forms of water usage in close relationship with their surrounding environment and conservation issues (Vila et al., 2018). Water regimes' disruptions are caused dominantly by climate change, unsustainable usage and unappropriated **spatial and soil** management that is particularly pronounced in the coastal areas (Papageorgiou, 2016). In general, **marine** PAs should strengthen environmental and visitor management systems that help to avoid or at least limit adverse effects of anthropogenic loads, caused mainly by high tourist seasonality (Martín-López et al., 2011). Intensified tourism activities can result in adverse impacts on **biological** resources, such as extraction of natural goods, hunting or harassment of animals (Oldekop et al., 2016). As an important part of the nature conservation agenda, **geodiversity management** plays a significant role, but a more formal platform is required to mainstream geoconservation within PA planning and management (Gordon et al., 2018). **Waste** management provides challenges and opportunities. In the case of poor waste management, tourism causes undesirable outcomes in the physical environment, but waste management in line with the EU Waste Management Hierarchy results in many benefits for tourism destinations (Jones & Wynn, 2019). As Zovko et al. (2021) explain, "overall waste reduction should be mainstreamed to better resource efficiency and with the shift towards more sustainable production and consumption patterns".

Decision-makers and experts must reconsider numerous **management** aspects of Pas, such as sustainable funding, capacity building, communication and cooperation, including education, monitoring and functional diversification with integration processes in the region. Although all these aspects are equally valuable, **funding** still plays a central role. It is generally accepted that tourism in Pas plays has a significant share in the economy of many countries. Simultaneously, Pas worldwide are underfunded. Limited public financial resources could be one of the main reasons for ineffective PAs' management, especially in a **crisis** (Mandić, 2020). It is reported that during the current pandemic, decreased numbers of visitors resulted in substantial PAs' economic losses for both PAs and local communities (Spenceley et al., 2021). Such situations should be avoided through a proactive response and recovery approach based on a "problem-solving action plan including prevention measures that combine all stakeholders' interests and activities" (Martens et al., 2016). In times of crisis, it is crucial to strengthen linkages between nature-based tourism and other **regional and local industrial sectors** (Romão et al., 2017). More precisely, scientists discuss systemic, integrative, and interdisciplinary approaches that should be used to increase **adaptive co-management** (Islam et al.,

2018), founded in **cooperation**, accountability, the rule of law, **monitoring**, transparency, and social learning (Kovács et al., 2021). **Functional diversification** of sustainable tourism also plays a significant role in creating more resilient nature-based tourism. Weidenfeld (2018) conclude that "diversification strategies result in more sustainable tourism", which pays considerable attention to "natural resources and value-added products and services". Additionally, many conflicting management issues and crises could be resolved through the **engagement** and **capacity building** of the local stakeholders. For a common and unambiguous understanding of emerging challenges and open questions, **education** is a fundamental prerequisite (Christie et al., 2017).

Establishing efficient benefit-sharing mechanisms of multiple objectives and stakeholders requires consideration of a wide range of **socio-demographic trends** associated with tourism market demands. It is understood that increased **visitor flow** results in adverse impacts on natural resources and, consequently, a decrease in the quality of experience (McCool et al., 2021). In this regard, it is imperative to establish effective **visitor flow management**. One of the most widely implemented methods is zoning, which determines the type of accommodation, transportation and tourist activities that should be adjusted to the size of the visitor groups (Roman et al., 2007). A significant challenge is developing and maintaining climate and environmentally **resilient tourism infrastructure**, which certainly implies the implementation of **digital solutions and "Smart" modes** for visitor flow management and environmental monitoring. The current pandemic has also raised the issue of timely and comprehensive **safety and health protocols** for visitors, employees and the local community. In line with the challenges above, measuring **visitor satisfaction** and the **development and diversification of tourism products** (Job et al., 2017; Snyman & Bricker, 2018); are crucial for the socio-economic resilience of each destination.

Income from partnership agreements, concessions, and private sector operators benefits conservation, habitat management, residents and visitors (Eagles et al., 2002). However, the **resilience of the local community** does not depend solely on adequate benefit distribution from tourism development but also on creating alternative sources **of income**. Such economic resilience significantly affects living standards and contributes to the resident's **quality of life**. In addition, the personal benefit from nature-based tourism fosters residents' motivation to improve environmental protection and nature conservation (Ali et al., 2020). Nevertheless, PAs' quality and proactive tourism management require a structured and responsive **monitoring** system designed to consolidate and process a wide range of diverse data.

Evaluation of existing systems – results and discussion

The main goal of this research was to define challenges in accordance with the modern needs of PA management and to investigate the extent to which these challenges could be monitored through existing indicator systems. Therefore, as a first step, it was important to determine whether the selected challenges were sufficiently represented through the existing indicator systems, based on the number of indicators and their quality determined according to the expert group. Research has unequivocally shown that the number of indicators does not necessarily imply clear and measurable evidence for a particular challenge. Therefore, the presentation and the discussion of the results of our analysis centre around three main themes, including: (1) *the challenges that are insufficiently represented through previously defined indicators systems*; (2) *challenges adequately represented by the number of previously defined indicators, but these indicators are not measurable, or the results of these indicators do not provide an adequate response*; and (3) *the extent to which each of the analysed five basic sets of indicators responds to the identified challenges and the criterion of data availability for the calculation of indicators.*

We have identified 27 challenges (Table 25.2), of which three (*Geodiversity management; Ecosystem service management and implementation of Smart approach for visitor flow management, design and experience monitoring*) were not correlated with indicators of five established international frameworks. As Gordon et al. (2018) stated, recognizing a wider range of geodiversity values should be the foundation of biodiversity management. In addition, the concept of ecosystem services has long been an essential topic in the literature related to PA management (Scolozzi et al., 2014), but still without a suitable monitoring tool. Features of digitalization and "Smart" solutions have become commonplace, but understanding their usefulness requires developing indicators and establishing monitoring.

Furthermore, for some of the mainstream challenges in contemporary scientific and public discourse, there is only a small number of (generic) indicators that cannot give us a clear picture of temporal change. This primarily refers to *Climate adoption* (4 indicators), *Socio-demographic trends and changes in the tourism demand market* (3 indicators), *Crisis management* (3 indicators) and *Functional diversification of the PA and strengthening of integration processes within the region* (2 indicators). These topics were not considered significant challenges when initial indicator systems were created (2005–2015), but today they certainly merit, particularly when it comes to responsible recovery amid the current pandemic and building resilience of complex socio-ecological systems.

A relatively high number of indicators could be associated with the following challenges, namely, *Education, employees, key stakeholders and visitors* (7), *Effective monitoring* (5), *Safety and health of visitors, employees and the community* (15) and *Local community resilience and adequate distribution benefit from tourism development* (11). However, our analysis showed that a very small share of such indicators (between 9.1 percent and 28.6 percent) is adequate to effectively monitor these challenges. The smallest share of appropriate indicators is recognized for the *Safety and health of visitors, employees and the community* (13.3 percent) and the *Local community resilience and adequate distribution benefit from tourism development* (9.1 percent). Given the global threats, such as climate change and pandemic crisis, the currently established frameworks need extensive revision and improvement.

Challenges with more than a third but less than half of the appropriate indicators have a slightly better status than the challenges mentioned above. Those challenges are *Climate change mitigation* (44.4 percent), *Energy and water resources management* (36.4 percent), *Agricultural and fisheries/aquaculture management* (33.3 percent) and *Creating alternative sources of income and linking them to the value chain* (42.9 percent), and it is evident that the related indicators also require significant revision.

An excellent example of a concise and well-affected list of indicators is undoubtedly related to *Forest and forest area management* since respondents rated both indicators as appropriate (Table 25.2). However, this status does not necessarily imply their sufficiency regarding the increasing loads on forest ecosystem services. Overall, the situation emphasizes the necessity to reconsider defined indicators in the context of the related challenges and their interdependences. Namely, it is not uncommon that indicators associated with one challenge contribute to the understanding of other challenges, as well.

It is important to point out that out of 168 analysed indicators, only 64 indicators have a score higher or equal to 3.5 in the case of both criteria and are thus considered suitable for use in PAs.

One of the critical sustainability issues is *Climate change mitigation* (Table 25.3). From 3 different indicator systems, 10 indicators are detected as appropriate for the quantification and

Table 25.2 Main challenges in relation to analysed indicators

Main category	*Sub-category*	*Challenge*	*Number of detected indicators*	*Number of indicators appropriate by both criteria (3.5 or more)*	*Average score indicator feasibility*	*Average score indicator usefulness*
CLIMATE CHANGES		*Climate change mitigation*	9	4	3.65	4.11
		Adaptation to climate change	4	1	3.69	4.09
EFFICIENT AND RESPONSIBLE MANAGEMENT	*Resource utilization*	*Biodiversity management*	10	7	3.91	4.18
		Energy and water resources management	11	4	3.56	4.58
		Space and soil management	10	8	4.31	4.18
		Geodiversity management	0	0		
		Forest and forest area management	2	2	4.75	4.88
		Agricultural and fisheries / aquaculture management	6	2	3.46	3.82
		Ecosystem service management	0	0		
		Waste reduction and promotion of sustainable circular economy practices	6	4	3.48	4.25
	PA management	*Sustainable funding*	6	5	4.19	4.4
		Capacity building, communication and cooperation (local community, tourism stakeholders, governing bodies)	5	3	3.43	4.13
		Education, employees, key stakeholders and visitors	7	2	3.39	4.04
		Crisis management	3	0	2.88	4.13
		Functional diversification of the PA and strengthening of integration processes within the region	2	1	3.38	4.25
		Effective monitoring	5	1	3.03	4.15
	Management of tourism development and visitor activities in the PA territory	*Visitor Flow Management (Zoning, CC)*	12	10	4.04	3.31
		Development and diversification of year-round tourism products	6	4	3.48	4.08
		Monitoring visitor satisfaction	10	6	3.75	4.19
		Implementation of digital solutions and "Smart" approach (visitor flow management, design and experience monitoring)	0	0		
		Socio-demographic trends and changes in the tourism demand market	3	2	3.33	4.25
		Maintenance and development of visitor infrastructure	7	5	4.11	4.61
		Safety and health of visitors, employees and the community	15	2	3.03	3.49

(*Continued*)

Table 25.2 (Continued)

Main category	*Sub-category*	*Challenge*	*Number of detected indicators*	*Number of indicators appropriate by both criteria (3.5 or more)*	*Average score indicator feasibility*	*Average score indicator usefulness*
STRENGTHENING THE LOCAL COMMUNITY AND THE ROLE IN PA MANAGEMENT		*Local community resilience and adequate distribution benefit from tourism development*	11	1	3.14	3.49
		Creating alternative sources of income and linking them to the value chain	7	3	3.43	3.98
		Quality of life and resilience of the local community (well-being)	8	4	3.55	3.97
		Monitoring the attitude of the local community on the development and effects of tourism development	6	0	3.31	4.23

Table 25.3 Challenge *Climate change mitigation* and associated indicators

Challenge	*Indicator*	*Source*	*Indicator feasibility*	*Indicator usefulness*
Climate change mitigation	Percentage of tourists and same-day visitors using local/soft mobility/public transport services to get around the destination	EIS	4.25	4.00
	Percentage of tourism enterprises involved in climate change mitigation schemes - such as CO_2 offset, low energy systems, etc.	ETIS	3.62	4.75
	Value of damage caused by extreme climatic events to the tourism sector	UNWTO	**2.50**	4.12
	A climate risk assessment, covering current and future risks - undertaken and made publicly available	GSTC	3.00	4.00
	The annual climate report, including monitoring and mitigation actions	GSTC	3.12	4.25
	% of the tourist area and infrastructure with sea defaces	UNWTO	3.87	3.87
	% of ski areas with snowmaking equipment	UNWTO	4.87	3
	The degree to which key tourist zones are covered by contingency or emergency planning (existence of plan)	UNWTO	**4.25**	**4.25**
	Total CO_2 produced due to the community's energy consumption	UNWTO	3.375	4.75

qualification of the causal climate-social phenomena. Based on the conducted research, only 4 offer a satisfactory response to the challenge, and it is reasonable to expect that they could be measured. Some of the indicators, such as the *Value of damage caused by extreme climatic events to the tourism sector* and the *Annual climate report, including monitoring and mitigation actions*, would essentially give a clear overview of the climate change issues. Unfortunately, it is not realistic to expect PAs worldwide (and even in developed countries) to have data on which the desired indication could be calculated. Simultaneously, indicators related to the implementation of documents related to climate change mitigation are more appropriate, but their usefulness is somewhat lower.

When considering the PA management, challenge *Capacity building, communication, and cooperation were detected* (Table 25.4) as essential for all sustainability dimensions. Among indicators detected as suitable for measuring this challenge, the existence of a community tourism plan (United Nations World Tourism Organization, 2004) was detected as of the highest feasibility and usefulness. The feasibility should be pretty high when dealing with plans and strategies within the PAs. In any case, the sole existence of these documents does not confirm the success of actual implementation on the territory. Also, the feasibility of indicators depends on PAs' location, the level of development and interconnections with the surrounding territory, modes of management and governance, and the fact that not all PAs can collect data in the same way. Therefore, even proved suitable, the data behind this indicator could blur the reality of the situation. All existing indicators (even those evaluated as suitable in this chapter) need to be re-examined from the data perspective. It is also interesting that the indicator *% of agencies incorporating sustainable tourism principles in their strategic planning processes* (United Nations World Tourism Organization, 2004) was the worst rated, given the insufficient contribution to monitoring the challenge and the issues arising during the data collection.

Concerning the challenge related to *monitoring the impact of agriculture and aquaculture* (Table 25.5), we can immediately notice the disparity of indicators related to agriculture (1) and aquaculture (5). Of the indicators shown, only two draw obvious conclusions and are easily measurable. Simultaneously, most indicators are not feasible, while, for example, *fishing failures* are a relatively simple indicator (readily available data) but offer little usefulness for tracking challenges.

Table 25.4 Challenge *Capacity building, communication and cooperation* and associated indicators

Challenge	*Indicator*	*Source*	*Indicator feasibility*	*Indicator usefulness*
Capacity building, communication and cooperation (local community, tourism stakeholders, governing bodies)	Existence of a community tourism plan	UNWTO	**4.13**	**4.13**
	% of agencies incorporating sustainable tourism principles into their strategic planning processes	UNWTO	**2.38**	**3.63**
	% of tourists participating in protection activities or contributing to conservation	UNWTO	3.75	4.38
	Capacity building for establishment and improvement of MSMEs: number of programs/events, level of participation	ETIS	3.13	4.63
	Evidence of stakeholder consultation, meetings in developing the plan	GSTC	3.75	3.88

Table 25.5 Challenge *Agricultural and fisheries/aquaculture management* and associated indicators

Challenge	*Indicator*	*Source*	*Indicator feasibility*	*Indicator usefulness*
Agricultural and fisheries/ aquaculture management	Number of berths and moorings for recreational boating	ETIS	**5.00**	**4.50**
	The area under organic farming	UN	**4.25**	**4.88**
	The proportion of fish stocks within their safe biological limits	UN	2.50	4.00
	Marine trophic index	UN	**2.13**	**3.47**
	Area of coral reef ecosystems and percentage of live cover	UN	3.00	3.49
	Fishing quotas	TSD	4.00	3.13

Of all the groups of challenges, the worst results were obtained in the main category *Strengthening the role of the local community*, with the challenge *Local community resilience and adequate distribution benefit from tourism development*, whose implications are far-reaching, especially in the COVID-19 era (Table 25.6). Of the 11 indicators analysed, the respondents recognized only *the Local unemployment rate in the off-season* as feasible and practical. One of the standard indicators for destinations, the *Relative contribution of tourism to the destination's economy (% GDP)*, in this case, is considered difficult to implement due to the complexity of PAs in economic terms.

The *UNWTO Indicator List* (United Nations World Tourism Organization, 2004) is the foundation of most indicator systems, and duplicate indicators were eliminated prior to analysis. From this research, we can draw general conclusions based on which the most innovations and deviations were shown by ETIS and GSTC lists. To some extent, it could be concluded that the UNWTO indicators are helpful to monitor the identified challenges within PAs. However, GSTC and ETIS lists of indicators enable the most effective monitoring, with 50 percent of valuable and feasible indicators (Figure 25.1).

Finally, one of the most critical questions is: how should indicators sustain PA management decisions to develop policies and mitigation activities? It is clear that no indicator or indicator system offers an answer in itself. The sound system should be grounded in a standard, acceptable limits (changes) and accompanied by adequate management strategies. One of the most significant issues that the existing (analysed) systems face is that many indicators are vague, almost immeasurable, or at least not measurable in a way that can ultimately support managerial decisions.

Ultimately, there is a clear need to develop a system of PA indicators to mitigate challenges, which will bring together all existing (and emerged) challenges. This novel system could only be defined by including the specific needs of diverse types of PAs (e.g. maritime, coastal, island, mountain river, wetlands). They, obviously, should be tested in diverse locations and verified by a third party. Finally, validated indicators should be easily communicated and justified as applicable for: (1) PAs adaptive management objectives, (2) effectiveness in PA management planning and (3) prevention and mitigation of the challenges. Thus, a system of PA indicators aimed at mitigating challenges is only a part of an integrated management model that should be reached in future PA management. Such a system should summarize the best world practices and specific management needs through goals, activities, monitoring and reevaluation of management.

Table 25.6 Challenge *Local community resilience and adequate distribution benefit from tourism development* and associated indicators

Challenge	*Indicator*	*Source*	*Indicator feasibility*	*Indicator usefulness*
Local community resilience and adequate distribution benefit from tourism development	Relative contribution of tourism to the destination's economy (% GDP)	ETIS	**2.88**	3.25
	Direct tourism employment as a percentage of total employment in the destination	ETIS	3.13	3.63
	Percentage of jobs in tourism that are seasonal	ETIS	3.38	4.13
	Number of social services available to the community (% which are attributable to tourism)	UNWTO	3.13	3.63
	% who believes that tourism has helped to bring new services or infrastructure	UNWTO	3.38	3.00
	% who believes that they or their family benefit from tourism	UNWTO	3.13	3.38
	The local unemployment rate in the off-season	TSD	**4.13**	**3.75**
	Value of receipts or invoices of funding for local groups (economic alternatives for local people to reduce exploitation of resources)	UNWTO	2.75	4.00
	Number (membership in local programs (lists of supported programs or lists of membership)	UNWTO	**3.25**	**2.00**
	Value of infrastructure investment by tourism enterprises	UNWTO	3.25	3.63
	Net tourism revenues accruing to the community	UNWTO	2.13	4.00

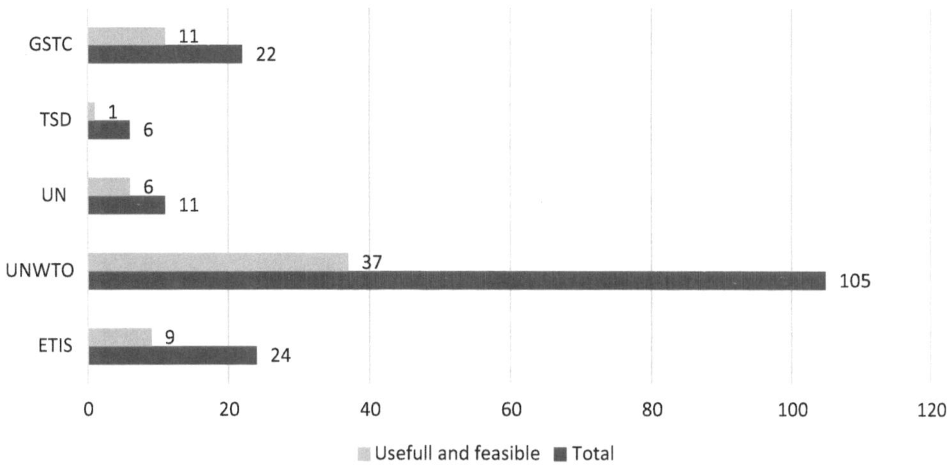

Figure 25.1 Comparison of the total number of analysed indicators of existing indicator systems and their feasibility and usefulness.

Conclusion

This chapter evaluates the usability of the established frameworks of indicators for monitoring the contemporary challenges in PAs as a starting point for a broader discussion related to three vital themes, including Climate change, the Efficient and resource utilization and Strengthening the local community (Table 25.2). The conducted analysis has demonstrated that the "Smart" solutions, Geodiversity management and Ecosystem services, are challenges that have not been addressed in existing monitoring frameworks. Additionally, the analysis pointed out the necessity of rapid refinement of indicators related to *Climate change adaptation and Crisis management, Functional diversification and integration in regional terms, including stakeholders' Health and safety issues, Resilience of the local community and Effective monitoring*. Judging by the numerous scientific and professional research, these are the emerging challenges, and paying greater attention to them in the post-pandemic period could support PAs' adaptive tourism management and resilience.

The proposed evaluation scheme with the main and sub-categories of challenges (Table 25.2) identified weaknesses, shortcomings and knowledge gaps in the existing monitoring frameworks. Their potentials and limitations must be emphasized, and the very logic behind the development of complex systems that are not globally operational or able to respond to current challenges. Furthermore, "slowly" adaptive indicator systems are the opposite of the dynamics of emerging challenges, influenced by urbanization, pandemics, climate change and the consequences of war events and disasters. This research also reveals that the strict division of the challenges into the three tiers of sustainability is insufficient, and the selection of appropriate indicators (according to indicator feasibility and data availability) is inadequate to a significant extent. As in previous revision processes, the future in-depth revision of existing indicators should focus on their relevance, legitimacy, feasibility, credibility, and availability of all necessary data and information. The revision of the established frameworks of indicators should consider the cross-cutting nature of the presented challenges with particular emphasis on clear, targeted variables.

It is widely recognized that adequate and timely available data are the fundamental prerequisite for successful management. This research indicates that many PAs are straggling with the availability of adequate data. In that context, collecting and manipulating real-time data is just one of the demanding future moves. New tools and methods, such as big data processing and forecasting models, should be considered in the planning and managerial processes designed to increase the harmony and resilience of the interaction between visitors and nature in PAs. The significance of new technical and technological solutions and the horizontal importance of modern information and communication technologies should also be emphasized to strengthen environmental monitoring in a short time frame. Although not discussed in this chapter, they certainly play a crucial role in addressing a wide range of emerging challenges arising from the climate, resource, management and socio-economic issues.

Finally, it is essential to emphasize the need for further research in line with the specifics of the challenges that arise in a wide range of PAs. The challenge's specifics will be reflected in related indicators, which will only then be adequate for monitoring the actual state of the environment and tourism development in the PAs.

References

Ali, M. B., Quaddus, M., Rabbanee, F. K., & Shanka, T. (2020). Community Participation and Quality of Life in Nature-Based Tourism: Exploring The Antecedents and Moderators. *Journal of Hospitality & Tourism Research, 46(3)*, 630–661. https://doi.org/10.1177/1096348020980094

Aliperti, G., Sandholz, S., Hagenlocher, M., Rizzi, F., Frey, M., & Garschagen, M. (2019). Tourism, Crisis, Disaster: An Interdisciplinary Approach. *Annals of Tourism Research, 79*, 102808. https://doi.org/10.1016/j.annals.2019.102808

Aretano, R., Semeraro, T., Petrosillo, I., De Marco, A., Pasimeni, M. R., & Zurlini, G. (2015). Mapping Ecological Vulnerability to Fire for Effective Conservation Management of Natural Protected Areas. *Ecological Modelling, 295*, 163–175. https://doi.org/10.1016/j.ecolmodel.2014.09.017

Bhuiyan, M. A. H., Siwar, C., & Ismail, S. M. (2016). Sustainability Measurement for Ecotourism Destination in Malaysia: A Study on Lake Kenyir, Terengganu. *Social Indicators Research, 128(3)*, 1029–1045. https://doi.org/10.1007/s11205-015-1068-5

Blancas, F. J., Lozano-Oyola, M., González, M., & Caballero, R. (2016). Sustainable Tourism Composite Indicators: A Dynamic Evaluation to Manage Changes in Sustainability. *Journal of Sustainable Tourism, 24*(10), 1403–1424. https://doi.org/10.1080/09669582.2015.1122014

Blancas, Francisco Javier, Lozano-Oyola, M., & González, M. (2015). A European Sustainable Tourism Labels proposal using a composite indicator. *Environmental Impact Assessment Review, 54*, 39–54. https://doi.org/10.1016/j.eiar.2015.05.001

Blancas, Francisco Javier, Lozano-Oyola, M., González, M., & Caballero, R. (2018). A Dynamic Sustainable Tourism Evaluation Using Multiple Benchmarks. *Journal of Cleaner Production, 174*, 1190–1203. https://doi.org/10.1016/j.jclepro.2017.10.295

Boluk, K. A., Cavaliere, C. T., & Higgins-Desbiolles, F. (2019). A Critical Framework for Interrogating the United Nations Sustainable Development Goals 2030 Agenda in Tourism. *Journal of Sustainable Tourism, 27*(7), 847–864. https://doi.org/10.1080/09669582.2019.1619748

Braat, L. C., & de Groot, R. S. (2012). The Ecosystem Services Agenda: Bridging the Worlds of Natural Science and Economics, Conservation and Development, and Public and Private Policy. *Ecosystem Services, 1(1)*, 4–15. https://doi.org/10.1016/j.ecoser.2012.07.011

Christie, P., Bennett, N. J., Gray, N. J., Wilhelm, T. A., Lewis, N., Parks, J., & Friedlander, A. M. (2017). Why People Matter in Ocean Governance: Incorporating Human Dimensions Into large-scale Marine Protected Areas. *Marine Policy, 84*, 273–284. https://doi.org/10.1016/j.marpol.2017.08.002

Dimitrakopoulos, P. G., & Jones, N. (2021). Protected Areas in Forest Conservation: Challenges and Opportunities. *Forests*, 12(*4*), 488. https://doi.org/10.3390/f12040488

Eagles, P., McCool, S., Haynes, C., & Phillips, A. (2002) *Sustainable Tourism in Protected Areas: Guidelines for Planning and Management.* IUCN Gland, Switzerland and Cambridge, UK. xv + 183 pp.

Eagles, P. F. J., Haynes, C. D., & McCool, S. F. (2018). Sustainable Tourism in Protected Areas: Guidelines for Planning and Management (Russian version). In *Sustainable Tourism in Protected Areas: Guidelines for Planning and Management (Russian Version)* (Issue 8). https://doi.org/10.2305/iucn.ch.2002.pag.8.ru

European Commission. (2016). The European Tourism Indicator System (Issue March). https://doi.org/10.2873/982144

European Commission. (2021). Commission Notice: Technical guidance on the climate proofing of infrastructure in the period 2021–2027. C373

Farinha, F., Jos, M., Silva, E. M. J., Lança, R., Pinheiro, M. D., & Miguel, C. (2019). *Selection Process of Sustainable Indicators for the Algarve Region – OBSERVE Project*. 1–24.

Font, X., Torres-Delgado, A., Crabolu, G., Palomo Martinez, J., Kantenbacher, J. & Miller, G. (2021) The Impact of Sustainable Tourism Indicators on Destination Competitiveness: The European Tourism Indicator System, *Journal of Sustainable Tourism*, https://doi.org/10.1080/09669582.2021.1910281

Forzieri, G., Girardello, M., Ceccherini, G., Spinoni, J., Feyen, L., Hartmann, H., Beck P. S. A., Camps-Valls, G., Chirici, G., Mauri, A., & Cescatti, A. (2021), Emergent Vulnerability to Climate-driven Disturbances in European Forests. *Nature Communications*, *12*, 1081. https://doi.org/10.1038/s41467-021-21399-7

Franzoni, S. (2015). Measuring the Sustainability Performance of the Tourism Sector. *Tourism Management Perspectives*, *16*, 22–27. https://doi.org/10.1016/j.tmp.2015.05.007

Global Sustainable Tourism Council (2019): GSTC Destination Criteria. Available at: https://www.gstcouncil.org/revising-gstc-destination-criteria/

Gordon, J. E., Crofts, R., Díaz-Martínez, & Woo, K. S. (2018). Enhancing the Role of Geoconservation in Protected Area Management and Nature Conservation. *Geoheritage*, *10*, 191–203. https://doi.org/10.1007/s12371-017-0240-5

Gössling, S. (2015). New Performance Indicators for Water Management in Tourism. *Tourism Management*, *46*, 233–244. https://doi.org/10.1016/j.tourman.2014.06.018

Gössling, S., & Lund-Durlacher, D. (2021). Tourist Accommodation, Climate Change and Mitigation: An Assessment for Austria, *Journal of Outdoor Recreation and Tourism*, *34*, 100367. https://doi.org/10.1016/j.jort.2021.100367

He, M. & Cliquet, A. (2020). Challenges for Protected Areas Management in China. *Sustainability*, *12*, 5879. https://doi.org/10.3390/su12155879

Higgins-Desbiolles, F. (2018). Sustainable Tourism: Sustaining Tourism Or Something More? *Tourism Management Perspectives*, *25*(October), 157–160. https://doi.org/10.1016/j.tmp.2017.11.017

Higgins-Desbiolles, F. (2020). The "War Over Tourism": Challenges to Sustainable Tourism in the Tourism Academy After COVID-19. *Journal of Sustainable Tourism*, 1–19. https://doi.org/10.1080/09669582.2020.1803334

Hobbs, R. J., Cole, D. N., Yung, L., Zavaleta, E. S., Aplet, G. H., Chapin, F. S., Landres, P. B. (2010). Guiding Concepts for Parks and Wilderness Stewardship in the Era of Global Environmental Change. *Frontieres in Ecelogy and Environment*, *8*(*9*), 483–490.

IPCC. (2007). Climate Change 2007: Synthesis Report. Contribution of Working Groups I, II and III to the Fourth Assessment Report of the Intergovernmental Panel on Climate Change [Core Writing Team, Pachauri, R.K. & Reisinger, A. (eds.)]. IPCC, Geneva, Switzerland, 104 pp.

Islam, M. W., Ruhanen, L., & Ritchie, B. W. (2018). Tourism Governance in Protected Areas: Investigating the Application of the Adaptive Co-management Approach. *Journal of Sustainable Tourism*, *26*(*11*), 1890–1908. https://doi.org/10.1080/09669582.2018.1526291

Job, H., Becken, S., & Lane, B. (2017). Protected Areas in a Neoliberal World and the Role of Tourism in Supporting Conservation and Sustainable Development: An Assessment of Strategic Planning, Zoning, Impact Monitoring, and Tourism Management at Natural World Heritage Sites. *Journal of Sustainable Tourism*, *25*(*12*), 1697–1718. https://doi.org/10.1080/09669582.2017.1377432

Jones, P., & Wynn, M. G. (2019). The circular Economy, Natural Capital and Resilience in Tourism and Hospitality. *International Journal of Contemporary Hospitality Management*, *31*(*6*), 2544–2563. https://doi.org/10.1108/IJCHM-05-2018-0370

Joppe, M. (2018). Tourism Policy and Governance: Quo vadis? *Tourism Management Perspectives*, *25*(October 2017), 201–204. https://doi.org/10.1016/j.tmp.2017.11.011

Jurigova, Z., & Lencsesova, Z. (2015). Monitoring System of Sustainable Development in Cultural and Mountain Tourism Destinations. *Journal of Competitiveness*, 7(*1*), 35–52. https://doi.org/10.7441/joc.2015.01.03

Klanjšček, J., Geček, S., Marn, N., Legović, T., & Klanjšček, T. (2018). Predicting Perceived Level of Disturbance of Visitors Due to Crowding in Protected Areas. *PLOS ONE 13*(*6*): e0197932. https://doi.org/10.1371/journal.pone.0197932

Kovács, E., Mile, M., Fabók, V., Margóczi, K., Kalóczkai, A., Kasza, V., Nagyné Grecs, A., Bankovics, A., & Mihók, B. (2021). Fostering Adaptive Co-management with Stakeholder Participation in the Surroundings of Soda Pans in Kiskunság, Hungary – An Assessment. *Land Use Policy*, *100*, 104894. https://doi.org/10.1016/j.landusepol.2020.104894

Leung, Y. F., Spenceley, A., Hvenegaard, G., & Buckley, R. (eds.) (2018) *Tourism and Visitor Management in Protected Areas: Guidelines for Sustainability*. Best Practice Protected Area Guidelines Series No. 27, Gland, Switzerland.

Lozano-Oyola, M., Blancas, F. J., González, M., & Caballero, R. (2012). Sustainable Tourism Indicators as Planning Tools in Cultural Destinations. *Ecological Indicators, 18*, 659–675. https://doi.org/10.1016/j.ecolind.2012.01.014

Lozano-Oyola, M., Blancas, F. J., González, M., & Caballero, R. (2019). Sustainable Tourism Tags to Reward Destination Management. *Journal of Environmental Management, 250*(July 2018), 109458. https://doi.org/10.1016/j.jenvman.2019.109458

Mandić, A. (2020). Structuring Challenges of Sustainable Tourism Development in Protected Natural Areas with Driving Force–Pressure–State–Impact–Response (DPSIR) Framework. *Environment Systems and Decisions, 40(4)*. https://doi.org/10.1007/s10669-020-09759-y

Mandić, A. (2022). Protected Area Tourism. In *Encyclopedia of Tourism Management and Marketing*. Edward Elgar Publishing, Cheltenham, UK. https://doi.org/10.4337/9781800377486.protected.area.tourism

Mandić A., & Marković Vukadin I. (2021) Managing Overtourism in Nature-Based Destinations. In Mandić, A., Petrić, L. (eds) *Mediterranean Protected Areas in the Era of Overtourism*. Springer, Cham. https://doi.org/10.1007/978-3-030-69193-6_3

Mandić A., & Petrić L. (2021) Summary and Outlook. In: Mandić, A., Petrić, L. (eds) *Mediterranean Protected Areas in the Era of Overtourism*. Springer, Cham. https://doi.org/10.1007/978-3-030-69193-6_20

Manning, R. E. (2010). *Studies in Outdoor Recreation: Search and Research for Satisfaction*. Corvallis, OR: Oregon State University Press, 1–468.

Marković Vukadin, I. (2017) Sustainability Issues in Management of Tourism in Protected Areas: Case Study of Plitvice Lakes National Park. In: Dwyer, L., Tomljenović, R., Čorak, S. (eds) *Evolution of Destination Planning and Strategy, The Rise of Tourism in Croatia*. Palgrave Macmillan, London, New York, str. 201–219.

Martens, H., Feldesz, K., & Merten, P. (2016). Crisis Management in Tourism – A Literature Based Approach on the Proactive Prediction of a Crisis and the Implementation of Prevention Measures. *Athens Journal of Tourism, 3*, 89–101. https://doi.org/10.30958/ajt.3-2-1

Martín-López, B., García-Llorente, M., Palomo, I., & Montes, C. (2011). The Conservation Against Development Paradigm in Protected Areas: Valuation of Ecosystem Services in the Doñana Social-ecological System (Southwestern Spain). *Ecological Economics, Elsevier, 70(8)*, 1481–1491. https://doi.org/10.1016/j.ecolecon.2011.03.009

McCool, S. F. (2016). *Reframing Sustainable Tourism* (Vol. 2). https://doi.org/10.1007/978-94-017-7209-9

McCool S. F., Eagles P. F. J., Skunca O., Vukadin V., Besancon C., Novosel A. (2021) Integrating Marketing and Management Planning for Outstanding Visitor Experiences in a Turbulent Era: The Case of Plitvice Lakes National Park. In: Mandić, A., Petrić, L. (eds) *Mediterranean Protected Areas in the Era of Overtourism*. Springer, Cham. https://doi.org/10.1007/978-3-030-69193-6_11

McCool, S. F., & Stankey, G. H. (2004). Indicators of Sustainability: Challenges and Opportunities at the Interface of Science and Policy. *Environmental Management, 33(3)*, 294–305. https://doi.org/10.1007/s00267-003-0084-4

McGinlay, J., Gkoumas, V., Holtvoeth, J., Fuertes R., Bazhenova, E., Benzon, A., Botsch, K., Martel, C. C., Sánchez, C. C., Cervera, I., Chaminade, G., Doerstel, J., García, C. J. F., Jones, A., Lammertz, M., Lotman, K., Odar, M., Pastor, T., Ritchie, C., Santi, S., Smolej, M., Rico, F. S., Waterman, H., Zwijacz-Kozica, T., Kontoleon, A., Dimitrakopoulos, P. G., & Jones, N. (2020). The Impact of COVID-19 on the Management of European Protected Areas and Policy Implications. *Forests, 11(11)*, 1214. https://doi.org/10.3390/f11111214

Múnera, C., & Kerkhoff, L. (2019). Diversifying Knowledge Governance for Climate Adaptation in Protected Areas in Colombia. *Environmental Science & Policy, 94*, 39–48. https://doi.org/10.1016/j.envsci.2019.01.004.

Newsome, D., & Lacroix, C. (2011). Changing Recreational Emphasis and the Loss of the ˝natural Experiances˝ in Protected Areas: An Issue that Deserves Consideration, Dialogue and Investigation. *Journal of Tourism and Laisure studies, 17* (2), 315–333.

OECD. (2008). *OECD Environmental Indicators: Towards Sustainable Development*. OECD. https://doi.org/10.1177/009155218501300304

Oldekop, J. A., Holmes, G., Harris, W. E., & Evans, K. L. (2016). A Global Assessment of the Social and Conservation Outcomes of Protected Areas. *Conservation Biology, 30*, 133–141. https://doi.org/10.1111/cobi.12568

Pan, S. Y., Gao, M., Kim, H., Shah, K. J., Pei, S. L., Chiang, P. C. (2018). Advances and Challenges in Sustainable Tourism Toward a Green Economy. *Science of the Total Environment, 635*, 452–469. https://doi.org/10.1016/j.scitotenv.2018.04.134

Papageorgiou, M. (2016). Coastal and Marine Tourism: A Challenging Factor in Marine Spatial Planning. *Ocean & Coastal Management, 129*, 44–48. https://doi.org/10.1016/j.ocecoaman.2016.05.006

Pérez, V., Guerrero, F., González, M., Pérez, F., & Caballero, R. (2013). Composite Indicator for the Assessment of Sustainability: The Case of Cuban Nature-based Tourism Destinations. *Ecological Indicators, 29*, 316–324. https://doi.org/10.1016/j.ecolind.2012.12.027

Petrić, L., & Mandić, A. (2014). Visitor Management Tools for Protected Aeeas Focused on Sustainable Tourism Development: The Croatian Expirience. *Environmental Engineering and Management Journal, 13(6)*, 1483–1495. http://omicron.ch.tuiasi.ro/EEMJ/

Plummer, R., & Fennell, D. A. (2009). Managing Protected Areas for Sustainable Tourism: Prospects for Adaptive Co-management. *Journal of Sustainable Tourism, 17(2)*, 149–168. https://doi.org/10.1080/09669580802359301

Rannow, S., Macgregor, N. A., Albrecht, J. et al. (2014). Managing Protected Areas Under Climate Change: Challenges and Priorities. *Environmental Management*, 54, 732–743. https://doi.org/10.1007/s00267-014-0271-5

Ristić, D., Vukoičić, D., & Milinčić, M. (2019). Tourism and Sustainable Development of Rural Settlements in Protected Areas - Example NP Kopaonik (Serbia). *Land Use Policy, 89*, 104231. https://doi.org/10.1016/j.landusepol.2019.104231

Roman, G., Dearden, P. & Rollins, R. (2007). Application of Zoning and "Limits of Acceptable Change" to Manage Snorkelling Tourism. *Environmental Management 39(6)*, 819–830. https://doi.org/10.1007/s00267-006-0145-6

Romão, J., Guerreiro, J., & Rodrigues, P. (2017). Territory and Sustainable Tourism Development: A Space-Time Analysis on European Regions. *Regions, 4*, 1–17. https://doi.org/10.18335/region.v4i3.142

Scolozzi, R., Schirpke, U. Morri, E., D'Amato, D., & Santolini, R. (2014) Ecosystem Services-based SWOT Analysis of Protected Areas for Conservation Strategies. *Journal of Environmental Management, 146*, 543–551. https://doi.org/10.1016/j.jenvman.2014.05.040

Sharpley, R. (2020). Tourism, Sustainable Development and the Theoretical Divide: 20 Years On. *Journal of Sustainable Tourism, 28(11)*, 1932–1946. https://doi.org/10.1080/09669582.2020.1779732

Sinclair, S. J., White, M. D., & Newell, G. R. (2010). How Useful are Species Distribution Models for Managing Biodiversity Under Future Climates? *Ecology and Society, 15(1)*. https://doi.org/10.5751/ES-03089-150108

Snyman, S., & Bricker, K. S. (2018). Living on the Edge: Benefit-sharing from Protected Area Tourism. *Journal of Sustainable Tourism, 27*, 705–719. https://doi.org/10.1080/09669582.2019.1615496

Spenceley, A., McCool, S., Newsome, D., Báez, A., Barborak, J. R., Blye, C-J, Bricker, K., Sigit Cahyadi, H., Corrigan, K., Halpenny, E., Hvenegaard, G., Malleret King, D., Leung, Y-F, Mandić, A., Naidoo, R., Rüede, D., Sano, J., Sarhan, M., Santamaria, V., Beraldo Sousa, T., & Zschiegner, A-K. (2021). Tourism in Protected and Conserved Areas Amid the COVID-19 Pandemic. *Parks 27(Special Issue)*. 103–118. https://doi.org/10.2305/IUCN.CH.2021.PARKS-27-SIAS.en

Squeo, F. A., Estévez, R. A., Stoll, A., Gaymer, C. F., Letelier, L., & Sierralta, L. (2012). Towards the Creation of an Integrated System of Protected Areas in Chile: Achievements and Challenges. *Plant Ecology & Diversity, 5(2)*, 233–243. https://doi.org/10.1080/17550874.2012.679012

Tanguay, G. A., Rajaonson, J., & Therrien, M. C. (2013). Sustainable Tourism Indicators: Selection Criteria for Policy Implementation and Scientific Recognition. *Journal of Sustainable Tourism, 21(6)*, 862–879. https://doi.org/10.1080/09669582.2012.742531

Thonfeld, F., Steinbach, S., Muro, J., Hentze, K., Games, I., Näschen, K., & Kauzeni, P. F. (2020). The Impact of Anthropogenic Land Use Change on the Protected Areas of the Kilombero Catchment, Tanzania. *ISPRS Journal of Photogrammetry and Remote Sensing, 168*, 41–55. https://doi.org/10.1016/j.isprsjprs.2020.07.019

Thongdejsri, M., & Nitivattananon, V. (2019). Assessing Impacts of Implementing Low-Carbon Tourism Program for Sustainable Tourism in a World Heritage City. *Tourism Review, 74(2)*, 138–156. https://doi.org/10.1108/TR-04-2017-0082

Torres-Delgado, A., & Palomeque, F. L. (2014). Measuring Sustainable Tourism at the Municipal Level. *Annals of Tourism Research, 49*, 122–137. https://doi.org/10.1016/j.annals.2014.09.003

Tverijonaite, E., Ólafsdóttir, R., & Thorsteinsson, T. (2018). Accessibility of Protected Areas and Visitor Behaviour: A Case Study from Iceland. *Journal of Outdoor Recreation and Tourism, 24*, 1–10. https://doi.org/10.1016/j.jort.2018.09.001

United Nations World Tourism Organization (2007). Indicators of Sustainable Development: Guidelines and Methodologies. In *New York* (Issue October). https://doi.org/10.1016/j.cirpj.2010.03.002

Vila, M., Afsordegan, A., Agell, N., Sánchez, M., & Costa, G. (2018). Influential Factors in Water Planning for Sustainable Tourism Destinations. *Journal of Sustainable Tourism, 26*(*7*), 1241–1256. https://doi.org/10.1080/09669582.2018.1433183

Weaver, D. B., & Lawton, L. J. (2017). A new Visitation Paradigm for Protected Areas. *Tourism Management, 60*, 140–146. https://doi.org/10.1016/j.tourman.2016.11.018

Weidenfeld, A. (2018). Tourism Diversification and Its Implications for Smart Specialisation. *Sustainability, 10*(*2*), 319. https://doi.org/10.3390/su10020319

Weng, L., He, B. J., Liu, L., Li, C., & Zhang, X. (2019). Sustainability Assessment of Cultural Heritage Tourism: Case Study of Pingyao Ancient City in China. *Sustainability (Switzerland), 11*(*5*). https://doi.org/10.3390/su11051392

United Nations World Tourism Organization. (2004). Indicators of Sustainable Development for Tourism Destinations. In *Tourism's potential as a sustainable development strategy. Proceedings from the 2004 WTO tourism policy forum at the George Washington University*, Washington, DC, USA, 18–20 October 2004. http://www.tourisk.org/content/projects/IndicatorsforTourismDevelopment.pdf

Zhong, L., Sun, S., Law, R., & Li, W. (2021). Tourism Crisis Management: Evidence from COVID-19. *Current Issues in Tourism 24*(*19*), 2671–2682. https://doi.org/10.1080/13683500.2021.1901866

Ziaabadi, M., Malakootian, M., Reza, M., Mehrjerdi, Z., Jalaee, S. A., & Mehrabi B. H. (2017). How to Use Composite Indicator and Linear Programming Model for Determine Sustainable Tourism. *Journal of Environmental Health Science and Engineering, 15*(*1*), 9–15. https://doi.org/10.1186/s40201-017-0271-5

Zovko, M., Melkić, S., & Marković Vukadin, I. (2021). Application of the DPSIR Framework to Assess Environmental Issues with an Emphasis On Waste Management Driven by Stationary Tourism in Adriatic Croatia. *Geoadria, 26*(*1*), 83–106. https://doi.org/10.15291/geoadria.3154

26

SUSTAINABLE MOUNTAIN TOURISM DEVELOPMENT IN THE FOOTHILLS OF HIMALAYA

A TISM-based approach to greater resilience of nature-based destinations

Suneel Kumar, Marco Valeri, Shekhar, Geetanjali Sageena and Ante Mandić

Introduction

The tourism industry is a significant driver of socio-economic development (Deshwal, 2015; Neto, 2003; World Tourism and Travel Council, 2019). However, due to the often-adverse impact on ecosystems, emphasizing sustainable tourism principles, particularly in developing countries experiencing higher growth rates and often over-tourism symptoms, is imperative (Kaul & Gupta, 2009; Ramdas & Mohamed, 2014). This undeniable importance led to sustainable tourism being a centre of debate between researchers and practitioners (Gossling, 2009; Liu, 2003; May, 1991; Nash & Butler, 1990; Ruhanen et al., 2015).

Indian mountain regions with abundant biodiversity have a vast potential for nature-based tourism development. Being home to the great Himalayas and its subranges in the northern and eastern parts and western and Eastern Ghats in the southern region, this area has a wide range of tourism possibilities. Many tourists are attracted to the mountains due to the adventure opportunity and unique scenery (Río-Rama et al., 2019; Saraswati & Ram, 2017). However, the impacts of mass tourism and negligence for carrying capacity has led to severe sustainability concerns among policymakers, residents, and visitors (Geneletti & Dawa, 2009; Singh et al., 2009). Although there is great potential to harness the mountains' tourism assets, there is an increasing understanding that the tourism development pathway going forward needs to be different and built on sustainable development principles (Gaur & Kotru, 2018).

Focusing on increasing the number of tourists for economic gains contradicts the notion of sustainability in nature-based destinations (López & Pardo, 2018; Mandić & Petrić, 2021; Petrić & Mandić, 2021). Tourism in the mountainous regions in India is developing fast, and the number of visitors increases continuously, posing pressure on local resources (Mbaiwa, 2003; Pignatti, 1993). The tradeoffs between nature conservation and the economic gains of tourism development are becoming more pronounced. According to Wang et al. (2019), the significant

DOI: 10.4324/9781003230748-31

challenges related to sustainability in mountain regions include weak governance, the overexploitation of natural resources, environmental degradation due to infrastructure development, and social degradation due to tourism. The outbreak of the current pandemic allowed the industry to implement sustainable practices in various sector activities (Mandić & Marković Vukadin, 2021; Romagosa, 2020; Seraphin, 2019). Many nature-based destinations, including secluded mountain areas, witnessed a dramatic decline in the numbers of international visitors, jeopardizing the decades of conservation efforts and sustainability of local communities (Mandić, 2021; Spenceley et al., 2021). This was particularly the case in developing and secluded spaces, where nature-based tourism is an essential source of income for communities and ecosystems (Newsome, 2020; Spenceley et al., 2021) and where the knowledge on how to develop tourism sustainably is often absent (Tosun, 2001). Simultaneously, there is an urgent need for a knowledge-based approach and knowledge transfer on integrating sustainability into nature-based tourism development planning. The practices involving reducing the number of tourists or increasing the average length of stay promoting the sustainability of the tourism industry (Gössling et al., 2020) have often been criticized as individualistic, ignoring and potentially harming the interests of stakeholders (Hunter, 1997). At the same time, Leung et al. (2018) propose limiting the number of visitors as a last resort to manage visitor flows in nature-based destinations and to instead employ the limits of acceptable change (LAC) or recreation opportunity spectrum (ROS) as frameworks to drive the long-term sustainability of tourism development.

The promotion of sustainability in nature-based destinations requires an understanding of the internal and external forces affecting the tourism development planning process (Mandić, 2019, 2020). Tourism development strategies have to be centred around visitors' satisfaction and conservation of the environment (Liu, 2003; McCool et al., 2021) and involve greater adaptability (flexibility) (McCool & Bosak, 2016; Zamfir & Corbos, 2015). Bearing that in mind, below, we first briefly review the literature on sustainable tourism with particular attention being paid to factors affecting the planning and development process in mountain destinations. Following, we introduce the study site and the empirical analysis. The discussion section is prepared to reflect on the primary constraints and opportunities to drive tourism sustainably in this mountainous and secluded Himalayan area.

Factors affecting the sustainable tourism development planning in mountain destinations

Recently, the sustainability of nature-based tourism development has been the subject of significant attention (Mandić & Petrić, 2021; McCool & Bosak, 2016). The complexity and uncertainty are driven, among others, by climate change population dynamics, the emergence of new technologies, and governance arrangements, but also changes in the public values, globalization, unprecedented tourism growth, destination management effectiveness, flexibility and adaptability of strategies and effectiveness of monitoring (Mandić, 2020; Mandić & McCool, 2022; McCool, 2016).

Tourism pressures infrastructure and degrades the quality of life in Himachal Pradesh (Shukla, 2017). Thus, transition toward a more sustainable development model is essential, and the development of nature-based tourism, paying greater attention to the inclusiveness, climate change, visitor pro-environmental behaviour, the contribution of tourism to the poverty reduction and the position of the indigenous communities, could play a crucial role in this process (Qian et al., 2018). However, the lack of social pressure to inspire positive change is apparent (Serenari et al., 2012). Studies have shown how the adaptation of sustainable tourism principles into the tourism development planning in mountain destinations is also affected by the stakeholders' culture, values, and interests (Thapa, 2012) and marketing and adaptation of the technologies (Tran, 2016).

In the mainstream tourism development and planning literature, sustainability is often discussed in the context of interrelationships between stakeholders with distinct political, environmental, economic, social, and cultural interests (McDonald, 2006). The most advocated Triple Bottom Line (TBL) development model emphasizes the need to reconcile economic prosperity with associated socio-cultural and environmental impacts (Stoddard et al., 2012). However, planning for sustainable tourism nowadays requires also strategic thinking about climate change, smart technologies, sustainable transportation, paying great attention to the participation of all the stakeholders, and raising their awareness and capacities (Aall, 2014; Bonzanigo et al., 2016; Cavallaro et al., 2017; Mandić & Garbin Praničević, 2019). Within the nature-based mountain destinations, often advocated collaborative planning (Pröbstl-Haider et al., 2018), involving the consideration of stakeholders' economic interests (Hammer & Pivo, 2017) and designing of memorable sustainable visitor experiences (Kuščer et al., 2017), could unsure more resilient communities, destinations, and ecosystems (Scuttari et al., 2019).

Sustainability in mountain tourism has received some attention over the previous decades. The studies have identified several factors that significantly influence a destination's resilience in the long run by triggering the sustainable behaviour of stakeholders (Table 26.1).

The studies presented in Table 26.1 demonstrate how sustainable tourism development in mountain destinations is affected/conditioned by several factors. For example, resource availability affects the development of niche tourism products, causing a change in the destination's marketing, which will undoubtedly impact the industry's long-term survival. Similar interrelationships could also be developed within other factors. As the development of strategies is constrained with limited time and financial resources, policymakers often cannot target each factor. Instead, policymakers need an adaptive/flexible hierarchical system to alter vital factors in order to inspire desired system change.

Table 26.1 Drivers of sustainable mountain tourism development

Drivers	*Characteristics*	*Relevant Studies*
Attitude of Planning/Nodal agencies	Being the nodal agency for policymaking, if they have a favourable attitude towards sustainability, only efforts will be made for the cause.	(Angelevska-Najdeska & Rakicevik, 2012; Berry & Ladkin, 1997; Buckley, 2012; Mahapatra et al., 2012)
Tourists' demand for sustainability	A product will not survive in the long run if the visitors do not desire it. Tourists are sometimes willing to pay more for sustainability. As people have become more environmentally conscious, the demand for sustainable development is rising in the tourism industry.	(Budeanu, 2007; Pulido-Fernández & López-Sánchez, 2016)
The attitude of other stakeholders	Pay a critical role in the long-term sustainability of the tourism industry. Stakeholders attitudes' are affected by the perception of the social and economic benefits and their dependence on tourism development.	(Chen, 2015; Knollenberg et al., 2016; Chen & Qiu, 2017; Zhang et al., 2017)

(Continued)

Table 26.1 (Continued)

Drivers	*Characteristics*	*Relevant Studies*
Business Ethics	Ethics significantly affect one's behaviour. Ethics includes the regard of the environment in which industry thrives. Whether or not every stakeholder works ethically will impact sustainable tourism development.	(Cassar, 2019; Lashley, 2016; Rushton, 2002; Wesarat et al., 2017)
Resource Availability	Availability of resources guides the behaviour of every stakeholder in the tourism industry. Limited and scarce resources affect the destination's performance and the success of the tourism industry. If the resources start depleting, the demand for sustainability rises. Although infrastructure development has a favourable economic impact on a destination, rapid urbanization will harm sustainability.	(Sati, 2004; Schilling& Chiang, 2011)
Disaster Management System (DMS)	It would mitigate the risk of a disaster. An effective DMS utilizes social learning to promote sustainability in the tourism industry.	(Nguyen et al., 2016; Petrović et al., 2017; Seraphin, 2019)
Marketing of destination	Marketing agendas are generally not compatible with sustainability concerns. However, marketing does play a pivotal role in a destination's success. Too much focus on attracting tourists and popularising a destination hurts a sustainable tourism mission. At the same time, the marketing of relatively under-explored destinations will contribute positively to sustainable development.	(Choo, 2011; Kemper et al., 2019; Kumar et al., 2012)
Niche Tourism Development	Innovative tourism products, which could act as an alternative to mass tourism, should be developed to reduce the negative impacts of mass tourism.	(Kumar et al., 2019)
Digitalization	The use of the technologies can help reduce the pressure on the environment, attract new tourists segments and can be an effective way to promote sustainability.	(Kumar & Shekhar, 2020)
Resilience in the Long run	The sustainable behaviour of the stakeholders could also be inspired by the desire to ensure the long term survival and the resilience of the communities, particularly those highly dependent on nature-based tourism development.	New factor proposed by authors

This chapter explains how identified factors interact, thereby broadening the understanding of the causes and consequences, enabling more efficient sustainable tourism development and planning in secluded mountain destinations.

Study site and methods

The case study region

Himachal Pradesh is located in the northern part of India. It consists of 12 districts and 109 tehsils and sub-tehsils, covering a total area of 55,673 square kilometers (Ramachandra et al., 2012). Himachal Pradesh is home to mountains that range in height from a few hundred meters to more than 6000 meters (NCERT, 2002). The state's highest peak is Reo Purgyil (6816 meters), located in the Kinnaur district (Department of Tourism and Civil Aviation, 2019). The terrain in the state is primarily hilly areas (Figure 26.1). The most renowned hill station destinations in Himachal Pradesh are Kasol, Shimla, Kullu, Kufri, Khajjar, Tirthan Valley, Chail, Palampur, Dharamshala, and Barog (Sen, 2019). The tourism industry is among the key contributors to the economic development of Himachal Pradesh. The three areas of Kullu, Kangra, and Shimla account for about 90% of the tourists (Kumar et al., 2018). Kullu (average altitude 1279 meters), Kangra (average altitude 733 meters), and Shimla (average altitude 2206 meters) are located at high altitudes, meaning that most of the tourism activities are located in the hills. Adventure tourism sites such as Triund trek (Kangra), ecotourism in Dharamshala (Kangra) and Kasauli (Solan), and even religious tourism sites such as the Shikari Mata temple (Mandi) and the Naina Devi Temple (Bilaspur) are located in the hills.

Figure 26.1 Hilly area in Himachal Pradesh.

Source: http://himachalpradeshtravel.com

Within this mountain area, the challenges associated with tourism development are different from those faced by other states (Nag, 2013). For example, limited accessibility and resource scarcity are common across the districts, and sustainability is an increasing concern. Although this area certainly has a lot to attract tourists, it failed to exploit the opportunities because of its weaknesses and external threats. To some extent, this also reflects the reluctance and position of the central state toward tourism development. Simultaneously, continuously increasing visitor numbers pose pressure on the ecosystem, while local policymakers, due to their limited knowledge of tourism development planning and their understanding of the global tourism context, are having a hard time finding solutions to inspire sustainable and inclusive tourism growth and destinations resilience.

Method

Total Interpretive Structural Modeling (TISM), developed by Sushil (2012), creates the foundation for understanding a structural relationship between phenomena based on people's judgments on how things are related. To develop a strategy, one must know about key areas that generate concerns and areas that could prove to be significantly beneficial. TISM helps in the identification of such areas (Jena et al., 2017). Researchers have widely used TISM to explain and develop the theory for sustainable development (Agrawal & Vinodh, 2019; Manjunatheshwara & Vinodh, 2018; Mohanty & Shankar, 2017; Shibin et al., 2017). The method is considered interpretive because it relies on the attitudes of a group of people and a modelling technique as the results are represented in a digraph model. Figure 26.2 displays the flowchart of the preparation of the TISM model.

The first step is to identify factors from the literature review and consultations with experts. Following the scoring, insignificant factors are removed and the final list is obtained. Once the factors are identified and coded, the next step is preparing the self-structural interpretation matrix (SSIM). The researcher converts SSIM into a binary form of zeroes and ones. All relationships between factors that are coded *V* and *X* are converted into *1*, and those coded *A* and *O* are converted into *0*. The matrix obtained from this is known as the initial reachability matrix. The next step is to incorporate transitive links into the matrix from the initial reachability matrix.

The transitivity concept implies that if the A factor leads to B and the B factor leads to C, then the A factor will lead to the C factor. After incorporating transitivity, the resultant matrix obtained is the final reach ability matrix, which further leads to the calculation of the driving power and the dependence level of each factor. The next step is to determine the partition level of each factor in the final ISM digraph. This step requires the use of a reachability set, antecedent set, and intersection set for each factor. The reachability set of a factor includes those factors, *including itself, driven by the factor*, and the antecedent set includes all those factors, *including itself, on which the factor depends*. The intersection set comprises factors present in the reachability and antecedent set. If the reachability set and the intersection set of a factor are the same, then that factor is eliminated, and its partition level is determined. This iteration is carried out until the partition level of each factor is determined. Once the partition levels are determined, the next step is preparing the digraph and establishing the links between these factors. Next, an interpretation matrix is prepared, which explains the interrelationship between the factors. The interpretation matrix inputs are incorporated in the digraph to prepare the final chart in the last step.

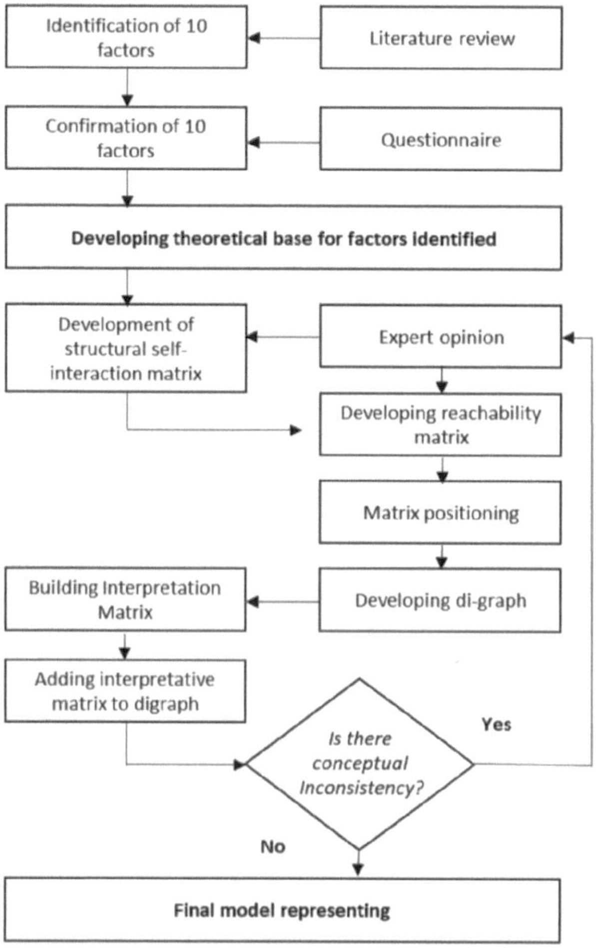

Figure 26.2 Flowchart of the solution methodology of the study.

Source: Adapted from Jena et al. (2017)

MICMAC analysis

Matrix of Cross-Impact Multiplications Applied to Classification (MICMAC) is an analysis used to determine the driving power and dependence of different elements used in the model (Jena et al., 2017). Adding every entry in the corresponding row in the final reachability matrix (drawn in TISM) gives a factor's driving power. On the other hand, adding entries in the corresponding column gives the dependence level. The driving power and dependence level calculation helps classify factors into *autonomous factors*, *linkage factors*, *driving factors*, and *dependent factors*. The autonomous factors have low dependence levels and low driving forces and neither depend on other factors nor cause significant influence. Driving factors have high driving power and low dependence level and are the framework's change bringers. The linkage factors have high driving power as well as dependence level. These factors are responsible for transferring the change in driving forces to the dependent forces, thereby acting as a linkage.

Dependent factors have a high dependence level and a low driving power. These factors absorb the changes from all other factors and help achieve the objectives. Categorizing factors (from the MICMAC analysis) are essential as they help identify controllable and uncontrollable factors influencing the system.

Survey instruments and measurement

To obtain the data, we have used a structured questionnaire containing the questions about the accuracy of the identified factors in influencing sustainable mountain tourism, the nature of the relationship between factors, its rationale. Respondents also had the opportunity to propose additional factors. The questionnaire involved the demographic variables of the respondents. We employed a five-point Likert scale to rate factors, indicating 1 for minimum and 5 for maximum. Respondents were also asked to fill up Self-Structuring Interaction Matrix (SSIM) used in studies applying TISM. The respondents replied in *V-A-X-O* coding, where *V* indicated a factor driving the other factor, *A* indicated a factor dependent on other factors, *X* indicated the interdependency of two factors, and *O* the existence of no relationship between the factors. The responses concerning the questions related to the authenticity of identified factors were coded as "yes," "no," and "cannot say."

The questionnaire items were validated by five scholars, translated into English and Hindi and distributed online using email. The data were collected between January and March 2020 using snowball sampling, and the final sample involved 88 representatives of the tourism industry and academia.

Results

The sample was dominated by research experts (69%), while the tourism industry representatives accounted for 31% of the sample. Males accounted for 54% of the sample, with the majority (62.5%) being between 30 and 40 years old. 55.68% of respondents hold a PhD, while 20% had a professional degree, and 14% had an MPhil. Only 10% of respondents had a graduation degree.

Examining the interrelationship between factors

We have calculated the average score for each factor based on the expert opinions and scores (Table 26.2). The highest score (4.65) was related to the factor *The attitude of nodal agency scored*. This was followed by the *Disaster Management Plan* (4.61) and *Demand of Tourist for sustainability* (4.52). *Digitalization* obtained the lowest score (2.55). All the other factors managed to score more than 2.50 (a cut-off score pre-decided by the author), so they were included in the study.

The SSIM (Table 26.3) shows the relationship identified by the experts among the factors. The planning agency's attitude drives the disaster management system, ensures resilience in the long run, marketing the destination, niche tourism development, and digitalization. Tourists' demand for sustainability drives planning agencies' sustainability, resilience, destination marketing, niche tourism development, and digitalization. Other stakeholders' attitude drives planning agencies' attitude, resilience, in the long run, niche tourism development, and digitalization in the industry. All the factors in the system drive resilience in the long run. Business ethics drive the attitude of planning agencies and other stakeholders and the marketing of destinations. Resource availability drives other stakeholders' attitudes, planning agencies' attitudes, tourists' demand for sustainability, resilience in the long run, destination

Table 26.2 Average scores obtained by the factors (left)

Factor No.	*Factor*	*Average Score*
1	The attitude of the planning agency	4.65
2	The demand of tourists for sustainability	4.52
3	The attitude of other stakeholders	2.84
4	Resilience in the long run	2.67
5	Business ethics	3.04
6	Resource availability	4.36
7	Disaster management system	4.61
8	Marketing of destination	3.84
9	Niche tourism development	3.10
10	Digitalization	2.55

Source: Authors

Table 26.3 Structural Self-Interaction Matrix (right)

Code	*1*	*2*	*3*	*4*	*5*	*6*	*7*	*8*	*9*	*10*
1	-	A	A	V	A	A	V	V	V	V
2	V	-	O	V	O	A	O	V	V	V
3	V	O	-	V	A	A	O	O	V	V
4	A	A	A	-	O	A	A	A	A	O
5	V	O	V	O	-	O	O	V	O	O
6	V	V	V	V	O	-	O	V	V	O
7	A	O	O	V	O	O	-	V	O	O
8	A	A	O	V	A	A	A	-	A	A
9	A	A	A	V	O	A	O	V	-	A
10	A	A	A	O	O	O	O	V	V	-

Source: Authors

marketing, niche tourism development, and digitalization. Disaster management systems, like niche tourism development, drive the marketing of destinations and resilience in the long run. Marketing of destination drives resilience in the long run. Digitalization in the industry drives niche tourism development. The SSIM was converted in binary form to have an initial reach ability matrix (Table 26.4).

Transitivity helps derive the new relationship between the factors (Table 26.5). For example, from the analysis above, one could conclude that tourists' demand for sustainability also drives disaster management systems, while the attitude of other stakeholders drives a disaster management system and the marketing of the destination. Business ethics also drive long-run resilience, disaster management system, niche tourism development, and digitalization. Resource availability has transitive links with disaster management systems and digitalization in the industry. The driving power and dependence level were calculated for each factor.

Driving power and dependence level enable determining the hierarchy level for each factor. For example, *resource availability* has a driving power of *9* and a *dependence* of *1*. This means that resource availability drives nine other factors and is dependent on one factor. The iteration is carried out until the partition level for each factor is determined. The partition level determines the positioning of the factor in the digraph (Table 26.6). Factors with low driving

Table 26.4 Initial Reachability Matrix (left)

Code	*1*	*2*	*3*	*4*	*5*	*6*	*7*	*8*	*9*	*10*
1	1	0	0	1	0	0	1	1	1	1
2	1	1	0	1	0	0	0	1	1	1
3	1	0	1	1	0	0	0	0	1	1
4	0	0	0	1	0	0	0	0	0	0
5	1	0	1	0	1	0	0	1	0	0
6	1	1	1	1	0	1	0	1	1	0
7	0	0	0	1	0	0	1	1	0	0
8	0	0	0	1	0	0	0	1	0	0
9	0	0	0	1	0	0	0	1	1	0
10	0	0	0	0	0	0	0	1	1	1

Source: Authors

Table 26.5 Final Reachability Matrix (Post Transitivity) (right)

Code	*1*	*2*	*3*	*4*	*5*	*6*	*7*	*8*	*9*	*10*	*Driving Power*
1	1	0	0	1	0	0	1	1	1	1	6
2	1	1	0	1	0	0	1*	1	1	1	7
3	1	0	1	1	0	0	1*	1*	1	1	7
4	0	0	0	1	0	0	0	0	0	0	1
5	1	0	1	1*	1	0	1*	1	1*	1*	8
6	1	1	1	1	0	1	1*	1	1	1*	9
7	0	0	0	1	0	0	1	1	0	0	3
8	0	0	0	1	0	0	0	1	0	0	2
9	0	0	0	1	0	0	0	1	1	0	3
10	0	0	0	0	0	0	0	1	1	1	3
Dependence	5	2	3	9	1	1	6	9	7	6	

Source: Authors

Table 26.6 Partition level of each factor

Factor Code	*Factor Name*	*Partition Level*
01	Attitude of Planning/Nodal agency	Fifth
02	Demand of tourists for sustainability	Sixth
03	Attitude of other stakeholders	Sixth
04	Survival in the Long run	First
05	Business Ethics	Seven
06	Availability of resources	Seven
07	Disaster Management System	Third
08	Marketing of Destination	Second
09	Niche Tourism Development	Third
10	Digitalization	Fourth

Source: Authors

power are eliminated initially and are identified at the early stages of the process. Results show that resilience, in the long run, is placed at level one. Marketing of destination occupies level two. Disaster management systems and niche tourism development are placed at level three. Digitalization is placed at level four, and the planning/nodal agency attitude is placed at level five. Tourists' demand for sustainability and the attitude of other stakeholders are placed at level six. Finally, business ethics and resource availability occupy level seven.

The initial digraph (Figure 26.3) presents the links between the factors. The placement of factors in the digraph follows increasing order, i.e., factors with level 1 are placed at the top, followed by factors at level 2, 3, and up to level 7. Factors identified at the same level are placed at the same level, i.e., horizontally adjacent.

The next step was to build an Interpretation Matrix (Table 26.7), explaining the relationship between the factors involved. The table shows that planning agencies' attitude drives the disaster management system to frame disaster management policies. An effective disaster management policy would lead to safety from, and recovery after, a disaster, thereby ensuring resilience. This interpretation matrix facilitated the development of the TISM digraph from the ISM digraph.

The final ISM digraph (Figure 26.4) demonstrates seven levels in the hierarchical structure of factors, with those below driving the factors at the upper level. At the bottom of the hierarchy are business ethics and resource availability. These factors determine and guide sustainable

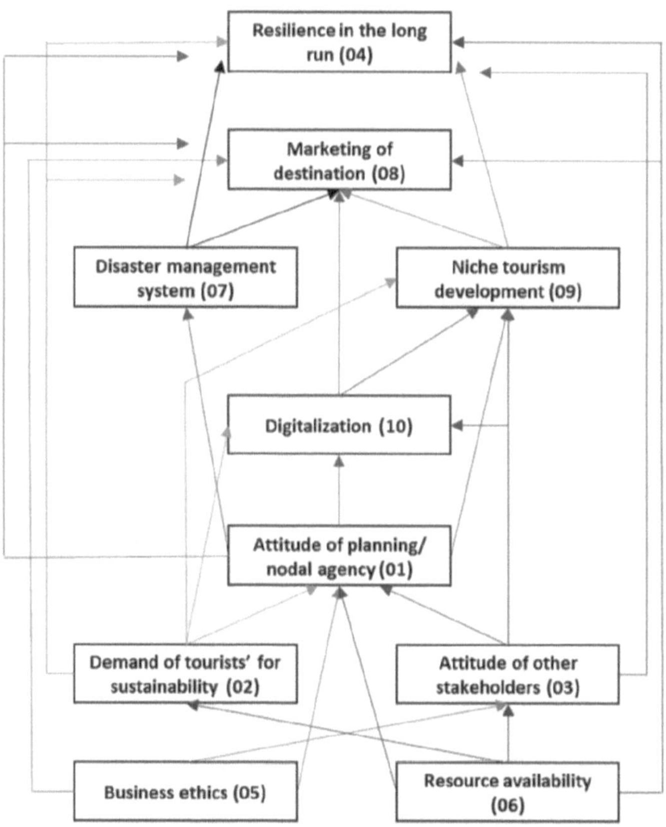

Figure 26.3 Inter-relationship among factors through TISM.

Source: Authors

Table 26.7 Interpretation Matrix

	The attitude of the Planning/ Nodal agency	*The demand of tourists for sustainability*	*The attitude of other stakeholders*	*Resilience in the Long run*	*Business Ethics*	*Availability of resources*	*Disaster Management System*	*Marketing of Destination*	*Niche Tourism Development*	*Digitalization*
The attitude of the Planning/ Nodal agency				The attitude of the planning agencies guides the policies formulation. Tourism development policies are vital for and determine the resilience of the destination.			The planning agencies hold responsibility for crisis management and establishing an effective disaster management system.	Planning agencies are responsible for developing an effective marketing program for a destination.	Planning agencies are responsible for developing new tourism products.	The attitude of planning agencies towards digitalization will determine the use of the technologies to facilitate destination management and inform planning.
The demand of tourists for sustainability	Demands of tourists for sustainability will impact the attitudes and actions of policymakers.			Visitors pro-environmental behaviour will certainly inspire destinations sustainability and resilience.				Visitors growing demand for sustainability will affect how tourism experiences are shaped, delivered and promoted.	Greater demand for sustainability creates an opportunity for innovative tourism products and experience development.	New technologies facilitate more sustainable tourism product development and delivery.
The attitude of other stakeholders	The actions of the planning agency are significantly affected by the attitudes of key stakeholders in tourism destinations, including communities, tourism businesses and NGOs.			The resilience of the tourism system requires joint efforts and commitments of all relevant stakeholders within the destination.					Tourism businesses and local communities have a central role in developing innovative visitor experiences.	The development of smart hospitality ecosystem requires a commitment of tourism businesses.

Resilience in the Long run								
Business Ethics	Ethics is the foundation for policy formulation and conduct of the nodal agency.		Ethics can also affect the behaviour of other stakeholders. Commitment to ethics relates to corporate social responsibility and sustainable production behaviour.				Ethics also affect our approach to destination marketing, the way the destination presents itself and communicates with emissive markets.	
Availability of resources	Availability of resources guides the decision making by the nodal agency.	Availability of resources at the destination can impact the behaviour of tourists and inspire actions.	Resource availability will decide what other will affect the decisions of other stakeholders, including the one on further investment in tourism.	Resource availability and environmental sustainability (natural environment condition) are vital aspects of resilience.			The state and the availability of resources can affect destination marketing—destinations employ de-marketing to reduce the pressure on depleted resources.	Resource availability will determine, among others, the development of niche tourism products, particularly nature-based tourism experiences.
Disaster Management System				An effective DRM would enable efficient mitigation and adaptation to global change and different kinds of crises.			Effective DMS requires efficient and timely communication and promotion activities, which are an essential part of destination marketing.	

(*Continued*)

Table 26.7 (Continued)

	The attitude of the Planning/ Nodal agency	*The demand of tourists for sustainability*	*The attitude of other stakeholders*	*Resilience in the Long run*	*Business Ethics*	*Availability of resources*	*Disaster Management System*	*Marketing of Destination*	*Niche Tourism Development*	*Digitalization*
Marketing of Destination				Destination resilience and marketing are two sides of the same coin, therefore interrelated. The way destinations approach marketing can build or deteriorate its resilience.						
Niche Tourism Development				Small scale, high-value tourism experiences can facilitate the destination's resilience.				Niche tourism products facilitate destinations authenticity and competitiveness.		
Digitalization								Innovative technologies transformed all aspects of destination marketing.	Digitalization facilitated the development of niche and innovative visitor experiences.	

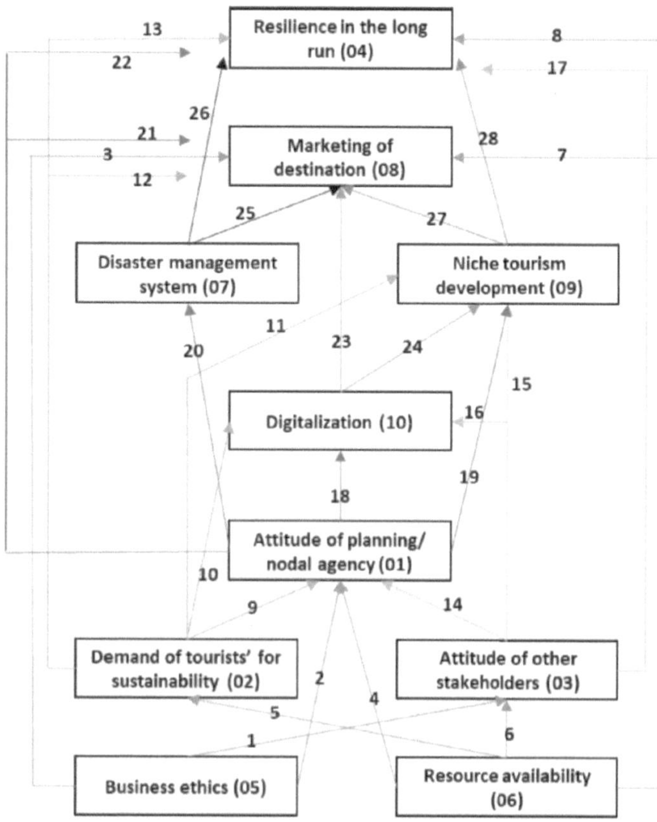

1	Affects stakeholders behaviour	11	Inspires innovative product development	21	Affects the development of effective marketing programs
2	Guides policymaking	12	Affect the destination marketing	22	Enables effective policy formulation
3	Affects destination marketing	13	Fosters sustainable use of destinations resources	23	Enables effective and innovative destination marketing
4	Guides policymaking	14	Affects policy formulation	24	Enables innovative experiences and products development
5	Inspires visitors pro-environmental behaviour	15	Foster innovative experiences and products development	25	Affects communication and promotion policies
6	Affects attitudes toward tourism	16	Increases competitiveness	26	Enables mitigation and adaptation to global change and crisis
7	Affects policy formulation	17	Inspired joint effort	27	Facilitates destination competitiveness and authenticity
8	Affects resilience	18	Enables adaptation of technologies	28	Reduces the pressure on the ecosystem
9	Affects policy formulation	19	Inspires innovative product development		
10	Increases reliance on technologies	20	Enables development of efficient DRM		

Figure 26.4 Final TISM digraph depicting interrelationship between factors.

Source: Authors

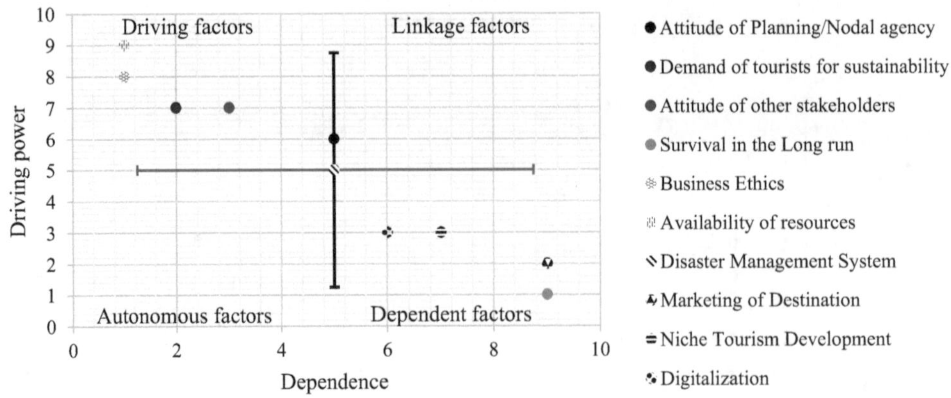

Figure 26.5 Driving and dependence power of the factors.

Source: Author's elaboration

mountain tourism development. They are perceived to inspire and affect tourists' demand for sustainability, other stakeholders' attitudes, planning agencies, niche tourism development, business ethics, destination marketing, disaster management system, and destinations resilience. The digraph can be divided into two parts—one below the planning agencies' attitude and one above it. The attitude of planning agencies acts as a linkage factor in the system. The inputs from the interpretation matrix express the interrelationship. All the factors above attitude of planning agencies drive less and are dependent on more factors.

MICMAC analysis (Figure 26.5) shows that *other stakeholders' attitudes, tourists' demand for sustainability, business ethics, availability of resources,* and *planning agencies' attitudes* are the drivers of sustainable tourism development in this mountain area. It is consistent with the digraph developed with the TISM. On the other hand, the *marketing of destination, digitalization, niche tourism development, disaster management system,* and *resilience in the long run,* are the dependent factors that support and are the outcomes of sustainable tourism development. The results confirm the validity of the conclusions drawn from the digraph.

Discussion

The analysis highlighted two pillars of sustainable tourism development in this mountain region. The first refers to the reliance on business ethics to help the industry build a social movement, implement corporate social responsibility, and, ultimately, strive toward greater sustainability and resilience of the destination. The second pillar, referring to the availability of resources, accentuates the common challenge in developing counties lacking strategic planning and governmental control over the effects of tourism development. Both pillars initially impact different dimensions of sustainable tourism but eventually converge by influencing planning agencies' behaviour. The nodal agency's attitude and behaviour are crucial, as it sets a course for tourism development and drives stakeholders' actions.

The final TISM digraph places resources and pro-environmental and responsible stakeholders' behaviour at the foundation of sustainable tourism development in this region. The factors placed above create the foundation of adaptive destination management, while the interlinkages between the factors at lower and upper hierarchical levels further enhance this adaptability. For example, the analysis demonstrated how the destination's resilience, in the long run, is the function of several variables. It is affected by the state of the resources, the attitude and behaviour of the stakeholders and the nodal agency, the type of tourism (product) development, and the strategic planning. The fail of one factor to contribute to achieving this goal, in the short run, could be compensated with other factors. However, the destination should strive to achieve the balance, i.e., all expected factors contribute to the desired goals. For example, contribution to sustainability and resilience will probably not be the initial goal of many tourism businesses and visitors. While in the short term, this can be compensated by some mitigation-focused activities, in the long term, tourism planning agencies must take responsibility for knowledge development, capacity building and awareness-raising so as to change the desired behaviour.

Greater sustainability is conditioned with the change in the mindset of stakeholders, including policymakers, tourism businesses and tourists. It must be inspired by those responsible for destination governance, reflected in tourism development policies, and must not compromise the tourist experience but rather be an intrinsic part of it (Taheri et al., 2018). Stakeholders' ethical and sustainable behaviour within the destination could inform visitors' actions and reduce the adverse influences on the ecosystem. The studies have already demonstrated that tourists' growing ecocentrism reflects the negative anthropocentric influences associated with tourism growth (Mandić & Vuković, 2021), so there is a need for capacity-building activities and education on implementing sustainable tourism practices and stimulating desired visitor behaviour on the destination scale (Gianna Moscardo & Murphy, 2015).

The availability and condition of resources often inspire the promotion of sustainability practices. The recent report "*Destinations at Risk: The Invisible Burden of Tourism*", published jointly by the Travel Foundation, Cornell University's Center for Sustainable Global Enterprise and Epler Wood International, suggests how destinations must uncover and account for tourism's hidden costs related to energy and GhG, water, solid waste, wastewater, natural and social capital referred to as the "invisible burden" (Epler Wood et al., 2019). According to the same report, the most vulnerable destinations are those with a high risk of climate change impacts, high economic dependence on tourism, poor destination governance and a rapidly growing population. Himalayan mountain destinations are a perfect example of developing spaces, where invisible burdens of tourism present costs that often outpace tourism annual revenue growth. Handling the invisible burden in their context requires innovative thoughts about detecting the state, e.g. using risk assessments and analysis, understanding the trajectory of change, e.g. using forecasting and scenario analysis, and, finally, employing a (tourism) development master planning or even advancing the conception of tourism development to achieve greater levels of resilience.

High exposure to risks, including climate change at the Himalayan mountain tourism destinations, accentuates the importance of monitoring, environmental accounting, and effective disaster management planning. DMS is recognized as the factor driving destinations resilience within the proposed model. Development planning agencies are expected to understand the causes and consequences of climate change and potential disasters (Ritchie, 2008) and implement or inspire the implementation of adequate mitigation and adaptation strategies (Castellani & Sala, 2010).

Finally, this mountain tourism destination requires product diversification (Nare et al., 2017), and, if adequately managed, nature-based tourism can inspire more inclusive and sustainable growth. However, planning sustainable nature-based experiences is a complex process requiring the reconciliation between environmental goals and high levels of visitor satisfaction (Mandić & McCool, 2022; Mandić, 2022), while demand is dependent upon gender, age, education, ethnicity, and season (Bonadonna et al., 2017; Fredman, 2008). Recent developments amid the Covid-19 pandemic demonstrated the role of digitalization and new technologies in designing and providing innovative nature-based virtual visitor experiences (Spenceley et al., 2021). Himachal Pradesh is a destination in the early tourism development stage, so greater efforts are needed to determine the acceptable development trajectory. Adapting the model proposed within this chapter within the traditional participatory community-based tourism development planning could be an initial, but vital step in this process.

Conclusions and recommendations for future research

Strategic planning is the precondition for sustainable tourism development and the resilience of tourism destinations. Understanding the internal and external factors affecting the development of the industry is vital to develop efficient plans and facilitates decision-making. This study, focused on the Himachal Pradesh mountain in the Himalayan destination, to identify factors significantly determining tourism development within the region and discuss the causal linkages between them.

The analysis yielded ten factors influencing the tourism industry's sustainability in these mountain destinations. The TISM converted experts' perceptions into a structured map that deciphered the contextual relationships (driver-driven) among them. We confirmed the presence of multiple interrelated factors that affect the tourism industry's sustainable development and the nature of their interrelatedness. The factors identified as drivers of sustainable tourism development were *tourists' demand for sustainability, stakeholders' attitudes and behaviours, business ethics and resource availability*. TISM analysis allows going beyond the established notion of static tourism planning by discussing different scenarios in a search for optimum tourism development conception. The study advanced the understanding of the factors that impact sustainable tourism development discussed within previous studies by discussing how they intervene in a specific context of mountain destinations. It creates the foundation for further developing theoretical and practical models for sustainable tourism development in mountain destinations in developing countries.

The mountain tourism destinations could differ in size, carrying capacity, ecology, and tourism typology offered. Further researches could explore the relationship by identifying destination-specific characteristics. Research is also needed to identify the differences in the sustainability of tourism development between states in the Himalayan belt. Since the environment is dynamic, researchers should periodically revisit and confirm sustainable development challenges in mountain destinations. Future research should also increase the number of significant factors and employ the factor subcategories. This can help build robust models and facilitate scenario-based planning.

The main limitation of this study is the interpretative nature of TISM. For example, TISM assumes a binary relationship between the variables, and it does not consider "how much" a factor influences or depends on another factor. A potential solution to address this limitation is to use fuzzy MICMAC, where relationships identified are further segregated.

Declaration of conflicting interest

The author(s) declare no conflict of interest in the publication of this chapter.

References

Aall, C. (2014). Sustainable tourism in practice: Promoting or perverting the quest for a sustainable development? *Sustainability (Switzerland)*, *6*(5), 2562–2583. https://doi.org/10.3390/su6052562

Agrawal, R., & Vinodh, S. (2019). Application of total interpretive structural modelling (TISM) for analysis of factors influencing sustainable additive manufacturing: A case study. *Rapid Prototyping Journal, 25*(7), 1198–1223. https://doi.org/10.1108/RPJ-06-2018-0152

Angelevska-Najdeska, K., & Rakicevik, G. (2012). Planning of sustainable tourism development. *Procedia-Social and Behavioral Sciences, 44*, 210–220. https://doi.org/10.1016/j.sbspro.2012.05.022

Berry, S., & Ladkin, A. (1997). Sustainable tourism: A regional perspective. *Tourism Management, 18*(7), 433–440. https://doi.org/10.1016/S0261-5177(97)00053-8

Bonadonna, A., Giachino, C., & Truant, E. (2017). Sustainability and mountain tourism: The millennial's perspective. *Sustainability, 9*(7), 1219. https://doi.org/10.3390/su9071219

Bonzanigo, L., Giupponi, C., & Balbi, S. (2016). Sustainable tourism planning and climate change adaptation in the Alps: A case study of winter tourism in mountain communities in the Dolomites. *Journal of Sustainable Tourism, 24*(4), 637–652. https://doi.org/10.1080/09669582.2015.1122013

Buckley, R. (2012). Sustainable tourism: Research and reality. *Annals of Tourism Research, 39*(2), 528–546. https://doi.org/10.1016/j.annals.2012.02.003

Budeanu, A. (2007). Sustainable tourist behaviour – a discussion of opportunities for change. *International Journal of Consumer Studies, 31*(5), 499–508. https://doi.org/10.1111/j.1470-6431.2007.00606.x

Cassar, C. (2019). Business ethics and sustainable development. In Filho, W. L. (ed) *Encyclopedia of Sustainability in Higher Education* (Living Edition, pp. 1–12). Springer International Publishing. https://doi.org/10.1007/978-3-319-63951-2_39-1

Castellani, V., & Sala, S. (2010). Sustainable performance index for tourism policy development. *Tourism Management, 31*(6), 871–880. https://doi.org/10.1016/j.tourman.2009.10.001

Cavallaro, F., Ciari, F., Nocera, S., Prettenthaler, F., & Scuttari, A. (2017). The impacts of climate change on tourist mobility in mountain areas. *Journal of Sustainable Tourism, 25*(8), 1063–1083. https://doi.org/10.1080/09669582.2016.1253092

Chen, B. Xia, & Qiu, Z. (2017). Community attitudes toward ecotourism development and environmental conservation in nature reserve: A case of Fujian Wuyishan National Nature Reserve, China. *Journal of Mountain Science, 14*(7), 1405–1418. https://doi.org/10.1007/s11629-016-3983-6

Chen, J. S. (2015). Tourism stakeholders attitudes toward sustainable development: A case in the Arctic. *Journal of Retailing and Consumer Services, 22*, 225–230. https://doi.org/10.1016/j.jretconser.2014.08.003

Choo, H. (2011). Marketing innovations for sustainable destinations. *Tourism Management, 32*(4), 959–960. https://doi.org/10.1016/j.tourman.2010.05.007

Department of Tourism and Civil Aviation. (2019). Mountaineering & rock climbing - Himachal tourism official website. Himachal Tourism. https://himachaltourism.gov.in/destination/mountaineering-rock-climbing/

Deshwal, P. (2015). Tourism industry: An instrument of Indian economic growth through FDI. *International Journal of Scientific Research and Management, 3*(2), 2137–2140. https://doi.org/10.18535/IJSRM

Epler Wood, M., M. Milstein, and K. Ahamed-Broadhurst. 2019. Destinations at risk: The invisible burden of tourism. 1–64.

Fredman, P. (2008). Determinants of visitor expenditures in mountain tourism. *Tourism Economics, 14*(2), 297–311. https://doi.org/10.5367/000000008784460418

Gaur, V. S., & Kotru, R. (2018). *Sustainable tourism in the Indian Himalayan region.* https://niti.gov.in/writereaddata/files/document_publication/Doc2.pdf

Geneletti, D., & Dawa, D. (2009). Environmental impact assessment of mountain tourism in developing regions: A study in Ladakh, Indian Himalaya. *Environmental Impact Assessment Review, 29*(4), 229–242. https://doi.org/10.1016/j.eiar.2009.01.003

Gossling, S. (2009). Carbon neutral destinations: A conceptual analysis. *Journal of Sustainable Tourism, 17*(1), 17–37. https://doi.org/10.1080/09669580802276018

Gössling, S., Scott, D., & Hall, C. M. (2020). Pandemics, tourism and global change: A rapid assessment of COVID-19. *Journal of Sustainable Tourism, 1–20.* https://doi.org/10.1080/09669582.2020.1758708

Hammer, J., & Pivo, G. (2017). The Triple Bottom Line and sustainable economic development Theory and practice. *Economic Development Quarterly, 31*(1), 25–36. https://doi.org/10.1177/0891242416674808

Hunter, C. (1997). Sustainable tourism as an adaptive paradigm. *Annals of Tourism Research, 24*(4), 850–867. https://doi.org/10.1016/s0160-7383(97)00036-4

Jena, J., Sumati Sidharth, Lakshman S. Thakur, Devendra Kumar Pathak, and V. C. Pandey. 2017. Total Interpretive Structural Modeling (TISM): Approach and application. *Journal of Advances in Management Research, 14*(2):162–181.

Kaul, H., & Gupta, S. (2009). Sustainable tourism in India. *Worldwide Hospitality and Tourism Themes, 1*(1), 12–18. https://doi.org/10.1108/17554210910949841

Kemper, J., Hall, C., & Ballantine, P. (2019). Marketing and sustainability: Business as usual or changing worldviews? *Sustainability, 11*(3), 780. https://doi.org/10.3390/su11030780

Knollenberg, W., Hao, H., Long, P., & Fridgen, J. (2016). Stakeholders' attitudes towards sustainable tourism development in coastal communities. *Travel and Tourism Research Association: Advancing Tourism Research Globally*, 11.

Kumar, S., Attri, K., & Shekhar (2018). Strategies for holistic development of tourism industry in India: A case study of Himachal Pradesh. *Journal of IMS Group, 15*(2), 12–22.

Kumar, S., Shekhar, & Guleria, N. (2019). Understanding dynamics of niche tourism consumption through interpretive structure modeling. *Saaransh RKG Journal of Management, 11*(01), 40–48.

Kumar, S., & Shekhar. (2020). Digitalization: A strategic approach for development of tourism industry in India. *Paradigm, 24*(1), 93–108. https://doi.org/10.1177/0971890720914111

Kumar, V., Rahman, Z., Kazmi, A. A., & Goyal, P. (2012). Evolution of sustainability as marketing strategy: Beginning of new era. *Procedia - Social and Behavioral Sciences, 37*, 482–489. https://doi.org/10.1016/j.sbspro.2012.03.313

Kuščer, K., Mihalič, T., & Pechlaner, H. (2017). Innovation, sustainable tourism and environments in mountain destination development: A comparative analysis of Austria, Slovenia and Switzerland. *Journal of Sustainable Tourism, 25*(4), 489–504. https://doi.org/10.1080/09669582.2016.1223086

Lashley, C. (2016). Business ethics and sustainability. *Research in Hospitality Management, 6*(1), 1–7. https://doi.org/10.2989/rhm.2016.6.1.1.1289

Leung, Y. F., Spenceley, A., Hvenegaard, G., Buckley, R. (eds.) (2018). *Tourism and Visitor Management in Protected Areas: Guidelines for Sustainability. Best Practice Protected Area Guidelines*. Gland, Switzerland: IUCN.

Liu, Z. (2003). Sustainable tourism development: A critique. *Journal of Sustainable Tourism, 11*(6), 459–475. https://doi.org/10.1080/09669580308667216

López, I., & Pardo, M. (2018). Tourism versus nature conservation: Reconciliation of common interests and objectives — an analysis through Picos de Europa National Park. *Journal of Mountain Science, 15*(11), 2505–2516. https://doi.org/10.1007/s11629-018-4943-0

Mahapatra, P. S., Pandey, R., & Pradhan, S. (2012). River rafting in mountainous regions of Uttarakhand: Impacts, suggested mitigation measures and sustainability. *Journal of Mountain Science, 9*(4), 511–522. https://doi.org/10.1007/s11629-012-2234-8

Mandić, A. (2019). Nature-based solutions for sustainable tourism development in protected natural areas: A review. *Environment Systems and Decisions, 39*(3). https://doi.org/10.1007/s10669-019-09718-2

Mandić, A. (2020). Structuring challenges of sustainable tourism development in protected natural areas with driving force – pressure – state – impact – response (DPSIR) framework. *Environment Systems and Decisions, 40*, 560–576. https://doi.org/10.1007/s10669-020-09759-y

Mandić, A. (2021). Protected area management effectiveness and COVID-19: The case of Plitvice Lakes National Park, Croatia. *Journal of Outdoor Recreation and Tourism. In press.* https://doi.org/10.1016/j.jort.2021.100397

Mandić, A. (2022) Protected area tourism. In Buhalis, D. (ed) *Encyclopedia of Tourism Management and Marketing*. Edward Elgar. https://doi.org/10.4337/9781800377486.protected.area.tourism

Mandić, A. & Garbin Praničević, D. (2019). Progress on the role of ICTs in establishing destination appeal: Implications for smart tourism destination development. *Journal of Hospitality and Tourism Technology, 10*(4), 791–813. https://doi.org/10.1108/JHTT-06-2018-0047

Mandić A., Marković Vukadin I. (2021) Managing overtourism in nature-based destinations. In Mandić, A. & Petrić, L. (eds) *Mediterranean Protected Areas in the Era of Overtourism*. Cham: Springer. https://doi.org/10.1007/978-3-030-69193-6_3

Mandić, A. & McCool, S. F. (2022). A critical review and assessment of the last 15 years of experience design research in a nature-based tourism context. *Journal of Ecotourism*. Accepted for publishing. https://doi.org/10.1080/14724049.2022.2099877

Mandić, A. & Petrić, L. (eds) (2021). *Mediterranean Protected Areas in the Era of Overtourism: Challenges and Solutions*. Cham: Springer. https://doi.org/10.1007/978-3-030-69193-6

Mandić, A. & Vuković, M. (2021). Millennials attitudes, choices and behaviour - integrative analysis. *Journal of Ecotourism*. https://doi.org/10.1080/14724049.2021.1932925

Manjunatheshwara, K. J., & Vinodh, S. (2018). Application of TISM and MICMAC for analysis of influential factors of sustainable development of tablet devices: a case study. *International Journal of Sustainable Engineering, 11*(5), 353–364. https://doi.org/10.1080/19397038.2017.1387186

May, V. (1991). Tourism, environment and development. Values, sustainability and stewardship. *Tourism Management, 12*(2), 112–118. https://doi.org/10.1016/0261-5177(91)90065-2

Mbaiwa, J. E. (2003). The socio-economic and environmental impacts of tourism development on the Okavango Delta, north-western Botswana. *Journal of Arid Environments, 54*(2), 447–467. https://doi.org/10.1006/jare.2002.1101

McCool, S. F. (2016) The changing meanings of sustainable tourism. In Stephen F. McCool (ed) *Reframing sustainable tourism* (pp. 13–32). Cham: Springer.

McCool, S. F. & Bosak, K. (2016). *Reframing Sustainable Tourism.* Dordrecht: Springer. https://doi.org/10.1007/978-94-017-7209-9

McCool S. F., Eagles P. F. J., Skunca O., Vukadin V., Besancon C., Novosel A. (2021) Integrating marketing and management planning for outstanding visitor experiences in a turbulent era: The case of Plitvice Lakes National Park. In Mandić, A., & Petrić, L. (eds) *Mediterranean Protected Areas in the Era of Overtourism.* Cham: Springer. https://doi.org/10.1007/978-3-030-69193-6_11

McDonald, J. (2006). Understanding sustainable tourism development from a complex systems perspective: A case study of the Swan River, Western Australia [Australia (Thesis). Edith Cowan University]. In *Theses: Doctorates and Masters.* https://ro.ecu.edu.au/theses/82

Mohanty, M., & Shankar, R. (2017). Modelling uncertainty in sustainable integrated logistics using Fuzzy-TISM. *Transportation Research Part D: Transport and Environment, 53*, 471–491. https://doi.org/10.1016/j.trd.2017.04.034

Moscardo, Gianna, & Murphy, L. (2015). Educating destination communities for sustainability in tourism. In Moscardo, G. & Benckendorff, P. (eds) *Education for Sustainability in Tourism* (pp. 135–154). Springer. https://doi.org/10.1007/978-3-662-47470-9_9

Nag, A. (2013). A study of tourism industry of Himachal Pradesh with special reference to ecotourism. *Asia Pacific Journal of Marketing and Management Review, 2*(04), 89–106.

Nare, A. T., Musikavanhu, G. M., & Chiutsi, S. (2017). Tourism diversification in Botswana-a stakeholder perspective. *African Journal of Hospitality, Tourism and Leisure, 6*(3), 1–14. http://www.ajhtl.com

Nash, D., & Butler, R. (1990). Towards sustainable tourism. *Tourism Management, 11*(3), 263–264. https://doi.org/10.1016/0261-5177(90)90051-A

NCERT. (2002). *India: Physical Environment: Textbook of Geography for Class XI.* New Delhi: Secretary, National Council of Educational Research and Training.

Neto, F. (2003). A new approach to sustainable tourism development: Moving beyond environmental protection. *Natural Resources Forum, 27*(3), 212–222. https://doi.org/10.1111/1477-8947.00056

Newsome, D. (2020). The collapse of tourism and its impact on wildlife tourism destinations. *Journal of Tourism Futures.* https://doi.org/10.1108/JTF-04-2020-0053

Nguyen, D., Imamura, F., & Iuchi, K. (2016). Disaster management in coastal tourism destinations: The case for transactive planning and social learning. *International Review for Spatial Planning and Sustainable Development, 4*(2), 3–17. https://doi.org/10.14246/irspsd.4.2_3

Petrić L., Mandić A. (2021) Governance and management of protected natural areas in the era of overtourism. In Mandić, A. & Petrić, L. (eds) *Mediterranean Protected Areas in the Era of Overtourism.* Cham: Springer. https://doi.org/10.1007/978-3-030-69193-6_2

Petrović, N., Bošnjak, I., & Nedeljković, S. (2017). Disaster risk reduction for sustainable development goals. *European Project Management Journal*, 7(2), 27–35. www.epmj.org

Pignatti, S. (1993). Impact of tourism on the mountain landscape of central Italy. *Landscape and Urban Planning, 24* (1–4), 49–53. https://doi.org/10.1016/0169-2046(93)90082-O

Pröbstl-Haider, U., Lund-Durlacher, D., Antonschmidt, H., & Hödl, C. (2018). Mountain bike tourism in Austria and the Alpine region–towards a sustainable model for multi-stakeholder product development. *Journal of Sustainable Tourism, 26*(4), 567–582. https://doi.org/10.1080/09669582.2017.1361428

Pulido-Fernández, J., & López-Sánchez, Y. (2016). Are tourists really willing to pay more for sustainable destinations? *Sustainability, 8*(12), 1240. https://doi.org/10.3390/su8121240

Qian, J., Shen, H., & Law, R. (2018). Research in sustainable tourism: A longitudinal study of articles between 2008 and 2017. *Sustainability, 10*(3), 1–13. https://ideas.repec.org/a/gam/jsusta/v10y2018i3p590-d133424.html

Ramachandra, T. V., Krishnadas, G., & Jain, R. (2012). Solar potential in the Himalayan Landscape. *International Scholarly Research Notices, 2012*, 1–13. https://doi.org/10.5402/2012/203149

Ramdas, M., & Mohamed, B. (2014). Impacts of tourism on environmental attributes, environmental literacy and willingness to pay: A conceptual and theoretical review. *Procedia - Social and Behavioral Sciences, 144*, 378–391. https://doi.org/10.1016/j.sbspro.2014.07.307

Río-Rama, M., Maldonado-Erazo, C. P., Durán-Sánchez, A., & Álvarez-García, J. (2019). Mountain tourism research. A review. *European Journal of Tourism Research*, *22*, 130–150.

Ritchie, B. (2008). Tourism disaster planning and management: From response and recovery to reduction and readiness. *Current Issues in Tourism*, *11*(4), 315–348. https://doi.org/10.1080/13683500802140372

Romagosa, F. (2020) The COVID-19 crisis: Opportunities for sustainable and proximity tourism. *Tourism Geographies*, *22*(3), 690–694.

Ruhanen, L., Weiler, B., Moyle, B. D., & McLennan, C. lee J. (2015). Trends and patterns in sustainable tourism research: A 25-year bibliometric analysis. *Journal of Sustainable Tourism*, *23*(4), 517– 535. https://doi.org/10.1080/09669582.2014.978790

Rushton, K. (2002). Business ethics: A sustainable approach. *Business Ethics: A European Review*, *11*(2), 137–139. https://doi.org/10.1111/1467-8608.00269

Saraswati, A., & Ram, P. (2017). Hill tourism and sustainable development of the destinations: A situational analysis of Kumaon region of Uttarakhand. *International Journal of Marketing & Financial Management*, *5*(01), 70–84.

Sati, V. P. (2004). Resource utilization pattern and development in hills—A case for the pindar basin of Garhwal Himalaya, India, *Journal of Mountain Science*, *1*(2), 155–165. https://doi.org/10.1007/bf02919337

Schilling, M., & Chiang, L. (2011). The effect of natural resources on a sustainable development policy: The approach of non-sustainable externalities. *Energy Policy*, *39*(2), 990–998. https://doi.org/10.1016/j.enpol.2010.11.030

Scuttari, A., Orsi, F., & Bassani, R. (2019). Assessing the tourism-traffic paradox in mountain destinations. A stated preference survey on the Dolomites' passes (Italy). *Journal of Sustainable Tourism*, *27*(2), 241–257. https://doi.org/10.1080/09669582.2018.1428336

Sen, S. (2019). *30 Best Hill Stations In Himachal You Should Visit In 2021*. Travel Triangle. https://traveltriangle.com/blog/best-hill-stations-in-himachal/

Seraphin, H. (2019). Natural disaster and destination management: The case of the Caribbean and hurricane Irma. *Current Issues in Tourism*, *22*(1), 21–28. https://doi.org/10.1080/13683500.2017.1422483

Serenari, C., Leung, Y. F., Attarian, A., & Franck, C. (2012). Understanding environmentally significant behavior among whitewater rafting and trekking guides in the Garhwal Himalaya, India. *Journal of Sustainable Tourism*, *20*(5), 757–772. https://doi.org/10.1080/09669582.2011.638383

Shibin, K. T., Gunasekaran, A., & Dubey, R. (2017). Explaining sustainable supply chain performance using a total interpretive structural modeling approach. *Sustainable Production and Consumption*, *12*, 104–118. https://doi.org/10.1016/j.spc.2017.06.003

Shukla, A. (2017, August 21). *Himachal Pradesh's model of tourism Is not sustainable*. Hill Post. https://hillpost.in/2017/08/himachal-mass-tourism-not-sustainable/109286/

Singh, R. B., Mal, S., & Kala, C. P. (2009). Community responses to mountain tourism: A case in Bhyundar Valley, Indian Himalaya. *Journal of Mountain Science*, *6*(4), 394–404. https://doi.org/10.1007/s11629-009-1054-y

Spenceley, A., McCool, S., Newsome, D., Báez, A., Barborak, J. R., Blye, C. J., Bricker, K., Sigit Cahyadi, H., Corrigan, K., Halpenny, E., & Hvenegaard, G. (2021). Tourism in protected and conserved areas amid the COVID-19 pandemic. *Parks*, *27*(March), 103–118, 10.2305/IUCN.CH.2021.PARKS-27-SIAS.en

Stoddard, J. E., Pollard, C. E., & Evans, M. R. (2012). The Triple Bottom Line: A framework for sustainable tourism development. *International Journal of Hospitality and Tourism Administration*, *13*(3), 233–258. https://doi.org/10.1080/15256480.2012.698173

Sushil, S. (2012). Interpreting the interpretive structural model. *Global Journal of Flexible Systems Management*, *13*(2), 87–106. https://doi.org/10.1007/S40171-012-0008-3

Taheri, B., Farrington, T., Curran, R., & O'Gorman, K. (2018). Sustainability and the authentic experience. Harnessing brand heritage–a study from Japan. *Journal of Sustainable Tourism*, *26*(1), 49–67. https://doi.org/10.1080/09669582.2017.1310867

Thapa, M. B. (2012). *Tourism and sustainable community development in Nepal* [Keski-Pohjanmaan ammattikorkeakoulu]. https://www.theseus.fi/bitstream/handle/10024/44290/Thapa_Mek.pdf?sequence=1#:~:text=The Nepal Tourism Board will, the local awareness towards tourism.&text=For the conservation of physical, backward community will be focused.

Tosun, C. (2001). Challenges of sustainable tourism development in the developing world: The case of Turkey. *Tourism Management*, *22*(3), 289–303. https://doi.org/10.1016/S0261-5177(00)00060-1

Tran, H. T. G. (2016). *An Exploration of the Potential for Sustainable Tourism Development in Selected Vietnamese Islands*. Lismore, NSW: Southern Cross University. https://epubs.scu.edu.au/cgi/viewcontent.cgi?article=1557&context=theses

Wang, Y., Wu, N., Kunze, C., Long, R., & Perlik, M. (2019). Drivers of change to mountain sustainability in the Hindu Kush Himalaya. In Wester, P., Mishra, A., Mukherji, A., & Shrestha, A. B. (eds) *The Hindu Kush Himalaya Assessment* (pp. 17–56). Springer International Publishing. https://doi.org/10.1007/978-3-319-92288-1_2

Wesarat, P., Sharif, M. Y., & Abdul Majid, A. H. (2017). Role of organizational ethics in sustainable development: A conceptual framework. *International Journal of Sustainable Future for Human Security*, *5*(1), 67–76. https://doi.org/10.24910/jsustwain/5.1/6776

World Tourism and Travel Council. (2019). *Travel and tourism economic impact 2018 world*. https://wttc.org/Research/Economic-Impact

Zamfir, A., & Corbos, R.-A. (2015). Towards sustainable tourism development in urban areas: Case study on Bucharest as tourist destination. *Sustainability*, 7(9), 12709. https://ideas.repec.org/a/gam/jsusta/v7y2015i9p12709-12722d55942.html

Zhang, Y. L., Zhang, J., Zhang, H. O., Zhang, R. Y., Wang, Y., Guo, Y. R., & Wei, Z. C. (2017). Residents' environmental conservation behaviour in the mountain tourism destinations in China: Case studies of Jiuzhaigou and Mount Qingcheng. *Journal of Mountain Science*, *14*(12), 2555–2567. https://doi.org/10.1007/s11629-016-4183-0

27

HUNTING TOURISTS' BEHAVIOURS ON PACKAGE TOURS

The case of Konya Province, Turkey

Vedat Acar, Nuray Yıldız and Abdullah Tanrisevdi

Introduction

2020 witnessed a decrease of international arrivals by 73 percent compared with 2019. The global tourism revenues in 2021 were higher than in 2020, but remained below the pre-pandemic (UNWTO, 2022). It is not surprising that Turkish tourism faced the same problems as global tourism. However, although the COVID-19 pandemic had adverse effects on the global tourism industry, it should be taken into consideration that such crises are temporary, and the growth trend will most likely be re-established. Because tourism has self-renewing characteristics, it has a capacity and potential to respond to visitors' changing needs, wishes, and expectations by diversifying different tourism products. Hunting tourism may be shown to be among those products, and destination managers who want to contribute to rural development should do their best to improve this product. Apart from mass tourism, Turkey has created a stable market for hunting tourism with different business models, such as private hunting lands (Şafak, 2003). In 2020, it was estimated that since 2000, some 21,496 foreign and 6735 domestic hunters have taken part in the hunting activities, and the total income from hunting tourism have been approximately US$4.8 million. Furthermore, of the total animals allowed to be hunted last decade, the rate of the wild boars was noted as 83 percent (Republic of Turkish Ministry of Agriculture and Forestry, 2020).

Relatively few studies of hunting tourism have investigated characteristics, motivations, expectations, and values of hunting tourists (Komppula & Suni, 2013), the differentiation of second home owners' intention to purchase hunting products according to motivation and demographic characteristics (Tangeland, Vennesland & Nybakk, 2013), the behavioural intentions of deer hunters in terms of planned behaviour theory (Shrestha et al., 2012), the socio-economic benefits of hunting tourism (Ege, 1996; Şafak, 2003; Alkan, 2014), underlying motivations for hunting tourism (Olcay, Dağlıoğlu & Sürme, 2019), and hunting tourism as a tourist product (Ulusoy, 2015). These studies have used a variety of data collection tools, including scales and questionnaires (e.g., Shrestha et al., 2012; Tangeland et al., 2013; Alkan, 2014, Akkuş & Şimşek, 2019), interviews (Komppula & Suni, 2013; Olcay et al., 2019), and secondary data comprising legislation and statistics (Ege, 1996; Şafak, 2003). The investigation of tourist behaviour may contribute to the marketing activities of both researchers and

 DOI: 10.4324/9781003230748-32

travel agency officials. Thus, this study aimed to reveal the behaviours (interacting, bargaining, shopping, timing, tipping, food and beverage preferences, complaining) of four hunting tourists (two German and two Portuguese hunters) who participated in a guided package tour in Konya, Turkey. The researchers, firstly, wanted the hunters who took part in the tours of travel agency X between September 1 and October 5 2021, to fill in open-ended question forms. However, the participants were unwilling to provide in-depth information. Therefore, the researchers turned to participant observation. The failure in data collection was the main reason why the researchers chose participant observation as the data collection tool. Based on the related literature (Pizam & Sussmann, 1995; Pizam & Jeong, 1996; Pizam & Reichel, 1996; Acar, 2018; Acar & Tanrısevdi, 2019a; Acar & Tanrisevdi, 2019b; Acar & Tanrısevdi, 2021), the main research question was "How do hunting tourists behave on package tours?" The following sub-questions were then identified for investigation:

- How do hunting tourists interact with agency staff, local people, and other tourists?
- What kinds of services/products do hunting tourists buy?
- Do hunting tourists tip?
- To what extent do hunting tourists comply with the tour schedule?
- What kind of food and beverages do hunting tourists prefer?
- How do hunting tourists react to the treats offered?
- To what extent do hunting tourists tend to take photographs?
- What do hunting tourists complain about?

Literature review

Komppula and Suni (2013) identified three types of hunting tourists (responsible, adaptable, and achievement-oriented). They found that hunters' attitudes toward shooting, game farming, and social relationships during the hunting trip were major distinguishing factors, while the willingness to experience new things associated with hunting tourism was the most important hunting tourism motivation (Komppula & Suni, 2013). According to Radder and Bech-Larsen (2008), male identity was particularly important for biltong hunters, and socialization was an important motivation. The study conducted by Olcay et al. (2019) show that hunting tourists are motivated by relieving daily stress, living healthily, and experiencing nature. Alkan (2014) found that almost half of the domestic hunting tourists in Antalya, Turkey, are working-class tradesmen, mostly aged between 35 and 49 and with education up to primary school level. These domestic hunting tourists hunt about once or twice a week. Similarly, Akkuş and Şimşek (2019) showed that hunting tourists in Sinop-Boyabat, Turkey, are mostly males aged 36–65, married with children. Domestic tourists hunt 1–3 prey each day in the season, whereas foreign tourists hunt between 2 and 3, with their primary target being wild boars. While most of the foreign tourists' monthly salaries are between 5,000-7499 US Dollars, the domestic tourists' monthly salaries are no more than 5,000 Turkish Liras (Akkuş & Şimşek, 2019). According to Ulusoy (2015), hunters are likely to spend on accommodation, shopping, and various other activities.

Many studies have investigated tourist behaviours from an intercultural perspective (Pizam & Sussmann, 1995; Pizam & Reichel, 1996; Pizam & Jeong, 1996; Pizam et al., 1997; Özdemir, 2014; Acar, 2018), although few have specifically considered hunting tourists. Therefore, based on the findings from previous studies, the literature review addressed the following seven tourist behaviour dimensions: interaction, shopping, timing, bargaining, tipping, food and beverage preference, and complaining.

Interaction

Interaction can be considered in terms of three fundamental forms: tourist–tourist, tourist–attendant, and tourist–local people. Recent research indicates clear intercultural differences. For example, Americans score high in terms of social interaction (Pizam & Sussmann, 1995; Pizam & Jeong, 1996; Pizam & Reichel, 1996; Pizam et al., 1997; Özdemir, 2014), whereas French and Italians tend to interact with people of their own nationalities (Pizam et al., 1997) as do Pakistani tourists visiting Istanbul (Acar & Tanrisevdi, 2019b).

According to Reisinger and Turner (2003), American, European, Asian, and Australian societies display controversial cultural tendencies in social interaction characteristics. For instance, American culture approves of being yourself, open, friendly, direct, confident, outspoken, and informal, whereas those characteristics are not generally approved of by Asian societies which can tend to perceive Americans as aggressive, lacking grace, manners, and awareness of Asian cultures (Reisinger & Turner, 2003). One reason for non-interaction with other nationalities is the language barrier, as occurs, for example, with Korean and Japanese tourists (Pizam & Jeong, 1996). Similarly, Cho (1998) found that Korean tourists do not wish to revisit Australia if they experience communication obstacles, while Kim and Prideaux (1999) found that most Koreans, particularly women, are shy about speaking in a foreign language and find it impolite to do so with new acquaintances. Heo, Jogaratnam and Buchanan (2004) reported that just 10 percent of Japanese and Korean travellers visiting New York City hotels could speak English fluently. Similarly, a group of South Korean package tourists tried to communicate with the tour staff and the other people, although only one of them could speak English (Acar & Tanrisevdi, 2021).

Farahani and Mohamed (2013) found that Iranian tourists who can speak English are more likely to communicate with other tourists and local people than Saudi tourists, although Acar and Tanrisevdi (2019a) found that Iranians on shopping tours in Turkey avoid interacting, either with other tourists in their group or with other people.

Shopping behaviour

Shopping is regarded as one of the significant aspects of tourist behaviours. Shopping is described as an exchange process with two parties: one side is the buyer, and the other is the seller (Acar & Tanrisevdi, 2021). Studies of tourists' shopping behaviours indicate that there are various behavioural differences between cultures. Pizam and Sussmann (1995), for example, found that Japanese and Americans are seen as likely to buy souvenirs, whereas the French visitors are the least. Similarly, American, Japanese, and Italian tourists are perceived as eager shoppers, whereas the French exhibit a neutral attitude (Pizam et al., 1997). According to Turkish tour guides, Japanese and Americans are most likely to shop, whereas German, French, and British tourists tend not to (Özdemir, 2014).

Japanese tourists are particularly interested in luxury brands (Timothy, 2005; Rosenbaum & Spears, 2006a; Rosenbaum & Spears, 2006b); on the other hand, Iranian tourists' primary motivation is value for money (Asadi & Daryaei, 2011a, 2011b, 2011c; Asgari & Borzooei, 2014; Duman, 2016; Fard & Saberi, 2015), and they prefer shopping for clothes from outlet stores (Acar, 2018). Generally, tourists from developing countries prefer malls and other modern shopping outlets (Egresi & Arslan, 2016). A study of Pakistani package tourists showed that they mainly bought clothes and Turkish delight during a three-day tour (Acar & Tanrisevdi, 2019b) and these findings support the statements of Egresi and Arslan (2016). Participant observations of South Korean package tourists suggest that they are highly interested in optional activities (e.g., balloon trips, jeep safaris, Turkish nights, and additional dinners) (Acar & Tanrisevdi, 2021). Similarly, tourist guides in Turkey state that Chinese tourists also like hot-air balloon trips, jeep safaris, and parachute jumps (Acar, 2022).

Timing

Scheduling is an essential aspect of package tours. To ensure that a tour flows smoothly throughout, the group members must observe the set timings. However, eastern and western cultures differ substantially in attitudes to time-keeping (Pizam & Sussmann, 1995; Pizam & Jeong, 1996; Acar, 2018). For example, participant observations of Kuşadası, Istanbul, Antalya, and Cappadocia showed that Japanese tourists are most sensitive about timing (Acar, 2018), whereas such behaviour is rarely observed in Near Eastern or Middle Eastern tourists. According to Avcıkurt (2007), time means money and punctuality for Western society but uncertainty for Eastern cultures. Similarly, discipline is perceived as a characteristic of Germans (Avcıkurt, 2007).

On the other hand, Farahani and Mohamed (2013) found that Arabic and Iranian tourists have a less strict understanding of timing than western nationalities. Acar and Tanrisevdi (2019b) reported half-hour delays on two of the days of a three-day package because one of the Pakistani tour members could not wake up on time. A similar delay was encountered on daily tours for Iranian tourists. In one case, the Iranian tour leader failed to appear at the specified time, thereby delaying a shopping trip (Acar, 2018). While there were partial delays for South Korean package tourists, this did not jeopardize the overall tour flow (Acar & Tanrisevdi, 2021).

Bargaining

Bargaining is a negotiation process involving intense interaction between buyer and seller. According to the Turkish Language Institution (TDK, 2022), bargaining is a discussion between buyer and seller to impose the most favourable price in their interest on the other party. As with the other behaviours, there are some intercultural differences with regard to this practice. For example, according to British Guild of Guides representatives working in London, Italians generally bargain, whereas Japanese and Americans pay the offered price (Pizam & Sussmann, 1995). Similarly, according to Dutch tourist guides, Italians bargain the most, whereas Japanese do this the least (Pizam et al., 1997). Pizam and Jeong (1996) also found that American tourists tend to pay the offered prices, according to Korean tourist guides. Lee (2000) found that Chinese tourists are more competitive bargainers than Americans. Similarly, Israeli tourist guides report that British and German tourists tend to pay the offered prices while American and French tourists indulge in some modest bargaining (Pizam & Reichel, 1996). Turkish tourist guides find that Spanish, Italian, Japanese, French, and German tourists are more likely to bargain, whereas American and British tourists are less likely to do so (Özdemir, 2014). According to Turkish shopkeepers, South Korean tourists tend to bargain (Tanrisevdi & Akyurt Kurnaz, 2018). Finally, according to Farahani and Mohamed (2013), Iranian service providers consider Saudi tourists to be "bargain hunters". However, this may be because sellers generally quote high opening prices with Saudi tourists, meaning that they are likely to negotiate a more realistic price (Farahani & Mohamed, 2013).

While cultural differences play an important role in tourists' bargaining behaviours, the behaviour of sellers in the countries visited also has an impact. For instance, British tourists believe that bargaining is more appropriate in Middle Eastern, African and Asian countries (Kozak, 2016). According to a Japanese-speaking tourist guide in Turkey, Japanese tourists tend to bargain when travelling abroad, which is behaviour they would never observe in their home country. This may partly be because tourist guides provide information that motivates Japanese tourists to bargain while shopping in Turkey (Acar, 2018). Another tourist guide suggested Pakistani tourists visiting Istanbul's Grand Bazaar to try to reduce the offered price by 30 percent (Acar & Tanrisevdi, 2019b). Participant observations of South Korean tourists showed that a leather shop sales assistant offered a 40 percent discount at first, before reducing it further because of the 'tourists' indecision (Acar & Tanrisevdi, 2021). Similarly, a British tourist in Turkey paid 10 Turkish Lira for two T-shirts, having initially offered some 50 Turkish Lira for them (Kozak, 2016). One of the main problems facing the Turkish tourism industry is that prices are not consistent; they tend to vary according to the tourists' home countries and nationalities (Acar et al., 2020). The bargaining behaviour of tourists in Turkey is also influenced by guidebooks (Acar, 2018) and also online shop reviews. For example, according to an English-speaking tourist guide with package tour experience, Chinese tourists read these comments and mostly avoid shopping or else ask for significant reductions of 50–70 percent (Acar, 2022).

Tipping

Lynn (2010) describes tipping as a voluntary payment made to a service provider in exchange for the service received. In other words, tipping provides material gratification to a service provider that reflects the level of the customer's satisfaction. However, tipping is a sensitive issue with meanings that vary across cultures, ethnic backgrounds, gender, and even region (Abraham, 2014). For example, Hofstede, Hofstede & Minkow (2010) argue that Far Eastern societies may consider tipping to be a humiliating experience or even the equivalent of bribery in complete contrast to western societies.

However, in western societies, this situation is reversed. Lynn (2010) notes that tipping amount equivalent to 15 to 20 percent of the overall check, in American restaurants, is a tradition, whereas Reisinger (2009) emphasizes that it is common practice in Europe to add the same amount of service charge into the check. Thus, a European visitor in the U.S. needs to be informed about this cultural difference in order to prevent potential misunderstandings.

Kim, Prideaux and Kim (2002) found that Japanese and South Koreans are more likely to tip croupiers in casinos in Seoul, Korea, after winning heavily, whereas Chinese visitors do not. Another cultural difference relates to obligatory tipping practices. Some travel agencies in Turkey require package tourists to pay service providers a fixed tip before their daily or weekly tour activities, as revealed by the study on South Korean (Acar & Tanrisevdi, 2021) and Chinese (Acar, 2022) package tourists. Tourists respond to these demands in different ways. For example, Bowie and Chang (2005) found that few tourists visiting Scandinavia leave tips despite information in the tour brochure about tipping, while some have even been known to use the rear door of the tour bus in order to avoid leaving tips. Similarly, participant observations in Turkey show that Pakistani package tourists tend not to leave tips (Acar & Tanrisevdi, 2019b). In contrast, Iranian tourists responded willingly to the tour leader's request for tips at the end of a shopping tour in Turkey (Acar & Tarisevdi, 2019a). Similarly, participant observations showed that South Korean package tourists in Turkey willingly tipped the tour leader and participant observer at the tour's end, despite paying predetermined tips during the tour (Acar & Tanrısevdi, 2021).

Food and beverage preferences

While all humans need food and drink to survive, the ways in which they satisfy this need vary from person to person. Regarding tourism specifically, novelty-seeking in food and beverage preferences also differs across cultures. For example, Americans are moderately interested in local foods, whereas Italian, French, and Japanese tourists show only a low level of interest (Pizam et al., 1997). Similarly, English, Israeli, Korean and Dutch tour guides reported that Americans have a slight preference for local dishes (Pizam, 1999), although Israeli tour guides stated that American, English, and French tourists are warier of trying local dishes than Germans (Pizam & Reichel, 1996). Similarly, Cömert and Sökmen (2017) found that German tourists are more inclined to try Turkish cuisine than British tourists, while Özdemir and Kınay (2004) reported that German tourists are more satisfied with Turkish foods and beverages than Russians. In another study, Avcıkurt (2007) has interestingly argued that Germans are interested in experiencing unfamiliar foods and beverages because traditional dishes in their culture lack variety.

In contrast, tourists from societies characterized by uncertainty avoidance, such as Japan (Pizam & Sussmann, 1995), may consider it risky to try unfamiliar local foods and beverages that do not appeal to their tastes. Similarly, Turkish tourist guides reported that Spanish, French, American, German, English, and Italian tourists had strong local food and beverage preferences, in contrast to Japanese tourists (Özdemir, 2014). More generally, tourists coming from Far Eastern countries may be dissatisfied with the food and beverages offered to them because of the radical differences between Eastern and Western cuisines. For example, Chinese tourists are dissatisfied with dishes served in Finland because they contain too many dairy products (Laitinen, 2004) while a Korean-speaking tour guide in Turkey reported that Chinese tourists bring their own noodle supply with them (Acar, 2018). In response to these different eating and drinking habits, package tour agencies in Turkey take both Chinese (Acar, 2022) and South Korean (Acar & Tanrisevdi, 2021) tourists several times to restaurants serving Far Eastern food.

Complaint behaviour

When their expectations are not satisfied, people express their feelings as complaints, and, thus, expectations are the focus of complaining behaviour. According to Doğan (2004), tourism providers need to avoid complaints because disappointed tourists may leave the destination earlier than planned, decide not to revisit the same destination, or even broadcast negative comments about the destination.

There are various causes of complaints. For example, Iranian tourists visiting Van, a province in eastern Turkey (Duman, 2016) and Cyprus (Bouzari, 2012) complained about the lack of shopping malls. Acar (2018) found that international tourists in Turkey complain about the high price for extra services offered during package tours, such as minibars in hotels and drinks in restaurants. Similarly, Farahani and Mohamed (2013) reported that Saudi tourists visiting Malaysia complained that they were charged higher prices than tourists from other countries.

Complaints can also arise from cultural differences in tourists' lifestyles (Doğan, 2004). For example, Chinese tourists are accustomed to drinking hot water and tea, so they are dissatisfied if there is no kettle in their hotel room (Laitinen, 2004), whereas Iranian tourists visiting Turkey complain if their hotel room lacks slippers, toothbrushes, and razors (Acar, 2018). Cultural differences also determine the manner in which tourists express their complaints. According to Reisinger and Turner (2000), tourists from eastern cultures, especially the Japanese, tend to keep their criticism private to avoid conflict or loss of face, which Ziff-Levine

(1990) describes as "polite inexplicitness". He emphasizes that Japanese respond indirectly not to humiliate, offend, or disturb the harmony of the group instead of responding directly to questions. Similarly, Acar (2018) found that Japanese tourists mostly express their complaints after returning home.

Methodology

Exploratory qualitative research involves collecting data from observations and interviews, whereas quantitative research uses questionnaires. An exploratory approach may be more appropriate when the researcher wants to comprehend the nature of the problem (Sekaran, 2003). The current study was exploratory as it aimed to identify hunting tourists' behaviours in the guided package tours in Turkey during the COVID-19 pandemic.

The main and sub-research questions were explored through case study design. According to Yin (2017), a case situation can be a person, institution, event, or behavioural condition. Here, the situation was the behaviours (such as bargaining, interaction, shopping, tipping, food, beverage preferences) of tourists participating in a hunting package tour in Turkey, created by a specialized travel agency. In this current study, participant observation was used as a data collection tool. Just such a method has been widely used for tourism research in recent years (Acar, 2018; Acar & Tanrisevdi, 2019b; Acar & Tanrısevdi, 2019a, 2021; Bowen, 2002; Chang et al., 2011; Kaaristo & Rhoden, 2017; Konijn et al., 2016; Mackellar, 2013; Verbeek et al., 2011). According to Tracy (2013), participant observation has a unique power for understanding gestalt meanings. In particular, researchers may not gain a realistic understanding of human behaviours unless they are examined in a real-world setting through the eyes of an outsider or insider (Yıldırım & Şimşek, 2013). Participant observation enables a cultural researcher to obtain all kinds of data, reduces reactivity, provides an intuitive understanding of what is going on in culture, and allows the researcher to discuss the meaning of the data confidentially (Bernard, 2006). According to Jorgensen (1989), participant observation is appropriate under six conditions. The appropriacy for the present study is indicated in brackets for each one (Jorgensen, 1989: 13):

- The research problem is related to human meanings and interactions and evaluated through the insider's eyes (*the observer had four years' experience of hunting tourism*)
- The investigated phenomenon is observable in an everyday life situation (*the observer witnessed shopping, tourist-tourist interactions, tourist-official interactions, tourist-local people interactions, tipping, timing, complaining, and food and beverage behaviours*).
- The researcher can access an appropriate setting (*the observer participated in the tour as a translator*).
- The phenomenon is sufficiently limited to be studied as a case (*the observation was limited to the behaviours of hunting tourists visiting Konya in Turkey*).
- The research questions are appropriate for a case study.
- The research problem can be addressed with qualitative data from direct observations and other appropriate methods for the field setting.

Given that these conditions were met, participant observation was chosen as the data collection tool. The participant observations took place in Turkey's central Anatolian province of Konya between 13 and 18 October 2021. Konya was preferred because it has several areas designated for hunting. The second author who has four years of experience in hunting tourism, carried out the observations. She was also the translator for the tourist group.

The tourist participants were not informed about the research, whereas the managers of travel agency X were informed in detail about the research process and provided written permission for the study. The rationale for this approach was that participants might not behave naturally if they knew they were being observed. In their study of volunteer tourism in Rwanda, for example, Barbieri et al. (2012) did not inform either the volunteers or the community members in order to avoid the observer effect, which is when participants become suspicious if the observer asks questions and/or takes notes (Barbieri et al., 2012). According to Veal (1997), it may not be possible to take notes or use a tape recorder if the researcher's identity needs to be concealed. Even if the researcher has a reasonable false identity, using devices may hinder the natural relationship they are trying to establish. Acar (2018), for example, studied international tourists visiting Turkey in 2016. He received permission from a travel agency to observe Iranian tourists' behaviours on a guided shopping tour as a participant but without explaining to the group. However, while he was writing in his observation notebook, one of the Iranian participants saw him and called the travel agency's manager to ask what Acar was doing. Therefore, in order to present such problems in subsequent participant observation studies, Acar introduced himself at the start as an "apprentice guide" (Acar, 2018; Acar & Tanrisevdi, 2019b; Acar & Tanrisevdi, 2021).

Travel agency X was chosen in accordance with the following criteria:

- The second author participated as a translator in the hunting tours, which the travel agency had been running for several years.
- Travel agency X had 23 years of experience in tourism.
- Travel agency X was allowed to export the tourists' hunting trophies due to its trophy centre, visited by the second author. According to information from the authorities, the meat and skins of hunted animals have been stored in freezers until they are exported.

The participant observations were conducted using the participant observation form developed by Acar (2018), derived from a scale measuring twenty typical tourist behaviours created by Pizam and Sussmann (1995). To evaluate the form's applicability (whether the items were observable on guided tours), pilot studies were carried out on daily tours in July and August 2016. After removing irrelevant items, the revised form was applied to tourists from various countries visiting Kuşadası, İstanbul, Antalya, and Cappadocia (Acar, 2018). The form has also been used in several other studies (Acar & Tanrısevdi, 2019b; Acar & Tanrısevdi, 2021).

Finally, participant observations were carried out during a six-day package tour with four experienced hunting tourists (two Portuguese and two German). The researcher first took notes on her phone as there was no opportunity to use a notebook before transferring them to a computer once the group returned to their hotel. The researcher tried to note every behaviour, as the following example shows:

> 13.10.2021, 13.30, after chatting for about half an hour in the lobby of the hotel, the two Portuguese hunters stated that they were hungry and added that hygiene was important for them. Firstly, they washed their hands and then they spent about twenty minutes choosing dishes. Finally, 2P wanted to eat fish but sea fish, not lake fish. 1P also ordered the same. Both wanted the fish to be grilled. However, they had to choose a different menu since the grill section of the restaurant will not be active before 16.00.

The data were analyzed item by item in the participant observation form, with each participant being assigned a code for the content analysis (Portuguese = 1P, 2P; Germans = 3P, 4P).

Whereas quantitative research should ensure internal validity, qualitative research relies on credibility. Various methods can be used to ensure credibility, including prolonged engagement, persistent observation, triangulation, and peer debriefing (Lincoln & Guba, 1985; Erlandson et al. 1993 cited by Yıldırım & Şimşek, 2013). As the second author participated throughout the package tour, prolonged engagement contributed to the credibility of the findings. According to Yıldırım and Şimşek (2013), as there is an increase in the number and duration of observations, there is a corresponding rise in the probability of the observed process reflecting its natural environment. Therefore, the second author participated as the group's appointed translator throughout the tour. Since the observer had contact with the tourist group from the beginning of the tour, she was better accepted by participants. In addition, the observer's prior experience in hunting tourism and the first author's role as a professional tourist guide increased the reliability of the data.

Research findings

This section presents the findings from the participant observations. Table 27.1 shows that the participants were mostly highly experienced male hunters (with an average of 23.25 years' experience).

Interaction behaviours of the hunting tourists

On the first day of the tour, the Portuguese hunters tried to interact with a waiter during lunch as they wanted to learn how "Thank you" and "Bon appétit" are expressed in Turkish. They then thanked the waiter. On the second day, at 16.20, they waved and smiled at an old peasant man they saw on the road on the way to the hunting ground. Then they talked with another villager for a while. On the fourth day, the hunters said to the translator that they (both the Portuguese and the Germans) had talked to each other while drinking in the lobby after returning to the hotel the previous evening. However, this was the only occasion when all four tourists socialized together as the tour officials took care not to bring them together to eat in the restaurant. On the fifth day, however, the four tourists and other tour officials (the guides and translators, a forester, and the agency owner) ate together. While the tourists were eating, they also talked with one another.

Novelty-seeking tendencies of hunting tourists

The tourists' preferences for novelty or familiarity could only be observed partially. On the first day, the translator said that she was living in Kuşadası, so the Portuguese hunters searched online for information about the location and asked about cruise ships. The translator then explained about Turkish coffee, so 2P wanted to try it. They also wanted to eat Turkish delight with the coffee, but had chocolate instead since Turkish delight was unavailable. While drinking their coffee, 1P video called his wife and showed her his coffee.

Table 27.1 Demographics of the participants

Participant (P)	*Gender*	*Age*	*Experience (Years)*	*Occupation*	*Country*
1	Male	46	28	Electrical technician	Portugal
2	Male	47	30	Farmer	Portugal
3	Male	65	30	Businessman	Germany
4	Male	27	5	Engineer	Germany

Photographing behaviours of hunting tourists

2P liked to show the others photos of previous hunts. At 20.32, 2P took several photos of a 200-kilo wild boar he had killed on the first day. He then shot another wild boar of 280 kilograms at 20.57 and asked the guide to clean the tusks while taking photos for ten minutes. On the way back, 2P photographed a tea glass during a break at the coffee house. On the second day, 1P also killed a wild boar at 23.40 and asked the guide to clean the tusks while taking photos. He also wanted to return to take more photos the following day. On the third day, the two Portuguese hunters wanted to take photos of the sunset by the lake while one of the Germans (3P) photographed the wild boar he had killed after its tusks were cleaned at 20.30. The hunters later shot another boar, which escaped, before being found and killed at 22.00. After cleaning the tusks, 3P was photographed with the corpse. Around 00.30, they shot another boar, and the tusks were cleaned and photographed. The group then returned to the hotel. Thus, the hunters mostly photographed the wild boars they hunted, rather than taking a wider interest in nature. Although it was not possible to observe if they shared the photos online, 2P stated on the first day that they did so. Similarly, 1P and 2P showed the translator the hunting photos in their social media profiles.

Food and beverage preferences of hunting tourists

Throughout the tour the hunters were provided with half-board hotel accommodation. Figure 27.1 lists the foods and beverages consumed by each hunter. On the third day, 3P (German hunter) poured hot spice from the bottle he had brought with him into his soup at dinner. Although he offered it to the guides, they did not like the taste. On the fourth day, when 1P (Portuguese hunter) saw sweets, such as Turkish fairy floss, Turkish delight, and cezerye, in a petrol station market, he wanted to return after hunting to buy them because he was already familiar with their taste from a visit two years earlier. The guide offered to take him to another shop with a greater variety.

As Figure 27.1 shows, the Portuguese hunters (1P, 2P) preferred cold starters and drank Turkish tea almost daily and beer with dinner. They also chose soup as the starter, as was the Turkish custom. On the other hand, they did not like the desserts. The food and beverage preferences of the two Germans could not be observed in as much detail.

Timing behaviours of hunting tourists

There were no delays. After the guide asked them to be ready for hunting at 17.00 each evening, the hunters were ready to depart. Having drunk Turkish tea and chatted with the guide for a while, the two Portuguese left the hotel at 17.35. On the second day, they followed the guide's request to be ready in the hotel lobby at 10.15. On the third day, at 12.00 p.m., the guide informed the hunters about the lunch, dinner, and hunting times (feeding and searching for the injured wild boar). At around 19.00, the guide gave ten minutes to prepare for hunting, and there was no delay. On the fourth day, at 17.00 p.m., the guide stated that they would depart for hunting between 18.30 and 19.00 p.m. All of the hunters were ready on time.

Hunting tourists' approach toward promotions/offers

The guide wanted to be rewarded with a whisky after 2P shot the second boar on the first day, and 2P agreed to treat the guide with a whisky after returning to the hotel. Both Portuguese hunters immensely enjoyed the Turkish teas offered to them. On the third day, complementary

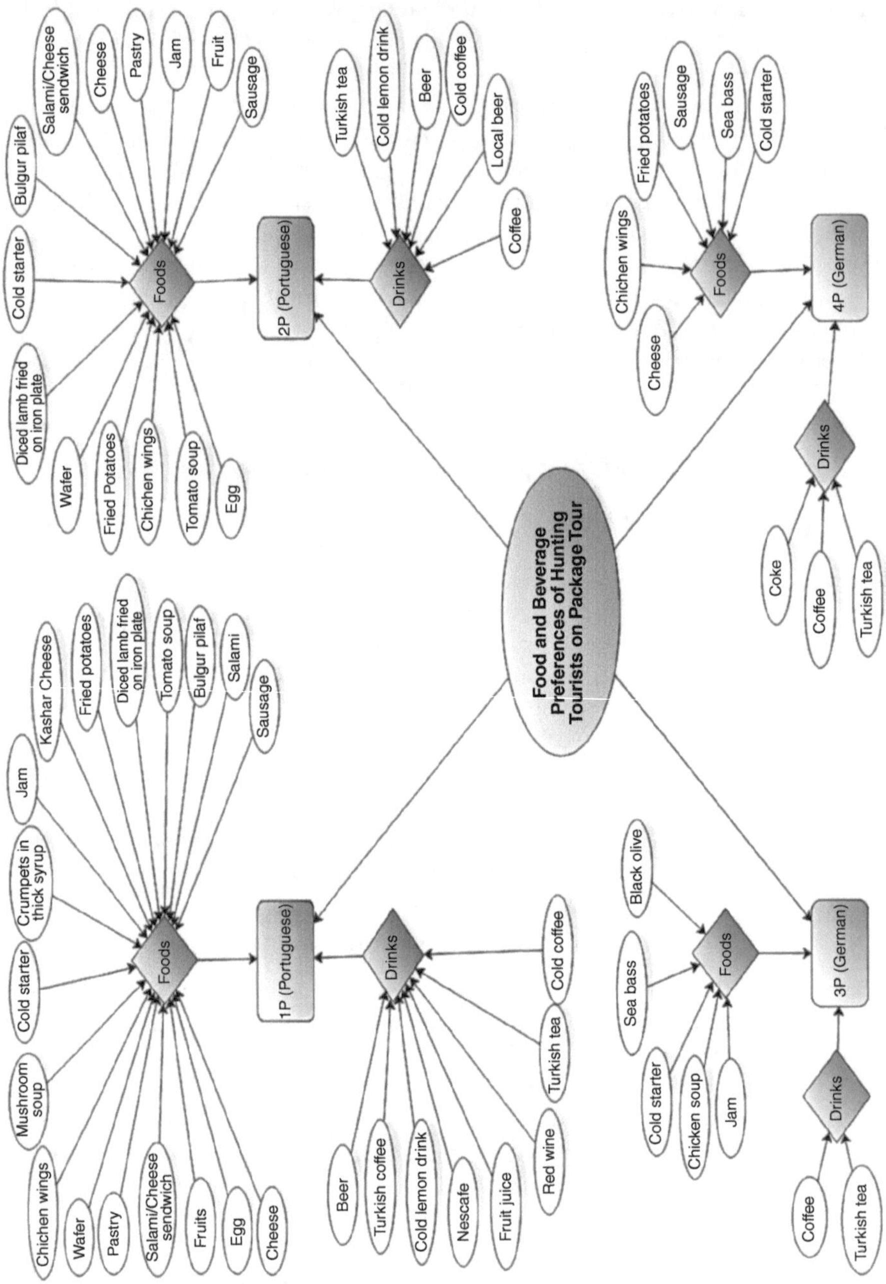

Figure 27.1 Hunters' food and beverage preferences.

Turkish tea was offered to them after lunch and during a break at a coffee house at 22.40 p.m. On the second day, at 12.50 p.m., the two Portuguese hunters asked the translator whether she wanted to eat the apples they had brought with them. On the third day, 3P (German) gave the guide one of the hunting photos as a gift. The guides also told the translator that 3P had offered them cigars and gifted one of the guides with a knife.

Hunting tourists' interaction with the staff

The Portuguese hunters were very willing to interact with the tour staff. On the first day, 1P showed the translator a video about how machines were used to prepare tomatoes in the processing factory. On the second day during the meal, 2P asked the translator to greet the video calling person. At 17.30, the guide and the Portuguese hunters spoke in English. On the third day, during the meal, after the waiter asked the translator whether the two tourists were Portuguese, 1P started to talk about Portuguese footballers. At 02.30 on the sixth day, 1P told the guide, translator, and forester about famous Portuguese dishes in Portugal and how to cook pork while eating soup after hunting.

Tipping behaviour of hunting tourists

No tipping was observed during the tour. However, on the fourth day, 2P told the guide that "good hunting means a good tip; that is not the kind of thing to be left to chance". On the final day, although the translator could not observe the participants' tipping behaviours having left the hotel before the participants checked out, she was informed that the other translator received some tips at the airport while the guides received tips at the hunting ground.

Complaining behaviour of hunting tourists

The only complaints occurred on the fourth day. At 13.10, the two Portuguese asked for their chicken to be cooked more as it seemed undercooked. There were no observed arguments during the five days of the tour. However, on the fourth day, 2P asked to go hunting with the German hunters' guides because he claimed to know the location of larger boars. After dinner, the guide invited 2P and the translator to the hotel's terrace of the hotel to resolve the disagreement. The guide reminded him that hunting depended on luck, although he did his best. He also wanted the translator to ask 2P why he had not shot the big wild boars before, although the guide had shown him where they were. After some discussion, it was decided that 2P would join the German hunters' guides just for that day. Then, 2P pointed at the German hunters, saying he wanted to hunt big boars like them. Eventually, both sides reached an agreement.

Shopping behaviours of hunting tourists

With the exception of one day, none of the participants purchased traditional Turkish products. On the fourth day, the guide and 1P bought cold coffee and wafers, which 1P said he liked a lot and had no hesitation in purchasing, during a break at the petrol station after they failed to find the injured boar shot the previous day.

Although the observation form did not ask for a record of the numbers of wild boars shot, the observer did note how many were hunted and injured. Table 27.2 shows the number of hours the tourists spent hunting and how many boars they either killed or injured.

Table 27.2 The wild boars which were hunted during the observations

Day	*Hunting time*	*Participants*	*Killed Boars*	*Injured Boars*
13-10-2021	17:35- 00.00	2P (Portuguese)	3	-
14-10-2021	18.22- 04.20	1P (Portuguese)	2	1
15-10-2021	19.40- 01.10	3P (German)	3	-
16-10-2021	20.00- 02.00	2P (Portuguese)	-	1
		3P (German)	5	-
		4P (German)	2	-
17.10.2021	19.20- 02.30	1P (Portuguese)	-	-
TOTAL			**15**	**2**

Table 27.3 Unobserved behaviours

Behaviours
Souvenir purchasing
Trust in tradesman
Shopping
Attitudes toward salespeople
Requesting a sales contract while shopping
Consulting with each other while shopping
Bargaining
Wanting to see original products
Purchasing postcards
Getting information about the destination
Being interested in local values
Applauding the guide

The tourists' concern for hygiene could not be observed. However, the Portuguese hunters claimed that hygiene was critical for them, and they washed their hands before the meal on the first day of the tour. They did not, however, pay sufficient attention to COVID-19 measures (such as wearing a mask).

Tourists who participate in hunting tours can be described as active rather than passive. On the other hand, whether the participants wanted their close friends and relatives to join such tours were not observed. However, 2P mentioned that he had two daughters but preferred to join tours with his friend rather than his family on the first day. Table 27.3 lists other behaviours that were not observed during the tour.

Discussion

The participants in the package tour were all males. In terms of gender, this current study supports the related studies (Radder & Bech-Larsen, 2008; Matilainen et al., 2016; Kupren & Hakuc-Blazowska, 2021), emphasizing that hunting tourists were primarily males. As stated above, the average hunting experience of the participants was 23.25 years. This finding also supports the study conducted by Willebrand (2009), revealing that most of the hunting tourists had an experience of more than ten years in the field of hunting.

The hunters in this study took part in the package tour with their friends, not their families. The study conducted in East Lapland by Matilainen et al. (2016) showed that most hunters travelled with their friends. According to Brotherton and Himmetoğlu (1997), most special interest tourists tend to participate in their preferred activities with potentially like-minded people. Another significant finding was that the Portuguese hunters were more likely to be novelty seekers. As shown in Figure 27.1, they drank Turkish tea almost daily and tried both Turkish coffee and local beer. This behaviour is evidence of "the explorer" tourist type (Cohen, 1972). According to Brotherton and Himmetoğlu (1997), most special interest tourists tended to be circumspect while just a minority of them were in search of new activities. With this respect, the current findings support the related literature.

According to the hunting activities report of The General Directorate of Nature Conservation and National Parks in Turkey. The highest hunting quotas were given to wild boars between 2000-2020. Furthermore, foreign hunters are permitted to hunt in the moonlight on seven consecutive days between April 1, 2021, and March 31, 2022 (Republic of Turkish Ministry of Agriculture and Forestry, 2020). On five of the days of the six-day package tours, the participants went hunting, so it is possible to state that the tour schedule was implemented following the regulations. Furthermore, the number of wild boars hunted is essential for satisfaction. It was observed that 17 wild boars were hunted (15 killed and two injured) totally. However, it is unclear whether that number met all the participants' expectations before taking part in the tour. The only one who probably achieved his goal was 3P (German hunter), who said that he did not want to hunt on the last day as he had already reached his target for the trip.

Kupren and Hakuc-Blazowska (2021) report that hunting-related expenditures are lower in Poland than in European countries. According to Ulusoy (2015), hunting tourists go hunting and tend to shop. On the other hand, the study by Matilainen et al. (2016) revealed that accommodation and transportation were the most significant expenses of hunting tourists. However, the participants of this current study showed no tendency to shop. In terms of shopping, the research findings partially support the related literature.

The study conducted by Akdu and Cengiz (2020) in the 5-star hotels in Antalya reveal that apologizing, one of the recovery strategies, positively affects customer's service quality perceptions. Türker and Özaltın Türker (2017) suggest that the most preferred compensation strategies used by travel agents are making explanations, apologizing, refunding, discounting, and gift-giving, respectively. Furthermore, Dalgıç et al. (2021) reveal that gift-giving and refunds/partial wage refunds are the compensation strategies that travel agents use. However, in this current study, there were almost no conflicts with the guide except for one Portuguese participant's request on the third day to change his guide. The travel agency officials fulfilled this request; thus, it is possible to state that the officials used a different compensation strategy compared to the literature.

Clothing and eating are among the physiological needs which are at the bottom of Maslow's five-tier model of human needs (McLeod, 2007); thus, all human beings want to eat something when they get hungry. However, cultural differences determine what, when, and how to eat (Rızaoğlu, 2012). Furthermore, tourists may also tend to eat or drink the things they are accustomed to during their trips. This current study revealed that one of the German hunters added hot spice he had brought with him to his soup, which is similar to behaviour noted by tour guides with South Koreans, who bring dried seafood and coffee from their country (Acar, 2018), and Pakistani package tourists who request English tea (Acar & Tanrisevdi, 2019b). They show a tendency to maintain their home habits while travelling. With this regard, this research supports the related literature.

Regarding tipping, the hunting tourists were not very generous, although one of the Portuguese hunters said that "good hunting means good tip; that is not the kind of thing to be left to chance". This suggestion may be explained by the win–win negotiation strategy.

Furthermore, there were no predetermined tips to be paid by the participants. Considering tipping behaviour, the findings of this current study differ from the studies emphasizing the predetermined tips that Chinese (Acar, 2022) and South Korean (Acar & Tanrisevdi, 2021) tourists have to pay upon joining the tour. The hunters mostly photographed the wild boars they had shot instead of natural scenery in terms of photography. This contrasts with Japanese package tourists, who photograph almost everything during the guided tours (Acar, 2018). Thus, the hunters' focused behaviour seems appropriate given the nature of hunting tourism. One of the significant extra findings was that the hunting tourists were satisfied with their treats. In addition, they also offered various treats to the tour officials (e.g., translator, guide). This finding supports the study with South Korean tourists (Acar & Tanrisevdi, 2021).

Conclusion

This study investigated the behaviours of hunting tourists on a package tour during the COVID-19 pandemic in Konya, a central Anatolian province of Turkey. The researchers used the observation form created by Acar (2018), based on the scale developed by Pizam and Sussmann (1995). The data obtained from participant observation were analyzed in terms of the parameters of the observation form. This current study provided a better understanding of why the officials of the travel agency X tried not to bring the participants together. Like in this package tour, bringing different tourist groups together may affect the tour flows negatively. Because hunting is an activity that mostly requires individual skills, ensuring the limited contacts of the participants with each other as much as possible will contribute to the elimination of the potential problems that travel agencies may face.

Shopping and optional tours are essential for travel agencies in the Turkish tourism industry. However, this study showed that the participants did not tend to buy souvenirs or optional tours during their visits. Because they spent most of their time hunting in the evenings and resting in the daytime, therefore, extra shopping tours except for hunting time may create added value for travel agencies.

The only one who probably achieved his goal was 3P (German hunter), who said that he did not want to hunt on the last day as he had already reached his target for the trip. Performance indicators can be important for special interest tourists, such as the number and height of mountains summited for mountaineers, the depth descended for a scuba diver, or the number of birds observed and recorded for a birdwatcher. Therefore, a performance indicator could be useful in future research into hunting tourism.

Regarding timing, the participants did not hinder the tour schedule, partly because the program was flexible. Although the hunting times were critical, the hunters themselves decided when to finish hunting rather than the guides. Thus, the hunters must comply with the determined schedule if they want to maximize their satisfaction and meet their expectations from the tour. Therefore, it should be noted that tourists participating in hunting tourism are more conscious tourists compared to the ones participating in mass tourism.

Limitations

As is common for every study, the present study had a series of limitations. Firstly, the hunting tourists haven't hunted together due to the tour agency's policy. Consequently, some sub-questions could not be fully addressed. A second limitation was that only a very few tourists' behaviours over a limited period could be observed. Third, several dimensions beyond the predetermined research sub-questions (satisfaction status, performance goals) only became apparent during the observations. Fourth, the observer was not able to go hunting on the fourth day of the tour because of the bad weather conditions but she learned how many animals were hunted from the guides. According to Karasar (2009), there should be at least two observers to make the right inference. However, there was only one observer in this current study during the tour. Finally, it is unknown whether any participants complained to the travel agency after returning to their country or made any negative electronic word of mouth (eWOM) about the country and/or package tour on social media. More valid and reliable results can be achieved if these limitations are eliminated while applying mixed research methods (both qualitative and quantitative) in the future.

References

Abraham, Z. (2014). The influence of national culture on tipping behavior. *UNLV Theses, Dissertations, Professional Papers, and Capstones*. 2626. https://digitalscholarship.unlv.edu/thesesdissertations/2626/

Acar, V. (2018). *Investigating international tourists' behaviors: A study on tourists taking part in guided tours in Turkey* [Unpublished doctoral dissertation]. Aydin Adnan Menderes University.

Acar, V. (2022). Chinese tourists' behavior on package tours in Turkey. In Wen, J. & Kozak, M. (Eds.), *Chinese Outbound Tourist Behaviour: An International Perspective*. London: Routledge.

Acar, V., Karaosmanoğlu, K., & Şenbahar, Y. (2020). Turistik halı ve deri mağazaları ile seyahat acentaları arasındaki güncel sorunları belirlemeye yönelik nitel bir araştırma: Kuşadası ve Selçuk örneği. *Marmara Üniversitesi Öneri Dergisi, 15*(54), 610–643. doi: 10.14783/maruoneri.771838

Acar, V., & Tanrisevdi, A. (2019a). Participant observations on Iranian tourists taking part in guided excursions, the case of Kusadasi, Turkey. *Manas Journal of Social Studies, 8*(2), 2187–2213. doi: 10.33206/mjss.502494

Acar, V., & Tanrisevdi, A. (2019b). Paket turlarda turist davranışlarının incelenmesi: İstanbul'u ziyaret eden Pakistanlı turistler üzerinde nitel bir araştırma. *Manas Sosyal Araştırmalar Dergisi, 8*(4), 3931–3957. doi: 10.33206/mjss.513298

Acar, V., & Tanrisevdi, A. (2021). Paket tur katılımcısı turistlerin davranışlarının incelenmesi: Güney Koreli turistlere yönelik bir katılımlı gözlem çalışması. *Marmara Üniversitesi Öneri Dergisi, 16*(56), 792–824. doi:10.14783/maruoneri.815611

Akdu, S. & Cengiz, E. (2020). Hizmet hatası telafi stratejileri, hizmet kalitesi algısı ve kurumsal imaj algısı arasındaki ilişki: Konaklama işletmeleri örneği. *İşletme Araştırmaları Dergisi, 12*(4), 4413–4428 doi: 10.20491/isarder.2020.1113

Akkuş, G. & Şimşek, A. (2019). Sinop-Boyabat'a Gelen Av Turistleri Tipolojisi. *8th International Vocational Schools Symposium, UMYOS' 19 Sinop*, 11–13 June.

Alkan, U. M. (2014). *Antalya bölgesinde yaban hayatı açısından av turizmi üzerine incelemeler.* [Unpublished master's thesis]. Kahramanmaraş Sütçü İmam University.

Asadi, R. & Daryaei, M. (2011a). Prioritization of pull factors of Malaysia as a destination for Iranian tourists. *European Journal of Social Sciences, 26*(2), 287–296.

Asadi, R. & Daryaei, M. (2011b). Prioritization of pull factors of Turkey as a destination for Iranian tourists. *Australian Journal of Basic and Applied Sciences, 5*(10), 1510–1515.

Asadi, R. & Daryaei, M. (2011c). Prioritization of pull factors of China as a destination for Iranian tourists. *Australian Journal of Basic and Applied Sciences, 5*(11), 1129–1135.

Asgari, M., & Borzooei, M. (2014). Evaluating the perception of Iranian students as educational tourists towards Malaysia: In-depth interviews. *Interdisciplinary Journal of Contemporary Research in Business, 5*(9), 81–109.

Avcıkurt, C. (2007). *Turizm sosyolojisi: Turist-yerel halk etkileşimi* (İkinci Baskı). Ankara: Detay Yayıncılık.

Barbieri, C., Santos, C.A. & Katsube, Y. (2012). Volunteer tourism: On-the ground observations from Rwanda. *Tourism Management, 33*(3), 509–516. doi: 10.1016/j.tourman.2011.05.009

Bernard, H.R. (2006). *Research methods in anthropology: Qualitative and quantitative approaches.* Lanham: AltaMira Press.

Bouzari, M. (2012). *Preliminary study on destination attributes of Northern Cyprus: Iranian travellers' perspective*, [Unpublished master's thesis] Eastern Mediterranean University.

Bowen, D. (2002). Research through participant observation in tourism: A creative solution to the measurement of consumer satisfaction/dissatisfaction (CS/D) among tourists. *Journal of Travel Research, 41*, 4–14. doi:10.1177/0047287502041001002

Bowie, D., & Chang, J.C. (2005). Tourist satisfaction: A view from a mixed international guided package tour. *Journal of Vacation Marketing, 11*(4), 303–322. doi:10.1177/1356766705056628

Brotherton, B. & Himmetoğlu, B. (1997). Beyond destinations-special interest tourism. *Anatolia, 8*(3), 11–30. doi: 10.1080/13032917.1997.9687118

Chang, R. C. Y., Kivela, J. & Mak, A. H. N. (2011). Attributes that influence the evaluation of travel dining experience: When East meets West. *Tourism Management, 32*(2), 307–316. doi:10.1016/j.tourman.2010.02.009

Cho, B.H. (1998). Assessing tourist satisfaction: An exploratory study of Korean youth toursts in Australia. *Tourism Recreation Research, 23*(1), 47–54 doi: 10.1080/02508281.1998.11014819

Cohen, E. (1972). Toward a sociology of international tourism. *Social Research, 39*(1), 164–182.

Cömert, M. & Sökmen, A. (2017). Türkiye'de gastronomi turizmi: Antalya'da konaklayan turistler üzerine bir araştırma. *Seyahat ve Otel İşletmeciliği Dergisi, 14*(3), 6–26 doi: 10.24010/soid.369897

Dalgıç, A., Toksöz, D., Ekici Çilgin, R. & Güvenç, A. (2021). Tur katılımcılarının şikayet unsurları ve izlenen şikayet telafi stratejiler: Turist rehberleri açısından bir değerlendirme. *Turist Rehberliği Nitel Araştırmalar Dergisi, 2*(2), 130–149.

Doğan, H.Z. (2004). *Turizmin sosyo-kültürel temelleri* (İkinci baskı). Ankara: Detay Yayıncılık.

Duman, M.Z. (2016). Van'a gelen İranlı turistlerin profilleri üzerine sosyolojik bir araştırma. *Uluslararası Sosyal Araştırmalar Dergisi, 9*(46), 399–408.

Ege, Z. (1996). *Bir rekreasyon türü olarak av turizmi ve Türkiye'de av turizminin geliştirilmesi ile ilgili bir model önerisi.* [Unpublished doctoral dissertation]. İzmir Dokuz Eylül University.

Egresi, I. & Arslan, S. (2016). Shopping and tourism in Turkey: The perfect combination. In Egresi, I. (Ed.), *Alternative Tourism In Turkey: Role, Potential Development and Sustainability* (pp. 211–228). Cham, Switzerland: Springer.

Farahani, B.M. & Mohamed, B. (2013). Relationship between nationality and tourists' behaviour: Case of Middle East tourists in Malaysia from tour guides perspective. *International Journal of Tourism Antropology, 3*(1), 43–57. doi:10.1504/IJTA.2013.054409

Fard, M.M. & Saberi, H. (2015). Studying the motivations of choosing foreign tourism destinations: Case study of Thailand. *Journal of Sustainable Development, 8*(9), 254–264.

Heo J. K., Jogaratnam, G., & Buchanan, P. (2004). Customer-focused adaptation in New York City hotels: exploring the perceptions of Japanese and Korean travelers. *International Journal of Hospitality Management, 23*(1), 39–53 doi:10.1016/j.ijhm.2003.07.003

Hofstede, G., Hofstede, G. J., & Minkov, M. (2010). *Cultures and Organizations: Software of the Mind.* (3th Edition). New York: McGraw Hill.

Jorgensen, D.L. (1989). *Participant Observation: A Methodology for Human Studies*, Applied Social Science Methods, 15, Newbury Park, CA: Sage Publication.

Kaaristo, M. & Rhoden, S. (2017). Everyday life and water tourism mobilities: Mundane aspects of canal travel, *Tourism Geographies, 19*(1), 78–95, doi: 10.1080/14616688.2016.1230647

Karasar, N. (2009). *Bilimsel araştırma yöntemi: kavramlar, ilkeler, teknikler* (20. Baskı), Ankara: Nobel Yayıncılık.

Kim, S. M., & Prideaux, B. (1999). Korean inbound tourism to Australia- a study of supply-side deficiencies. *Journal of Vacation Marketing, 5*(1), 66–81. doi:10.1177/135676679900500106

Kim, S. S., Prideaux, B., & Kim, S. H. (2002). A cross-cultural study on casino guests as perceived by casino employees. *Tourism Management, 23*(5), 511–520. doi:10.1016/S0261-5177(02)00008-0

Komppula, R. & Suni, J. (2013). Identifying hunting tourist types – an exploratory case study from Finland. *Tourism Review, 68*(1), 48–61 doi: 10.1108/16605371311310075

Konijn, E., Sluimer, N. & Mitas, O. (2016). Click to share: Patterns in tourist photography and sharing. *International Journal of Tourism Research, 18*(6), 525–535 doi: 10.1002/jtr.2069

Kozak, M. (2016). Bargaining behavior and the shopping experiences of British tourists on vacation. *Journal of Travel & Tourism Marketing, 33*(3), 313–325. doi:10.1080/10548408.2015.1051199

Kupren, K. & Hakuc-Blazowska (2021). Profile of a modern hunter and the socio-economic significance of hunting in Poland as compared to European data. *Land, 10*(11), 1178 doi: 10.3390/land10111178

Laitinen, K. (2004). *The perceptions of tour guides of Finland as a travel destination for Chinese tourists* [Unpublished master's thesis]. Bournemouth University.

Lee, D. Y. (2000). Retail bargaining behaviour of American and Chinese customers. *European Journal of Marketing, 34*(1/2), 190–206. doi:10.1108/03090560010306287

Lincoln, Y. S., & Guba, E. G. (1985). *Naturalistic Inquiry*. Beverly Hills: Sage Publications.

Lynn, W. M. M. (2010). Tipping. In Pizam, A. (Ed.), *International Encyclopedia of Hospitality Management* (pp. 651–652). Oxford: Butterworth Heinemann.

Mackellar, J. (2013). Participant observation at events: theory, practice and potential. *International Journal of Event and Festival Management*, *4*(1), 56–65. doi:10.1108/17582951311307511

Matilainen, A., Keskinarkaus, S. & Törmä, H. (2016). The economic significance of hunting tourism in East Lapland, Finland. *Human Dimensions of Wildlife*, *21*(3), 203–222. doi: 10.1080/10871209.2016.1129652

McLeod, S. (2007). Maslow's hierarchy of needs. Retrieved March 1, 2022, from https://www.simplypsychology.org/maslow.html

Olcay, A., Dağlıoğlu, T., & Sürme, M. (2019). Turistleri av turizmine yönelten nedenlerin belirlenmesine yönelik araştırma. *Journal of Recreation and Tourism Research*, *6*(3), 285–293. doi: 10.31771/jrtr.2019.31

Özdemir, B. & Kınay, F. (2004). Yabancı ziyaretçilerin Türk mutfağına ilişkin görüşleri: Antalya'yı ziyaret eden Alman ve Rus turistler üzerine bir araştırma. *Gazi Üniversitesi Ticaret ve Turizm Eğitim Fakültesi Dergisi*, *2*, 5–28.

Özdemir, C. (2014). *Kültürlerarası turist davranışı: Turistlerin tur esnasındaki davranışlarının belirlenmesi üzerine bir araştırma.* [Unpublished master's thesis]. Eskişehir Anadolu University.

Pizam, A. (1999). The American group tourist as viewed by British, Israeli, Korean, and Dutch tour guides. *Journal of Travel Research*, *38*, 119–126. doi:10.1177/004728759903800204

Pizam, A., Jansen-Verbeke, M., & Steel, L. (1997). Are all tourists alike, regardless of nationality? The perceptions of Dutch tour guides. *Journal of International Hospitality Leisure and Tourism Management*, *1*(1), 19–38. doi:10.1300/J268v01n01_03

Pizam, A. & Jeong, G.H. (1996). Cross-cultural tourist behavior: Peerceptions of Korean tour-guides. *Tourism Management*, *17*(4), 277–286. doi:10.1016/0261-5177(96)00019-2

Pizam, A. & Reichel, A. (1996). The effect of nationality on tourist behavior: Israeli tour-guides' perceptions. *Journal of Hospitality & Leisure Marketing*, *4*(1), 23–49. doi:10.1300/J150v04n01_03

Pizam, A. & Sussmann, S. (1995). Does nationality affect tourist behavior? *Annals of Tourist Research*, *22*(4), 901–917. doi:10.1016/0160-7383(95)00023-5

Radder, L. & Bech-Larsen, T. (2008). Hunters' motivations and values: A South African perspective. *Human Dimensions of Wildlife: An International Journal*, *13*(4), 252–262. doi: 10.1080/10871200801986739

Reisinger, Y. (2009). *International Tourism: Cultures and Behaviours*. Oxford: Butterworth-Heinemann.

Reisinger, Y. & Turner, L. (2000). Japanese tourism satisfaction: Gold Coast versus Hawaii. *Journal of Vacation Marketing*, *6*(4), 299–317. doi:10.1177/135676670000600402

Reisinger, Y., & Turner, L. W. (2003). *Cross-cultural Behaviour in Tourism: Concepts and Analysis*. Oxford: Butterworth Heinemann.

Republic of Turkey Ministry of Agriculture and Forestry (2020). *Hunting activities, 2000–2020*, Retrieved March 1, 2022, from https://www.tarimorman.gov.tr/DKMP/Menu/19/Yaban-Hayati-Istatistikleri#

Rızaoğlu, B. (2012). *Turizm davranışı* (3.baskı), Ankara: Detay Yayıncılık.

Rosenbaum, M. S., & Spears, D.L. (2006a). Who buys what? Who does that? The case of Golden Week in Hawaii. *Journal of Vacation Marketing*, *12*(3), 246–255. doi:10.1177/1356766706064621

Rosenbaum, M. S., & Spears, D. L. (2006b). An exploration of spending behaviors among Japanese tourists. *Journal of Travel Research*, *44*(4), 467–473. doi:10.1177/0047287505282949

Şafak, İ. (2003). Türkiye'deki av turizmi uygulamalarının özel avlak işletmelerine etkileri. *Süleyman Demirel Üniversitesi Orman Fakültesi Dergisi*, *4*(2), 133–148.

Sekaran, U. (2003). *Research Methods for Business: A Skill Bulding Approach.* (4th Edition). New York: John Wiley & Sons.

Shrestha, S. K., Burns, R. C., Pierskalla, C. D. & Selin, S. (2012). Predicting deer hunting intentions using the theory of planned behavior: A survey of Oregon big game hunters. *Human Dimensions of Wildlife*, *17*(2), 129–140 doi:10.1080/10871209.2012.649885

Tangeland, T., Vennesland, B. & Nybakk, E. (2013). Second-home owners' intention to purchase nature-based tourism activity products- A Norwegian case study. *Tourism Management*, *36*, 364–376, doi:10.1016/j.tourman.2012.10.006

Tanrısevdi, A., & Akyurt Kurnaz, H. (2018). Hediyelik eşya satın alma davranışında kültürel özellikler: Hediyelik eşya satıcıları açısından bir inceleme. *Journal of Travel and Tourism Research*, *13*, 1–18.

Timothy, D. J. (2005). *Shopping Tourism, Retailing, and Leisure.* Clevedon: Channel View Publications.

Tracy, S.J. (2013). *Qualitative Research Methods: Collecting Evidence, Crafting Analysis, Communicating Impact.* Wiley-Blackwell: West Sussex.

Türk Dil Kurumu. (2022). Pazarlık. Retrieved February 15, 2021, from https://sozluk.gov.tr/

Türker, A. & Özaltın Türker, G. (2017). Seyahat acentalarının müşteri şikayetlerine yaklaşımı: İstanbul, Ankara ve Antalya karşılaştırması, *Researcher: Social Science Studies*, *5*(10), 23–35.

Ulusoy, H. (2015). Av turizminin kırsal turizm açısından kırsal kalkınma üzerindeki etkisinin irdelenmesi. *Türk Bilimsel Derlemeler Dergisi*, *8*(2), 74–80.

UNWTO (2022). Tourism grows 4% in 2021 but remains far below pre-pandemic levels. Retrieved March, 6, 2022, from https://www.unwto.org/news/tourism-grows-4-in-2021-but-remains-far-below-pre-pandemic-levels

Veal, A. J. (1997). *Research Methods for Leisure and Tourism: A Practical Guide* (2nd Edition). London: Pitman Publishing.

Verbeek, D. H. P. Bargeman, A. & Mommaas, J. T. (2011). A sustainable tourism mobility passage. *Tourism Review*, *66*(4), 45–53.

Willebrand, T. (2009). Promoting hunting tourism in North Sweden: Opinions of local hunters. *European Journal of Wildlife Research*, *55*(3), 209–216. doi:10.1007/s10344-008-0235-2

Yıldırım, A. & Şimşek, H. (2013). *Sosyal bilimlerde nitel araştırma yöntemleri* (9. Genişletilmiş baskı), Ankara: Seçkin Yayıncılık.

Yin, R. K. (2017). *Durum çalışması araştırması uygulamaları*. 3. Basımdan çeviri. (Günbayı, İ. Çev.). Ankara: Nobel Akademik Yayıncılık.

Ziff-Levine, W. (1990). The cultural logic gap: A Japanese tourism research experience, *Tourism Management*, *11*(2), 105–110. doi:10.1016/0261-5177(90)90025-5

28

THE FIRST INSIGHTS INTO THE IMPACTS ASSOCIATED WITH THE DEVELOPMENT OF HIKING TOURISM IN MOUKHTARA, LEBANON

Viana Hassan and Georges Bellos

Introduction

The rising number of tourists in some regions may negatively impact local populations and the natural environment, given that development is not adequately controlled (Bosch et al., 2017). Such impacts initiated an urgent need to manage tourism activities better, ensuring movement towards more sustainable tourism practices (UNWTO, 2020a, b).

Hiking is termed as a recreational activity whereby an individual or a group walks across several regions in a country, typically within rural areas, to explore the natural world. It is one of the most often-conducted activities in Moukhtara, where the number of tourists and hikers visiting the site is increasing (Kenny & Brown, 2018). Recent studies have established a link between hiking and sustainability. The authors claim that hiking activities and hiking tours can affect the different aspects of local ecosystems or the sustainability of ecosystems as they induce economic, environmental, and socio-cultural impacts based on arguments regarding sustainability (Bosch etal., 2017; Kiper, 2013). Considering the environmental aspect, hiking can either stimulate conservation or induce the deterioration of the local ecosystem (Khalil, 2009). It can establish friendships from a social standpoint, yet it can also create problems among hikers, visitors, and communities (Pozzato, 2017–2018). While the economic impact is also essential as tourism activities can stimulate the local community's economic development (Bellos, 2021; Bellos & Knio, 2021).

This study examines the impacts of hiking on the environment and the community in Moukhtara and its relation to sustainable development. This is defined as the conscious actions taken to preserve natural resources and meet the needs of present generations to ensure that future generations are not jeopardized (Kamran, 2020). Within this study, we analyse how hiking supports sustainable development activities and in what way this contributes to nature conservation and the development of local communities. Several specific research questions have been posed: *What is the effect of hiking on the environment and sustainable development in Moukhtara? Does hiking affect wildlife communities in Moukhtara*? If yes, *What are the recommendations to be sustainable on a hike and keep the hiking activity in Moukhtara as healthy as possible*? Throughout

DOI: 10.4324/9781003230748-33

the study, extensive answers will be provided to obtain insightful information on these matters, following Törn, Tolvanen, Norokorpi, Tervo and Siikamaki (2009) and Pickering, Rossi and Barros (2011) and Babí, Yuba and Soler (2019). Furthermore, the specific objective of this study is to analyse the effects of hiking on the ecological, social, and economic dimensions of sustainable practice, promoting trail races as a small step toward sustainability in Moukhtara and showing how hiking tourism can inspire the sustainable tourism development of local communities, through establishing proper ways in achieving this.

Sustainable tourism development

Previous studies discussing the concepts about sustainable tourism development focus on the UNWTO definition, depicted in Figure 28.1 (Babí, Yuba, & Soler, 2019; Hall, 1998; Higgins-Desbiolles, 2018; Kiper, 2013; Kuhlman, & Farrington, 2010; Pickering, Rossi, & Barros, 2011; Sharpley, 2020; Törn, 2007; Weaver, 2006). The proposed definition states that: "*Tourism leads to management of all resources in such a way that economic, social and aesthetic needs can be fulfilled while maintaining cultural integrity, essential ecological processes, biological diversity and life support systems*" (UNWTO, 2020a, b).

Sustainable tourism development reflects the conscious actions taken to preserve natural (and cultural) resources and meet the needs of present generations to ensure that future generations can sustain their modes of life. Failing to do so can create unprecedented adverse effects, especially in the current case in Lebanon (Al-Sawi, 2018).Such adverse impacts can be reversed with the implementation of appropriate action, which aims to defend and conserve the natural

Figure 28.1 The different categories of sustainable tourism (Kamran, 2020).

world for future generations. On the other hand, in the context of this case study, hiking has positively impacted the economy by generating revenue for small businesses, including shops and restaurants. Additionally, it raised awareness regarding the importance of nature for current and future generations. Exploring the impacts associated with tourism activities, as hiking in remote communities is essential to inform future decisions and policy development for the benefit of communities and visitors (see Khalil, 2015).

Literature review

Nature-based tourism (NBT) is based on natural attractions that conserve the environment and improve local people's welfare (Kamran, 2020). Hiking is recognized both as one of the nature-based activities but also as niche tourism (Bichler, & Peters, 2021). As an activity, hiking involves following paths on foot, which may or may not be signposted, for sporting and cultural purposes (e.g. Gomez-Martin, 2019; Hall, 2019; Maria Raya, Martínez-Garcia, & Celma, 2018). Although with important regional variations, hiking was traditionally linked to mountaineering and rambling clubs in each country practising such tourism activities (see: Babí, Yuba, & Soler, 2019). Hiking is often defined as a kind of sport that takes place in the scenic natural environment and tracks, and a multiphase exercise conducted in sporting places, touristic locations and natural areas (Carvalhinho, Rosa, & Gomes, 2015).

Many sports and outdoor activities are currently being redefined in terms of specialized applications or specific types of participants. Hiking today is further identified through activities such as trekking (which involves slow, laborious travel often in mountain environments), rambling (which occurs in more remote or often rural areas as through the British countryside where the term originates), heli-hiking (which begins by flying by helicopter to distant or inaccessible areas such as glaciers), and animal-assisted backpacking (which uses pack animals including horses, donkeys or llamas) as Bichler and Peters (2021) discussed. Recently, Gómez-Martín (2019) discussed hiking as an exercise that includes walking on determined paths that could be signposted for sports or heritage causes. Types of hiking vary from one region to another depending on the landscape, weather and environment. Hiking can also take on various forms: it may be a short walk between two designated short points; or it can be a hike that can last all day and which can expand through the night or even overnight (McNamara & Prideaux, 2011).

Hikers seek to immerse themselves in the history, culture and ambience while requiring developed services and amenities (Bichler, & Peters, 2021; Gomez-Martin, 2019; Hall, 2019). The impacts associated with hiking as activity and niche tourism are well researched (Dayananda, 2016; Kenny & Brown, 2018; Kiper, 2013; McNamara & Prideaux, 2011; Pickering, Rossi, & Barros, 2011; Törn, 2007; Törn, Norokorpi, Tervo, & Siikamaki, 2009).

Spain, for example, among the most appealing hiking destinations, provides a unique scenery and distinctive experience, and receives substantial volumes of annual tourists. Gomez-Martin (2019) asserts that hiking, in Spain, amidst the 1990s has exceeded its primary label of being a sporting activity and has become a form of tourism and leisure due to being renowned and popular among travellers. This sets the tone to discuss our case of Moukhtara village in the Baakline area as similar activities were attested in the Chouf region, depending on the localities (see: Khalil, 2009).

Moukhtara is a small village located in the Chouf area of Mount Lebanon. According to Fattouh (2017), Jacques Cauvin excavated the locality in 1963; since then, it has remained an ancient archaeological site that many people enjoy exploring. Research in valorizing rural sites for tourism activities like the ones discussed by Saule (2004) and Khalil (2015) present similar cases to the one currently under study. Moukhtara's main prominence lies in its contents: the old traditional houses, old oil presses, and two El Saydeh churches, also referred to as 'Our Lady' churches (see: Cauvin, 1963; Cauvin & Cauvin, 1968). The fact that hiking is a growing activity in Moukhtara supporting the local community's wellbeing and development (Khalil, 2009) emphasizes the need to understand better the long-term impacts associated with this activity on the destination scale. This creates the foundation for tailoring the effective policies at the destination scale and the visitor experiences.

Methodology

This research is expected to broaden the understanding of the effects of hiking on all three tiers of sustainable tourism development. It will lay the foundation for tailoring specific policies and promoting trail races as a small step toward the destination's sustainability (Figure 28.5). We relied on critical realism, abduction, surveying, cross-sectional time horizons and sampling techniques to cover qualitative and quantitative data collection strategies (Figure 28.2).

Edson, Henning and Sankaran (2017) commend the researchers' use of adductive methods of inquiry through missed methods, utilizing critical realists' viewpoints, as the realities seen in the studies must be questioned. Brynman (2012) and McLeod (2018) pointed out that the selected research methods were necessary for working with (semi-)structured questionnaires. As such, the devices used to research a particular subject and collect data from the participants were helpful for the qualitative part of the study. Following Creswell (2012), and after administering a pilot study, since we noticed that the secondary data we investigated were inconclusive (Figure 28.4), so, we designed the quantitative survey whose results are shown in Figure 28.3. Sileyew (2019) confirmed that the research design was dictated by the use of one

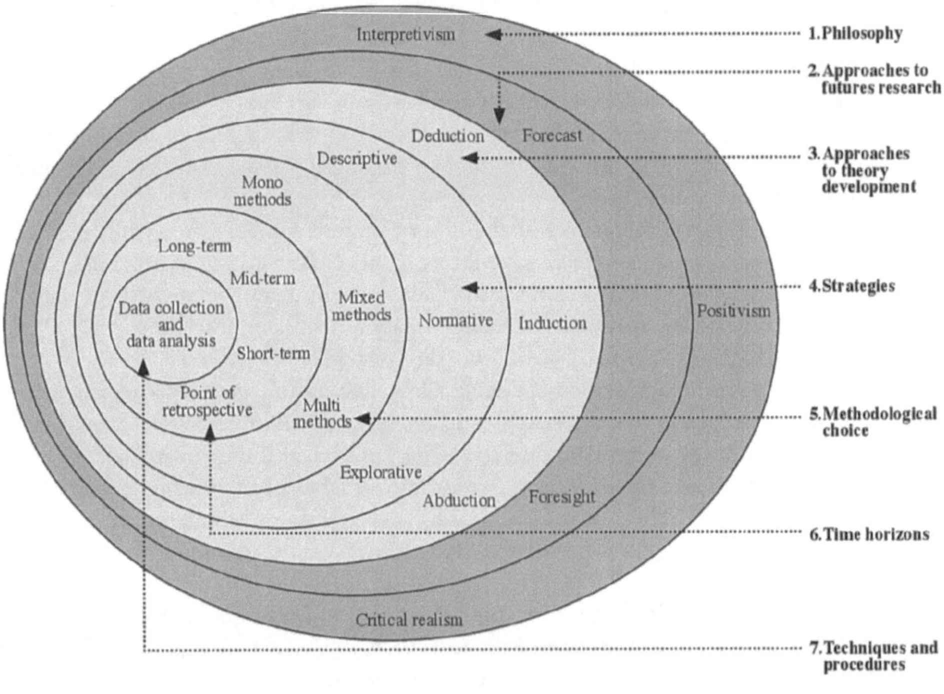

Figure 28.2 Research design (Saunders, Lewis & Thornhill, 2019: 130).

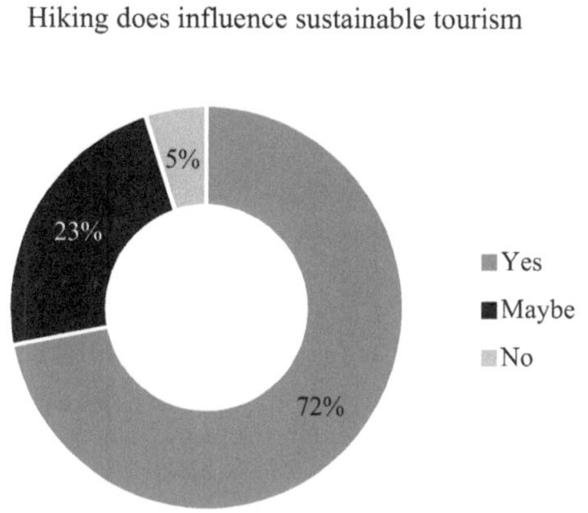

Figure 28.3 Statistical pie chart to show that our survey results agree with the pilot study. (left).

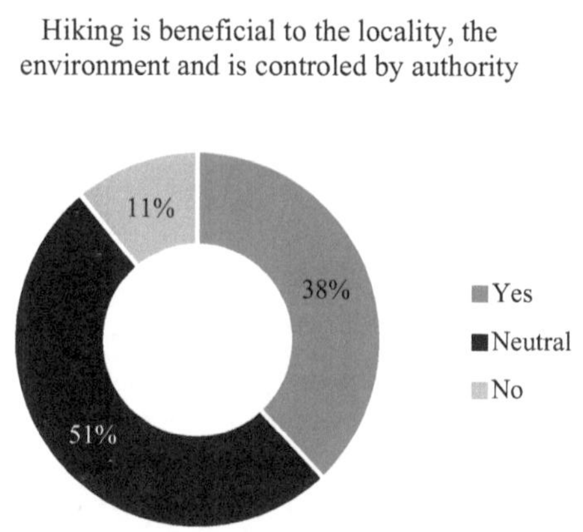

Figure 28.4 Overall results showing the link with hiking and sustainability, based on interpreting secondary data collected since Spring 2020 during the literature review phase. (right).

quantitative tool (survey), as well as two qualitative data collection tools (interview and survey). This complex mixed-method approach also relied on both the collection and interpretation of both qualitative primary and secondary data sources and observations.

Following the suggestion by Sekaran and Bougie (2016), quantitative (probability random and cluster) and non-quantitative (judgmental) sampling strategies were used. The two distinct sets of questionnaires were distributed in the areas surrounding Moukhtara upon initial research conduction in Spring 2020, yet, due to COVID-19, the study took longer to complete (see Gössling, Scott, & Hall, 2020; McLeod, 2018). So, in order to emphasize the multiple perspectives of the phenomenon under investigation, we decided to target tourists and hikers independently from locals via two distinct surveys and an interview addressed to an expert in the domain, as well as observations (Altinay & Paraskevas, 2008; Lucas-Alfieri, 2015; Rakic & Chambers, 2012).

The sampling strategy was tailored accordingly to collecting data through in-depth interviews and through using semi-structured questionnaires, following a judgmental sampling method (Boyce & Neale, 2006; Brynman, 2012; Edson, Henning, & Sankaran, 2017; Khalil,

2009; Saunders, Lewis & Thornhill, 2019; Sekaran & Bougie, 2016). The quantitative sampling was undertaken through a combination of cluster and random sampling (Creswell, 2012). Due to pandemics, some of the data were hard to reach or collect (Creswell, 2012; Gössling, Scott, & Hall, 2020). Overall, the listed resources helped gather information to form coherent research through mixed methods (Altinay & Paraskevas, 2008; Denzin & Lincoln, 2005; Pechlaner & Dal Bò, 2012). The research permitted an in-depth investigation of the phenomena under study while seeing the bigger picture, aiming to understand the issue holistically (Denzin & Lincoln, 2005).

Thus, the investigation commenced in Spring 2020 when secondary data were collected, amid the pandemic to investigate the current issue, through a qualitative method, using two sets of questionnaires for data collection, and an expert interview in Summer 2021. The quantitative survey was mainly utilized to test the selected research design through a pilot study as well as to generalize the results. The first survey was addressed to tourists and hikers while covering issues pertaining to sustainable tourism development and pro-environmental behaviour; the second survey targeted local residents. For this purpose, both questionnaires and the interviews with an organizer of a hiking group composed the qualitative data collection. Quantitatively speaking, the sample was generalized to the population. In terms of quantitative sampling, simple random and cluster sampling were utilized and distributed online to both tourists and local residents, of whom some were hikers, to gather data to assist in the research component (see Saunders, Lewis & Thornhill, 2019).

Findings and results

The findings comprised two qualitative surveys (interviews and semi-structured questionnaires) targeting 150 participants and numerical surveys on a randomly selected sample comprising 40 subjects. Geographically, the qualitative sample comprised hikers and tourists from all over Lebanon and from local inhabitants from Moukhtara, while the 40 people recruited for the quantitative study were given equal chances to participate were randomly selected from the Lebanese population. The survey was disseminated through *Survey Monkey*.

Qualitative data

Below, we present the interview results and our initial survey results that corroborate the current findings in Figure 28.3. Information on the Lebanese International Hikes website also seems to concur.

The first question focused on the *effect of hiking on the environment and sustainable development in Moukhtara*, and indicated the positive and negative effects of this activity in the Moukhtara area. The interviewee and survey subjects responded in detail and somehow agreed that hiking does affect sustainable tourism in the area. This is in line with Figure 28.3, and information on the Lebanese International Hikes website seems to corroborate this. All participants replied positively to *whether hiking affects Wildlife communities in Moukhtara. However, none of them recommended how to keep Moukhtara as healthy as possible. When discussing if hiking in Moukhtara valley could be developed in a more sustainable and eco-friendly manner*, over 70 percent of the respondents agreed.

Referring to the *link between hiking and sustainability*, 65 percent of respondents were neutral, suggesting that the community still has not developed its attitude toward hiking and its associated impacts. Simultaneously, only 9 percent of respondents agreed, while 26 percent considered how hiking could not prevent adverse environmental impacts.

To some extent, these answers already revealed what could be done in the context of strategic planning and capacity building to optimize the outcomes of the hiking activities. The lack of tourism education could be a result of the incapacity of the authorities in preserving a locality, or by the majority of people not being aware of the steps to take, or why the site was not appropriately valorized (Al-Sawi, 2018; Bichler & Peters, 2021; Bou-Melhem, 2021; Gomez-Marin, 2019; Khalil, 2015; Saule, 2004).

The respondents were asked if hiking helps preserve a locality in the next question. 64 percent of the participants were "*neutral about the restoration of biological diversity and about overusing environmental resources*", while 23 percent of them responded as being "*not sure if hiking can cause a decrease in the species types within nature*". Similarly, only 18 percent agreed and added that "*hiking does help preserve nature for the future*", demonstrating the lack of tourism education (Bou-Melhem, 2021). When asked about the *impacts of hiking*, 29 percent of the respondents perceived positive social effects associated with *enthusiasm*. Simultaneously, 53 percent of them show neutral attributes, as they were unsure if *hiking could create new jobs, and about contributing environmental protection and conservation, and the responsible use of natural resources*. Only the remaining 18 percent show the negative attributes since they claim that the "*increase in the number of hikers will lead to friction between them and residents*".

The qualitative data gave first insight but stressed that more research is needed, targeting larger samples and properly sampling the population. For example, we targeted tourists and locals in two distinct surveys. The current interpretation of these data is solely based on analysing the available raw data for triangulation purposes, while the final interpretation will be shown graphically. The below section presents the findings of the quantitative data, while the following one deals with the interpretations of our data collection following field observations.

Findings of the pilot study and quantitative surveys

The pilot study was conducted to confirm our initial survey results in Summer 2021. The actual survey was well correlated as the Cronbach Alpha value was above 0.95 and is therefore reliable (Table 28.1), following Cronbach (1951). Excel was utilized for interpreting quantitative data. From our research, and based on the pilot study results, we noticed that 72 percent of the responses agreed to the *influences of hiking on sustainable tourism*, 23 percent of the responses showed a neutral stance, while the remaining 5 percent of the responses showed to disagree. Figure 28.3 shows the overall results of our survey.

Table 28.1 Cronbach Alpha value (Cronbach, 1951)

Cronbach's Alpha	*0.987732969*
Split-Half (odd-even) Correlation	0.999558804
Split-Half with Spearman-Brown adjustment	0.999779354
Mean for Test	3.333333333
The standard deviation for Test	2.866422813
KR21 (use only 0 and 1 to enter data for this)	0.810597274
KR21 (use only 0 and 1 to enter data for this)	0.810597274

The claim that "*Hiking supports the achievement of SDGs, or contributes to nature conservation, development of local communities*" was agreed by 73 percent of the respondents, while 27 percent were neutral. The same results were observed regarding whether hiking activities infer *sustainable development in Moukhtara,* if the *characteristics and behaviour of hikers have changed over the years,* and if it is *vital to conserve the natural world and greenery within Moukhtara to attract local and foreign hikers to the site.* The results were similar to the one displayed in Figure 28.3.

Furthermore, the majority of respondents (73 percent) agreed that hiking affects the environment and sustainable development in Moukhtara and that it impacts upon its flora and fauna, while only 9 percent of respondents considered the opposite. The vast majority of respondents (91 percent) spoke of the importance of *drawing attention to the tourism industry for assistance in advocating hiking activity.* Hiking is predominately (73 percent of respondents) recognized, from a social standpoint, as an activity that can foster the establishment of friendships; however, it can also create problems among hikers, visitors and communities. When asked about the importance of the economic impacts and the role of tourism as a driver of the local community development, 82 percent of respondents expressed a confirmative attitude. It is essential to emphasize that only 45 percent of respondents observed the effects associated with excessive tourism development.

Having finished the presentation of findings from the range of qualitative and quantitative data, we now deliver the interpretations in the form of the discussion of the links between our field observations and the results of the empirical analysis.

Interpretations

Our analysis demonstrated how hiking is an integral part of the tourism activity in Moukhtara, while the hikers who are de facto seeking NBT activities are usually amateurs. Locals and tourists who practice or seek such activities in Moukhtara are considered as being concerned for the environment. However, we must emphasize how this information is conflicting with the secondary data interpretation, revealing that most people are not sure of the link between hiking and sustainability. When it comes to the impacts, indeed hiking in the Moukhtara region is predominately associated with positive social and economic impacts, which support the finding of the previous research (Al Sawi, 2018; Bellos, 2019, 2021; Bellos & Knio, 2021; Bosch et al., 2017; Gomez-Martin, 2019; Khalil, 2009, 2015; Knio & Bellos, 2018; Pozzato, 2017–2018; Saule, 2004).

Our analysis points out that valorizing natural localities, such as Moukhtara, actually benefit the area's tourism activity and, eventually, the entire country to ultimately enhance its economic activity (Khalil, 2015). So, in this case, as Al Sawi (2018) denotes, both natural and cultural heritages need to be valorized to benefit tourism activity, while Khalil (2009) shows, in some cases, that the Moukhtara municipality demonstrated efforts and in others not. Indeed, if properly managed the development of hiking could contribute to the preservation of the Moukhtara ecosystem. Referring to Gomez-Martin's (2019) arguments about Spain, the current researchers see a similarity with Khalil (2009) in terms of hiking as being central to sustainable tourism development. In cases, scenic nature and the history, culture, and ambience are central drivers of visitor experience (see Khalil, 2015).

In terms of cleanliness, what Bichler and Peters (2021) believes to be a clean hiking path is, in fact, well maintained in terms of natural preservation and the sentiment of the community. This means that the locals wish to instruct the hikers and visitors to keep the surrounding environment clean and avoid pollution, similar to the Lebanese Mountain trail path guidelines (cf. Babí, Yuba, & Soler, 2019). The historical and cultural identity of Moukhtara is dictated by the archaeological vestiges remaining in the area, which provide this locality with its uniqueness as a touristic destination (Al Sawi, 2018). So, the established trail discussed in this study may be one of those ancient vestiges (see Khalil, 2009). The local municipality works to enhance tourism activities to valorize the landscape to sustain the local ecosystem, creating alternative revenue streams for the local community, and supporting the sustainable recovery post-pandemic (UNWTO, 2020a, b). The authors wish to show how responsible hikers can preserve their localities' ecosystems. As Khalil (2015) and Gomez-Martin (2019) argue, these areas' historical, cultural, and social ambiences depend on the activities these localities wish to propose by attracting the right audiences and addressing the proper messages.

Conclusions and recommendations

In this section, we deliver brief conclusions and main recommendations. Before proceeding, we would like to emphasize how the conclusions drawn in this case study were constrained by the challenges associated with data collection due to the influence of the COVID-19 pandemic.

Conclusions

Several major conclusions could be drawn from the current analysis. Hiking is leaving negative impacts on the environment and the wildlife communities. Due to the aggressive stomping on the soil, there is a gradual degradation of the flora. The soil is at risk because of the stress and the pressure of hiking activity. Similarly, hikers can pollute the environment, which means that there will be a reduced growth of greenery. These impacts affect the flora, and wildlife is also at risk. When these settings are not being adequately cared for, animals tend to become discouraged; they resort to isolating themselves as a means of coping with the ongoing damage being enforced onto their habitat. Through the brief reflection on the perception of impacts associated with hiking development, we conclude how, if properly managed, hiking could be seen as a central activity within sustainable nature-based tourism discourse.

When it comes to the environmental impacts associated with hiking in the Moukhtara region, both positive and adverse, have been perceived. Residents perceive the positive impacts on the locality because it draws attention toward district areas. They also believe that hikers have a positive impact on the economy because they make purchases in local businesses – the money spent in these stores generates revenue to improve the economy. Promoting trail races in Moukhtara, as stated by individuals, acts as a small step toward sustainability within the region (Carvalhinho, Rosa, & Gomes, 2015). Thus, hiking can be considered a sustainable form of tourism as it has been proven that hiking allows for interactions with nature, and it helps individuals appreciate the natural world.

Finally, and in the conclusion of this research paper, the study reveals that hiking influences sustainable tourism and could be beneficial to the locality and the environment if the authorities supervise it. Such findings support the conclusions drawn by Al-Sawi (2018) and Bellos (2019, 2020), who respectively discussed how Tyre and Beirut also were deemed too to have somehow mismanaged their cultural and their natural resources. We also suggest considering tourism education as a potential solution for increasing awareness in terms of tourist site conservation and municipal involvement. Since most people are still unaware of the effects of hiking on the environment, and those sceptical about the effectiveness of the authorities to enhance the valorization of their cultural and natural resources, properly valorizing sites will benefit communities (Khalil, 2015) Figure 28.5 proposes a way in which localities can do this, and spared awareness to all stakeholders of that given locality.

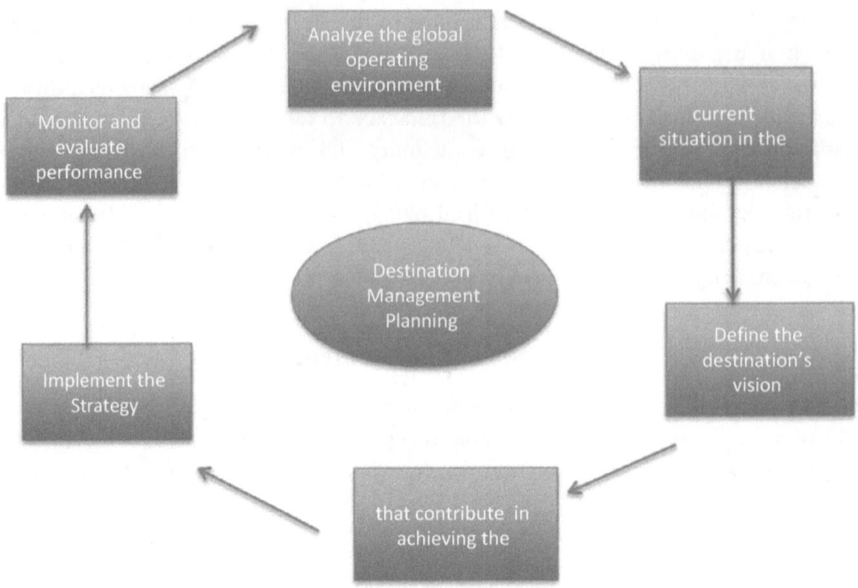

Figure 28.5 Successful destination management plan chart.

Recommendations

In summary, the substantial changes are always community-driven. Firstly, Moukhtara's municipality may combat growing issues that affect the environment by taking further action to eliminate soil compaction, greenery degradation, and violating wildlife animals. Implementing effective strategies will certainly foster sustainable tourism development in Moukhtara, as Figure 28.5 suggests. To reduce the impact of hiking, cautionary measures must be considered. These strategies include, among others, limiting the number of individuals who wish to access the site through hiking. This will enable reductions in harsh soil trampling since large groups of visitors will not be permitted onsite.

Secondly, proposing entry fees into the site to further limit the number of individuals who wish to perform hiking activities. Organizers within the region can reinforce these measures, which will allow the message to spread further and quicker. Following, placing recycling bins around the site with informative signs indicating how one must dispose of waste will educate people while, at the same time, significantly reducing the pollution within the protected site; At the same time the entry fees could be used for investing in maintaining the site, and thus contribute to sustainable tourism.

Thirdly, pre-established trails are tracks within the wilderness that are set up in advance to create a passage for hikers to walk on (Carvalhinho, Rosa, & Gomes, 2015). These trails allow for a great experience of any natural setting, and they also ensure that flora and fauna and the greenery are protected at all times. Pamphlets and information brochures must be distributed amongst hikers before entering the area and organizing a qualified tour guide to provide greater detail about hiking in accordance with these trails (see: Carvalhinho, Rosa, & Gomes, 2015; Khalil, 2015).

Fourthly, the municipality must consider hiring rangers to monitor hikers' activity and behaviour within Moukhtara. Rangers can observe whether hikers are adhering to rules, for example, keeping their fires small and ensuring individuals are not polluting lakes and streams.

There is also a need to inform hikers to be mindful of the animals and respect their space by not interacting with them.

Finally, overall, these recommendations successfully eliminate arising issues in the Moukhtara natural area and could allow for an improvement in soil condition, pollution prevention and wildlife living standards. This, together with increasing the awareness of the people visiting this site, or in encouraging them in participating in its preservation, and thus contributing to the participative approach some sustainable tourism strategies propose, whereby the people themselves become encouraged to preserve the sites.

References

Al-Sawi, M. (2018). "The Effective Valorization of Resources for the Economic Development of Tyre". *MBA Project presented at the Lebanese International University, Spring 2018*, Saida, Lebanon.

Altinay, L., & Paraskevas, A. (2008). *Planning Research in Hospitality and Tourism*. Elsevier Ltd.

Babí, J., Yuba, E. I., & Soler, S. (2019). "Trail traces in protected mountain areas and their effects on sustainable development",*eco.mont - Journal on Protected Mountain AreasResearch*, *11*(2), 18–26.

Bellos, G. S. G. (2019). "Estimating the Impact of the Effective Valorization of Cultural and Natural Resources for the Economic Development of Beirut". *MBA Thesis Submitted to the Lebanese International University on Sept. 2nd 2019*, Lebanon.

Bellos, G. S. G. (2020). *La valorisation efficace des ressources culturelles et naturelles du Liban pour son développement économique: Un aperçu*. Retrieved from: https://libnanews.com/la-valorisation-efficace-des-ressources-culturelles-et-naturelles-du-liban-pour-son-developpement-economique-un-apercu/

Bellos, G. S. G. (2021). "Estimating the impact of the SustainableDevelopment Goals on the Lebanese Society", *Project Thesis submitted to the Lebanese International University, School of Education, Beirut Campus*, June, 2021, Beirut, Lebanon.

Bellos, G. S. G., & Knio, M. S.(2021). "Estimating the impact of the effective valorization of cultural and natural resources for the economic development of Beirut", *Journal of Transnational Management*, 26(2), 90–117. https://doi.org/10.1080/15475778.2021.1915566

Bichler, B. F., & Peters, M. (2021), "Soft adventure motivation: An exploratory study of *hiking* tourism", *Tourism Review*, 76(2). https://doi.org/10.1108/TR-10-2019-0403

Bosch, P., Jongeneel, S., Rovers, V., Neumann, H.M., Airaksinen, M., & Huovila, A. (2017). "CITYkeys indicators for smart city projects and smart cities". *s.l.:European Commission*.

Bou-Melhem, J. (2021). "Implementing the culture of tourism in the secondary stage curriculum". *Project Thesis submitted to the Lebanese International University, School of Education, Beirut Campus, June, 2021*, Beirut, Lebanon.

Boyce, C., & Neale, P. (2006). *Conducting In-Depth Interviews: A Guide for Designing and Conducting In-Depth Interviews for Evaluation Input*. Watertown: Pathfinder International.

Brynman, A., 2012. *Social Research Methods*(2nd Ed.),New York: Oxford University Press.

Carvalhinho, L., Rosa, P., & Gomes, F. (2015). "Hiking trails evaluation in the natural park of Serras de Aire E Candeeiros, Portugal", *European Journal of Tourism, Hospitality and Recreation*, 6(2), 139–156.

Cauvin, J., (1963). "The Neolithic Moukhtara (South Lebanon)",*L'Anthropologie*, 67, 489–511.

Cauvin, J. et Cauvin, M.-C., (1968). *Des ateliers campigniens au Liban, in Mélanges R. Vaufrey, La préhistoire, problèmes et tendances*. Paris: Éditions du CNRS, 103–116.

Creswell, J.-W. (2012). *Educational Research: Planning, Conducting, and Evaluating Quantitative and Qualitative Research*(4th Ed.), Boston, MA: Pearson.

Cronbach, L. J. (1951). "Coefficient alpha and the internal structure of tests", *Psychometrika*, 16, 297–334.

Dayananda, K. (2016). "Tourism and its Impact on Indian Economy", *IOSR Journal of Humanities And Social Science* (IOSR-JHSS), 21(6), 24–28.

Denzin, N. K., & Lincoln, Y. S.(2005). *The SAGE Handbook of Qualitative Research*. California: Sage Publications.

Edson, M.-C., Henning, P.-B.& Sankaran, S. (2017). *A Guide to System Research* (2nd Ed.), New York, NY: Springer.

Fattouh, M. S. (2017). *Moukhtara (Arabic: المختارة, "The Chosen"), a small town in the Chouf...*, Retrieved from: https://www.lebanoninapicture.com/pictures/moukhtara-arabic-D8%A7%D9%84%D9%85%D8%AE%D8%AA%D8%A7%D8%B1%D8%A9-the-chosen-a-small-town-in-th

Gómez-Martín, M. B. (2019). "Hiking tourism in Spain: Origins, issues and transformations", *Sustainability*, 11(13), 3619. Repositori Institucional. ISSN: 2071-1050.

Gössling, S., Scott, D., & Hall, C.-M. (2020). "Pandemics, tourism and global change: Arapid assessment of COVID-19", *Journal of Sustainable Tourism*, 28, 1–20.

Hall, C.-M. (1998). "Historical antecedents on sustainable development: New labels on old bottles?" In C.-M. Hall and A.-A. Lew (eds.), *Sustainable Tourism: A Geographical Perspective*. London: Longman.

Hall, C.-M. (2019). "Constructing sustainable tourism development: The 2030 agenda and the managerial ecology of sustainable tourism", *Journal of Sustainable Tourism*, 27(7), 1044–1060. https://doi.org/10.1080/09669582.2018.1560456

Higgins-Desbiolles, F. (2018). "Sustainable tourism: Sustaining tourism or something more?" *Tourism Management Perspectives*, 25, 157–160.

Kamran, M. (2020). "Global sustainability overview and role of policy instruments for sustainable tourism management in Pakistan", *PhD Thesis submitted to the Faculty of Geography and Tourism, Rovira I Virgili University*, Tarragona, Spain.

Kenny, N., & Brown, R. (2018). "Effects of recreational use impacts on hiking experiences in natural areas", *Landscape and Urban Planning*, 64(1–2), 77–87.

Khalil, W. (2009). "Prospection dans le Liban méridional: étude du développement de la zone montagneuse de l'époque hellénistique à la fin de l'époque médiévale, 323 av. J.-C. -1516 ap. J.-C.", Ph.D. Thesis, Université Paris 1 Panthéon-Sorbonne, France.

Khalil, W. (2015) *Qab Elyes: Étude archéologique et de plan de valorisation touristique*, Rapport présenté à la municipalité de Qab Elyes.

Kiper, T. (2013). "Role of ecotourism in sustainable development", In M. Ozyavuz (ed.), *Advances in Landscape Architecture*. http://dx.doi.org/10.5772/55749

Knio, M. S., & Bellos, G. S. G., (2018). "The effective valorization of resources for the economic development of tyre", *Account and Financial Management Journal*, 3(11), 1794–1800. https://doi.org/10.31142/afmj/v3i11.04

Kuhlman, T., & Farrington, J. (2010). "What is sustainability?", *Sustainability*, 2(11), 3436–3448. https://doi.org/10.3390/su2113436

Lucas-Alfieri, D. (2015). *Marketing the 21st Century Library*. Amsterdam, the Netherlands: Chandos Publishing.

Maria Raya, J., Martínez-Garcia, E., & Celma, D.(2018). « Economic and social yield of investing in hiking tourism: The case of Berguedà, Spain". *Journal of Travel & Tourism Marketing*, 35(2), 148–161.

McLeod, S. (2018). *Questionnaire: Definition, Examples, Design and Types*. Retrieved from:https://www.simplypsychology.org/questionnaires.html

McNamara, K. E., & Prideaux, B. (2011). "Planning Nature-based Hiking Trails in a Tropical Rainforest Setting", *Asia Pacific Journal of Tourism Research*, 16(3), 289–305.

Pechlaner, H., & Dal Bò, G. (2012). "What makes tourism an attractive industry for new minority entrepreneurs: Results from an exploratory qualitative study?" In: *Tourism Review: The Official Journal of the AIEST*, 67(1), 11–22. ISSN 1660-5373.

Pickering, C. M., Rossi, S., & Barros, A. (2011). "Assessing the impacts of mountain biking and hiking on subalpine grassland in Australia using an experimental protocol", *Journal of Environmental Management*, 92 (12), 3049–3057.

Pozzato, A.(2017–2018). Smart cities: An innovative solution for the management of global risks. *MA thesis submitted for the Master's Degree programme In Innovation and Marketing to the Ca'Foscari University of Venice*, Italy.

Rakic, T., & Chambers, D. (2012). *An Introduction to Visual Research Methods in Tourism*. Abingdon: Routledge.

Saule, M. (2004). "*Rural Tourism as a tool for Local Development, Diagnosis of the French area of Haute-Corrèze". Master's thesis submitted to the Bournemouth University*, UK.

Saunders, M. N. K., Lewis, A., & Thornhill, P. (2019). *Research Methods for Business Students* (8th Ed.), New York: Pearson.

Sekaran, U., & Bougie, R. (2016). *Research Methods for Business: A Skill-building Approach* (7th Ed.), New York, NY: John Wiley & Sons.

Sharpley, R. (2020). "Tourism, sustainable development and the theoretical divide: 20 years on", *Journal of Sustainable Tourism*, 28(11), 1932–1946. https://doi.org/10.1080/09669582.2020.1779732

Sileyew, K. J. (2019). *Research Design and Methodology*. Retrieved from: https://www.intechopen.com/online-first/research-design-and-methodology

Törn, A. (2007). "Sustainability of nature-based tourism", *Acta Universitatis Ouluensis, Series A*, 298.
Törn, A., Tolvanen, A., Norokorpi, Y., Tervo, R., & Siikamaki, P. (2009). "Comparing the impacts of hiking, skiing and horse riding on trail and vegetation in different types of forest", *Journal of Environmental Management*, 90(3), 1427–1434. https://doi.org/10.1016/j.jenvman.2008.08.014
UNWTO (2020a). *Impact assessment of the covid-19 outbreak on international tourism.* Retrieved from: https://www.unwto.org/impact-assessment-of-the-covid-19-outbreak-on-international-tourism
UNWTO (2020b). *International tourism growth continues to outpace the global economy*, Retrieved from: https://www.unwto.org/international-tourism-growth-continues-to-outpace-the-economy
Weaver, D. (2006). *Sustainable Tourism. Theory and Practice.* Butterworth-Heinemann, Amsterdam: Elsevier.

29

A TOURISM ACCOMMODATION PERSPECTIVE IN A POST-PANDEMIC NATURE-BASED TOURISM DEVELOPMENT CONTEXT

Günay Erol, Ayşe Deliormanli and Ebru Düşmezkalender

Introduction

Tourism and the environment are interdependent phenomena and cannot be considered separately. When these two notions are evaluated independently, natural attractions, which are fundamental components of tourism, suffer (Sunnetcıoglu and Yeşildağ, 2021). For example, levels of tourist arrivals which are above destinations' capacity can harm nature and impact the lives of local people (Garau-Vadell et al., 2018). Rapid development of the tourism industry causes environmental problems, including pollution, noise, biodiversity loss, and the destruction of valuable living things such as coral reefs (Das and Chatterjee, 2015). Environmental and natural resources' rapid consumption may cause the exhaustion of such locations. Thus, it is necessary to diversify tourism and manage it sustainably without doing any harm to the environment (Karadağ and Marangoz, 2017). Travelling to other places increases the carbon footprint, causing global warming and increasing greenhouse gases (Akoglan Kozak, 2012). And climate change due to global warming made people more sensitive to the environment (Steg and Vlek, 2009). In addition to the adverse effects of tourism, its positive effects are also observed. Undoubtedly, the most important of these is the foreign exchange effect on the economy. However, although its negative effects on the environment are taken into account, it also has significant effects, especially on a regional and national scale (Tovar and Lockwood, 2008).

Nature-based tourism can be defined as recreational activities where interaction with plants and animals is incidental or travel to a natural destination (Wolf et al., 2019). Such activities could benefit nature conservation (Yılmaz and Akman, 2020) if recreational behaviours are conducted in an environmentally responsible manner (Formica and Uysal, 2001). This type of tourism, covering travel naturally and especially to relatively unspoiled natural areas, is interconnected with ecotourism (Winter et al., 2020). The number of people participating in recreation and tourism activities in parks, wilderness, and protected areas worldwide is on a continuously rising trend (Balmford et al., 2015), and this growth has been additionally accentuated with

DOI: 10.4324/9781003230748-34

the current pandemic. Although nature-based tourism includes activities that are unlikely to damage the environment, it should be remembered that it impacts animal habitats. For example, diving during last-chance tourism activities to see endangered sea corals can also prove a nuisance to other species or the little reptiles that live in the wild (Geffroy et al., 2015).

Due to the need to protect the environment, mass tourism has begun to give way to alternative tourism forms. Nature-based tourism, ecotourism, wildlife tourism, and rural tourism are conspicuous examples (Batman and Demirel, 2015; Mandic, 2020). The growth in interest in nature triggered by the pandemic has changed people's travel expectations, and the demand for such leisure activities has increased as the offer has expanded and diversified globally. With the relaxation of restrictions, the quest for cleanliness, social distance, and a location free of crowds has begun to push individuals to seek a more secluded experience. Nature-based tourism generally fulfils these requirements. This conceptual research aims to explain the potential future developments of nature-based accommodation options in the aftermath of the pandemic.

Nature-based tourism in the scope of alternative tourism

Tourism is one of the world's largest industries regarding employment and expenditure. Every year, a significant portion of the world's population takes part in national and international travels. Thus, tourism supports airlines, highway, rail systems, accommodation enterprises, agricultural production, and water facilities. On a more personal level, it also offers individuals with opportunities to escape from the tediousness of daily life and experience the new and unknown (Kuenzi and McNeely, 2008). This change and development pave the way for alternative tourism types. Most of the so-called new alternative tourism types aims to improve the local environment, social and economic conditions through ecotourism, heritage tourism, religious tourism, social tourism and community-based tourism (Conway and Timms, 2010). Alternative tourism types based on nature strive to ensure the protection of nature simultaneously. Furthermore, it is a powerful instrument in the transmission of intangible cultural heritage, such as traditions and customs, from generation to generation and local 'people's involvement in decision-making (Agaliotou, 2015). Tourists that engage in alternative tourism are often environmentally and culturally conscious. They are, in general, well-educated and respectful of local customs and traditions; they interact with local communities, and are eager to learn about their culture. Nature-based tourism, which is among the alternative tourism types, refers to experiencing the wildlife in the natural environment, observing the local values in the ecosystem and visiting areas open to tourism (Ankre et al., 2016). Nature-based tourism benefits from the natural and cultural attractions of a destination. Visits to these attraction centres sometimes take the form of trips to watch wildlife and learn about cultural values from local communities (Divinagracia et al., 2012).

One of the fastest-growing tourism industry segments is nature-based travel (Christ et al., 2003). Nature-based travel has been dealt with extensively for 20 years (Luo and Deng, 2008). Although there is no generally agreed-upon definition of nature-based tourism, contemporary literature associates it with recreation and adventure (Laarman and Durst, 1987), perceptions of pristine nature (Valentine, 1992), protected natural areas and destinations. The first definition of nature-based tourism was proposed by Valentine (1992) as "*direct use of relatively undisturbed natural resources*". Laarman and Durst (1987) used the term "nature travel" (or sometimes "nature-oriented tourism") to refer to a style of tourism that *combines education, recreation, and often an adventure*. Lucas (1985) defined nature tourism as "*tourism based on exploiting natural areas and observing nature*" and claimed that this type of tourism has a low environmental impact, is labour-intensive, and contributes socially and economically.

Elements of nature and protection of nature (Hall and Boyd, 2004), activities directly related to the natural environment (Mehmetoglu, 2007), spending time in natural areas outside the home environment (Fredman et al., 2012), and living in areas rich in genuine opportunities are activities directly related to tourism and natural environments. Accordingly, nature-based tourism supply can be considered a specific service space where the physical environment plays a vital role in 'visitors' exchanges (Arnould et al., 1998; Buckley et al., 2015; Hultman et al., 2015). In a nature-based service environment, activities such as scenery, relaxation, adventure, escape from civilization are sold as tourism commodities (Fredman et al., 2012).

Nature-based tourism involves recreational activities in exceptionally preserved areas that are carried out as sustainable and responsible nature-based trips (Batman and Demirel, 2015; Valentine, 1992). Over the course of the past decade or so, traditional activities such as relaxation and vacationing with family and friends have been replaced by nature-based, technology-focused, more active and adventurous economic activities (Batman and Demirel, 2015).

Nature-based accommodation

Tourism plays an essential role in the world economy with a market share that has recently grown and diversified. However, this growth brings with it some unsustainable practices. Along with tourism's rapid growth, variables such as global warming, climate change and natural resources' endangerment, tourism concepts, nature and sustainability have all come on to the agenda as essential issues. So against this growing awareness tourism industry is evolving into a tourism approach that values nature and tries to understand consumers preferring eco-friendly products (Trang et al., 2019). Nature-based solutions (NBS) are observed worldwide in many areas with rapidly increasing interest (Mandic, 2019). It has a positive effect on people's behaviour and quality of life by being implemented in many places from houses to workplaces, from public areas to buildings. Green and eco-friendly environments are no longer sought just in permanent residences. Tourists search for these places when choosing their vacations, destinations, and hotels. Therefore, a green and eco-friendly environment positively impact guest satisfaction (Han et al., 2020). Eco-friendly endeavours provide a double benefit for both tourists and businesses. It improves tourist satisfaction with their vacation experience while also providing a marketing opportunity for service suppliers.

With nature-based tourism, tourism, instead of creating an economic contribution by consuming natural resources, supports the development of alternative revenue streams that could be used to support the development of communities and nature conservation (Stradas, 2006). Tourists who enjoy nature, want to see and experience natural landscapes and are ready to pay a high price account for a significant market share. In addition to their experiences, tourists also experience a distinct satisfaction by contributing to nature protection (De, 2013). Nature-based tourism activities were considered outdoor activities and a low-cost type. Today, when it comes to nature-based tourism, a market with a significant income share and compromising different consumer behaviours comes to mind (Fredman et al., 2012). Therefore, tourism enterprises need to improve their services by analysing the tourist behaviours in nature-based tourism (Mandic and Vukovic, 2021; Tangeland, 2011). From this perspective, nature-based tourism includes nature-based outdoor activities and other critical components of tourism offer, including accommodation. Although accommodation is not a motivation to visit a particular location, it is a significant part of the tourist experience, whether they choose to sleep in a hotel, a campsite, or somewhere else.

Tourism travels inspired by nature usually involve responsible travellers who share a sense of responsibility towards the local community and ecosystem. They are based on natural attractions such as birdwatching, photography, stargazing, camping, and hiking (Monz et al., 2013). On these trips, tourists are interested in various natural and cultural resources they wish to experience the authentic ones. Nature-based tourism also uses a sustainable approach for wildlife and nature conservation (Texas Park and Wildlife Foundation, 2022).

Nature-based tourism is the most common use of protected areas. They areas may have a range of beneficial and detrimental effects. Managers of protected areas must develop long- and short-term objectives and policies to define conservation and recreation activities and priorities in order to manage this process responsibly (Mandic, 2021). Many operators working in protected areas have based their business models on the principles of sustainable development, and they have actively contributed to safeguarding biodiversity and the local economy.

In nature-based tourism, camping is one of the most enjoyed outdoor activities, evolved from tent camping to caravans, recreational vehicles and luxury amenities (Brooker and Joppe, 2013). Nature-based accommodation enables tourists to be intertwined with nature and natural life interactively and enriches their experience with nature. In addition, nature-based accommodation enterprises provide a natural, realistic and authentic experience to ensure tourists interact with the people living in that nature and environment. Thus, tourists' experience with the local people and other businesses in their accommodation neighbourhood contributes to nature-based experiences and improves the destination's brand image. For example, those who stay in treehouses in the forest; or who engage with nature with eco-activities. These activities may include walking in the forest, bird sounds, meals prepared with fresh food, night walks, and nature views created by waterfalls and streams. This contributes to ecological learning and ecologic experience (Brochado, 2019).

Cheng et al. (2016) stated that variables such as nature, local culture, authenticity for tourists seeking relaxation, spiritual tranquillity and different satisfactions have begun to be used in hotel services. Even some international hotels have incorporated these variables into their service processes. Today, hotels are built as interbedded with nature, and materials and artworks are preferred in hotels (such as paintings, handicrafts, keys, trinkets, doors, upholstery) that emphasize nature. Thus, guests are more satisfied by having a nature-based experience. Hotels are built into the unspoiled wilderness; window views make the rooms feel like in nature. Moreover, the room's interior design and furniture aim to provide guests with a nature-based experience. This might be an authentic house or a modern-looking building constructed in a tranquil environment with nature-based products being incorporated in its interior design. Because according to Wang, Tsai and Tang's (2018) research, nature-based service environments in hospitality businesses attract more customers' attention than built-based servicescapes, empower marketing image, and further increase customers' behavioural intentions. As supported in the research by Han et al. (2020), improving visitors' mental health by incorporating natural elements into the hotel's design improves their satisfaction and encourages them to return.

As tourists have different behaviours and desires, their accommodation expectations change accordingly. It is well known that there are significant changes in the health and wellbeing of people living in interaction with nature (for instance, physical health, emotional health, behavioural issues, social interaction, cognitive function, quality of life, spiritual wellbeing) (Barret et al., 2019). Sometimes, for the guest staying in a nature-based accommodation facility, the hotel's facilities may be more important than the hotel's harmony with nature. For instance, even if thermal hotels are described as a facility with natural resources, the primary preference for guests could be a spa and bath-related activities. In this case, guests can highlight their activity needs rather than nature-based opportunities. In this case, the nature-based "nature" of the hotel may be an accidental and secondary situation. Therefore, even if the hotels are nature-based, guests' expectations can change their perceptions and demands (Chang, 2016).

As Valentine (1992) stated, the relationship between nature and experience changes according to the role of nature. These changes offer three different approaches: (1) experiences (activities) entirely dependent on nature; (2) experiences empowered and increased by nature; and (3) nature or natural setting is incidental in experiences.

For example, the person who wants to stay in a treehouse needs a tree. In this case, nature can be an integral part of the experience and accommodation wholly dependent on nature. Second, as Valentine (1992) states, camping is generally an experience enhanced by nature. In other words, the individual can camp in the garden of their house or wherever they want. Still, camping in nature enhances this accommodation experience even more. Third, as in the thermal example above, if the person's primary purpose is a bath or spa, it is coincidental that the accommodation business is built with thermal resources or nature-based.

The fact that accommodation establishments are only nature-based or built in harmony with nature does not make them directly nature-based. It depends on how a nature-based accommodation business protects and sustains the ecosystem with its practices. Nature-based tourism activities are handled as both commercial and non-commercial activities (Acuner and Tayfun, 2016). Commercial ones are based on natural resources for a fee and act in the natural environment (Tangeland and Aas, 2011). Non-commercials are, primarily doing in nature, dependent on or influenced by the natural environment and free of charge (Tangeland et al., 2013). A hotel built in natural settings should be responsible for the hotel's land and be able to participate in projects that protect and support nature. For example, one accommodation business established on the island of Mustique in the Caribbean has contributed to the planting of more than 7,000 new coral reefs by supporting the coral reef restoration program. And this practice increased corals and fish diversity and water quality. A notable portion of the income from the snorkelling service offered to the guests was used to support this project. This caught the guests' attention and increased awareness (Hohenlohe, 2021). In this case, the materials to be used in hotel construction, its harmony for the green building system and the other green areas, and the construction site's ecosystem and biodiversity should all be taken into consideration (Legrand, 2021).

Although the relationship between tourism and nature is significant, it encompasses many variables, from local people to local nature. The physical impact of an overabundance of visitors can harm the local community and wildlife. In addition, renewable energy and ecologically friendly materials obtained from sustainable sources, and variables such as reduced reliance on fossil fuels, are also relevant in terms of understanding green tourism. This is not just concerned with the construction and design of hotels. At the same time, it can be made possible by raising awareness, and informing and training the visitors. Furthermore, a tourism understanding that covers those practices that protect nature by local and national governments is also essential. Thus, reducing the harmful effects of leisure industry practices is critical for the sustainability of natural, rural, and agricultural regions and urban life (Padma et al., 2019).

In addition to the hotels' design, the construction must be done not to harm nature and biodiversity. Further, it is crucial to consider sustainability principles, aid the local community, and prevent resource depletion and biodiversity conservation while establishing a new hotel. Hotel location, its construction structure, the size of the construction area and the resources utilized are also crucial in this case. In addition, using sustainable food and raw materials in hotel restaurants is a sign of a hotel's environmental consciousness. For example, excluding endangered species of fish or agricultural items that harm local agriculture from a hotel's menu demonstrates a hotel's environmental awareness. Therefore, a nature-based hotel should not be limited to the existence of a hotel built into nature. On the contrary, it should be considered *in toto*, including the processes from the hotel construction to the management techniques, local people, suppliers, customers, public and local organizations (IUCN, 2008). Some of the most popular nature-based accommodation types are discussed below.

Camping and caravanning

The most popular form of nature-based accommodation is camping. Global trends in camping indicate that this recreational activity is currently emerging in Asia. It is defined as follows: "Camping is a kind of recreational activity connected with the leisure time, relaxation, and the entertainment of holiday activities" (Wong et al., 2021). Camping, considered as a recreational activity, provides dynamic and flexible content and accommodation intertwined with nature and away from the permanent living area (Bilim and Özer, 2021). According to another description, camping is a popular outdoor activity for refreshing and revitalizing the visitor, and also socializing with family, friends, and other campers. Although it emerged as a low-cost temporary accommodation option in a remote area, it has now evolved into a niche market with multiple segments. For the traditionalist, camping is synonymous with tenting and outdoor activities. However, for increasing numbers of tourists, it incorporates at least one night in a recreational vehicle, caravan, cabin, or home comfort shelter.

Consequently, there is now a wide range of camping activities and accommodations, from tents to luxury cabins, which can accommodate all types of visitors (Brooker and Joppe, 2014). It is far less expensive than staying in a hotel, and guests can bring their food and cook it. Fresh air, stargazing, and appreciating nature are all regarded as essential aspects of camping. Compared to the hotel's impersonal ambience, individuals in the campsites form close relationships and may share equipment. As a result, they have a higher level of social interaction (Brooker and Joppe, 2013). Camping has a unique significance for each person. It may be carried out for a variety of reasons, including low-cost accommodation, a type of lifestyle, an environmentalist worldview, forming relationships with individuals who share that worldview, and exploring new locations.

Caravan tourism is considered a subset of camping tourism as a recreational vehicle or mobile home. It is a type of recreation activity that occurs in natural settings and, like camping, uses environmental assets. Therefore, caravan and camping trips require an awareness of the environment. Caravanning is comparable to camping, but with the added benefit of mobility and access to certain domestic utilities (toilet, water, kitchen, bed). The tendency to see caravanning as a subset of camping is also a common one (Bilim and Özer, 2021).

Camping and caravan tourism are seen as excellent ways to get away from the crowds. The desire to 'escape' and enjoy a secluded vacation has grown increasingly popular, particularly during the COVID-19 epidemic. With the sharp rise in the popularity of camping and caravanning to enjoy isolated holidays, campgrounds and caravan parks have become increasingly overcrowded. Even though camping and caravanning are often seen as environmentally friendly forms of tourism, the current overcrowding condition is in direct contrast to these ideals.

Glamping

The basic camping offer has recently been advanced by the development of luxury facilities such as those associated with glamping. Glamping is a new form of accommodation which combines comfortable accommodation with outdoor activities (Brochado and Pereira, 2017). An increasing interest in nature-based accommodation has started to attract tourists from communities. This situation has led to the growth of the glamping market, addressing the preferences of the tourist group seeking luxury and comfort. Glamping, a combination of "glamorous" and "camping", is an outdoor activity presenting a comfortable holiday experience to tourists. The concept has recently been seen as a prevalent nature-based accommodation option and outdoor activity. Glamping, which has grown in popularity worldwide, is seen as an experience that allows travellers to enjoy nature without compromising their luxury. There is no need to pitch a tent, unpack a sleeping bag, or build a fire at these facilities (Brochado and Brochado, 2019). Combining the best aspects of the hotel and camping sectors and providing outdoor accommodations with the most significant degree of amenities, glamping supports the new era of tourist growth (Milohnic et al., 2019).

Glamping is considered more convenient than accommodation options such as hotels that consume more energy and are less environmentally friendly. Like all other forms of nature-based accommodations, glamping accommodation appeals, in particular, to environmental visitors. Even though ecotourists are supposed to have low expectations for nature-based accommodations, comfort remains a priority, in contrast to other forms of accommodation. On the other hand, glamping tourists are conscious that being in nature implies that all resources cannot be obtained immediately and that a vacation should be planned with restricted resources.

Eco-lodge

A growing number of tourists worldwide have become more aware of their impact on the environment and are therefore seeking environmentally friendly accommodation options. As a result of this circumstance, "eco-lodge" accommodation facilities have become a popular choice for travellers seeking eco-friendly and wildlife-friendly accommodation options (Ghaith et al., 2019). Parallel to the growth in the number of environmentally conscious travellers, the accommodation industry has seen the emergence of ecological hotels and accommodation options. One of these, eco-lodges, has much interest, particularly among nature-based tourists. The term "eco-lodge" was coined during the International Ecolodge Forum and Field Seminar in 1994. For eco-lodge designers and developers, the proceedings were published as a reference book (Gawad, 2020). Eco-lodges are low-impact buildings developed in line with the ecotourism concept and sustainability principles while at the same time delivering economic benefits to host communities (Sumanapala et al., 2015). Costa Rica, a popular tourist destination, was the site of the first eco-lodge. While it was originally developed in locations with wildlife, it has now extended to rural and urban areas (Holiable, 2022). A set of norms and frameworks has been established to design eco-lodges. However, there is still uncertainty regarding which facilities would be classified as eco-lodges (Sumanapala et al., 2015). Nevertheless, the following characteristics of an eco-lodge accommodation facility should be considered (Holiable, 2022):

- It must protect both the natural and the cultural environments.
- During installation, it should have a minimal effect on the environment.
- It should strive to work with local communities.
- It should use sustainable water extraction technologies and save water.
- It must ensure that the waste is properly handled.
- Waste management must be implemented.
- Through training programs, it should contribute to the development of locals.
- It should pay attention to the landscaping and consider the local architecture in the arrangement.

To some extent, the eco-lodge's design, construction and infrastructure investments are unique compared to other traditional tourism projects. For example, it is usually established in remote and isolated tourist areas, it may be difficult to reach the supports provided by the state such as electricity, potable water and roads. This circumstance poses a significant challenge to eco-lodging owners and managers (Gawad, 2020). However, eco-lodges are spreading rapidly across the globe. Accommodation investors are abandoning conventional projects and investments, favouring new, nature-based solutions as a global trend. In this regard, eco-lodge designs are more profitable, less complex, and could be favoured because of their reduced environmental effect. In addition to these favourable impacts, eco-lodges have a negligible impact throughout the development phase (Ghaith et al., 2019).

The future of nature-based tourism

Most tourism research understandably prefers to take a long-term view of where the industry is headed and focuses on current trends and problems. Moving ahead from this point, it is the intention of this study to give an idea about nature-based tourism's development and the prediction of its future. Considering the pressures that may influence the supply of nature-based tourism (for example, climate change, and the clearing of the rainforests), the social and cultural changes of nature-based tourism destinations or regions are significant. It is seen as a potential source of transformation for people living the existential forms of nature-based experiences, authenticity (Wang, 1999), and the busy pace of the urban lifestyle. When we look at the literature, even though there is no unique net definition, it emphasizes nature-based tourism as activities that are regarded as being respectful to nature. For nature-friendly activities to occur, nature-based tourists must be made up of people who have environmental consciousness. To talk about the continuity, development, and future of something, the assets owned must be protected and guarded. Silverberg et al. (1996) characterized nature tourists as persons who desire to see natural events, be physically present in nature, learn about nature, and travel with others who share their interests. Similarly, Köroglu and Karaman (2014) define nature tourists as explorers who wish to experience the marvels of nature while also learning about the local culture's conservation efforts. In addition, the demographic characteristics of tourists participating in nature-based tourism activities differ from those participating in other tourism activities. The way of utilizing natural resources has a significant impact on shaping nature-based tourism's future. Results of people's perspectives are valuable factors and tourism activities. For this reason, the purpose and form of use of nature are essential in continuing its existence and protecting the habitats of other living things. Similarly, Köroglu and Karaman (2014) concluded that methods used to protect natural resources and nature from past to present are insufficient in terms of protection, and this goal could only be achieved with an integrated program involving the whole society. If society-based approaches in natural resource management are considered, it will significantly contribute to the development of nature-based tourism activities.

The concept of authenticity is evaluated by what people perceive as authentic rather than based on objective measures in the postmodern period (Cohen, 1988). It sometimes represents an individual's desire to find their real identity (Wang, 1999). A return to nature allows people to feel a strong sense of self and understand themselves better, becoming more popular to ensure transformation in lives. Moreover, a growing perception that natural heritage is part of our cultural heritage. It also helps people to protect and connect with nature by creating the perception that natural heritage is a part of cultural heritage. This may contribute to the formation of a sense of identity towards nature in the public (Frost et al., 2014). Balmford et al. (2015) stated that the popularity of natural areas estimates that there have been more than eight thousand million visits annually to global protected areas. Tourism and recreation parks evaluated within the visit to protected areas began to be considered a component of a range of cultural ecosystem services provided by protected areas (Guo et al., 2010; Tenerelli et al., 2016). It is widely accepted that nature-based tourism can promote the conservation of protected areas (Kasangaki et al., 2012). Moreover, tourism generally benefits the middle and upper classes or tour companies (generally international). With nature-based tourism, the benefit of local people living around potential natural tourism destinations will be increased. For example, locations of the houses recognized as tiny houses today and whose images are commonly reproduced in the bungalow style are located in more forested and mountainous living areas. The transformation of lifestyles to this type of accommodation remains popular today, and it is possible that this will continue into the future. Nature-based tourism close to such areas will allow residents to generate income from the sustainable use of forests and alternative jobs for farming and timber harvesting (Schreckenberg and Luttrell, 2009). By 2050, Asia's highly urbanized and technologically driven population will continue to seek experiences that will help them relax, relieve stress, and connect with the natural world (Frost et al., 2014).

Nature-oriented recreational activities have gained significant attention with the COVID-19 pandemic (Fredman and Margaryan, 2021). According to data from Norway, during the curfew at the peak of the pandemic, there was a 291 percent increase in outdoor activities (Venter et al., 2020). A similar trend was experienced in Swedish national parks. Such figures were not recorded even in the high season before the pandemic (Hansson, 2020), emphasizing the importance of nature-based activities. The local people have suffered just as much as the visitors due to the limitations of nature-based tourism (Spenceley, 2021). Research shows that people prioritized health, hygiene, and social distancing measures during the pandemic, and that they planned their travel and vacation preferences accordingly. Tour operators and destinations must address the rising demand for ecologically and socially responsible tourist experiences. It is vital to promote a shared understanding of sustainable tourism to overcome obstacles when support cannot be supplied (Spenceley, 2021).

Additionally, these studies indicate that sustainable tourism, social welfare, economic development, adventure, and travel to natural areas have all garnered increased interest (Bloom, 2020). Although the influence of COVID-19 is still being felt and was significant in its early days, the tourism sector relies on a number of factors to recover and return to its prior levels of growth (Bremner, 2020), with sustainability, climate action and digital transformation being the primary drivers (Smith, 2020a, b).

Individuals are searching for natural areas, preferring to escape to quieter locations rather than busy ones, and wishing to appreciate nature. As a result, travellers from all over the globe are concentrating their trips on nature-based destinations (UNWTO, 2020). Tour operators' attempts to develop nature-based activities reflect this development. These travel businesses conducted virtual tours throughout the pandemic period, allowing individuals to learn about future destinations. As soon as the limits on virtual tours are relaxed, tourists will be able to use the information they have gathered from these excursions to organize future trips to real-world natural attractions (Spenceley, 2021). These projects inspire people and catalyse developing and implementing nature-based tourist activities.

The 2020 coronavirus pandemic has helped reveal unprecedented changes in mobility, economic activity and the associated environmental footprint (Muhammad et al., 2020). Although the interest in nature-based tourism continues, the same interest, if it cannot be controlled and managed well, will lead to substantial habitat losses. On this issue, studies emphasize that global challenges threaten the natural environment (many habitats and wildlife species, climate change, biodiversity decline, unsustainable resource consumption, and other anthropogenic factors) (Blumstein et al., 2017; Øian et al., 2018; Wearing and Schweinsberg, 2018).

Conclusion

Tourism has grown rapidly in recent years and is expected to grow further in the near future. However, over the past couple of years the sector has been greatly impacted by COVID-19, which has had many adverse effects in economic, social and cultural terms. Fear and anxiety brought on by the pandemic has caused some changes in tourist decisions. The desire to live a sustainable life has increased the interest in a return to nature, increased levels of solo travel, escapes to non-crowded destinations, and an upsurge in interest in nature-based travels and accommodation. Nature options, such as ecological hotels, camping, caravans, and glamping, have become popular in this context and these will become more important sectors in the upcoming years. In fact, according to Türker's (2020) study, interest in nature tourism will increase within the scope of holidays away from the crowd.

At the same time increasing interest in nature-based tourism and accommodation options is considered important in terms of protecting resources and ensuring sustainability. Besides the importance of assuming risk by those responsible for any adverse effects, it should be borne in mind that the benefits to be provided are equally distributed. All stakeholders who have a stake in managing the risks and sharing the benefits are expected to be self-sacrificing in carrying out the process effectively and efficiently.

While nature-based tourism continues to expand and garner attention, the impact of destination pull factors on visitor motivation is equally critical. Accommodation, transportation, and other attractions, along with nature-based tourist sites, are all different forms of pull factors. In nature-based tourism, pull motivation is essential in influencing travellers' recommendations to others (Xu and Chan, 2016). In this situation, the destination's nature-based tourism attractions influence tourists' future behaviour and may also affect tourists' willingness to travel. There are a variety of reasons why nature-based tourists travel. As Brochado (2019) mentioned, guests staying in treehouses place a high value on nature-based resources such as nature, experience, wildlife, rainforest, experience, trip, and the hotel's service quality.

Furthermore, previous research revealed that guests who stay in nature-based hotels had a high propensity to recommend such experiences to other people. In nature-based accommodation, there are a variety of motivations and expectations that can be highlighted. While some visitors emphasized the aspects of relaxation, tranquillity, proximity, spontaneity, and natural life, others may have placed a high value on entertainment, activity, and service. In nature-based tourism, these motivations and expectations influence customer satisfaction and perceived service quality, and there is a strong connection between experience and post-purchase behaviour (Sotiriadis, 2017). The quality of service provided by businesses that offer nature-based tourism services may influence tourist behaviour both before and after the purchase. Thus, the destination's attractiveness, design, motivational factors, and service are all shown to be crucial in nature-based tourism. According to a study conducted by Meng et al. (2008), guests at nature-based resorts place a higher value on service and accommodations than they do on food and location. Overall, the size of nature-based accommodation facilities, their opportunities, and the service content and location all have a significant impact on the visitors' choice and satisfaction.

The link between nature and tourism raises specific issues. Climate change and its consequences on tourism are two of the major future environmental concerns. Dube and Nhamo (2020) stress that tourism and nature have a complicated connection, and they offer a discussion of how the two impact one another. It is shown that nature and tourism are both interconnected and determinative. For example, as a result of climate change, nature is harmed, diminishing the attraction of nature-based tourism locations and leading to declining visitor numbers and hotel occupancy rates. It is therefore necessary to introduce sustainable and nature-friendly policies. Promoting green construction for hotels and lodges in natural regions, for example, may help to mitigate the effects of climate change (Dube and Nhamo, 2020). Because of this, eco-friendly practices should be given precedence in the planning, design, and building of any nature-based tourist destination. This requirement concerns not only nature but also the long-term viability of human reliance on natural resources. Nature offers tangible resources for humans, but it also has both spiritual and psychological advantages. As a result, green buildings may also be found in cities and workplaces, positively impacting people's mental and behavioural health. Eco-friendly hotels are beneficial to people's mental wellbeing and positively influence their purchase behaviour, satisfaction, and re-visitation (Han et al., 2020). Thus, in this research nature-based tourism can be regarded as a kind of tourism that both contributes economically to tourist mobility and also enhances visitors' social and psychological wellbeing as well as their awareness of sustainability.

It is well known that previous pandemic outbreaks have had a detrimental impact on tourist mobility and the economy. The influence of COVID-19 has both good and bad consequences, particularly in areas that rely on nature-based tourism. Among African countries that depend on nature tourism, the loss of nature protection funds and the rise in wildlife poaching are seen as negative consequences (Hambira et al., 2022). There has also been a decline in the number of tourists and revenues generated by tourism, the loss of jobs and the closure of tourist attractions, and a rise in illegal activity and damage to tourism features in natural regions (Mandic, 2021). Apart from these negative social and economic reflections, the decrease in the demand for nature, the self-renewal of natural resources and the reduction of damage to natural tourism attractions are also considered favorable outcomes (Bhuiyan et al., 2020). Also, COVID-19 may have some effects on nature tourism travel habits. As a result, people's perceptions of COVID-19 risk can positively impact their environmental responsibility, environmental concerns, and even their readiness to spend more money to protect the environment. Therefore, albeit indirectly, COVID-19 may affect future nature-related travel behaviour (O'Connor and Assaker, 2021). Sustainable environment-based tourism may benefit from global climate change and pandemic issues such as COVID-19 and discussions about tourism's damaging impact on nature. Thus, nature-based tourism resources may be integrated into tourism mobility as a sustainable tourism product. People embrace a nature-centred strategy because the pressure and human effect on nature are at a level that can alter tourists' actions, choices, and motivations. This issue demonstrates the ecocentrism inherent in nature-based tourist activities and tourism development while emphasizing the significance of tourism development in such a setting (Mandic and Vukovic, 2021).

Nature-based tourism destination management is critical for the implementation of sustainable tourism policies and mitigating the effects of social and economic losses during times of crisis. The drop in the number of travellers and tourism income, particularly during economic downturns, epidemic such as COVID-19 or other crisis conditions, has produced numerous challenges, particularly in places heavily reliant on tourism. In this scenario, the need for proactive and recovery protection strategies is highlighted (Mandic, 2021). As a result, it is critical to promote the development of alternative strategies by minimizing the damage to natural tourism resources in such circumstances. Additionally, alternative revenue streams to support nature-based tourist resources in a climate where incomes are declining and redistributing revenue streams to nature conservation are suggested. Otherwise, nature conservation plans may lose their function since their economics and viability rely entirely on nature-based tourist income. As Mandic (2021) points out, crisis scenarios, such as COVID-19, where the long-term impact is uncertain, highlight the need for studies on nature-based tourism and other activities. However, societies that depend on sectors with a fragile structure, such as nature-based tourism, need policies that foster collaboration and support among the state, local populations, businesses, and other stakeholders. In addition to the devastating effects of COVID-19 on nature-based tourism activities, there has been an increasing interest in sustainable tourism products. Even marketing activities have focused on nature-oriented tourism (such as health, escape, nature, open space, and rejuvenation), and new standards have been implemented. With the increase in the demands of travelers for sustainable and natural tourism products, investments in this field are also increasing. Therefore, the demands for sustainable and nature-based tourism products are variables that should be considered for businesses, society and tourism planners (Spenceley, 2021). As a result, nature-based tourism businesses and destinations must engage in future-oriented promotional activities that emphasize environmental sustainability and human health.

References

Acuner, E., & Tayfun, A. (2016). Nature-based accessible tourism: Case of Rize. *Journal of Tourism and Gastronomy Studies* 4(2), 63–63.

Agaliotou, C. (2015). Reutilization of industrial buildings and sites in Greece can act as a lever for the development of special interest/alternative tourism. *Procedia-Social and Behavioral Sciences*, 175, 291–298.

Akoglan Kozak, M. (2012). *General tourism information* (Ed. N. Kozak). Eskişehir Anatolian University Publications.

Ankre, R., Fredman, P., & Lindhagen, A. (2016). Managers' experiences of visitor monitoring in Swedish outdoor recreational areas. *Journal of Outdoor Recreation and Tourism*, 14, 35–40.

Arnould, E. J., Price, L. L., & Tierney, P. (1998). Communicative staging of the wilderness servicescape. *Service Industries Journal*, 18(3), 90–115.

Balmford, A., Green, J. M., Anderson, M., Beresford, J., Huang, C., Naidoo, R., … & Manica, A. (2015). Walk on the wild side: Estimating the global magnitude of visits to protected areas. *PLoS Biology*, 13(2), 1–6.

Batman, Z. P., & Demirel, Ö. (2015). Natured based tourism activity in the route of Altindere valley Meryemana creek route: Paragliding. *İnönü University Journal of Art and Design*, 5(11), 13–26.

Bhuiyan, M. A. H., Hassan, S., Darda, M. A., & Habib, M. W. (2020). Aspects of sustainable tourism development and Covıd-19 pandemic. Preprints, 2020080418.

Bilim, Y. & Özer, Ö. (2021). No 'over', yes 'minimal'! camp and caravan tourism. *Revista Anais Brasileiros De Estudos Turísticos*, 11, 1–13.

Bloom, L. B. (2020). 'Coronavirus travel: This country wants to charge you $3000 to visit including the cost of your potential 'funeral'. *Forbes*. 24 June 2020. Available at: https://www.forbes.com/sites/laurabegleybloom/2020/06/24/coronavirus-travel-cambodia-charging-visitfuneral/#8f37bbc2b7d8 (Accessed: 02 March 2022).

Blumstein, D. T., Geffroy, B., Samia, D. S., & Bessa, E. (2017). Creating a research-based agenda to reduce ecotourism impacts on wildlife. In D. T. Blumstein, B. Geffroy, D. Samin, & E. Bessa (Eds.), *Ecotourism's promise and peril*. Cham: Springer, 180–185. doi: 10.1007/978-3-319-58331-0_11

Bremner, C. (2020). Travel 2040: Sustainability and digital transformation as recovery drivers. Euromonitor International. Available at: https://go.euromonitor.com/wb-videotravel 2020-travel-2040.html (Accessed: 02 March 2022).

Brochado, A. (2019). Nature-based experiences in tree houses: Guests' online reviews. *Tourism Review*, 74(3), 310–326.

Brochado, A. and Brochado, F. (2019). What makes a glamping experience great?. *Journal of Hospitality and Tourism Technology*, 10(1), 15–27.

Brochado, A. & Pereira, C. (2017). Comfortable experiences in nature accommodation: Perceived servicequality in glamping. *Journal of Outdoor Recreation and Tourism*, 17, 77–83.

Brooker, E., & Joppe, M. (2013). Trends in camping and outdoor hospitality-An international review. *Journal of Outdoor Recreation and Tourism*, 3(4), 1–6.

Brooker, E. & Joppe, M. (2014). Entrepreneurial approaches to rural tourism in the Netherlands: Distinct differences, *Tourism Planning & Development*, 11(3), 343–353.

Buckley, R., Gretzel, U., Scott, D., Weaver, D., & Becken, S. (2015). Tourism megatrends. *Tourism Recreation Research*, 40(1), 59–70.

Chang, K. C. (2016). Effect of servicescape on customer behavioral intentions: Moderating roles of service climate and employee engagement. *International Journal of Hospitality Management*, 53, 116–128. http://dx.doi.org/10.1016/j.ijhm.2015.12.003

Cheng, J. S., Tang, T. W., Shih, H. Y., & Wang, T. C. (2016). Designing lifestyle hotels. *International Journal of Hospitality Management*, 58, 95–106. http://dx.doi.org/10.1016/j.ijhm.2016.06.010

Christ, C., Hillel, O., Matus, S., & Sweeting, J. (2003). Tourism and biodiversity – mapping tourism's global footprint, conservation ınternational: Washington 2003. Available at: https://www.unep.org/resources/report/tourism-and-biodiversity-mapping-tourisms-global-footprint

Cohen, E. (1988). Authenticity and commoditisation in tourism. *Annals of Tourism Research*, 15(3), 371–386.

Conway, D., & Timms, B. F. (2010). Re-branding alternative tourism in the Caribbean: The case for 'slow tourism'. *Tourism and Hospitality Research*, 10(4), 329–344.

Das, M., & Chatterjee, B. (2015). Ecotourism: A panacea or a predicament? *Tourism Management Perspectives*, 14, 3–16.

De, U. K. (2013). Sustainable nature-based tourism, involvement of indigenous women and development: A case of North-East India. *Tourism Recreation Research*, 38(3), 311–324. http://dx.doi.org/10.1080/02508281.2013.11081756

Divinagracia, L. A., Divinagracia, M. R. G., & Divinagracia, D. G. (2012). Digital media-induced tourism: The case of nature-based tourism (NBT) at East Java, Indonesia. *Procedia-Social and Behavioral Sciences*, 57, 85–94.

Dube, K. & Nhamo, G. (2020). Vulnerability of nature-based tourism to climate variability and change: Case of Kariba resort town, Zimbabwe. *Journal of Outdoor Recreation and Tourism*, 29, 100281.

Formica, S., & Uysal, M. (2001). Segmentation of travelers based on environmental attitudes. *Journal of Hospitality & Leisure Marketing*, 9(3–4), 35–49.

Fredman, P., & Margaryan, L. (2021). 20 years of Nordic nature-based tourism research: A review and future research agenda. *Scandinavian Journal of Hospitality and Tourism*, 21(1), 14–25.

Fredman, P., Wall-Reinius, S., & Grundén, A. (2012). The nature of nature in nature-based tourism. *Scandinavian Journal of Hospitality and Tourism*, 12(4), 289–309.

Frost, W., Laing, J., & Beeton, S. (2014). The future of nature-based tourism in the Asia-Pacific region. *Journal of Travel Research*, 53(6), 721–732.

Garau-Vadell, J. B., Gutierrez-Taño, D., & Diaz-Armas, R. (2018). Economic crisis and residents' perception of the impacts of tourism in mass tourism destinations. *Journal of Destination Marketing & Management*, 7, 68–75.

Gawad, I. O. (2020). Eco-lodge design and architectural education: A new approach for design studios. *International Journal of Engineering Research and Technology*, 13(11), 3877–3892.

Geffroy, B., Samia, D. S., Bessa, E., & Blumstein, D. T. (2015). How nature-based tourism might increase prey vulnerability to predators. *Trends in Ecology & Evolution*, 30(12), 755–765.

Ghaith, A., Abdel-Wahab, M., Abdel-ate, M. A., & Qoura, O. (2019). Profiling of Egyptian eco-lodge guests. *International Journal of Heritage, Tourism and Hospitality*, 13(2), 52–67.

Guo, Z., Zhang, L., & Li, Y. (2010). Increased dependence of humans on ecosystem services and biodiversity. *PloS One*, 5(10), e13113.

Hall, C. M., & Boyd, S. (2004). Nature-based tourism in peripheral areas: Introduction. In C. Michael Hall & Stephen W. Boyd (Eds.), *Nature-based tourism in peripheral areas: Development or disaster?* Bristol, Blue Ridge Summit: Channel View Publications, 3–18. https://doi.org/10.21832/9781845410025-003

Hambira, W. L., Stone, L. L., & Pagiwa, V. (2022). Botswana nature-based tourism and COVID-19: Transformational implications for the future. *Development Southern Africa*. http://dx.doi.org/10.1080/0376835X.2021.1955661

Han, H., Yu, J. & Hyun, S. S. (2020). Effects of naturebased solutions (NBS) on eco-friendly hotel guests' mental health perceptions, satisfaction, switching barriers, and revisit intentions. *Journal of Hospitality Marketing & Management*, 29(5), 592–611. http://dx.doi.org/10.1080/19368623.2019.1660751

Hansson, E. (2020, Apr. 21). Coronakrisen har lett till kraftig ökning av naturutflykter. Natursidan. Available at: https://www.natursidan.se/nyheter/sa-klarar-naturomradena-det-okade-besokstrycket/ (Accessed: 04 July 2020).

Hohenlohe, X. Z. (2021). Hotels located in natural settings (e.g. forest, coastline): What actions can be undertaken to maintain or restore the ecosystems? Available at: https://www.hospitalitynet.org/panel/125000113/126001127.html

Holiable. (2022). Digital platform involved in sustainable and responsible travel. Available at: https://www.holiable.com/lodging/eco-lodge/ (Accessed: 04 March 2022).

Hultman, M., Kazeminia, A., & Ghasemi, V. (2015). Intention to visit and willingness to pay premium for ecotourism: The impact of attitude, materialism, and motivation. *Journal of Business Research*, 68(9), 1854–1861.

IUCN (2008). *Biodiversity: My hotel in action A guide to sustainable use of biological resources*, Gland, Switzerland: IUCN (International Union for Conservation of Nature), 128 pp.

Karadağ, L., & Marangoz, M. (2017). The importance of alternative tourism for sustainable tourism: Marmaris sample. *Tourism Academic Journal*, 4(2), 29–40.

Kasangaki, A., Kanyamibwa, S., Burgess, N. D., Baghabati, N., Olwero, N., Anderson, M., ... & Maritim, Z. (2012). Capturing the Benefits of Ecosystem Services to Guide Decision-Making in the Greater Virungas Landscape of the Albertine Rift Region. ARCOS, University of Cambridge and WWF-US (Project Technical Report to MacArthur Foundation).

Köroglu, Ö., & Karaman, S. (2014). The importance of community-based natural resource management in the development of nature-based tourism activities. *KMU Journal of Social and Economic Research*, 2014(1), 95–106.

Kuenzi, C. & McNeely, J. (2008). Chapter 8. Nature based tourism. In O. Renn & K. Walker (Eds.), *Global risk governance: Concept in practice using the IRGC framework*. Netherlands: Springer, 155–178.

Laarman, J. G., & Durst, P. B. (1987). Nature travel in the tropics. *Journal of Forestry*, 85(5), 43–46.

Legrand, W. (2021). Nature-based solutions for urban hotel real estate. White Paper Series on Sustainable Hospitality, June, 2021.

Lucas, P. H. C. (1985). How protected areas can help meet society's evolving needs. In National parks, conservation, and development. The role of protected areas in sustaining society. Proceedings of the *World Congress on National Parks*, Bali, Indonesia, 11–12 October 1982 Smithsonian Institution Press, 72–77.

Luo, Y., & Deng, J. (2008). The new environmental paradigm and nature-based tourism motivation. *Journal of Travel Research*, 46(4), 392–402.

Mandic, A. (2019). Nature-based solutions for sustainable tourism development in protected natural areas: A review. *Environment Systems and Decisions*, 39(3), 249–268. https://doi.org/10.1007/s10669-019-09718-2

Mandic, A. (2020). Structuring challenges of sustainable tourism development in protected natural areas with driving force–pressure–state–impact–response (DPSIR) framework. *Environment Systems and Decisions*, 40(4), 560–576.

Mandic, A. (2021). Protected area management effectiveness and COVID-19: The case of Plitvice Lakes National Park, Croatia. *Journal of Outdoor Recreation and Tourism*, https://doi.org/10.1016/j.jort.2021.100397

Mandic, A., & Vukovic, M. (2021). Millennials' attitudes, choices and behaviour – integrative analysis. *Journal of Ecotourism*, 21(1), 87–97.

Mehmetoglu, M. (2007). Typologising nature-based tourists by activity – Theoretical and practical implications. *Tourism Management*, 28(3), 651–660.

Meng, F., Tepanon, Y. & Uysal, M. (2008). Measuring tourist satisfaction by attribute and motivation: The case of a nature-based resort. *Journal of Vacation Marketing*, 14(1), 41–56.

Milohnic, I., Bonifacic, J. C. & Licul, I. (2019). Transformation of camping into glamping – trends and perspectives. *Tourism in Southern and Eastern Europe*, 5, 457–473.

Monz, C. A., Pickering, C. M., & Hadwen, W. L. (2013). Recent advances in recreation ecology and the implications of different relationships between recreation use and ecological impacts. *Frontiers in Ecology and the Environment*, 11(8), 441–446.

Muhammad, S., Long, X., & Salman, M. (2020). COVID-19 pandemic and environmental pollution: A blessing in disguise?. *Science of The Total Environment*, 728, 138820.

O'Connor, P.,& Assaker, G. (2021). COVID-19's effects on future pro-environmental traveler behavior: An empirical examination using norm activation, economic sacrifices, and risk perception theories. *Journal of Sustainable Tourism*. http://dx.doi.org/10.1080/09669582.2021.1879821

Øian, H., Fredman, P., Sandell, K., Sæþórsdóttir, A. D., Tyrväinen, L., & Jensen, F. S. (2018). Tourism, nature and sustainability: A review of policy instruments in the Nordic countries.

Padma, P., Ramakrishna, S., & Rasoolimanesh, S. M. (2019). Nature-based solutions in tourism: A review of the literature and conceptualisation. *Journal of Hospitality & Tourism Research*, Published Online. https://doi.org/10.1177/1096348019890052

Schreckenberg, K., & Luttrell, C. (2009). Participatory forest management: A route to poverty reduction?. *International Forestry Review*, 11(2), 221–238.

Silverberg, K. E., Backman, S. J., & Backman, K. F. (1996). A preliminary investigation into the psychographics of nature-based travelers to the Southeastern United States. *Journal of Travel Research*, 35(2), 19–28.

Smith, J. (2020a). Will 2020 define the future of tourism?. Wtm Global Hub. Available at: https://hub.wtm.com/will-2020-define-the-future-of-tourism/, (Accessed: 02 March 2022).

Smith, J. (2020b). Which trends should guide ' 'tourism's recovery?. Wtm Global Hub. Available at: https://hub.wtm.com/what-trends-should-guide-tourisms-recovery/ (Accessed: 02 March 2022).

Sotiriadis, M. (2017). Experiential dimensions and their influence on behavioral intentions within the context of nature-based tourism. *Tourism and Hospitality Management*, 23(1), 35–50.

Spenceley, A. (2021). *The future of nature-based tourism: Impacts of covid-19 and paths to sustainability*. Luc Hoffmann Institute. Available at: https://luchoffmanninstitute.org/wp-content/uploads/2021/04/luchoffmanninstitute-future-nature-based-tourism-report-2021.pdf

Steg, L., & Vlek, C. (2009). Encouraging pro-environmental behaviour: An integrative review and research agenda. *Journal of Environmental Psychology*, 29(3), 309–317.

Sumanapala, H. D. P., Perera, P. K. P., Kotagama, S. W., Silva, D. A. C. S. (2015). Eco-lodge patrons' characteristics: The Sri Lankan perspective. *International Research Journal of Social Sciences*, 5(2), 509–525.

Sunnetcioglu, S., & Yesildag, B. (2021). Ecotourism and environment relation. In Aslı Ozgen Cigdemli (Ed.), *Understanding ecotourism*. Ankara: Detay Publications, 12–23.

Tangeland, T. (2011). Why do people purchase nature-based tourism activity products? A Norwegian case study of outdoor recreation. *Scandinavian Journal of Hospitality and Tourism*, 11(4), 435–456. http://dx.doi.org/10.1080/15022250.2011.619843

Tangeland, T., & Aas, Ø. (2011). Household composition and the importance of experience attributes of nature based tourism activity products e A Norwegian case study of outdoor recreationists. *Tourism Management*, 32, 822–832.

Tangeland, T., Vennesland, B., & Nybakk, E. (2013). Second-home owners' intention to purchase nature-based tourism activity products–A Norwegian case study. *Tourism Management*, 36, 364–376.

Tenerelli, P., Demšar, U., & Luque, S. (2016). Crowdsourcing indicators for cultural ecosystem services: A geographically weighted approach for mountain landscapes. *Ecological Indicators*, 64, 237–248.

Texas Park & Wildlife Foundation (2022). What is nature tourism? Available at: https://tpwd.texas.gov/landwater/land/programs/tourism/what_is/

Tovar, C., & Lockwood, M. (2008). Social impacts of tourism: An Australian regional case study. *International Journal of Tourism Research*, 10(4), 365–378.

Trang, H. L. T., Lee, J. S., & Han, H. (2019). How do green attributes elicit pro-environmental behaviors in guests? The case of green hotels in Vietnam. *Journal of Travel & Tourism Marketing*, 36(1), 14–28. http://dx.doi.org/10.1080/10548408.2018.1486782

Türker, G. Ö. (2020). How does the Covid-19 outbreak affect the tourism industry? An evaluation from the perspective of tourism academics. *International Journal of Social Sciences and Education Researh*, 6(2), 207–224.

UNWTO (2020). Understanding Domestic Tourism and seizing its opportunities, UNWTO Briefing Note-Tourism and Covid-19, issue 3. Available at: https://www.e-unwto.org/doi/epdf/10.18111/9789284422111 (Accessed: 02 March 2022).

Valentine, P. S. (1992). Nature-based tourism. In B. Weiler & C. M. Hall (Eds.), *Special interest tourism*. London: Belhaven. Available at: https://researchonline.jcu.edu.au/1632/1/Nature-based_tourism.pdf

Venter, Z. S., Barton, D. N., Gundersen, V., Figari, H., & Nowell, M. (2020). Urban nature in a time of crisis: Recreational use of green space increases during the COVID-19 outbreak in Oslo, Norway. *Environmental Research Letters*, 15(10), 104075.

Wang, N. (1999). Rethinking authenticity in tourism experience. *Annals of Tourism Research*, 26(2), 349–370.

Wang, T. C., Tsai, C. L., & Tang, T. W. (2018). Exploring advertising effectiveness of tourist hotels' marketing ımages containing nature and performing arts: An eye-tracking analysis. *Sustainability*, 10(9), 3038, 1–13. http://dx.doi.org/10.3390/su10093038

Wearing, S., & Schweinsberg, S. (2018). *Ecotourism: Transitioning to the 22nd century*. Routledge.

Winter, P. L., Selin, S., Cerveny, L., & Bricker, K. (2020). Outdoor recreation, nature-based tourism, and sustainability. *Sustainability*, 12(1), 81.

Wolf, T., Debut, L., Sanh, V., Chaumond, J., Delangue, C., Moi, A., … & Rush, A. M. (2019). Huggingface's transformers: State-of-the-art natural language processing. arXiv preprint arXiv:1910.03771.

Wong, J.-Y., Hsiung, M.-L., Lee, S.-J., & ChouHuang, C.-Y. (2021). The relationship between endurance involvement and travel behavior in camping and the moderating effect of place attachment. *Sustainability*, 13, 5016. https://doi.org/10.3390/su13095016

Xu, J. B., & Chan, S. (2016). A new nature-based tourism motivation model: Testing the moderating effects of the push motivation. *Tourism Management Perspectives*, 18, 107–110.

Yılmaz, G., & Akman, S. (2020). The concept of symbioticism and its relationship with gastronomy tourism. *Aydın Gastronomy*, 4(2), 125–134.

30

NATURE-BASED TOURISM AND NATURE CONSERVATION

Exploring the elements and links

Jeetesh Kumar, Priyakrushna Mohanty and Neethiahnanthan Ari Ragavan

Introduction

Many studies have focused on nature-based tourism (NBT) from the supply side, particularly on destination development, lifestyle entrepreneurship, and the role of tour guides (Fredman & Margaryan, 2021, p. 529). NBT is defined by Margaryan (2017): as "activities by humans occurring when visiting nature areas outside the person's ordinary neighbourhood". As such, the sector is heavily dependent on natural resources in its efforts to meet the needs of nature tourists. NBT is mainly manifested as nature tourism, environmental tourism, ecotourism, geographic tourism, bicycle tourism, adventure tourism, wildlife tourism, cruise tourism, outdoor tourism and other forms of tourism. NBT is one of the fastest-growing segments in the tourism industry, with numerous public lands and protected regions serving as popular tourist attractions (Winter, Selin, Cerveny, & Bricker, 2020). Chung, Dietz, and Liu (2018) find that the higher the biodiversity, the older the history, the larger the area, the more accessible the city, and the higher the altitude, the higher the level of NBT. However, rapid growth in NBT can cause complex changes in local communities, have an impact on local infrastructure and facilities, trigger competition for scarce resources, result in unwelcome marketing of indigenous products and traditions, and result in an influx of non-native workers, business owners, and amenity immigrants (Jones, McGinlay, & Dimitrakopoulos, 2017). In recent years, tourism has been highly vulnerable to local phenomena, in particular, the impact of nature-based tourism, such as earthquakes, forest fires, volcanic eruptions, tsunamis or floods, as well as global events such as disease pandemics (Ma, Chiu, Tian, Zhang, & Guo, 2020; Rogerson & Rogerson, 2020). This chapter explores the essential elements of NBT, and examples from the region will help illustrate its diversity of form. The work also focuses on providing a context for analysing nature-based tourism and discussing the relative merits of options.

Nature and tourism – global context

From a global perspective, "nature-based tourism is huge," as Blumstein, Geffroy, Samia, and Bessa (2017) noted. Many natural ecosystems and wildlife species are under threat from global challenges such as climate change, biodiversity decline, unsustainable resource usage, and other human factors (Wearing & Schweinsberg, 2018). While NBT can help broaden the audience

 DOI: 10.4324/9781003230748-35

for biodiversity and conservation efforts, increased recreational and tourist access can impact the natural and social environment. Large numbers of NBT visitors may put additional pressure on natural regions, as any tourism activity may have an impact on the environment (Øian et al., 2018), and the lack of monitoring of spatial patterns and characteristics of visitors in the protected area (Kim, Lee, & Kim, 2020).

Nature and tourism – regional context

The Asiatic region has emerged as the top performer globally in terms of growth (Mohanty, Dhoundiyal, & Thomas, 2022). This growth has been primarily spearheaded by NBT destinations across the region (Jones, Apollo, & Bui, 2021). China, for example, has rich tourism resources with its vast geographical area, presenting a variety of tourism types. The most prominent is natural and cultural tourism, which is relatively unaffected by the seasons due to the climate and cultural differences between the provinces in China. In recent years, nature tourism has brought economic benefits to local communities, but has also a number of ecological problems in China (Wang et al., 2021). Therefore, the quality of nature tourism sites may be negatively affected by these ecological problems, and this deterioration, in itself, is detrimental to the sustainable development of nature tourism. According to the summary report of "Green Shield 2017", China has found more than 20,000 examples of ecological problems relating to national nature reserves. Most of these ecological problems are caused by irrational resource development, including, but not limited to, irrational tourism development (Yuxi & Linsheng, 2020).

Japan's 35 national parks are rich in biodiversity, ranging from snow-capped mountains to coral reefs. Together with well-designed seasons, abundant resources and opportunities enrich multiple types of NBT, including everyday activities such as hiking and sightseeing and more adventurous activities such as skiing and mountain biking (Jones & Nguyen, 2021). However, Japan's NBT industry has stagnated or shrunk since the 1990s (Jones, 2012). This is mainly reflected in the fact that the domestic NBT tourists are elderly tourists with moderate education levels, and the possibility of NBT tourists staying overnight is gradually decreasing (Jones & Nguyen, 2021; Suganuma, Hibino, & Morichi, 2011).

In Sri Lanka, tourism has long been the third-largest source of foreign exchange, accounting for more than 13 percent of the country's GDP (Ranasinghe & Sugandhika, 2018). In the most recent year for which there are records, more than 2.7 million international tourists visited wildlife tourism destinations, including about 173,000 who visited coastal wetlands and marine parks (Sri Lanka Tourism Development Authority, SLTDA, 2019). However, the spatial concentration of tourists in a few famous destinations is a common but problematic phenomenon in Sri Lanka's NBT industry (Prakash, Perera, Newsome, Kusuminda, & Walker, 2019). For example, almost 65 percent of national park visitors came from the four popular national parks in 2018 (SLTDA, 2019). The high level of tourism access may be unsustainable, leading to the degradation of nature-based destinations and dissatisfaction among visitors (Prakash et al., 2019; Newsome, Rodger, Pearce, & Chan, 2019).

Similarly, in Indonesia, forest reserves contain 54 national parks, of which nine are mainly marine parks, six are UNESCO World Heritage sites, and five are Ramsar wetlands. The current 54 national parks are protected areas with indigenous ecosystems managed by zoning to promote research, education, biodiversity and tourism goals. However, tourism resource protection and tourist behaviour control are necessary for the sustainable development of national parks. In Bromo and other mountain national parks, unregulated nature tourism negatively impacts the environment, such as air and water pollution, habitat deterioration, solid waste pollution, erosion of volcanic edges, vandalism and illegal plant collection (Pamungkas & Jones, 2021).

In Vietnam, NBT is increasingly recognized as playing an essential role in national development, especially in the national tourism industry, supporting the socio-economic development of rural areas around the reserve. Vietnam's National Parks (NPs) are protected areas in the National System of Special Use Forests (SUFs), designed to protect nature. Other goals include preserving landscapes and cultural and historical sites and providing recreation and tourism. However, Vietnam's NBT suffers from a lack of planning or strategies on the national level (Markowski, Bartos, Rzenca, & Namiecinski, 2019). National planning is essential for developing NBT destinations and improving the country's tourist attractions, such as its natural surroundings, biodiversity, cultures, and history.

Diversity in nature-based tourism

The spectacular natural assets of the Nordic area have long intrigued and attracted travellers, and their tourism value has only grown over time (Fredman & Margaryan, 2021). Its natural assets are highlighted in the diversity of outdoor recreational activities available, ranging from relatively common activities such as animal-based activities (fishing and equestrian) or skiing (remote areas) to more niche activities such as glacier travel and mountain biking and Viking tourism. However, Nordic tourism destinations are vulnerable to climate change, such as changing precipitation patterns, reduced winter snowfall and seasonal changes (Landauer, Goodsite, & Juhola, 2018). In particular, the climate vulnerability of winter tourism depends not only on the activity itself but also on the region where the activity occurs. About three-quarters of winter tourism businesses have been affected by weather events, such as extreme cold or high winds leading to cancellations. The future of winter tourism in Finland and similar areas may be more sensitive to changes in the frequency and severity of extreme weather than to changes in the length of seasons (Tervo, 2008) (Figure 30.1).

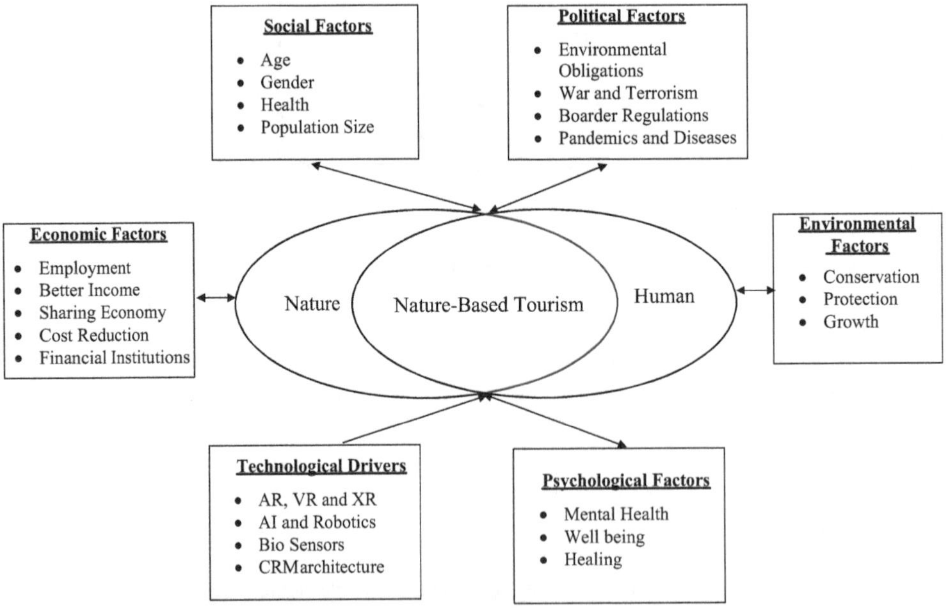

Figure 30.1 A framework for NBT.

Source: Authors' Own Work

Natural resources evaluation

In the initial stage of regional development studies, the problems and potential of the region are diagnosed. Evaluating the natural resource base is the basis of any development planning and project formulation that offers baseline information to help develop strategies and identify projects. Priskin (2001) indicates that the first stage in effective planning is systematically identifying and evaluating the resource base for its potential development. The critical point of rural ecotourism development is to choose characteristic tourism resource points. China's *Tourism resources Classification, survey and Evaluation* (GB/T18972-2003) is the national criteria to determine the level of natural tourism resources (Xiang & Yin, 2020). The NBT development potential of the national parks and protected areas should be evaluated before developing and promoting NBT destinations. For example, ecotourism is a burgeoning niche sector that can help protected places thrive sustainably; ecotourism development requires a preliminary evaluation of destination resources (Bekele, Teshome, & Asteray, 2017; Kunasekaran & Kumar, 2021).

Xiang and Yin (2020) construct the evaluation system of rural ecotourism resources, which consists of three levels, target level, factor level and evaluation index level, and the study indicates that for rural ecotourism resources, human and natural conditions are the primary elements; of these, the natural landscape and social-cultural objects are the most important. Africa's natural resources, wildlife, national parks, and landscapes lay the foundation for its tourism development (Bekele et al., 2017). For example, East African countries are known for wildlife diversity, an abundance of wildlife species, and leading tourism destinations. Bekele et al. (2017) find that Maze National Park has high ecotourism potential regarding natural resources but moderate potential in site infrastructure provision and human resources. Local communities living near national parks have a low potential for tangible cultural resources but have various intangible cultural resources that attract visitors. Therefore, Maze National Park has a high potential for eco-tourism development.

Destination attractiveness

NBT is defined as travel to enjoy undeveloped natural regions, and almost all NBT destinations are located in protected areas (Markowski et al., 2019). With the increase in the urban population, rural tourism is becoming more and more popular (Zhou, 2014). Natural and cultural attractiveness are significant factors in the development of NBT (Sadowski & Wojcieszak, 2019). The attractiveness of a tourism destination is a crucial determinant of its competitiveness and becomes an essential factor in driving the local tourism economy development. The fundamental attractiveness of a tourism destination consists of various destination appeals, namely natural resource attractiveness, cultural heritage attractiveness and social appeals (Ndivo, Waudo, & Waswa, 2012).

Destination attractiveness is a measure of the potential of a particular destination to provide travellers with a pleasant experience and ensure a sense of happiness during the vacation (Jin, Weber, & Bauer, 2012). Attractiveness is related to tourists' perception of the potential of a place to meet their tourism expectations (Reitsamer, Brunner-Sperdin, & Stokburger-Sauer, 2016). Chin, Lo, Razak, Pasbakhsh, and Mohamad (2020) find that destination attractiveness is the top concern for tourists in Sarawak's semi-rural and rural tourism destinations. From the global perspective, tourism destinations are no longer regarded as various natural, social-cultural and ecological resources but as an attractive tourism product of a destination (Reitsamer et al., 2016). Destination attractiveness is one of the critical factors in selecting a tourism destination. The success of a tourism destination is determined primarily by its tourism advantage in terms of unique attractiveness (Jin et al., 2012). The more attractive a destination, the greater the chance tourists will choose that destination (Kim & Perdue, 2011). Image is also a critical factor that influences destination attractiveness. The better a destination's reputation and image, the more attractive it will be. When tourists are faced with the complexity of tourism offers selection, their choice is mainly affected by the resources the destination possesses (Puška, Šadić, Maksimović, & Stojanović, 2020). The diversity of tourism resources determines the uniqueness and attractiveness of the destination; tourists' knowledge of these resources can affect the attractiveness of destinations (Guan & Jones, 2015). However, Zawadka (2019) points out that the problem with rural tourism's low attractiveness is tourist stereotypes and their erroneous assumption that rural regions are unattractive to tourists.

When a traveller visits a destination, the destination must either meet or exceed their expectations (Lee, 2016). Before going on holiday, tourists have certain expectations about the destination they will visit. Their experience of the destination will either confirm or deny their expectations. Therefore, if a destination can meet or exceed the expectations of tourists' needs, it automatically becomes a more attractive destination (Ma, Hsiao, & Gao, 2018). One popular approach is to associate specific criteria with destinations (Lee, 2016). These standards attract visitors to a given location (Puška et al., 2020), such as tourist infrastructure, culture and tradition, and natural and ecological resources. For example, the rural area of Poland is regarded as an attractive rural tourism destination, characterized by the diversity of tourist offerings, a beacon of culture and tradition, and a place to preserve unique natural and landscape values (Sadowski & Wojcieszak, 2019).

Accessibility

In recent years, accessibility has significantly affected regional tourism flows and global tourism industry performance. However, tourism encompasses many sectors, and accessibility measurements may not fully incorporate the simultaneous accessibility of corresponding destinations (Yen, Chen, & Ho, 2021). Tourism has a network structure that includes primary suppliers, tourism intermediaries and tourism production activities within the tourism system. In addition, both major suppliers (such as airlines) and travel agents (such as tour operators) constitute channels that offer opportunities for travellers to interact with destinations, with different channels associated with their forms of accessibility (Romero & Tejada, 2011).

Several studies are interested in travellers' perceptions that evaluate destination accessibility. Ceccato, Deflorio, Diana, and Pirra (2020) show that accessibility can be defined as high geographical connectivity between regions, provided by a good travel experience and efficient modes of transportation, and indicators are designed based on tourists' perception of accessibility to travel time. In general, accessibility is divided into two categories, including individual accessibility (such as health condition, psychological condition, opportunity, economic factor, and social factor) and destination accessibility (such as driving attributes and destination attributes) (Wirajaya, Rahmafitria, Nurazizah, & Jamin, 2021). Some studies show that low individual accessibility awareness negatively impacts visits, while high destination accessibility awareness positively impacts visits (Wirajaya et al., 2021).

Destination accessibility refers to the accessibility of a particular destination and is related to the facilities and infrastructure of the destination and the time and effort required to get there (Gehrke, Akhavan, Furth, Wang, & Reardon, 2020; Zhu & Diao, 2020). Destination accessibility is generally considered a significant factor in driving tourists' travel decisions (Park, Kim, Kim, & Park, 2019). For example, tourist service is based on assistive technology that can enable hiking using wheelchairs that can improve the accessibility of tourist destinations and its attraction (Nanchen, Margot-Cattin, Ramseyer, & Schegg, 2020). However, the natural and cultural environment in the Greek islands is the main competitive advantage of tourism there, whereas the absence of ground transportation limits the accessibility of the islands, especially in winter (Paraschi, 2020). The critical question for tourism destinations and countries is regarding and measuring accessibility and its performance. Therefore, understanding destination accessibility can assist destination managers and governments identify tourism opportunities and implementing suitable strategies (Yen et al., 2021).

Activities (exploration, participative)

The projects and features of the natural environment are inherent in NBT. When people leave their homes for recreation in nature, the experience they get from outdoor activities often depends on specific natural resources and environments (Tyrväinen, Uusitalo, Silvennoinen, & Hasu, 2014). For example, Brčko District is located in a superior location and rich in natural resources. It can be used as a base for carrying out hunting, leisure, ecological, mountain, agricultural, cultural, health tourism, folk, religious, historical, food and other NBT branches (Puška et al., 2020). With the increasing popularity of NBT environments, the relationship between leisure activities and natural resources requires further attention (Margaryan, 2018). For example, environments and resources suitable for fishing activities excepted rivers and fish include shelters, beautiful scenery, pleasant weather, campfire sites, and space and time for social interaction with owners and other anglers.

Similarly, ski trips require more than good weather and suitable snow conditions, iconic destinations and magnificent environments, negotiable terrain, and safe routes (Fossgard & Fredman, 2019). Providing visitors with a range of particular resources can enhance their final activity experience. Most NBT activities can be classified as sports and adventure (Buckley, 2006). For example, with nature as the backdrop, rock climbing and bike park as stage entertainment; Kayaking on calm seas and skiing on untouched snow are real experiences that connect with nature; Kayaking in breaking waves and skiing in steep canyons (Fossgard & Fredman, 2019). To participate in and experience the immersion of nature, when faced with a wide variety of outdoor activities, tourists can choose from simple hiking or cycling trips staying in hotels and more intense NBT experiences such as camping (Filipe, Barbosa, & Santos, 2020).

In the 1990s, Lindberg and Lindberg (1991) divided nature tourists into four categories based on different motivations and activities: (1) Core nature tourists, including those with educational purposes; (2) Dedicated nature tourists, primarily those seeking the natural and cultural history of protected areas; (3) Mainstream nature tourists, for unusual travel activities; and (4) Casual nature tourists; tourists who experience nature by chance. There are a wide range of destinations for NBT, but natural areas such as national parks and protected areas are the most significant component. Landscape experience, sightseeing and nature photography are the main activities in terms of preference and participation in NBT activities by different foreign visitors to Norway in summer. Still, they tend to be associated with sports activities, walking and hiking in mountains and forest areas being the most common (Dybedal, Haukeland, & Stemmer, 2021). Everyday NBT activities in Western Australia include bushwalking, backpacking, camping, wildlife watching, and fishing. More adventurous activities in natural areas include off-road driving, scuba diving and rock climbing. The central coastal area of Western Australia has open space for various natural resources in both marine and terrestrial environments (Priskin, 2001).

Supportive service (other infrastructure)

Tourism infrastructure is divided into two categories, namely soft infrastructure and complex infrastructure. Soft infrastructure refers to training, while hard infrastructure refers to tangible elements such as transportation facilities, tourists' access methods, and information facilities (Kumar, 2016). In addition to tourism infrastructure, service quality has always been a primary concern for service providers in today's modern, globalized and competitive environment (Ramseook-Munhurrun, Naidoo, Seebaluck, & Pillai, 2016). In this context service refers to non-physical goods provided in exchange for money in a business context (Davidoff, 1994). Service quality is considered one of the critical factors in developing destination marketing activities (Yaru, Liu, & Jing, 2018). In the development of rural tourism in the Rostov region, the number of enterprises in the hospitality sector is increasing, and they are providing more and more additional services, such as distinctive reception services, the development of rural tourism new routes, and cultural leisure activities (Provotorina, Kazmina, Petrenko, & Ekinil, 2020). Accommodation facilities are a significant component of tourism infrastructure because accommodation services are vital for tourism products. Tinakhat, Viriyachaikul, and Vorasingha (2018) believe that the service quality of homestay providers plays a crucial role in the success or failure of the rural tourism industry. Al-Ababneh (2013) points out that the quality of services provided and experienced by travellers "will determine their level of happiness, and the better the quality, the happier the tourists will be" that is a key aspect of building visitor satisfaction.

Tourism infrastructure has become one of the primary factors that affect tourists' decision-making on tourism destinations (Zaei & Zaei, 2013), which will inevitably enhance its competitiveness and contribute to the successful development of tourism destinations. The elements of destination attractiveness and tourism infrastructure affect the development of rural tourism destinations' competitiveness (Owiyo, 2018). In rural tourism, natural resources or facilities have become the critical input of rural tourism products (Nooripoor, Khosrowjerdi, Rastegari, Sharifi, & Bijani, 2021). Therefore, it is proposed that good tourism infrastructure will become an essential promotional tool for the marketing of rural tourism destinations (Karalkova, 2016). Infrastructure assets are essential for ecotourism development (Parker & Khare, 2005). Proximity to cultural and historical sites and roads and rivers are considered critical natural attractions for ecotourism development (Abrehe, Girma, & Nigusse, 2021). As a result of infrastructure development, tourism resources are being developed, and the attractiveness and accessibility to tourists are increasing, along with the tourism capacity of the territory (Provotorina et al., 2020). Almeida and Machado (2021) indicate that investment in infrastructure has helped Madeira's rural tourism sector attract more visitors. Infrastructure development is crucial for improving destination accessibility (Wirajaya et al., 2021). Moric (2013) points out that the efficiency of tourism infrastructure plays a crucial role in attracting tourists to destinations, especially rural tourism destinations (Goffi, 2013). Davidson (2016) emphasizes the effectiveness of tourist facilities and services to provide unforgettable destination experiences, thereby further improving tourist satisfaction.

Local community

In a particular destination, community participation is necessary for increasing tourism efficiency (Su, & Wall, 2014). Community participation can be spontaneous, induced, or forced, depending on the nature of tourism development (Rasoolimanesh, Ringle, Jaafar, & Ramayah, 2017). Effective community participation requires that community residents have the right knowledge and skills for conservation and tourism development. The community understands environmental protection and participates in various communication processes at tourism destinations, such as attending training on environmental issues, making crafts with natural ingredients, and serving food and beverages. During the process, the community residents price their products and services, thereby sustaining their livelihood (Ali, Quaddus, Rabbanee, & Shanka, 2020). Mogomotsi, Stone, Mogomotsi, and Dube (2020) find that When local livelihoods and wildlife conservation come into conflict, it is essential to understand communities' attitudes close to wildlife and explore how to respond to wildlife conservation. Human interaction with wildlife can only be improved by understanding the experiences of local communities (Tessema, Lilieholm, Ashenafi, & Leader-Williams, 2010). According to the sustainable-use approach hypothesis, local community user rights promote natural resource conservation (Mogomotsi, Mogomotsi, Gondo, & Madigele, 2018).

In existing NBT literature, community participation has received substantial attention (Ali et al., 2020; Mogomotsi et al., 2020). Community participation refers to the participation of communities in tourism activities for social, environmental, and economic benefits, which influences the nature and level of participation (Hung, Sirakaya-Turk, & Ingram, 2011). Ecotourism, as one of the aspects of NBT, can encourage local communities to participate in conservation (Abrehe et al., 2021). Community participation depends on the attractiveness of nature and the motivation of people to participate in ecotourism activities, and it affects the quality of life of local community residents. Political instability and government policy significantly moderated the relationship between ecotourism motivation and community participation and also between community participation and quality of life (Ali et al., 2020). Community residents are the essential participants in the success of the NBT industry, and community participation is crucial to the development of NBT in a destination by promoting positive relationships among various stakeholders in the community. Participation use and utilization of community resources in the planning and decision-making process increase community awareness of the benefits of tourism initiatives (Rasoolimanesh et al., 2017).

Ecological degradation

Tourism plays a vital role in the socio-cultural transformation of countries and is also one of the fastest-growing industries in the world (Geary, 2018). With the growth of per capita income worldwide, tourists have increased rapidly. This tourism development is reflected in the associated energy use and subsequent environmental degradation (Kalayci, 2019; Tang, Bai, Shi, Liu, & Li, 2018). For example, it was estimated that in 2013 8 percent of global carbon emissions from greenhouse gas emissions were generated by global tourism (Lenzen et al., 2018). The emergence of ecotourism is one of the main reasons for the migration of animal habitats and the faster depletion of natural resources in their destinations (Fletcher, 2019).

Measures such as greenhouse gas (GHG) emissions, carbon dioxide (CO_2) emissions, and ecological footprint (EFP) can be used to access the level of environmental degradation (Hassan, Xia, Khan, & Shah, 2019). For example, the growth of tourism in China is complemented by growth in the transport sector, which is an essential contributor to tourism's carbon emissions. In addition, tourist accommodation also accelerates the consumption of natural resources and produces waste gas, which leads to environmental degradation (Bianco, 2020; Wang, Du, Li, He, & Xu, 2019; Wang, Zhou, Chen, & Rong, 2019). Increased tourism activities and solid waste generation will negatively impact vegetation cover, landscaping and overall ecosystem health if rational management measures are not taken for fragile ecosystems (Maldonado-Oré & Custodio, 2020).

Some tourism-oriented sports and other recreational activities have led to the migration of animal habitats, the depletion of natural resources and the generation of greenhouse gases (Chen, Thapa, & Yan, 2018). With tourism development, related industries formed by tourism destinations also aggravate this ecological dilemma (Sharif et al., 2020). The Greek Islands show a fragile ecosystem, such as drought, sea-level rise and land erosion. Fish stocks are limited because of overfishing, and many islands suffer from water shortages, which worsen in the summer as tourist flows increase (Paraschi, 2020). Geotourism, which is an aspect of culturally and environmentally responsible tourism, is particularly important in the Andaman and Nicobar Islands, which are environmentally fragile and are already facing challenges such as biodiversity loss, ecosystem destruction, water scarcity and loss of indigenous populations and seismic activities (Arora, Rajput, & Anand, 2020). Mountain tourism often occurs in protected regions and fragile ecosystems, and activities such as skiing and trekking are endangering the sustainability of mountain tourism (Seraphin & Dosquet, 2020). Although mountains are fragile ecosystems and threatened by tourism, tourism economic activity is necessary for the survival of rural regions due to their low population density and the lack of integration of extensive economic networks (Luthe, Wyss, & Schuckert, 2012).

Discussion and conclusion

Over the past decade, NBT has been gathering momentum as a distinct field of study (Qiu, Sha, & Scott, 2021). While various facets of nature-based tourism have been studied across the different regions of the world, many new trends, such as the use of technology, stakeholder participation, and visitor experience management continue to emerge, widening the understanding of nature-based tourism. In all these developments, sustainability has remained the bottom line of the discussions. The critical tourism literature has also contributed to the development of the field by highlighting the negative impacts of the NBT and providing cautionary marks (Hall & Boyd, 2005). In this chapter, attempts have been made to shed light on the various aspects of NBT and discuss the future directions of this emerging doctrine of study. For better understanding, a framework for NBT has been developed by the authors.

The concept of NBT originated from the environmental movements of the 1970s and 1980s and primarily focused on minimizing the negative impacts of tourism on the environment (Mehmetoglu, 2007). With no standardized definition in the literature, Buckley and Coghlan (2012) define NBT as the wasteful and non-consumptive forms of tourism occurring in a relatively undisturbed natural setting. Here, the diversity of the NBT is one of the key factors to recognize. Since its inception, the concept of NBT has evolved as a hugely debated concept, with many forms of tourism coming under the umbrella term. Thereby, a host of new forms of NBT tourism has evolved at different periods and assisted in the growth of the concept in terms of both length and breadth (Buckley & Coghlan, 2012). It is also argued that a host of tourists demand nature-based products and products that fulfil the needs of nature-based tourists.

Following its inception, NBT was initially studied as a form of tourism, which aimed to educate the tourists about the negative impacts and propagate the forms of tourism with minimal impacts on the environment. During this period, the study of NBT was majorly examined from the conservation of environment aspect. Gradually, research on NBT was extended to community development (Fredman, Wall-Reinius, & Grundén, 2012) and host–guest interactions (Carmody, 2008). Next, the visitor satisfaction component was given due attention in the context of NBT. As time progressed, NBTs gained momentum as a form of tourism that benefits the hosts, the visitors, and the environment (O'Neill, Riscinto-Kozub, & Van Hyfte, 2010). However, the advocacy approach to NBT has not been reflected all across. Critics like

Duffy (2015, p. 529) noted that NBT follows "a process of neoliberalism nature, which cuts the threads that bind ecosystems together so that the constituent parts can be transformed into new commodities". Thus, NBT was blamed for destroying the very resources that it aimed to protect. Further, NBT was critiqued as a form of tourism that commoditizes flora and fauna of any particular destination providing shelter to neoliberal beliefs (Butt, 2012). However, despite all these criticisms, the concept of NBT kept growing steadily mainly due to its deep affiliations to the sustainability aspect, which emerged as the most significant topic of discussion in the 2000s.

NBT is studied as a multifaceted concept that influences, and is influenced by, social, political, economical, environmental, and technological factors (Elmahdy, Haukeland, & Fredman, 2017). The primary aim of NBT has remained the same: to achieve environmental goals of conservation, protection and growth (Jones et al., 2021). In society, NBT has been associated with gender, age, population size, culture, and harmony in the community (Zhang, Chan, Liu, & Zhu, 2020). Similarly, when observed from the political angle, NBT is characterized as a geopolitical phenomenon covering many issues concerning both developed and developing countries and power imbalances (Fredman, Haukeland, Tyrväinen, Stensland, & Wall-Reinius, 2021). NBT is studied as a form of tourism that generates, and can generate, substantial economic benefits, especially in developing and Least Developed Countries (Cisneros-Montemayor et al., 2020). Further, trends suggest that NBT has excellent potential for being an agent of psychological wellbeing and wellness (Kim, Lee, Uysal, Kim, & Ahn, 2015). This emphasizes the psychological benefits of NBT, which will drive the future research agenda in the doctrine of tourism development.

Finally, similar to other forms of tourism, technology has emerged as the most critical factor driving the future of NBT. The applications of technology in the context of NBT consist of travel planning (Silva, Marques, & Abrantes, 2020), visitor experience management (Albrecht, 2021), new product development (Coghlan & Carter, 2020), knowledge mobilization (Coghlan, 2020), environmental protection (Wheaton et al., 2016), safe payments (Nagle & Vidon, 2021), and awareness generation (da Mota & Pickering, 2020). Therefore, in the framework, technological inputs like AR, VR, XR, AI, Robotics, and CRM architecture have been depicted as the drivers of NBT.

Overall, NBT has come a long way from being a form of tourism that emphasizes environmental protection to being an exclusive segment of tourism that impacts all aspects of life. The current chapter provides a holistic view of the different facets of NBT and the trajectory of movement since its inception. With environmental protection and conservation emerging as one of the most significant global agendas, discussions, and debates about NBT will only intensify. Lastly, through proper planning and execution, NBT can act as one of the critical factors driving the future of humanity.

References

Abrehe, S., Girma, A., & Nigusse, A. G. (2021). Potential eco-tourism site suitability evaluation for sustainable natural resource management in Kafta Sheraro National Park (KSNP), north-western Tigray, Ethiopia. *Journal of Ecotourism, 20*(4), 341–370.

Al-Ababneh, M. (2013). Service quality and its impact on tourist satisfaction in Jordan. *An International Journal, 4*(12), 164–177.

Albrecht, J. N. (2021). The future of visitor experiences in nature-based tourism. In J. N. Albrecht (Ed.), *Managing visitor experiences in nature-based tourism* (pp. 205). London, UK: CABI.

Ali, M. B., Quaddus, M., Rabbanee, F. K., & Shanka, T. (2020). Community participation and quality of life in nature-based tourism: Exploring the antecedents and moderators. *Journal of Hospitality & Tourism Research*, 1096348020980094.

Almeida, A., & Machado, L. P. (2021). Rural development and rural tourism: The impact of infrastructure investments. In R. A. Castanho, G. Couto, & R. Santos (Eds.), *Peripheral territories, tourism, and regional development*. IntechOpen. https://doi.org/10.5772/intechopen.95610

Arora, K., Rajput, S., & Anand, R. R. (2020). Geomorphosites Assessment for the Development of Scientific Geo-tourism in North and Middle Andaman's, India. *Geojournal of Tourism and Geosites, 32*(4), 1244–1251.

Bekele, H., Teshome, E. & Asteray, M. (2017). Assessing protected areas for eco-tourism development: The case of Maze National Park, Ethiopia. *Journal of Hospitality Management and Tourism, 8*(3), 25–31.

Bianco, V. (2020). Analysis of electricity consumption in the tourism sector. A decomposition approach. *Journal of Cleaner Production, 248*, 119286.

Blumstein, D. T., Geffroy, B., Samia, D. S. M., Bessa, E. (2017). Introduction: Eco-tourism's promise and peril. In D. Blumstein, B. Geffroy, D. Samia, & E. Bessa (Eds.), *Eco-tourism's promise and peril*. Cham: Springer. https://doi.org/10.1007/978-3-319-58331-0_1

Buckley, R. (2006). *Adventure tourism*. CABI.

Buckley, R., & Coghlan, A. (2012). Nature-based tourism in breadth and depth. In Tej Vir Singh (Ed.), *Critical debates in tourism* (vol. 57, pp. 304–307). Griffith University, UK, USA and Canada: Channel View Publications.

Butt, B. (2012). Commoditising the safari and making space for conflict: Place, identity and parks in East Africa. *Political Geography, 31*(2), 104–113.

Carmody, J. (2008). Profiling specialist accommodation operators in the nature-based tourism sector of North Queensland. In Scott Richardson, Liz Fredline, Anoop Patiar, & Megan Ternel (Eds.), *CAUTHE 2008: Tourism and hospitality research, training and practice; "Where the 'Bloody Hell' Are We"*, Vol. 847. Gold Coast, Qld: Griffith University.

Ceccato, R., Deflorio, F., Diana, M., & Pirra, M. (2020). Measure of urban accessibility provided by transport services in Turin: A traveller perspective through a mobility survey. *Transportation Research Procedia, 45*(1), 301–308.

Chen, L., Thapa, B., & Yan, W. (2018). The relationship between tourism, carbon dioxide emissions, and economic growth in the Yangtze River Delta, China. *Sustainability, 10*(7), 2118.

Chin, C. H., Lo, M. C., Razak, Z. B., Pasbakhsh, P. & Mohamad, A. A. (2020). Resources confirmation for tourism destinations marketing efforts using PLS-MGA: The moderating impact of semirural and rural tourism destination. *Sustainability, 12*(17), 6787.

Chung, M. G., Dietz, T., & Liu, J. (2018). Global relationships between biodiversity and nature-based tourism in protected areas. *Ecosystem Services, 34*(1), 11–23.

Cisneros-Montemayor, A. M., Townsel, A., Gonzales, C. M., Haas, A. R., Navarro-Holm, E. E., Salorio-Zuñiga, T., & Johnson, A. F. (2020). *Nature-based marine tourism in the Gulf of California and Baja California Peninsula: Economic benefits and key species*. Paper presented at *the Natural Resources Forum*.

Coghlan, A. (2020). Mobilising knowledge: An action design research case study of using technology to safeguard an endangered nature-based tourism attraction. *Journal of Sustainable Tourism*, [ahead-of-print], 1–20.

Coghlan, A., & Carter, L. (2020). New product design: Creating a digital VR game to promote the conservation of nature-based tourism attractions. In A. Hassan & A. Sharma (Eds.), *The emerald handbook of ICT in tourism and hospitality*. Bingley, UK: Emerald Publishers.

da Mota, V. T., & Pickering, C. (2020). Using social media to assess nature-based tourism: Current research and future trends. *Journal of Outdoor Recreation and Tourism, 30*, 100295.

Davidoff, D. M. (1994). *Contact: Customer service in the hospitality and tourism industry*. Prentice-Hall Career and Technology.

Davidson, R. (2016). *Harnessing local expertise and knowledge: A new concept in business tourism destination marketing*. Paper presented at the *Book of Proceedings Edited by Milovan Stanišić in Singidunum International Tourism Conference*.

Duffy, R. (2015). Nature-based tourism and neoliberalism: Concealing contradictions. *Tourism Geographies, 17*(4), 529–543.

Dybedal, P., Haukeland, J. V., & Stemmer, K. (2021). Characteristics of different nature-based tourism activity markets. In Peter Fredman & Jan V. Haukeland (Eds.), *Nordic Perspectives on Nature-based Tourism*. Northampton, UK, USA: Edward Elgar Publishing.

Elmahdy, Y. M., Haukeland, J. V., & Fredman, P. (2017). *Tourism megatrends: A literature review focused on nature-based tourism* (2535–2806). Retrieved from https://nmbu.brage.unit.no/nmbu-xmlui/handle/11250/2648159

Filipe, S., Barbosa, B., & Santos, C. A. (2020). Nature-based tourism: A study on consumer behavior when participating in camping activities. In Devkant Kala & Satish Chandra Bagri (Eds.), *Global opportunities and challenges for rural and mountain tourism* (pp. 112–130). Hershey, USA: IGI Global.

Fletcher, R. (2019). Eco-tourism after nature: Anthropocene tourism as a new capitalist "fix". *Journal of Sustainable Tourism, 27*(4), 522–535.

Fossgard, K., & Fredman, P. (2019). Dimensions in the nature-based tourism experiencescape: An explorative analysis. *Journal of Outdoor Recreation and Tourism, 28*, 100219.

Fredman, P., Haukeland, J. V., Tyrväinen, L., Stensland, S., & Wall-Reinius, S. (2021). Nature-based tourism in a Nordic context. In Peter Fredman & Jan V. Haukeland (Eds.), *Nordic perspectives on nature-based tourism*. UK, USA: Edward Elgar Publishing.

Fredman, P., & Margaryan, L. (2021). 20 years of Nordic nature-based tourism research: A review and future research agenda. *Scandinavian Journal of Hospitality and Tourism, 21*(1), 14–25.

Fredman, P., Wall-Reinius, S., & Grundén, A. (2012). The nature of nature in nature-based tourism. *Scandinavian Journal of Hospitality and Tourism, 12*(4), 289–309.

Geary, A. (2018). *Marketing in the modern world: Tourism after terrorist attacks*. Laramie, Wyoming: University of Wyoming. Libraries.

Gehrke, S. R., Akhavan, A., Furth, P. G., Wang, Q., & Reardon, T. G. (2020). A cycling-focused accessibility tool to support regional bike network connectivity. *Transportation Research Part D: Transport and Environment, 85*, 102388.

Goffi, G. (2013). A model of tourism destination competitiveness: The case of the italian destinations of excellence (Un Modelo De Destino Turístico Competitivo: El Caso De Los Destinos Italianos De Excelencia). *Anuario turismo y sociedad, 14*, 121–147.

Guan, J., & Jones, D. L. (2015). The contribution of local cuisine to destination attractiveness: An analysis involving Chinese tourists' heterogeneous preferences. *Asia Pacific Journal of Tourism Research, 20*(4), 416–434.

Hall, C. M., & Boyd, S. W. (2005). *Nature-based tourism in peripheral areas: Development or disaster?* (Vol. 21). Toronto: Channel View Publications.

Hassan, S. T., Xia, E., Khan, N. H., & Shah, S. M. A. (2019). Economic growth, natural resources, and ecological footprints: Evidence from Pakistan. *Environmental Science and Pollution Research, 26*(3), 2929–2938.

Hung, K., Sirakaya-Turk, E., & Ingram, L. J. (2011). Testing the efficacy of an integrative model for community participation. *Journal of Travel Research, 50*(3), 276–288.

Jin, X., Weber, K., & Bauer, T. (2012). Impact of clusters on exhibition destination attractiveness: Evidence from Mainland China. *Tourism Management, 33*(6), 1429–1439.

Jones, N., McGinlay, J., & Dimitrakopoulos, P. G. (2017). Improving social impact assessment of protected areas: A review of the literature and directions for future research. *Environmental Impact Assessment Review, 64*(1), 1–7.

Jones, T. (2012). A life cycle analysis of nature-based tourism policy in Japan. *Journal of the German Institute for Japanese Studies Tokyo, 24*(2), 179–211.

Jones, T. E., Apollo, M., & Bui, H. T. (2021). Mountainous protected areas & nature-based tourism in Asia. In T. E. Jones, H. T. Bui, & M. Apollo (Eds.), *Nature-based tourism in Asia's mountainous protected areas* (pp. 3–25). Singapore: Springer.

Jones, T. E., & Nguyen, M. H. (2021). Nature-based tourism motivations and visit profiles of domestic and international segments to a Japanese National Park. *Quaestiones Geographicae, 40*(2), 77–92.

Kalayci, C. (2019). The impact of economic globalisation on CO2 emissions: The case of NAFTA countries. *International Journal of Energy Economics and Policy, 9*(1), 356.

Karalkova, Y. (2016). *Rural tourism destination competitiveness: Portugal vs Belarus* [Master's Thesis] Instituto Politécnico de Bragança.

Kim, D., & Perdue, R. R. (2011). The influence of image on destination attractiveness. *Journal of Travel & Tourism Marketing, 28*(3), 225–239.

Kim, H., Lee, S., Uysal, M., Kim, J., & Ahn, K. (2015). Nature-based tourism: Motivation and subjective well-being. *Journal of Travel & Tourism Marketing, 32*(1), S76–S96.

Kim, Y. J., Lee, D. K., & Kim, C. K. (2020). Spatial tradeoff between biodiversity and nature-based tourism: Considering mobile phone-driven visitation pattern. *Global Ecology and Conservation, 21*, e00899.

Kumar, S. (2016). Developing India as smart tourism destination–A Sap-Lap analysis. *South Asian Journal of Tourism and Heritage, 9*, 124–136.

Kunasekaran, P. & Kumar, J. (2021). Rethinking engagements with indigenous communities through tourism. In S. K. Walia (Ed.), *Community based tourism management: Concepts, issues & implications*. London, UK: Routledge- Taylor & Francis Group.

Landauer, M., Goodsite, M. E., & Juhola, S. (2018). Nordic national climate adaptation and tourism strategies–(how) are they interlinked? *Scandinavian Journal of Hospitality and Tourism, 18*(1), S75–S86.

Lee, C. F. (2016). An investigation of factors determining industrial tourism attractiveness. *Tourism and Hospitality Research, 16*(2), 184–197.

Lenzen, M., Sun, Y.-Y., Faturay, F., Ting, Y.-P., Geschke, A., & Malik, A. (2018). The carbon footprint of global tourism. *Nature Climate Change, 8*(6), 522–528.

Lindberg, K., & Lindberg, K. (1991). *Policies for maximising nature tourism's ecological and economic benefits*. Washington, DC: World Resources Institute.

Luthe, T., Wyss, R., & Schuckert, M. (2012). Network governance and regional resilience to climate change: empirical evidence from mountain tourism communities in the Swiss Gotthard region. *Regional Environmental Change, 12*(4), 839–854.

Ma, E., Hsiao, A., & Gao, J. (2018). Destination attractiveness and travel intention: The case of Chinese and Indian students in Queensland, Australia. *Asia Pacific Journal of Tourism Research, 23*(2), 200–215.

Ma, H., Chiu, Y.-H., Tian, X., Zhang, J., & Guo, Q. (2020). Safety or travel: Which is more important? The impact of disaster events on tourism. *Sustainability, 12*(7), 3038.

Maldonado-Oré, E. M., & Custodio, M. (2020). Visitor environmental impact on protected natural areas: An evaluation of the Huaytapallana Regional Conservation Area in Peru. *Journal of Outdoor Recreation and Tourism, 31*, 100298.

Margaryan, L. (2017). *Commercialisation of nature through tourism*. Sweden: Mid Sweden University.

Margaryan, L. (2018). Nature as a commercial setting: The case of nature-based tourism providers in Sweden. *Current Issues in Tourism, 21*(16), 1893–1911.

Markowski, J., Bartos, M., Rzenca, A., & Namiecinski, P. (2019). An evaluation of destination attractiveness for nature-based tourism: Recommendations for the management of national parks in Vietnam. *Nature Conservation, 32*, 51.

Mehmetoglu, M. (2007). Nature-based tourism: A contrast to everyday life. *Journal of Ecotourism, 6*(2), 111–126.

Mogomotsi, G. E. J., Mogomotsi, P. K., Gondo, R., & Madigele, T. J. (2018). Community participation in cultural heritage and environmental policy formulation in Botswana. *Chinese Journal of Population Resources and Environment, 16*(2), 171–180.

Mogomotsi, P. K., Stone, L. S., Mogomotsi, G. E. J., & Dube, N. (2020). Factors influencing community participation in wildlife conservation. *Human Dimensions of Wildlife, 25*(4), 372–386.

Mohanty, P., Dhoundiyal, H., & Thomas, A. (2022). Technological innovations in Asian tourism. In Azizul Hassan (Ed.), *Technology application in tourism in Asia* (pp. 69–80). Singapore: Springer.

Moric, I. (2013). Clusters as a factor of rural tourism competitiveness: Montenegro experiences. *Business Systems Research: International journal of the Society for Advancing Innovation and Research in Economy, 4*(2), 94–107.

Nagle, D. S., & Vidon, E. S. (2021). Purchasing protection: Outdoor companies and the authentication of technology use in nature-based tourism. *Journal of Sustainable Tourism, 29*(8), 1253–1269.

Nanchen, B., Margot-Cattin, P., Ramseyer, R., & Schegg, R. (2020). *Improving the accessibility of touristic destinations with an assistive technology for hiking*. Paper presented at the *Proceedings of the 17th International conference on computers helping people with special needs*.

Ndivo, R., Waudo, J., & Waswa, F. (2012). Examining Kenya's tourist destinations' appeal: The perspectives of domestic tourism market. *Journal of Tourism and Hospitality, 1*(5), 1–7.

Newsome, D., Rodger, K., Pearce, J., & Chan, K. (2019). Visitor satisfaction with a key wildlife tourism destination within the context of a damaged landscape. *Current Issues in Tourism, 22*(6), 729–746.

Nooripoor, M., Khosrowjerdi, M., Rastegari, H., Sharifi, Z., & Bijani, M. (2021). The role of tourism in rural development: Evidence from Iran. *GeoJournal, 86*(4), 1705–1719.

O'Neill, M. A., Riscinto-Kozub, K. A., & Van Hyfte, M. (2010). Defining visitor satisfaction in the context of camping oriented nature-based tourism—the driving force of quality! *Journal of Vacation Marketing, 16*(2), 141–156.

Øian, H., Fredman, P., Sandell, K., Sæþórsdóttir, A. D., Tyrväinen, L., & Jensen, F. S. (2018). Tourism, nature and sustainability: A review of policy instruments in the Nordic countries. Denmark: Nordisk Ministerråd.

Owiyo, V. (2018). *Strategic determinants of destination competitiveness: A case of Western tourist circuit, Kenya*. Rongo, Kenya: Rongo University.

Pamungkas, W., & Jones, T. E. (2021). Indonesia's mountainous protected areas: National parks and nature-based tourism. In Thomas E. Jones, Huong T. Bui, & Michal Apollo (Eds.), *Nature-based tourism in Asia's mountainous protected areas* (pp. 111–131). Cham, Switzerland: Springer.

Paraschi, E. P. (2020). Accessibility, tourism and social welfare: Covid19 and a new quality-of-life tourism model for the Greek islands. *International Journal of Cultural and Digital Tourism, 6*, 10–21.

Park, D., Kim, J., Kim, W. G., & Park, H. (2019). Does distance matter? Examining the distance effect on tourists' multi-attraction travel behaviors. *Journal of Travel & Tourism Marketing, 36*(6), 692–709.

Parker, S., & Khare, A. (2005). Understanding success factors for ensuring sustainability in eco-tourism development in southern Africa. *Journal of Ecotourism, 4*(1), 32–46.

Prakash, S. L., Perera, P., Newsome, D., Kusuminda, T., & Walker, O. (2019). Reasons for visitor dissatisfaction with wildlife tourism experiences at highly visited national parks in Sri Lanka. *Journal of Outdoor Recreation and Tourism, 25*, 102–112.

Priskin, J. (2001). Assessment of natural resources for nature-based tourism: The case of the Central Coast Region of Western Australia. *Tourism Management, 22*(6), 637–648.

Provotorina, V., Kazmina, L., Petrenko, A., & Ekinil, G. (2020). *Organisation and functioning of accommodation facilities as a component of rural tourism infrastructure in the Rostov Region.* Paper presented at *the E3S Web of Conferences.*

Puška, A., Šadić, S., Maksimović, A., & Stojanović, I. (2020). Decision support model in the determination of rural touristic destination attractiveness in the Brčko District of Bosnia and Herzegovina. *Tourism and Hospitality Research, 20*(4), 387–405.

Qiu, M., Sha, J., & Scott, N. (2021). Restoration of visitors through nature-based tourism: A systematic review, conceptual framework, and future research directions. *International Journal of Environmental Research and Public Health, 18*(5), 2299.

Ramseook-Munhurrun, P., Naidoo, P., Seebaluck, V., & Pillai, P. (2016). *The impact of destination service quality on tourist satisfaction and loyalty: Evidence from Mauritius.* Paper presented at the Proceedings of the *International Academic Research Conference on Marketing & Tourism* (MTCI16 Paris Conference).

Ranasinghe, R., & Sugandhika, M. (2018). The contribution of tourism income for the economic growth of Sri Lanka. *Journal of Management and Tourism Research, 1*, 67–84.

Rasoolimanesh, S. M., Ringle, C. M., Jaafar, M., & Ramayah, T. (2017). Urban vs. rural destinations: Residents' perceptions, community participation and support for tourism development. *Tourism Management, 60*, 147–158.

Reitsamer, B. F., Brunner-Sperdin, A., & Stokburger-Sauer, N. E. (2016). Destination attractiveness and destination attachment: The mediating role of tourists' attitude. *Tourism Management Perspectives, 19*, 93–101.

Rogerson, C. M., & Rogerson, J. M. (2020). COVID-19 and tourism spaces of vulnerability in South Africa. *African Journal of Hospitality, Tourism and Leisure, 9*(4), 382–401.

Romero, I., & Tejada, P. (2011). A multi-level approach to the study of production chains in the tourism sector. *Tourism Management, 32*(2), 297–306.

Sadowski, A., & Wojcieszak, M. M. (2019). Geographic differentiation of agritourism activities in Poland vs. cultural and natural attractiveness of destinations at district level. *PloS one, 14*(9), e0222576.

Seraphin, H., & Dosquet, F. (2020). Mountain tourism and second home tourism as post COVID-19 lockdown placebo? *Worldwide hospitality and tourism themes.*

Sharif, A., Godil, D. I., Xu, B., Sinha, A., Khan, S. A. R., & Jermsittiparsert, K. (2020). Revisiting the role of tourism and globalisation in environmental degradation in China: Fresh insights from the quantile ARDL approach. *Journal of Cleaner Production, 272*, 122906.

Silva, C., Marques, J., & Abrantes, J. L. (2020). *Nature-based tourism travel planning: A Generational approach.* Paper presented at the *International Conference on Tourism, Technology and Systems.*

SLTDA (2019). Sri Lanka Tourism Development Authority. Tourism Research and Statistics. Retrieved from http://www.sltda.lk/statistics

Su, M. M., & Wall, G. (2014). Community participation in tourism at a world heritage site: Mutianyu Great Wall, Beijing, China. *International Journal of Tourism Research, 16*(2), 146–156.

Suganuma, Y., Hibino, N., & Morichi, S. (2011). Trend analysis of domestic tourist travel in Japan based on individual data from tourism statistics. *Journal of the Eastern Asia Society for Transportation Studies, 9*, 826–841.

Tang, Z., Bai, S., Shi, C., Liu, L., & Li, X. (2018). Tourism-related CO2 emission and its decoupling effects in China: A spatiotemporal perspective. *Advances in Meteorology, 2018*, 1–9. https://doi.org/10.1155/2018/1473184

Tervo, K. (2008). The operational and regional vulnerability of winter tourism to climate variability and change: The case of the Finnish nature-based tourism entrepreneurs. *Scandinavian Journal of Hospitality and Tourism*, *8*(4), 317–332.

Tessema, M. E., Lilieholm, R. J., Ashenafi, Z. T., & Leader-Williams, N. (2010). Community attitudes toward wildlife and protected areas in Ethiopia. *Society and Natural Resources*, *23*(6), 489–506.

Tinakhat, P., Viriyachaikul, V., & Vorasingha, Y. (2018). The international guests' satisfaction about service quality of guesthouse in Sukhothai. *Journal of Tourism Theory and Research*, *4*(2), 57–69.

Tyrväinen, L., Uusitalo, M., Silvennoinen, H., & Hasu, E. (2014). Towards sustainable growth in nature-based tourism destinations: Clients' views of land use options in Finnish Lapland. *Landscape and Urban Planning*, *122*, 1–15.

Wang, S., Du, J., Li, S., He, H., & Xu, W. (2019). Impact of tourism activities on glacial changes based on the tourism heat footprint (THF) method. *Journal of Cleaner Production*, *215*, 845–853.

Wang, Y., Wu, C., Wang, F., Sun, Q., Wang, X., & Guo, S. (2021). Comprehensive evaluation and prediction of tourism ecological security in droughty area national parks—a case study of Qilian Mountain of Zhangye section, China. *Environmental Science and Pollution Research*, *28*(13), 16816–16829.

Wang, Y., Zhou, T., Chen, H., & Rong, Z. (2019). Environmental homogenisation or heterogenisation? The effects of globalisation on carbon dioxide emissions, 1970–2014. *Sustainability*, *11*(10), 2752.

Wearing, S., & Schweinsberg, S. (2018). *Eco-tourism: Transitioning to the 22nd century*. Routledge.

Wheaton, M., Ardoin, N. M., Hunt, C., Schuh, J. S., Kresse, M., Menke, C., & Durham, W. (2016). Using web and mobile technology to motivate pro-environmental action after a nature-based tourism experience. *Journal of Sustainable Tourism*, *24*(4), 594–615.

Winter, P. L., Selin, S., Cerveny, L., & Bricker, K. (2020). Outdoor recreation, nature-based tourism, and sustainability. *Sustainability*, *12*(1), 81.

Wirajaya, I., Rahmafitria, F., Nurazizah, G., & Jamin, A. (2021). *The effect of individual and destination accessibility on willingness to visit: Nature-based tourism destination*. Paper presented at the *Promoting Creative Tourism: Current Issues in Tourism Research: Proceedings of the 4th International Seminar on Tourism (ISOT 2020)*, November 4–5, 2020, Bandung, Indonesia.

Xiang, C., & Yin, L. (2020). Study on the rural eco-tourism resource evaluation system. *Environmental Technology & Innovation*, *20*, 101131.

Yaru, L., Liu, X., & Jing, M. (2018). Study on the Quality of Service in Rural Homestay-Taking Shanli Lohas as an Example. *J Tourism Hospit*, 7(4), 370.

Yen, H. P., Chen, P. C., & Ho, K. C. (2021). Analysing Destination Accessibility From the Perspective of Efficiency Among Tourism Origin Countries. *SAGE Open*, *11*(2). https://journals.sagepub.com/doi/10.1177/21582440211005752

Yuxi, Z., & Linsheng, Z. (2020). Identifying conflicts tendency between nature-based tourism development and ecological protection in China. *Ecological indicators*, *109*, 105791.

Zaei, M. E., & Zaei, M. E. (2013). The impacts of tourism industry on host community. *European Journal of Tourism Hospitality and Research*, *1*(2), 12–21.

Zawadka, J. (2019). Agritourism as a way of spending free time of urban families with children. *Roczniki (Annals), 2019*(1230-2020-782).

Zhang, S., Chan, C.-S., Liu, J., & Zhu, H. (2020). Different stakeholders' perceptions and asymmetric influencing factors towards nature-based tourism in Hong Kong. *Asia Pacific Journal of Tourism Research*, *25*(8), 889–901.

Zhou, L. (2014). Online rural destination images: Tourism and rurality. *Journal of destination marketing & management*, *3*(4), 227–240.

Zhu, Y., & Diao, M. (2020). Crowdsourcing-data-based dynamic measures of accessibility to business establishments and individual destination choices. *Transportation Research Part D: Transport and Environment*, *87*, 102382.

PART V

The future of nature-based tourism

31

NATURE-BASED TOURISM IN LIGHT OF COVID-19

Anna Spenceley

Impacts of the COVID-19 pandemic on the nature-based tourism sector

Prior to the COVID-19 pandemic, protected areas received roughly 8 billion visits annually, and generated approximately US$600 billion per year in direct in-country expenditures and US$250 billion per year in consumer surplus (Balmford et al., 2015). Furthermore, wildlife tourism, according to the World Travel & Tourism Council (WTTC), contributed 4.2 percent of direct in-country expenditure and 6.8 percent of employment globally (WTTC, 2019). In 2014, the entrance, gate and park fees from all types of protected areas in 14 sub-Saharan African countries were estimated by the UNWTO at US$142 million per year (UNWTO, 2015). Safari tourism in Kenya alone generated an estimated US$1.1 billion in 2019 (ALUSWC, 2020). Revenues from tourism – whether through entrance fees paid by tourists, or from concession and lease fees paid by tourism operators – generate a substantial proportion of budgets of conservation areas in some countries. Certain protected areas retain tourism revenues for local conservation and management activities; for example, areas under co-management agreements, conservancies and private protected areas. Other protected areas' tourism revenues go to a central budget, and the protected areas receive government budget allocations, particularly state-operated protected areas. Many tourism companies make further contributions to conservation, including species and habitat conservation, directly through their operations, or by channelling donations towards them (Snyman and Spenceley, 2019).

Local people also benefit from economic linkages with nature-based tourism through employment, by selling products and services that tourism companies and tourists need, or by owning tourism businesses, or through revenue-sharing systems (World Bank, 2020). The direct economic effects can be substantial at both local and national levels. To illustrate, the total economic effects of tourism in South Luangwa National Park in Zambia comes to US$34.8 million, which includes US$7.4 million in local wages and 1,759 jobs (Souza et al., 2021). Aside from direct financial benefits, local communities may benefit from infrastructure development, telecommunications, education, training and healthcare, as well as strengthening communities through skills development, empowerment and improved governance (Spenceley, 2021a).

However, many of these benefits are at risk due to dramatic reductions in wildlife populations and natural habitats, coupled with increases in wildlife crime and habitat conversion and

DOI: 10.4324/9781003230748-37

climate change. The world is facing a triple planetary crisis relating to the climate, pollution, and nature. Therefore, it is becoming increasingly important for governments and businesses to shift toward more sustainable and circular models in the tourism sector (Spenceley, 2022):

- **Climate**: Tourism is an ever-increasing contributor to climate change. Pre-pandemic, it accounted for about 8–11 percent of global GHG emissions, with 3.9–5.4 billion tons of CO_2 emissions in 2019 (WTTC, 2021). The majority of tourism emissions are exerted by and in high-income countries for transport and accommodation (UNWTO and UNEP, 2008). Tourism is also highly vulnerable to climate change, including extreme weather events, loss of biodiversity, damages to assets and increasing insurance costs and safety concerns.
- **Pollution**: Tourism contributes to air, water and land pollution: the transport and electricity generated can cause air pollution, while water contamination may be caused by sewage, solid waste, petrol and toxins entering the environment from tourism facilities or discarded by tourists. Solid waste, particularly plastic waste, can harm wildlife and create toxic environments and unsightly beaches, oceans and landscapes. Every year, humans produce 300 million tons of plastic waste, including 11 million tons that end up in the ocean (UNEP, 2020). As 80 percent of all tourism takes place in coastal areas, plastic from the sector can be a large contributor to this pollution (e.g. marine litter can increase by up to 40 percent during the peak tourist season in the Mediterranean region) (ibid.).
- **Nature**: Biodiversity is vital for tourism; coasts, mountains, rivers and forests and their wildlife are major tourist attractions across the world. However, there have been severe reductions in wildlife populations and natural habitats globally, coupled with increases in wildlife crime and habitat conversion (Almond et al., 2020). Tourism can have both positive and negative impacts on nature. Negative impacts include the degradation of air and water quality, increased water use, changes in landforms through infrastructure development, mineral and energy consumption, disturbance or destruction of wildlife habitat, behavioural change in animals, increased pressure on endangered species, introduction of invasive alien species (including for hotel gardens), land-based pollution, and direct mortality (e.g. from vehicle collisions, hunting, fishing or disease transmission). Indeed, the origin of zoonotic disease pandemics and epidemics, such as COVID-19, lie in our unsustainable exploitation of nature (Patz et al., 2004) – emphasizing how interlinked humans are with their environment (Hockings et al., 2021).

Much of the industry has focused on volumes of visitors and returns on investment rather than long-term sustainable destination management. One of the few reliable indications of sustainable practices is independent certification, but this is only used by a minority of accommodations, tour operators and destinations (Spenceley, 2018). Collectively, these problems threaten human health and the health of the planet (Spenceley, 2022).

In 2020 the COVID-19 pandemic impacted millions of human lives, people's wellbeing, and also the global tourism sector. Protected areas globally closed due to travel restrictions in efforts to slow the spread of the virus while scientists sought treatment and established guidance on preventative measures. The collapse of nature-based tourism globally – particularly between March and May 2020[1] – led to lower salaries for tourism employees, and reduced incomes for entrepreneurs and small businesses that provide products and services within the tourism value chain (Spenceley, 2021a). Many analyses emerged over time, considering the social, economic, and environmental impacts of the pandemic at global, continental, and destination level, and its implications for sustainable nature-based tourism and particularly implications for poverty reduction in developing countries (e.g. Spenceley, 2021a, 2021b; Spenceley, McCool, Newsome et al., 2021; Spenceley, 2022). This chapter provides an overview of some of the documented responses by NBT operators and destinations, tools and resources to adjust management and support sustainable practices in light of the pandemic, and identifies future research priorities and necessary interventions.

Responses by operators and destinations

As tourism came to a standstill at the start of the pandemic, companies innovated in order to survive. Those that already had diverse products and services were able to galvanize alternative income sources. For example, Gibbs Farm in Tanzania offers a luxury lodge on an organic farm and coffee estate on the edge of the Ngorongoro Crater. Despite a drop in tourism, its coffee harvest continued and revenues from the farm were used to feed staff and their families (O'Brien, 2020). In addition, donations were sought to support organizations and staff that relied on tourism income, including through crowd-funding, with some calls generating thousands of dollars of support (Spenceley, 2001a).

Then, as domestic and international travel restrictions lifted, NBT operations could adapt the products and services provided to emphasis COVID-19 preventative measures, and also adjust their source markets to adjust to changes in demand (ATTA, 2020). For example, safari operators that had formerly relied on international clients began providing packages and experiences for domestic clients while also stimulating new interest in conservation and nature-based tourism locally (Spenceley, 2021a). Tourist accommodation was also offered to local people to safeguard or quarantine vulnerable members of society. For example, Bulungula Lodge on South Africa's Wild Coast used its premises as a place for local community members to quarantine, particularly the elderly and those with pre-existing conditions that are at most risk from the virus (Ateino, 2020). Artisans that usually generated income from selling craft to tourists adapted too – for example, the Ku-Humelala Craft Group in South Africa began making fabric face masks to protect people from COVID-19 (and Beyond, 2020). Technology also played a key role, and virtual tours set up by protected areas and nature-based operators received considerable traction and interest across the world. While some of these are free and used to sustain interest (and lift our spirits), others have been fee-based and used to sustain conservation revenues and pay local staff salaries (Spenceley, 2021c). Couch safaris, virtual classrooms, and other innovations provided a means for would-be-tourists and new generations of travellers to experience natural attractions from their homes and inspire future travel plans (Spenceley, 2021c, Table 31.1).

However, as travel restrictions lifted, many natural destinations in North America, Europe and parts of Asia were flooded with new cohorts of domestic visitors (who were often still restricted from international travel), and these areas were sometimes overwhelmed by crowding and waste left behind. For example, in Germany's Black Forest National Park, there was an increase of 100,000 visits between April and June 2020 compared with the same period in 2019 (a 50 percent increase), coupled with challenges where new visitors have not complied with park rules (Spenceley, McCool, Newsome et al., 2021). Also, in the UK's Devon and Snowdonia National Parks – with increased camping, wood collection, makeshift toilets, anti-social behaviour and litter – birds like the nightjar vanished (Fyfe, 2020) and as a result emergency powers were used to temporarily ban camping in August 2020 (Barkham, 2020). Also in the UK, visitors disturbed nesting birds and seal rookeries in the Holkham Estate nature

Table 31.1 Examples of innovative nature-based tourism product resilience and diversification amid COVID-19

Category	*Examples*
Agriculture and food products	Gibbs Farm offers a luxury lodge on an organic farm and coffee estate on the edge of the Ngorongoro Crater in Tanzania. Despite a drop in tourism, the coffee harvest continued and produce from the farm fed staff and their families (O'Brien, 2020). Tutka Bay Lodge on the edge of Kachemak Bay State Park in Alaska grows most of the produce used in the kitchen in greenhouses and gardens on site. While rooms were empty, staff made sausage, pickled and shrimped, learned new skills and developed new offerings (O'Brien, 2020). A drop in customers for community-based tourism, such as homestays, in the Spiti Valley in India were offset by a focus on agriculture as a main source of income for local communities (Equations, 2020).
Community support	Bulungula Lodge on South Africa's Wild Coast used its premises as a place for local community members to quarantine, particularly the elderly and those with pre-existing conditions that are at most risk from the virus (Ateino, 2020). Tourism in Need collected donations to support organizations that rely on income from tourism. They include social businesses that help youth, community development and wildlife conservation (Tourism in Need, 2020).
Competitions	The Ol Petjeja Conservancy in Kenya, through its Art of Survival fund, organized an art competition that people pay to enter and can win a fully paid trip to the conservancy once the pandemic is over (DHDNR, 2020).
Craft products	The Ku-Humelala Craft Group in South Africa pivoted from making crafts to sell to tourists to making fabric face masks to protect people from COVID-19 (and Beyond, 2020). The Vukukhanya development project supported women's groups around iSimangaliso Wetland Park and Tembe Elephant Park in South Africa, where they made masks for community members and planted vegetable gardens for food security (Vukukhanye Development Project, 2020). Launched in January 2020, the Sijwa Project has re-purposed all recyclable waste from the African Monarch lodges and worked with the local community to transform them into exquisite, saleable crafts and art. The project has trained glass recycling artisans to transform bottles – that often end up in rubbish pits or littering escarpments – into beautiful jewellery. A sewing workshop has been established in collaboration with The Collective Boutiques and Myeisha Namibia to create shopping bags (Spenceley, 2021f).
Source markets	Mbazi Safaris found innovative ways to tap into the South African diaspora and previous visitors to the Greater Kruger National Park. They were invited to sponsor a game drive experience for neighbouring communities once lockdowns were lifted, showing how domestic tourism could lead recovery (Boundless Southern Africa, 2020) Camping Khapela created packages and payment terms for the hugely untapped township tourism market to Transfrontier Conservation Areas and other conservation landscapes in Southern Africa (Boundless Southern Africa, 2020)

(*Continued*)

Table 31.1 (Continued)

Category	*Examples*
Virtual tours	Virtual 'visits' have been promoted to the Grand Canyon, Yosemite, Zion national parks in the US, Mount Everest in Nepal, Phong Nha-Ke Bang National Park in Vietnam, the Perito Moreno glacier in Argentina, and the Giants Causeway in Northern Ireland (Wilson, 2020; Spenceley, 2021c). Virtual underwater experiences have been created by dive and snorkel tour operators in New Zealand (Booking.com, 2020). In Namibia, Na-Beat hosted a virtual festival from local Nama communities near the Richtersveld World Heritage site (Boundless Southern Africa, 2020). An increase in remote access to Mexico's forests expanded access to nature-based tourism experiences to new audiences, providing immersive conservation experiences (Jack of all trades, 2020).

Source: Drawn from Spenceley, 2021a.

reserve, nitrous oxide canisters were left on Marsend Moor, illegal boating took place in the Ore estuary, and huge quantities of human excrement have been left in the Lake District as visitors avoided public toilets due to coronavirus fears (Barkham, 2020).

The openings of National Parks and associated crowding have led to some visitors being turned away, even after travelling for hundreds of miles to reach natural areas (Marsh, 2020). To illustrate, in October 2020, Wales introduced a ban on people travelling to the region from hotspots in England, Scotland and Northern Ireland (Morris, 2020). A partial reopening of Yellowstone National Park in the US in May saw thousands of domestic visitors descend on the park after lockdown (Gilbert, 2020). And at Rinjani Geopark in Indonesia the pre-COVID-19 challenges of overcrowding re-emerged as parks reopened, despite the continued presence of the virus (Spenceley, McCool, Newsome et al., 2021).

With these re-openings and resurgence of crowds, there have been concerns about the potential to spread COVID-19 in protected areas and into gateway communities, particularly where social distancing is not observed and face coverings are not worn (Gilbert, 2020), or where vaccination rollouts have been slow. There has been a clear need to prevent the trend of going from undertourism back to overtourism (not least because of the health threat of the coronavirus, but also for social well-being of local populations), and to find better ways to disperse tourists within and between destinations (Berge, 2020; Tripadvisor, 2020).

Tools and resources to adjust management and support sustainability

The pause in tourism during 2020 gave many tourism and conservation professionals an opportunity to reflect on whether the sector will inevitably go back to the way it was, or if it will build back in a more responsible, sustainable and regenerative way. Numerous international policies, guidance documents and online discussions push for a more sustainable future. Leaders within the tourism sector play an active role in this drive toward a better future, and active partnerships have arisen or been strengthened as stakeholders seek to galvanize and mainstream increased demand for sustainable travel (see Spenceley, 2021a).

With the backdrop of existing resources to support sustainable nature-based tourism, such as the IUCN's Best Practice Guidelines on Tourism and Visitor Management in Protected areas (Leung, Spenceley, Hvengaard and Buckley, 2018) and the World Bank's Tools and Resources for Nature-Based Tourism (World Bank, 2020), numerous new guidance materials emerged to support protected areas and nature-based tourism operators emerge from the pandemic (e.g. see Spenceley, 2020, for over 1500 online resources).

Perhaps the one of the most notable resources to emerge was a synthesis of materials published by the European Union (EU), following international peer review within the IUCN's World Commission on Protected Areas, and international agencies including UNESCO and WWF. The EU's 'Tourism and Visitation to protected areas amid COVID-19: Guidance for protected area authorities and managers' (Spenceley, 2021d). This arose from a recognition that protected area managers were under pressure to provide facilities safely to visitors as destinations lifted their travel restrictions, but that there were challenges to doing so given rapidly changing conditions and new health and safety requirements related to the virus. Providing pragmatic guidance to protected area managers and authorities on operating tourism safely amid the COVID-19 crisis, the resource is accompanied by supplementary hyperlinks to reputable online sources and a range of examples. The material includes links to background on the virus itself from the World Health Organization and Centres for Disease Control and Prevention, and contextual data on the impacts of the pandemic on the tourism sector and implications for protected areas. Then, it shares examples of protected areas that had closed and re-opened during the pandemic, with associated guidance for doing so safely. A checklist of actions protected area managers could undertake was shared (see Box 31.1), and accompanied by examples of protocols and guidance from protected area agencies across the world, including

Box 31.1 Checklist for protected area managers re-opening amid COVID-19 (Spenceley, 2021d)

- Coordinate with other relevant agencies on policy and information exchange.
- Assessment and feasibility of screening visitors for evidence of vaccination or recovery from COVID-19, and for possible signs of COVID-19 infection.
- Inform the public that the protected area or trail could be closed again if visitors fail to follow the rules or local infections have increased.
- Limit the total number of visitors in sensitive or heavily visited parks, or attractions within parks.
- Disperse visitors away from heavily visited parks to alternative sites.
- Limit the size of informal groups, e.g. to single households or specific numbers.
- Allow customary activities of formal groups (e.g. scouts, birders and school classes, when their leaders agree to enforce the rules).
- Require social distancing of at least 2 metres/6 feet. Require face coverings.
- Protect sources of drinking water or require visitors to bring their own.
- Toilets: leave open and keep clean.
- Admission, rentals and other charges: suspend or make cashless.
- Trails: limit the number of hikers, bikers, and equestrians on a trail. Close narrow trails temporarily or make them one-way
- Beaches: limit to active land and water recreation.
- Rivers and lakes: regulate boating, swimming, etc., as required.

Australia, Belgium, Brazil, Costa Rica, Egypt, France, Italy, Korea, South Africa, the UK, and the USA. The guidance further included links to risk assessment tools and standards, and also how to manage interactions between wildlife and people (recognizing that most zoonotic diseases arise from unsustainable exploitation of nature) (Jones et al., 2008), and more frequent contact makes it more likely that potential pathogens will jump from wildlife to humans (and, in some circumstances, from humans to wildlife) (Hockings et al., 2020).

While nature-based tourism is well-placed to provide open-air, socially distanced recreational experiences, the material appreciates the risks of overcrowding as the public emerge from lockdowns and quarantines, and seek out natural spaces *en masse*. Therefore, it provides links to visitor management guidance, including the IUCN Best Practice Guidelines (Leung et al., 2018) and the Visitor Use Management Framework (IVUMC, 2019), along with advice on how to undertake consultation and communicate with local people and stakeholders remotely as responses are formulated. One of the most critical elements of a COVID-19 response strategy is communication with visitors, and therefore information and examples consider information required before people arrive (e.g. where and when they can visit, how to make reservations) and during their visits (e.g. what to expect from screening or proof of vaccination, paperless payments, health and safety considerations, and signage). Different types of tourist activities pose different risks and require different options to reduce the transmission of the virus. These relate to the equipment used, where the activities take place, and whether they involve groups of people (Spenceley, 2021d). Therefore, materials on tours and activities, accommodation, food and drink and workforce considerations are also shared. Pragmatically, there are also links to financing support and other options of generating income for conservation and local livelihoods where protected areas are not yet fully open (such as virtual safaris). At the time of writing, this material was available in 7 languages enabling broad dissemination.

In light of the pandemic, the IUCN's WCPA Tourism and Protected Areas collaborated with many collaborators to produce an urgent Motion to the IUCN World Conservation Congress in 2021, on 'Strengthening sustainable tourism's role in biodiversity conservation and community resilience' (IUCN, 2021). IUCN members passed the motion by a substantial majority, recognizing that more effective implementation of sustainable nature-based tourism strategies and benefit-sharing programmes were an integral part of community recovery to strengthen resilience during future adverse events and to build a more sustainable future for people and wildlife. They committed the IUCN to put nature-based tourism firmly in its recovery agenda, and for members to: (a) support development of diversified sustainable livelihood activities, skill-training programmes and alternative protein-sourcing markets in tourism-dependent communities to bolster community resilience against current and future adverse events; (b) establish enterprise-based partnerships to incorporate conservation and biodiversity monitoring across the tourism supply chain; and (c) establish more sustainable financing campaigns, including endowment funds, to support key biodiversity assets during tourism industry recessions (IUCN, 2021).

Investing in sustainable ventures makes sense, and investors will allocate capital where they see long-term returns, for example, by investing in companies with great staff relations, sustainable practices and solid climate strategies (Spenceley, 2021e). Many tourism operators see becoming sustainable as too hard, but in reality, it is not difficult, and if we are unable to remove the climate crisis, all other sustainability elements are irrelevant (WTTC, 2020). Others simply do not understand the essential elements of sustainability or know where to begin. And therefore the 'Handbook for sustainable tourism practitioners: The essential toolbox' was produced during the pandemic to provide better guidance for operators, destinations, and governments on the step-by-step processes needed (Spenceley, 2021f). Like this Handbook, the toolbox forms part of the Edward Elgar series, and focusses on many elements of protected area and nature-based tourism, the handbook shares expertise from more than 50 expert practitioners on: sustainable planning and design; enhancing the sustainability of existing tourism; balancing overtourism and undertourism through sound visitor management; and monitoring and evaluation. Specifically, the monitoring of economic impacts of tourism in the long term, Visitors Count! can be used by destinations to establish how poor, local people, and broader economies benefit from tourism (Spenceley, Schägner, Engels, et al., 2021). The guidelines outline how demonstrating the positive impact of protected areas can lead to greater buy-in and ownership of conservation practices, less poaching and land encroachment, and help offset human–wildlife conflict where it occurs. It is clear that once managers understand the number and behaviour of visitors they host, and the revenues and costs they generate, informed decisions on management plans and tourism strategies can be made. Such economic analyses have been used to clearly illustrate visitor spending effects, jobs, labour income, value added, and economic output – and how the pandemic dramatically affected these indicators in 2020 (see Figure 31.1).

Spenceley, McCool, Newsome et al. (2021) recommend that planning for tourism should become more holistic, inclusive, equitable and adaptable and focused on the question of what tourism can sustain. Many of the communities involved are particularly vulnerable to change

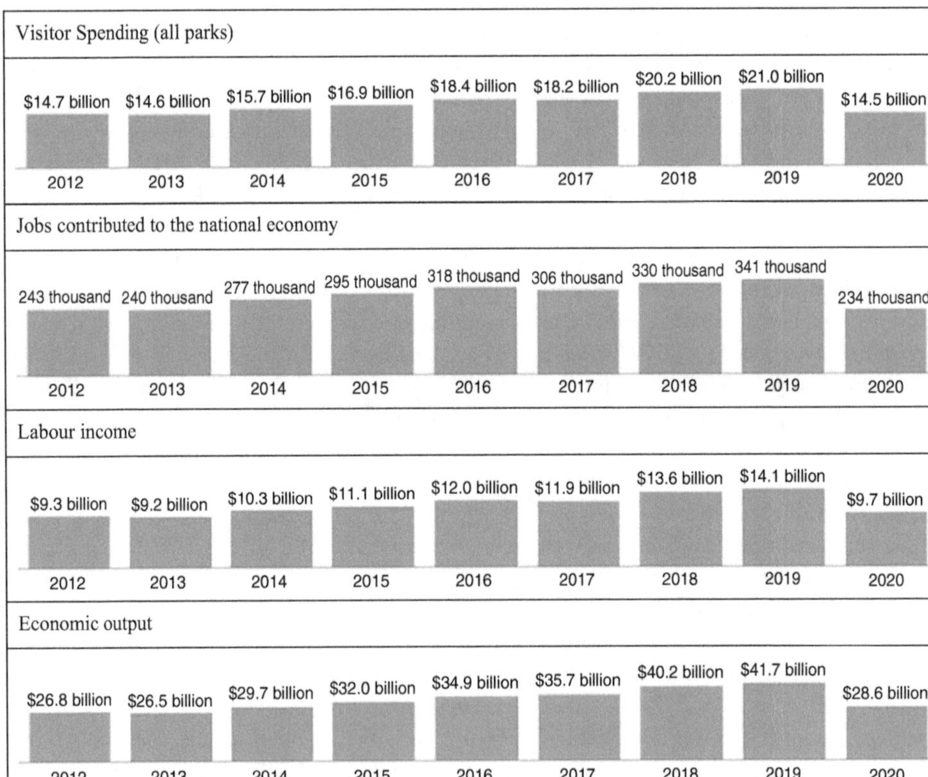

Figure 31.1 US National Parks Service economic evaluations of visitor spending by millions of visitors, and transition over the pandemic period (2012-2020).

Source: (National Parks Service, 2022).

because of distances from markets and their dependence on natural resources for livelihoods. The way forward for tourism could have five dimensions: (1) fostering openness to change, with a willingness to embrace new ways of thinking and acting; (2) developing a vision for the tourism offer of the future; (3) protecting biodiversity for its importance to the ecology of the area and peoples' dependency on tourism; (4) recovering and rebuilding local livelihoods and the health of residents and visitors; and (5) reframing tourism, including the resources it uses, to achieve productive and healthy livelihoods without degrading the biodiversity upon which it depends.

Spenceley, McCool, Newsome, et al. (2021) further reflect that the COVID-19 pandemic has revealed how dependent some conservation areas and many local communities are on tourism, and also the physical and mental health benefits of nature for visitors. Furthermore, it has shown how vulnerable tourism is to forces beyond its control. Even once we have overcome the immense health challenges of the coronavirus, tourism may still remain vulnerable to uncertainties and risks of subsequent health, security or economic shocks. Surely, the only way to make it more robust is to plan for a different type of tourism that is less exploitative, more sustainable and more in tune with the long-term needs of nature, the communities that depend on it, and the tourists themselves (ibid.). Tourism will only thrive if it is adaptable and functions as an essential environmental and social service, without undermining the natural and human resource base that it relies upon. Therefore, it must fully integrate the principles of sustainable development and focuses on equity, inclusiveness and integration (ibid.).

A lot has been learned during the pandemic. COVID-19 has forced the world to face its own fragility, stop and reflect, and begin to innovate and accelerate sustainable futures. One scenario might be to take this opportunity to restructure existing systems to be more inclusive and equitable. Seeds of a wildlife economy less reliant on tourism, such as platforms that link donors directly to local communities, payments for ecosystem services, or investments in measures that strengthen local sovereignty, may even begin to take root and evolve beyond initial small-scale experiments and pilot projects. COVID-19 has also opened a window of opportunity for transformative economic and social change, for change in how nature is conserved, what economic models it depends on, and how funders choose to invest for resilience and regeneration. Understanding the system is one of the first steps to systemic change, and this report seeks to contribute to that understanding. Furthermore, we have learned that evidence-informed systems analyses such as this report play a key role in informing dialogue, bridging perspectives and allowing systemic change to be imagined and created (Spenceley, 2021a).

We have a unique opportunity to establish a more resilient nature-based tourism sector for the future: one where communities participate and benefit meaningfully, where accommodation and tours are better for the environment, society and their clients, where a diversified and more robust sector can establish improved biodiversity conservation, and where the climate emergency is resolved. The rapid recovery of this sector should help to alleviate financial pressures on conservation areas, nature-based tourism operators and their workforce and help them survive and thrive (Spenceley, 2021a).

Acknowledgement

This essay is largely drawn from Spenceley, A. (2021) *The future of nature-based tourism: Impacts of COVID-19 and paths to sustainability*, Luc Hoffmann Institute, https://luchoffmanninstitute.org/future-nature-based-tourism-report

Note

1 See Our World In Data (2022) Parks and outdoor spaces: How did the number of visitors change since the beginning of the pandemic? Available at: https://ourworldindata.org/grapher/change-visitors-parks-covid?time=2020-05-01 (Accessed 2 February 2022).

References

Adventure Travel Trade Association (ATTA) (2020) Financial strategies and tactics for Adventure travel businesses during economic uncertainty. Available at: https://www.adventuretravel.biz/research/financial-strategies-and-tactics-for-adventure-travel-businesses-during-economic-uncertainty/ (Accessed: 1 October 2020).

African Leadership University School of Wildlife Conservation (ALUSWC) (2020) The State of the Wildlife Economy in Kenya. Country case study for the State of the Wildlife Economy in Africa report. 27 October 2020. Available at: https://sowc.alueducation.com/state-wildlife-economy-africa-report-kenya-country-case-study-published/ (Accessed: 12 November 2020).

Almond, R. E. A., Grooten, M. and Petersen, T., eds, (2020) *Living planet report 2020: Bending the curve of biodiversity loss* (Gland, World Wide Fund for Nature (WWF). Available at https://f.hubspotusercontent20.net/hubfs/4783129/LPR/PDFs/ENGLISH-FULL.pdf; Christopher H Trisos, Cory Merow and Alex L. Pigot The projected timing of abrupt ecological disruption from climate change. *Nature*, 580 (April 2020). https://doi.org/10.1038/s41586-020-2189-9

and Beyond (2020) From crafts to masks: South Africa. What do you do when there are no guests to buy your crafts? Turn your talents to a COVID necessity… Available at: https://www.andbeyond.com/impact/our-impact-platforms/africa-foundation/from-crafts-to-masks/?utm_campaign=339779_Indirect_No%20room%20for%20I%20in%20COVID-19_13May2020&utm_medium=email&utm_source=Dotdigital&dm_i=5BIT,7A6B,26Z1XH,S701,1 (Accessed: 9 October 2020).

Ateino, L. (2020) Covid-19: Fear of job losses could make tourism indifferent to wellbeing. *African Arguments*. 1 April 2020. Available at: https://africanarguments.org/2020/04/01/covid-19-fear-of-job-losses-could-make-tourism-indifferent-to-wellbeing/

Balmford, A., Green, J. M. H., Anderson, M., Beresford, J., Huang, C., Naidoo, R., Walpole, M and Manica, A. (2015) Walk on the wild side: estimating the global magnitude of visits to protected areas. *PLoS Biol* 13(2): e1002074

Barkham, P. (2020) 'The worst of human nature': UK staycationers' trail of destruction. *The Guardian*. 14 August 2020. Available at: https://www.theguardian.com/environment/2020/aug/14/uk-staycations-countryside-coast-visitors-trail-of-destruction (Accessed: 1 October 2020).

Berge, C. (2020) How can we be sustainable post-Covid-19. *BBC Travel*. 16 April 2020. Available at: http://www.bbc.com/travel/story/20200415-how-can-we-be-sustainable-post-covid-19 (Accessed: 9 October 2020).

Booking.com (2020) Travalyst coalition announces development of new sustainability frameworks to help travellers find sustainable travel and tourism options. 21 February 2020. Available at: https://globalnews.booking.com/travalyst-coalition-announces-development-of-new-sustainability-frameworks-to-help-travelers-find-sustainable-travel-and-tourism-options/ (Accessed: 21 October 2020).

Boundless Southern Africa (2020) 11 Healing solutions for tourism from Southern Africa. *Department of Environmental Affairs, South Africa*.

Department of Human Dimensions of Natural Resources (DHDNR) (2020) Protected areas for all. *Webinar series part 1: Session 1A Protected area tourism beyond 2020*. (Webinar). 26 August 2020. Available at: https://www.youtube.com/watch?feature=youtu.be&v=lOKS8M35cNE&app=desktop (Accessed: 1 October 2020).

Equations (2020) Is tourism a viable livelihood option? 7 May 2020. (Webinar). Available at: https://www.youtube.com/watch?v=HDnJ4_YZ5hk&app=desktop (Accessed: 9 October 2020).

Fyfe, W. (2020) 'COVID-19': Dark days; for tourism in Wales, but some 'silver lining'. *BBC News*. 17 October 2020. Available at: https://www.bbc.com/news/uk-wales-54562701 (Accessed: 19 October 2020).

Gilbert, S. (2020) Not a mask in sight': Thousands flock to Yellowstone as park reopens. *The Guardian*. 20 May 2020. Available at: https://www.theguardian.com/environment/2020/may/19/yellowstone-coronavirus-reopening-grand-teton-covid-19 (Accessed: 14 October 2020).

Hockings, M., Dudley, N., Elliott, W., Napolitano Ferreira, M., MacKinnon, K., Pasha, MKS., Phillips, A., Stolton, S., Woodley, S., Appleton, M., Chassot, O., Fitzsimons, J., Galliers, C., Golden Kroner, R., Goodrich, J., Hopkins, J., Jackson, W., Jonas, H., Long, B., Mumba, M., Parrish, J., Paxton, M., Phua, C., Plowright, R., Rao, R., Redford, K., Robinson, J., Rodríguez, C. M., Sandwith, T., Spenceley, A., Stevens, C., Tabor, G., Troëng, S., Willmore, W. and Yang, A. (2020) Editorial essay: COVID-19 and protected and conserved areas. *Parks*, 26.1: 7–24. https://doi.org/10.2305/IUCN.CH.2020.PARKS-26-1MH.en

Interagency Visitor Use Management Council (2019) Visitor capacity guidebook: managing the amounts and types of visitor use to achieve desired conditions, Edition 1, Available at: https://visitorusemanagement.nps.gov/Content/documents/IVUMC_Visitor_Capacity_Guidebook_newFINAL_highres.pdf

IUCN (2021) 130 - Strengthening sustainable tourism's role in biodiversity conservation and community resilience, IUCN /world Conservation Congress, Marseille, Available at: https://www.iucncongress2020.org/motion/130

Jack of all trades (2020) In Mexico's vibrant forests, locals adapt to a year without tourists. Life euphoria. 9 December 2020. Available at: https://www.nationalgeographic.com/travel/article/mexico-forests-adapts-to-a-coronavirus-year-without-tourists (Accessed: 18 February 2021).

Jones, K. E., Patel, N. G., Levy, M. A., Storeygard, A., Balk, D., Gittleman, J. L. and Daszak, P. (2008) Global trends in emerging infectious diseases. *Nature*, 451: 990–993.

Leung, Y-F., Spenceley, A., Hvenegaard, G., and Buckley, R. (eds) (2018) *Tourism and visitor management in protected areas: Guidelines for sustainability*, Best Practice Protected Area Guideline Series No. 27, IUCN, Geneva.

Marsh, S. (2020) Tourist hotspots busy despite coronavirus warnings. *The Guardian*. 17 May 2020. Available at: https://www.theguardian.com/world/2020/may/16/tourist-hotspots-busy-despite-coronavirus-warnings (Accessed: 14 October 2020).

Morris, S. (2020), "Crazy, isn't it?" Devon resort reports no shortage of visitors'. *The Guardian*. 16 October 2020. Available at: https://www.theguardian.com/uk-news/2020/oct/16/crazy-isnt-it-devon-resort-reports-no-shortage-of-visitors (Accessed: 19 October 2020).

National Parks Service (2022) Visitor spending effects – economic contributions of National Park visitor spending, Available at: https://www.nps.gov/subjects/socialscience/vse.htm

O'Brien, M. (2020) These 5 eco-friendly resorts around the world have survived the pandemic without guests by focussing on sustainability, caring for their orchards and farms, and delivering fresh food to locals. *Business Insider Australia*. 19 July 2020. Available at: https://www.businessinsider.com.au/how-eco-friendly-resorts-around-world-survived-pandemic-without-guests-2020-7?r=US&IR=T (Accessed: 5 October 2020).

Patz, Jonathan A. and others. 2004. Unhealthy landscapes: Policy recommendations on land use change and infectious disease emergence. *Environmental Health Perspectives*, 112(10) (April). https://doi.org/10.1289/ehp.6877

Snyman, S. and Spenceley, A. (2019) *Private sector tourism in conservation areas in Africa*. Oxford, UK; Boston, MA: CABI.

Souza, T., Chidakel, A., Child, B., Chang, W-H., Gorsevski, V. (2021) Economic effects assessment approaches: Tourism Economic Model for Protected Areas (TEMPA) for developing countries. In Spenceley, A. (ed) *Handbook for sustainable tourism practitioners: The essential toolbox*. Cheltenham, UK; Northampton, USA: Edward Elgar.

Spenceley, A. (2018) Sustainable tourism certification in the African hotel sector. *Tourism Review*. https://doi.org/10.1108/TR-09-2017-0145

Spenceley, A. (2020) COVID-19 and sustainable tourism: Information resources and links. Available at: https://annaspenceley.wordpress.com/2020/04/14/eu-study-on-covid-19-and-protected-area-tourism/

Spenceley, A. (2021a) The future of nature-based tourism: Impacts of COVID-19 and paths to sustainability. *Luc Hoffmann Institute*. Available at: https://luchoffmanninstitute.org/future-nature-based-tourism-report/

Spenceley, A. (2021b) COVID-19 and protected area tourism: A spotlight on impacts and options in Africa. *European Union*. https://doi.org/10.2841/33171

Spenceley, A. (2021c) Virtual protected area experiences in Africa: Status and potential for post-COVID-19 resilience. *European Union*. doi (TBC).

Spenceley, A (2021d) Tourism and Visitation to protected areas amid COVID-19: Guidance for protected area authorities and managers, *European Union*.

Spenceley, A. (2021e) How YOU can help make travel more sustainable – COVID-19 lessons, Africa Geographic, 20 December 2021. Available at: https://africageographic.com/stories/demand-for-sustainable-travel-what-travellers-can-do-to-help-build-back-better-from-covid-19/

Spenceley, A. (ed) (2021f) *Handbook for Sustainable Tourism Practitioners: The Essential Toolbox*, Edward Elgar, Available at: https://www.e-elgar.com/shop/gbp/handbook-for-sustainable-tourism-practitioners-9781839100888.html

Spenceley, A. (2022) *Transforming tourism in the pan-European region for a resilient and sustainable post-COVID world*, UNEP, Paris. Available at: https://wedocs.unep.org/handle/20.500.11822/39782;jsessionid=C10FC9F677DDF6B394BD75C729A561C5

Spenceley, A., McCool, S., Newsome, D., Baez, A., Barborak, J., Blye, C-J., Bricker, K., Cahyadi, H., Corrigan, K., Halpenny, E., Hvenegaard, G., King, D., Leung, L., Mandic, A., Naidoo, R., Ruede, D., Sano, J., Sarhan, M., Santamaria, V., Sousa, T. B., and Zschiegner, A-K. (2021) Tourism in protected areas amid the COVID-19 pandemic. *PARKS*, 27 (Special issue): 103–118. https://doi.org/10.2305/IUCN.CH.2021PARKS-27SI.en

Spenceley, A., Schägner J. P., Engels, B., Engelbauer, M., Erkkonen, J., Job, H., Kajala, L., Majewski, L., Metzler, D., Mayer, M., Rylance, R., Scheder, N., Smith-Christensen, C., Beraldo Souza, T., Cullinane Thomas, C., and Woltering, W. (2021) Visitors count! Guidance for protected areas on the economic analysis of visitation. UNESCO, BfN, EU JRC (eds). Paris/Bonn/Ispra.

Tourism in Need (2020) Tourism in Need. Available at: https://www.tourisminneed.com (Accessed: 6 October 2020).

Tripadvisor (2020) From the editors desk: COVID-19's impact on the travel industry and what likes ahead. Tripadvisor talk. (Webinar). Available at: https://www.youtube.com/watch?v=gQvRupGIW14&list=PL4xxeqMpxEYJtanXcO54jvJ9o9DeW3DNp&index=6 (Accessed: 5 October 2020).

UNEP Tourism to tackle plastic pollution with new commitment, 29 January 2020. Available at: https://www.unep.org/news-and-stories/story/tourism-tackle-plastic-pollution-new-commitment

UNWTO (2015) Towards Measuring the Economic Value of Wildlife Watching Tourism in Africa – Briefing Paper, UNWTO, Madrid. Available at: https://www.e-unwto.org/doi/book/10.18111/9789284416752 (Accessed: 20 October 2020).

UNWTO and UNEP (2008) *Climate change and tourism – Responding to global challenges*. UNWTO, Madrid. https://doi.org/10.18111/9789284412341

Vukukhanye Development Project (2020) Vukuhanye Development Project. Available at: https://www.facebook.com/16project/ (Accessed: 20 October 2020).

Wilson, A. (2020) 10 of the best virtual tours of the world's natural wonders. *The Guardian*. 6 April 2020. Available at: https://www.theguardian.com/travel/2020/apr/06/10-best-virtual-tours-of-worlds-natural-wonders-everest-patagonia-grand-canyon-yosemite (Accessed: 6 October 2020).

World Bank (2020) *Tools and resources for nature-based tourism*. World Bank, Washington, DC. Available at: https://openknowledge.worldbank.org/handle/10986/34433 (Accessed: 20 October 2020).

World Travel & Tourism Council (WTTC) (2021) *A net zero roadmap for travel & tourism: Proposing a new target framework for the travel & tourism sector* London. Available at: https://wttc.org/Portals/0/Documents/Reports/2021/WTTC_Net_Zero_Roadmap.pdf

World Travel and Tourism Council (WTTC) (2019) The economic impact of global wildlife tourism - Travel and tourism as an economic tool for the protection of wildlife. August 2019.

WTTC (2020) One Voice – accelerating the sustainability agenda (webinar), 24 September 2020. Available at: https://www.youtube.com/watch?v=B748cs3wpIA&feature=youtu.be (Accessed: 5 October 2020).

32

AN EVALUATION OF TURKEY'S NATURE-BASED TOURISM AGENDA

Policy, planning, and management

Mehmet Şeremet, Emine Cihangir and Kübra Cihangir-Çamur

Introduction

The complex inter-related system that links nature and human society need to be managed, governed, and understood by stakeholders (Folke, 2006). Recently, this issue has become integral to the United Nations' Sustainable Development Goals 2030 project (SDGs 2030) (Cihangir & Şeremet, 2021). However, given the importance of tourism activities to the economic development of many developing countries, the degree to which sustainability in nature-based tourism (NBT) destinations can be achieved remains a matter for discussion. This is the case not least because of the degree to which natural resources and environmental quality may be sacrificed in the cause of socio-economic development for tourism and its marketing strategies (Yu-Fai et al., 2018).

Any NBT destination is a multi-functional location with its unique animal species,-biodiversity and cultural heritage sites (e.g. castles, monuments, and archaeological sites) and, of course, its environmental repositories (lakes, valleys, waterfalls and many others) and intangible cultural assets (foods, dance, rituals, festivals) (Worboys et al., 2015). It is also usually possible to participate in adventure activities in the natural environment around the NBT destinations, including trekking, snow skiing, rafting, bird-watching, climbing, scuba diving, water skiing, sailing and camping. Therefore, many studies in travel literature have been focused on the concept of nature-based tourism. These have largely paid attention to local perceptions, tourist motivation and experience, community development and conservation policy, and the management and planning of NBT (Worboys et al., 2015; Alaeddinoğlu & Şeremet, 2016; Giddy & Webb, 2018; Gordon et al., 2018; Sagala et al., 2018; Chikuta et al., 2019; Mitsui et al., 2020; Şeremet et al., 2021a). Given that all of the marketing and design processes are related to a place's growth, planning, and management (Gordon et al., 2018), this chapter will analyse the Turkish setting in relation to these three issues, which have received slightly less attention.

With regard to the evaluation framework in this chapter, three themes have been used as follows: policy, planning and management (PPM). The rationale of the selection of these concepts in the framework has been based principally on an axiom that decision-making in NBT PPM has been intrinsically complex and to be scrutinized by a constructivist perspective which is beyond the positivist and modern paradigm of political and management approaches. Hall

DOI: 10.4324/9781003230748-38

(2008) argued that policy is simply the way the government adopt the public sphere. Therefore, public policy is the reflection of the government's practical actions, where many different factors (economic, social and cultural, ideology, values, identities) have been driven. Dredge and Jenkins (2011) conceived of a framework in tourism policy that can be understood as being based on an "institutional framework" which underlines the understanding of policymaking, which is a cascade from the factors (economic, cultural and social) driving state's role, practices, structures as well as historical context and agencies, actors, and their attributes. In the final step of decision-making, the way institutions, agencies and organizations have been cooperated and collaborated with the remaining stakeholders is important in that "policy dialogues" can be regarded as intermediary agents. A caveat should be noted here: this framework has been created based on the developed country context and can be critically scrutinized.

Conceptual framework

Many implementations of the sustainable management of NBT destinations around the globe have initially been driven by the political process and its sub-mechanisms, such as regulations, interventions, and incentives. Therefore, governmental agencies and institutions have been largely involved in the process in which nature-based tourism has been managed and planned. Institutional framework and arrangements such as normative/informal rules and inter-institutional relationships are also relevant for the destination planning and management of NBT, which is again a relevant subject for the destination where socio-ecological resilience might be fostered by the local individuals who have the necessary skills and knowledge of the local socio-natural values (Worboys et al., 2015). However, NBT destination management is not confined to these concepts. It has been noted that the country's tourism policy has always been largely affected by the sector's lobbying activities which often create conflicts between the national tourism strategies of the country and local destination planning (Hill & Hill, 2011). Buckley (2012) asserted, in a meta-analysis of sustainability literature, that one reason for the adoption of sustainability principles is mainly related to the extent of the sector's response pace to scientific outcomes and advocates. Therefore, power, knowledge and wider policy discourses also have implications for the planning and management of NBT destinations which has been saturated with the socio-ecological system (Willis, 2015). For example, Zhang and Xiao's (2014) study of the Chinese context asserted that the local institutions and their attributes have played an emerging role in the development of an NBT destination. This was in contrast to the approaches in developed countries where the central government incentives and sector ventures are dominant (p. 214). Although the sector holds a dominant role in some changes of political decision, the wider policy focuses of the government, such as internationalism and geopolitics discourses, have affected the local management of nature-based destinations in which the juxtaposing of geopolitics and national tourism policy have been practised. Nevertheless, some critical issues in the management of NBT destinations remain elusive.

The association of planning with tourism policy is also of interest to many studies, where the regional (district, cluster) and local (e.g. urban, national park, marine) practices have been the foci (Jones et al., 2003; Esfehani & Albrecht, 2019; Willis, 2015). Hall (2008) asserted some tenets about tourism planning which is not confined to the activities of governmental institutions. Its wider definition gaining prominence in the literature is that the conceptualization of planning within the tourism is more related to the factors driving the development of tourism in a destination to amalgamate social, economic, political and environmental factors (Hall & Page, 2006). Although the protected area strategy widely adopted approach to "spin-off" the sustainable development of NBT, Pröbstl -Haider and Haider (2014) asserted that their effect on the tourist's destination choice is relatively small. While protected areas (PA) have been discussed widely within the extant planning and policy literature (Pieraccini, 2015), exterior nature-based tourism destinations have received relatively little attention. Gaining the importance of both "stakeholder" and "collaboration" approaches in spatial planning fosters the sustainability agenda in NBT destinations (Healey, 2006; Borrie et al., 2020; Yrza & Filimonau, 2021). Therefore, a strategic planning approach might prove relevant in discussing the blend of policy and management in nature-based tourism destinations (Healey, 2004; Albrecht, 2017). This is not least the case as the socio-cultural and policy contexts in tourism planning have been regarded as increasingly important (Higham et al., 2016). In several types of tourism, the advocated approach toward planning and destination management were focused on the collaboration of both local and regional authorities and the interplay of local values and knowledge with the planning policy and strategies (Jones et al., 2003; Esfehani & Albrecht, 2019). Yet, although nature-based touristic destinations deserve community-oriented local management and integrated planning strategies, this has proven to be a relatively uneven practice for many emerging economies such as China, Peru, Chile, and Turkey where some contrasts might be existing between the political approaches and planning strategies (Marzuki et al., 2014; Cheng et al., 2017; Outeiro et al., 2018; Borrie et al., 2020).

The term 'management' has created a divergent conceptualization in the geographical scale of touristic destinations, with some concepts being used in a vast array of literature, such as environmental management (Fraser et al., 2006), PA area management (Borrie et al., 2020) and destination management (Fyall & Garrod, 2020). Although tourism destinations have different management models which are again evolving towards a community-driven approach in which the way is deeply embedded in a socio-cultural construct. This premise is also advocated by those scholars who are inclined to support the value of local socio-cultural context for NBT areas (Mancini et al., 2017). Thus, stakeholder theory and collaboration management perspectives have gained importance in destination management studies (Yrza & Filimonau, 2021). Nevertheless, given that developing countries are inclined to be involved in more central approaches to the management process, the onus of local communities in this process is still unclear. Even in the developed world perspectives, there are still some flaws in the management of protected areas (Priskin, 2003; Pieraccini, 2015; Peroff et al., 2017). Therefore, the extent of local people's involvement in the process might be varied in the socio-cultural context, depending upon the social capital, deprivation, economic capital, skills and vulnerability, as there is no "one-size-fits-all" solution. Taken as a whole, the literature suggests that policy, planning, and management are not mutually exclusive in the local context and need to be critically evaluated for practical implications.

Having set the scene for a critical understanding of the extant literature on NBT development and identifying research priorities and gaps, the second part of the chapter analyzes the changes in political approaches and practices in NBT destinations. This will be followed by the planning approaches in the third part. The penultimate section deals with the management of NBT destinations, including the local management approaches and their practical consequences. The chapter will be finalized by a synopsis and suggestions that might allow the readers to draw some broader lessons from the Turkish experience.

Policy practice in Turkey's nature-based touristic destinations

In history, the policy of economic liberalization was introduced in the early 1980s under the government of Mr Ozal; Turkey's tourism profile was initially based on the "sea-sun-sand" style of tourism, which has largely been offered by the estuary of the Mediterranean region of the country. From the second half of 2005 onward, tourism has spread across the rest of the country. For example, the eastern Black Sea region devoted itself to nature-based "high-land" tourism which today receives the majority of tourists from the countries of the Middle East. From the perspective of political modernization theory, this stage can be seen as reflecting the "neoliberalism" of the AKP Government era. Adopting a critical perspective of the country's political economy, they mostly refer to this term as "authoritarian neoliberalism" since they see the government-driven tourism incentives and development as a disguised tool for transcending the conservation rules and over-use of natural resources (İnal & Turhan, 2019). This was the case not least because although the country has now 47 National Parks, 35 of them had been legitimized in the pre-conservative government era. The government's main policy has been slightly toward drawing the border of developments areas and then declaring the specific conservation fields within the developing regions. Additionally, in this term, the status of two of the National Parks (Gallipoli Historical and Goreme National Parks) has changed to "site management", which is a recent policy of the government to draw the borders of cultural tourism destinations in the country (Cihangir-Çamur & Bozkaya-Yetkin, 2019). In the meantime, the Nature Park and Natural Monument protection status has increased in importance as its field size and conditions are more manageable and open to the development of tourism and recreational activities.

With regard to the policy of internationalism, Turkey's ratification with the international agreement on environment conservation could herald the signal of PA's settlement into the country's environmental policy. The first agreement the country joined was the EU's Bern convention, which brings the conservation of animal wildlife and their habitat areas. This was followed by the Ramsar Agreement in the mid-1990s and subsequently by the UN's Biodiversity agreement in Rio De Janeiro. For example, the Agenda-21 initiative, which was mainly supported by the Social Democratic Party's municipalities in the eastern part of Turkey, has triggered the transformation of small-scale NBT destinations within the form of a 'slow-city' concept. Today, the country has been a participant of thirty international agreements and conventions with the recent effort to participate in the adaptation and mitigation of climate change effects. Some 17 wetlands sites are now under the protection of the Ramsar Convention. The section will also analyze how the environmental policy discourses affect the planning and management strategies, as well as its local practice.

The policy documents, including tourism development strategies and mandate duties for the governmental organization, were mainly placed within the national development plans. These development strategies were into practice in the middle of the 1960s. NBT was involved in these policy documents with the 7th planning stage in the middle of the 1990s, when the focus was on the development of NBT. This continued with the production of the 8th and 9th policy documents by 2013 (Çiçekdaği, 2020). In the 9th term, the emphasis has been changed to the promotion of 'ecotourism' alongside NBT. Yet, since then, the policy with regard to the development of NBT has been replaced with other tourism types. Within these policy documents, so-called 'eco'-tourism was often connected with NBT. Yet, these are often faced by the "prima facie" in local practice due to the lack of a defined and cohesive strategy about the roles of local communities and NGOs in the region as well as the incentives provided to the local population. The major incentives in the policy document of TRTS2023 are by no means comprehensive in terms of enhancing the local community's social and economic capital. There might be a reason that some destinations to be rejuvenated by the new tourism types, as the leading destinations in NBT has begun to enter the stagnation stage with the "over-tourism" and become "mundane" to many international tourists. Probably, at this time was coincided with the loss of western tourists' attention towards the country's NBT destinations as well as with the country's new geopolitics of "neo-ottomanism".

Nevertheless, since the beginning of 2007, the Ministry of Tourism has been involved in the policy and strategies production process that the first and extensive tourism strategies of Turkey have been disseminated as a part of strategic documents, at both the destination and regional level (which now refers to TRTS2023) (MoCaT, 2007a). With this strategic policy and planning document, both cultural and natural development areas were underlined with the steps to be taken to achieve the transformation of destination development as well as the action plans of rehabilitation and rejuvenation within the current destinations. Although these policies and their action plans have allowed a period of six years for the achievement of these targets and aim for Turkey's 2023 tourism agenda, many destinations and tourism types are still lagging behind their major targets. Not least, because many destinations' life cycles have been built on the policy experience of "massive" tourism which is based on sea-sun-and-sand tourism in the southern part of Turkey. In most cases, the local and cultural context has been avoided and protection rules and policies are most often relaxed in order to be able to support the sector's investment into the regions.

The scale and nature of Protected Areas in Turkey

Table 32.1 shows the number of people visiting protected areas (PAs) in Turkey between the years 2002 and 2021. Between 2002 and 2017, the total number of visitors increased by 20.83 times in 15 years. Unlike the 37 per cent decrease due to pandemic restrictions and closures in 2020, there has been a significant increase in visitor numbers and percentages in all years. Additionally, despite the ongoing pandemic conditions of 2021, the total number of visitors exceeded the pre-COVID period of 2019. In sum, between 2002 and 2021, the total number of visitors increased by 77.24 times. This suggests that the demand for PAs will continue to rise in the future.

Although both the lack of institutional framework and arrangements compelling the rules and regulations of NBT destinations was a big challenge (albeit an attempt was unsuccessful in 2012), national tourism policy showed that the creation of PAs is the prevailing approach across the country. Apart from the NBT destinations in the south and the south-eastern part, the remaining destinations have been declared either as natural parks (NAP), natural protection sites or national parks (NP). As can be seen from Tables 32.2 and 32.3, seven national parks were declared, two before 1960 and five in the period between 1960 and 1980. Accordingly, the long-term protection plans of these national parks could only be made in the second half

Table 32.1 Number of visitors to protected areas in Turkey (2002–2021)

Turkey	*2002*	*2017*	*2018*	*2019*	*2020*	*2021*	*Total (2002–2021)*
Number of Visitors	**670.015**	**13.958.512**	**37.578.613**	**50.970.084**	**32.016.460**	**51.756.443**	**186.950.127**
Annual Change Rate (%)	**1983**	**20,83**	**169,21**	**35,64**	**COVID-19 Pandemic Year −37,18**	**61,65**	**7625**

Source: Ministry of Agriculture and Forestry

Table 32.2 Detailed information about the top-visited NBT Destinations of Turkey in National Park Areas

Turkey		*National Park*				*Resource Value*				
*NUTS**	*Province*	*NP Site Name*	*Area (Hectare)*	*Proclamation Date*	*Long-term Development Plan*	*Natural*	*Wetlands*	*Geological*	*Landscape*	*Recreational*
TR723	Yozgat	Yozgat Pine Grove NP	267	5.02.1958	16.05.2011	Forest				Day Trip/Excursion-Trekking-Mountain Biking Trail-Fauna and Flora Observation
TR221	Balıkesir	Kuş Cenneti NP	17058	27.07.1959	6.05.2019	Pelecanus crispus -Oxyura leucocephala	Lake		Natural Landscape	Bird Watching-Hiking-Photography-Wildlife Watching-Rock Climbing
TR411	Bursa	Uludağ NP	13024	20.09.1961	10.12.2009	Forest		Glacial Lakes	Natural Landscape	Winter Tourism-Cycling-Picnic-Hiking-Photography-Wildlife Watching-Mountaineering-Skiing-Tent Camping-Highland Tourism-Orienteering-Rock Climbing-Bird Watching-Botanical Trip-Caravan Camping-Mountain Biking-Accommodation-Nature Tourism-ATV Safari-Outdoor Sports-Volleyball-Ropeway-Landscape Viewing-Day trip

TR424	Bolu	Yedigöller NP	1623	29.04.1965	28.01.2016	Forest	Landslide lakes	Waterfalls	Natural Landscape	Bungalows-Tent and Caravan-Picnic-Camping-Nature walks-Sport fishing
TR321	Aydın	Dilek Peninsula-Buyuk Menderes Delta NP	27598	19.05.1966	31.12.2019	Anatolian chestnut (Castanea sativa) - Anatolian leopard (Panthera pardus tulliana)	Lagoons		Natural and Cultural Landscape	Visitors- Angling-Cycling-Picnic-Hiking-Photography-Wildlife-Observing-Bird Watching-Swimming-Landscape-Day Trip
TR611	Antalya	Köprülü Canyon NP	47473	12.12.1973	30.05.2014	Kardelenler (Galanthus) - Red Spotted Trout (Salmo spp.)		Canyon	Cultural Landscape	Nature Walks-Rafting- Angling-Cycling- -Hiking-Photography-Wildlife Observing-Tent Camping-Highland Tourism-Orienteering-Birdwatching-Horseback-Rafting-Canyoning-Caravan Camping-Canoeing-Mountain Biking
TR821 TR822	Kastamonu+ Çankırı	Ilgaz Mountain NP	1118	2.06.1976	25.12.2009	Forest			Forest Landscape	Winter Tourism-Ski Activities-Smokeless Picnic (Summer-Winter)-Nature Walks-Fotasafari-Botanical Tours-Wildlife Observation-Bird Watching

* Nomenclature of territorial units for statistics
Source: Ministry of Agriculture and Forestry

Table 32.3 An detailed information about the top-visited NBT Destinations of Turkey (cont.)

Turkey		*National Park*				*Resource Value*				
*NUTS**	*Province*	*NP Site Name*	*Area (Hectare)*	*Proclamation Date*	*Long-term Development Plan*	*Natural*	*Wetlands*	*Geological*	*Landscape*	*Recreational*
TR901	Trabzon	Altındere Valley NP	4468	9.09.1987	17.04.2014	Kuş Üvezi (Sorbus aucuparia)	Rivers		Natural Landscape	Mountaineering-Water Sports-Trekking-Camping-Climbing-Sport Fishing-Nature Research-Scientific Studies
TR221	Balıkesir	Kazdağı NP	20935	17.04.1994	17.06.1995	Forest			Natural Landscape	Nature Tours-Cycling-Picnic-Hiking-Photography-Wildlife Watching-Mountaineering-Tent Camping-Rock Climbing-Bird Watching-Botanical Trip-Caravan Camping-Mountain Biking
TR905	Artvin	Hatila Valley NP	16944	31.08.1994	28.10.2014	Forest		Topography	Natural Landscape	Nature Tours-Photography-Hiking-Photography-Rock Climbing-Bird Watching
TR905	Artvin	Karagöl-Sahara NP	3251	31.08.1994	8.11.2016	Forest	Lake		Natural and Cultural Landscape	Bird Watching-Camping-Picnic-Hiking-Photography-Rock Climbing

TR905 TR904 TRA11	Rize+ Artvin+ Erzurum	Kaçkar Mountains NP	52970	31.08.1994	8.11.2016	Forest		Glacial Valleys	Natural and Cultural Landscape	Nature Tours-Camping-Mountaineering-Trekking-Agro tourism
TR813 TR821	Bartın+ Kastamonu	Küre Mountains NP	37753	7.07.2000	3.12.2012	Forest		Giant Canyons - Caves	Natural Landscape	Nature Tours - Accommodation-Angling-Cycling-Picnic-Hiking-Photography-Wildlife Observation-Mountaineering-Tent Camping-Highland Tourism-Orienteering-Rock Climbing-Bird Watching-Cave Research
TRC34	Siirt	Botan Valley	11 358	15.08.2019		Rocky-Steppe		River	Natural Landscape	Ecotourism-Photography Paragliding-Rafting
TRB24	Hakkari	Cilo & Sat Mountains NP	27 500	26.09.2020			Glacial Lakes		Natural Landscape	Ecotourism-Photography-Mountaineering-Trekking-Climbing-Camping

* Nomenclature of territorial units for statistics

Source: Ministry of Agriculture and Forestry

of the 2000s, after a 20–40-year gap. These long gaps between declaration and planning were, and still are, major problems in Turkey.

This was also further tempered by the mining activities in the region. For example, the national parks in the Artvin region were opened to mining activities which were operated by international ventures. This suggests, in particular, that the government's conservation policy and laws are not necessarily strict and coercive for everyone. For example, the recent law of exploitation from the natural sites allows the entrepreneurs to facilitate new fashion tourism types and their business infrastructure within the protected areas. Additionally, vesting of regulations within the PAs adhered to the Ministry of Tourism. Although these are not fixed settlements, the experience of the country in Kaçkar Mountain and Uzungöl showed that the extractive development pathways putting habitable environments at risk are closely intertwined with infrastructural politics regulations for the development of tourism, which might harm again the NBT destinations' natural quality and the feature of habitat as well as the wild animal life.

Given that the country's geopolitical discourses towards the NBT destinations are relevant matters for discussion, while the southern regions have been recently dominated by Russian tourists, the NBT destinations (i.e., Black sea region, Bolu, Sakarya, Yalova) were also most visited by Muslim tourists coming from both the Middle Eastern and South Asian countries (e.g. Indonesia, Malaysia, Singapore). The Middle Eastern tourists have principally come from Israel, Kuwait, Saudi Arabia, Qatar, Jordan, Lebanon, Libya and the UAE (Figures 32.1–32.4). This shows that Turkey's goodwill with Middle East countries in the last decade has increased the influx of tourists from these countries. Although there is no study on the host–guest relationship, the personal experience of the researchers across the country and informal interviews with the hosts revealed that the local people are not satisfied with the tourists either coming from Middle Eastern or Gulf countries. Given that tourist is not only visitors spending a short period in the destination, as elucidated in eastern cases, those people are also empowering the local's conservation attributes towards nature and culture. Therefore, in the past, the 'high-brown' tourists who are mainly coming from either EU countries or the US used to make a valuable contribution to increasing local people's self-awareness toward environmental conservation, as the local and central government's interventions toward the protection of natural resources.

Figure 32.1 A (a)pre- (2001) and (b)post-picture (2015) of Trabzon's Uzungol Destination (NAP).

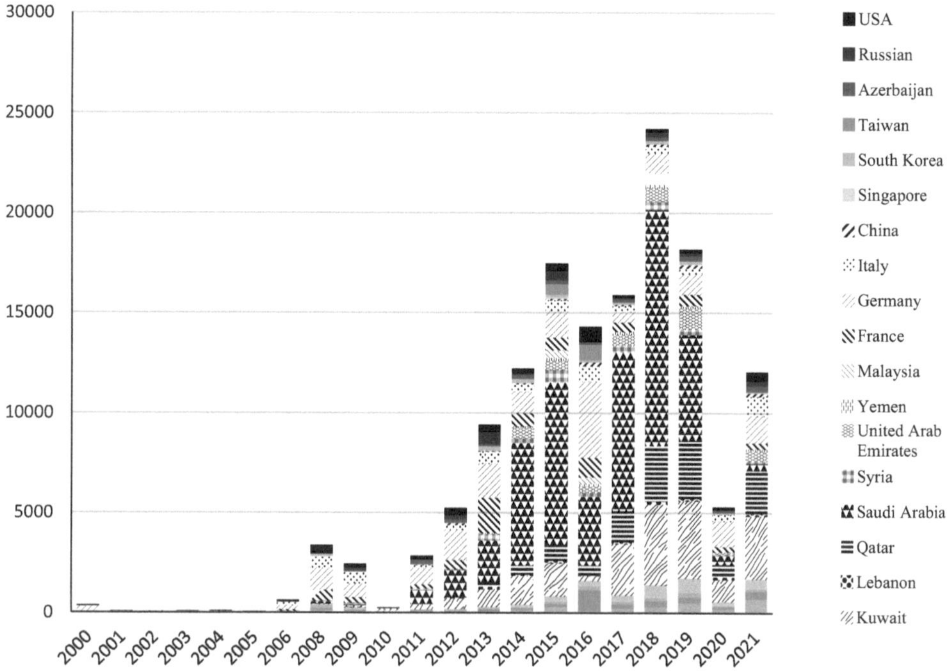

Figure 32.2 A graph of accommodation statistics of international tourists from lake Sapanca destination.

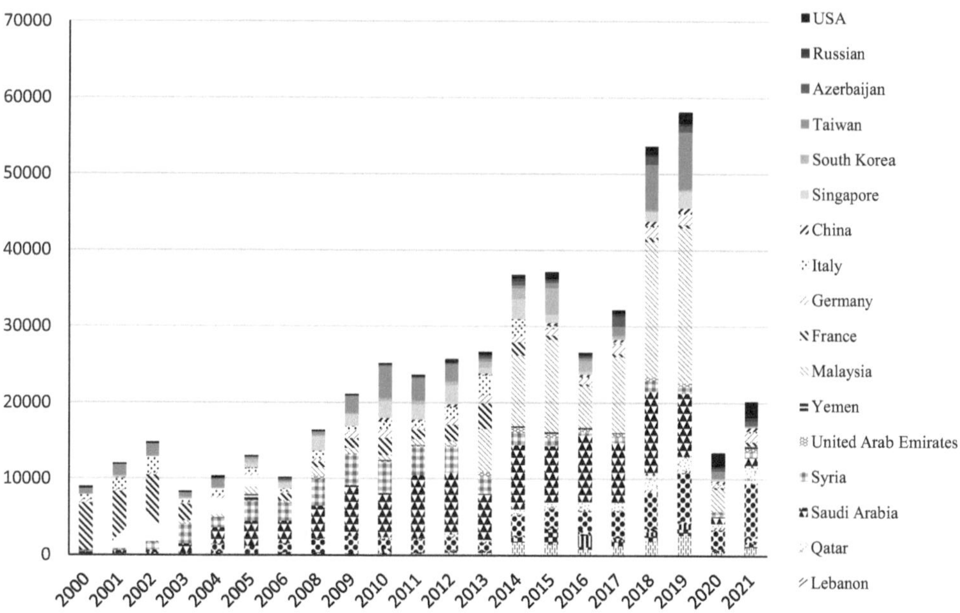

Figure 32.3 A graph of accommodation statistics of international tourists from lake Abant destination.

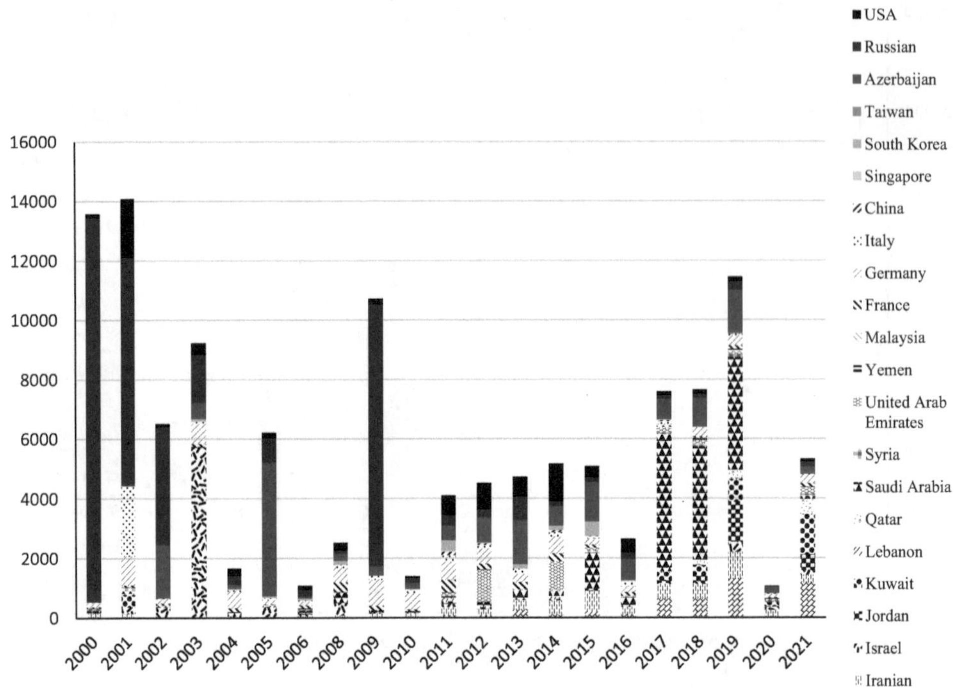

Figure 32.4 A graph of accommodation statistics of international tourists from the Kackar Mountains ("highland") destination

Unlike the NPs, the NAPs are another set of NBT destinations receiving a high number of visitors from both either domestic or international destinations. For example, the Uzungöl NAP has received over 2.5 million visitors in 2021. Turkey's new geopolitics alignments with the Middle Eastern countries were operationalized by the country's flag-carrier company Turkish Airlines and charter flights from Gulf and Middle East countries toward the vicinity airports in NBT destinations (e.g. İstanbul, Bursa, Trabzon and, more recently, Van). Cihangir et al. (2022) argued that Turkey's numerous destinations also attract Arabic people who are more inclined to be in cold and less humid climates alongside luxury shopping and local food. Therefore, the tourists coming from the Gulf countries (Saudi Arabia, Qatar, Kuwait, Oman) commerce to the guided-tour from Trabzon (Uzungöl) towards the north-eastern border of Turkey with Georgia (Figure 32.4). In addition, one further present-day destination for Arab tourists is the Bolu (Lake Abant, Golcuk NAPs) district, which bears similar characteristics to those found in the Black Sea region. Although they visit the Black Sea destinations, they prefer to stay in Trabzon (Yomra and centre provinces). Both of these destinations attract more than a half-million visitors each year. Most tourist agents agree that the visitors to NBT destinations of Turkey doing a kind of "Insta" and "mass" tourism. As seen in Figure 32.5, Artvin destination is losing its popularity for the international visitor as it offers more experience-based tourism activities to international tourists.

After all, although the governmental policy and rules were well set up in the protection regime of NBT destinations, there is a prima-facie approach that many destinations even protected under the National Park framework suffers from "over-tourism" and "over-capacity use" of nature. Many governmental policies support the incentives of tourism development, which often involve the construction of tourist amenities expand across unspoiled ecosystems.

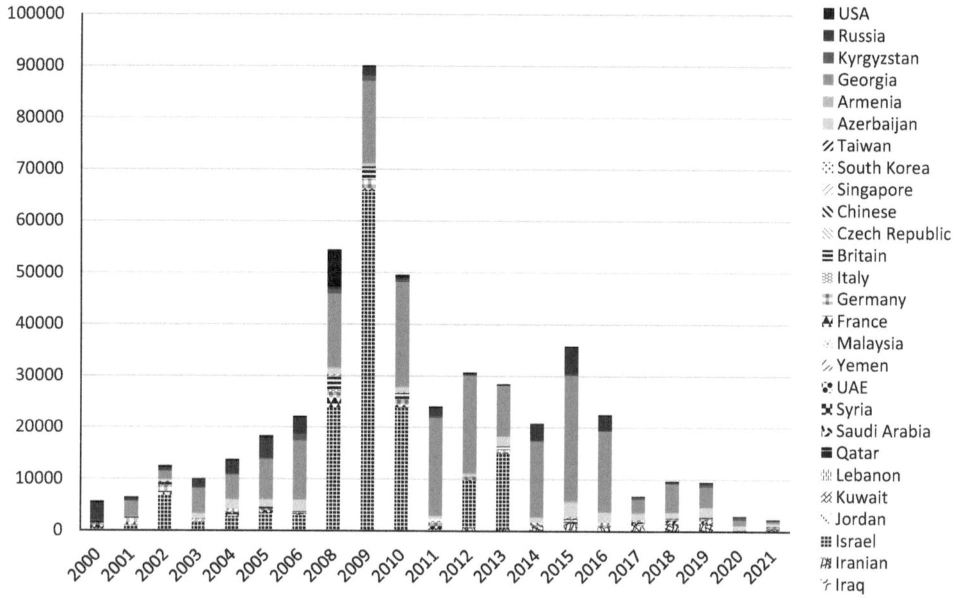

Figure 32.5 A graph of accommodation statistics of international tourists from Artvin ("valley" and highland") destination.

Planning history and approaches in the Turkish context

TRTS2023 is a very significant document in the recent history of Turkey; the goal of "planned development of nature-based tourism" under the strategy of "eco-tourism zones" was clearly stated in the white paper. The actions necessary to achieve this objective are likely to influence both natural and cultural influences. Drawing on this issue, holistic planning approaches become essential. In this section, the development of the tourism sector, the emergence of the nature-based tourism phenomenon and the relationship between these two and the planning in the background are all evaluated with their mutual interactions based on the breaking points in Turkey's economic, political, and social processes.

As elucidated in the earlier section, balancing the pressures and providing full protection where necessary, "legislation that defines and guides the planning and planning processes and constitutes the system of legal norms" appears as an intervention tool. In the case of Turkey, the most important problem is the inefficiency and never-ending change of both the planning authorities and the legislation regarding NBT-related protected areas and tourism. Therefore, a periodization effort was employed to describe the emerging circumstances in the planning of tourism in Turkey, in which three-term underscoring the changes in the economic, political, legal-administrative and social processes of the country were provided as follows.

- *The First Period of Before 1980 "state-controlled development period, and tourism as a newly emerging economic sector in the young term of the republic":*

This period also reflects a state-controlled, closed economic structure and its reflections on tourism. Within this period, the most important step toward planning was the establishment of the State Planning Organization (SPO) in 1961 as a new institutional body that is a candidate to manage Five Year National Development Plans that have been prepared to plan all economic and social sectors since then. These studies paved the way for sectoral planning, and accordingly tourism-related steps and new structuring were initiated. Alongside the planned development term, one further important development in tourism was the establishment of the Ministry of Tourism and Promotion in 1963 as a specialized state organization to plan, direct, support, coordinate, and manage the tourism sector. Until this time, the planning studies in the country were focused on only big city centres, and there was no planning study for tourism development. The first "Tourism Development Fund" was established in 1962 by the USA International Development Agency to provide funds and support for the construction of accommodation facilities and roads in the touristic areas (TTB-Turizm ve Tanıtma Bakanlığı, 1965). In the 1970s, the Association of Türkiye Travel Agencies (TÜRSAB) was established in 1972 to support travel agencies to provide professional and qualified service, to carry out activities in the country and abroad for the development of tourism, and to work in coordination with the Ministry of Tourism (TÜRSAB, 1997).

In addition to the PAs, the awareness of the conservation dimension of this period was reflected in the legislation, and the monuments and antiquities board, which constitutes the beginning of today's conservation boards, was also established in this period. Seven NPs were declared; two before 1960 and five between 1960 and 1980 (Table 32.1). It can be observed that the long-term protection plans of these national parks could only be made in the second half of the 2000s, after a 20–40-year gap (Table 32.1). These long gaps between declaration and planning remain major problems in Turkey.

The Southern Antalya Tourism Development and Infrastructure Management Association (GATAB) project was begun in 1974 in line with the development plans' predictions to realize the infrastructure investments for the development of the tourism accommodation potential in the priority regions by the state (Inskeep, 1993). The project was defined in the 1/25 000-scale Environmental Plan approved in 1976, on the 80-kilometre long coastline. The project, realized under the umbrella of a single project in terms of planning, programming, financing and implementation, has the most holistic planning approach in tourism development in Turkey. It is an accomplished project in terms of regarding the planning, programming, financing, implementation processes as a single project (Bozkurt, 2019). Within the framework of the project, infrastructure and superstructure applications were carried out in line with the Southwest Antalya Environmental Plan, which was prepared with the aim of integrating tourism development with the economic and social structure of the region, and protecting the forest and coastal structure.

Although GATAB set an example of a pioneering and successful holistic planning and implementation approach for Turkey in the 1970s, its autonomy was ended by joining the Antalya Water and Waste Administration General Directorate in 2014 (https://www.kemerhaber.com/kemer-guncel/ve-gatab-kapandi.html). It should also be underlined that Turkey still does not have an integrated country physical planning document.

- *The Second Period of 1980–2002 "growing tourism sector through laissez-faire approach on the axis of neoliberal and planning policies with fragmented authority period"*:

In this term, MoCaT was gained more power to determine tourism development zones in spatial dimensions, such as tourism regions, tourism areas and tourism centres, which will be approved by the Council of Ministers. 1/5000-scale planning authority was given to the MoCaT in the announced regions, the process for the allocation of treasury lands was simplified and all allocation authority was given to the MoCaT in all tourism regions and centres.

This period also saw the enactment, in 1983, of the Law on the Protection of Cultural and Natural Assets No. 2863, which determined the protection and application conditions for the protection of cultural and natural assets and environmental values (Law No. 2863, 1983), and also the Law on National Parks No. 2783 (Law No.2783, 183) and the Law on Environment No. 2872 (Law No. 2872, 1983). This process was the most vital step towards the practical implementation of the MoCaT's policies for sustainable tourism development and PAs after the Law on Tourism Encouragement No. 2634 (Law No. 2634, 1982).

Culture and tourism were separated in 1989 and two separate ministries were established (these two ministries were merged again in 2003). This year, the Ministry of Tourism put into effect the ATAK project (the Mediterranean, Aegean Tourism Infrastructure Coastal Management Project). The financing of this project was covered by a World Bank loan. With this project, drinking water and sewage are projected to be delivered to the entire project area about 2000 km. of coastline. It aimed to bring solutions until 2020, via a 1/25,000-scale masterplan, feasibility studies and the creation of corporate and financial structure issues (ER-GE, n.d.).

Given that Ayder highland and Uzungöl were declared as tourism development zones by the decision of the Council of Ministers, the lack of planning and monitoring, a large number of illegal constructions suddenly emerged on the natural areas that were subject to high-conservation regulations. The major could be seen as a lack of a holistic masterplan and infrastructure planning. The effects of mass tourism, which showed rapid development after the incentive and support process that started with Law no. 2634 in 1982 (Law no. 2634, 1982), on the sustainability of natural resources and the negativities brought by the dependence on the season began to be discussed again in the 2000s (Ministry of Tourism, 2002). The 2001 economic crisis marked the beginning of the Justice and Development Party (AKP) government and the construction-oriented conservative re-structuring period in Türkiye.

- *The Third Period–Part 1, 2002–2016 the Rise of JDP Period "single party conservative government and increased capital dominance-decreased planning intervention through construction led economic approach" and Part 2, After 2016–Beyond "the stagnation and regression process of JDP with increasing conservative policies and excessive centralization trends especially in the design of economics and planning authority parallel to power loss"*

This period was led by neoliberal economic growth policies that were driven by the JDP government's construction-oriented economic growth. The legislative changes began with the reunification of the Ministry of Culture and Ministry of Tourism in 2003, and the gathering of the tourism planning authority under a single ministry marked the beginning of a new era for the public sector. Given the burgeoning trend in sustainable tourism, TRTS2023 and its Action Plan for 2007–2013 document were prepared by the MoCaT in 2007 (MoCaT, 2007b).

Encouraging the demand for natural areas was also begin in this period. For example, with the TRTS2023 of creating a "highland" (*yayla*) tourism and connection of all potential areas with each other has been taken into action (Figure 32.6). However, the necessary changes were not made in the organizational structure and planning legislation within the framework of the targets of the TRTS2023. Although the most influential example of conservation-oriented development in the field of planning and protection is the establishment of the "Ministry of Environment and Urbanization" with Decree No. 644, which entered into force in 2011, and the establishment of the "Ministry of Forestry and Water Affairs" with Decree No. 645, the resegregation of the two ministries were merged before the last election. Although the planning authorities seem to be gathered under a single ministry at the stage of separation, the planning

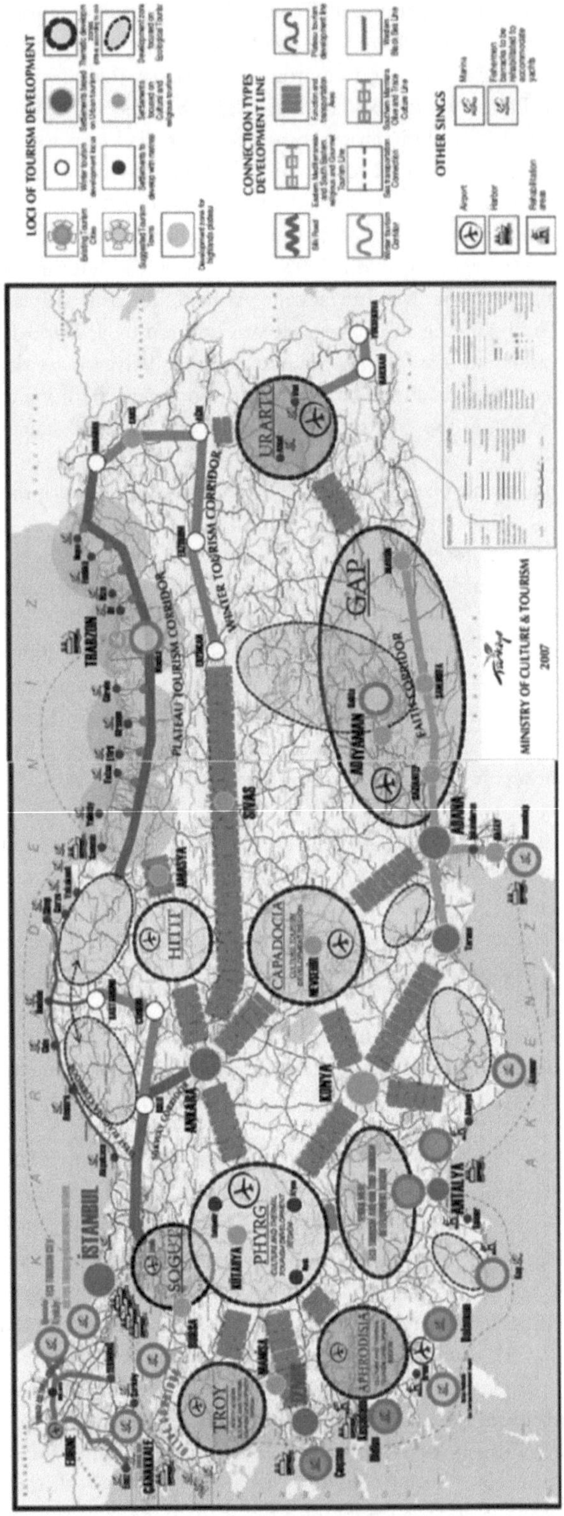

Figure 32.6 Turkey's Tourism Strategy Conceptual Action Plan Chart.

Source: Türkiye's 2023 Tourism Strategy (MoCaT, 2007a)

authorities for tourism and protection have been shared by the different institutions, so that the possibility of ensuring the integrity of the plan in tourism centres and development regions have been lost. At the same time, the complete protection of conflicting upper-scale plan types separately for each institution showed that interventions to protected areas will continue when needed. For example, according to the results of Serter's (2020) comparative analysis of the quantitative and qualitative changes of protected areas in Turkey after 2002, the criterion of compliance with international national park standards is underlined in Article 6 of the National Parks Regulation was ignored. In most of the post-2002 period studies, to increase the quantity, areas that can sometimes be measured in square meters began to be declared as nature parks, and this procedure caused the quality to decrease.

Consequently, the review of the three main periods of recent Turkish history reveals the need for strong and comprehensive institutionalization and specialization, and the importance of all assets in nature for the holistic protection of ecosystems which is very vital and required for overall sustainability and sustainable NBT. However, the ongoing transfer of authority and the changes in the authorized territories among the institutions raise several problems. This fragmented and disorganized structure in policy, planning, and management in protected areas appears to be the main obstacle. Consequently, the current structure is far from producing solutions to the erosion that occurs in protected areas and to the sustainable development of tourism as well as NBT, as it has focused on short-term interests and construction-led economics. It is necessary to develop content for a holistic and systematic structure in protected natural areas (Cihangir-Çamur et al., 2021) and to redefine the relationship between policy, plan, and practice by aiming to establish the protection-utilization balance.

Governance and management practice of Turkey in NBT destinations

An NBT destination is a relatively more functional and diverse place regarding its socio-cultural structures and attributes as well as landscape diversity; therefore, its management process is expected to be slightly different than for the urban areas at national and regional levels (Caldwell & Freire, 2004). Given that the TRTS2023 white paper was provided with no attempt to the establishment of an institutional framework, this was also echoed by "deliberative" theory which is simply a premise of democratic management. Pieraccini (2015) looks through the PA in the context of Britain with this lens to provide a critical view that the PA has become subject to the transmissions toward 'new' commons.

Bingeman et al. (2004) argued that an institution is an important frame, which features both formal and informal constraints that might hinder the structure from the interplay of local people and managers. TRTS2023 agenda refers mainly to the national tourism councils and its subdivision such as ministerial offices (environment, tourism, development etc.) and the local authorities (municipalities, sector-driven NGOs, and local development agencies). Most importantly, the way in which the local communities are involved in the management process is not even placed within the documents. Although it emphasizes including public views in the planning process and creating masterplans, neither the national nor the local committees include local voices. This might be related to cultural context. For instance, volunteer activities in the autonomous NBT destination development initiatives (e.g., Geoparks) have resulted in a thriving management process and the pursuit of sustainability goals (collaboration with women-driven NGOs and training of women as field guides). Yet, there is still some shortcoming in the community-driven approach to the management process. While the management practice is based more on the elite board, including the business sector and the major managers, one of the weaknesses of this approach is the exclusion of the local community and its voices from the overall management process.

In today's sustainable tourism development goals, NBT destinations have been receiving great attention from the growing numbers of tourists that avoiding the "over-tourism" might be a challenge in the management (Şeremet et al., 2021b). Therefore, over-tourism has been raised as an issue not only for urban districts, but also in terms of its threat to the sustainable future of NBT destinations. There is no evidence that Turkey's national parks or protected areas have regulations designing the capacity level for the visitor numbers or indeed any appointment system, which is at present becoming an increasingly common approach to many protected areas offering NBT experiences around the world. In Turkey, some well-known NBT destinations have been placed within the protected areas. These include Winter Sky Centers (Uludağ, Sarıkamış, Ilgaz), Highlands (Ayder Recreational Area), and Lakes (Lake Abant, Lake Sapanca, Lake Karagöl, Lake Salda). Finally, there is also a UNESCO-approved Geopark, which is located in the Kula province of Manisa. Of all these destinations, Uludağ receives the highest number of annual visitors, with nearly two million. The accommodation statistics have shown that most of these are daily-recreation seeking visitors.

World Tourism Organization's (UNWTO) seminal guidance (2007) is so far a very valuable source for destination management for many different tourism types. Additionally, a specimen case for NBT was identified by Tayler (1997), who provided a framework as to how to set up management and planning processes for the NBT-oriented protected areas. Although their framework suggests that earlier works to be undertaken are mainly related to quality and impact assessment of natural assets in the PA at the earlier stage of the management phase. The latter steps focus on the dimension of a "well-organized" 'visitor's experience. In Turkey, although many visitors have to pay an administration fee to National Park Authorities at the entrance, the service and conditions offered to visitors are comparatively limited. For example, guidance, regulations, and relevant materials are not always provided to the visitors and neither is professional guidance. In many instances, the local people and the nation's tour guides are offering guiding services to the visitors. Although many nation-wide tour companies work with professional guides to employ and lead the tour, for example, in Kackar and Cilo-Sat Mountains NPs can be exemplified for such cases, some local people and climbers offer guide services to visitors or "closed-group" visitors who are more inclined to make trekking and camping expeditions in glacial lakes. This might be seen as the interplay of NBT destinations with the local people. For example, with the efforts of volunteers in the independent NBT destination development projects (e.g. Geoparks) and even there is a flourishing management process and sustainability goals targeted (collaboration with women-driven NGOs and training of women as field guides.

Although NBT destination management needs to be achieved through institutional arrangements which are a set of agreements on the division of the respective responsibilities of agencies involved in protecting, offering training, managing, and doing research. While NBT has not necessarily adopted this management practice, in the Geopark case, networking, and sharing roles and responsibilities have been achieved by a volunteer committee under the support of the local municipalities from which the budget was transferred. In the certificated Geopark case, there is a tendency toward the locally-driven guidance and providing service (e.g. accommodation, food and beverage) with the local partnerships. As the parking area covers nearby villages and includes local communities, the professional management system in the Kula Geopark case suggests that offering training to vulnerable people from the local community might help to facilitate the empowerment of local people who may be from a disadvantaged background. Additionally, they do consider that TURSAB-certificated nationwide tour guides are not necessarily experts and have relevant knowledge about NBT areas such as the Geopark. Compared to the remaining cases, this is a slightly better approach of exploitation of nature areas towards the empowerment of the local community, who can be seen as the custodian of the natural and cultural heritage sites.

Additionally, Jenkins (2006) argued that the implementation of tourism strategies at the local level was hindered by financial constraints. In Turkey's case, this seems to be slightly overcome by the support of local municipalities and DA. As the municipalities have more independence in financial decisions and setting up companies, the involvement of municipalities into DMOs have also been represented by a few further cases, such as Kayseri Erciyes Ski Centre and Şanlıurfa. Both of which have independent professional companies that have their financial management system and allow to create a budget. Kayseri's Winter tourism centre in Erciyes mountain was financed and initiated by the metropolitan municipality and DA. This might be an example of thematic management organization in tourism, while the Şanlıurfa case is the integrated approach of cultural tourism destinations. To date, there is no evidence that NBT destinations will be managed based on neither a "thematic" nor a "holistic" approach. Yet the practice of DAs and their management plan provides us with some cruel estimation about the fact that the government's tourism policy expects them to establish an integrated DMO.

However, as many scholars (Worboys et al., 2015; Albrecht, 2017; Fyall & Garrod, 2020) argued that the sustainability of local implementation and the management of destination requires a holistic way to integrate local support which the resident communities should be part of the DMOs. Leadership and advocacy have also emerged as paramount matters by Albrecht (2021), who underscored the importance of NGO-informed management practice in destination management. Therefore, she put forward a "local champions" role in the local communities. Particularly, the action plan of DA in the Western Black Sea region is on the way to the development of a DMO which might manage the NBT destinations and culture-oriented tourism destinations. However, this 'holistic' way might not be helpful, as the skills of managers and the level of protection do not have equal weight in the management of eclectic touristic attractions. So, each requires different skill sets and experience in the management process. An integrated approach aligning with the National Tourism strategy documents seems to be the most agreeable way of establishing DMOs. Yet their management plans for the NBT destinations remains elusive.

Interestingly, in many cases, the local or DA have played an important role in the commerce of destination management offices. Although Turkey has no common practice of Destination Management Organizations (DMOs), many touristic destinations' tourism development plans (feasibility study, strategic and masterplans) and marketing analyses (e.g. tourist experience and motivation surveys) for the early development have been undertaken by the local RA offices. Yet these organizations are principally undertaking these duties and carrying the roles with the support of consultancy companies and occasionally with the support of universities and internal experts. Therefore, it does not provide continuous support and management practice for the touristic destinations as the NBT destinations deserve the introduction of an innovative management process for sustainable development, which requires experience quality, training, conservation and involvement. In Turkey's NBT sites, it is too complicated as most areas placed within the PA border, the overlapping roles of the local agents and institutions are often more apparent, not least because there are a few authorities have that a right to control the area and often their roles and regulations collide with one another. For example, Kackar National Park has been principally controlled by the local Forest Institution which is based in the city (Rize) and connected to the central government, but the destination is more close to the province of Çamlıhemşin which has both municipality and province governance (Alaeddinoğlu & Şeremet,

2016). Their roles to tourism management are not clear, often complicated as the branch of Forestry and Agriculture ministry mainly follows the regulation of environmental conservation, municipality deals with the infrastructure/settlement (dealing with illegal dwellings) and sub-governorship involves in more security and administration issues.

One final matter to be raised is the adaptation of technology into NBT destinations, Fyall and Garrod (2020) suggested that technology needs to be considered for the future of destination management. From this perspective, a GIS-based application of the PA directorate of Forestry and Agriculture Ministry has set up a Web-GIS-based application to offer a wide range of place-based information about where to go and what to experience types of tool for the NBT destinations. All of which have detailed information and activity types and route plans. Yet its mobile application and transformation to mobile devices are not compatible with the end-users. Nevertheless, it might prove beneficial for visitors who would like to experience NBT in a different part of the country.

Synopsis and conclusion

In this chapter, the panorama of Turkey's nature-based destinations has been evaluated in the context of policy, planning, and management. Turkey's high potential of NBT has been underscored regarding spatial and temporal distribution across the country. The role of national and international policy staggering the planning process was found to have mainly dominated the tourism development and destination management issues. While the planning approach has been mainly "holistic" and "deintegrated" in the early terms of the Turkish Republic. The conservative government era has resulted in slight changes and the tourism ministry takes the leading role in making the strategic decision which mainly underlines the three main principles of strategic planning such as where to reach and what to do in the middle of the 2000s. This shows that NBT destinations have been classified as an aspect of tourism development across the country. As there is no specific institutional arrangement for NBT, defining the responsibilities, interests, and values of stakeholders and agencies at the destination level might be viewed as an intermediary obligation for a portion of these activities.

The tourism policy supports strategic planning, yet it places a greater emphasis on principally a "holistic" approach to allow for a regional tourism agency or management organization. However, the lack of a community-driven approach in planning seems to be the main challenge of achieving the sustainable development goals in NBT destinations. With the introduction of the recent law, the tourism ministry has a coercive role in the exploitation of natural areas which are even to be found in PAs. The recent trends in tourism, such as glamping, camping and caravanning, have been reflected in the new post-COVID tourism policy of the government (Cihangir-Çamur et al., 2022). However, the lack of local management organization seems to be the main obstacle standing in the way of the exploitation from the natural areas. This might result in the NBT locations of the country in the near future confronting the difficulty of becoming ordinary and overcrowded for many European and even local tourists.

Given that the government's management of the country is becoming 'authoritative neo-liberalist', the local people's raising awareness towards the natural areas is not very welcomed by the formal institutions, this situation mainly hinders the local people from participating in the decision-making process. Again, the recent tourism strategy of the country suggests that the stakeholder approach should be operationalized, which is probably mainly refers to the increasing interplay of government and sector. Therefore, while the International Union For Nature Conservation (IUCN) guidance (Worboys et al., 2015; Yu-Fai et al., 2018) advocates public involvement in PA and visitor management, the current practices loom large in "massive tourism" practice in NBT destinations, which is again creating an over-usage effect on the nature and wildlife of the country. The widening of tourism development in less developed regions might be considered as a disguise in revitalizing the sector rather than the local people and community.

Finally, the last issue is the interaction of hosts and tourists in NBT destinations. While the country has experienced a change in geopolitics, its local practice now seems to be attracting more tourists from the Middle East countries. Although the tour companies and chain hotels are satisfied with this outstanding trend, the local people's expectations seem to be receiving western-oriented tourists who are seeking more experiential activities and culture-based experiences. As the community's wellbeing is also another important factor, the interplay of local people and tourists might contribute to the sustainable future of the destination as the locals are the "gatekeepers" of the NBT destinations. Taken as a whole, NBT destinations now appear to be managed arbitrarily, making it difficult to accurately estimate visitor numbers or assess their expectations. Further, destinations struggle to keep up with new trends or to be managed and protected sustainably. This might also reflect to weak policy dialogues with the local communities and agents. The concentration of power in central organizations and the lack of incentives to local vulnerable groups may help to explain barriers to the effectiveness of collaboration strategies (Selin, 1998).

Given that many NBT destinations have been rejuvenated with the government's latest initiative in geopolitics to receive more tourists from the Middle East countries, this situation has been mainly organized and operationalized by the sector. The local people are most receptive and do not involve in any decision to change international tourism policy. The majority of the visitor experience at NBT sites is planned by tour firms, which solely provide food and beverage service. As a result, the majority of international visitors do not engage in outdoor activities at NBT areas; rather, they only observe them. The insufficient organization of the tours and the lack of estimation of tourist numbers, mean that in the high season (e.g. summer) destinations suffer from over-tourism (e.g traffic jams, crowded spaces, and busy service in the catering sectors) which principally results in lowering tourism satisfaction and leaving many tourists' expectations at a very basic level As tourist expectations vary, the sustainable development of NBT destinations should be facilitated by allowing the local people to play an active role in the management and design of visitor experiences.

References

Alaeddinoğlu, F., & Şeremet, M. (2016). Nature-based tourism in Turkey: The yayla in Turkey's eastern black sea region. In *Alternative Tourism in Turkey* (pp. 71–86). Cham: Springer.

Albrecht, J. N. (2017). Challenges in national-level tourism strategy implementation–a long-term perspective on the new zealand tourism strategy 2015. *International Journal of Tourism Research, 19*(3), 329–338.

Albrecht, J. N. (2021). Destination management for sustainability. In A. Correia & S. Dolnicar (Eds), *Women's Voices in Tourism Research Contributions to Knowledge and Letters to Future Generations* (pp. 18–20). Brisbane: The University of Queensland. https://uq.pressbooks.pub/tourismknowledge. https://doi.org/10.14264/817f87d

Bingeman, K., Berkes, F., & Gardner, J. S. (2004). Institutional responses to development pressures: resilience of social–ecological systems in Himachal Pradesh, India. *International Journal of Sustainable Development and World Ecology 11*, 99–115.

Borrie, W. T., Gale, T., & Bosak, K. (2020). Privately protected areas in increasingly turbulent social contexts: Strategic roles, extent, and governance. *Journal of Sustainable Tourism*, 1–18. https://doi.org/10.1080/09669582.2020.1845709

Bozkurt, G. (2019). *Yerel ve mekansal dinamiklerin turizm temelli girişimcilik üzerine etkileri: marmaris yerleşmesindeki girişimcilik yapısı ve girişimciliği etkileyen faktörlerin.* [Unpublished doctoral dissertation]. Gazi University, FBE, Department of City and Regional Planning.

Buckley, R. (2012). Sustainable tourism: Research and reality. *Annals of Tourism Research, 39*(2), 528–546.

Caldwell, N., & Freire, J. R. (2004). The differences between branding a country, a region and a city: Applying the Brand Box Model. *Journal of Brand Management, 12*(1), 50–61.

Cheng, C., Robinson, B. E., Xiao, Y., Ouyang, Z., & Rao, E. (2017). Increasing the value of China's environment for recreation: The case of Jiuzhaigou, Sichuan. *Environmental Engineering & Management Journal (EEMJ), 16*(12), 2665–2672.

Chikuta, O., du Plessis, E., & Saayman, M. (2019). Accessibility expectations of tourists with disabilities in national parks. *Tourism Planning & Development, 16*(1), 75–92.

Çiçekdaği, M. (2020). Kalkinma Planlarinda Turizm: İçerik Analizi Yöntemi İle Türkiye Kalkinma Planlarinin Analizi. *ÇATALHÖYÜK Uluslararası Turizm ve Sosyal Araştırmalar Dergisi, 5*, 1–11.

Cihangir, E., & Şeremet, M. (2021). Education for Sustainable Development (ESD) in tourism- oriented vocational degree programmes in Turkey. In E. Pénzesné Kónya, J. Křeček, & M. Haigh (Eds.), *Environmental Sustainability Education for a Changing World* (pp. 185–203). Springer. https://doi.org/10.1007/978-3-030-66384-1_11

Cihangir, E., Şeremet, M., & Cihangir-Çamur, K. (2022). Turkey at the crossroads: A study of geopolitics and tourism re-alignment. *Geography, 107*(3), 145–152. https://doi.org/10.1080/00167487.2022.2114165

Cihangir-Çamur, K, & Bozkaya-Yetkin, F. (2019). Arkeolojik Sit Alanlarında Yönetim Planlamasının Turizme Etkisinin Değerlendirilmesi; Şanlıurfa Örneği *Harran Üniversitesi Mühendislik Dergisi, 4*(3), 38–48.

Cihangir-Çamur, K., Cihangir, E., & Şeremet, M. (2022). Turkey's tourism recovery process during COVID-19: Policy, planning and management. In J. Kennell, P. Mohanty, A. Sharma, & A. Hassan (Eds.), *Crisis Management, Destination Recovery and Sustainability* (pp. 97–111). London: Routledge. https://doi.org/10.4324/9781003295839-13

Cihangir-Çamur, K., Şeremet, M., Cihangir, E., & Özer B. (2021). Kış Destinasyonlarında Başarı İçin Stratejik Planlama: Kayseri-Erciyes Kayak Merkezi Hikayesinden Çıkarımlar. In Aydın Ünal (Ed.), *Destinasyon Konulu Güncel Araştırmalar* (pp. 493–507). Çizgi Kitabevi. https://www.cizgikitabevi.com/kitap/1264-destinasyon-konulu-guncel-arastirmalar-i

Dredge, D., & Jenkins, J. (Eds.). (2011). *Stories of Practice: Tourism Policy and Planning*. Surrey, England: Ashgate Publishing, Ltd.

ER-GE (n.d.). *Mediterranean and aegean cost touristic infrastructures and coastal management project (ATAK)*. Retrieved February, 10, 2022 from https://www.ergeproje.com/en/master-plan-feasibility-projects-and-eia-environmental-impact-assessment-reports/mediterranean-and-aegean-cost-touristic-infrastructures-and-coastal-management-project-atak

Esfehani, M. H., & Albrecht, J. N. (2019). Planning for intangible cultural heritage in tourism: Challenges and implications. *Journal of Hospitality & Tourism Research, 43*(7), 980–1001.

Folke, C. (2006). Resilience: The emergence of a perspective for social–ecological systems analyses. *Global Environmental Change, 16*(3), 253–267. https://doi.org/10.1016/j.gloenvcha.2006.04.002

Fraser, E. D., Dougill, A. J., Mabee, W. E., Reed, M., & McAlpine, P. (2006). Bottom up and top down: Analysis of participatory processes for sustainability indicator identification as a pathway to community empowerment and sustainable environmental management. *Journal of Environmental Management, 78*(2), 114–127.

Fyall, A. and Garrod, B. (2020), Destination management: A perspective article, *Tourism Review, 75*(1), 165–169. https://doi.org/10.1108/TR-07-2019-0311

Giddy, J. K., & Webb, N. L. (2018). The influence of the environment on adventure tourism: From motivations to experiences. *Current Issues in Ttourism, 21*(18), 2124–2138.

Gordon, J. E., Crofts, R., Díaz-Martínez, E., & Woo, K. S. (2018). Enhancing the role of geoconservation in protected area management and nature conservation. *Geoheritage, 10*(2), 191–203.

Hall, C. M. (2008). *Tourism Planning: Policies, Processes and Relationships*. Harlow, UK: Pearson Education.

Hall, C. M., & Page, S. (2006). *The Geography of Tourism and Recreation: Space, Place and Environment*, 3rd edn., London: Routledge.

Healey, P. (2004). The treatment of space and place in the new strategic planning in Europe. *International Journal of Urban and Regional Research, 28*(1), 45–67.

Healey, P. (2006). *Collaborative Planning: Shaping Places in Fragmented Societies*. New York: Palgrave Macmillan.

Higham, J., Vidar Haukeland, J., Hopkins, D., Inge Vistad, O., Lindberg, K., & Daugstad, K. (2016). National parks policy and planning: a comparative analysis of friluftsliv (Norway) and the dual mandate (New Zealand). *Journal of Policy Research in Tourism, Leisure and Events, 8*(2), 146–175.

Hill, J. L., & Hill, R. A. (2011). Ecotourism in Amazonian Peru: uniting tourism, conservation and community development. *Geography*, *96*(2), 75–85.

İnal, O., & Turhan, E. (Eds.). (2019). *Transforming Socio-Natures in Turkey: Landscapes, State and Environmental Movements*. Oxon and New York: Routledge.

Inskeep, E. (1993). Tourism planning: An integrated and sustainable development approach. *Journal of Travel Research*, 31(4), 70–71. https://doi.org/10.1177/004728759303100459

Jenkins, C. L. (2006). Tourism development: Policy, planning and implementation issues in developing countries. In J. Cuckier (Ed.), *Tourism Research: Policy, Planning and Prospects* (pp. 21–30). Waterloo: Department of Geography.

Jones, T., Shone, M. C., & Memon, P. A. (2003). *Emerging Tourism Planning Processes and Practices in New Zealand: A Local and Regional Perspective* Canterbury, NZ: Lincoln University.

Law No. 2634. 1982. *Turizmi teşvik kanunu*. https://www.mevzuat.gov.tr/mevzuatmetin/1.5.2634.pdf. Accessed: December 2021.

Law No. 2783. 1983. *Milli parklar kanunu*. https://www.mevzuat.gov.tr/MevzuatMetin/1.5.2873.pdf. Accessed: December 2021.

Law No. 2863. 1983. *Kültür ve tabiat varlıklarını koruma kanunu*, https://www.mevzuat.gov.tr/mevzuatmetin/1.5.2863.pdf. Accessed: December 2021.

Law No. 2872. 1983. *Çevre kanunu*. https://www.mevzuat.gov.tr/mevzuatmetin/1.5.2872.pdf. December 2021.

Mancini, F., Coghill, G. M., & Lusseau, D. (2017). Using qualitative models to define sustainable management for the commons in data poor conditions. *Environmental Science & Policy*, *67*, 52–60. https://doi.org/10.1016/j.envsci.2016.11.002

Marzuki, A., Rofe, M., & Mohd Hashim, N. A. (2014). Disputes on nature-based tourism development in northern peninsular Malaysia. *Tourism Analysis*, *19*(4), 525–530.

Ministry of Tourism. (2002). *II. Tourism Council Sub-Commission Reports*. General Assembly Discussions and II. Tourism Council Decisions. 12–14 April: 2002.

Mitsui, S., Kubo, T., & Shoji, Y. (2020). Understanding residents' perceptions of nature and local economic activities using an open-ended question before protected area designation in Amami Islands, Japan. *Journal for Nature Conservation*, *56*, 125857.

MoCaT (2007a). *2023 Türkiye Turizm Stratejisi*. Ankara: Kültür ve Turizm Bakanlığı.

MoCaT (2007b). *Türkiye Turizm Stratejisi Eylem Planı (2007 – 2013)*. Ankara: Kültür ve Turizm Bakanlığı.

Outeiro, L., Villasante, S., & Oyarzo, H. (2018). The interplay between fish farming and nature based recreation-tourism in Southern Chile: A perception approach. *Ecosystem Services*, *32*, 90–100.

Peroff, D. M., Deason, G. G., Seekamp, E., & Iyengar, J. (2017). Integrating frameworks for evaluating tourism partnerships: An exploration of success within the life cycle of a collaborative ecotourism development effort. *Journal of Outdoor Recreation and Tourism*, *17*, 100–111.

Pieraccini, M. (2015). Democratic legitimacy and new commons: Examples from English protected areas. *International Journal of the Commons*, *9*(2), 552–572.

Priskin, J. (2003). Issues and opportunities in planning and managing nature-based tourism in the central Coast region of Western Australia. *Australian Geographical Studies*, *41*(3), 270–286.

Pröbstl-Haider, U., & Haider, W. (2014). The role of protected areas in destination choice in the European Alps. *Zeitschrift für Wirtschaftsgeographie*, *58*(1), 144–163.

Sagala, S., Rosyidie, A., Sasongko, M. A., & Syahbid, M. M. (2018, May). Who gets the benefits of geopark establishment? A study of Batur Geopark Area, Bali Province, Indonesia. In *IOP Conference Series: Earth and Environmental Science* (Vol. 158, No. 1, p. 012034). IOP Publishing.

Selin, S. (1998). The promise and pitfalls of collaborating. *Trends*, *35*(1), 9–13.

Şeremet, M., Alaeddinoğlu, F., & Okudum, R. (2021a). Rural community development through ecotourism in Turkey: Challenges and opportunities in The Eastern Black Sea region. In V. Kunjuraman, C. R. Aziz, & N. Z. A. Baka (Eds.), *Ecotourism and Rural Community Development*. Malaysia: Penerbit Umk.

Şeremet M., Haigh M., & Cihangir E. (2021b). Fostering constructive thinking about the 'wicked problems' of team-work and decision-making in tourism and geography, *Journal of Geography in Higher Education*, *45*(4), 517–537. https://doi.org/10.1080/03098265.2020.1869924

Serter, G. (2020). 2002 sonrası milli parklar kanunu kapsamında ilan edilen doğa koruma alanlarına eleştirel bir bakış. *Planlama*, *30*(2), 242–256. https://doi.org/10.14744/planlama.2020.63625

TTB (Turizm ve Tanıtma Bakanlığı). (1965). *Tourism in Turkey*. İstanbul: Ajans-Türk Matbaasi.

TÜRSAB (1997). 25. yılında TÜRSAB, İstanbul: TÜRSAB.

Willis, C. (2015). The contribution of cultural ecosystem services to understanding the tourism–nature–wellbeing nexus. *Journal of Outdoor Recreation and Tourism*, *10*, 38–43.

Worboys, G. L., Lockwood, M., Kothari A., Feary S., & Pulsford I. (eds) (2015). *Protected Area Governance and Management*, Canberra: ANU Press.

World Tourism Organization (UNWTO) (2007). A Practical Guide to Tourism Destination Management: 21.06.2021.

Yrza, B., & Filimonau, V. (2021). Integrating sustainability in destination management plans and policies of a post-Soviet state. *Tourism Planning & Development*, 1–23. https://doi.org/10.1080/21568316.2021.1875037

Yu-Fai L., Anna S., Glen H., & Ralf B. (eds.) (2018). *Tourism and Visitor Management in Protected Areas: Guidelines for Sustainability*. Best Practice Protected Area Guidelines Series No. 27, Gland, Switzerland: IUCN, xii + 120 pp.

Zhang, C., & Xiao, H. (2014). Destination development in China: owards an effective model of explanation. *Journal of Sustainable Tourism*, *22*(2), 214–233.

33

SUSTAINABLE PRODUCT DEVELOPMENT FOR VISITOR EXPERIENCES IN NATURE-BASED TOURISM

Applying and extending the Double Diamond Design Process

Julia N. Albrecht and Marco Haid

Introduction

Nature-based tourism products are resource-based, and the related visitor experiences rely on the quality of the environment in which they take place. Nature-based tourism destinations are thus vulnerable to visitation impacts. They need to protect the resource on which they depend, at the same time as offering high-quality visitor experiences. Where tourism damages or depletes these natural resources, visitor experiences are impacted, and the resulting reputational damage can be severe (Weber et al., 2019). It is consequently crucial for tourism businesses who plan for (new) tourism products and experiences to consider the environments in which these will take place. Although many industries have developed and applied frameworks for sustainable production, the tourism literature still largely draws on broad sustainability frameworks that address the destination as a whole rather than individual or sets of products (e.g., Roxas et al., 2020, in their application of systems thinking or Calgaro et al., 2014, in their application of the destination sustainability framework in Thailand). Whilst such holistic approaches are necessary and helpful, they generally fail to identify and outline implications for tourism businesses, tourism product development and delivery. Our chapter thus clearly addresses the *development* of products and experiences, not the marketing thereof. As such, and in line with Tukamushaba et al.'s (2016) differentiation of experience as product and marketing, we prioritize the product component of the overall visitor experience here. Even though the possible usefulness of product design concepts in a tourism context has been demonstrated (Haid & Albrecht, 2021), models that specifically posit pathways toward sustainable product development for visitor experiences are still lacking.

This chapter addresses this gap in the literature by positing a framework for sustainable tourism product development for high-quality nature-based visitor experiences. It takes into account two important, previously established notions that are relevant to tourism product development as well as visitor management: First, that typical, linear processes in tourism

DOI: 10.4324/9781003230748-39

product and experience development fail to take into account sustainability needs, paired with the increasingly established view in the literature outside of tourism that products for sustainability must be integrated with their context and supply chain (Ceschin & Gaziulusoy, 2016); and second that any visitor (impact) management intervention will have a positive or negative impact on the visitor experience (Albrecht, 2017), which suggests the requirement for at least one feedback loop in related product development. This chapter illustrates the resulting usefulness of a design thinking approach by applying the Double Diamond Design Process (3DP) to product development for sustainable tourism. 3DP is the outcome of a large field study of the design processes in a range of global companies led by the Design Council (Ball, 2019; the approach is explained in detail in the section "The Double Diamond Design Process"). In applying 3DP to sustainable product development for visitor experiences, the chapter demonstrates how existing linear conceptualizations of tourism product and experience development can be transcended by current design thinking, with a view of achieving improved results not only for destination resources and relevant managing agencies but also for business owners and operators, as well as visitors to the nature-based tourism destination.

This chapter provides context for the application and adaptation of the 3DP model to the tourism context by reviewing relevant literature in the areas of visitor (impact) management and visitor experiences with a focus on nature-based tourism, sustainable (tourism) product development and design, and design for sustainability in other fields (including the original 3DP). Then, our own adaptation, the extended 3DP framework, is presented and explained, and conceptual implications and practical recommendations for tourism practitioners are provided.

Visitor (impact) management and visitor experiences

The management and attempted avoidance of negative impacts from visitation as well as the related visitor management for nature-based tourism is typically associated with the public sector (Mason, 2015). This is because nature-based tourism often (but not exclusively) draws on protected or conserved areas, which tend to be public lands. Government organizations with mandates in environmental management or conservation develop and implement legislation and mandatory frameworks that regulate the utilization of natural areas in and for tourism (Larsen & Brockington, 2018). The Recreation Opportunity Spectrum (ROS; Clark, 1982) and the Limits of Acceptable Change (LAC; McCool et al., 2007) are examples of resource-/supply-driven approaches that define and regulate any visitor experience design, product development, or related stakeholder activities within the nature-based destination. National or Regional Tourism Organizations or Destination Management Organizations may have the opportunity to contribute to the development of such regulation, for example, in the form of a formal submissions process, but they are rarely represented in these processes directly. The private sector in nature-based tourism is even further removed; businesses are typically presented with a fixed structure within which activities can be developed, offered, and run (Gajdošík et al., 2017).

Reasons for the sometimes restricted frameworks in which nature-based tourism can occur lie in the responsibility and moral obligations on the part of tourism businesses in terms of common resource preservation, restoration or depletion, as well as acknowledged shortcomings and failures in previous and current relevant destination management (Frost & Laing, 2018; Marzuki et al., 2014). Indeed, the businesses' vested interests in protecting the resources that their offerings depend can sometimes be overlooked (Albrecht, 2017). Thus, there are several lines of argument culminating in the conclusion that product development for nature-based tourism is an important activity requiring care for the environment, attention to detail, (local) knowledge, as well as knowledge of visitor expectations and experiences (Albrecht, 2021; Weber et al., 2019).

As such, tourism product development needs to balance and negotiate relevant visitor (impact) management interventions and visitor experiences (Leung et al., 2018). Although there is a small, but growing number of destinations that collaborate and pool their knowledge for sustainable tourism product development (e.g., DestiMED for ecotourism in the Mediterranean; Interreg Mediterranean, 2019), the resulting approaches can be linear, one-off, and destination-specific. Given these complexities, it is surprising that tourism has not yet sought to derive benefit from existing knowledge of and approaches to product development for sustainability.

Sustainable (tourism) product development/design

Tourism products enable people to travel and experience outside their usual environment (Smith, 1994). They can be divided into primary and secondary tourism products. The former attracts tourists and act as the (primary) motivation to visit the destination, while the latter have less specific appeal, such as accommodation and catering (Benur & Bramwell, 2015). Due to the complex interplay of tangible and intangible elements, primary tourism products, in particular, are geographically and thematically very wide-ranging (Benur & Bramwell, 2015; Smith, 1994), making their sustainable design particularly important.

Sustainably designed tourism products derive benefits without compromising the environment, community cultures or tourism development (Machado, 2003; WTO, 2005). In terms of strategic product development, sustainable tourism products go beyond traditional resource-based approaches in their design and incorporate destination-specific values and core competencies to achieve sustainability (Eckert & Pechlaner, 2019). At the destination level, the adaptation of supply chains and distribution channels is posited as an approach to achieve sustainable tourism products (Schwartz et al., 2008). It follows that cooperation and communication among stakeholders are crucial for the development and implementation of sustainable tourism products. In addition, sustainable product development approaches are encouraged by the increasing demand for 'green' products (Font & McCabe, 2017). In doing so, it is important to distinguish genuine product development for sustainable tourism from 'greenwashing' (Hall, 2015). Greenwashing is an instance where a product or experience suggests environmental sustainability to enhance its image, or the image of the business offering said product or experience, generally with a view of increasing sales. We suggest that a comprehensive, iterative design process can help providers of tourism experiences to transcend deliberate or inadvertent greenwashing by weaving decision-making for sustainable tourism into the collaborative product or experience design process (refer also Weaver & Lawton, 2017).

Design for sustainability in other fields

Existing design concepts for sustainable products have significantly expanded in scope and focus (Vezzoli et al., 2018), and they can provide valuable insights and approaches in the tourism context (Haid & Albrecht, 2021). In this regard, design concepts have increasingly evolved from insular, very limited to more holistic perspectives, especially including the environment and the context of the product (Burall, 1991; Mackenzie, 1997). Aiming to foster social justice and cohesion, these more holistic approaches incorporate all pillars of sustainability: (1) people, (2) profit, and (3) planet (Clark et al., 2009; Crul & Diehl, 2008; Vezzoli et al., 2018; Karlsson & Luttropp, 2006) and are closely linked to concepts of circular economy and regenerative systems (Geissdoerfer et al., 2017; Kirchherr et al., 2017).

Although these approaches already provide valuable insights and can be useful as a concept for sustainable tourism product development (Haid & Albrecht, 2021), their focus on the level of innovation (Ceschin & Gaziulusoy, 2016), and the impact on the three sustainability pillars: (1) People (social sustainability), (2) Profit (economic sustainability) and (3) Planet (environmental sustainability) (Clark et al., 2009; Crul & Diehl, 2008; Haid & Albrecht, 2021) can be especially helpful in a nature-based tourism context where there is, by definition, a strong focus on the area and resource(s). The level of innovation refers to the degree of systemic nature of the design and development approaches, ranging from insular to integrated and systemic solutions (Ceschin & Gaziulusoy, 2016). This involves the levels "Product Innovation" (innovations relate to the product itself), "Product-Service System Innovation" (the focus of the innovation is not only the product but the service or utility behind it), "Spatio-Social" (innovations involve social units and include entire communities) and "Socio-Technical" (innovation involves technology, social issues, organizations and institutions) (Ceschin & Gaziulusoy, 2016).

In our adaptation of the 3DP model, we bring together the design thinking approach, the sustainability considerations as explained above, and the aspiration to achieve high levels of innovation in sustainable product development in a nature-based tourism context. Notwithstanding the broader and more holistic view compared to previous perspectives, tourism, in particular, requires the involvement of further factors to promote effective approaches to sustainable product development. In this sense, the circular implementation dynamics (Haid et al., 2021), the diversity and heterogeneity of stakeholders and their interests and requirements (Beritelli, 2011, McAreavey, 2006; Strobl & Peters, 2013) are particularly important in tourism. We can then introduce current design thinking and connect applicable approaches (such as the 3DP at the focus of this chapter), identify and explain how and why these hold potential in (nature-based) tourism, with emphasis on products for sustainability which are integrated with their context and supply chain (Ceschin & Gaziulusoy, 2016).

The Double Diamond Design Process (3DP)

Though its intellectual origins can be traced back to Banathy (1996), the 3DP has been formally introduced by the Design Council in 2005 (Ball, 2019). It was the outcome of a field study of design processes employed in innovative global companies such as Alessi, Microsoft, Sony, Lego, and Starbucks, among others (Design Council, 2007). The study deliberately involved companies engaged in both product and service innovation. In a nutshell, the model (Figure 33.1) posits four stages in a design process: A discovery phase that allows designers to familiarize themselves with the problem(s) at hand, for example, by conducting market or user research; a definition phase where the information ascertained in stage 1 is synthesized to form a problem definition; a development phase that explores possible solutions; and a delivery phase, which also includes evaluation and feedback loops. Stages 1 and 2 form the first diamond; stages 3 and 4 the second. The diamond-shape exemplifies how for each set of stages, the designers' thinking must first diverge to take into account a wide range of information before converging again in order to zero in on possible solution strategies. To illustrate further, the diamond on the left, which is primarily associated with problem definition, has light-heartedly been labelled as "designing the right thing", whereas the solution-oriented diamond on the right was named "designing things right". It is vital to note that these phases do not occur in a linear manner but allow ample space for iteration and reflection (Design Council, 2007). In fact, it is its non-linearity that sets the 3DP model apart from its predecessors in the design canon. It is further important to note that unlike in colloquial language, where the term design often refers to the appearance of function of an object, design in a professional context refers to all thought processes that shape the appearance, functions, and processes related to a product or service.

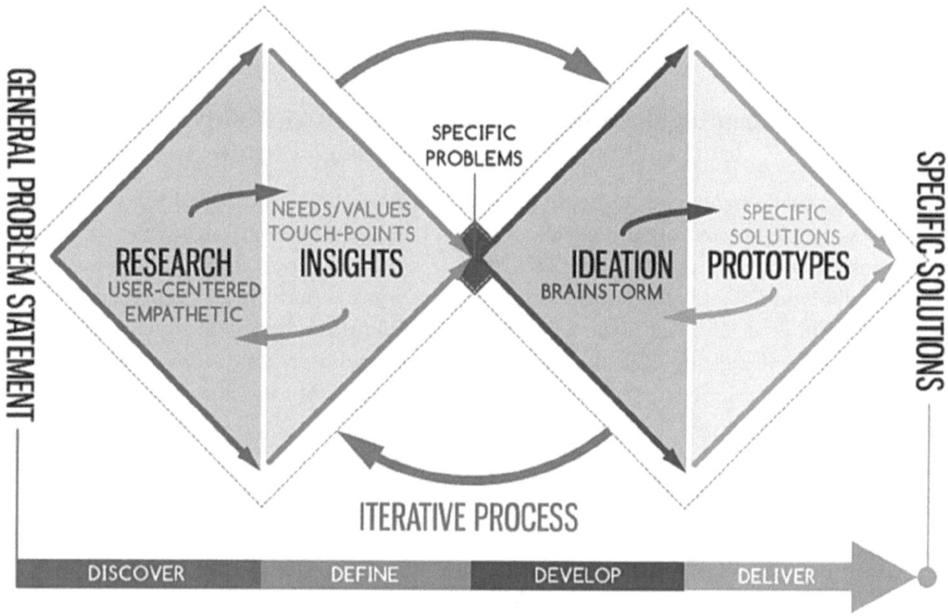

Service Design Vancouver
building innovation • value for businesses and people

Figure 33.1 Double Diamond Design Process (3DP).

Source: (from https://servicedesignvancouver.ca/wp-content/uploads/2014/11/SDV-DoubleDiamond.pdf; accessed 30 October 2021)

3DP and nature-based tourism

3DP was initially developed for product design, but its suitability and applicability in service contexts was soon realized (Design Council, 2015). In cases where 3DP is applied to service development, the relevant methods are user-focused for the first diamond (user mapping, diaries, shadowing/observation, creation of user personas) and process-focused for the second diamond (service blueprinting, experience prototyping, scenarios). It is no coincidence that these methods are familiar to tourism researchers and practitioners with interests in experience and product design as well as tourism planning more broadly. Evidence of their potential usefulness is scattered throughout the tourism literatures on tourism (product) development, planning and management (Albrecht, 2014; Gunn & Var, 2020; McLoughlin & Hanrahan, 2021; Volgger et al., 2021), tourism design (Fesenmaier & Xiang, 2017; Scuttari et al., 2021), and even design for transformative tourism experiences (Sheldon, 2020). It is important to acknowledge the strong theoretical underpinning of design science in and for tourism which lends credibility to the integration of design concepts and tools in a tourism context.

This chapter focuses on nature-based tourism and posits a design approach specific to its context, characteristics and requirements. As such, it is necessary to emphasize the necessity of sustainability in any related tourism, product, and experience design, development, and planning. The following section demonstrates how we integrate sustainability in a 3DP framework for nature-based tourism.

Extending the framework to address sustainability in tourism: the Triple Diamond Design Process

The 3DP is a useful framework addressing the design process. However, as it does not explicitly consider sustainability and other related innovation considerations, extensions are required for the design of sustainable products and, in particular, sustainable tourism products. Specifically, these are discretion of and communication among tourism stakeholders (as empirically demonstrated in Haid et al., 2021). Sustainability is added as the third diamond in Figure 33.2.

In addition, the existing 3DP design framework assumes a predominantly research-driven development. However, research results show the importance of ideas and initiating problems

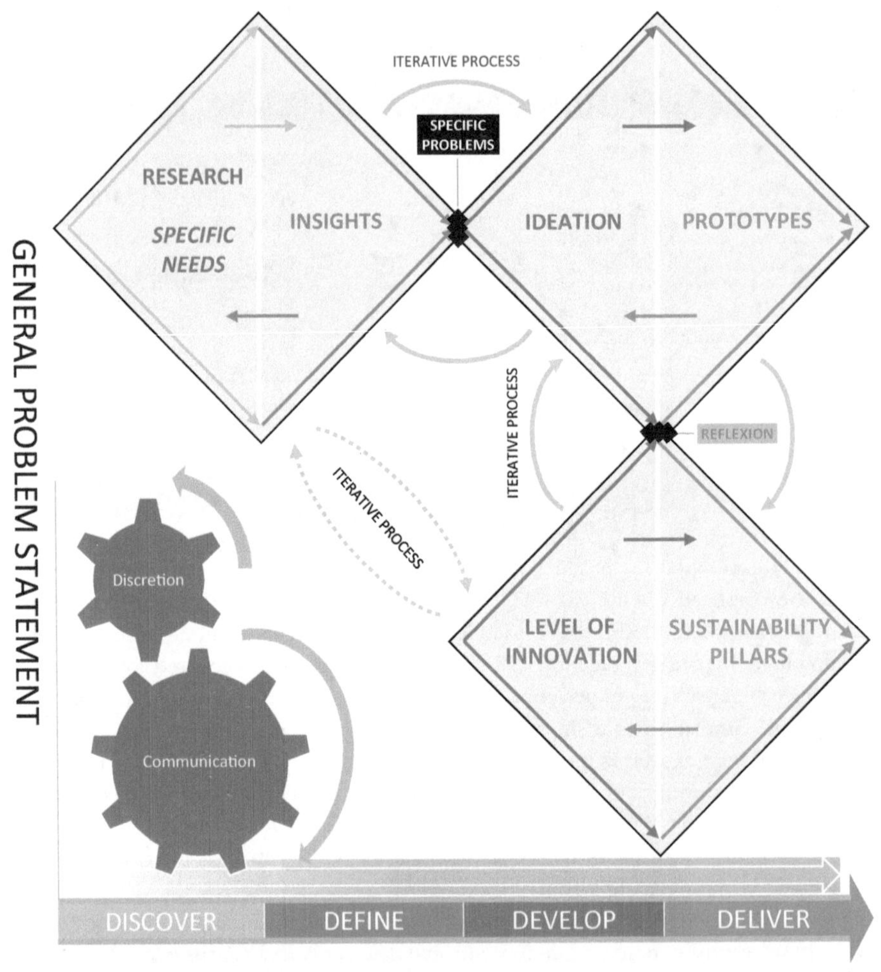

Figure 33.2 Triple Diamond Design Process.

Source: authors, after Ball, 2019

for the start of implementation and development processes (e.g., Thomann et al., 2018; Burke, 1990), as well as stakeholder communication and discretion as driving factors. Given these identified gaps, the 3DP model is extended, as illustrated in Figure 33.2.

The extensions of the 3DP (Ball, 2019) are both generic and specific to tourism. The extended model can be used for all sustainable design processes regardless of context or sector, although special features for (nature-based) tourism, in particular, have been included. Consequently, the framework in Figure 33.2 offers an avenue to address all possible, generic and destination- and area-specific concerns regarding sustainability as part of a formal set of processes.

From a generic point of view, in addition to "Research", "Specific Needs" in the sense of thought-provoking ideas, problems, and initiatives of individuals and groups were added to the 3DP in the first diamond. Especially in the mostly small-structured tourism context, individuals or individual problems often initiate a design process. Thus, the model is expanded in that the (early stage) problem can already exist at the beginning and then become specific, especially through the second step "Insights".

The extension to include the third diamond ("Level of Innovation" and "Sustainability Pillars") and the associated interface of reflection on the level of innovation and sustainability is also generic. This turns the existing 3DP into a sustainable design model that includes the (systemic) level of innovation. While the sustainability pillars are basically applicable in any context across sectors and domains (Shmelev & Brook, 2021), the level of innovation is more context-specific and case-dependent (Ostrom, 2009, puts forward a systems approach that distinguishes resources, the related management system, the governance system, and resource-users). For example, in the context of nature-based tourism, there needs to be specific attention on the resources and governance system which differs according to the type of natural area and its possible protection status and determines the level and type of possible innovation in product and experience development as well as the composition of the relevant stakeholder groups. The additional diamond and thereby generated reflection loop further promotes the iterative nature of the model and can reduce possible remaining linearity, contributing to a more thoughtful and reflective design process. Another generic extension is the addition of driving factors for the design process, which can differ from one context and sector to another.

A particularly tourism-specific extension is the concrete addition of the driving factors communication and discretion, which are particularly important in implementation and development processes in (nature-based) tourism (Haid et al., 2021). The former refers to the collaboration and exchange of stakeholders and the resulting innovative potential (Colebatch, 2006), the latter to the power of individual representatives in innovation processes (Thomann et al., 2018). The existing and extended 3DP, with the iterative procedures therein, are particularly suitable for tourism's prevailing circularity of development and implementation processes (Haid et al., 2021; Wray, 2009). With the expansion of the third diamond and thus the inclusion of sustainability and innovation concerns, the framework is of great importance, especially in nature-based tourism, as particularly in these (vulnerable) regions, sustainable tourism products are necessary to reduce negative impacts of tourism on ecosystems and communities (Visbeck et al., 2014).

Implications for tourism practitioners

The extended 3DP model, the Triple Diamond Design Process, for sustainable product development provides several practical recommendations for destination managers and other practitioners in (nature-based) tourism. First, we know that a pluralistic and multi-perspective view and the representation of all interests can be beneficial for overall destination development and management (Benur & Bramwell, 2015), and that view can be reasonably extended to development in the destination more broadly, including product development (Smith, 1994). Implementing an approach such as the posited Triple Diamond Design Process will thus benefit from involving all relevant stakeholders in the destination, including those not immediately involved with the business managing a product. In doing so, the relevant management structures for any protected, conserved, or otherwise publicly managed resources for the nature-based tourism offering in question will determine which other stakeholders to consult. Secondly, the consideration and reflection of sustainability and the level of innovation are complex. Not only are the concepts used and prioritized in different ways (Albrecht et al., 2021), but also a comprehensive knowledge of the tourism destination's objectives in this regard is necessary. The nomination and involvement of competent individuals who act as critical reflection staff (third diamond) will be valuable for a sustainable (systemic) innovative design process. Thirdly, initiatives in the form of initial problems and ideas are very important to start design processes. Especially in centralized procedures, measures should be taken to encourage individuals and groups to submit problems and ideas. Fourthly, communication should be promoted due to its high importance as a driving factor (Haid et al., 2021). Therefore, regular formal and informal exchanges and cooperation (Beritelli, 2011), in the form of events, joint meetings, excursions, or regulars' tables, are highly recommended.

Critical assessment, limitations

There is some linearity remaining within the model, which counteracts its basic notion of iteration and can thus lead to a less reflective and considered design process. However, the extension of the 3DP with the third diamond and thus the additional reflection loop regarding sustainability and innovation level can counteract this effect and foster a more iterative procedure. Although the extension of the model with the third diamond is an important step for sustainable design, it is worth mentioning that sustainability as a vague and inconsistent concept is interpreted and prioritized differently (Albrecht et al., 2021; Salas-Zapata & Ortiz-Munoz, 2019). The existing model addresses this problem only insofar as the pillars of sustainability (Clark et al., 2009; Crul & Diehl, 2008; Vezzoli et al., 2018; Karlsson & Luttropp, 2006) are mentioned, and thus at least a first guidance is given. However, this does not entirely address the lack of clarity of the term, which can be problematic for implementation and reflection.

The implementation of the model can be particularly resource- and time-consuming in the case of isolated economic entities, such as individual companies, and long-established product development procedures, as the assumptions and processes in the model can be alienating and counter-inductive for the otherwise familiar heuristic approaches, such as "learning by doing" (i.e., fine-tuning the product/service during the (active) offer), "trial and error" or the imitation of existing products. Therefore, the model should first be implemented in a centralized manner for integrated products and services, for example, by the Destination Management Organizations, with the involvement of all stakeholders. In this way, the process can be (better) guided, and the people involved can be made aware of the handling and opportunities of the model. This has the potential for the model to be used in other economic entities in the destination as well as for more isolated products.

Regarding the time after completing the design process, the model does not include feedback and reflection loops. This can be problematic, especially for products and services in nature-based tourism, which takes place in particularly vulnerable regions and areas, in that possible negative impacts may only become apparent after a certain time. Predictions about potential problems and negative effects are hardly or not possible immediately after completion of the design process. In this respect, the existing model should be expanded to protect particularly vulnerable regions and areas.

As some aspects of the Triple Diamond Design Process can be seen as generic and intrinsic to design processes (especially in protected areas), we emphasize that its contribution is more at the process and organizational levels rather than at the content level. As such, the model can contribute to the transparency of processes, influencing factors and critical phases.

Conclusion, future research

This chapter presents a framework for the development of sustainable tourism products, building on the need for sustainable management of the limited and vulnerable resources in nature-based tourism. To address the problems that any intervention in visitor management has an impact on the visitor experience (Albrecht, 2017), that sustainability requirements are hardly considered in current linear tourism product developments, and that many stakeholders and thus different interests exist (Beritelli, 2011), current design thinking approaches were brought into the tourism context. The 3DP was used and adapted for this purpose. Its iterative, problem-centred and open-inclusive character makes it possible to overcome linear, less reflective processes and involve all relevant stakeholders and perspectives. In addition, it places the user in the centre of the considerations, and thus, in the tourism context, both the perspectives of the tourists and their visitor experience as well as the providers and local communities can be included and considered. In order to take into account the need to protect the sensitive ecosystems in nature-based tourism, the 3DP was extended by another iterative reflection loop on sustainability to promote a more holistic design process addressing all areas of sustainability and thus responsible and sustainable products. This requires clear communication and a common understanding of the term sustainability in the design process (Albrecht et al., 2020). In addition, proper leadership and guidance of the design process are needed to optimally utilize the innovation potential of stakeholder communication and collaboration (Haid et al., 2021) and avoid conflicts and thus negative development tendencies (McAreavey, 2006). Positive results are anticipated in several ways: The user-centred, evidence-based approach to product development aims to ensure a close fit between visitor expectations/abilities and values and product and experience offerings which will benefit businesses as well as visitors. The feedback loops between ideation and the piloting of products will serve to increase these benefits. Bringing together innovation and sustainability requirements in both development and delivery of the tourism products in the form of iterative rather than linear processes will help carry through and consolidate the link between the evidence-based approach and its evolving outcomes.

The adapted and extended framework model for the design of sustainable tourism products has great potential, especially in nature-based tourism. How and to what extent it is or could be applied needs to be addressed in further research. This could provide interesting insights into implementation, realization, and possible barriers. Further research is also needed to extend the model after the design is completed in terms of introducing ongoing reflection and monitoring of the impact of the sustainable products.

References

Albrecht, J. N. (2014). Micro-mobility patterns and service blueprints as foundations for visitor management planning. *Journal of Sustainable Tourism*, 22(7), 1052–1070.

Albrecht, J. N. (2017). Introduction to visitor management in tourism destinations. In J. N. Albrecht (Ed.), *Visitor Management in Tourism Destinations* (pp. 3–8). CABI.

Albrecht, J. N. (2021). Current issues in managing visitor experiences in nature-based tourism. In J. N. Albrecht (Ed.), *Managing Visitor Experiences in Nature-based Tourism* (pp. 1–8). CABI.
Albrecht, J. N., Haid, M., Finkler, W., & Heimerl, P. (2020). What's in a name? The meaning of sustainability to destination managers. *Journal of Sustainable Tourism*, 30(1), 32–51.
Ball, J. (2019). The Double Diamond: A universally accepted depiction of the design process. *Design Council*. https://www.designcouncil.org.uk/news-opinion/double-diamond-universally-accepted-depiction-design-process (accessed 29 October 2021).
Banathy, B. H. (1996). *Designing Social Systems in a Changing World*. Springer.
Benur, A. M., & Bramwell, B. (2015). Tourism product development and product diversification in destinations. *Tourism management*, 50, 213–224.
Beritelli, P. (2011). Cooperation among prominent actors in a tourist destination. *Annals of Tourism Research*, 38(2), 607–629.
Burall, P. (1991). *Green design*. Gower Publishing.
Burke, J.P., (1990). Policy implementation and the responsible exercise of discretion. In Palumbo, D.J., Calista, D.J. (Eds.), *Implementation and the Policy Process: Opening up the Black Box* (pp. 133–148). Greenwood Press.
Calgaro, E., Dominey-Howes, D., & Lloyd, K. (2014). Application of the destination sustainability framework (DSF) to explore the drivers of vulnerability in Thailand following the 2004 Indian Ocean tsunami. *Journal of Sustainable Tourism*, 22(3), 361–383.
Ceschin, F., & Gaziulusoy, I. (2016). Evolution of design for sustainability: From product design to design for system innovations and transitions. *Design studies*, 47, 118–163.
Clark, G., Kosoris, J., Hong, L. N., & Crul, M. (2009). Design for sustainability: current trends in sustainable product design and development. *Sustainability*, 1(3), 409–424.
Clark, R. (1982). Promises and pitfalls of the ROS in Resource Management. *Australian Parks and Recreation*, May, 9–13.
Colebatch, H. K. (Ed.). (2006). *Beyond the Policy Cycle: The Policy Process in Australia*. Allen and Unwin.
Crul, M., & Diehl, J. C. (2008, October). Design for sustainability (D4S): Manual and tools for developing countries. In *Proceedings of the 7th annual ASEE global colloquium on engineering education, Cape Town* (pp. 19–23).
Design Council (2007). *Eleven Lessons: Managing Design in Eleven Global Companies*. Design Council.
Design Council (2015). *Design Methods for Developing Services*. Design Council. https://www.designcouncil.org.uk/sites/default/files/asset/document/Design%20methods%20for%20developing%20services.pdf (accessed 29 October 2021).
Eckert, C., & Pechlaner, H. (2019). Alternative product development as strategy towards sustainability in tourism: The case of Lanzarote. *Sustainability*, 11(13), 3588.
Fesenmaier, D. R., & Xiang, Z. (Eds.). (2017). *Design Science in Tourism*. Springer.
Font, X., & McCabe, S. (2017). Sustainability and marketing in tourism: Its contexts, paradoxes, approaches, challenges and potential. *Journal of Sustainable Tourism*, 25(7), 869–883.
Frost, W., & Laing, J. (2018). Public–private partnerships for nature-based tourist attractions: the failure of Seal Rocks. *Journal of Sustainable Tourism*, 26(6), 942–956.
Gajdošík, T., Gajdošíková, Z., Maráková, V., & Flagestad, A. (2017). Destination structure revisited in view of the community and corporate model. *Tourism Management Perspectives*, 24, 54–63.
Geissdoerfer, M., Savaget, P., Bocken, N. M., & Hultink, E. J. (2017). The Circular Economy–A new sustainability paradigm? *Journal of Cleaner Production*, 143, 757–768.
Gunn, C. A., & Var, T. (2020). *Tourism Planning - Basics Concepts Cases* (4th ed.). Routledge.
Haid, M. & Albrecht, J.N. (2021) Sustainable tourism product development: An application of product design concepts. *Sustainability*, 13, 7957.
Haid, M., Albrecht, J. N., & Finkler, W. (2021). Sustainability implementation in destination management. *Journal of Cleaner Production*, 127718.
Hall, C. M. (2015). Economic greenwash: On the absurdity of tourism and green growth. In M. V. Reddy & K. Wilkes (Eds.), *Tourism in the Green Economy* (pp. 339–358). Routledge.
Interreg Mediterranean. (2019). *How protected areas & local stakeholders can work together to create sustainable tourism products*. Retrieved 25 January 2022 from https://destimed.interreg-med.eu/news-events/news/detail/actualites/how-protected-areas-local-stakeholders-can-work-together-to-create-sustainable-tourism-products/
Karlsson, R., & Luttropp, C. (2006). EcoDesign: what's happening? An overview of the subject area of EcoDesign and of the papers in this special issue. *Journal of Cleaner Production*, 14(15–16), 1291–1298.

Kirchherr, J., Reike, D., & Hekkert, M. (2017). Conceptualizing the circular economy: An analysis of 114 definitions. *Resources, Conservation and Recycling*, 127, 221–232.

Larsen, P. B., & Brockington, D. (2018). Introduction: Rethinking the Boundaries of Conservation NGOs. In P. B. Larsen & D. Brockington (Eds.), *The Anthropology of Conservation NGOs: Rethinking the Boundaries* (pp. 1–15). Springer International Publishing.

Leung, Y.-F., Spenceley, A., Hvenegaard, G., & Buckley, R. (2018). *Tourism and Visitor Management in Protected Areas, Guidelines for Sustainability*. IUCN, International Union for Conservation of Nature.

Machado, A. (2003) *Tourism and Sustainable Development, Capacity Building for Tourism Development in Vietnam*. Available online: http://www.antoniomachado.net/wp-content/uploads/pdf/tourism/2003-tourism-and-sustainable-development.pdf (accessed on 20 July 2021).

Mackenzie, D. (1997). *Green Design: Design for Environment Laurence King*. Laurence King Publishing.

Marzuki, A., Rofe, M., Hashim, M., & Arbaayah, N. (2014). Disputes on nature-based tourism development in Northern Peninsular Malaysia. *Tourism Analysis*, 19(4), 525–530.

Mason, P. (2015). *Tourism Impacts, Planning and Management* (3rd, Ed.). Routledge.

McAreavey, R. (2006). Getting close to the action: the micro-politics of rural development. *Sociologia Ruralis*, 46(2), 85–103.

McCool, S., Clark, R., & Stankey, G. (2007). *An assessment of frameworks useful for public land recreation planning*. Gen. Tech Rep. PNW-GTR-705. U.S. Department of Agriculture, Forest Service, Pacific Northwest Research Station.

McLoughlin, E., & Hanrahan, J. (2021). Evidence-informed planning for tourism. *Journal of Policy Research in Tourism, Leisure and Events*. https://doi.org/10.1080/19407963.2021.1931257

Ostrom, E. (2009). A general framework for analyzing sustainability of social-ecological systems. *Science*, 325(5939), 419–422.

Roxas, F. M. Y., Rivera, J. P. R., & Gutierrez, E. L. M. (2020). Framework for creating sustainable tourism using systems thinking. *Current Issues in Tourism*, 23(3), 280–296.

Salas-Zapata, W., & Ortiz-Munoz, S. (2019). Analysis of meanings of the concept of sustainability. *Sustainable Development*, 27, 153–161.

Schwartz, K., Tapper, R., & Font, X. (2008). A sustainable supply chain management framework for tour operators. *Journal of Sustainable Tourism*, 16(3), 298–314.

Scuttari, A., Pechlaner, H., & Erschbamer, G. (2021). Destination design: A heuristic case study approach to sustainability-oriented innovation. *Annals of Tourism Research*, 86, 103068. https://doi.org/10.1016/j.annals.2020.103068

Sheldon, P. (2020). Designing tourism experiences for inner transformation. *Annals of Tourism Research*, 83. https://doi.org/10.1016/j.annals.2020.102935

Shmelev, S., & Brook, H. R. (2021). Macro sustainability across countries: Key sector environmentally extended input-output analysis. *Sustainability*, 13(21), 11657.

Smith, S. L. (1994). The tourism product. *Annals of Tourism Research*, 21(3), 582–595.

Strobl, A., & Peters, M. (2013). Entrepreneurial reputation in destination networks. *Annals of Tourism Research*, 40, 59–82.

Thomann, E., van Engen, N., Tummers, L., 2018. The necessity of discretion: a behavioral evaluation of bottom-up implementation theory. *Journal of Public Administration Research and Theory*, 28 (4), 583–601. https://doi.org/10.1093/jopart/muy024

Tukamushaba, E. K., Xiao, H., & Ladkin, A. (2016). The effect of tourists' perceptions of a tourism product on memorable travel experience: Implications for destination branding. *European Journal of Tourism, Hospitality and Recreation*, 7(1), 2–12.

Vezzoli, C., Ceschin, F., Osanjo, L., M'Rithaa, M. K., Moalosi, R., Nakazibwe, V., & Diehl, J. C. (2018). *Designing Sustainable Energy for All: Sustainable Product-service System Design Applied to Distributed Renewable Energy*. Springer Nature.

Visbeck, M., Kronfeld-Goharani, U., Neumann, B., Rickels, W., Schmidt, J., van Doorn, E., Matz-Lück, N., Ott, K., & Quaas, M. (2014). Securing blue wealth: The need for a special sustainable development goal for the ocean and coasts. *Marine Policy*, 48, 184–191.

Volgger, M., Erschbamer, G., & Harald Pechlaner. (2021). Destination design: New perspectives for tourism destination development. *Journal of Destination Marketing & Management*, 19, 100561. https://doi.org/10.1016/j.jdmm.2021.100561

Weaver, D. B., & Lawton, L. J. (2017). A new visitation paradigm for protected areas. *Tourism Management*, 60, 140–146.

Weber, F., Crameri, U., Stettler, J., Gunzinger, T. L., Eggli, F., & Barth, M. (2019). *Measuring Overtourism.* World Tourism Forum Lucerne.

Wray, M. (2009). Policy communities, networks and issue cycles in tourism destination systems. *Journal of Sustainable Tourism*, 17(6), 673–690.

WTO; UNEP (2005). *Making Tourism More Sustainable–A Guide for Policy Makers.* UNEP & WTO: Paris, France; Madrid, Spain.

34

TOWARD AN ERA OF MULTISPECIES LIVELIHOODS IN COASTAL AND MARINE COMMUNITY-BASED TOURISM

Samuel R. Fennell, Kellen Copeland, Bastian Thomsen, Shelby Copeland and Max Duggan

Introduction

The COVID-19 pandemic has led to tourism industries becoming financially suffocated by the phenomenon of under-tourism. This largely effects the communities that have become dependent upon the tourism industry, including many rural communities across the globe. But what if this was an opportunity to rethink how humans treat one another within the industry and the communities which we operate? This chapter discusses certain integral aspects of the sustainable tourism paradigm to focus on community-based tourism (CBT). It considers how to remediate the effects of over-tourism for (rural) economies as they transition into living with COVID-19. As this unprecedented era of under-tourism persists, it provides humanity a rare opportunity to critically analyze how we operate on a quotidian basis. Not only should community-based tourism focus on the rights, welfare, and agency of local receiving human cultures (Thomsen et al., 2020), they should also extend to the nonhuman animals that are so frequently used in tourism (see Moorhouse et al., 2017). This can be achieved through emerging wildlife-human paradigms such as multispecies livelihoods (Thomsen et al., 2021a). Thomsen et al. define multispecies livelihoods:

> the right for human and nonhuman animal species to not only exist but to secure the necessities of life in a manner that does not infringe on another species' right to live, except for sustenance hunting or legitimate safety concerns to foster optimal conditions for wildlife-human coexistence.
>
> (2021a, p. 4)

Can humanity leverage this opportunity to foreground local cultures and nonhuman animals in an era of COVID-19? This chapter discusses animal labour in community-based tourism development, and then presents a brief case study on coastal marine tourism in Oregon that provides insights into the complexities of tourism policy, degrowth, and education. We conclude with

DOI: 10.4324/9781003230748-40

thoughts on how to reconsider a multispecies livelihoods perspective in community-based tourism as humanity learns to live with COVID-19.

Sustainable tourism

Zhenhua Liu's (2003) critique of sustainable tourism development outlined six issues in tourism research: the nature of tourism resources, the role of tourism demand, the imperative of intra-generational equity, the measurement of sustainability, the role of tourism in promoting sociocultural progress, and various forms of sustainable development. This interdisciplinary approach is essential when reviewing a particular tourist destination(s), or entire sanction of the tourism industry's policy. As policy within the sustainable tourism sector continues to update and improve (e.g., animal welfare) (Winter, 2020), the process of developing sustainable and culturally appropriate livelihoods through tourism has gained traction (Liu, 2003; Thomsen et al., 2021a). By accounting for both the theoretical and practical bases of sustainability, local cultures and communities will continue to benefit from sustainable tourism as it strives to give due account to their national wealth, ecological diversity, and local communities (Moyle et al., 2020; Thomsen et al., 2021b).

Cultural tourism is "the movement of persons to cultural attractions away from their normal place of residence, with the intention to gather new information and experiences to satisfy their cultural needs" (Csapó, 2012, p. 205). Cultural tourism and community-based tourism (CBT) possess a grounding and support base within the current literature, as they are either indirectly or directly linked to the many benefits tourism can have for a local destination (cultural, ecological, etc.). Among the total number of global tourists, those who are either wholly or partially motivated by culture are those who can be placed into the category of a cultural tourist (Csapó, 2012). The social benefits found when systems of cultures connect between communities through the medium of tourism possess high importance. The indirect effects cultural tourism has on a local culture are much higher than the direct ones, due to its multiplicator effects. As an example, the 'cultural tourist' has been found to spend one-third more money on average than other types of tourists (Csapó, 2012). Community-based tourism maximizes the tourist's participation and engagement with local cultures to maximize the authenticity of one's visit. However, for any community development project to be truly sustainable, it must embrace local perspectives, culture, and desires (Thomsen et al., 2020). The following sections review sustainable tourism development in conjunction with CBT in directing a post-pandemic industry that embraces multispecies livelihoods.

US policy and the COVID-19 pandemic

Tourism's historical significance and cultural erudition enables the industry to face challenges posed by global economic conditions, geopolitical turmoil, and natural disasters with remarkable resilience. Tourism remains one of the largest and fastest-growing sectors in the world economy and a valuable source of job creation, economic growth, export revenue, and domestic value added (Steinbrink, 2012). The COVID-19 crisis has led to a collapse in international travel, which has devastated world markets (Mandić, 2021). According to the World Tourism Organization (WTO), international tourist arrivals declined globally by 74 percent in 2020 with one billion fewer travelers compared to 2019. The WTO estimates that this shift has impacted between 100 and 120 million direct tourism jobs (WTO, 2022). This has negatively affected international revenue for tourism-dependent economies, exports of travel services, and contributed to the decline in exports of transport services. Despite compounding challenges, tourism countries with strong centralized tourism bodies and a national tourism policy are remediating losses faster than those who do not; Spain, France, and Canada among the top countries (Bork et al., 2020).

In the United States, tourism is a major contributor to the national economy, supplying 2.9 percent of the GDP and accounting for 10 percent of net exports, more than agriculture, mining, and the utility sector (OECD US, 2018). The tourism and travel industry are one of the nation's largest employers, subsidizing 16.8 million jobs in 2019 (OECD US, 2020). According to the 2019 World Population Review, the USA ranked third in international visits, accounting for roughly 79.3 million tourists, directly behind France (1) and Spain (2). However, the US market share in international tourism is currently declining due to the lack of a national tourism policy. The National Travel and Tourism Office (NTTO) has become embedded within the subcommittee of the International Trade Administration (ITA). The US federal government does not regulate travel and tourism as a distinct industry, and tourism marketing has not been a focal point of recent administrations. The National Tourism Policy Act of 1981 was effectively disbanded by the Clinton era in 1996 which further decentralized tourism initiatives in the USA. This has made it more difficult to compete on a global market due to its fragmented nature. A highly centralized tourism policy that transects government bodies (economic, trade and commerce, foreign affairs, transport) and private sectors (airlines, hotels, and car rentals), is needed to promote tourism as an economic mainstay and a possible tool for economic diplomacy with other nations. Sustainable tourism development, community-based tourism, and cultural tourism initiatives are believed to provide solutions to place-based resilience by tending to the consequences of over-tourism through strategic management.

Community-based tourism

Community-based tourism (CBT) is the integration of involved populations with the tourism industry, becoming a part of the tourism system in the relevant area. The host population is involved with all processes of the tourism industry, allowing the residents control over operations with outside influence on practices minimized. The local population's involvement in tourism operations can facilitate sustainable practices, improving on waste oversight and reducing strains within the industry's systems. Sustainability can be improved through further integration of the local resources and the tourism system to create new systems that complement the existing tourism industry. The intent of CBT is to improve the livelihoods of the host community through opportunities in the tourism industry while improving existing tourism models. It is a strategic investment for communities to participate in the broader tourism sector. CBT places heavy reliance on local communities, where planning and process both are dictated by the host population.

The extent to which the host population is involved will affect the effectiveness of the CBT system as well as any other measures tied to the CBT efforts. Sustainability measures will be especially reliant on host connection; one cannot thrive without the other's success. Guidelines and systems must be agreed upon and planned out before beginning CBT to allow for productive, real sustainable measures, evaluated by the community as well as the non-host organizations. Adherence to this plan will give way to more sustainable outcomes in the host community that can be recognized and improved upon with future CBT development. Planning also plays a key role in the future developments, where host populations are especially susceptible to rising land prices and higher costs of living. CBT plays a role in giving these host populations a voice for upcoming tourism improvements and pathways for least impact to their communities.

A community participation approach, according to Okazaki (2008), has been identified as integral within the process of sustainable tourism development. The proportion of active tourists continues to increase, more and more destinations not previously utilized for tourism are able to afford authentic and culturally diverse experiences to those who are willing to engage and learn (Okazaki, 2008). A mutually beneficial option for tourism has come about at around the turn of the 21st century, which assists in socio-economic development for particular regions that are in need (Dangi and Jamal, 2016). It is also complementary to traditional practices based solely on agriculture, livestock farming, and fishing. It is essential that the development of policies for the purpose of exploiting natural and cultural resources at a particular location must be done by the local communities themselves. Similar to other forms of tourism, CBT also allows for the creation of resources such as health services, infrastructure, and education. For these locals, the level of involvement they have with CBT, too, must be entirely in their hands (López-Guzmán et al., 2011). Finally, for a tourist to 'participate' with a local culture, beyond simply obtaining material resources, they must engage in activities of knowledge and cultural exchange with locals (Okazaki, 2008). The following section considers coastal and marine community-based tourism, that is often local and mixed with sometimes competing industrial priorities when considering the rights, welfare, and agency of nonhumans.

Coastal and marine community-based tourism

Marine wildlife tourism can be understood as any tourist activity that has the primary purpose of studying, watching, or enjoying wildlife in a marine habitat. Zeppel and Muloin (2008) outlined that this form of tourism includes:

> Marine wildlife-watching holidays, wildlife boat trips in marine or estuarine areas, guided island or coastal walks, observing marine life from land viewpoints, visiting marine or coastal nature reserves, participating in a marine life study tour or conservation holiday, and visiting marine wildlife visitor centres and marine aquaria.
>
> (p. 20)

Coastal and marine tourism is considered a mixed bag of both 'wicked' and 'super wicked' problems that emphasize the responsibility of tourism researchers and experts in communicating issues to policymakers (Miller et al., 2017). There is an increasing focus on the types of questions and research goals in answering issues related to the Anthropocene (Arias-Maldonado, 2016). This is the geological epoch succeeding the Holocene, marked by significant anthropogenic disturbance to Earth's climate, oceans, land, and biosphere (Waters et al., 2016).

Flora and fauna that live in the coastal and maritime zone and are dependent upon resources in a particular marine environment, are also either directly or indirectly involved within marine wildlife tourism. Animals or *nonhumans* can be gazed upon in entirely wild habitats, coastal habitats (habituated to humans), or to provisioned i.e., feed habitats. Numerous ways for tourists to interact with marine wildlife have emerged. At sea or near-shore tourism vantage points include shore-based, boat-based, onshore, vantage points, interactive driving, etc. (Zeppel and Muloin, 2008). When considering marine wildlife tourism from the perspective of the human consumer, various benefits include physiological, economic, environmental psychological and social outcomes. Those who seek this form of tourism for biocentric reasons (i.e., all life deserves equal moral standing and consideration) may seek to enjoy a form of non-consumptive activities such as cetacean viewing in opposition to partaking in activities involved in whale killing, fishing, etc. It is through an understanding of the benefits sought by tourists which mitigates both how marine wildlife tourist attractions are marketed and how they are managed. (Zeppel and Muloin, 2008). However, anthropocentric tendencies have dominated policy decisions in marine and coastal tourism. This effects tourism planning, which can result in the violation of animal ethics, rights, and welfare (Leposa, 2020).

Cater and Cater (2008) discuss the economic impacts of coastal and marine tourism are encapsulated by a series of three multiplier effects. These effects are direct use of commodities by tourists, indirect effects arising from companies who receive money from tourists buying goods and services, and induced effects understood as locals who spend their money on local goods and services. Although limited by leakage, these positive multiplier effects can be found to be omnipresent within tourism, with coastal and marine tourism without exception. Marine tourism often results in benefits for tourists looking for particular social, cultural, and health-related advantages. These can include various participatory (sea bathing, sea sports, etc.), health care (seas often possess metabolism-beneficial properties), and heterogeneous cultural (marine culture itself is often regarded for its openness, innovativeness, and overall extroversion) (Wang and Zhang, 2019).

Grounding animal ethics in/of animal labour

Bohn et al. (2018) explain that since animals have become a very important part of tourism and leisure experiences around the world, they can be seen in captivity, as entertainers, in the wild, and as part of tourism activities themselves (Moorhouse et al., 2017; Bohn et al., 2018; Copeland, 2021). Online marketing campaigns from Sweden, Finland, and Norway portray photographs that contain reindeer and huskies engaging in activities performed by tourists, with sledding being a common trend. Tourism companies around the world are akin to these Nordic destinations that utilize the advertising strategy of depicting human–animal encounters more frequently. Destinations without a possibility of offering animal encounters have become an exception within the tourism industry (Bohn et al., 2018). Fennell (2015) stresses the importance of paying attention to the animal ethics literature, as tourism theorists and practitioners must contain an understanding of the rightness and wrongness of a range of practices involving animals. Some of the major perspectives in this line of thought – animal rights, animal welfare, utilitarianism, ecofeminism, suffering, and ecocentrism – offer unique positions on if or how animals ought to be used for the purpose of entertainment (Fennell, 2015; Fennell and Thomsen, 2021).

When looking at animal ethics and tourism from a cultural anthropological perspective, von Essen et al. (2020) brought forth the issue behind imposing universal standards of animal welfare on other cultures. Although they vary in form, tourism activities that involve animals are indeed typically authentic, multisensory, and tactile in nature. Additionally, the demographics associated with nonhuman animal tourism tends to be relatively consistent. They are typically urban clients who have a willingness to pay a premium price in order to 'get away' from their home community and rediscover the wilderness. Due to these trends and the degree of demand for this particular area of tourism, the emergence of animals used at tourist destinations for the purpose of entertainment, self-fulfilment, and edutainment to tourist-consumers is increasing. It can be said that animals become labourers in a global capitalist economy (Copeland, 2020; von Essen et al., 2020). Wadiwel (2018) explained that humans' understanding of 'domesticated' animals is unstable within a variety of production processes. In opposition to 'harvested' animals, such as chickens, the author contends:

> There is thus a curious instability in how we understand the concept of the "domesticated animal" who, it is assumed through generations of training, habituation, body

> modification, reproductive controls, and enclosure, occupies some sort of position of docility or beneficent relationality with respect to humans and thus, through a kind of failure of resistance, is imagined as "not wild" or "not belonging in the wild.
>
> (p. 528)

Consistent with Karl Marx, anthropocentric views toward animal labour provide an ideology which holds that when animals labour under capitalism (i.e., *part* of the working class) they are seen to share the structural condition of humans in the production processes. Critical tourism studies have begun to focus on unearthing what is distinctive about animal labour to reconcile the growing phenomenon (Wadiwel, 2018). Ecotourism, a relatively new form of tourism, is arguably amongst the most ethical forms of leisure travel. Fennell (2020) explains that certain activities such as hunting, fishing, sled dogging are constantly criticized as not conforming to the parameters of ecotourism due to their consumptive nature (see also Thomsen et al., 2021a).

Other human–animal interactions that can be as brief as taking selfies with animals in the background are also contentious in a way such that these instances pose serious welfare risks to the animal because of the interaction. An example of this can be found when a baby dolphin was killed while being passed from tourist to tourist on an Argentinian beach (O'Neil, 2016). It was reported that the tourists were convinced that the calf was at ease during the handing-off process as it had a 'smile' on its face. It was only until someone noticed the calf had passed away when they understood something was wrong (Fennell, 2020). The anthropocentric nature of how humans understand nonhuman beings becomes a greater concern in sustainable tourism development, and CBT as animal labour is increasingly deployed.

Cui and Xu (2019) indicated that a non-anthropocentrism model which favors animal rights ethics does not have any practicality in Thailand due to spatial, temporal, social, and economic constraints. They concluded that there cannot be a single set of ethics which can solve all problems and be sustainable in all contexts. In response, von Essen et al. (2020) presented two approaches that can be used to practice compassionate animal-based tourism, and thus combat the various forms of labour that animals can be subject to in the tourism industry. First, the three R's: "Replacement of animals with alternative methods, Reduction of the number of used animals and Refinement of the methods, including housing and care, to mitigate suffering and promote animal welfare" (von Essen et al., 2020, p. 11). This model challenges humans to ask questions regarding whether it is worth using animals for a given tourist destination, or whether the cost of labour on the animal is worth the interest garnered by the tourists. Thomsen (2021) evaluates the role of individual animals in wildlife tourist attractions (WTAs) from a posthumanist lens. He posits that while noncaptive settings are ideal for nonhumans, the individual animal's circumstances (e.g., what conditions did they live in prior to living in the WTA) must be considered when evaluating the efficacy of using animals in tourism. The second approach underlines the need to inform tourists of animals' welfare, rights, and agency in WTAs. Keeping tourists well informed may help them make animal-welfare-friendly and compassionate decisions when traveling (Fennell and Sheppard, 2021).

Balon (2000) uses moral philosophy to argue against fishing for pleasure, such as in tournaments and for fishing for sport. Garrod (2007) identified three ethical issues emended to the utilization of marine wildlife for tourism stemming from the continual evolution of moral consideration humans hold towards animals: 1) a conservationist approach to the ethical treatment of animals, 2) the welfare of individual animals, and 3) the rights of animals. Ecotourism and marine wildlife tourism possess a strong affinity with one another by incorporating human well-being from nature-based activities, management according to sustainable development principles, benefits of local communities, and a source for conservation-related resources, among others.

However, there are many practical applications within marine wildlife tourism that do not adopt the characteristics of ecotourism. As a result, there have been numerous accounts of poorly managed marine wildlife tourism destinations (Garrod, 2007) which poses the issue around ethical dilemmas within marine wildlife tourism because of unjust animal labour (Crater et al., 2015). Only by embracing pro-animal, non-consumptive practices can marine wildlife tourism transcend to marine wildlife *ecotourism*. The following case study briefly highlights some of the complexities of marine based wildlife tourism, particularly in the USA during the COVID-19 pandemic.

A brief case study of Oregon coastal tourism

In Oregon, USA whale watching tours, science centres, and coastal aquariums serve to educate the public on issues of plastic pollution, ocean acidification, and marine life in general (Nalven, 2019; see also Baechler et al., 2020). Tourism management and planning has examined the relationship between resident attitudes and agency trust in the development of new 'Marine Protected Areas' and 'Marine Reserves' (Perry et al., 2017). Outdoor recreation activities such as surfing and fishing have been utilized to understand both residents and tourists' perceptions of current environmental issues (i.e., record-breaking high temperatures, drought, unusual oceanographic conditions, morbidity/mortality of typically harvested aquatic species) and long-term environmental issues (sea level rise, ocean acidification, increased storm intensity, and coastal erosion rates) across the Oregon Coast (Hoelting and Burkardt, 2017). Hoelting and Burkardt (2017) identify several challenges for the growing tourism industry in Oregon's coastal (rural) townships that include increased presence of tropical aquatic species such as Opah, Marlin, and Plankton; warmer temperatures linked to illnesses and die-offs of marine animals, and increased conditions for toxic algae blooms throughout the region (see also Miller, 2015; McCabe et al., 2016). Many coastal communities struggle to remediate environmental effects due to economic losses associated by declines of the timber industry, an influx of retirees, and fewer fishing opportunities (Ackerman et al., 2016). Tourism growth, declining extractive industries (e.g., timber harvesting), and climate change impacts have increased the need to develop marine tourism and conservation along the Oregon Coast.

The outdoor recreation economy accounts for 2.1 percent of the national GDP in the US and 2.9 percent of Oregon's economy – rivaling sectors of mining, agriculture, and big pharma (BEA, 2020). Revenue growth has exceeded the national average by 60 percent, generating 5.2 million jobs through $788 billion in economic output (BEA, 2020). It is characterized as an emerging economy consisting of innovators and early adopters with major players undergoing significant disruption. The COVID-19 pandemic has resulted in 91 percent of industry supply chains experiencing difficulty with production and distribution while 47 percent of businesses have laid off or furloughed a portion of their workforce (ORR, 2021). There is an opportunity to assist the sector in the promotion of sustainability, education, and conservation throughout the state.

Oregon State University has recently launched the Center for the Outdoor Recreation Economy for industry and cross-university partnerships (Branam, 2021). Key questions remain in how these investments will be distributed and whom they aim to serve. *Travel Oregon* is the State's "semi-independent agency that works to enhance Oregon's economy by developing world-class visitor experiences and implementing sales and marketing campaigns that inspire travel and convey the exceptional quality of Oregon as a destination" (Travel Oregon, 2022). It aims to advance sustainable forestry and fishing, ecotourism initiatives, and marine tourism but are at risk of becoming fragmented without a national tourism policy or substantial statewide regulatory body.

As of 2020, community volunteers had long been acting as independent stewards of *Otter Rock Marine Reserve* (ORMR) and formalized their efforts with the creation of the non-profit *Friends of Otter Rock* (FoOR). Consensus for collaborative efforts that promote marine conservation have been established where outreach and interpretive programs alongside the Oregon Department of Fish and Wildlife (marine reserve managers) are emphasized. The objectives are to engage visitors, decision-makers, scientists, and the local community in enhancing protection of the Otter Rock Marine Reserve. FoOR exists to strengthen a stewardship ethic toward protection of the tangible and intangible assets of the ORMR and achieve this by raising awareness of wildlife and the ORMR, connecting visitors and community members with the area, and promoting responsible visitation practices. Visitor demand at ORMR between April and October exceeds the reserve's carrying capacity and problems include disturbance of nesting birds, damage to marine gardens, tide pools, visitor safety on cliffs, and congested traffic (Friends of Otter Rock, 2021).

iNaturalist and *Bioblitz* are two initiatives to promote citizen science, engage tourists and local citizens, and enhance understanding of wildlife within the area by collecting data through wildlife photography with the use of an iPhone. FoOR aims to seek funds from the local municipality for these programs but has been met in opposition from city council (Brock, 2021). Whilst the development of the reserves was to establish goals for conservation, research, and community enrichment, the MRs' lack of size and competing interests hinder their ecological value toward the conservation of marine protection. They were developed in conjunction with the stronghold of the consumptive-based fishing industry and are therefore undersized and placed in areas where fishing did not typically occur. The case of coastal and marine tourism in Oregon, like tourism studies themselves, is a mixed bag of competing claims where scaling conservation initiatives with the benefits of community-based tourism is representative of settler interests who favor industry and privatization over community organizing and conservation.

Toward a 'post'-pandemic future of coastal & marine CBT

There is a surging interest in tourism research, and amongst practitioners who are looking to manage and valorize new ethos of tourism impacts with the transformational affordance of COVID-19. Hall et al. (2020) contend that the reduction of tourism spending and travel has created enough space between profit and industry to suggest that the crisis can be a trigger of change. Tourism transformation must focus on equity by implementing a resilient post-pandemic tourism industry that addresses economic and environmental gaps by using a multidisciplinary lens (Sigala, 2020). Transforming tourist consumerism to a holistic system of responsible tourism that supports and protects cultural heritage, fragile ecosystems, and the socio-economic position of locals is imperative in remediating the effects of over-tourism and its exploitative nature (Gretzel et al., 2020).

Sustainable tourism strategies such as community-based tourism promote wellbeing, environmental stewardship, and cultural unity and allyship (Hunt and Harbor, 2019; Dwyer, 2020). In Tanzania, it has been reported that the inflow of affluent tourists "can be instrumental in helping communities to have more realistic expectations about tourism development… people in developing countries like Tanzania have a tendency to essentialize tourists and the mythical developed world they come from" (Salazar, 2012, p. 19). An authentic and holistic experience provided by CBT can evoke feelings of attachment and empathy toward developing cultures (Salazar, 2012). Since the USA does not have a centralized federal policy, implementing one now with responsible tourism initiatives is essential to curb the climate crisis, mitigate socio-economic problems, and protect fragile ecological zones. However, in the current ecological crisis, nonhuman rights, welfare, and agency must be foregrounded as humanity learns how to emerge from the COVID-19 lockdowns and embrace pent-up demand for a tourism experience. A multispecies livelihoods approach affords community-based tourism operators the opportunity to embrace local and culturally appropriate approaches to human–animal relations that foreground the rights, welfare, and agency of all individuals, no matter the species (see Thomsen et al., 2021a). This is particularly important in coastal and marine areas that are confronted with competing industry and identity demands.

Conclusion

The focus of this chapter was to draw attention to emerging debates and converging phenomena of the (eco)tourism sector as the world emerges from lockdowns and learns to live with COVID-19. Amongst the seemingly endless strife that COVID-19 has caused, its 'silver lining' may be that it allows humans to reconsider how we treat all living creatures. By embracing a multispecies livelihoods perspective to community-based tourism, and in particular coastal and marine-based CBT, a new era of wildlife–human coexistence could emerge to present an ethic of care and decency.

References

Ackerman, R., Neuenfeldt, R., Eggermont, T., Burbidge, M., Lehrman, J., Wells, N., & Chen, X. (2016). *Resilience of Oregon coastal communities in response to external stressors*. Master's Thesis. University of Michigan.

Arias-Maldonado, M. (2016). The anthropocenic turn: Theorizing sustainability in a postnatural age. *Sustainability*, *8*(1), 10.

Baechler, B. R., Granek, E. F., Hunter, M. V., & Conn, K. E. (2020). Microplastic concentrations in two Oregon bivalve species: Spatial, temporal, and species variability. *Limnology and Oceanography Letters*, *5*(1), 54–65.

Balon, E. K. (2000). Defending fishes against recreational fishing: An old problem to be solved in the new millennium. *Environmental Biology of Fishes*, *57*(1), 1–8.

Bohn, D., García-Rosell, J. C., & Äijälä, M. (2018). Animal-based tourism services in Lapland. *University of Lapland / Multidimensional Tourism Institute*. 1–14.

Bork, S., Geerts, W., Wang, H., de Vries, J., Scully, E., Seely, J., & Krishnan, V. (2020, September). *The travel industry turned upside down - McKinsey & Company*. The Travel Industry Turned Upside Down. Retrieved February 2022, from https://www.mckinsey.com

Branam, C. (2021). New OSU center for the outdoor recreation economy will support $788 billion U.S. Industry. Newsroom. https://today.oregonstate.edu/news/new-osu-center-outdoor-recreation-economy-will-support-788-billion-us-industry

Brock, M. (2021). Marine reserve group asks depoe bay for support. *Newport News Times*. Retrieved 1 March 2022, from https://www.newportnewstimes.com/news/marine-reserve-group-asks-depoe-bay-for-support/article_0d8f8c32-32bb-11ec-90d1-f3c6513b011f.html

Cater, C., & Cater, E. (2008). The economic impacts of marine wildlife tourism. *Marine Wildlife and Tourism Management: Insights from the Natural and Social Sciences*, 145–162. https://doi.org/10.1079/9781845933456.0145

Copeland, K. (2020). Commodifying biodiversity: Socio-economic approaches to wildlife-human coexistence. *Encyclopedia of the UN Sustainable Development Goals, Partnerships for the Goals*, *1*, 1–12.

Copeland, K. (2021). Reimagining innovation for "social" entrepreneurship: Nonhuman spaces for the SDGs. *Journal of the International Council for Small Business*, *2*(2), 134–146.

Csapó, J. (2012). The role and importance of cultural tourism in modern tourism industry. *Strategies for Tourism Industry - Micro and Macro Perspectives*. https://doi.org/10.5772/38693

Cui, Q., & Xu, H. (2019). Situating animal ethics in Thai elephant tourism. *Asia Pacific Viewpoint*, *60*(3), 267–279. https://doi.org/10.1111/apv.12221

Dangi, T., & Jamal, T. (2016). An integrated approach to "Sustainable community-based tourism." *Sustainability*, *8*(5), 475. https://doi.org/10.3390/su8050475

Department of Resources, Energy and Tourism. (n.d.). *Tourism 2020 - Australia*. Tourism 2020 Australia. Retrieved February 2022, from https://www.tourism.australia.com/content/dam/assets/document/1/6/w/u/3/2002107.pdf

Dwyer, L. (2020). Tourism development and sustainable well-being: A Beyond GDP perspective. *Journal of Sustainable Tourism*, 1–18.

Fennell, D. A. (2015). Ethics in tourism. In S.O. Idowu & R. Schmidpeter (Ed.), *Education for sustainability in tourism* (pp. 45–57). Springer, Berlin, Heidelberg.

Fennell, D. A. (2020). *Ecotourism*. Routledge, Milton Park, Abingdon, Oxon; New York, NY.

Fennell, D. A., & Sheppard, V. (2021). Tourism, animals and the scales of justice. *Journal of Sustainable Tourism*, *29*(2–3), 314–335.

Fennell, D. A., & Thomsen, B. (2021). Tourism & animal suffering: Mapping the future. *Annals of Tourism Research*, *91*, 103317.

Friends of Otter Rock. (2021). *Work Plan for 2022*. https://legistarweb-production.s3.amazonaws.com/uploads/attachment/pdf/1066971/2022_Work_Plan_Friends-of-Otter-Rock.pdf

Garrod, B. (2007). Marine wildlife tourism and ethics. *Marine Wildlife and Tourism Management: Insights from the Natural and Social Sciences*, 257–271. https://doi.org/10.1079/9781845933456.0257

Gretzel, U., Fuchs, M., Baggio, R., Hoepken, W., Law, R., Neidhardt, J., Pesonen, J., Zanker, M., & Xiang, Z. (2020). e-Tourism beyond COVID-19: A call for transformative research. *Journal of Information Technology & Tourism*, *22*(2), 187–203. http://link.springer.com/10.1007/s40558-020-00181-3

Hall, C. M., Scott, D., & Gössling, S. (2020). Pandemics, transformations, and tourism: Be careful what you wish for. *Tourism Geographies*. https://doi.org/10.1080/14616688.2020.1759131

Hoelting, K., & Burkardt, N. (2017). *Human dimensions of climate change in coastal Oregon*. US Dept. of the Interior, Bureau of Ocean Energy Management, Headquarters, Washington, DC. (OCS Study BOEM 2017-052).

Hunt, C. A., & Harbor, L. C. (2019). Pro-environmental tourism: Lessons from adventure, wellness and eco-tourism (AWE) in Costa Rica. *Journal of Outdoor Recreation and Tourism*, *28*, 1–7.

Leposa, N. (2020). Problematic Blue Growth: A thematic synthesis of social sustainability problems related to growth in the marine and coastal tourism. *Sustainability Science*, *15*(4), 1233–1244. https://doi.org/10.1007/s11625-020-00796-9

Liu, Z. (2003). Sustainable tourism development: A Critique. *Journal of Sustainable Tourism*, *11*(6), 459–475. https://doi.org/10.1080/09669580308667216

López-Guzmán, Sánchez-Cañizares, & Pavón. (2011). Community - based tourism in developing coiuntries: A case study. *Tourismos: An International Multiduisciplinary Journal of Tourism*, *6*(1), 69–84.

Mandić, A. (2021). Protected area management effectiveness and COVID-19: The case of Plitvice Lakes National Park, Croatia. *Journal of Outdoor Recreation and Tourism*, *100397*, 1–11.

McCabe, R. M., Hickey, B. M., Kudela, R. M., Lefebvre, K. A., Adams, N. G., Bill, B. D., … & Trainer, V. L. (2016). An unprecedented coastwide toxic algal bloom linked to anomalous ocean conditions. *Geophysical Research Letters*, *43*(19), 10–366.

Miller H. 2015. Bizarre catches bring new meaning to the term 'hot fishing'. *Statesman Journal*. Retrieved December 7, 2016, from http://www.statesmanjournal.com/story/travel/outdoors/huntingfishing/2015/08/31/bizarre-catches-bring-new-meaning-term-hot-fishing/71485872/

Miller, M. L., Auyong, J., Szuster, B. W., Needham, M. D., Orams, M. B., Wilks, J., & Lück, M. (2017). A field takes stock: Papers from the 8th International Congress on Coastal and Marine Tourism. *Tourism in Marine Environments*, *12*(3–4), 139–153.

Moorhouse, T., D'Cruze, N. C., & Macdonald, D. W. (2017). Unethical use of wildlife in tourism: What's the problem, who is responsible, and what can be done?. *Journal of Sustainable Tourism*, *25*(4), 505–516.

Moyle, B., Moyle, C. -lee, Ruhanen, L., Weaver, D., & Hadinejad, A. (2020). Are we really progressing sustainable tourism research? A bibliometric analysis. *Journal of Sustainable Tourism*, *29*(1), 106–122. https://doi.org/10.1080/09669582.2020.1817048

Nalven, K. B. (2019). Adapting the Community Based Social Marketing Framework to Create Actionable Messages About Plastic Pollution at the Oregon Coast Aquarium.

O'Neil, L. (2016). Selfie-taking tourists kill rare dolphin calf on Argentina beach | CBC News. *CBC*. https://www.cbc.ca/news/trending/dolphin-calf-dies-after-tourists-use-it-for-selfies-1.3456188

OECD. (2018). OECD Tourism Trends and Policies 2018, Publishing Paris. https://www.oecd-ilibrary.org/urban-rural-and-regional-development/oecd-tourism-trends-and-policies-2018_tour-2018-en

OECD. (2020). OECD Economic Outlook, Volume 2020 Issue 2, OECD Publishing, Paris. https://doi.org/10.1787/39a88ab1-en

Okazaki, E. (2008). A community-based tourism model: Its conception and use. *Journal of Sustainable Tourism*, *16*(5), 511–529. https://doi.org/10.1080/09669580802159594

Outdoor Recreation Roundtable. 2021. Covid-19 impacts on the outdoor recreation economy. Retrieved August 4, 2021, from https://recreationroundtable.org/covid-19-impacts-survey/

Perry, E. E., Needham, M. D., & Cramer, L. A. (2017). Coastal resident trust, similarity, attitudes, and intentions regarding new marine reserves in Oregon. *Society & Natural Resources*, *30*(3), 315–330.

Salazar, N. B. (2012). Community-based cultural tourism: Issues, threats and opportunities. *Journal of Sustainable Tourism*, *20*(1), 9–22. https://doi.org/10.1080/09669582.2011.596279

Sigala, M. (2020). Tourism and COVID-19: Impacts and implications for advancing and resetting industry and research. *Journal of Business Research*, *117*, 312–321.

Statista Research Department, & 3, F. (2022, February 3). Travel and tourism: Share of global GDP 2020. Statista. Retrieved February 2022, from https://www.statista.com/statistics/1099933/travel-and-tourism-share-of-gdp/

Steinbrink, M. (2012). 'We did the slum!' – urban poverty tourism in historical perspective. *Tourism Geographies*, *14*(2), 213–234. https://doi.org/10.1080/14616688.2012.633216

Thomsen, B. (2021). Wolf ecotourism: A posthumanist approach to wildlife ecotourism. In D. Fennell (Ed.), *Routledge handbook of ecotourism* (pp. 117–131). United Kingdom: Routledge.

Thomsen, B., Muurlink, O., Best, T., Thomsen, J., & Copeland, K. (2020). Transcultural development. *Human Organization*, *79*(1), 43–56.

Thomsen, B., Thomsen, J., Cipollone, M., & Coose, S. (2021b). Let's Save the Bear: A multispecies livelihoods approach to wildlife conservation and achieving the SDGs. *Journal of the International Council for Small Business*, *2*(2), 114–124. https://doi.org/10.1016/j.tmp.2012.05.001

Thomsen, B., Thomsen, J., Copeland, K., Coose, S., Arnold, E., Bryan, H., … & Chalich, G. (2021a). Multispecies livelihoods: A posthumanist approach to wildlife ecotourism that promotes animal ethics. *Journal of Sustainable Tourism*, 1–19. https://doi.org/10.1080/09669582.2021.1942893

Travel Oregon. (2022). 2021–2023 strategic plan. Travel Oregon. Retrieved March 1, 2022, from https://industry.traveloregon.com/about/strategic-plan/2021-2023-strategic-planning/

U.S. Bureau of Economic Analysis (BEA). (2020). Outdoor recreation satellite account, U.S. and States, 2019 https://www.bea.gov/news/2020/outdoor-recreation-satellite-account-us-and-states-2019

von Essen, E., Lindsjö, J., & Berg, C. (2020). Instagranimal: Animal welfare and animal ethics challenges of animal-based tourism. *Animals*, *10*(10), 1830. https://doi.org/10.3390/ani10101830

Wadiwel, D. (2018). Chicken harvesting machine. *South Atlantic Quarterly*, *117*(3), 527–549. https://doi.org/10.1215/00382876-6942135

Wang, L., & Zhang, H. (2019). The impact of Marine tourism resources development on sustainable development of marine economy. *Journal of Coastal Research*, *94*(sp1), 589. https://doi.org/10.2112/si94-117.1

Waters, C. N., Zalasiewicz, J., Summerhayes, C., Barnosky, A. D., Poirier, C., Gałuszka, A., … & Wolfe, A. P. (2016). The anthropocene is functionally and stratigraphically distinct from the Holocene. *Science*, *351*(6269), aad2622.

Winter, C. (2020). A review of research into animal ethics in tourism: Launching the annals of tourism research curated collection on animal ethics in tourism. *Annals of Tourism Research*, *84*, 102989.

WTO. (2022). 2020: Worst Year in Tourism History with 1 Billion Fewer International Arrivals. https://www.unwto.org/news/2020-worst-year-in-tourism-history-with-1-billion-fewer-international-arrivals. Accessed 22 November 2022.

Zeppel, H., & Muloin, S. (2008). Marine wildlife tours: Benefits for participants. *Marine Wildlife and Tourism Management: Insights from the Natural and Social Sciences*, 19–48. https://doi.org/10.1079/9781845933456.0019

INDEX